env- environmental

~~60.00~~ 44.00

ENVIRONMENTAL FIELD TESTING AND ANALYSIS READY REFERENCE HANDBOOK

ENVIRONMENTAL FIELD TESTING AND ANALYSIS READY REFERENCE HANDBOOK

Donald A. Drum, B.S., M.S., Ph.D., Professor Emeritus

Shari L. Bauman, B.S., M.S., Geologist and Environmental Science

Gershon J. Shugar, B.S., M.S., Ph.D., Professor Emeritus

McGRAW-HILL, INC.

New York San Francisco Washington, D.C. Auckland Bogotá
Caracas Lisbon London Madrid Mexico City Milan
Montreal New Delhi San Juan Singapore
Sydney Tokyo Toronto

Library of Congress Cataloging-in-Publication Data

Shugar, Gershon J.
Environmental Field Testing and Analysis Ready Reference Handbook/
Gershon J. Shugar, Donald A. Drum, Shari L. Bauman
p. cm.
Includes index.
ISBN 0-07-135964-8
1. Pollutants—Analysis—Handbooks, manuals, etc.
2. Environmental Sampling—Handbooks, manuals, etc. I. Drum, Donald A. II. Bauman, Shari L. III. Shugar, Gershon J., 1918–
IV. Title.
TD193 .D78 2000
628.5′028′7—dc21 00-062449

McGraw-Hill

A Division of The ***McGraw·Hill*** *Companies*

1 2 3 4 5 6 7 8 9 0 DOC/DOC 9 0 9 8 7 6 5 4 3 2 1 0 9

ISBN 0-07-135964-8

The sponsoring editor for this book was Ken McCombs, the editing supervisor was Reina Zatylny, and the production supervisor was Sherri Souffrance. It was set in Times Roman by Pro-Image Corporation.

Printed and bound by R. R. Donnelley & Sons Company.

This book is printed on recycled, acid-free paper containing a minimum of 50% recycled, de-inked fiber.

CONTENTS

CONTRIBUTORS

Pete Ambrose, Columbia-Greene Community College, Hudson, N.Y.
Deborah Bowers, Data Entry, Jacksonville, FL.
Denise Budzinski, Mine Safety Appliance, Inc., Cranberry Township, PA.
Cathi Christianson,* HACH Company, Loveland, CO.
James Cudahy, Focus Environmental, Inc., Knoxville, TN.
Mary Jo Dawson, Bacharach, Pittsburgh, PA.
Irene DeGraff, SUPELCO, Bellefonte, PA.
Mary Ann Detlefsen, Environmental Chemist, Latham, N.Y.
Mike Denez, Ward's Natural Science Establishment, Inc., Rochester, N.Y.
Bob Esposito,* McGraw-Hill, New York, N.Y.
Bill Gorsuch, LaMotte Company, Chestertown, MD.
Scott Grillo, McGraw-Hill, New York, N.Y.
Bob Harrison, Cape Technologies, South Portland, ME.
Floyd Hasselriis, P.E., Consultant, Forest Hills, N.Y.
Greg Haughey, Callery Chemical Co., Evans City, PA.
Inga Henderson, Timberline Instruments, Boulder, CO.
Margaret Hill, LaMotte Company, Chestertown, MD.
Andrew Jolin, GESS Environmental, Bayside, CA.
Ken Keppel, Dräger, Pittsburgh, PA.
Richard LaMotte, LaMotte Company, Chestertown, MD.
Jack Lauber, P.E., Consultant, Latham, N.Y.
Bernie J. Lerner, Beco Engineering Co., Oakmont, PA.
Ed Ligus, Dräger, Pittsburgh, PA.
Ken McCombs, McGraw-Hill, New York, N.Y.
Cathy McDonald, Mine Safety Appliance, Inc., Cranberry Township, PA.
Mary McGinley, Mine Safety Appliance, Inc., Cranberry Township, PA.
Tim Parent, LaMotte Company, Chestertown, MD.
Ben Pierson, New York State Department of Health, Troy, N.Y.
Jon & Sue Powell, Columbia-Greene Community College, Hudson, N.Y.
Christopher Rappe, Ph.D., University of Umea, Umea, Sweden
Ron Roberson, Sensidyne Inc., Clearwater, FL.
Carlene Roberts, HACH Company, Loveland, CO.
Dave Ruane,* HACH Company, Loveland, CO.
Ryan Rudebusch, HACH Company, Loveland, CO.
Edward Schut, LaMotte Company, Chestertown, MD.
Alan Seideman, Zero Waste, La Jolla, CA.
Denton Slovacek, HACH Company, Loveland, CO.
Amy Steigerwald, LaMotte Company, Chestertown, MD.
Ward Stone, Del Mar Wildlife Laboratories, Del Mar, N.Y.
Eric Umlreit, HACH Company, Loveland, CO.
Ed White, State Soil Scientist, Harrisburg, PA.

*No longer with the Company as of August 2000.

Contributing Editors:

Ronald A. Shugar, B.S., M.D.; Rose Shugar Bauman, B.S.; Lawrence Bauman, B.S., D.D.S.; Ellen Shugar, B.S., M.A., R.H.; Allan J. Bauman, B.S., M.D.; Karen Bauman, B.S., M.S.; Pamela A. Drum, B.S.; Deborah C. Bowers, B.S., M.Sc.

ACKNOWLEDGMENTS

The authors wish to express their grateful appreciation to the following Environmental Field Testing and Analysis Companies for their total and complete cooperation by supplying us with literature, brochures, flyers, manuals, information packets, and telephone communications without reservations. Without their help and assistance, this ready reference could not have been completed to the level of excellence that the authors wish to achieve. They have indicated that they are willing to answer any and all environmental field testing and analysis questions without reservation.

If we have inadvertently omitted listing any of the environmental field testing and analysis suppliers, we apologize. We would appreciate any corrections that we should make on the next addition. We would be grateful for any and all suggestions regarding inclusions of missing material for the next addition. We, the authors, feel that this is the first complete, in all respects, Ready Reference Environmental Field Testing and Analysis Handbook.

Company	Address	
Drager Safety, Inc.	101 Technology Drive	Pittsburgh, PA. 15725
GESS Environmental	2575 Old Arcata Rd.	Bayside, CA. 95524
HACH Company	P.O. Box 389	Loveland, CO. 80539
LaMotte Company	P.O. Box 329	Chestertown, MD. 21620
Sensidyne, Inc.	16333 Bay Vista Drive	Clearwater, FL. 34620
Strategic Diagnostics	111 Pencader Drive	Newark, DE. 19702
Ward's Natural Science Establishment, Inc.	5100 West Henrietta Road	Rochester, NY. 14692

As authors, we express our appreciation to each person for their contributions and efforts. We express our personal regrets if we have not included any name that we have personally contacted about information for the text. Thank you for your contribution(s) to the text.

A special thank you is extended to Deborah Bowers, Ed Ligas, Denton Slovacek, Jack Lauber, Mary Ann Detlefsen, Amy Steigerwald and Mike Denez for their extensive assistance in gathering and preparing materials for the text. Each of these individuals contributed a great deal of thought and advice during preparation of materials.

The authors wish to express our appreciation to Scott Gillo, Editor-in-Chief, and Ken McCombs, Acquisition Editor, Professional and Reference Division, McGraw-Hill, New York for their encouragement and invaluable guidance during the successful completion of the Environmental Field Testing and Analysis Ready Reference Handbook.

This text was an attempt to bring factual information and experiments together into one document. In many cases, the description of a specific pollutant is the most recent information available on the topic. Some of the information has been published during the summer of 2000. For more information, please review the reference section after each experiment for a list of papers and texts.

Donald A. Drum, Ph.D.
Gershon Shugar, Ph.D.
Shari L. Bauman

ENVIRONMENTAL FIELD TESTING AND ANALYSIS READY REFERENCE HANDBOOK

CHAPTER 1
OVERVIEW OF ENVIRONMENTAL TESTING

ABBREVIATIONS

Abbreviation*	Definition
°C	Degree(s) Celsius (Centigrade)
°F	Degree(s) Fahrenheit
ACS	American Chemical Society reagent grade purity
ASPHA Standard Methods	*Standard Methods for the Examination of Water and Wastewater,* published jointly by the American Public Health Association (APHA), the American Water Works Association (AWWA), and the Water Environment Federal (WEF). Purchase the text from Hach requesting Cat. No. 22708-00 or from the Publication Office of the American Public Health Association. This book is the standard reference work for water analysis. Many procedures contained in this manual are based on *Standard Methods.*
AV	AccuVac
Bien	bicinchoninate
Conc	concentrated
CFR	Code of Federal Regulations
EPA	Environmental Protection Agency
FAU	Formazin Attenuation Units. Turbidity unit of measure based on a Formazin stock suspension.
g	grams
gr/gal	grains per gallon (1 gr/gal = 17.12 mg/L)
HR	high range
kg/ha	kilograms per hectare
L	Liter. Volume equal to one cubic decimeter (dm^3)
lbs/Ac	pounds per acre
LR	low range
MDL	Method detection limit
mg/L	milligrams per liter (ppm)
μg/L	micrograms per liter (ppb)
mL	(milliliter)-approximately the same as a cubic centimeter (cc) or 1/1000 of a liter. Also known as a "cc."
MR	medium range
NIPDWR	National Interim Primary Drinking Water Regulations
NPDES	National Pollutant Discharge Elimination System
P	Phosphorus
PCB	Polychlorinated biphenyl

Abbreviation*	Definition
PPB	parts per billion
PPM	parts per million
TPH	Total petroleum hydrocarbons
TPTZ	2,4,6-Tri-(2-Pyridyl)-1,3,5-Triazine
USEPA	United States Environmental Protection Agency

* The abbreviations are used in several sections of the text. Compliments of HACH Company.

STANDARDS

Accuracy and Precision*

Accuracy is the nearness of a test result to the true value. Precision is how closely repeated measurements agree with each other. Although good precision suggests good accuracy, precise results can be inaccurate. The following paragraphs describe how to improve accuracy and precision of analysis by using Standard Additions.

USEPA Accepted

The USEPA has reviewed and accepted certain procedures for reporting purposes. A few of these methods are referenced to the equivalent USEPA method in the procedure.

Standard Additions**

Standard Additions is a common technique for checking test results. The technique can test interferences, bad reagents, faulty instruments, and incorrect procedures.

Perform standard additions by adding a small amount of a standard solution to your sample and repeating the test. Use the same reagents, equipment and technique. You should get about 100% recovery. If not, you have an identifiable problem.

If Standard Additions works for your test, a Standard Additions Method section will be in the procedure after each experiment under *Accuracy Check* in the *HACH Water Analysis Handbook,* 3rd Edition, 1997. Follow the detailed instructions given. If you get 100% for each addition, everything is working right and your results are correct.

If you don't get 100% recovery for each addition, follow the procedure listed on the bottom of p. 26 and the checklist on pp. 27–37 of the *HACH Water Analysis Handbook,* 3rd Edition, 1997.

Method Performance**

Estimated Detection Limit, Method Detection Limit (MDL), Precision, and Estimating Precision are provided for the HACH experiments.

* Compliments of HACH Company.

** *HACH Water Analysis Handbook,* 3rd Edition, 1997.

CALIBRATION

Calibration Curves*

1. Prepare five or more standards of known concentration that covers the expected range of the test. Run tests as described in the procedure on each prepared standard. Then pour the customary volume of each known solution into a separate clean sample cell of the type specified for your instrument.
Note: Unknown sample results should fall within the range of the measured standards.
2. Select the proper wavelength. Standardize (zero) the instrument using an untreated water sample or a reagent blank, whichever the procedure instructs you to perform.
Note: Check to see if the sample cell and reference cell match by filling each with water and running an absorbance reading. Both cells should have the same reading.
3. Measure and record the %T or absorbance of the known solutions.
4. Follow the **Absorbance vs Concentration Calibration** or the ***%T* vs Concentration Calibration** procedure provided below or in the HACH Water Analysis Handbook, 3rd Edition, pp. 42–43.

Absorbance Versus Concentration Calibration

If absorbance values are measured, plot the results from standard solutions on linear graph paper. Plot the absorbance value on the vertical axis and the concentration on the horizontal axis. Plot increasing absorbance values from bottom to top on the y axis. Also, plot increasing concentration values from left to right on the horizontal (x) axis. If interferences and mistakes are avoided, the plot of the data from the standard solution dilutions should extrapolate through the points [0,0] on the graph.

A calibration table can be extrapolated from the plotted line or the concentration values can be read directly from the graph. Also, an equation for the line using the slope and the y-intercept can be developed.

Alternatively, plot *%T* (vertical axis) versus concentration (horizontal x axis) on semilogarithmic graph paper. The concentration can be read directly from the plot or a calibration table can be developed for interpolation purposes. For additional information, refer to pp. 42–43 of the HACH Water Analysis Handbook, 3rd Edition, 1997.

PROCEDURES AND EQUIPMENT

Adapting Procedures and DR/2010 Instrument Operational Procedures*

1. Adapting procedures to other spectrophotometers is not a problem if calibration curves that convert absorbance or %T to concentration are plotted.
2. To select the best wavelength on the DR/2010, follow the instructions on pp. 45–46 of the HACH Water Analysis Handbook, 3rd Edition, 1997. This information also provides the basic operational procedures for the DR/2010 instrument.
3. A buret can be adapted for use with a Digital Titrator.
Note: The Digital Titrator dispenses 1 mL per 800 digits on the counter.

Interferences*

Substances in the sample may interfere with a measurement. Common interferences are usually mentioned in the test procedures. The reagent formulations eliminate many interfer-

* *HACH Water Analysis Handbook,* 3rd Edition, 1997.

ences. Other interferences can be removed with sample pretreatments described in the experiment procedures.

If an unusual answer is given, a color is unexpected, or an unusual odor or turbidity is observed in the experimental test, the result may be wrong. Repeat the test on a sample diluted with deionized water and compare the result (corrected for the dilution) with the result of the original test.

pH Interference*

Many of the procedures are effective within a certain pH range. Reagents normally contain buffers to adjust the pH of the typical sample to the correct pH range. The pH of the sample may have to be adjusted before testing. If so, follow the procedure listed on p. 50 of the HACH Water Analysis Handbook, 3rd Edition, 1997. The *Sampling and Storage* section of each procedure gives the proper pH range for the sample.

Sample Collection, Preservation and Storage*

Correct sampling and storage are critical for accurate measuring. For greatest accuracy, thoroughly clean sampling devices and containers to prevent problems from previous samples. Preserve the sample properly. Each procedure has information about sample preservation.

1. Avoid soft glass containers for metals in the microgram/L range.
2. Store samples for silver determination in amber containers or in light-absorbing containers.
3. Thoroughly clean sample containers.
4. Preservation may impact sample analysis. Analyze samples as soon as possible after collection.
5. Preservation methods include pH control, chemical addition, refrigeration and freezing. See Table 10 on pp. 52–54 of the HACH Water Analysis Handbook, 3rd Edition, 1997, for additional information.

Collecting Water Samples*

Obtain the best sample by careful collection. In general, collect samples near the center of the vessel or duct and below the surface. Use only clean containers (bottles, beakers, etc.). Rinse the container several times with the water to be collected.

Take samples as close as possible to the source of the supply. Collect water samples from wells after the pump has run long enough to deliver water representative of the groundwater feeding the well. Let the water run long enough to flush the system. Fill sample containers slowly with a gentle stream to avoid turbulence and air bubbles. Make certain that the nozzle or spigot delivering the water supply does not contaminate the water.

Try to collect representative samples of the water source. Best results are obtained by testing several different samples at identifiable sites.

---

* *HACH Water Analysis Handbook,* 3rd Edition, 1997.

Test the sample as soon as possible after collection. Prevent natural interferences such as organic growths and loss or gain of dissolved gases.

Acid Washing Bottles*

If a procedure suggests acid washing, the listed procedure should be followed. Phosphate-free detergent is recommended for washing glassware. Containers should be rinsed with tap water, 1:1 hydrochloric acid solution or 1:1 nitric acid solution. The nitric acid rinse is important for testing lead. For ammonia and Kjeldahl nitrogen, rinse with ammonia-free water. Rinse several times with deionized water.

Correcting for Volume Additions*

If a large volume of preservative is used, correct for the volume of preservative added. This accounts for dilution due to the acid added to preserve the sample and the base used to adjust the pH to the range of the procedure. This correction is made as follows:

1. Determine the volume of initial sample, the volume of acid and base added, and the total final volume of the sample.
2. Divide the total volume by the initial volume.
3. Multiply the test result by this factor.

Boiling Aids

Bumping in the heated reaction flask can be avoided by adding boiling chips. Glass beads are recommended. The boiling chips must not contaminate the sample.

Sample Filtration*

Filtering separates particles causing turbidity and precipitates from solution. Gravity filtration uses gravity to pull the sample through filter paper in a funnel. Vacuum filtration uses suction by an aspirator or vacuum pump to move the sample rapidly through a sintered glass filter or some other appropriate filtering device.

Reagent and Standard Stability*

Chemicals supplied with the DR/2010 Spectrophotometer and the DREL/2010 Portable Laboratories have an indefinite shelf life when stored under average room conditions unless designated otherwise. Notations on product labels specify any special storage conditions required. Generally, reagents should be stored in a cool, dry, and dark place for maximum life. Date the chemicals upon receipt and use the older supplies first. If in doubt about the reagent shelf life, run a standard to check reagent effectiveness.

Reagent Blank*

The term "reagent blank" refers to that portion of the test result contributed solely by the reagent and not the sample. In several of the tests, the reagent blank is of such magnitude

---

* *HACH Water Analysis Handbook,* 3rd Edition, 1997.

that compensation is made each time the test is performed. This is done by zeroing the instrument on deionized water and reagents. A note is included in the appropriate procedures describing this activity.

SAMPLE CELLS

Orientation of Sample Cells

The DR/2010 uses two types of matched sample cells; a matched pair of taller 25 mL sample cells and a shorter matched pair of 10 mL sample cells. Both types are matched with the spectrophotometer light beam passing through the side with the fill mark and the opposite side.

To minimize variability of measurements using a particular cell, always place the cell into the cell holder with the same orientation. The cells are placed in the instrument with the fill marks facing left (viewer's left). In addition to proper orientation, the sides of the cells must be free of smudges, fingerprints, spilled liquid, etc. to ensure accurate readings. Wipe the side of the cells with a soft cloth or Kimwipe to clean the surface before taking measurements.

Care of Sample Cells

Empty and clean cells after completing analyses of the sample. A few rinses of deionized water will usually complete the cleaning procedure. Individual procedures often recommend specific cleaning methods for special circumstances. Try to avoid touching the cells on the sides that the light passes through by grasping the top and bottom of the sample bottle or holding the bottle by the top.

Cleaning Sample Cells

Most laboratory detergents can be used at recommended concentrations. If possible, use distilled water to rinse the cell. Be aware that many soaps and detergents contain carbonates and phosphates which may interfere with the field test.

Sample Cell Matching*

Sample cells and reference cells may develop nicks and scratches with handling. A procedure for checking the optical matching of cells is provided. If the cells do not match optically, a new cell may have to be purchased and used in the experiments.

DILUTION TECHNIQUES

Sample Dilution Techniques (HACH Equipment)*

Volumes of 10 mL and 25 mL samples are used for most colorimetric tests. However, the color may be too intense to be measured. Also, unexpected colors may develop in other

* *HACH Water Analysis Handbook,* 3rd Edition, 1997.

tests. In both cases, dilute the sample and rerun the sample to determine if interfering substances are present.

To dilute the sample easily, pipet the chosen sample portion into a clean graduated cylinder (or volumetric flask for greater accuracy). Fill the cylinder (flask) to the desired volume with distilled water. Mix well. Use the diluted sample when running the test.

Table 1.1 titled "**Sample Dilution Volumes**" displays the amount of sample volume (mL) used, the amount of deionized water used to bring the volume up to 25 mL, and the multiplication factor. The concentration of the sample is equal to the diluted sample reading multiplied by the multiplication factor.

TABLE 1.1 Sample Dilution Volumes

Sample volume*(mL)	mL deionized water used to bring the volume to 25 mL	Multiplication factor
25.0	0.0	1
12.5	12.5	2
10.0	15.0	2.5
5.0	20.0	5
2.5	22.5	10
1.0	24.0	25
0.250	24.75	100

*For sample volumes of 10 mL or less, use a pipet to measure the sample into the graduated cylinder or volumetric flask.

More accurate dilutions can be performed with a pipet and a 100 mL volumetric flask (See Table 1.2 titled **Multiplication Factors for Diluting to 100 mL**). Pipet the sample volume (mL) into a 100 mL volumetric flask and dilute the sample with distilled water to the 100 mL volume. Swirl the sample to mix. After using this diluted sample for the test, the sample concentration is equal to the diluted sample multiplied by the appropriate multiplication factor.

TABLE 1.2 Multiplication Factors for Diluting to 100 mL

Sample volume (mL)	Multiplication factor
1	100
2	50
5	20
10	10
25	4
50	2

Sample Dilution Techniques with Lamotte Equipment*

In some cases, the concentration of the analyte in the sample is too high to be measured by the colorimetric instrument in the LaMotte experiments. Reference guidelines on dilutions of various proportions are listed in the table below. All dilutions are based on a 10 mL volume final volume of sample. Graduated pipets or volumetric pipets should be used for all dilutions.

To dilute the sample easily, pipet the selected sample into a clean small graduated cylinder (or small volumetric flask). Then, continue filling the container to the 10 mL volume with deionized water. Mix the sample well before testing. The diluted solution will be used for the test.

Size of sample	Deionized water to bring volume to 10 mL	Multiplication factor
10 mL	0 mL	1
5 mL	5 mL	2
2.5 mL	7.5 mL	4
1 mL	9 mL	10
0.5 mL	9.5 mL	20

Sample Dilution and Interfering Substances*

Sample dilution may influence the level at which a substance may interfere. The effect of the interferences decreases as the dilution increases.

MEASUREMENTS AND SAMPLING

Temperature Considerations

For best results, the tests should be performed with sample temperatures between 20° C (68° F) and 25° C (77° F).

Using Pipets and Graduated Cylinders**

1. Rinse all glassware two or three times with the sample to be tested before testing. Pour the rinse on the ground. **Never add** the rinse back in the sample to be tested.
2. Use a pipet filler or pipet bulb to draw samples into the pipet. Never pipet chemicals or the sample by mouth.
3. Measure liquids with a graduated cylinder, graduated pipet (serological pipet) or a volumetric pipet.

*Compliments of LaMotte Company

** *HACH Water Analysis Handbook,* 3rd Edition, 1997.

4. Read the volume of the sample by holding the graduated cylinder and pipets vertically and read the scale on the graduated glassware at the bottom of the meniscus. See Fig. 8 on page 64 of the HACH Water Analysis Handbook, 3rd Edition, 1997.

Using AccuVac Ampuls*

AccuVac ampuls contain pre-measured powder or liquid to be used in some of the experimental procedures. Follow the procedures below:

1. After collecting the sample in an open beaker, place the ampul tip well below the sample surface and break the tip off against the beaker wall or with an AccuVac Breaker.
2. Invert the ampul several times to dissolve the reagent. Do not place your finger over the broken end; the liquid will stay in the ampul when inverted.
3. Wipe the ampul with a towel or KimWipes to remove fingerprint and for drying purposes.
4. Insert the ampul into the AccuVac adapter in the DR/2010 and read the results directly.
5. Collect and discard the used ampul appropriately.

Using Reagent Powder Pillows and PermaChem Pillows*

Dry powdered reagents are used to minimize leakage problems and deterioration problems.

1. Open the powder pillow by tapping one end of the pillow on a hard surface to collect the powdered reagent in the bottom.
2. Cut the powder pillow with finger nail clippers (tear the PermaChem Pillow) holding the pillow away from your face.
3. (Only for the PermaChem Pillow.) Using two hands, push both sides toward each other to form a spout.
4. Carefully pour the pillow contents into the sample cell and continue the procedure as directed.
5. Collect and properly discard the powder pillows.

Mixing Water Samples and Testing Agents*

Specific procedures are recommended when mixing chemicals in a sample cell or in a graduated cylinder.

1. When mixing sample in a square sample cell, swirl or rotate the cell by holding the top (neck) of the cell with the thumb and the index finger of one hand while resting the bottom on the tip of the index finger of the other hand. Do not touch the sides of the sample cell to avoid bad readings.
2. For a titration flask or a graduated cylinder, grasp the top of the graduated cylinder or titration flask with three fingers. Hold the sample container at a 45° angle. Swirl the container to mix the sample. Avoid spilling the sample.

---

* *HACH Water Analysis Handbook,* 3rd Edition, 1997.

Using the Pour-Thru Cell*

The DR/2010 uses the 25 mL cell to perform measurements. *Do not use the Pour-Thru Cell with organic solvents such as acetone, chloroform, toluene or cyclohexanone.*

Volume Measurement Accuracy*

Do not use the 10 mL and 25 mL sample cells to measure volumes of liquids for sample dilutions. Use a pipet, buret or another accurate measuring device.

Distillation and Speciality Glassware*

Distillation is an effective way of separating chemical components in water and wastewater. Several experiments require distillation procedures. Some experiments require speciality glassware with threaded connectors for ease and safety during the experiment. See pp. 69–70 of the HACH Water Analysis Handbook, 3rd Edition, 1997, for more details.

INSTRUMENT OPERATIONS

Measurement Equipment

HACH Company, LaMotte Company and WARD's Natural Science Establishment, Inc. prepare field kits which utilize comparator methods, electronic instruments, and simple titration methods. Most of the companies prepare kits to measure a developing color from a variety of chemical reactions by both visual methods and by electronic colorimeters. Some kits use visual methods to compare the color of an "unknown" sample with a set of color standards to determine the concentration of the unknown. More sophisticated kits use electronic colorimeters to measure the amount of light which travels through the reacted sample. This measurement is converted to an analog or digital reading as ppm, absorbance, or % transmittance.

Titration Procedures

Titration methods are based on adding a titrant (a solution of known strength) to a specific volume of a sample in the presence of an indicator. Usually, the color forming indicator is an intermediate complex in the reaction or an indicator that is added directly to the reaction. The indicator produces a color change indicating the reaction is complete. For field work, titration methods include automatic burets, direct reading titrators, dropper pipets, dropper bottles, multi-channel pipettors, Digital Titrators, Tensette Pipets, calibrated eye droppers, graduated cylinders, and additional items.

The Direct Reading Titrator from LaMotte provides a good quality hands-on method of titrating samples. The equipment provides reliable and fairly accurate results during general studies of pollution. The Titrator equipment is a good tool used for field and laboratory studies.

---

* *HACH Water Analysis Handbook,* 3rd Edition, 1997.

Titration experiments using the Digital Titrator (HACH) provides more accuracy and precision during the measurements than other available titration methods. A volume of 1.0 mL of titrant is equivalent to 800 digits on the Digital Titrator. If a Digital Titrator is not available, then a TenSette Pipet, a Direct Reading Titrator, a calibrated pipet or a buret may be used to titrate the sample. For less accurate results, a calibrated eye dropper may be used for titration purposes. The eye dropper should be calibrated before assuming 1 mL = 20 drops. When using the different titration devices, the titrant will have to be available and accessible in an appropriate container.

Direct Reading Titrator Instructions (LaMotte Company)

1. Fill the titration tube to the line with the water sample.
2. Add the reagents as specified in the instructions for the individual test method. Cap the tube with the special titration tube cap. Mix by swirling gently.
3. Depress the plunger of the Titrator to expel air.
4. Insert the Titrator into the plastic fitting of the titrating solution bottle.
5. To fill the Titrator, invert the bottle and slowly withdraw the plunger until the bottom of the plunger is opposite the zero mark on the scale.
   Note: When filling the Titrator from a container not fitted with a special plug, submerge the tip of the Titrator below the surface of the solution and withdraw the plunger. Always use the same Titrator and be sure that the Titrator is clean and dry if it has not been used for a period of time.
   Note: A small air bubble may appear in the Titrator barrel. Expel the bubble by partially filling the barrel and pumping the titration solution back into the inverted reagent container. Repeat this pumping action until the bubble disappears.

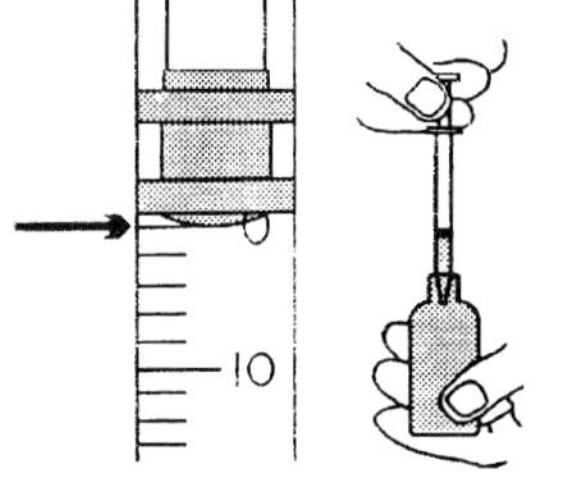

6. Turn the bottle right-side-up and remove the Titrator.
7. Insert the tip of the Titrator into the opening of the titrator tube cap. Slowly depress the plunger to dispense the titrating solution. Gently swirl the tube to mix.

8. Continue adding the titrating solution until the specific color change occurs. If no color change occurs by the time the plunger tip reaches the bottom of the scale, refill the Titrator to the zero mark. Continue the titration. Include both titration amounts in the final test result.

9. Read the test result directly from the scale opposite the bottom of the plunger tip.
Note: The Titrator illustrated is an **example only**. Refer to individual kit instructions for actual graduation range and increments.

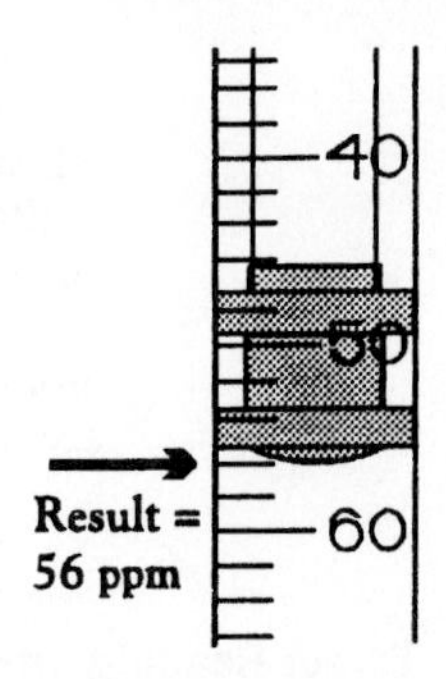

10. If no additional tests are to be made, discard the titrating solution in the Titrator. Thoroughly rinse the Titrator and the titration tube.
Note: For extended life, the plunger tip should periodically be coated with silicone grease and stored apart from the barrel.

HACH Digital Titrator Instructions*

1. Select a sample volume and titration cartridge corresponding to the expected sample concentration from the table given in each procedure. If the expected sample concentration is not known, start with one of the smaller volumes and determine its approximate concentration. Retest with the appropriate sample size.
2. Slide the cartridge into the titrator receptacle and lock in position with a slight turn.
3. Remove the polyethylene cap and insert a clean delivery tube into the end of the cartridge. Use a straight tube with a hook at the end for hand-held titrations. **Do not insert the tube past the cartridge extension.** See the illustration on p. 73 of the HACH Water Analysis Handbook.
4. For stationary setups, use a 90° tube with a hook at the end. See the discussion and illustrations on pp. 73–74 of the HACH Water Analysis Handbook.
5. To start the titrant flowing and to flush the delivery tube, hold the tip of the cartridge up. Push the plunger release button in and toward the cartridge to engage the piston with the cartridge. Turn the delivery knob until air is expelled and several drops of solution flow from the tip. Use the counter reset knob to turn the digital counter back to zero and wipe the tip. If desired, the tip can be rinsed with deionized water.
6. Use the smallest appropriate graduated cylinder or pipet to measure the sample volume in the experiment. Transfer the sample into a 125-mL or 250-mL erlenmeyer flask. Dilute to the appropriate total volume with deionized water if necessary.
Note: Sample dilutions must be made accurately.
7. Add the necessary reagents to the sample and swirl to mix.
8. Immerse the delivery tube tip in the solution and swirl the flask while holding the top of the flask under the neck. Titrate by turning the delivery knob. Keep turning the knob and swirling the sample until the end point is reached. Record the number of digits that appear in the digital counter window.
Note: The number of digits required will usually range from 100 to 400. If the digits

* *HACH Water Analysis Handbook,* 3rd Edition, 1997.

required are less than 100 or more than 400, an alternate sample volume or titrant cartridge should be used.
Note: Inaccurate results will occur if the delivery tube tip is held out of the solution rather than under the solution surface.

9. Calculate the concentration of your sample by using the following formula:

$$\text{Digits Required} \times \text{Digit Multiplier} = \text{Sample Concentration}$$

Where:
Digits Required = the number that appeared in the digital counter in Step 8.
Digit Multiplier = the number from the table given in the procedure. This number takes into account the sample dilution and titrant strength.

10. After completing the testing for the day, press the plunger release button and manually retract the plunger into the body of the titrator. Remove the cartridge. Remove the delivery tube and reseal the cartridge with the polyethylene cap.
11. Discard or clean the delivery tube immediately after use. To clean, force water, then air, into the tube opening with a syringe or wash bottle.

HACH TenSette Pipet Instructions*

1. For best results use a new pipet tip each time you pipet. After several uses, the pipet tip may retain some liquid, causing inaccurate delivery.
2. If the pipet does not operate smoothly, disassemble and coat the piston and retainer with high quality stopcock grease.
3. Never lay the pipet down with the liquid in the top. Solution could leak into the pipet and cause corrosion.

TenSette Pipet Operations*

1. Attach a clean tip by holding the pipet body in one hand and gently pressing the large end of the pipet tip onto the tapered end of the pipet.
2. Turn the turret cap to align the desired volume with the mark on the pipet body.
3. Using a smooth motion, press down on the turret cap until it reaches the stop. Immerse the tip about 5 mm (1/4 inch) below the solution surface to avoid drawing air into the pipet. Do not insert the tip any deeper or the delivery volume may be affected.
4. While maintaining a constant pressure, allow the turret to return slowly to the extended position. A rapid return may affect the delivery volume.
5. With the turret tip up, take the tip out of the solution and move it to the receiving vessel. Do not press on the turret cap while moving the pipet.
6. Use the thumb and forefinger to twist the turret cap to the next higher volume position to ensure quantitative transfer of the sample titrant. The "F" position provides full blow-out.
7. With the tip in contact with the side of the receiving vessel, slowly and smoothly press down on the turret cap until it reaches the stop and the solution is completely discharged.

---

* *HACH Water Analysis Handbook,* 3rd Edition, 1997.

Note: If a TenSette pipet is not available, a graduated pipette, a buret or a calibrated eye dropper can perform the titration. Some accuracy may be lost.

Operation of the DR/2010 Spectrophotometer and DREL/2010 Portable Laboratories

Specific sample volumes, reagents, sample cells, and even timing intervals for a given procedure can vary depending on which spectrophotometer is being used. Basic laboratory procedures are listed in a **Portable Laboratory Manual** contained in the DREL/2010 Portable Laboratories Kits. The kits are self-contained laboratories on water quality, soil and irrigation water, aquaculture, and water conditioning.

Each instrument must be properly calibrated before measurements. If using the HACH DR/2010 Spectrophotometer, read the Instrument Manual, last published during 1999, for the experimental details about measurements, operational procedures, setup menu details, zeroing the spectrophotometer, measuring the prepared sample, cleaning procedures, battery installation for field operations, lamp replacement and calibration, and any instructions required for proper operation of the instrument. For field measurements using the HACH DR/2010 Spectrophotometer, basic operator procedures are also listed on p. 80 of the HACH Water Analysis Handbook.

While performing field studies and battery power is required, the submenu option **LAMP: Momentary** may become an important issue. To change the lamp status, press **enter** (if not, press the down arrow key). Use the **arrow** key to toggle between Momentary and Constant On. Select **Momentary** by pressing **enter** when the desired choice is flashing. **Always use Momentary mode when operating on battery power**. The lamp is on a short time only, saving battery power on the HACH instrument.

Operation of the DR/800 Series Field Colorimeters (DR/820, DR/850, and DR/890)

The HACH DR/820, DR/850 and DR/890 Portable Datalogging Colorimeter Instrument Manual provides instrument operation information, battery operations, setup menu, recall menu, sample preparation, analysis procedures, standard curve information, user-entered programs, recalling and erasing stored data, measurement techniques and many other details. The Instrument Manual provides operational and care information to the user.

The DR/850 Colorimeter Procedures Manual and the DR/890 Colorimeter Procedures Manual provide information for many of the field kit experiments on sample preparation, dilutions, measurement techniques, and additional information. Most of the experiments in Chapter 2 can be performed on the LED-based DR/890 or DR/850 instruments and are included in the field kit. For field work, these instruments are portable, reliable, adequate and easy to operate for most industrial and environmental measurements.

Operation of Pocket Colorimeters™ Analysis Systems

Pocket Colorimeters are fairly accurate instruments. The instruments are used primarily in the field and in industrial plants to perform a number of tests. The instrument operates on four AA alkaline batteries and provides a direct LCD readout in concentration units.

Data Collection and Data Transfer Tools

The TREKKER™ instrument is a portable data collection device which operates on a 9 v AC power adapter or on two AA batteries for field applications. The instrument performs a variety of experiments such as pH, turbidity, conductivity and color measurements. Various probes can measure samples at several different rates and record as much as 190 data points per probe. The TREKKER™ can be connected to a calculator or computer for graphing, printing, and analyzing collected data using a spreadsheet program or available software packages. The equipment is manufactured by TEK-GEAR and is sold by both LaMotte and Ward's Companies. Additional computer software is also available through both sources.

The DR/2010 Spectrophotometer (HACH) has data logging capabilities with a portable printer for field measurements and printouts. Collected data can be transferred into most spreadsheet programs for storage and reporting. The DR/800 Colorimeters store measurements in memory. Field readings can be recalled, displayed or downloaded at a later time. A Data Transfer Adapter (DTA) interfaces with the DR/800 Colorimeters to transfer stored field measurements to be displayed or downloaded into a microcomputer for analyses. This equipment is available through HACH.

DIGESTION PROCEDURES

A number of procedures require sample digestion for reporting purposes or because of sample interferences. Digestion uses chemicals and heat to break down substances into quantities that can be analyzed. For USEPA reporting, USEPA-approved digestions are required. USEPA approves the mild and vigorous digestions for metal analysis. Additional digestion procedures are required for phosphorus and total kjeldahl nitrogen (TKN). The HACH Digesdahl system can be used for the determination of metals, total phosphorus and total kjeldahl nitrogen.

EPA Mild Digestion with Hot Plate for Metal Analysis Only*

1. Acidify the entire sample at the time of collection with concentrated nitric acid by adding 5 mL of acid per liter of sample.
2. Transfer 100 mL of well-mixed sample to a beaker or flask. Add 5 mL of distilled 1:1 hydrochloric (HCL) solution.
3. Heat using a hot plate or steam bath until the volume has been reduced to 15 to 20 mL. Make certain the solution does not boil.
4. After this treatment, the sample may be filtered to remove any insoluble material.
5. Adjust the digested sample to pH 4 by drop-wise addition of 5.0 N Sodium Hydroxide Standard Solution. Mix thoroughly and check the pH after each addition.
6. Quantitatively transfer the sample with demineralized water to a 100-mL volumetric flask and dilute to volume with demineralized water. Continue with the procedure. The reagent blank should also be carried through the digestion and measurement procedures.

EPA Vigorous Digestion with Hot Plate for Metal Analysis Only*

A vigorous digestion is required to ensure all organic-metallic bonds are broken.

1. Acidify the entire sample with redistilled 1:1 Nitric Acid Solution to a pH of less than two. Do not filter the sample before digestion.

* *HACH Water Analysis Handbook,* 3rd Edition, 1997.

2. Transfer an appropriate sample volume (see Table 1.3 below) into a beaker and add 3 mL of concentrated redistilled nitric acid.
3. Place the beaker on a hot plate and evaporate to near dryness, making certain the sample does not boil.
4. Cover the beaker with a watch glass and return it to the hot plate. Increase the temperature of the hot plate so that a gentle reflux occurs. Add additional acid, if necessary, until the digestion is complete (generally indicated when the digestate is light in color or does not change color or appearance with continued refluxing.)
5. Again, evaporate to near dryness (do not bake) and cool the beaker. If any residue or precipitate results from the evaporation, add redistilled 1:1 hydrochloric acid (5 mL per 100 mL of final volume). See Table 1.3.
6. Warm the beaker. Add 5 mL of 5.0 N Sodium Hydroxide and quantitatively transfer the sample with demineralized water to a volumetric flask. See Table 1.3 below for the suggested final volume.
7. Adjust the sample to pH 4 by dropwise addition of 5.0 N Sodium Hydroxide Standard Solution; mix thoroughly and check the pH after each addition. Dilute to volume with demineralized water. Multiply the result by the correction factor in Table 1.3. A reagent blank also should be carried through the digestion and measurement procedures.

Digesdahl Digestion

For details about using the apparatus and safety information, read the Digesdahl Digestion Manual and the HACH Water Analysis Handbook, 3rd Edition, 1997, pp. 88–101. Digestion guidelines of specific sample types and some properties of kjeldahl nitrogen standards are provided in Tables 15 and 16. A pH adjustment for a metal procedure and a Colorimetric Method for TKN is available on page 96–98. A Digesdahl Digestion procedure for aqueous liquids (pp. 102–104), for oils (pp. 112–114) and for solids (pp. 122–124) is provided for review.

DIGESTION PROCEDURE FOR AQUEOUS LIQUIDS

pH Adjustment for Metals Method*

Note: If analyzing aliquots smaller than 0.5 ml, pH adjustment is not necessary.

1. Pipet the appropriate analysis volume into the appropriate mixing graduate cylinder. See Sample and Analysis Volume Tables for Aqueous Liquids following this procedure to

TABLE 1.3 Vigorous Digestion Volumes

Expected metal concentration	Suggested sample Vol. for digestion	Suggested volume of 1:1 HCL	Suggested final volume after digestion	Corrector factor
1 mg/L	50 mL	10 mL	200 mL	4
10 mg/L	5 mL	10 mL	200 mL	40
100 mg/L	1 mL	25 mL	500 mL	500

* *HACH Water Analysis Handbook,* 3rd Edition, 1997.

determine the analysis volume.
Note: Some methods require pipetting into a volumetric flask or a graduate cylinder.

2. Dilute to about 20 mL with deionized water.
3. Add one drop of 2,4-Dinitrophenol Indicator Solution.
4. Add one drop of 8 N Potassium Hydroxide (KOH) Standard Solution, swirling between each addition, until the first flash of yellow appears (pH 3). Do not use a pH meter if analyzing for silver.
5. Add one drop of 1 N KOH. Stopper the cylinder and invert several times to mix.
Note: Use pH paper to insure the pH is 3. If it is higher than 4, do not readjust with acid; start over with a fresh aliquot volume.
6. Continue to add 1 N KOH in this manner until the first permanent yellow color appears (pH 3.5–4.0).
Note: High iron content will cause precipitation (brown cloud) which will coprecipitate other metals. Repeat this procedure with a smaller aliquot volume.
7. Add deionized water to the volume indicated in the colorimetric procedure for the parameter you are analyzing. Fill a second graduated cylinder to the same volume with deionized water.
8. Continue with the colorimetric procedure for the parameter you are analyzing.

pH Adjustment for Total Kjeldahl Nitrogen (TKN), Colorimetric Methods

To complete the TKN analysis, consult the spectrophotometer or colorimeter procedure. Consult the procedure on p. 106 in the HACH Water Analysis Handbook, 3rd Edition, 1997, if a guide is not available.

*SAMPLE AND ANALYSIS VOLUME TABLES FOR AQUEOUS LIQUIDS**

Digestion procedures for beverages (wines, soda, etc,), wastewater (influent, effluent, waste tanks, etc.) and potable water, surface water and ground water are provided on a previous page or in the HACH Water Analysis Handbook.

Aluminum, Aluminon (Method 8012)*

Expected AL conc. (mg/L)	Sample amount (mL)	Analysis volume (mL)	Dilute to
0.1–6	40.0	20.0	50.0 mL
0.5–25	20.0	10.0	50.0 mL
2.0–95	10.0	5.00	50.0 mL
20–950	5.00	1.00	50.0 mL
200–9500	1.00	0.50	50.0 mL

*For Total Aluminum Analysis.

* *HACH Water Analysis Handbook,* 3rd Edition, 1997.

$$\text{Total mg/L Al} = \frac{A \times 5000}{B \times C}$$

A = mg/L reading from the instrument

B = mL (g) sample amount from the table

C = mL analysis volume from table

Cadmium, Dithizone (Method 8017)*

Expected Cd conc. (mg/L)	Sample amount (mL)	Analysis volume (mL)	Dilute to
0.05–5	40.0	20.0	250.0 mL
0.2–20	20.0	10.0	250.0 mL
1–80	10.0	5.00	250.0 mL
10–800	5.00	1.00	250.0 mL
100–8000	1.00	0.500	250.0 mL

*For Total Cadmium Analysis.

$$\text{Total mg/l Cd} = \frac{A \times 25}{B \times C}$$

A = μg/L reading from the instrument

B = mL (g) sample amount from the table

C = mL analysis volume from table

Chromium, Total (Method 8024)*

Expected Cr conc. (mg/L)	Sample amount (mL)	Analysis volume (mL)	Dilute to
0.05–2.0	40.0	20.0	25.0 mL
0.20–8	20.0	10.0	25.0 mL
0.75–33	10.0	5.00	25.0 mL
7.5–330	5.00	1.00	25.0 mL
75–3300	1.00	0.500	25.0 mL

*Digestion required if high amounts of organic material are present.

$$\text{Total mg/L Cr} = \frac{A \times 2500}{B \times C}$$

A = mg/L reading from the instrument

B = mL (g) sample amount from the table

C = mL analysis volume from table

Cobalt (Method 8078)*

Expected Co conc. (mg/L)	Sample amount (mL)	Analysis volume (mL)	Dilute to
0.1–6.0	40.0	20.0	25.0 mL
0.5–25	20.0	10.0	25.0 mL
2.0–100	10.0	5.00	25.0 mL
20–1000	5.00	1.00	25.0 mL
200–10000	1.00	0.500	25.0 mL

*For Total Recoverable Cobalt and if EDTA is present in the sample.

$$\text{Total mg/L Co} = \frac{A \times 2500}{B \times C}$$

A = mg/L reading from the instrument

B = mL (g) sample amount from the table

C = mL analysis volume from table

Copper, Bicinchoninate (Method 8026)*

Expected Co conc. (mg/L)	Sample amount (mL)	Analysis volume (mL)	Dilute to
0.25–15	40.0	20.0	25.0 mL
1–60	20.0	10.0	25.0 mL
4–240	10.0	5.00	25.0 mL
40–2400	5.00	1.00	25.0 mL
400–24000	1.00	0.500	25.0 mL

*For Total Copper Analysis. Digestion required for USEPA accepted reporting for wastewater analysis.

$$\text{Total mg/L Cu} = \frac{A \times 2500}{B \times C}$$

A = mg/L reading from the instrument

B = mL (g) sample amount from the table

C = mL analysis volume from table

Iron, Total (Method 8008)*

Expected Fe conc. (mg/L)	Sample amount (mL)	Analysis volume (mL)	Dilute to
0.15–8	40.0	20.0	25.0 mL
0.6–35	20.0	10.0	25.0 mL
2.5–125	10.0	5.00	25.0 mL
25–1250	5.00	1.00	25.0 mL
250–12500	1.00	0.500	25.0 mL

*For Total Iron Analysis of samples containing iron oxides, iron sulfides, etc. Digestion required for USEPA accepted reporting of wastewater analysis.

$$\text{Total mg/L Fe} = \frac{A \times 2500}{B \times C}$$

A = mg/L reading from the instrument

B = mL (g) sample amount from the table

C = mL analysis volume from table

Lead, Dithizone (Method 8033)*

Expected Pb conc. (mg/L)	Sample amount (mL)	Analysis volume (mL)	Dilute to
0.1–9.0	40.0	20.0	250.0 mL
0.4–35	20.0	10.0	250.0 mL
1.5–140	10.0	5.00	250.0 mL
15–1400	5.00	1.00	250.0 mL
150–14000	1.00	0.500	250.0 mL

*Digestion is required for USEPA reporting. Samples of water, wastewater, paints and other substances containing lead may require digestion.

$$\text{Total mg/L Pb} = \frac{A \times 25}{B \times C}$$

A = μg/L reading from the instrument

B = mL (g) sample amount from the table

C = mL analysis volume from table

Manganese, PAN (Method 8149)*

Expected Mn conc. (mg/L)	Sample amount (mL)	Analysis volume (mL)	Dilute to
0.05–2.2	40.0	20.0	25.0 mL
0.2–10	20.0	10.0	25.0 mL
0.8–35	10.0	5.00	25.0 mL
8–350	5.00	1.00	25.0 mL
80–3500	1.00	0.500	25.0 mL

*For Total Manganese Analysis.

$$\text{Total mg/L Mn} = \frac{A \times 2500}{B \times C}$$

A = mg/L reading from the instrument

B = mL (g) sample amount from the table

C = mL analysis volume from table

Nickel, PAN (Method 8150)*

Expected Ni conc. (mg/L)	Sample amount (mL)	Analysis volume (mL)	Dilute to
0.05–3	40.0	20.0	25.0 mL
0.2–12	20.0	10.0	25.0 mL
0.8–47	10.0	5.00	25.0 mL
8–470	5.00	1.00	25.0 mL
80–4700	1.00	0.500	25.0 mL

*Requires digestion if chelating agents (EDTA) are present.

$$\text{Total mg/L Ni} = \frac{A \times 2500}{B \times C}$$

A = mg/L reading from the instrument

B = mL (g) sample amount from the table

C = mL analysis volume from table

Nitrogen, Total Kjeldahl (Method 8075)**

Expected nitrogen conc. (mg/L)	Sample amount (mL)	Analysis volume (mL)	Dilute to
0.5–35	40.0	20.0*	25.0 mL*
2–140	20.0	10.0*	25.0 mL*
11–700	10.0	5.00*	25.0 mL*
45–2800	5.00	1.00*	25.0 mL*
425–28000	1.00	0.500*	25.0 mL*

*These are guidelines only. See the spectrophotometric procedure manual.
**Nessler Method requires digestion for organically-bound nitrogen.

$$\text{Total mg/L TKN} = \frac{A \times 75}{B \times C}$$

A = mg/L reading from the instrument

B = mL (g) sample amount from the table

C = mL analysis volume from table

Phosphorus, Ascorbic Acid (Method 8048)*

Expected PO_4 conc. (mg/L)	Sample amount (mL)	Analysis volume (mL)	Dilute to
0.12–6	40.0	20.0	25.0 mL
0.5–23	20.0	10.0	25.0 mL
2–90	10.0	5.00	25.0 mL
20–900	5.00	1.00	25.0 mL
200–9000	1.00	0.500	25.0 mL

*Requires Digestion for Total Phosphorus.

$$\text{Total mg/L phosphate} = \frac{A \times 2500}{B \times C}$$

A = mg/L reading from the instrument

B = mL (g) sample amount from the table

C = mL analysis volume from table

Silver (Method 8120)*

Expected Ag conc. (mg/L)	Sample amount (mL)	Analysis volume (mL)	Dilute to
0.08–4	40.0	20.0	50.0 mL
0.3–16	20.0	10.0	50.0 mL
1.0–65	10.0	5.00	50.0 mL
12–650	5.00	1.00	50.0 mL
120–6500	1.00	0.500	50.0 mL

*Requires digestion if the sample contains cyanide, organic matter or thiosulfate.

$$\text{Total mg/L Ag} = \frac{A \times 5000}{B \times C}$$

A = mg/L reading from the instrument

B = mL (g) sample amount from the table

C = mL analysis volume from table

Zinc (Method 8009)*

Expected Zn conc. (mg/L)	Sample amount (mL)	Analysis volume (mL)	Dilute to
0.2–15	40.0	20.0	50.0 mL
0.8–60	20.0	10.0	50.0 mL
3.0–250	10.0	5.00	50.0 mL
30–2500	5.00	1.00	50.0 mL
300–25000	1.00	0.500	50.0 mL

*Digestion required for reporting of USEPA approved wastewater analysis.

$$\text{Total mg/L Zn} = \frac{A \times 5000}{B \times C}$$

A = mg/L reading from the instrument

B = mL (g) sample amount from the table

C = mL analysis volume from table

DIGESTION GUIDELINES OF SPECIFIC SAMPLE TYPES*

Sample type	Sample weight	Vol. of acid	Preheat time (Step 5)	Vol. of peroxide	Special instructions
Plant Tissue	0.25 to 0.5 g	4 mL	4 min.	10 mL	Use Nitrogen-free paper to weigh samples.
Meat & Poultry	0.5 g or predigestion	4 mL or as in predigest	4 min.	10 mL	—
Fluid Fertilizers	0.1 to 0.25 g	4 mL	4 min.	10 mL	Add 0.4 g Kjeldahl Reduction Powder to flask before adding sulfuric acid. Place the flask in 80°C oven 15 min. before digestion. Use N-free paper to weigh samples.
Feed & Forage	0.25 g	4 mL	4 min.	10 mL	—
Dairy	0.25 g to 2.0 g	4 mL	4 min.	10 mL	—
Cereal	0.25 g to 0.5 g	4 mL	4 min.	10 mL	Use Nitrogen-free paper to weigh samples.
Beverage	about 5 g (pipet into funnel)	4 mL	1 min.	10 mL	Preheat acid for 1 min. then add sample through funnel. Heat flask for 30 sec. after sample is in the flask.
Sludge	<2.5 g wet sludge <0.5 g dried sludge	4 mL	3–5 min.	10 mL or increase in 5 mL increments	Heat the diluted digest for 15 min. and filter.
Water & Wastewater	Not more than 0.5 g solid (mL = 40/C; C = % solids)	3 mL	until acid is refluxing	10 mL or increase in 5 mL increments	Water must evaporate before acid will reflux. **Boiling chips required.**
Bath Solutions	0.3 to 10 mL	4 mL	4 min.	10 mL	Water must evaporate before acid will reflux. **Boiling chips required.**
Edible Oils	0.25 to 5.0 g	4–6 mL	4 min.	5 mL immediately and 5 mL later	Weigh samples into flask and record exact weight.
Ion Exchange Resins	equivalent of 0.25 g dry resin	10–15 mL	12 min.	20 mL	Digest will be clear with particles on bottom if metal oxides are not soluble in H_2SO_4. Add aqua regia or suitable solvent to dissolve particles. If particles are floating, start again using 15 mL H_2SO_4 and longer char time.

Sample type	Sample weight	Vol. of acid	Preheat time (Step 5)	Vol. of peroxide	Special instructions
Soil	0.25 to 0.5 g	6 mL	4 min.	10–20 mL	—
Fuels	0.25 to 0.5 g	6 mL	4 min.	20 mL	Heat the diluted digest for 15 min. and filter. Lower heater temperature if foaming or burning occurs.

*Compliments of HACH Company.

GUIDELINES TO MINIMIZE TOXIC EXPOSURE

An analyst can minimize exposure of toxic substances by following some general guidelines:*

1. Know the chemical and toxic properties of all materials involved by reviewing the Material Safety Data Sheets MSDS).
2. Substitute a less toxic substance whenever possible. For example, rinse glassware with hexane instead of benzene or other more toxic substances.
3. When indoors, always use a fume hood and periodically test its efficiency. When testing samples outdoors, perform tests with the wind to your back if possible.
4. Know that toxic exposure can occur through respiration, ingestion, or through the skin. Always use personal protective equipment (gloves, masks, spatulas, etc.)
5. *Never* taste any chemicals and *avoid* smelling chemicals directly. If you have to smell a chemical, position a test tube containing the chemical about 8 (or more) inches from your nose and waft your hand over the container so that you smell a little bit of the vapors. Minimize breathing of chemical vapors.
6. Avoid excessive exposure to all chemicals, even those considered not to be high on any toxicity listing.
7. Do not drink alcoholic beverages on the job. This practice is never permissible for obvious reasons. Furthermore, ethanol (the active ingredient in alcoholic beverages) has a synergistic effect with many other solvents and should be avoided.
8. Routinely monitor the laboratory atmosphere for specific contaminants and their concentration levels. This can be done quickly with a gas analyzer. A gas analyzer manually draws air through some specific colorimetric reagents contained in a calibrated glass tube. Another way to monitor the laboratory for contaminants is to simply wear a *vapor monitor badge*.

Toxic Exposure Levels

A number of substances are potentially harmful in the workplace and in the environment. Enclosed structures containing gaseous vapors often present life threatening problems. Fires,

*Compliments of Dr. Gershon Shugar, Chemical Technicians' Ready Reference Handbook.

explosions, leaks, and spills provide the greatest opportunity to exposure to air contaminates and chemical contact.

A number of countries have few, if any, standards for air contaminates. The workplace should be monitored to ensure long-term and short-term maximum exposure limits are not exceeded. Therefore, systematic air analysis is necessary as a preventive measure.

Germany has established MAK-values for **MAK Peak Exposure Limitation Categories***. In addition, TRGS 900 standards have been developed for the carcinogenic potential of a substance. According to TRGS 900, TRK-values are provided as shift mean values for an 8-hour workday and a 40-hour workweek.

Great Britain enforces two types of Occupational Exposure Limit values: the **Occupational Exposure Standard (OES)** and the **Maximum Exposure Limit (MEL)***. Both types usually have a long-term Time Weighted Average (TWA) and a Short-Term Exposure Limit (STEL). The OES time-weighted average values and short-term exposure limit vales are very similar to the ACGIH Threshold Limit Values (TLV) of the USA.

The American Conference of Governmental Industrial Hygienists (ACGIH) has established **Threshold Limit Values (TLVs).*** TLVs establish guidelines for airborne concentrations of substances and represent conditions under which it is believed that nearly all workers may be repeatedly exposed day after day without adverse health effects.* The list below includes a variety of definitions including the three categories of Threshold Limit Values (TLVs) with the two categories of carcinogens.

Definitions*

***BEI*—**Biological Exposure Indices

***TLV*—**Threshold Limit Value

***Threshold Limit Value-time-weighted average (TLV-TWA)*—**The time-weighted average concentration for a normal 8-hour workday or 40-hour workweek, to which nearly all workers may be repeatedly exposed, day after day, without adverse effects.

***Threshold Limit Value-short-term exposure (TLV-STEL)*—**The maximum concentration to which workers can be exposed for a period of up to 15 minutes continuously without suffering from (1) irritation, (2) chronic or irreversible tissue change, (3) narcosis of sufficient degree to increase the likelihood of accidental injury, impair self-rescue or materially reduce work efficiency, and provided that the daily TLV-TWA is not exceeded. The STEL should not exceed 15 minutes and should not occur more than four times per day, with at least 60 minutes between successive exposures.

Threshold Limit Value-Ceiling (TLV-C) or **Ceiling Limits (C)—**a definite boundary that concentrations should not be permitted to exceed. The concentration that should not be exceeded even instantaneously. The definitions of two categories are the following:
A 1: confirmed human carcinogen
A 2: suspected human carcinogen

***Permissible Exposure Level (PEL)*—**The highest allowable concentration of a substance to which an employee may be exposed over an 8-hour time-weighted average.

***Immediately dangerous to life to health (IDLH)*—**An atmosphere that poses an immediate hazard to life or produces immediate irreversible health effects.

***Oxygen-deficient atmosphere*—**An atmosphere of less than 19.5% oxygen by volume measured at sea level.

*Compliments of the Dräger-Tube Handbook, 11th Edition, 1998.

Time-Weighted Average (TWA)—The calculation of average concentration of exposure to a substance during a workweek rather than a workday.

GUIDE TO CHEMICAL HAZARDS

The **American Conference of Governmental Industrial Hygienists (ACGIH)** provides a comprehensive listing of potentially hazardous chemical substances and physical agents in the work environment.* These *Threshold Limit Values* (TLVs) refer to airborne concentrations of substances and conditions under which nearly all workers may be repeatedly exposed without adverse effects. The TLV for a gaseous contaminant is reported in parts per million (ppm). Particulate matter is normally reported in milligrams per cubic meter (mg/m^3).

The table below is limited by listing potentially hazardous substances with TWA and STEL/C values that are normally recorded below 50 ppm and 10 mg/m^3. Also, chemicals that provide irritation below a concentration level of 50 ppm are included in a second listing of chemicals titled "**Substance Irritation.**" In many cases, chemicals in the water and/or air will be present at relatively low concentrations. Where dangerous concentrations of chemicals exist, the person performing the experiments (collection of sample, testing procedures, waste disposal) must follow appropriate procedures.

If the investigator is in doubt about the danger of a specific substance, check the most recent edition of the **American Conference of Governmental Industrial Hygienists (ACGIH)** in the TLV/BEI booklet. The updated edition and complete listing of the **Threshold Limit Values for Chemical Substances in the Work Environment** have been published recently.

TWA AND STEL VALUES*

Substance	TWA ppm or mg/m^3	STEL/C ppm or mg/m^3	General effects**
Acrolein	—	C 0.1 ppm	Irritation, pulmonary edema
Acrylic Acid	2 ppm	—	Irritation, reproductive
Adipic Acid	5 mg/m^3	—	Irritation, neurotoxicity, GI†
Aniline	2 ppm	—	Anoxia
Arsenic (metal & compounds)	0.01 mg/m^3	—	Cancer (lung & skin)
Benzo[b]fluoranthene	—	—	Cancer (suspected)
Benzo[a]pyrene	—	—	Cancer (suspected)
Cadmium (metal & compounds)	0.01–0.002 mg/m^3	—	Cancer (suspected)
Chromate compounds (Various compounds)	0.001–0.01 mg/m^3 Cr & above	—	Cancer (suspected), cancer (lung), reproductive, CVS†, etc.
Carbon Disulfide	10 ppm	—	CVS, CNS†, neuropathy
Carbon Monoxide	25 ppm	—	Anoxia, CVS, CNS, reproductive
Carbon Tetrachloride	5 ppm	10 ppm	Liver, cancer
Chlorine	0.5 ppm	1 ppm	Irritation, potential death
Chlorine Dioxide	0.1 ppm	0.3 ppm	Irritation, bronchitis, potential death
Chlorobenzene	10 ppm	—	Liver
Chloroform (Trichloromethane)	10 ppm	—	CVS, liver, kidney, CNS, reproductive
Chromium (VI) (metal & compounds)	0.01–0.05 mg/m^3	—	Cancer, liver, kidney
Cobalt (metal and compounds)	0.02 mg/m^3	—	Asthma, lung, CVS
Cyanogen Chloride	—	C 0.3 ppm	Irritation, pulmonary function

Substance	TWA ppm or mg/m³	STEL/C ppm or mg/m³	General effects**
DDT	1 mg/m³	—	Seizures, liver
Dichloromethane (Methylene Chloride)	50 ppm	—	CNS, anoxia
Dimethylformamide or DMF	10 ppm	—	Liver
Epichlorohydrin	0.5 ppm	—	Irritation, liver, kidney
Ethylene	—	—	Asphyxiation
Ethylene Oxide	1 ppm	—	Lung, liver, kidney, blood, CNS, cancer (suspected)
Fluorides (anionic)	2.5 mg/m³	—	Irritation, bone, fluorosis
Formamide	10 ppm	—	Irritation, liver
Hydrazine	0.01 ppm	—	Irritation, liver
Hydrogen	—	—	Asphyxiation
Hydrogen Chloride	—	C 5 ppm	Irritation, corrosion
Hydrogen Cyanide and cyanide salts	—	C 4.7 ppm & C 5 mg/m³	CNS, irritation, anoxia, lung, thyroid
Hydrogen Fluoride	—	C 3 ppm	Irritation, burns, bone, teeth, fluorosis
Hydrogen Peroxide	1 ppm	—	Irritation, pulmonary edema, CNS
Hydrogen Sulfide	(10 ppm)	(15 ppm)	Sudden death, irritation, CNS
Hydroquinone	2 mg/m³	—	CNS, dermatitis, ocular
Iodoform	0.6 ppm	—	CNS, liver, kidney, CVS
Lead (metal and compounds)	0.05 mg/m³	—	CNS, GI, blood, kidney, reproductive
Manganese (metal and compounds)	0.2 mg/m³	—	lung, reproductive, CNS
Mercury (metal and compounds)	0.01 to 0.1 mg/m³	(0.03 mg/m³)	CNS, kidney, vision, neuropathy, etc.
Methyl Bromide	1 ppm	—	CNS, neurotoxicity, pulmonary edema
Methyl Iodide	2 ppm	—	CNS, irritation
Nickel (insoluble compounds)	0.2 mg/m³	—	Lung cancer, irritation, dermatitis
Nickel (metal & soluble compounds)	0.1–1.5 mg/m³	—	Dermatitis, kidney, CNS, etc.
Nitric Acid	2 ppm	4 ppm	Irritation, corrosion, pulmonary edema
Nitric Oxide	25 ppm	—	Anoxia, irritation, cyanosis
Nitrogen Dioxide	3 ppm	—	Irritation, pulmonary edema
Nitrous Oxide	50 ppm	—	Blood, neuropathy, reproductive, etc.
Oxalic Acid	1 mg/m³	2 mg/m³	Irritation, burns
Ozone (workload & time dependent)	0.05–0.2 ppm	—	pulmonary function, headache, etc.
Perchloroethylene/ Tetrachloroethylene	25 ppm	100 ppm	Irritation, CNS
Persulfates (Group IA & Ammonium)	0.1 mg/m³	—	Irritation, dermatitis
Phenol	5 ppm	—	Irritation, CNS, blood
Phosgene	0.1 ppm	—	Irritation, pulmonary edema, anoxia
Phosphine	0.3 ppm	1 ppm	Irritation, CNS, GI
Polychlorobiphenyls (42%-54% Cl)	0.5–1 mg/m³	—	Irritation, chloracne, liver
Potassium Hydroxide	—	C 2 mg/m³	Irritation, corrosion
Pyridine	5 ppm	—	Irritation, CNS, liver, kidney, blood
Resorcinol	10 ppm	20 ppm	Irritation, dermatitis, blood
Silver (metal & soluble compounds)	0.01–0.1 mg/m³	—	Argyria (skin, eyes, etc.)
Sodium Azide	—	C 0.29 mg/m³	CNS, CVS, lung
Styrene, monomer	20 ppm	40 ppm	Neurotoxicity, irritation, CNS
Sulfuric Acid	1 mg/m³	3 mg/m³	Irritation, cancer (larynx)
Sulfuryl Fluoride	5 ppm	10 ppm	Irritation, CNS
Terephthalic Acid	10 mg/m³	—	Lung
1,1,2,2-Tetrachloroethane	1 ppm	—	Liver, CNS, GI

Substance	TWA ppm or mg/m^3	STEL/C ppm or mg/m^3	General effects**
Tin (metal, inorganic) (organic	2 mg/m^3	—	Stannosis
compounds)	0.1 mg/m^3	0.2 mg/m^3	CNS, immunotoxicity, irritation
Toluene	50 ppm	—	CNS
Toluene-2,4-diisocyanate	0.005 ppm	0.02 ppm	Irritation, sensitization
Toluidine (ortho, meta, para)	2 ppm	—	Anoxia, kidney
1,1,2- Trichloroethane	10 ppm	—	CNS, liver
Trichloroethylene	50 ppm	100 ppm	CNS, headache, liver
Vinyl Chloride	1 ppm	—	CNS, liver, cancer, CVS, etc.
Xylidine (mixed isomers)	0.5 ppm	—	Anoxia, liver, kidney

* The list of chemicals may not include all chemicals to be analyzed or all of the chemicals used in the experiments.
** Items in this column have been expanded. Some gases can cause death at higher levels of human exposure.
† Definitions: CNS–central nervous system; CVS–cardiovascular system; and GI–gastrointestinal.

Substance Irritation

The following substances are irritants at concentrations below 50 ppm or below 10 mg/m^3: Acetaldehyde, Acetic Acid, Acetic Anhydride, Aluminum Salts, Ammonia, Bromine, Chromium (III), Cyclohexylamine, Dimethyl Sulfate, Fluorine, Formic Acid, Iodine, Iron Salts (soluble), Methyl Acrylate, Phosphoric Acid, Propionic Acid, Selenium, Sodium Hydroxide, Sodium Bisulfite, Sulfur Dioxide, Sulfur Fluoride Compounds, Thionyl Chloride, and Triethanolamine. This list of chemicals may not include all chemicals to be analyzed or all of the chemicals used in the experiments.

HARDNESS FACTORS

The table below lists the factors for converting one unit of measure for hardness to another unit of measure. For example, to convert mg/L $CaCO_3$ to German hardness parts/100,000 CaO, multiply the HACH experimental value in mg/L by 0.056.

Hardness Conversion Factors*

Units of measure	mg/L $CaCO_3$	British gr/gal (Imperial) $CaCO_3$	American gr/gal (US) $CaCO_3$	French parts/ 100,000 $CaCO_3$	German parts/ 100,000 CaO	meq/L[1]	g/L CaO	lbs./cu ft $CaCO_3$
mg/L $CaCO_3$	1.0	0.07	0.058	0.1	0.056	0.02	5.6×10^{-4}	6.23×10^{-5}
English gr/gal $CaCO_3$	14.3	1.0	0.83	1.43	0.83	0.286	8.0×10^{-3}	8.9×10^{-4}

* Compliments of HACH Company.

Units of measure	mg/L $CaCO_3$	British gr/gal (Imperial) $CaCO_3$	American gr/gal (US) $CaCO_3$	French parts/ 100,000 $CaCO_3$	German parts/ 100,000 CaO	meq/L[1]	g/L CaO	lbs./cu ft $CaCo_3$
US gr/gal $CaCO_3$	17.1	1.2	1.0	1.72	0.96	0.343	9.66×10^{-3}	1.07×10^{-3}
Fr. p/ 100,000 $CaCO_3$	10.0	0.7	0.58	1.0	0.56	0.2	5.6×10^{-3}	6.23×10^{-4}
Ger. p/ 100,000 CaO	17.0	1.25	1.04	1.79	1.0	0.358	1×10^{-2}	1.12×10^{-3}
meq/L	50.0	3.5	2.9	5.0	2.8	1.0	2.8×10^{-2}	3.11×10^{-2}
g/L CaO	1790.0	125.0	104.2	179.0	100.0	35.8	1.0	0.112
lbs./cu ft $CaCO_3$	16,100.0	1,123.0	935.0	1,610.0	900.0	321.0	9.0	1.0

[1] "epm/L," or "mval/L." Note: 1 meq/L = 1 N/1000

WASTE MANAGEMENT PRACTICES

Waste management regulations change regularly and additional state and local laws may apply to your waste. Each waste generator is responsible for knowing and obeying the laws governing waste management. Generally, the USEPA, state agencies (DER and DEP), and local boards have regulations which control hazardous waste management practices.

Waste Minimization

Each person performing field tests must be responsible for used chemicals and throw away items. Every attempt should be made to minimize the waste stream and recover the waste stream rather than discarding it carelessly.

Hazardous Waste Regulations and Requirements

Title 40 Code of Federal Regulations (CFR) part 260 provides the Federal waste disposal regulations with emphasis on controlling hazardous waste disposal. The waste regulations depend on the amount of hazardous waste generated in one month. The USEPA, under 40 CFR 261.20–261.33, has defined hazardous wastes by a set of codes such as DOO1 (Ignitability), DOO2 (Corrosivity), DOO3 (Reactivity), DOO4 (Barium) through DO11 (Silver). However, many very toxic compounds are not regulated by the hazardous waste definition. These compounds are regulated by CERCLA (Superfund) regulations or common law tortes.

Hazardous wastes must be managed and disposed of according to federal, state and local regulations. The waste generator is responsible for making the determination whether a waste is or is not a hazardous waste. For additional information, contact the USEPA or your State Department of Environmental Resources (Protection). The HACH Water Analysis Handbook, 3rd Edition, 1997, pp. 139–150 discusses a number of issues related to hazardous waste.

Material Safety Data Sheets

Material Data Safety Sheets must be available to the laboratory technicians and any person performing the tests. The sheets should be readily available in the field kits or in an accessible file with the field instruments. The MSDS will provide the important data on product manufacturer and phone numbers, as well as product or chemical name(s), CAS number, information on exposure limits (TLV, PEL, Hazards), physical properties, data on flash point and flammable limits (Fire, Explosion Hazard and Reactivity Data), health hazards, first aid, spill and disposal procedures, protective equipment and additional details. Before performing the field tests or laboratory procedures, the analyst must be familiar with the precautions necessary to perform the experiment safely. Each person using the kits or performing any experiment should read the MSDS included in the kits or stored in the laboratory.

*INDEX TO TEST METHODS**

The "Index to EPA Test Methods" may be obtained by internet address:

http://www.epa.gov/epa home/index/

The procedures listed in the "Index to EPA Test Methods" are approved by EPA for use in the field and in the laboratory. The programs below are hyperlinked to "Netscape" or "Explorer" for reference. The programs include the following:

Introduction to Index (FAQ section)
How to Interpret Information in the Columns
Chemical or Name Index: A-D, E-M, N-Se, or Sh-Z
Sources of EPA Test Methods
Key to Obtaining Sources & Vendor Contact Information
Related Sources of Methods Information
Electronic Sources of Methods
Links to Sources of Test Methods on the Internet

The Index to EPA Test Methods, 11/98 Edition, is also available in Portable Document Format (PDF). For further information on changes and updates of material contact the following:

EPA Region 1 Library
JFK Federal Building
Boston, MA 02203
Fax: 617-565-9067

*Compliments of Mary Ann Detlefsen, Ph. D., Environmental Chemist, Latham, N.Y.

COMPARISON OF INTERNATIONAL DRINKING WATER AND FDA BOTTLED WATER GUIDELINES*

Parameter[1]	USEPA[2] maximum contamination level (MCL)	Canada[3] maximum acceptable concentration	EEC[4] maximum admissable concentration	Japan[5] maximum admissable concentration	WHO[6] guideline	Bottled water U.S. Federal Drug Administration level
Aluminum	0.05-0.2 mg/L[7]		0.2 mg/L	0.2 mg/L	0.2 mg/L	
Ammonium			0.5 mg/L	No standard	1.5 mg/L	
Antimony	0.006 mg/L		0.01 mg/L	0.002 mg/L[8]	0.005 mg/L	
Arsenic	0.05 mg/L	0.025 mg/L	0.05 mg/L	0.01 mg/L	0.01 mg/L	0.05 mg/L
Barium	2.0 mg/L	1.0 mg/L	No standard	No standard	0.7 mg/L	2.0 mg/L
Boron		5.0 mg/L	1.0 mg/L	0.2 mg/L[8]	0.3 mg/L	
Cadmium	0.005 mg/L	0.005 mg/L	0.005 mg/L	0.01 mg/L	0.003 mg/L	0.005 mg/L
Chloride	250 mg/L[7]	250 mg/L	250 mg/L	200 mg/L	250 mg/L	
Chromium	0.1 mg/L	0.05 mg/L	0.05 mg/L	0.05 mg/L	0.05 mg/L	0.1 mg/L
Coliforms, total Organisms/100 mL	≤ % positive	0	0 or MPN ≤ 1	0	0	≤ 1 MF
Coliforms (*E. coli*) Organisms/100 mL	0	0	0	0	0	
Color	15 cu[7]	15 cu	20 mg Pt-Co/L	5 cu	15 cu	< 15 cu
Copper	1.3 mg/L[7]	1.0 mg/L	2.0 mg/L	1.0 mg/L	1–2 mg/L	1.0 mg/L
Cyanides	0.2 mg/L	0.1 mg/L	0.05 mg/L	0.01 mg/L	0.07 mg/L	
Fluoride	2.0-4.0 mg/L[7]	1.5 mg/L	0.7-1.5 mg/L	0.8 mg/L	1.5 mg/L	
Hardness			50 mg/L	300 mg/L		
Iron	0.3 mg/L[7]	0.3 mg/L	0.2 mg/L	0.3 mg/L	0.3 mg/L	
Lead	0.015 mg/L	0.01 mg/L	0.01 mg/L	0.05 mg/L	0.01 mg/L	0.005 mg/L
Manganese	0.05 mg/L	0.05 mg/L	0.05 mg/L	0.01–0.05 mg/L	0.1–0.5 mg/L	
Mercury	0.002 mg/L	0.001 mg/L	0.001 mg/L	0.0005 mg/L	0.001 mg/L	0.002 mg/L
Molybdenum				0.07 mg/L	0.07 mg/L	

COMPARISON OF INTERNATIONAL DRINKING WATER AND FDA BOTTLED WATER GUIDELINES* (*Continued*)

Parameter[1]	USEPA[2] maximum contamination level (MCL)	Canada[3] maximum acceptable concentration	EEC[4] maximum admissable concentration	Japan[5] maximum admissable concentration	WHO[6] guideline	Bottled water U.S. Federal Drug Administration level
Nickel	0.1 mg/L		0.02 mg/L	0.01 mg/L[8]	0.02 mg/L	
Nitrate/Nitrite, total	10.0 mg/L as N			10.0 mg/L as N		10 mg/L as N
Nitrates	10.0 mg/L as N	10.0 mg/L as N	50 mg/L	10.0 mg/L as N	50 mg/L as NO_3^-	
Nitrites	1 mg/L as N	3.2 mg/L	0.1 mg/L	10 mg/L	3 mg/L as NO_2^-	1 mg/L as N
Odor	3 TON[9]		2 dilution no. @ 12 °C; 3 dilution no. @ 25 °C.	3 TON[9]		
pH	6.5–8.5	6.5–8.5	6.5–9.5	5.8–8.6	6.5–8.5	
Phosphorus			5 mg/L	No standard		
Phenols		0.002 mg/L	0.5 μg/L C_6H_5OH	0.005 mg/L		
Potassium			12 mg/L	No standard		
Selenium	0.05 mg/L	0.01 mg/L	0.01 mg/L	0.01 mg/L	0.01 mg/L	0.05 mg/L
Silica Dioxide			10 mg/L	No standard		
Silver	0.1 mg/L[7]	0.05 mg/L	0.01 mg/L	No standard	No standard	
Solids, total dissolved	500 mg/L[7]	500 mg/L	No standard	500 mg/L	1000 mg/L	
Sodium			75–150 mg/L	200 mg/L	200 mg/L	
Sulfate	250 mg/L[7]	500 mg/L	250 mg/L	No standard	250 mg/L	
Turbidity	0.5–5 NTU	1 NTU	4 JTU	1–2 units	5 NTU	
Zinc	5 mg/L[7]	5.0 mg/L	No standard	1.0 mg/L	3.0 mg/L	

*Compliments of HACH Company.
[1] To our knowledge, data in this table were accurate and current at the publication date. Contact the regulatory agency in your area for the most current information.
[2] United States Environmental Protection Agency
[3] These limits are established by Health Canada
[4] In the EEC (European Economic Community), these limits are set by the European Committee for Environmental Legislation.
[5] In Japan, these limits are established by the Ministry of Health and Welfare.
[6] World Health Organization
[7] U.S. Secondary MCL
[8] Identified as a parameter to be regulated in the future
[9] Threshold Odor Number

CHAPTER 2
WATER ANALYSES

ACIDITY

The measurement of acidity plays an important role in determining the quality of a water source. Acidity can be caused by natural substances or industrial pollution. Acids can be significantly corrosive and influence biological development, decay and decomposition processes, and chemical reaction rates. Federal and State regulations require that different types of acid used by industry must be neutralized before being dumped into lakes, streams and oceans. Additional acid wastes are manifested for burial purposes or for recovery by other industries.

Acidity is a measure of the ionized and the unionized forms of acids. Acidity is attributed to both strong acids, such as hydrochloric acid, sulfuric acid and nitric acid, and weak acids including acetic acid, tannic acid, hydrogen sulfide and carbonic acid (H_2CO_3). Carbonic acid is formed by carbon dioxide dissolving in water and is the most common source of acidity in water. In addition, salts of specific metals such as aluminum tend to hydrolyze and contribute to the acidity of aqueous solutions.

Total acidity includes acidity attributed to mineral acids, weak organic acids, hydrolyzed metals, and carbon dioxide in water. In most natural water bodies, the phenolphthalein acidity is equal to the total acidity due to the fact that carbon dioxide is usually the primary cause of acidity in natural waters. The neutralization of carbonic acid to bicarbonate is determined by titrating a sample of water to a phenolphthalein end point at pH of 8.3. Acidity can be determined by kit or in the laboratory.

Before testing for acidity, samples of industrial wastes, acid mine drainage or other solutions that contain appreciable amounts of hydrolyzable metal ions such as iron, aluminum or manganese are treated with hydrogen peroxide and boiled to ensure complete oxidation of the metals and to hasten hydrolysis of the metals. Waste samples may be subject to microbial action and to loss or gain of CO_2 or other gases when exposed to air. Thus, acidity of a sample can change if oxidation, hydrolysis or microbial action occurs over a period of time in a sample.

In this experiment, there are two major problem areas. Dissolved gases contributing to acidity, such as CO_2 and H_2S, may be lost or gained during sampling, storage, or titration. These effects can be minimized by avoiding vigorous shaking or mixing of the sample and protecting the sample from heat, and by immediately titrating the sample to the end point after collecting the sample or promptly after opening the sample container. In addition, residual free available chlorine in the sample may bleach the indicator. This source of interference can be eliminated by adding 1 drop of 0.1 M Sodium Thiosulfate Solution to the solution before testing.

ACID-BASE (1–4000 meq/L)

Method 8200*
Acid Determination with the Digital Titrator

1. Select the sample volume corresponding to the expected acid concentration in milliequivalents (meq)/L or normality (N) from Table 2.1.
Note: See Sampling and Storage following these steps.

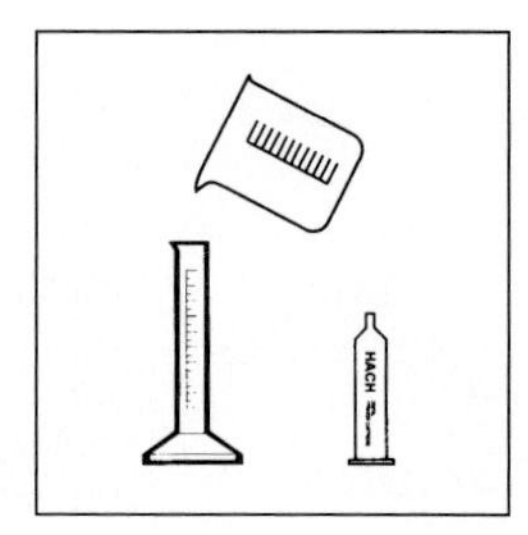

TABLE 2.1

Range meq/L	Range N	Sample Volume (mL)	Titration Cartridge	Catalog Number	Digit Multiplier
1–4	0.001–0.004	100	1.6 N NaOH	14379-01	0.02
			1.6 N H_2SO_4	14389-01	
4–10	0.004–0.01	50	1.6 N NaOH	14379-01	0.04
			1.6 N H_2SO_4	14389-01	
10–40	0.01–0.04	100	8 N NaOH	14381-01	0.1
			8 N H_2SO_4	14391-01	
			8 N HCl	14390-01	
20–80	0.02–0.08	50	8 N NaOH	14381-01	0.2
			8 N H_2SO_4	14391-01	
			8 N HCl	14390-01	
50–200	0.05–0.2	20	8 N NaOH	14381-01	0.5
			8 N H_2SO_4	14391-01	
			8 N HCl	14390-01	
100–400	0.1–0.4	10	8 N NaOH	14381-01	1.0
			8 N H_2SO_4	14391-01	
			8 N HCl	14390-01	
200–800	0.2–0.8	5	8 N NaOH	14381-01	2.0
			8 N H_2SO_4	14391-01	
			8 N HCl	14390-01	5.0
500–2000	0.5–2	2	8 N NaOH	14381-01	
			8 N H_2SO_4	14391-01	
			8 N HCl	14390-01	
1000–4000	1–4	1	8 N NaOH	14381-01	10.0
			8 N H_2SO_4	14391-01	
			8 N HCl	14390-01	

*Compliments of HACH Company.

2. Insert a clean delivery tube into the appropriate Sodium Hydroxide Titration Cartridge. Attach the cartridge to the titrator body.

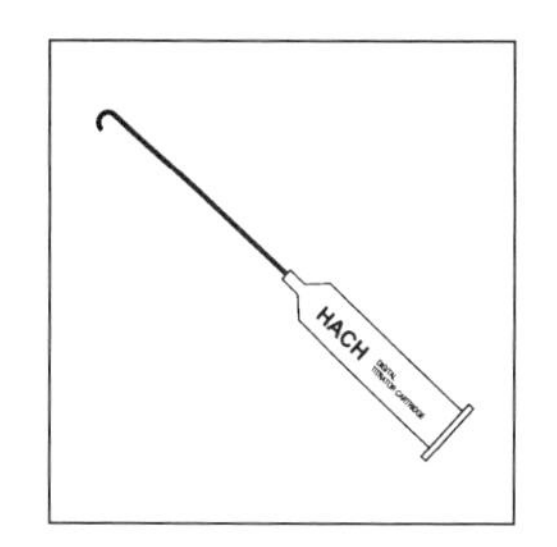

3. Flush the delivery tube by turning the delivery knob to eject a few drops of titrant. Reset the counter to zero and wipe the tip.
Note: For added convenience use the TitraStir stirring apparatus.

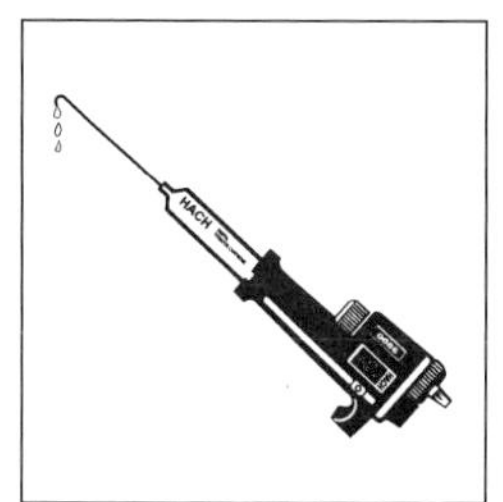

4. Use a graduated cylinder or pipet to measure the sample volume from Table 2.1. Transfer the sample into a clean 250-mL erlenmeyer flask. Dilute to about the 100-mL mark with demineralized water, if necessary.

5. Add the contents of one Phenolphthalein Indicator Powder Pillow and swirl to mix. The solution should be colorless.
Note: Four drops of Phenolphthalein Indicator Solution may be substituted for the Phenolphthalein Indicator Powder Pillow.

6. Place the delivery tube tip into the solution and swirl the flask while titrating with sodium hydroxide until a light pink color forms and persists for 30 seconds. Record the number of digits required.

7. Calculate: Digits Required × Digit Multiplier = Millequivalents per Liter of Acid
Note: To determine the normality of the sample, divide the millequivalents per liter obtained by 1000.

Digits Required x Digit Multiplier
= Milliequivalents/L of Acid

SAMPLING AND STORAGE

Collect samples in clean plastic or glass bottles. Fill completely and cap tightly. Minimize agitation or prolonged exposure to air. Sample may be stored at least 24 hours by cooling to 4°C (39°F) or below if they cannot be analyzed immediately. Warm to room temperature before analyzing.

ACCURACY CHECK

See p. 156 of the HACH Water Analysis Handbook, 3rd Ed., 1997.

INTERFERENCES

Highly colored or turbid samples may mask the color change at the end point. Use a pH meter for these samples.

ACIDITY

Method 8202*
Phenolphthalein (total) Acidity

1. Measure a second portion of the sample selected from Table 2.2 into a clean 250-mL erlenmeyer flask. Dilute to about the 100-mL mark with demineralized water, if necessary.

*Compliments of HACH Company.

TABLE 2.2

Range (mg/L as $CaCO_3$)	Sample Volume (mL)	Titration Cartridge (N NaOH)	Catalog Number	Digit Multiplier
10–40	100	0.1600	14377-01	0.1
40–160	25	0.1600	14377-01	0.4
100–400	100	1.600	14379-01	1.0
200–800	50	1.600	14379-01	2.0
500–2000	20	1.600	14379-01	5.0
1000–4000	10	1.600	14379-01	10.0

2. Add the contents of one Phenolphthalein Indicator Powder Pillow and swirl to mix.
Note: Four drops of Phenolphthalein Indicator Solution may be substituted for the Phenolphthalein Indicator Powder Pillow.

3. Titrate with sodium hydroxide from colorless to a light pink color that persists for 30 seconds. Record the number of digits required.
Note: A solution of one pH 8.3 Buffer Powder Pillow and one Phenolphthalein Powder Pillow in 50 mL of demineralized water is recommended as a comparison for determining the proper end-point color.

4. Calculate: Digits Required × Digit Multiplier = mg/L as $CaCO_3$ Phenolphthalein Acidity.

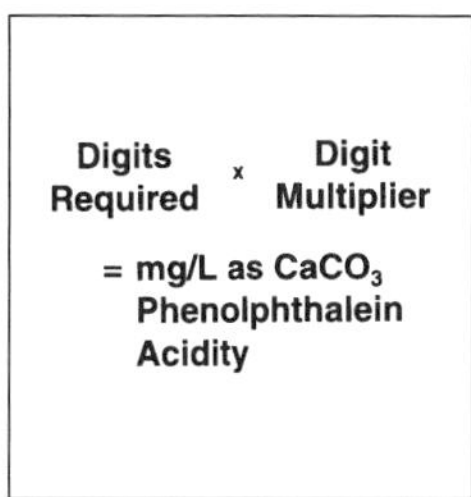

SAMPLING AND STORAGE

Collect samples in clean plastic or glass bottles. Fill completely and cap tightly. Minimize agitation or prolonged exposure to air. Sample may be stored at least 24 hours by cooling to 4°C (39°F) or below if they cannot be analyzed immediately. Warm to room temperature before analyzing.

ACCURACY CHECK

See p. 167 of the HACH Water Analysis Handbook, 3rd Ed., 1997.

INTERFERENCES

- Highly colored or turbid samples may mask the color change at the end point. Use a pH meter for these samples.
- Chlorine may interfere with the indicators. Add one drop of 0.1 N Sodium Thiosulfate to eliminate this effect.
- To determine the phenolphthalein acidity of samples containing hydrolyzable metals such as iron, manganese or aluminum, use the following procedure:
  a) Adjust the sample in Step 1 for phenolphthalein acidity to pH 4.0 or less (if necessary) by using the Digital Titrator with a titration cartridge identical to the Sodium Hydroxide Titration Cartridge used. Record the number of digits of acid added to lower the pH.
  b) Add 5 drops of 30% Hydrogen Peroxide Solution and boil the solution for 2 to 5 minutes.
  c) Cool to room temperature. Titrate following the Phenolphthalein Procedure Step 2 and 3 above. Subtract the number of digits of acid added to lower the pH from the number of digits required in Step 3 of the Phenolphthalein Procedure. Continue with Step 4.

ACIDITY, PHENOLPHTHALEIN

Method 8010 for water, wastewater and seawater*
Using Sodium Hydroxide with a Buret**
USEPA Accepted

1. Select a sample volume corresponding to the expected acidity concentration in mg/L as calcium carbonate ($CaCO_3$) from Table 2.3.

See Table 2.3

TABLE 2.3

Range (mg/L as $CaCO_3$)	Sample Volume (mL)	Standard Titrant Solution (N)	HACH Catalog Number	Multiplier
0–1000	50	0.020	193-53	20
800–2000	25	0.020	193-53	40
2000–5000	10	0.020	193-53	100
4000–10,000	5	0.020	193-53	200

* Compliments of HACH Company.
** Adapted from *Standard Methods for the Examination of Water and Wastewater, 2320 B.*

2. Use a graduated cylinder or pipet to measure the sample volume.
Note: If samples cannot be analyzed immediately, see Sampling and Storage following these steps.

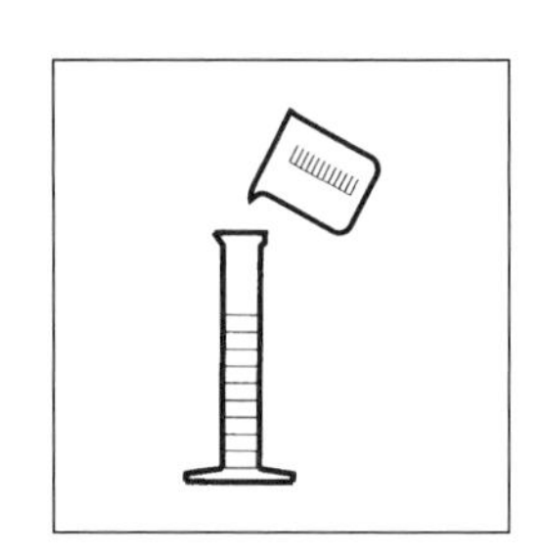

3. Transfer the sample into a 250-mL erlenmeyer flask. Dilute to about 50 mL with deionized water if necessary.
Note: Excessive agitation of the sample should be avoided to prevent loss of dissolved gases such as carbon dioxide, hydrogen sulfide or ammonia.

4. Add the contents of one Phenolphthalein Indicator Powder Pillow. Swirl to mix.
Note: Omit this step if a pH meter is used.
Note: Six drops of Phenolphthalein Indicator Solution may be substituted for the Phenolphthalein Indicator Powder Pillow.

5. Fill a 50 mL buret to the zero mark with 0.020 N Sodium Hydroxide Standard Solution.

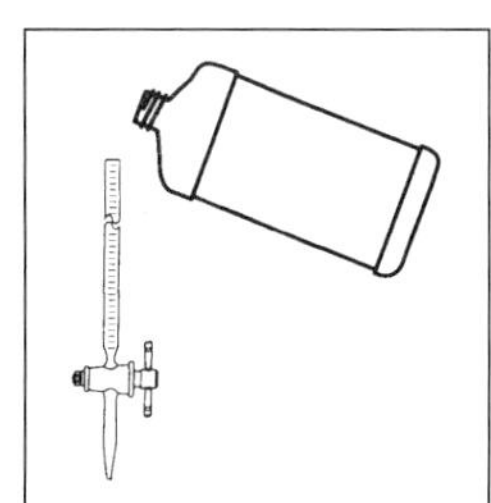

6. While swirling the flask, titrate the sample with 0.020 N Sodium Hydroxide Standard Solution until the solution color changes from colorless to a light pink that persists for 30 seconds.
Note: The pH meter end point is pH 8.3.

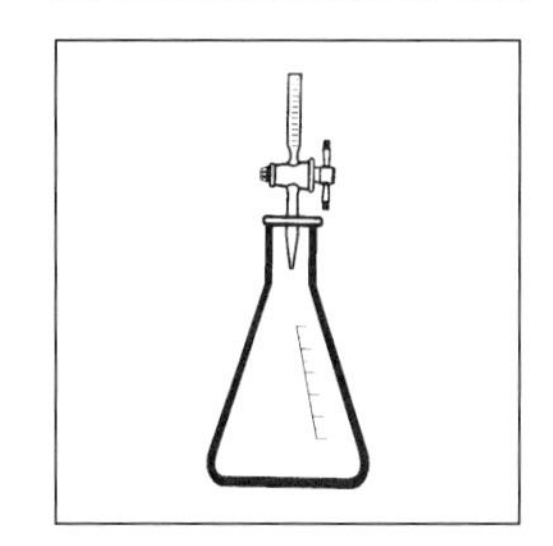

7. Calculate:
   mL Titrant × Multiplier Used = mg/L phenolphthalein acidity as $CaCO_3$
   Note: To convert from mg/L phenolphthalein acidity to grains per gallon equivalent, multiply by 0.0584.

mL Titrant Used x Multiplier = mg/L phenolphthalein acidity as $CaCO_3$

SAMPLING AND STORAGE

Collect samples in clean plastic or glass bottles. Fill completely and cap tightly. Avoid excessive agitation and prolonged exposure to air. Samples should be analyzed as soon as possible after collection but can be stored at least 24 hours by cooling to 4°C (39°F) or below. Warm to room temperature before analyzing.

ACCURACY CHECK

Sodium Hydroxide Standard Solution slowly absorbs carbon dioxide when exposed to air, causing a partial loss of strength. This solution should be checked frequently (monthly) by titrating 50 mL of Potassium Acid Phthalate Standard Solution and using a Phenolphthalein Indicator Powder Pillow. The titration should require 5.68 mL of Sodium Hydroxide Standard Solution. If the volume required for the titration is greater than 5.88 mL, the solution should be discarded and replaced with a fresh supply.

INTERFERENCES

Adding a drop of 0.1 N Sodium Thiosulfate Standard Solution will remove residual chlorine which may interfere with the indicator.

Samples containing hydrolyzable metal ions such as iron, aluminum or manganese in significant amounts must be pretreated as follows:

a) Adjust the sample taken in Step 1 to pH 4.0 or less (if necessary) by adding 5.0-mL increments of 0.020 N Sulfuric Acid Standard Solution. Record the amount of acid added. Use a pH meter or appropriate indicator to determine when pH 4.0 is reached. Remove the pH electrodes after completing this step if a pH meter is used.

b) Using a 1-mL glass serological pipet and pipet filler, add 5 drops of 30% Hydrogen Peroxide Solution.

c) Boil the solution for 2 to 5 minutes.

d) Cool to room temperature and titrate to pH 8.3 as described in the procedure beginning with Step 3.

e) Subtract the number of milliliters of 0.020 N Sulfuric Acid Standard Solution added from the number of milliliters of 0.020 N Sodium Hydroxide Standard Solution used in the titration before multiplying the volume of sodium hydroxide by the multiplier used in Step 7.

END POINT CONFIRMATION AND STANDARD ADDITIONS METHOD

For more details on these issues, see ACIDITY, PHENOLPHTHALEIN on p. 178 of HACH Water Analysis Handbook, 3rd Ed.

INTERFERENCES

Adding a drop of 0.1 N Sodium Thiosulfate Standard Solution will remove residual chlorine which may interfere with the indicator.

Samples containing hydrolyzable metal ions such as iron, aluminum or manganese in significant amounts must be pretreated by the procedure indicated under Phenophthalein Acidity in the HACH Water Analysis Handbook, 3rd Ed., pp. 178–179.

QUESTIONS AND ANSWERS

1. What does acidity measure?
Acidity is the measure of both the ionized and the unionized forms of acid.

2. Why is excessive sample agitation avoided before phenolphthalein indicator is added in this experiment?
Excessive agitation of sample is avoided to prevent loss of dissolved gases such as carbon dioxide, hydrogen sulfide, etc. that contributes to sample acidity.

3. Why should a relatively fresh 0.020 N Sodium Hydroxide Standard Solution be used to titrate the sample for acidity?
Sodium Hydroxide Standard Solution slowly absorbs carbon dioxide when exposed to air, causing a partial loss of strength. Therefore, the 0.020 N NaOH Solution should be checked monthly.

4. How does residual chlorine in water impact the phenolphthalein acidity test?
Residual chlorine may interfere with the phenolphthalein indicator during the acidity test. A drop of 0.1 N Sodium Thiosulfate will remove the residual chlorine.

REFERENCES

1. *Standard Methods for the Examination of Water and Wastewater:* ACIDITY, 16th Ed., A. E. Greenberg, R. R. Trussell, L. S. Clesceri and Mary Ann H. Franson (eds.), pp. 265–269, Washington, DC, American Public Health Association, 1985.
2. *HACH Water Analysis Handbook,* 3rd ed., (Loveland , CO.: HACH Company, 1997).
3. *Hawley's Condensed Chemical Dictionary,* 11th Ed., rev'd by N. Irving Sax and Richard J. Lewis (New York, N.Y.: Van Nostrand Reinhold, 1987).

ALGAE IN WATER

Algae can be named by several different characterizations. Some descriptions of algae are the taste and odor algae, the filter clogging algae, polluted water algae, clean water algae, surface water algae, and algae growing on reservoir walls. Each category contains algae of different shapes, design patterns and sizes with unique characteristics. The different types of algae are best observed and identified by using a microscope and slides. Algae color plates (A through F) are displayed in the text titled: Standard Methods for the Examination of Water and Wastewater, 16th Ed., 1985.

Anchovies and other small fish feed on toxic algae containing a poisonous acidic substance. Sea lions and other sea life are sometimes poisoned by feeding on the anchovies and small fish. Toxic algae can contribute to the death of sea lions, cormorants (a type of pelican), and several additional species. A similar type of algae sickened people and killed marine life a few years earlier.

A unique filtration procedure determines if an unacceptable level of algae in water is present. The algae cells are filtered out of the water by means of a syringe with filter attachment. A yellow-green color on the glass fiber filter indicates algae may be present. The algae collected on the filter is subjected to a chlorophyll extraction. A yellow to green color in the extract solution indicates an unacceptable level of algae in the water.

ALGAE*

Filtration Assembly

1. Unscrew the cover of the filter holder and install a filter disc in the holder. Position the disc carefully to avoid by-passing the filter. If a membrane filter is used, install a support pad in back of the membrane disc. Replace the cover.

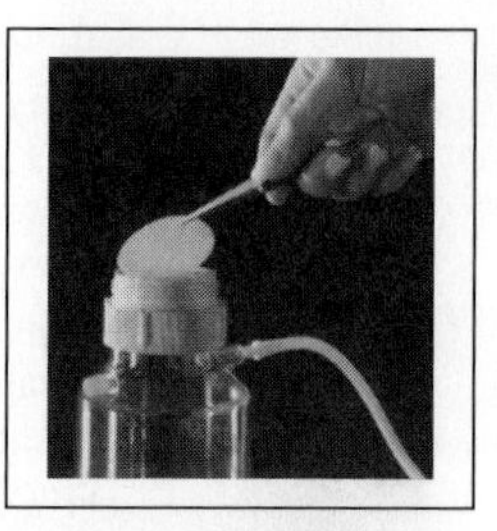

2. Insert the end of the check valve that does not have threads and does have small tabs into the Luer tip of the syringe. Insert the opposite end of the check valve with the threads into the larger opening in the filter holder.

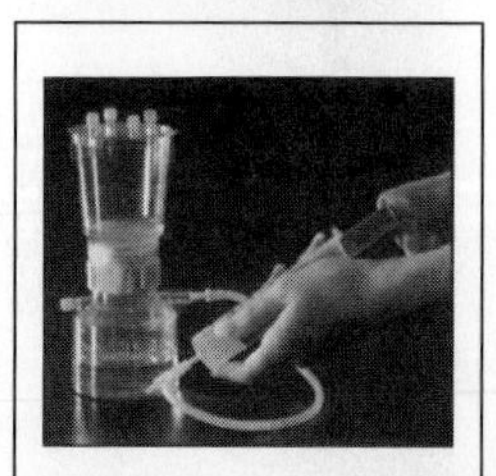

3. Depress the syringe plunger to the "0" position. Attach the plastic tubing to the side arm of the check valve. Submerge the free end of the tubing in the sample water.

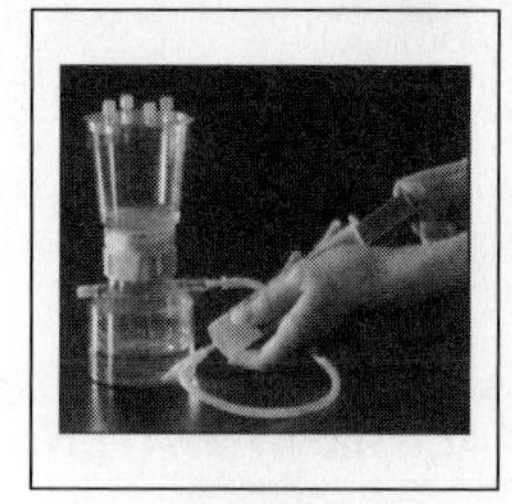

*Compliments of Ward's Natural Science Establishment, Inc. and LaMotte Company

PROCEDURE

1. After the syringe filter device has been prepared according to the above procedure, complete one stroke of the syringe by pulling the plunger and drawing water into the syringe through the plastic tubing. Fill the barrel to the 50 cc mark. Slowly depress the plunger to expel the water through the filter holder. Repeat at least 4 times.
   Note: When excessive pressure is exerted upon the filter disc by depressing the plunger too rapidly, there is a tendency for the filter disc to rupture. If this happens, replace the filter disc and repeat the procedure.

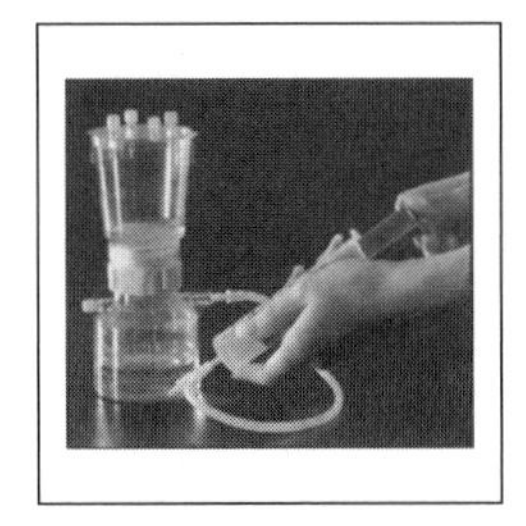

2. Disconnect the filter holder from the syringe. Unscrew the filter holder and carefully remove the filter disc. The presence of a green-yellow color on the filter disc indicates algae may be present.

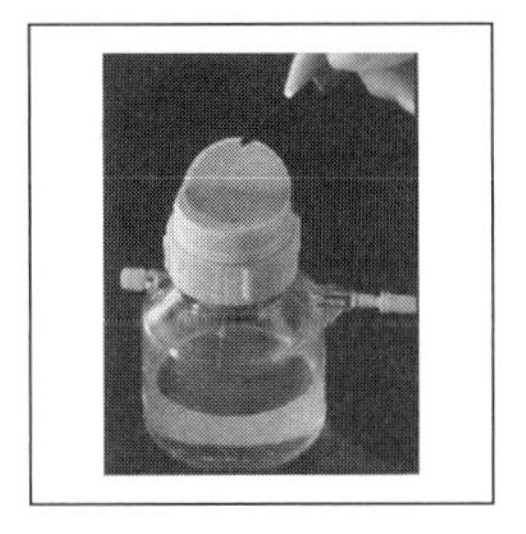

3. Place the filter disc in the calibrated vial. Fill to 5 ml mark with methyl alcohol. Cap and shake vigorously for approximately 2 minutes. This will extract the green chlorophyll from the algae cells and disintegrate the disc. Filter the disintegrated disc out of the solution.

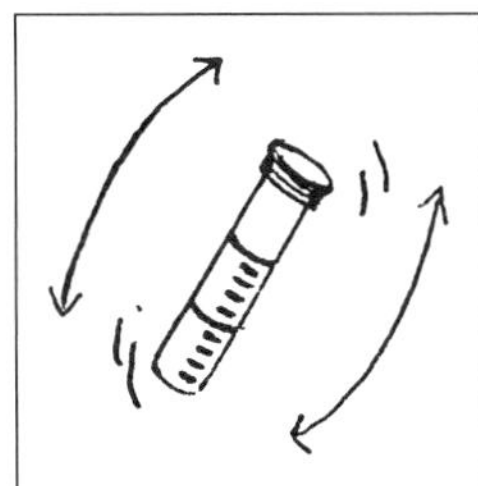

4. Place a new filter disc in the filter holder.

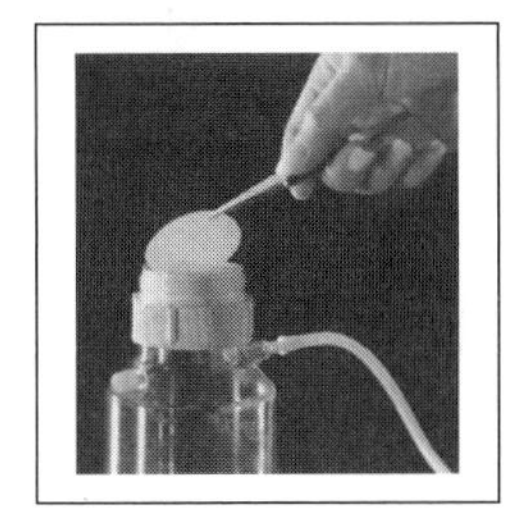

5. Remove the check valve from the syringe. Attach the filter holder *directly* to the syringe.
6. Remove the plunger from the syringe and pour the contents of the test vial into the syringe barrel. Place the outlet of the filter holder into the test tube, replace the plunger in the syringe, and slowly depress until all of the solution has been collected in the test tube.
7. View across the diameter of the test tube or down through the solution. A clear yellow to green color indicates algae present.

CAUTION AND SAFETY INFORMATION

1. Read the MSDS for methyl alcohol before performing this test procedure.
2. When the test kit is used as directed, no chemical reagent will come in contact with the user. Should contact occur, flush skin and eyes with water. Call a physician, should any reagent be swallowed.
3. The person performing the tests should always wear safety glasses.

QUESTIONS AND ANSWERS

1. How do you distinguish between algae and a chemical, which is also yellow-green?
   Minerals or chemicals (yellow-green) will not be trapped on the filter. Algae will be trapped on the filter.
2. What chemical is used to extract chlorophyll?
   Methyl alcohol is used to extract chlorophyll.
3. Which substance in the experiment is considered to be hazardous?
   Methyl alcohol is considered to be a hazardous substance.
4. How does toxic algae affect sea life?
   Toxic algae consists of a substance called domoic acid, which attacks the brains of sea lions, cormorants, and other sea life. The substance does not appear to sicken anchovies and other small fish.

REFERENCES

1. *Algae in Water Test Kit,* LaMotte Test Kit, WARD's Natural Science Establishment, Inc., Rochester, N.Y.
2. *Algae in Water Test Kit,* LaMotte Test Kit, LaMotte Company, Chestertown, Md.
3. *Standard Methods for the Examination of Water and Wastewater:* ALGAE COLOR PLATES (A-F), 16th Ed., A. E. Greenberg, R. R. Trussell, L. S. Clesceri and Mary Ann H. Franson (eds.), pp. 1201 *ff*, Washington, DC, American Public Health Association, 1985.
4. Pete Ambrose, Columbia-Greene Community College, Hudson, N.Y.

ALKALINITY

The alkalinity of most surface waters is attributed to carbonate, bicarbonate and hydroxide concentrations in water. Substances such as borates, phosphates, silicates and organic bases may also contribute to the alkalinity of water.

A number of industries neutralize aqueous wastes to a pH of 7.0. These wastes are usually used by another industry or allowed to empty into streams, rivers and other waterways depending on the level of other pollutants. A few industries manifest wastes for additional treatment purposes or for burial.

Bicarbonate is the primary contributor to alkalinity problems. Carbonates and hydroxides are usually less important contributors to water alkalinity. These substances can present major problems in boilers, water-cooling towers and in industrial water and wastewater treatment systems. Alkalinity tests are often used to identify the problem of water pipe scaling in homes and in many different types of industry.

Both phenolphthalein alkalinity and total alkalinity are determined by titration with a sulfuric acid standard solution to an end point evidenced by a color change of an indicator solution. Alternatively, the end point can be determined by a pH meter. The phenolphthalein alkalinity is determined by titration to a pH of 8.3 and measures the total hydroxide and one-half the carbonate present. The total alkalinity depends on the total concentration of the carbonate, bicarbonate and hydroxide ions in solution. Depending on the alkalinity concentration in the sample, the total alkalinity is determined by titration of the sample to a pH of 5.1, 4.8, 4.5, or 3.7. Industrial wastes and other interferences can impact the pH of the solution when testing for total alkalinity.

ALKALINITY (10-4000 mg/L as $CaCO_3$)

Method 8203*
Digital Titration Method

1. Select the sample volume and Sulfuric Acid (H_2SO_4) Titration Cartridge corresponding to the expected alkalinity concentration in mg/L as calcium carbonate ($CaCO_3$) from Table 2.4.
Note: See Sampling and Storage following these steps.

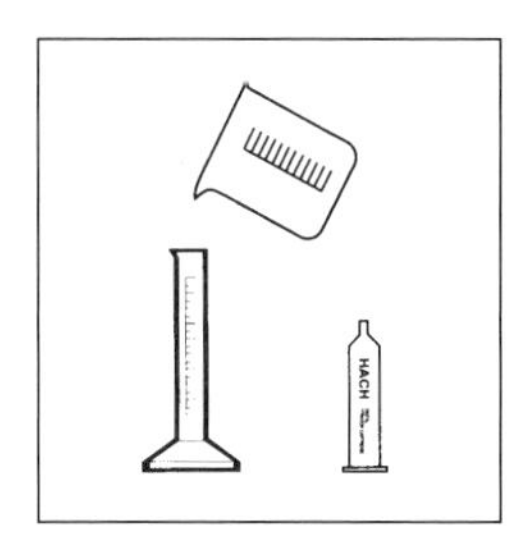

2. Insert a clean delivery tube into the titration cartridge. Attach the cartridge to the titrator body.

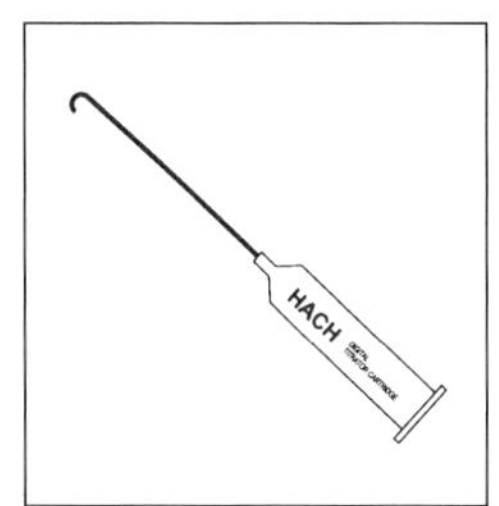

TABLE 2.4

Range mg/L as $CaCO_3$)	Sample Volume (mL)	Titration Cartridge N H_2SO_4)	Catalog Number	Digit Multiplier
10–40	100	0.1600	14388-01	0.1
40–160	25	0.1600	14388-01	0.4
100–400	100	1.600	14389-01	1.0
200–800	50	1.600	14389-01	2.0
500–2000	20	1.600	14389-01	5.0
1000–4000	10	1.600	14389-01	10.0

*Compliments of HACH Company.

3. Turn the delivery knob to eject a few drops of titrant. Reset the counter to zero and wipe the tip.
Note: For added convenience use the TitraStir stirring apparatus.

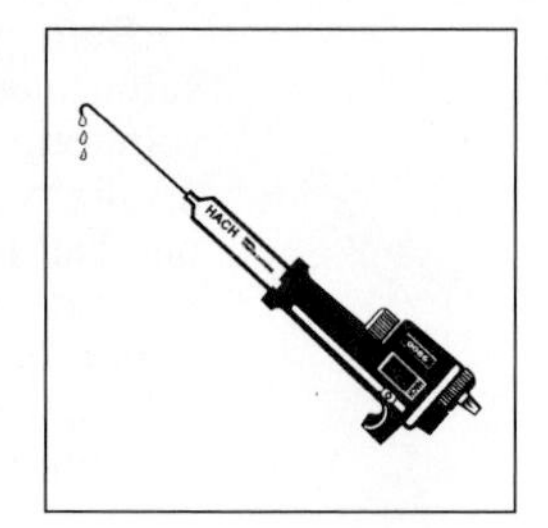

4. Use a graduated cylinder or pipet to measure the sample volume from Table 2.4. Transfer the sample into a clean 250-mL erlenmeyer flask. Dilute to about the 100-mL mark with demineralized water, if necessary.

5. Add the contents of one Phenolphthalein Indicator Powder Pillow and swirl to mix.
Note: A solution of one pH 8.3 Buffer Powder Pillow and one Phenolphthalein Powder Pillow in 50 mL of demineralized water is recommended as a comparison for determining the proper end-point color.
Note: Four drops of Phenolphthalein Indicator Solution may be substituted for the Phenolphthalein Indicator Powder Pillow.

6. If the solution turns pink, titrate to a colorless end point. Place the delivery tube tip into the solution and swirl the flask while titrating with sulfuric acid. Record the number of digits required.
Note: If the solution is colorless before titrating with Sulfuric acid, the Phenolphthalein (P) alkalinity is zero; proceed with Step 8.

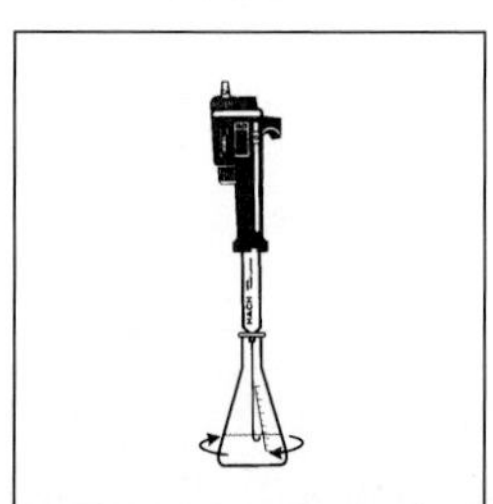

7. Calculate: Digits Required × Digit Multiplier = mg/L $CaCO_3$ P Alkalinity.

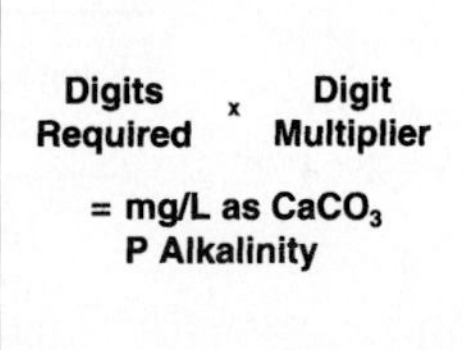

8. Add the contents of one Bromcresol Green-Methyl Red Indicator Powder Pillow to the flask and swirl to mix.
Note: Four drops of Methyl Purple Indicator Solution may be substituted for the Bromcresol Green-Methyl Red Indicator Powder Pillow. Titrate from green to a gray end point (pH 5.1).
Note: Four drops of Bromcresol Green-Methyl Red Indicator Solution may be substituted for the Bromcresol Green-Methyl Red Indicator Powder Pillow.

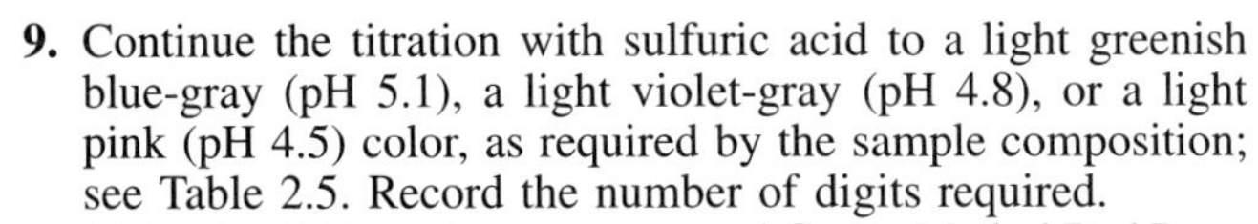

9. Continue the titration with sulfuric acid to a light greenish blue-gray (pH 5.1), a light violet-gray (pH 4.8), or a light pink (pH 4.5) color, as required by the sample composition; see Table 2.5. Record the number of digits required.
Note: A solution of one Bromcresol Green-Methyl Red Powder Pillow and one pillow of the appropriate pH buffer in 50 mL of demineralized water is recommended as a comparison for judging the proper end-point color. If the pH 3.7 end point is used, use a Bromphenol Blue Powder Pillow instead of a Bromcresol Green-Methyl Red and titrate to a green end point.

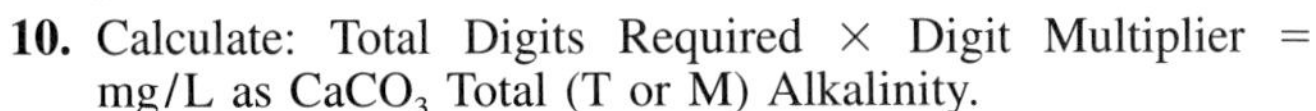

10. Calculate: Total Digits Required × Digit Multiplier = mg/L as $CaCO_3$ Total (T or M) Alkalinity.
Note: Carbonate, bicarbonate, and hydroxide concentrations may be expressed individually using the relationships shown in Table 2.6 below.
Note: meq/L Alkalinity = mg/L as $CaCO_3$ ÷ 50.

Total Digits Required x Digit Multiplier
= mg/L as $CaCO_3$
Total (T or M) Alkalinity

TABLE 2.5

Sample Comparison	End Point
Alkalinity about 30 mg/L	pH 5.1
Alkalinity about 150 mg/L	pH 4.8
Alkalinity about 500 mg/L	ph 4.5
Silicates or Phosphates present	pH 4.5
Industrial waste or complex system	pH 3.7

ALKALINITY RELATIONSHIP TABLE

Total alkalinity primarily includes hydroxide, carbonate, and bicarbonate alkalinities. The concentration of these alkalinities in a sample may be determined when the phenolphthalein and total alkalinities are known (see Table 2.6).

To use this table, follow these steps:

a) Does the phenolphthalein alkalinity equal zero? If yes, use Row 1.

b) Does the phenolphthalein alkalinity equal total alkalinity? If yes, use Row 2.

TABLE 2.6

Row	Result of Titration	Hydroxide Alkalinity is equal to:	Carbonate Alkalinity is equal to:	Bicarbonate Alkalinity is equal to:
1	Phenolphthalein Alkalinity = 0	0	0	Total Alkalinity
2	Phenolphthalein Alkalinity equal to Total Alkalinity	Total Alkalinity	0	0
3	Phenolphthalein Alkalinity less than one-half of Total Alkalinity	0	2 times the Phenolphthalein Alkalinity	Total Alkalinity minus 2 times Phenolphthalein Alkalinity
4	Phenolphthalein Alkalinity equal to one-half of Total Alkalinity	0	Total Alkalinity	0
5	Phenolphthalein Alkalinity greater than one-half of Total Alkalinity	2 times the Phenolphthalein Alkalinity minus Total Alkalinity	2 times the difference between Total and Phenolphthalein Alkalinity	0

c) Multiply the phenolphthalein alkalinity by 2.

d) Select Row 3, 4, or 5 based on comparing the result of Step c with the total alkalinity.

e) Perform the required calculations in the appropriate row, if any.

f) Check your results. The sum of the three alkalinity types will equal the total alkalinity.

SAMPLING AND STORAGE

Collect samples in clean plastic or glass bottles. Fill completely and cap tightly. Avoid excessive agitation or prolonged exposure to air. Samples should be analyzed as soon as possible after collection but can be stored at least 24 hours by cooling to 4°C (39°F) or below if they cannot be analyzed immediately. Warm to room temperature before analyzing.

ACCURACY CHECK

See p. 192 of the HACH Water Analysis Handbook, 3rd ed., 1997.

INTERFERENCES

- Highly colored or turbid samples may mask the color change at the end point. Use a pH meter for these samples.
- Chlorine may interfere with the indicators. Add one drop of 0.1 N sodium Tiosulfate to eliminate this interference.

ALKALINITY

Method 8221for water, wastewater and seawater*
Buret Titration Method**
USEPA Accepted

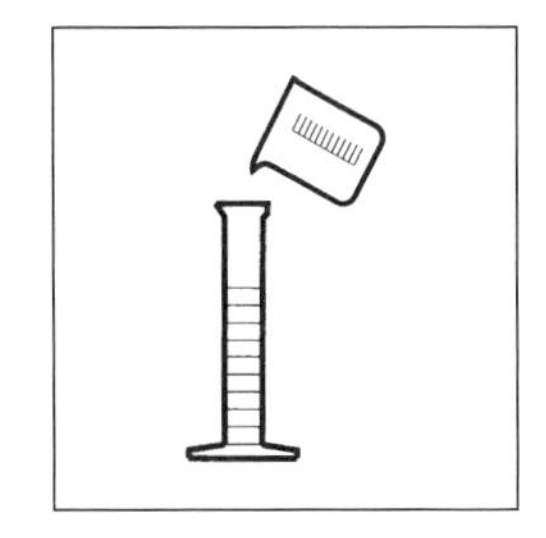

1. Select a sample size corresponding to the expected alkalinity concentration in mg/L as calcium carbonate ($CaCO_3$) from Table 2.7 below.
   Note: If samples cannot be analyzed immediately, see Sampling and Storage following these steps.

2. Use a graduated cylinder or pipet to measure the sample volume from Table 2.7. Transfer the sample into a 250 mL erlenmeyer flask. If necessary, dilute to 50 mL with deionized water.
   Note: For proof of accuracy, see HACH Water Analysis Handbook, Method 8221.

TABLE 2.7

Range (mg/L as $CaCO_3$)	Sample Volume (mL)	Standard Titrant Solution (N)	HACH Catalog Number	Multiplier
0–500	50	0.020	203-53	20
400–1000	25	0.020	203-53	40
1000–2500	10	0.020	203-53	100
2000–5000	5	0.020	203-53	200

Phenolphthalein Alkalinity

3. Add the contents of one Phenolphthalein Indicator Powder Pillow. Swirl to mix.
   Note: Six drops of Phenolphthalein Indicator Solution can be substituted for the Phenolphthalein Indicator Powder Pillow.
   Note: Omit this step if a pH meter is used. A pH meter is required for NPDES reporting and is recommended for best results.

*Compliments of HACH Company.
** Adapted from *Standard Methods for the Examination of Water and Wastewater, 2320 B.*

4. Fill a 25-mL buret to the zero mark with 0.020 N Sulfuric Acid Standard Solution.

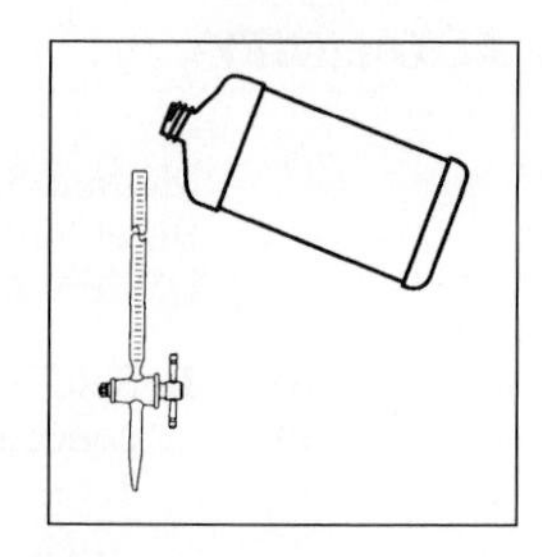

5. Titrate the sample while swirling the flask until the solution changes from pink to colorless.
Note: When using a pH meter, the end point is 8.3 pH.

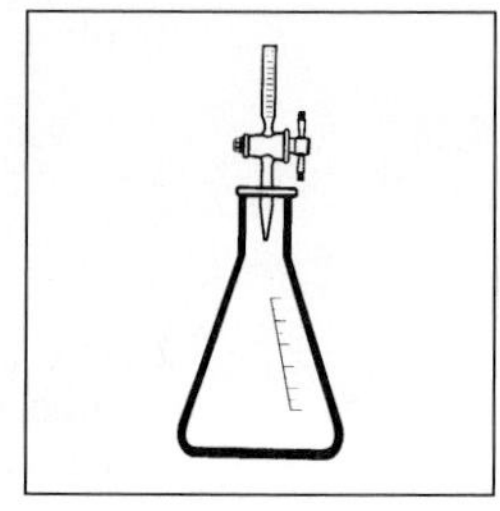

6. Calculate:
mL Titrant × Multiplier Used = mg/L Phenolphthalein Alkalinity as $CaCO_3$
Note: To convert to grains per gallon, divide mg/L value by 17.12.

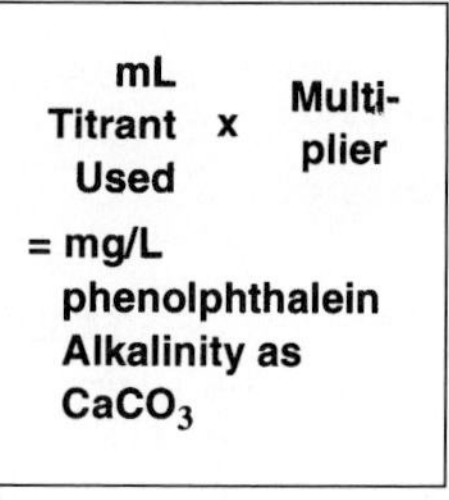

Total Alkalinity

7. Add the contents of one Bromcresol Green-Methyl Red Indicator Powder Pillow to the titrated sample. Swirl to mix.
Note: Six drops of Bromcresol Green-Methyl Red Indicator Solution can be substituted for the Bromcresol Green-Methyl Red Powder Pillow.
Note: See Total Alkalinity End Points in Table 2.8 for alternate end points.
Note: Do not add indicator if a pH meter is used. A pH meter is required for NPDES reporting and is recommended for best results.

8. Continue the titration until the appropriate end point is reached.
Calculate:
Total mL Titrant × Multiplier Used = mg/L Total Alkalinity as $CaCO_3$.
Note: To convert to grains per gallon, divide mg/L value by 17.12.

mL Titrant Used x Multi-plier
= mg/L Total Alkalinity as $CaCO_3$

TOTAL ALKALINITY END POINTS

The following end points are recommended for determining total alkalinity in water samples of various compositions and alkalinity concentrations. See Table 2.8 below.

TABLE 2.8

Sample Trait	End Point	
	pH	Color
Alkalinity approx. 30 mg/L	5.1	light green blue-gray
Alkalinity approx. 150 mg/L	4.8	light violet gray
Alkalinity approx. 500 mg/L	4.5	light pink
Silicates or phosphates known present or suspected	4.5	light pink
Industrial waste or complex system	3.7	Green*

*Use a Bromophenol Blue Indicator Powder Pillow instead of Bromcresol Green-Methyl Red.

SAMPLING AND STORAGE

Collect samples in clean plastic or glass bottles. Fill completely and cap tightly. Avoid excessive agitation and prolonged exposure to air. Samples should be analyzed as soon as possible after collection but can be stored at least 24 hours by cooling to 4°C (39°F) or below. Warm to room temperature before analyzing.

ACCURACY CHECK (END POINT CONFIRMATION AND STANDARD ADDITIONS METHOD)

See p. 185 of the HACH Water Analysis Handbook, 3rd Ed., 1997.

INTERFERENCES

Residual chlorine interference with the indicator may be removed by adding a drop of 0.01 N Sodium Thiosulfate Solution. Highly colored or turbid samples may mask the color change at the end point. Use a pH meter for these samples.

ALKALINITY RELATIONSHIP TABLE

Total alkalinity primarily includes hydroxide, carbonate and bicarbonate alkalinities. The concentration of these types in a sample may be determined when the phenolphthalein and total alkalinities are known; see Table 2.8a.

To use this table, follow these steps:

TABLE 2.8a

	Result of Titrations	Hydroxide Alkalinity is Equal to	Carbonate Alkalinity is Equal to	Bicarbonate Alkalinity is Equal to
1	Phenolphthalein Alkalinity equal to zero	0	0	Total Alkalinity
2	Phenolphthalein Alkalinity equal to Total Alkalinity	Total Alkalinity	0	0
3	2 times the Phenolphthalein Alkalinity less than Total Alkalinity	0	2 times the Phenolphthalein Alkalinity	Total Alkalinity minus 2 times Phenolphthalein Alkalinity
4	2 times the Phenolphthalein Alkalinity equal to Total Alkalinity	0	Total Alkalinity	0
5	2 times the Phenolphthalein Alkalinity greater than Total Alkalinity	2 times the Phenolphthalein Alkalinity minus Total Alkalinity	2 times the difference between Total and Phenolphthalein Alkalinity	0

a) Does the phenolphthalein alkalinity equal zero? If yes, use Row 1.
b) Does the phenolphthalein alkalinity equal total alkalinity? If yes, use Row 2.
c) Multiply the phenolphthalein alkalinity by 2.
d) Select Row 3, 4, or 5 based on comparing the result of Step c with the total alkalinity.
e) Perform the required calculations if any. Check your results. The sum of the three alkalinity types will equal the total alkalinity.

INTERFERENCES

Residual chlorine interference with the indicator may be removed by adding a drop of 0.01 N Sodium Thiosulfate Standard Solution.

Highly colored or turbid samples may mask the color change at the end point. Use a pH meter for these samples.

QUESTIONS AND ANSWERS

1. What does alkalinity measure?
Alkalinity is the measure of carbonate, bicarbonate and hydroxide concentrations in water. Other substances such as phosphates and organic bases do not usually contribute much to the alkalinity of the solution.

2. How does residual chlorine in water impact the alkalinity test?
Residual chlorine may interfere with the indicator during the test and may be removed by adding a drop of 0.1 N Sodium Thiosulfate Solution.

3. What changes in alkalinity procedure is required for NPDES reporting?
For NPDES reporting, a pH meter is required instead of indicators in Steps 3 and 7.

4. What is the difference between phenolphthalein alkalinity and total alkalinity?
Phenolphthalein alkalinity is determined by titration to a pH of 8.3 and measures the total hydroxide and one-half the carbonate present. The total alkalinity determines all the carbonate, bicarbonate, and hydroxide alkalinity.

REFERENCES

1. *Standard Methods for the Examination of Water and Wastewater*: ALKALINITY, 16th Ed., A. E. Greenberg, R. R. Trussell, L. S. Clesceri and Mary Ann H. Franson (eds.), pp. 269–273,Washington, DC, American Public Health Association, 1985.
2. *HACH Water Analysis Handbook,* 3rd Ed., (Loveland, CO.: HACH Company, 1997).
3. *Hawley's Condensed Chemical Dictionary,* 11th Ed., rev'd by Irving Sax and Richard J. Lewi, (New York, N.Y.: Van Nostrand Reinhold, 1987).

ALUMINUM

Aluminum is an abundant element occurring in rocks, minerals and clay. Natural water contains the metal as a soluble salt, a colloid, or as an insoluble compound. These soluble, colloidal, and insoluble compounds often appear in treated water and wastewater during alum coagulation. Alum flocculation is a common technique to remove suspended solids as interferences from samples being tested for dissolved oxygen and other chemicals in solutions.

Aluminum ions hydrolyze in water to form weakly acidic solutions. A drop in the pH tends to make aluminum more soluble in surface waters and thus increases the uptake of aluminum by plants. Aluminum is of particular concern because of its toxic effects on plants. Aluminum hydroxide and similar aluminum compounds are often used to concentrate viruses from small volumes of water, wastewater and other solvents. Viruses adsorb on the $Al(OH)_3$ precipitate and are collected by centrifugation and filtration procedures. The viruses can be isolated and used for further study or identification purposes.

Aluminum metal has extensive uses in cans, building materials, cooking ware, vehicles, electrical devices, household items, toys and many other devices. Aluminum, as a metal, can be easily recycled. Aluminum compounds have a number of practical uses in society such as the following: sewage and water purification, paper manufacturing, medical research, catalyst, nucleonics, anti-corrosion agent, antiperspirant, pigments, pharmaceuticals, cosmetics, special papers, photography, textiles, alums, dyeing, and foaming agents.

Some of the industries utilize closed loop systems or partial closed loop systems to recover aluminum wastes. Industries and metal recycling sites collect and recycle aluminum metal. Aluminum wastes are manifested for burial purposes and for recovery and reuse. A few aluminum compounds in water are dumped directly into waterways.

The Aluminon Method of determination has some interferences which can be eliminated by additional treatment of the solution. These interferences are alkalinity (>1000 mg/L as $CaCO_3$), iron (>20 mg/L), acidity (>300 mg/L as $CaCO_3$), and phosphate (>50 mg/L). Polyphosphate causes a negative interference at all levels and must be converted to orthophosphate by acid hydrolysis (see phosphorus procedures). Also, fluoride interferes at all levels by complexing with aluminum. Aluminum can be determined accurately by using the Fluoride Interference Graph when the fluoride concentration is known in the sample.

The Aluminon Method for determination of aluminum can be measured in the field by the DR/820, DR/850, and the DR/890 Colorimeters and the DR/2010 Spectrophotometer. Additional methods to determine aluminum by both colorimetric and atomic absorption methods are available in the literature.

ALUMINUM (0 to 0.80 mg/L)

Method 8012 for water and wastewater*
Aluminon Method**

1. Enter the stored program number for aluminum (A1). Press: **1 0 ENTER.** The display will show: **Dial nm to 522.**
Note: The Pour-Thru Cell can be used if rinsed well with deionized water between the blank and prepared sample.

2. Rotate the wavelength dial until the small display shows: **522 nm.** When the correct wavelength is dialed in, the display will quickly show: **Zero Sample,** then: **mg/L Al^{+3}.**
Note: Total aluminum determination needs a prior digestion; use the Digestion Procedure and the Sample and Analysis Volume Table (Aluminum) for Aqueous Liquids in Section 1.

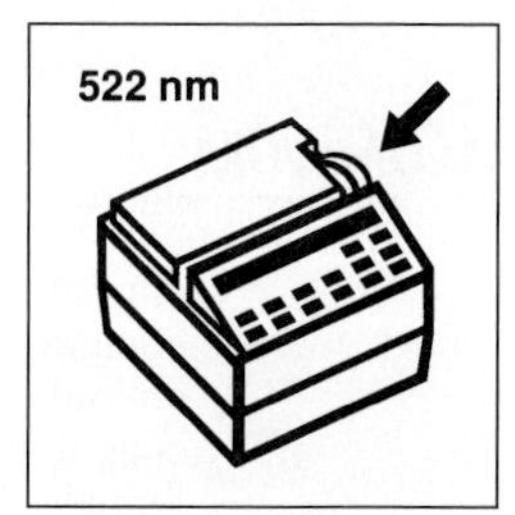

3. Fill a 50-mL graduated mixing cylinder to the 50-mL mark with sample.
Note: Rinse cylinder with 1:1 Hydrochloric Acid and deionized water before use to avoid errors due to contaminants absorbed on the glass.
Note: The sample temperature must be between 20 to 25°C (68 to 77°F) for accurate results.

4. Add the contents of one Ascorbic Acid Powder Pillow. Stopper. Invert several times to dissolve powder.

*Compliments of HACH Company. Analysis may be performed by DR/820, DR/850, and DR/890 Colorimeters and DR/2010 Field Spectrophotometer

** Adapted from *Standard Methods for the Examination of Water and Wastewater,* 12th Ed., p. 53.

ACCURACY CHECK

See p. 198 of the HACH Water Analysis Handbook, 3rd Ed., 1997.

INTERFERENCES

The following do not interfere up to the indicated concentrations.

TABLE 2.9

Alkalinity	1000 mg/L as $CaCO_3$
Iron	20 mg/L
Phosphate	50 mg/L

- Interferences from higher alkalinity concentrations can be eliminated by the following pretreatment:
  a) Add one drop of m-Nitrophenol Indicator Solution to the sample taken in Step 3. A yellow color indicates excessive alkalinity.
  b) Add one drop of 5.25 N Sulfuric Acid Standard Solution. Stopper the cylinder. Invert to mix. If the yellow color persists, repeat until the sample changes to colorless. Continue with the test.
- Polyphosphate causes a negative interference at all levels and must be absent. Before testing, polyphosphate must be converted to orthophosphate by acid hydrolysis as described under the phosphorus procedures.
- Acidity interferes at greater than 300 mg/L as $CaCO_3$. Samples with greater than 300 mg/L acidity as $CaCO_3$ must be treated as follows:
  a) Add one drop of m-Nitrophenol Indicator Solution to the sample taken in Step 3.
  b) Add one drop of 5.0 N Sodium Hydroxide Standard Solution. Stopper the cylinder. Invert to mix. Repeat as often as necessary until the color changes from colorless to yellow.
  c) Add one drop of Sulfuric Acid Standard Solution, 5.25 N, to change the solution from yellow back to colorless. Continue with the test.
- Calcium does not interfere.
- Fluoride interferes at all levels by complexing with aluminum. The actual aluminum concentration can be determined using the Fluoride Interference Graph (HACH Water Analysis Handbook, 3rd Ed., 1997, p. 200) when the fluoride concentration is known.

QUESTIONS AND ANSWERS

1. What are some of the problems with aluminum dissolved in water?
Aluminum dissolved in water makes the solution somewhat acid, which has a tendency to dissolve additional materials.

2. What are some of the interferences with aluminum?
Alkalinity interferes at 1000 mg/L as $CaCO_3$*. Some additional interferences are acidity*

(>300 mg/L as $CaCO_3$), iron (>20 mg/L) and phosphate (>50 mg/L). Polyphosphate causes a negative interference. Fluoride interferes at all levels and accurate aluminum results depend on the interpretation of the Fluoride Interference Graph.

2. Does iron interfere with the aluminum test?
Ascorbic acid is added to remove iron interference.

4. How stable is the AluVer 3 Aluminum reagent?
The reagent packaged in powder form is very stable.

REFERENCES

1. *Standard Methods for the Examination of Water and Wastewater:* ALUMINUM, 16th Ed., A. E. Greenberg, R. R. Trussell, L. S. Clesceri and Mary Ann H. Franson (eds.), pp. 182, 959–961, Washington, DC, American Public Health Association, 1985.
2. *HACH Water Analysis Handbook,* 3rd Ed., (Loveland, CO.: HACH Company, 1997).
3. *Hawley's Condensed Chemical Dictionary,* 11th Ed., rev'd by N. Irving Sax and Richard J. Lewis (New York, N.Y., Van Nostrand Reinhold, 1987).

ARSENIC

Some compounds of arsenic are highly toxic and may be carcinogenic. The accumulation of arsenic in the body occurs primarily through inhalation, ingestion, and skin contact. Chronic effects can appear from its accumulation in the body at low intake levels. Ingestion of as little as 100 mg of arsenic can lead to severe poisoning of a person.

Organic compounds and inorganic compounds of As (III) and As (V) can be found in varying limited quantities in waters and wastewaters. A number of arsenic compounds are increasingly soluble in water as the temperature of the water increases above 0°C. Some arsenic compounds decompose in water and solubilize arsenic metal. Most potable waters contain less than 10 micrograms per liter of arsenic. Arsenic can be found in water from insecticide applications, industrial discharges, and mineral dissolution. Some sources of arsenic and arsenic compounds are the following: insecticides, herbicide, dyeing and printing, metal adhesives, weed killing, colored glass, pharmaceuticals, ceramics, wood preservative, rodenticide, sheep and cattle dip, wood preservative, hide preservative, doping agent, poisons, leather industry, pyrotechnics, and additional applications.

Industries manufacturing and using many of the arsenic compounds must undertake special precautions to avoid contamination of effluents released to the environment. Several industries treat and detoxify arsenic-type wastes, while others implement a closed loop system to reuse the waste streams in the processes. Some facilities sell the arsenic wastes for uses in other applications at another site. Smaller industries tend to manifest wastes for burial purposes. A number of arsenic compounds of varying toxicities are discarded in the garbage. A few sources of arsenic waste and applications of arsenic compounds in our society still provide the opportunity for contamination in our environment.

The silver diethyldithiocarbamate method is applicable when interferences of certain metals (chromium, cobalt, copper, nickel, mercury, molybdenum, and platinum) are minimal in the sample being analyzed. The low concentration of these metals in water from the natural environment does not interfere significantly with the test method. To measure the total amount of arsenic in an aqueous sample, sample digestion (distillation) in a laboratory fume hood is required to solubilize particulate arsenic and to convert any organic arsenic compounds to inorganic arsenic. This procedure requiring distillation is USEPA accepted for reporting.

In a distillation apparatus, arsenic ions and compounds are reduced to arsine gas by a mixture of zinc, stannous chloride, potassium iodide, and hydrochloric acid. After arsine is passed through a scrubber containing lead acetate saturated on cotton, the arsenic reacts with silver diethyldithiocarbamate in pyridine to form a red complex which is measured colorimetrically at 520 nm.

ARSENIC (0 to 0.200 mg/L)

Method 8013 for water, wastewater, and seawater*
Silver Diethyldithiocarbamate Method**
USEPA accepted (distillation required)†

1. This procedure requires a user entered calibration before sample measurement. See the steps in the Initial Setup of Arsenic Program and the User Calibration of Arsenic Program below.

User Calibration

2. Enter the user stored program number for arsenic (As). Press **9 ? ?** then **ENTER.** The display will show: **Dial to 520 nm.** *Note:* The Pour-Thru Cell cannot be used.

3. Rotate the wavelength dial until the small display shows: **520 nm.** When the correct wavelength is dialed in the display will quickly show: **Zero Sample,** then **mg/L As.**

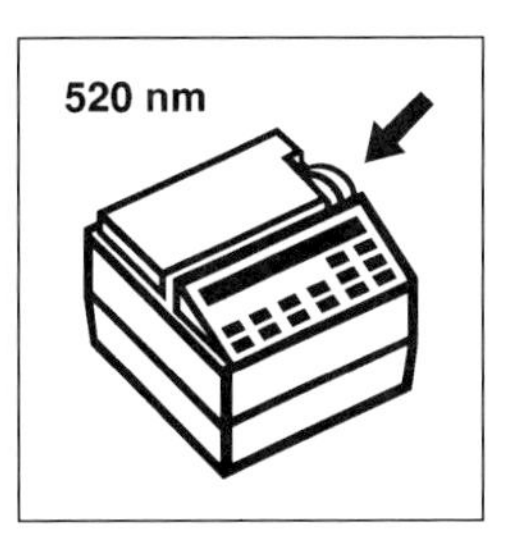

*Compliments of HACH Company.
**Adapted from *Standard Methods for the Examination of Water and Wastewater, 307 B.*
†Equivalent to USEPA Method 206.4 for wastewater and Standard Method 3500-As for drinking water.

4. Prepare the HACH distillation apparatus for arsenic recovery. Place it under a fume hood to vent toxic fumes.
Note: See the HACH Distillation Manual for assembly instructions.

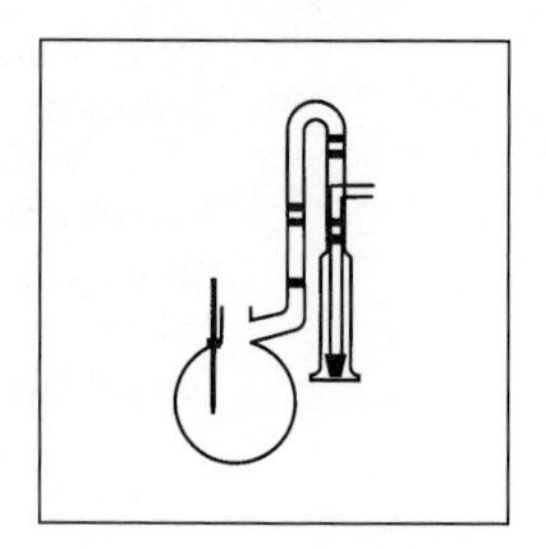

5. Dampen a cotton ball with 10% Lead Acetate Solution. Place it in the gas scrubber. Be certain the cotton seals against the glass.

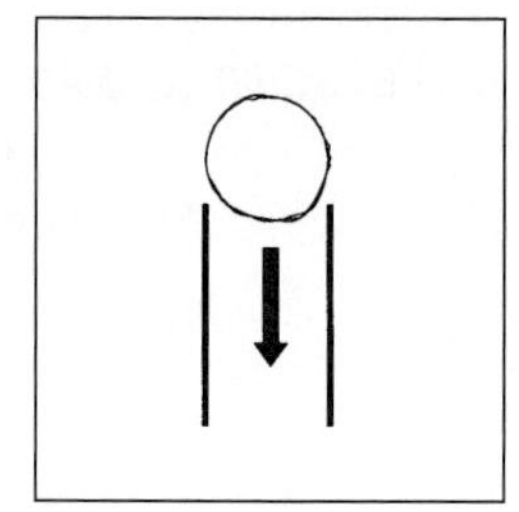

6. Measure 25 mL of prepared arsenic absorber solution into the cylinder/gas bubbler assembly with a graduated cylinder. Attach it to the distillation apparatus.
Note: Prepare the arsenic absorber solution as directed under Reagent Preparation below.

7. Measure 250 mL of sample into the distillation flask using a graduated cylinder.

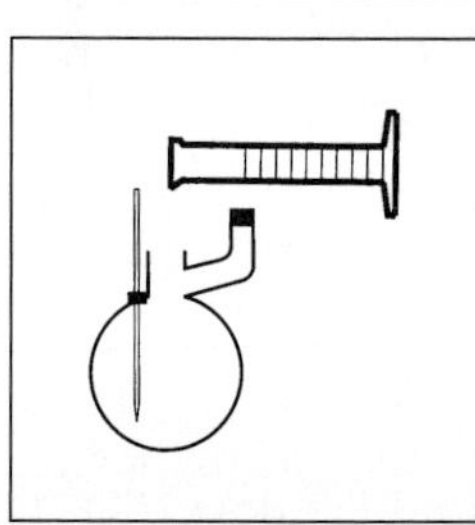

8. Turn on the power switch. Set the stir control to 5. Set the heat control to 0.

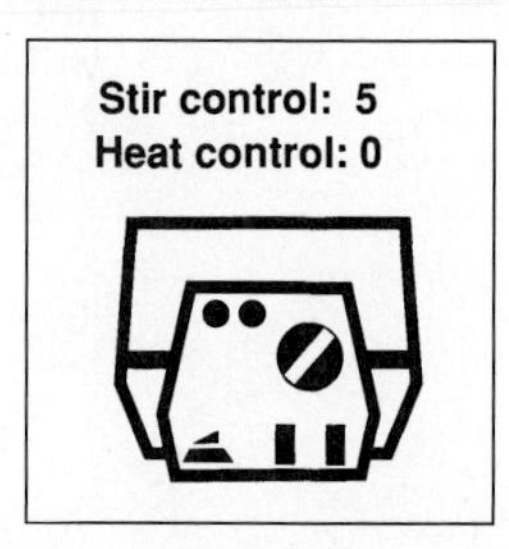

9. Measure 25 mL of hydrochloric acid, ACS, into the flask using a graduated cylinder.

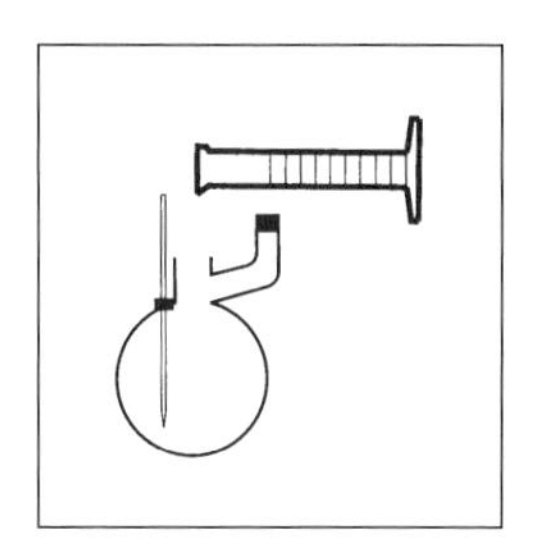

10. Measure 1 mL of Stannous Chloride Solution into the flask. ***Note:*** Use a clean serological pipet to measure the solution.

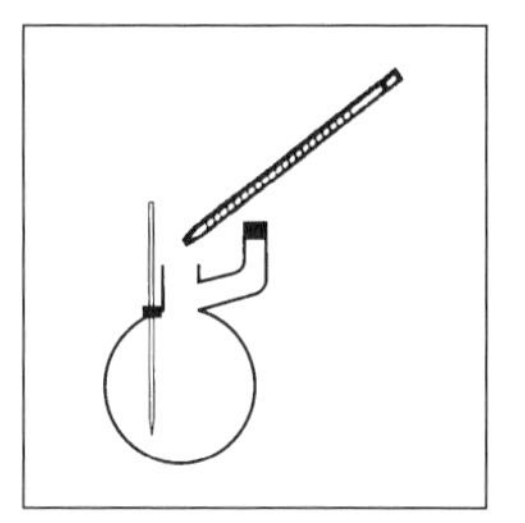

11. Add 3 mL of Potassium Iodide Solution to the flask. Cap. ***Note:*** Use a serological pipet to measure the solution.

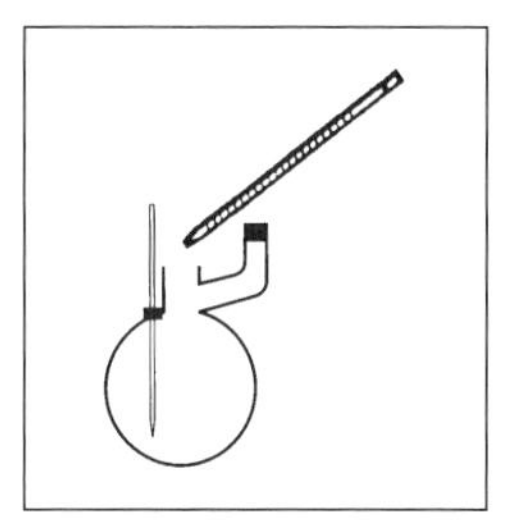

12. Press: **SHIFT TIMER.** A 15-minute reaction period will begin.

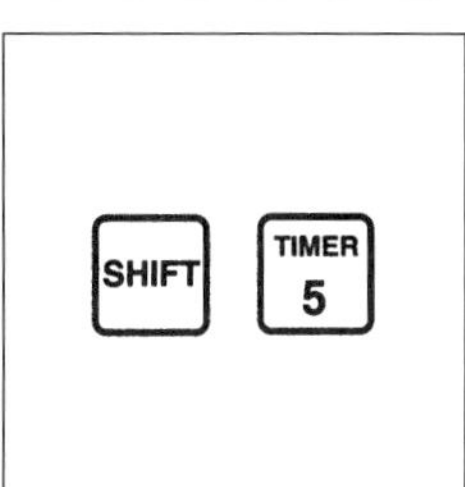

13. When the timer beeps, add 6.0 g of 20-mesh zinc to the flask. Cap immediately.

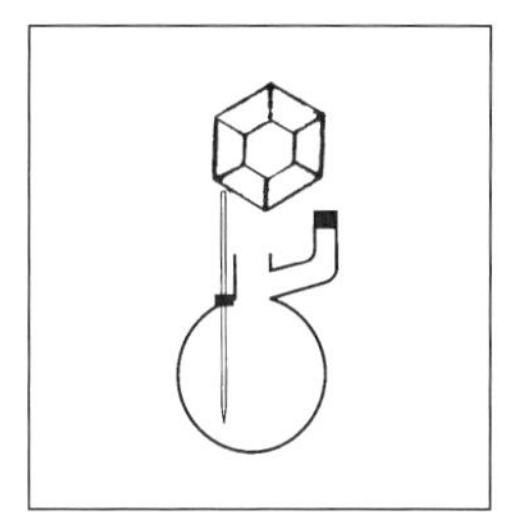

14. Set the heat control to 3. Press **SHIFT TIMER.** A second 15-minute reaction period will begin.

15. When the timer beeps, set the heat control to 1. Press: **SHIFT TIMER.** A third 15-minute reaction will begin.

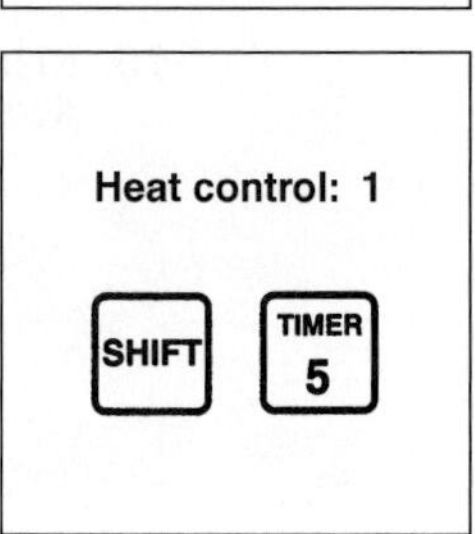

16. When the timer beeps, the display will show: **mg/L As.** Turn on the heater. Remove the cylinder/gas bubbler assembly as a unit.

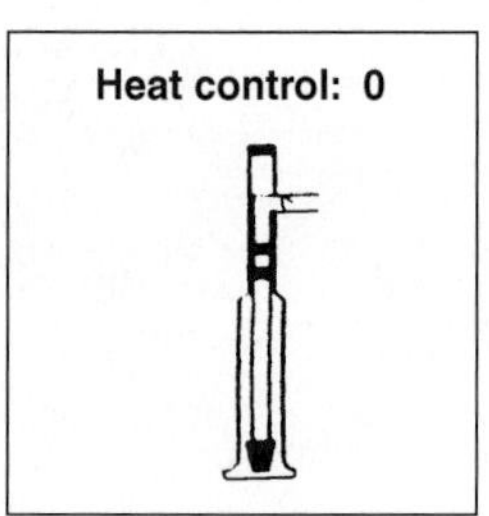

17. Rinse the gas bubbler by moving it up and down in the arsenic absorber solution.

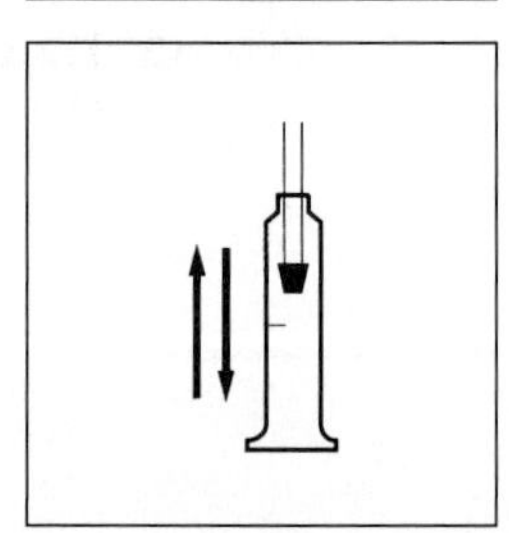

18. Fill a dry sample cell with unreacted arsenic absorber solution (the blank). Stopper. Place it into the cell holder.

19. Place the blank into the cell holder. Close the light shield.

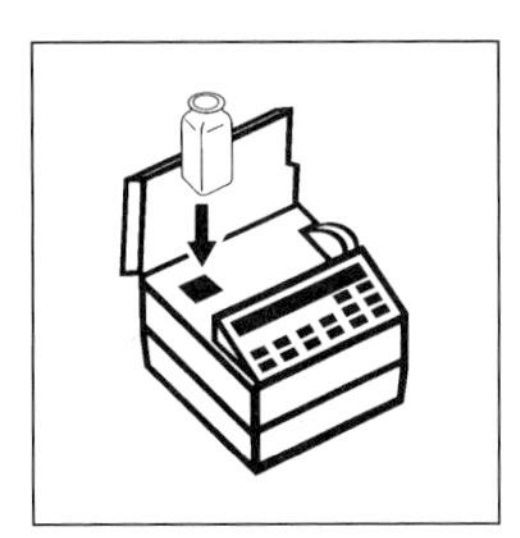

20. Press: **ZERO.** The display will show: **Zeroing ....,** then: **0.000 mg/L As.**

21. Pour the reacted arsenic absorber solution into a sample cell (the prepared sample). Stopper.
Note: If the solution volume is less than 25 mL, add pyridine to bring the volume to exactly the 25-mL mark. Swirl to mix.

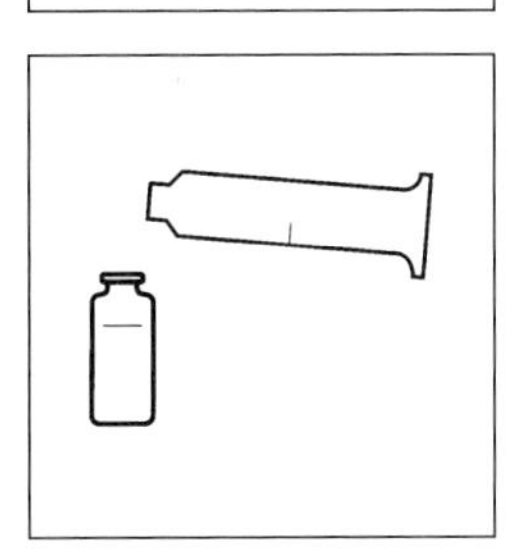

22. Place the prepared sample into the cell holder. Close the light shield.

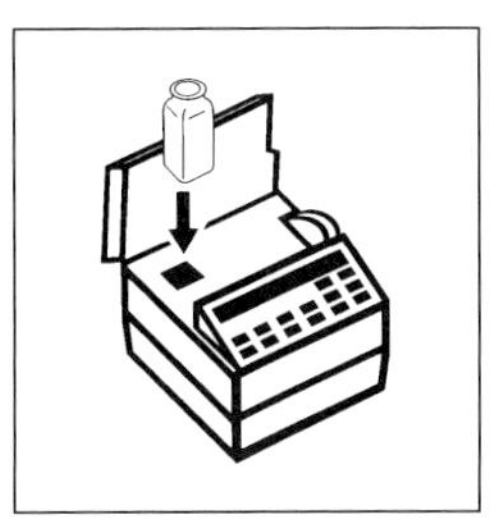

23. Press: **READ ...** The display will show: **Reading . . .** then the result in mg/L arsenic (As) will be displayed.

SAMPLING AND STORAGE

Collect samples in acid washed glass or plastic bottles. Adjust the pH to 2 or less with nitric acid. (about 2 mL per liter). Preserved samples may be stored up to six months at room temperature. Correct the test result for volume additions; see "Standard Additions" in Section 1.

REAGENT PREPARATION

Prepare the arsenic absorber solution as follows:

1. Weigh 1.00 g of silver diethyldithiocarbamate on an analytical balance.
2. Transfer the powder to a 200-mL volumetric flask. Dilute to volume with pyridine. (Use pyridine only in a fume hood).
3. Mix well to dissolve. Store the reagent, tightly sealed, in an amber bottle. The reagent is stable for one month if stored in this manner. Larger volumes of reagent can be prepared if the reagent is used within one month.

USER CALIBRATION (STANDARD PREPARATION)

1. Prepare a 10.0 mg/L arsenic working standard by pipetting 1.00 mL of Arsenic Standard Solution, 1000 mg/L As, into a 100-mL volumetric flask. Dilute to volume with deionized water.
2. Prepare standards of 0.04, 0.08, 0.12, and 0.16 mg/L arsenic by diluting 1.0, 2.0, 3.0, and 4.0 mL, respectively, of the working standard into four 250-mL volumetric flasks. Dilute to volume with deionized water.

INITIAL SETUP OF ARSENIC PROGRAM

A one-time setup of a program for arsenic is required. An arsenic program template is preprogrammed into memory to make the process easier. After the setup is complete, the calibration can be entered for each new lot of reagents used or as necessary.

a) Press **SHIFT USER PRGM.** Use the **UP** arrow key to scroll to **Copy Program.** Press **ENTER.**

b) Scroll to or enter the template number for arsenic (900). Press **ENTER.**

c) Scroll to or enter the desired user program number for arsenic (>950). Press **ENTER.** Record the program number for reference.

d) The display will show: **Program Copied.**

e) Press **EXIT.** The program is now ready to be calibrated.
Note: The templates within User Program cannot be run directly. They must be copied into a usable program number (greater than 950) as in Step c and d. Then, calibrate the program.

USER CALIBRATION OF ARSENIC PROGRAM

a) Use the test procedure to develop color in the standards just before recording the absorbance values for the calibration.

b) Press **SHIFT USER PRGM.** Use the **UP** arrow to scroll to **Edit Program.** Press **ENTER.**

c) Scroll to or enter the program number for arsenic (from Step c in Setup). Press **ENTER.**

d) Use the **DOWN** arrow to scroll down to **Calib Table: X** (X=denotes a number which indicates the number of data points in the table). Press **ENTER.**

e) The instrument will prompt **Zero Sample.** Place the blank solution (unreacted arsenic absorber) in the cell holder. Close the light shield. Press **ZERO.** The instrument will prompt you to adjust to the proper wavelength if necessary.

f) The first concentration point will be displayed. Press **ENTER** to display the stored absorbance value of the first concentration point.

g) Place the first developed standard solution (same concentration as the value displayed) in the cell holder. Close the light shield. Press **READ** to display the measured absorbance of the standard. Press **ENTER** to accept the displayed absorbance value.

h) The second concentration point will be displayed. Press **ENTER** to display the stored absorbance value of the second concentration. Place the second developed standard solution in the cell holder. Close the light shield. Press **READ** to display the measured absorbance value of the standard.

i) Press **ENTER** to accept the absorbance reading. The next concentration point will then be displayed.

j) Repeat Steps h and i as necessary for the remaining standards.

k) When you are finished reading the absorbance values of the standard, press **EXIT.** Scroll down to **Force Zero.** Press **ENTER** to change the setting. Change to **ON** by pressing the arrow key, then press **ENTER.**

l) Scroll down to **Calib Formula.** Press **ENTER** twice or until only the **0** in **F(0)** is flashing. Press **DOWN** arrow to select **F1** (linear calibration). Press **ENTER** to select **F1.**

m) Press **EXIT** twice. The display will show **Store Changes?.** Press **ENTER** to confirm.

n) Press **EXIT.** The program is now calibrated and ready for use. Start on step 2 of the iconed procedure.
Note: Other calibration fits may be used if appropriate.

INTERFERENCES

Antimony salts may interfere with color development.

QUESTIONS AND ANSWERS

1. How much arsenic can a human body tolerate?
Ingestion of a small amount (100 mg) of arsenic can cause severe poisoning to the human body. At low intake levels, chronic effects can appear from its accumulation in the body. Arsenic may also be carcinogenic to some degree.

2. How much arsenic can be found in potable waters?
Arsenic concentrations in potable waters usually are less than 10 micrograms per liter. In certain areas, the arsenic concentration in potable water has exceeded 90 to 100 micrograms per liter.

3. Why is digestion of the sample necessary during analysis?
Sample digestion is necessary to be sure that all the solubilized particulate arsenic and organic arsenic compounds are converted to inorganic arsenic, which can be ultimately converted to a soluble red complex after reaction with silver diethyldithiocarbamate in pyridine.

4. What are the interferences during the test for arsenic?
Antimony in the sample may interfere with the red color development. Metals (copper, nickel, cobalt, chromium, mercury, silver, platinum, and molybdenum) are sometimes present in significant concentrations in water and may interfere in the generation of arsine during the distillation process. Usually, the metal concentrations in water are low and do not impact arsenic results.

REFERENCES

1. *Standard Methods for the Examination of Water and Wastewater:* ARSENIC, 16th Ed., A. E. Greenberg, R. R. Trussell, L. S. Clesceri and Mary Ann H. Franson (eds.), pp. 165–166 and 187–189, Washington, DC, American Public Health Association, 1985.
2. *HACH Water Analysis Handbook, 3rd Ed.* (Loveland, CO.: HACH Company, 1997).
3. *Hawley's Condensed Chemical Dictionary,* 11th Ed., rev'd by N. Irving Sax and Richard J. Lewis (New York, N.Y.: Van Nostrand Reinhold, 1987).

BARIUM

Barium impacts the heart, blood vessels and nerves of the human body. Barium can get into the body by inhalation, consumption or absorption. A barium dose of 550 to 600 milligrams (mg) is considered fatal to human beings. Despite a relatively high amount of barium in nature, water contains a trace amount of barium. The barium concentration in drinking water averages slightly less than 50 micrograms per liter. Higher concentrations in excess of 900 micrograms of barium per liter of drinking water may suggest an industrial pollution problem.

In general, a number of barium compounds are increasingly soluble in water as the temperature of the water increases above 0°C. For example, 5.0 grams of barium nitrate is soluble in 100 mL of water at 0°C, whereas 7.2 grams of the material is soluble in 100 mL of water at 10°C. A container of barium material discarded in a water body could present greater problems as the temperature of the water increases.

Barium is sometimes found with radioactive substances (Radium 228 and Radium 226) in water and suspended matter. Radioactive barium is usually present with radioactive strontium in water. Barium compounds are often used in medical diagnosis applications and in medical research.

Barium has been used as a lubricant additive and as an alloy in vacuum tubes and spark plugs. Applications of barium compounds have been the following: paint and varnish driers, textile mordant, explosives and pyrotechnics, corrosion inhibitor, photographic compounds, optical glass, rodenticide, safety matches, paint pigment and stabilizer, metallurgy, x-ray screens, lubricant applications, electronics, insecticide, steel deoxidizers, ceramic insulation, plastic stabilizer, electroplating, paper manufacturing and rubber preparations.

Many industries try to implement a "semi-closed" loop for barium compounds where the more toxic compounds are used internally in the process. Additional industries attempt to minimize release of barium by precipitation and other methods. Some industries manifest the waste stream, while others may sell the "waste" stream to another industry for a different application. A few industries still dump barium wastes in diluted form into water bodies. Hospital and research facility waste streams often contain low level radioactive wastes which are buried or treated by other means. Often, barium-containing wastes are discarded in landfills.

In the test for barium, the BariVer 4 Barium Reagent Powder reacts with barium to form a barium sulfate precipitate. The barium precipitate is held in suspension by a protective colloid and forms a solution with a fine white dispersion of particles. The amount of turbidity caused by the particle dispersion is directly proportional to the amount of barium present. Strontium interferes at any level of barium. The total concentration of both strontium and barium may be expressed as PS (Precipitated by Sulfate) and will provide an accurate indication of scaling tendency.

BARIUM (0 to 100 mg/L)

Method 8014 for water, wastewater, oil-field water and seawater*
Turbidimetric Method**
Powder Pillow Method

1. Perform a User-Entered Calibration to obtain the most accurate results. See **User Calibration Section** following this procedure. Programs 20 and 25 can be used directly for process control or applications where a high degree of accuracy is not needed.
Note: Reagent lot variation and the nature of turbidity testing require user calibration for best results.

2. Enter the stored program number for barium (Ba). Press **2 0 ENTER** or **9 ? ? ENTER.** The display will show: **Dial to 450 nm.**
Note: The Pour-Thru Cell cannot be used with this procedure.

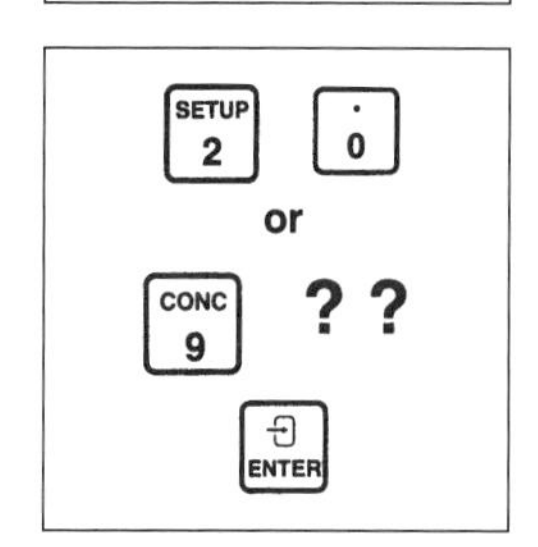

*Compliments of HACH Company.

**Adapted from Snell and Snell, *Colorimetric Methods of Analysis,* Vol. II, 769 (1959).

3. Rotate the wavelength dial until the small display shows: **450 nm.** When the correct wavelength is dialed in the display will quickly show: **Zero Sample,** then **mg/L Ba.**

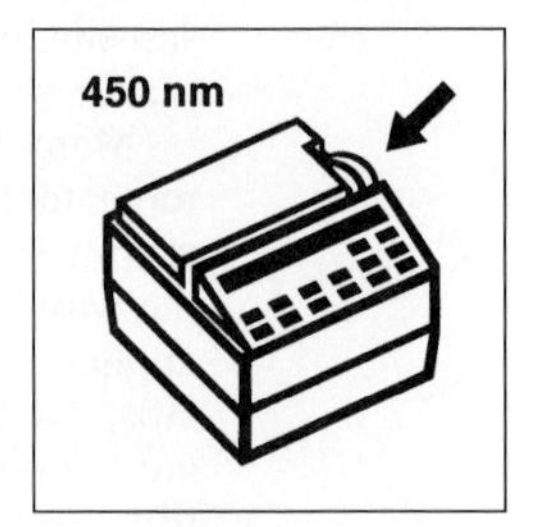

4. Fill a sample cell with 25 mL of sample.
Note: Filter highly colored or turbid samples. Large amounts of color or turbidity will cause high readings. Use the filtered sample in Steps 4 and 7.

5. Add the contents to one BariVer 4 Barium Reagent Powder Pillow to the cell (the prepared sample). Swirl to mix.
Note: A white turbidity will develop if barium is present.

6. Press: **SHIFT TIMER.** A 5-minute reaction period will begin.
Note: Do not disturb the sample during this period. If the BariVer 4 Barium Reagent does not dissolve, mix the reagent and sample in a 25-mL graduated cylinder before pouring it into the sample cell.

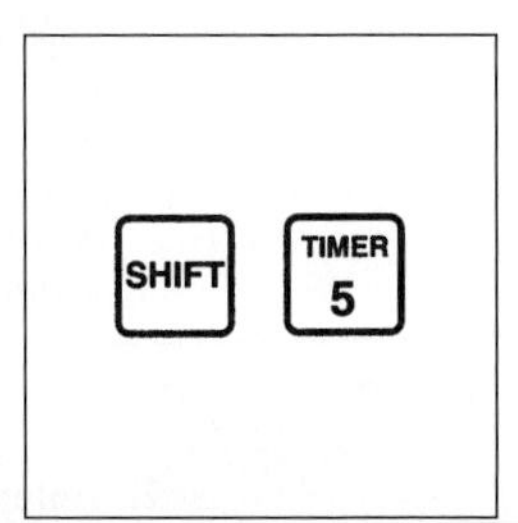

7. Fill another sample cell (the blank) with 25 mL of sample.
Note: Determine a reagent blank for each new lot of reagent by repeating Steps 4 through 11 using deionized water as the sample.

8. When the timer beeps, the display will show: **mg/L Ba.** Place the blank into the cell holder. Close the light shield.

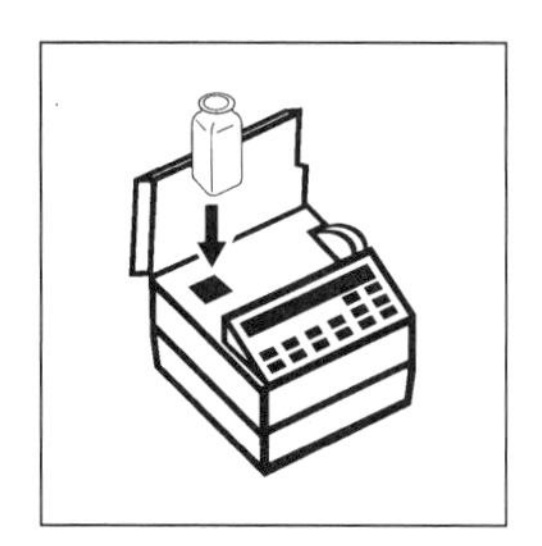

9. Press: **ZERO.** The display will show: **Zeroing ...** then **0.mg/L Ba.**

10. Within 5 minutes after the timer beeps, place the prepared sample into the cell holder. Close the light shield.

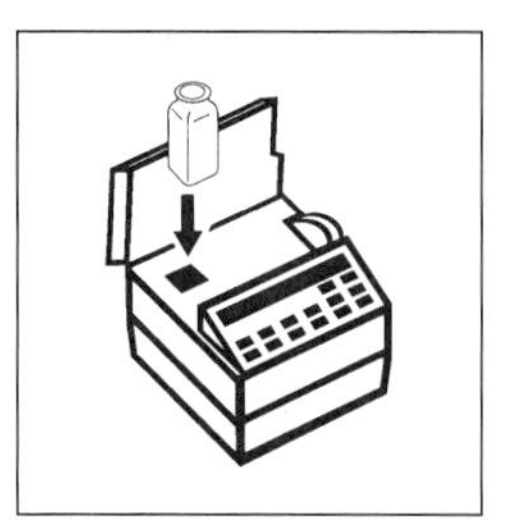

11. Press: **READ.** The display will show: **Reading ...** then the result in mg/L barium will be displayed.
Note: Clean the sample cell immediately after each test with soap, water, and a brush to prevent a film of barium sulfate from developing on the inside of the sample cell.

SAMPLING AND STORAGE

Collect samples in an acid cleaned glass or plastic container. Adjust the pH to 2 or less with nitric acid (about 2 mL per liter). Preserved samples can be stored up to six months at room temperature. Adjust the pH to 5 with 5.0 N Sodium Hydroxide before analysis. Correct the test result for volume additions; see "Correcting for Volume Additions" in Section 1.

ACCURACY CHECK

See HACH Water Analysis Handbook, 3rd Ed., 1997, p. 253.

USER CALIBRATION

For most accurate results, the use of a user-calibrated program is highly recommended. The HACH stored programs 20 and 25 are intended for process control samples or other applications where a high degree of accuracy is not necessary.

A one-time setup of a user program for barium is required. A barium program template is pre-programmed into memory to make the process easier. After the setup is complete, the calibration can be entered for each new lot of reagents or as necessary.

STANDARD PREPARATION

A new calibration should be performed for each new lot of BariVer 4 Barium Reagent as follows:

a) Prepare calibration standards of 10, 20, 30, 50, 80, 90 and 100 mg/L Ba by pipetting 1, 2, 3, 5, 8, 9, and 10 mL of the 1000 mg/L Barium Standard Solution into 100-mL volumetric flasks.

b) Dilute to the mark with deionized water. Mix thoroughly.

c) Use this stored program number in the Powder Pillow Procedure above. Prepare a new calibration for each new lot of reagent, using the same stored program number.

INITIAL SETUP OF BARIUM PROGRAM

a) Press **SHIFT USER-PRGM.** Use the **UP** arrow key to scroll to **Copy Program.** Press **ENTER.**

b) Scroll to or enter the template number for barium [901, 902 (AV)]. Press **ENTER.**

c) Scroll to or enter the desired user program number for barium (>950). Press **ENTER.** Record the program number for reference.

d) The display will show **Program Copied.** Press **EXIT.** The program is now ready to be calibrated.
Note: The templates within User Program cannot be run directly. They must be copied into a usable program number (greater than 950) as in Steps c and d. The program must then be calibrated.

USER CALIBRATION OF BARIUM PROGRAM

a) Use the test procedure to develop the turbidity in the standards just before recording the absorbance values for the calibration.

b) Press **SHIFT USER PRGM.** Use the **UP** arrow to scroll to **Edit Program.** Press **ENTER.**

c) Scroll to or enter the program number for barium (from Step c in Setup). Press **ENTER.**

d) Use the **DOWN** arrow to scroll down to **Calib Table:X** (X=denotes a number which indicates the number of data points in the table). Press **ENTER.**

e) The instrument will prompt **Zero Sample.** Place the blank solution in the cell holder. Close the light shield. Press **ZERO.** The instrument will prompt you to adjust to the proper wavelength if necessary.

f) The first concentration point will be displayed. Press **ENTER** to display the stored absorbance value of the first concentration point.

g) Place the first developed standard solution (same concentration as the value displayed) in the cell holder. Close the light shield. Press **READ** to display the measured absorbance of the standard. Press **ENTER** to accept the displayed absorbance value.

h) The second concentration point will be displayed. Press **ENTER** to display the stored absorbance value of the second concentration. Place the second developed standard solution in the cell holder. Close the light shield. Press **READ** to display the measured absorbance value of the standard.

i) Press **ENTER** to accept the absorbance reading. The next concentration point will then be displayed.

j) Repeat Steps h and i as necessary for the remaining standards.

k) When you are finished reading the absorbance values of the standards, press **EXIT.** Scroll down to **Force Zero.** Press **ENTER** to change the setting. Change to **ON** by pressing the arrow key, then press **ENTER.**

l) Scroll down to **Calib Formula.** Press **ENTER** twice or until only the **0** in **F(0)** is flashing. Press **DOWN** arrow to select **F3** (cubic calibration). Press **ENTER** to select **F3.**

m) Press **EXIT** twice. The display will show **Store Changes?** Press **ENTER** to confirm.

n) Press **EXIT.** The program is now calibrated and ready for use. Start on Step 2 of the iconed procedure.

INTERFERENCES

The following may interfere when present in concentrations exceeding those listed below:

Silica	500 mg/L
Sodium Chloride	130,000 mg/L as NaCl
Magnesium	100,000 mg/L as $CaCO_3$
Calcium	10,000 mg/L as $CaCO_3$
Strontium	Interferes at any level

If strontium is known to be present, the total concentration between barium and strontium may be expressed as a PS (Precipitated by Sulfate). While this does not distinguish between barium and strontium, it gives an accurate indication of scaling tendency.

Highly buffered samples or extreme sample pH may exceed the buffering capacity of the reagents and require sample pretreatment; see "pH Interference" in Section 1.

QUESTIONS AND ANSWERS

1. How is barium determined in this experiment?
The BariVer 4 Barium Reagent Powder combines with barium to form a barium sulfate precipitate, which is held in suspension by a protective colloid. The amount of turbidity present caused by the fine white dispersion of particles is directtly proportional to the amount of barium present in the sample.

2. What are some of the interferences when testing for barium in the environment?
Strontium interferes at any level in water. Low levels of radioactive barium and radioactive strontium are usually present at the same locations in soil and water. Also, barium is usually found in the presence of radium nuclides. If present in significant amounts, barium should be precipitated with sulfate from the radium and strontium nuclides before analysis.
3. Why is the sample cell cleaned immediately after taking the reading for barium?
The sample cell is cleaned immediately after taking the reading to prevent a film of barium sulfate from developing on the inside surface of the sample cell.
4. How much barium can a human body tolerate?
Humans can tolerate less than 550 to 600 milligrams of barium. Much less barium is required to impact the blood vessels, nerves and heart. Barium stimulates the heart muscle.

REFERENCES

1. *Standard Methods for the Examination of Water and Wastewater:* BARIUM, 16th Ed., A. E. Greenberg, R. R. Trussell, L. S. Clesceri and Mary Ann H. Franson (eds.), pp. 191, 646–647 and 667–668, Washington, DC, American Public Health Association, 1985.
2. *HACH Water Analysis Handbook,* 3rd Ed. (Loveland, CO.: HACH Company, 1997).
3. *Hawley's Condensed Chemical Dictionary,* 11th Ed., rev'd by N. Irving Sax and Richard J. Lewis (New York, N.Y.: Van Nostrand Reinhold, 1987).

BROMINE

Bromine is unstable in natural waters and easily forms bromide in water containing other chemicals. The concentration of bromine in natural water is essentially non-existent, whereas the concentration of bromide in drinking water seldom exceeds 1 mg/L. The reaction and decomposition of bromine in water is influenced by reactant types and concentrations, sunlight, salinity, pH and temperature.

The compound is a strong oxidizing agent and reacts rapidly with various inorganic compounds. Bromine slowly oxidizes many organic compounds and often reacts directly with specific organic compounds (phenols, alkoxyaromatics and aniline-type) to form brominated complexes in water at normal temperatures. Normally, the brominated organic compounds are insoluble in water and are more stable in water than the reactants. These reactions occur even with relatively low concentrations of bromine in aqueous industrial effluents.

Bromine is used in organic synthesis, bleaching, water purification, fumigant preparation, fire retardant for plastics, dyes, pharmaceuticals, photography, and shrink-proofing wool. Bromo-organic compounds are used heavily in industrial organic synthesis and as fumigants, germicides, and fungicides.

Bromine wastes from a number of sources are often flushed down the drain with water or buried as contaminated wastes with other chemicals. In specific cases, the bomine-contaminated waste stream is recycled in-house or utilized by an industrial site for another purpose.

Samples suspected of containing bromine should be analyzed immediately after collection to avoid changes of bromine to bromide. A number of interferences such as chlorine, iodine, ozone, and bromamines in the sample may react and show as bromine. Oxidized forms of manganese and chromium may require some additional sample pretreatment before tests for bromine. In the experiment, bromine reacts with DPD (N,N-diethyl-p-phenylenediamine) to form a magenta color which is proportional to the total bromine concentration. The concentration of the bromine in the sample may be determined in the field by the DR/820, DR/850 and DR/890 Colorimeters or the DR/2010 Spectrophotometer. Bromine can also be determined by titration methods.

BROMINE (0 to 4.50 mg/L)

DPD Method*
Powder Pillows
Method 8016 for water, wastewater, and seawater**

1. Enter the stored program number for bromine (Br_2) powder pillows. Press: **5 0 ENTER.** The display will show: **Dial nm to 530.**
Note: The Pour-Thru Cell can be used with 25 mL reagents only.

2. Rotate the wavelength dial until the small display shows: **530 nm.** When the correct wavelength is dialed in the display will quickly show: **Zero Sample,** then: **mg/L Br_2.**

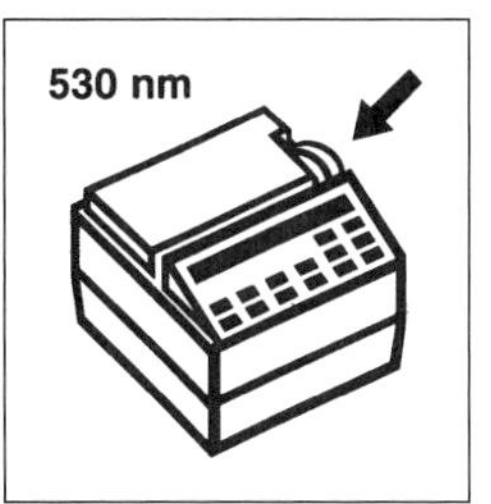

3. Insert the 10 mL Cell Riser into the cell compartment.

4. Fill a sample cell with 10 mL of sample.
Note: Samples must be analyzed immediately.

*Adapted from *Standard Methods for the Examination of Water and Wastewater.*
***Compliments of HACH Company. Analysis may be performed by DR/820, DR/850, and DR/890 Colorimeters and DR/2010 Spectrophotometer.*

5. Add the contents of one DPD Total Chlorine Powder Pillow to the sample cell (the prepared sample). Swirl to mix. ***Note:*** A pink color will develop if bromine is present.

6. Press: **SHIFT TIMER.** A 3-minute reaction period will begin.

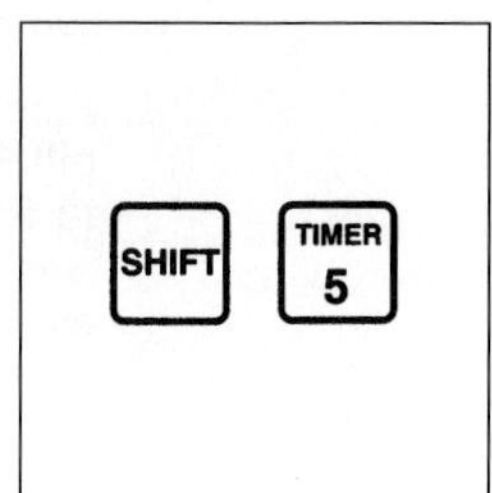

7. When the timer beeps, the display will show: **mg/L Br_2**. Fill a second sample cell (the blank) with 10 mL of sample.

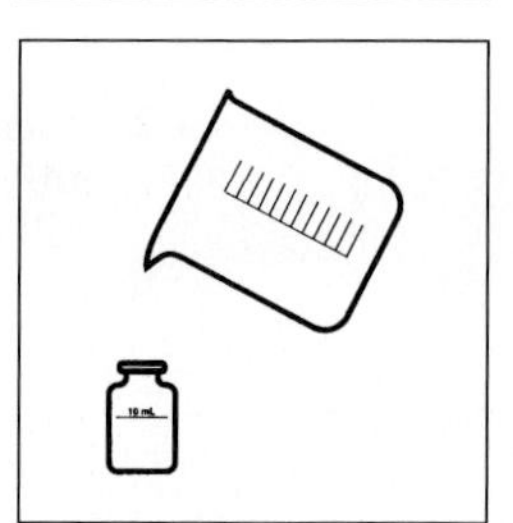

8. Place the blank into the cell holder. Close the light shield.

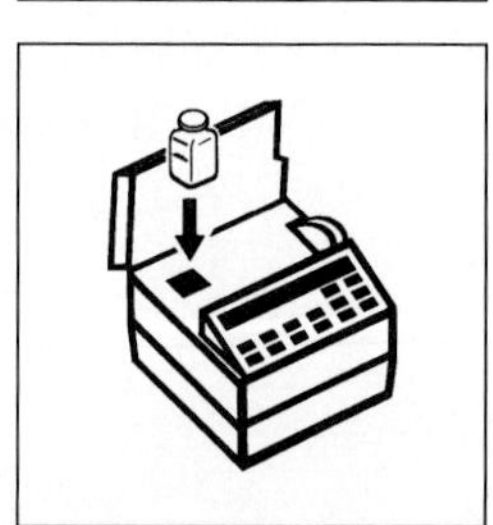

9. Press: **ZERO.** The display will show: **Zeroing ...** then **0.00 mg/L Br_2**.

10. Within 3 minutes after the timer beeps, place the prepared sample into the cellholder. Close the light shield.

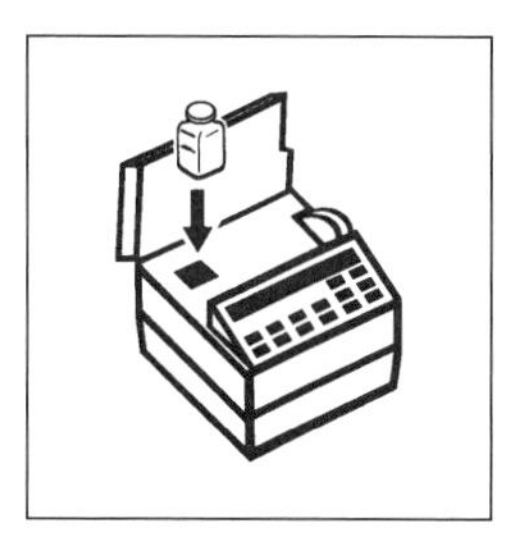

11. Press: **READ.** The display will show: **Reading ...** then the result in mg/L Br_2 will be displayed.
Note: If the sample temporarily turns yellow after reagent addition, or shows **OVER-RANGE,** dilute a fresh sample and repeat the test. A slight loss of bromine may occur during dilution. Multiply the result by the appropriate dilution factor; see "Sample Dilution Techniques" in Section 1.

SAMPLING AND STORAGE

Analyze samples for bromine **immediately** after collection. Bromine is a strong oxidizing agent, and it is unstable in natural waters. It reacts rapidly with various inorganic compounds and more slowly oxidizes organic compounds. Many factors, including reactant concentrations, sunlight, pH, temperature, and salinity influence decomposition of bromine in water.

Avoid plastic containers since these my have a large bromine demand. Pretreat **glass** sample containers to remove any bromine demand by soaking in a dilute bleach solution (1 mL commercial bleach to 1 liter of deionized water) for at least one hour. Rinse thoroughly with deionized or distilled water. If sample containers are rinsed thoroughly with deionized or distilled water after use, only occasional pre-treatment is necessary.

A common error in testing for bromine is introduced when a representative sample is not obtained. If sampling from a tap, let the water flow for at least 5 minutes to ensure a representative sample. Let the container overflow with the sample several times, then cap the sample container so there is no headspace (air) above the sample. If sampling with a sample cell, rinse the cell several times with the sample, then carefully fill to the 10 mL mark. Perform the bromine analysis immediately.

ACCURACY CHECK

See HACH Water Analysis Handbook, 3rd Ed., p. 279.

INTERFERENCES

Samples containing more than 300 mg/L alkalinity or 150 mg/L acidity as $CaCO_3$ may not develop the full amount of color, or it may instantly fade. Neutralize these samples to a pH of 6 to 7 with 1 N sulfuric acid or 1 N sodium hydroxide. Determine the amount required on a separate 10 mL sample. Add the same amount to the sample to be tested. Correct the test result for volume additions; see "Correction for Volume Additions" in Section 1 for more information.

Chlorine, iodine, ozone, bromamines, and oxidized forms of manganese and chromium also may react and show as bromine. Compensate for the effects of oxidized manganese or chromium by adjusting the pH to 6 to 7 as described above. Add 3 drops of 30 g/L potassium iodide to 25 mL of sample, mix, and wait one minute. Add 3 drops of 5 g/L Sodium Arsenite and mix. Analyze 10 mL of this sample as described above. (If chromium is present, allow exactly the same reaction period with DPD for both analyses.) Subtract the result of this test from the original analysis to obtain the correct bromine result.

DPD Total Chlorine Reagent Powder Pillows contain a buffer formulation which will withstand high (>1000 mg/L) levels of hardness without interference.

QUESTIONS AND ANSWERS

1. Where would the investigator find bromine?
Bromine could be found in industrial discharges, laboratory effluents or in discarded containers containing bromine solutions.

2. How would an investigator find evidence of bromine in water effluents or in streams?
Water with the most severe case of bromine contamination would have a reddish color. Since bromine easily oxidizes both inorganic compounds and some organic compounds, bromine would be present as bromide in water. Bromine in water can react with specific organic complexes to form bromo-organic compounds at normal temperatures.

3. Why should a sample suspected of containing bromine be tested immediately?
Bromine is easily reduced to bromide in aqueous solution containing contaminates. In water, bromine is a strong oxidizing agent and reacts easily at normal temperatures with metal cations, anions, and many organic complexes. Special care must be taken to avoid bromine demand.

REFERENCES

1. *HACH Water Analysis Handbook,* 3rd Ed., (Loveland, CO.: HACH Company, 1997).
2. Donald Pavia, Gary Lampman, and George Kriz, *Introduction to Organic Laboratory Techniques,* 3rd Ed., (New York, N.Y.: Saunders College Publishing and Harcourt Brace College Publishers, 1988) at 453.
3. *Hawley's Condensed Chemical Dictionary,* 11th Ed., rev'd by N. Irving Sax and Richard J. Lewis (New York, N.Y.: Van Nostrand Reinhold, 1987).
4. *Standard Methods for the Examination of Water and Wastewater:* BROMIDE, 16th Ed., A. E. Greenberg, R. R. Trussell, L. S. Clesceri and Mary Ann H. Franson (eds.), pp. 278, Washington, DC, American Public Health Association, 1985.

CADMIUM

Cadmium is highly toxic and has been implicated in some cases of food poisoning. Cadmium causes some types of cancers in laboratory animals and has been linked to human cancers. Small quantities of cadmium appear to produce adverse changes in arteries of human kidneys.

Levels of cadmium in water have been reported as high as 60 micrograms per liter with an average near 8 micrograms per liter. A cadmium concentration of nearly 200 micrograms per liter appears to be toxic to trout and to other specific types of fish. A number of cadmium compounds are increasingly soluble in water as the temperature of the water increases above 0°C. Water contamination from cadmium-containing items is a common occurrence.

Cadmium may enter water through industrial discharges and through the deterioration of galvanized pipes. Even though cadmium is fairly stable as a metal, cadmium metal exposed to acid, bases, and electrical current has a tendency to dissolve or deteriorate over a period of time. Cadmium metal has practical applications in metal coatings, alloys, batteries, transmission wire, enamels, pigments and glazes, fungicides, photography, photoelectric cells, and several electrical devices. As compounds, cadmium can be found in fungicides, pigments, ceramics semiconductors, reagent chemicals, photography, optical applications, mirror manufacturing, electroplating, lubricants, phosphors, catalysts, solar cells, and several additional applications.

Industries manufacturing cadmium and cadmium compounds must protect the water bodies by eliminating discharges of the metal and compounds to the environment. A number of industries have implemented closed loop systems to recover and reuse waste, while other facilities manifest waste for landfilling purposes. Some cadmium materials are sold to other companies to produce additional items. A number of materials containing cadmium is discarded in the garbage.

Some of the chemicals used in the determination of cadmium should be stored away from light and heat. A treatment procedure can eliminate the interferences caused by the presence of copper, bismuth, mercury and silver in the experiment. During the analysis, cadmium ions react with dithizone in basic solution to form a pink to red complex that is extracted with chloroform. The cadmium-dithizonate complex in chloroform is measured photometrically at 515 nm.

CADMIUM (0 to 80 μg/L)

Method 8017 for water and wastewater*
Dithizone Method**

1. Enter the stored program number for cadmium (Cd). Press: **6 0 ENTER.** The display will show: **Dial nm to 515.**
   Note: The Pour-Thru Cell cannot be used.
   Note: Clean all glassware with Nitric Acid Solution, 1:1. Rinse with deionized water.

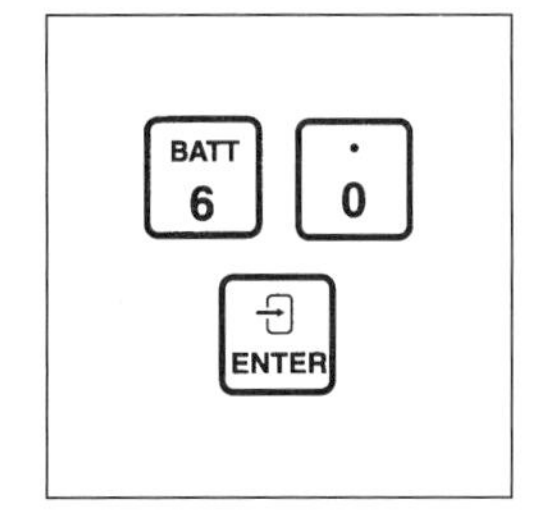

2. Rotate the wavelength dial until the small display shows: **515 nm.** When the correct wavelength is dialed in the display will quickly show: **Zero Sample,** then: **μg/L Cd.**
   Note: Total cadmium determination requires a digestion; see "Digestion Procedure and Sample and Analysis Volume Table for Aqueous Liquids" in Section 1.

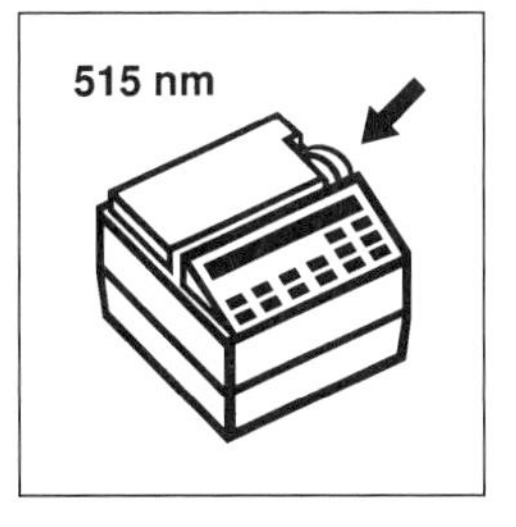

* Compliments of HACH Company.
** Adapted from *Standard Methods for the Examination of Water and Wastewater, 310B.*

3. Fill a 250-mL graduated cylinder to the 250-mL mark with sample. Pour the sample into a 500-mL separatory funnel.
Note: Cloudy and turbid samples may require filtering with a glass membrane filter before running test. Report results as μg/L soluble cadmium.

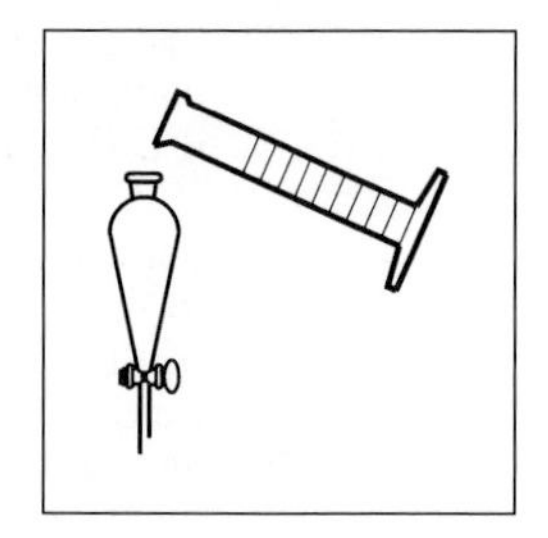

4. Add the contents of one Buffer Powder Pillow, citrate type for heavy metals. Stopper the funnel. Shake to dissolve.
Note: For best results, determine a reagent blank. Use deionized water in place of the sample. Subtract the reagent blank value from each sample reading. Repeat for each new lot of reagent.

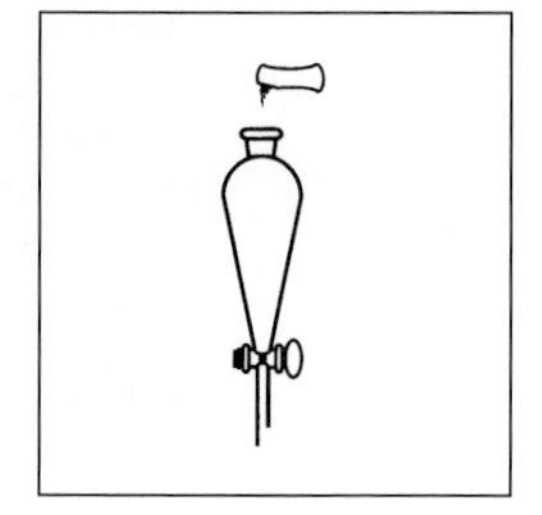

5. Add 30 mL of chloroform to a 50-mL mixing graduated cylinder. Add the contents of one DithiVer™ Metals Reagent Powder Pillow. Stopper. Invert repeatedly to mix (DithiVer Solution).
Note: Use adequate ventilation. The DithiVer powder will not all dissolve. See DithiVer Solution Preparation and Storage below.

6. Add 20 mL of 50% Sodium Hydroxide Solution and then a 0.1-g scoop of potassium cyanide to the funnel. Shake vigorously for 15 seconds. Remove the stopper and let stand for one minute.
Note: Spilled reagent will affect test accuracy and is hazardous.

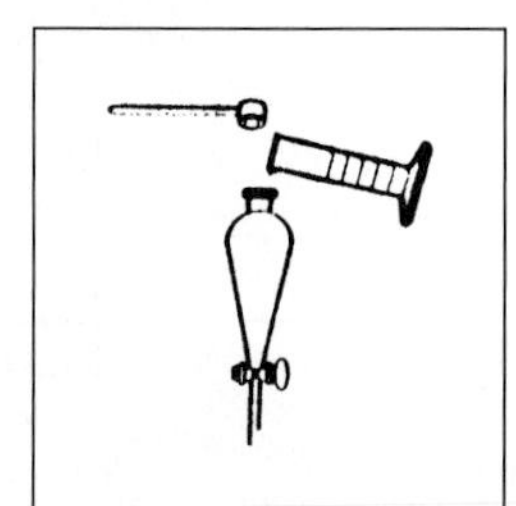

7. Add 30 mL of the DithiVer Solution to the 500-mL separatory funnel. Stopper, invert, and open stopcock to vent. Close the stopcock and shake funnel once or twice; vent again. Close the stopcock and shake the funnel vigorously for 60 seconds.

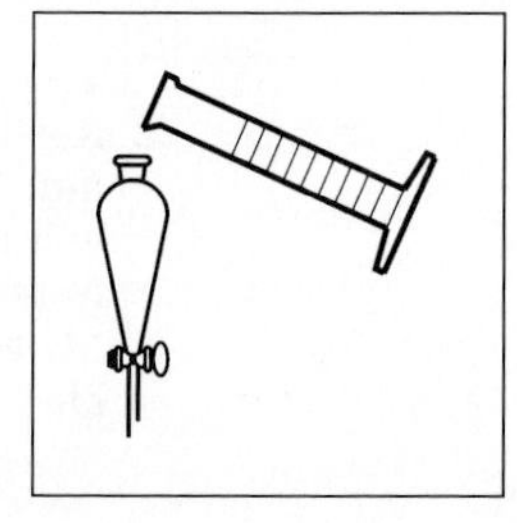

8. Let the funnel stand undisturbed for roughly one minute. ***Note:*** The bottom (chloroform) layer will be pink if cadmium is present.

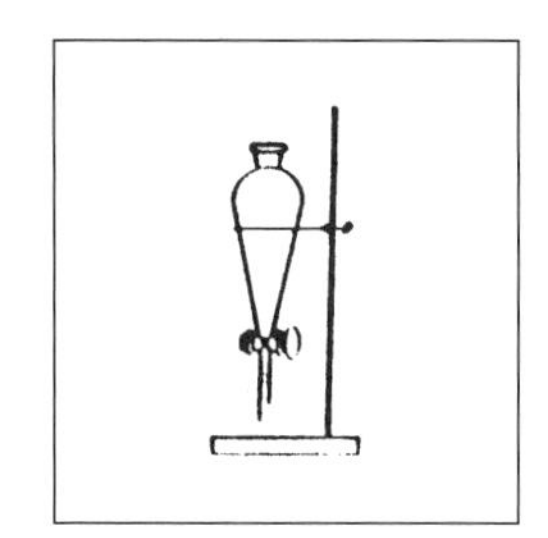

9. Insert a cotton plug the size of a pea into the delivery tube of the funnel and slowly drain the bottom (chloroform) layer into a dry 25-mL sample cell (the prepared sample). Cap. ***Note:*** The cadmium-dithizone complex is stable for hours if the sample cell is kept tightly capped and out of direct sunlight.

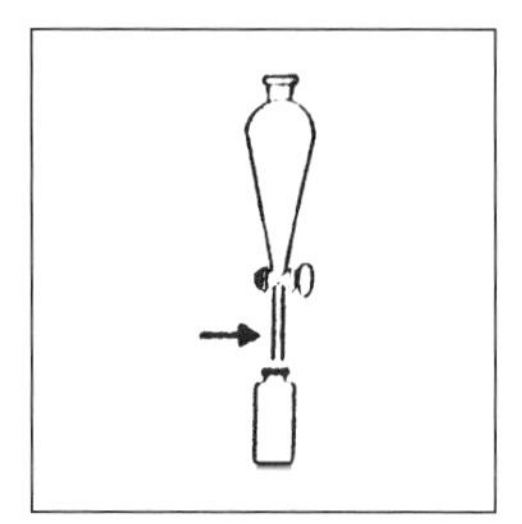

10. Fill a dry 25-mL sample cell with chloroform (the blank). Cap.

11. Place the blank into the cell holder. Close the light shield.

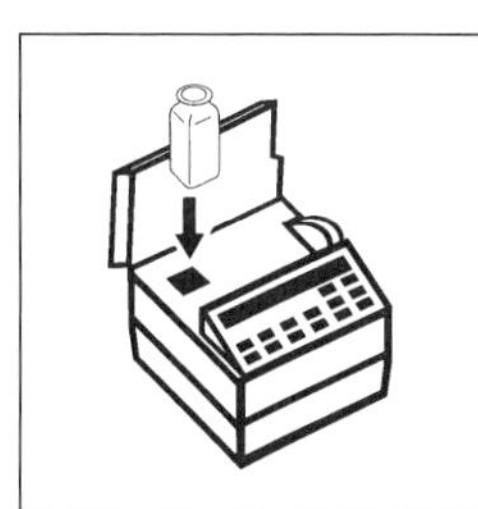

12. Press: **ZERO.** The display will show: **Zeroing ...,** then **0. μg/L Cd.**

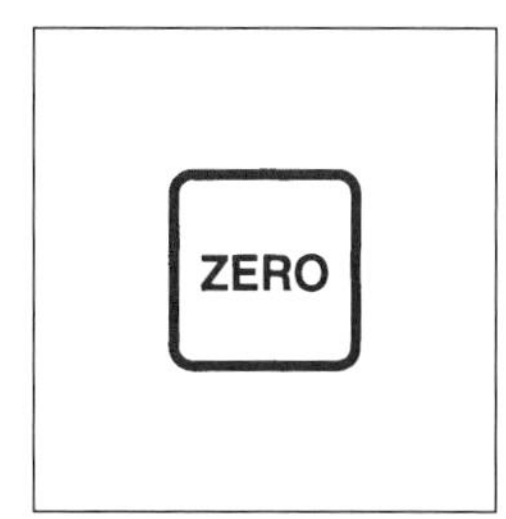

13. Place the prepared sample into the cell holder. Close the light shield.

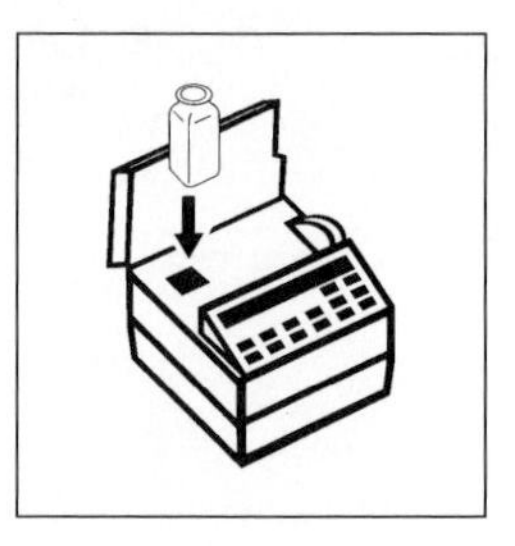

14. Press: **READ.** The display will show: **Reading ...** then the result in $\mu g/L$ cadmium will be displayed.

SAFETY PRECAUTIONS

Perform the entire procedure in a fume hood if possible. Using gloves that are resistant to chloroform solutions (i.e. VITON™) and safety goggles is recommended.

SAMPLING AND STORAGE

Collect samples in acid cleaned glass or plastic container. Adjust the pH to 2 or less with nitric acid. (about 2mL per liter). Store preserved samples up to six months at room temperature. Adjust the pH to 2.5 with 5.0 N Sodium Hydroxide before analysis. Correct the test result for volume additions; see "Correcting for Volume Additions" in Section 1.

DITHIVER SOLUTION PREPARATION AND STORAGE

Store DithiVer powder pillows away from heat and light. A convenient way to prepare this solution is to add the contents of 16 DithiVer Metals Reagent Powder Pillows to a pint bottle of chloroform and invert several times until well mixed (carrier powder may not dissolve). Store dithizone solution in an amber glass bottle. This solution is stable for 24 hours.

ACCURACY CHECK (STANDARD ADDITIONS METHOD AND STANDARD SOLUTION METHOD)

See HACH Water Analysis Handbook, 3rd Ed., 1997, p. 284–285.

INTERFERENCES

The following (Table 2.10) do not interfere:

TABLE 2.10

Aluminum	Lead
Antimony	Magnesium
Arsenic	Manganese
Calcium	Nickel
Chromium	Tin
Cobalt	Zinc
Iron	

The following interfere with cadmium causing high results when present in concentrations exceeding those listed below (Table 2.11):

TABLE 2.11

Copper	2 mg/L
Bismuth	80 mg/L
Mercury	all levels
Silver	2 mg/L

Eliminate interference from these metals by the following treatment, beginning after Step 5.

a) Measure about 5 mL of the DithiVer solution into the separatory funnel. Stopper the funnel, invert and open the stopcock to vent. Close the stopcock and shake the solution vigorously for 15 seconds. Allow the funnel to stand undisturbed until the layers separate (about 30 seconds). A yellow, red or bronze color in the bottom (chloroform) layer confirms the presence of interfering metals. Draw off and discard the bottom (chloroform) layer.

b) Repeat extraction with fresh 5-mL portions of the DithiVer Solution (discarding the bottom layer each time) until the bottom layer shows a pure dark green color for three successive extracts. Extractions can be repeated several times without appreciably affecting the amount of cadmium in the sample.

c) Extract the solution with several 2- or 3-mL portions of pure chloroform to remove any remaining DithiVer, again discarding the bottom layer each time.

d) Continue with Step 6.

e) In Step 7, substitute 28.5 mL of DithiVer Solution for the 30 mL.

f) Continue with Step 8.

Highly buffered samples or extreme sample pH may exceed the buffering capacity of the reagents and require sample pretreatment; see "pH Interference" in Section 1.

POLLUTION PREVENTION AND WASTE MANAGEMENT

Both chloroform (D002) and cyanide (D003) solutions are regulated as hazardous wastes by the Federal RCRA. Do **not** pour these solutions down the drain. Collect chloroform solutions and the cotton plugs used in the delivery tube of the separatory funnel for disposal with laboratory solvent waste. Be sure that cyanide solutions are stored in a caustic solution with a pH > 11 to prevent release of hydrogen cyanide gas.

QUESTIONS AND ANSWERS

1. Can the human body tolerate cadmium?
   Very small quantities of cadmium appear to produce adverse changes in arteries of human kidneys. Cadmium is highly toxic and has been linked epidemiologically with specific human cancers. Cadmium has also been responsible for food poisoning.
2. How does cadmium get into the water supply?
   Cadmium metal and ions tend to get into water through industrial discharges. Galvanized pipes containing cadmium metal tend to deteriorate with time and may allow small amounts of cadmium metal to dissolve in water, especially if the pipe is not properly grounded.
3. What interferences cause error in the measurement of cadmium?
   High results for cadmium are caused by the presence of mercury (all levels), copper (>2 mg/L), bismuth (>80 mg/L) and silver (> 2 mg/L) when using the dithizone method. Most metals do not cause error in the measurement of cadmium.
4. Why should the DithiVer Powder Pillows be stored in a cool and dark place in the laboratory?
   DithiVer Powder Pillows tend to decompose slowly in light and heat. Even chloroform solutions of the powder should be stored in an amber glass bottle.

REFERENCES

1. *Standard Methods for the Examination of Water and Wastewater:* Cadmium, 16th Ed., A. E. Greenberg, R. R. Trussell, L. S. Clesceri and Mary Ann H. Franson (eds.), pp. 193–196, Washington, DC, American Public Health Association, 1985.
2. *HACH Water Analysis Handbook,* 3rd Ed. (Loveland, CO.: HACH Company, 1997).
3. *Hawley's Condensed Chemical Dictionary,* 11th Ed., rev'd by N. Irving Sax and Richard J. Lewis (New York, N.Y.: Van Nostrand Reinhold, 1987).

CARBON DIOXIDE

Surface waters normally contain less than 10 mg of free carbon dioxide per liter, whereas ground waters often exceed that concentration. As the carbon dioxide concentration in air increases, the carbon dioxide in water should also increase depending on water temperature, flow rates, plant types and the level of contaminates. The carbon dioxide content in water may contribute to corrosion problems.

During water softening operations, recarbonation of a water supply is a common treatment procedure. Carbonation of soft drinks is enjoyed by most people. Carbonation processes are also used to package items to prevent air oxidation and growth of specific harmful species.

Solid carbon dioxide, dry ice, has many applications for dramatically cooling liquid substances and displaying clouds of vapors.

Metal ions (Al, Cu, Cr, Fe) that easily precipitate in solution by forming hydroxide complexes can contribute to high results. The presence of weak bases (such as ammonia and amines) and by salts of a few acids and bases (such as sulfide, nitrite, silicate, borate, and phosphate) in the sample can produce positive errors if the substances exceed 5% of the CO_2 concentration in the sample. Negative errors can be observed if total solids in the sample are too high or by the addition of excess indicator.

The phenolphthalein indicator test provides an approximation of free carbon dioxide in water in the range 10 to 1000mg/L as CO_2. This procedure is appropriate for field tests and for control and routine applications. However, this method of analysis is not applicable to samples of acid mine wastes. In the experiment, a water sample is titrated with a Sodium Hydroxide Standard Solution to a phenolphthalein end point at pH=8.3. Aeration of the sample by swirling or prolonged exposure to air must be avoided to prevent loss of carbon dioxide to the air.

CARBON DIOXIDE (10-1000 mg/L as CO_2)

Method 8205*
Digital Titrator Method using Sodium Hydroxide

1. Select a sample size and a Sodium Hydroxide (NaOH) Titration Cartridge corresponding to the expected carbon dioxide (CO_2) concentration; see Table 2.12.
Note: See Sampling and Storage of the HACH Water Analysis Handbook following these steps.

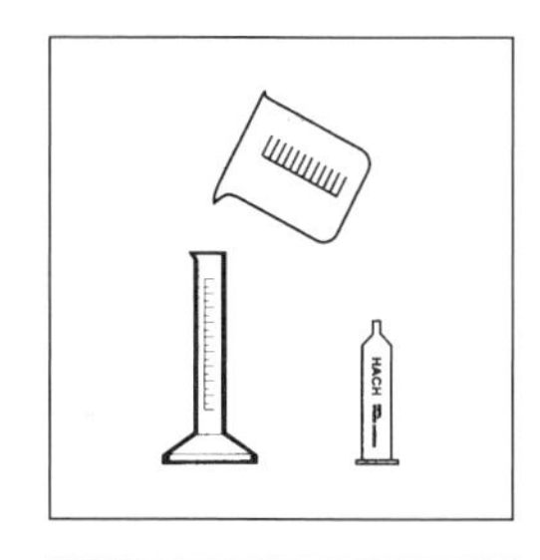

2. Insert a clean delivery tube into the titration cartridge. Attach the cartridge to the titrator body.

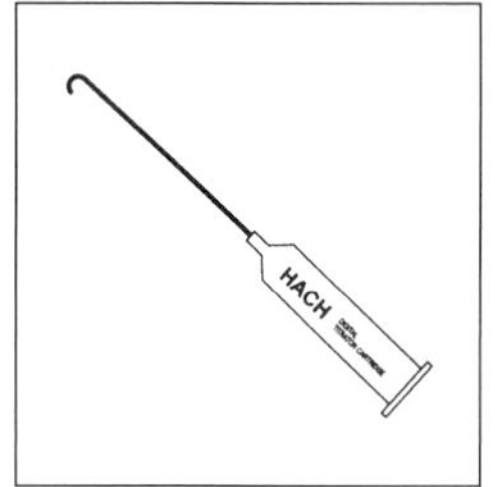

3. Turn the delivery knob to eject a few drops of titrant. Reset the counter to zero and wipe the tip.
Note: For added convenience use the TitraStir stirring apparatus.

*Compliments of HACH Company.

TABLE 2.12

Range (mg/L as CO_2)	Sample Volume (mL)	Titratiom Cartridge (N NaOH)	Catalog Number	Digit Multiplier
10–50	200	0.3636	14378-01	0.1
20–100	100	0.3636	14378-01	0.2
100–400	200	3.636	14380-01	1.0
200–1000	100	3.636	14380-01	2.0

4. Collect a water sample directly into the titration flask by filling to the appropriate mark.
Note: Minimize agitation because carbon dioxide may be lost.
Note: For most accurate results, check the calibration of the erlenmeyer flask by measuring the proper volume in a graduated cylinder. Mark the proper volume on the flask with a permanent marker.

5. Add the contents of one Phenolphthalein Indicator Powder Pillow and mix.
Note: Four drops of Phenolphthalein Indicator Solution may be substituted for the Phenolphthalein Indicator Powder Pillow.
Note: If a pink color forms, no carbon dioxide is present.

6. Place the delivery tube tip into the solution and swirl the flask gently while titrating with sodium hydroxide from colorless to a light pink color that persists for 30 seconds. Record the number of digits required.

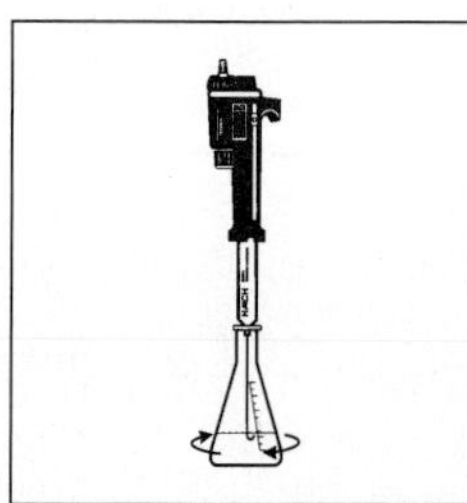

7. Calculate:
Total Digits Required × Digit Multiplier = mg/L as CO_2

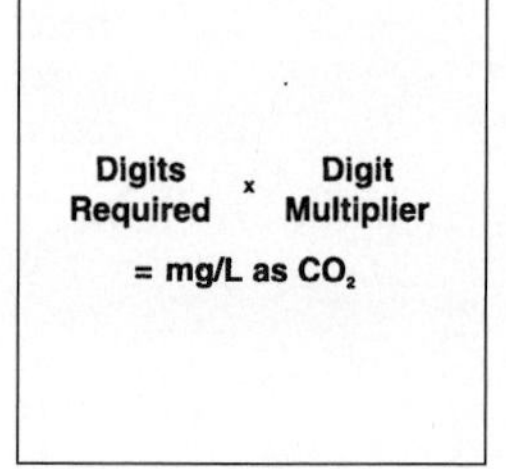

SAMPLING AND STORAGE

Collect samples in clean plastic or glass bottles. Fill completely and cap tightly. Avoid excessive agitation or prolonged exposure to air. Analyze samples as soon as possible after collection. If immediate analysis is not possible, the samples may be stored at least 24 hours by cooling to 4°C (39°F) or below. Before analysis, warm the samples to room temperature.

ACCURACY CHECK

See p. 294 of the HACH Water Analysis Handbook, 3rd Ed., 1997.

INTERFERENCES

- Other acid components in the sample will be titrated and interfere directly in this determination.
- Sodium Hydroxide Standard Solutions tend to lose strength slowly with age and should be checked periodically by titrating a known standard. Check the solution frequently (monthly) by titrating 50 mL of Potassium Acid Phthalate Standard Solution, 100 mg/L CO_2, using Phenolphthalein Indicator Solution. The titration should require 5.00 mL of titrant. If the volume required for this titration is greater than 5.25 mL, discard the sodium hydroxide and replace it with a fresh supply.

QUESTIONS AND ANSWERS

1. Why is excessive sample agitation avoided before phenolphthalein indicator is added in this experiment?
   Excessive agitation of sample is avoided to prevent loss of carbon dioxide.
2. Why should a relatively fresh 0.0227 N Sodium Hydroxide Standard Solution be used to titrate the sample for carbon dioxide?
   Sodium Hydroxide Standard Solution slowly absorbs carbon dioxide when exposed to air, causing a partial loss of strength. Therefore, the NaOH solution should be stored in an appropriate tightly sealed container and the solution should be checked periodically.
3. How accurate is this method of determination?
   This indicator method is satisfactory for field tests and for control and routine applications. The method provides only an approximation of free carbon dioxide in water.

REFERENCES

1. *Standard Methods for the Examination of Water and Wastewater:* CARBON DIOXIDE, 16th Ed., A. E. Greenberg, R. R. Trussell, L. S. Clesceri and Mary Ann H. Franson (eds.), pp. 279–285, Washington, DC, American Public Health Association, 1985.
2. *HACH Water Analysis Handbook,* 3rd Ed. (Loveland, CO.: HACH Company, 1997).

CHLORIDE

Chloride is a major inorganic anion in wastewater and water. Water containing about 250 mg/L sodium chloride per liter may have a detectable salty taste. Most common chloride compounds, such as sodium chloride, require no treatment. However, some of the organic chlorides and heavy metal chlorides require extensive handling and recovery efforts due primarily to the presence of highly toxic heavy metal ions and organic species.

Since sodium chloride in the diet passes through the digestive system, wastewater contains a higher chloride concentration than normal water. Zeolite-type water softeners contribute a large amount of chloride to sewage and wastewaters. Chloride content is greater along the seacoast due to ocean waters. Industrial processes and food cafeterias produce considerable quantities of chloride salts which are added to our waters. Road salts often contribute significant amounts of chloride to our waterways.

Many chloride compounds have a variety of uses during the preparation of foods, chemicals, and metals. A few uses for chloride compounds are the following: salt for food, deicing roads and pathways, pharmaceutical preparations, photography, sewage and industrial waste treatment, mordant in dyeing, etching agent, feed additive, medicine, printed circuitry, fire proofing, preservative, adhesives, embalming and taxidermists' fluid, adhesives, and many other applications. Metal chlorides and organic chlorides have numerous uses in our society.

The silver nitrate titration of chloride requires an indicator of potassium chromate. Substances at low concentrations in potable water do not interfere with the test. However, testing of water containing bromide, iodide, and cyanide can precipitate as silver compounds and register as equivalent chloride concentrations. Orthophosphate above 25mg/l precipitates as silver phosphate and interferes during the test for chloride. Sulfite, thiosulfate, and sulfide anions can be oxidized with hydrogen peroxide to prevent interference with the chloride test. In addition, high iron concentrations can mask the end point during the titration for chloride.

Silver Nitrate Standard Solution in the presence of potassium chromate is used to titrate the solution for chloride. Silver nitrate reacts with the chloride in solution to form an insoluble silver chloride. After all of the chloride has been precipitated, the silver ions react with the chromate to form a red-brown precipitate indicative of the end point.

CHLORIDE (10-10000 mg/L as Cl^-)

Method 8207*
Silver Nitrate Method

1. Select the sample volume and Silver Nitrate Titration Cartridge corresponding to the expected chloride concentration from Table 2.13 below.

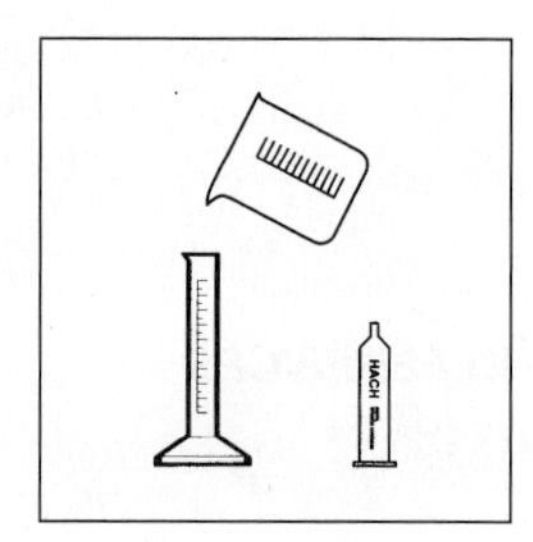

* Compliments of HACH Company.

TABLE 2.13

Range (mg/L as Cl^-)	Sample Volume (mL)	Titration Cartridge (N of $AgNO_3$)	HACH Catalog Number	Digit Multiplier
10–40	100	0.2256	14396-01	0.1
25–100	40	0.2256	14396-01	0.25
100–400	50	1.128	14397-01	1.0
250–1000	20	1.128	14397-01	2.5
1000–4000	5	1.128	14397-01	10.0
2500–10000	2	1.128	14397-01	25.0

2. Insert a clean delivery tube into the titration cartridge. Attach the cartridge to the titrator body.

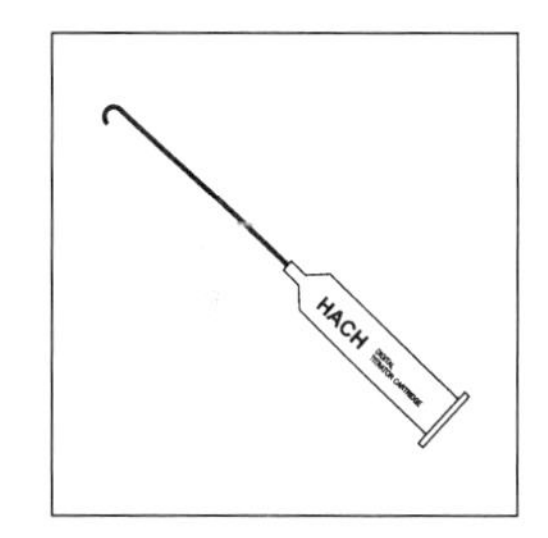

3. Turn the delivery knob to eject a few drops of titrant. Reset the counter to zero and wipe the tip.
Note: For added convenience use the TitraStir stirring apparatus.

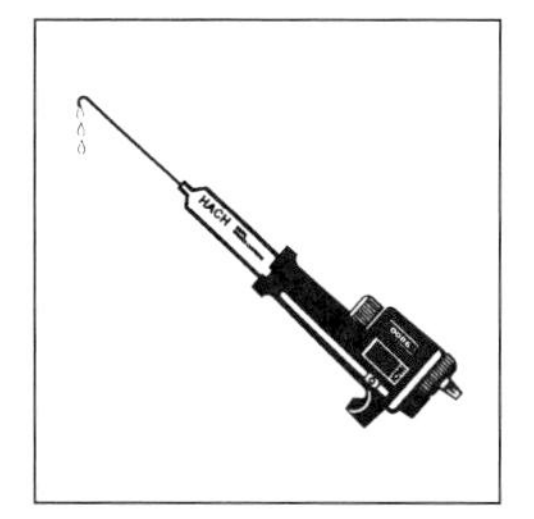

4. Use a graduated cylinder or pipet to measure the sample volume from Table 2.13. Transfer the sample into a clean 250-mL erlenmeyer flask. Dilute to about the 100 mL mark with deionized water, if necessary.
Note: See Sampling and Storage following these steps.

5. Add the contents of one Chloride 2 Indicator Powder Pillow and swirl to mix.
Note: Results will still be accurate if a small amount of the powder does not dissolve.

6. Place the delivery tube tip into the solution and swirl the flask while titrating with silver nitrate from a yellow to red-brown color. Record the number of digits required.

7. Calculate:
   Digits Required × Digit Multiplier = mg/L Chloride.
   Note: Results may be expressed as mg/L sodium chloride by multiplying the mg/L chloride by 1.65.
   Note: meq/L Chloride = mg/L Cl^- ÷ 35.45.

Digits Required x Digit Multiplier

= mg/L Chloride

SAMPLING AND STORAGE

Collect at least 100 to 200 mL of sample in a clean glass or polyethylene container. Samples may be stored up to 7 days before analysis.

ACCURACY CHECK

See p. 306 of the HACH Water Analysis Handbook, 3rd Ed., 1997.

INTERFERENCES USING THE SILVER NITRATE METHOD

- Iron in excess of 10 mg/L masks the end point.
- Orthophosphate in excess of 25 mg/L will precipitate the silver.
- Sulfite in excess of 10 mg/L interferes. Eliminate sulfite interference by adding 3 drops of 30% hydrogen peroxide in Step 4.
- Remove sulfide interference by adding the contents of one Sulfide Inhibitor Reagent Powder Pillow to about 125 mL of sample, mixing for one minute, and filtering through a folded paper.
- Cyanide, iodide, and bromide interfere directly and titrate as chloride.
- Neutralize strongly alkaline or acid samples to a pH of 2 to 7 with 5.25 N Sulfuric Acid Standard Solution or 5.0 N Sodium Hydroxide Standard Solution. Determine the amount of acid or base necessary in a separate sample because pH electrodes will introduce chloride into the sample.

CHLORIDE

Method 8225 for water, wastewater, and seawater*
Silver Nitrate Buret Titration Method**
USEPA Accepted†

1. Select the sample volume and standard titrant solution, that correspond to the expected chloride (Cl^-) concentration; see Table 2.14 below.

0.0141 N
or
0.141 N

2. Use a graduated cylinder or pipet to measure the sample volume from Table 2.14.
Note: If samples cannot be analyzed immediately, see Sampling and Storage following these steps.

3. Transfer the sample into a 250-mL erlenmeyer flask. Dilute to 100 mL with deionized water if necessary.
Note: Highly acidic or alkaline samples should be adjusted to a pH of 7 to 9 before testing. Use pH paper to measure the pH; using a pH meter may contaminate the sample. Test the deionized water for chloride by adding silver nitrate. If a white precipitate forms, use a chloride free deionized water.

TABLE 2.14

Range (mg/L as Cl^-)	Sample Volume (mL)	Standard Titrant Solution** (N of $AgNO_3$)	HACH Catalog Number	Multiplier
0–125	100	0.0141	316-53	5
100–250	50	0.0141	316-53	10
200–500	25	0.0141	316-53	20
500–1250	100	0.141	12551-49	50
1000–2500	50	0.141	12551-49	100
2500–10000	25	0.141	12551-49	200
5000–25000	10	0.141	12551-49	500

* Compliments of HACH Company.
** Adapted from *Standard Methods for the Examination of Water and Wastewater, 407A.*
† Silver Nitrate Standard Solution, 0.0141 N, is required for NPDES reporting purposes.

4. Add the contents of one Chloride 1 Indicator Powder Pillow. Swirl to mix (the prepared sample).

5. Fill a 25-mL buret to the zero mark with the appropriate Silver Nitrate Standard Solution.

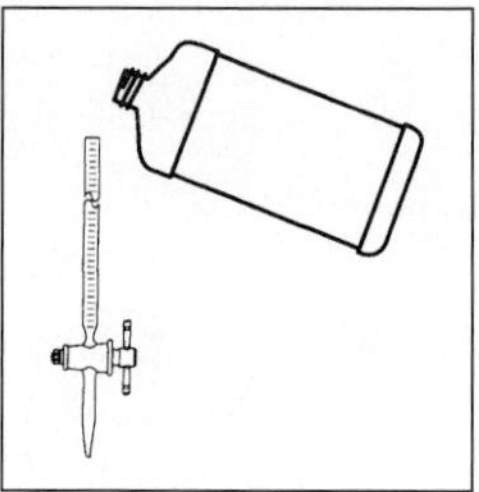

6. Titrate the prepared sample while swirling the flask until the color changes from yellow to red-brown.
Note: When the total amount of Silver Nitrate Standard Solution required exceeds 25 mL per titration, testing a smaller sample volume is recommended to minimize errors from multiple buret readings.

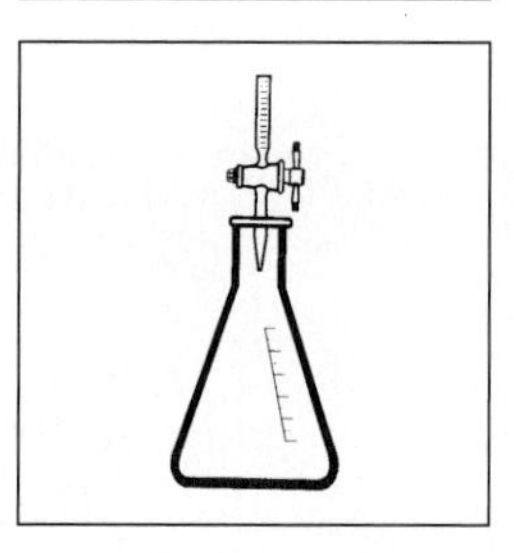

7. Calculate:
mL Titrant $\times$ Multiplier Used $=$ mg/L as Cl^-

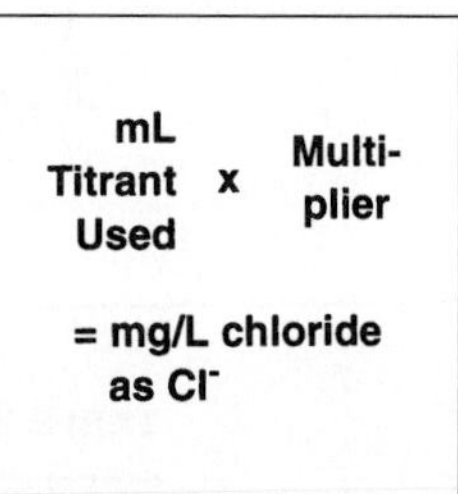

SAMPLING AND STORAGE

Collect samples in clean plastic or glass bottles. Samples can be stored in sealed containers.

ACCURACY CHECK

See HACH Water Analysis Handbook, 3rd Ed., 1997, p. 315.

SILVER NITRATE STANDARD SOLUTION CHECK

See HACH Water Analysis Handbook, 3rd Ed., 1997, p. 315–316.

INTERFERENCES

The effect of interferences increases as the sample size increases. When testing a 100-mL sample, the effect of interferences is twice as strong as when testing a 50-mL sample volume diluted to 100 mL.

For example, iron concentrations over 20 mg/L mask the end point when using a 100-mL sample. With a 50-mL sample, iron over 40 mg/L masks the end point.

When using a 100-mL sample volume, orthophosphate in excess of 25 mg/L will precipitate the silver. Interference from 10 mg/L or more sulfite can be eliminated with three drops of 30% hydrogen peroxide per 100 mL of sample before running the test. Sulfide interference can be removed by adding the contents of one Sulfide Inhibitor Reagent Powder Pillow to about 125 mL of the sample, mixing for one minute and filtering through a folded filter paper. Cyanide, bromide and iodide interfere directly and are titrated as chloride.

QUESTIONS AND ANSWERS

1. Why is the silver nitrate solution stored in dark bottles or away from light?
   All silver nitrate solutions decompose when exposed to light.
2. What are some of the interferences during the test for chloride?
   High iron concentrations can mask the end point during titration. Orthophosphate in excess of 25mg/L will precipitate silver. Sulfite, sulfide, and thiosulfate must be oxidized to prevent interference. Cyanide, bromide and iodide interfere directly and are titrated as chloride.
3. Identify the titrant and the indicator in the experiment.
   Silver nitrate is the titrant and potassium chromate is used as the indicator.
4. What harm can chloride salts cause?
   High concentrations of chloride salts in water will create an unacceptable taste and can cause physical problems in humans and animals.

REFERENCES

1. *HACH Water Analysis Handbook,* 3rd Ed. (Loveland, CO.: HACH Company, 1997).
2. *Hawley's Condensed Chemical Dictionary,* 11th Ed., rev'd by N. Irving Sax and Richard J. Lewis (New York, N.Y.: Van Nostrand Reinhold, 1987).
3. *Standard Methods for the Examination of Water and Wastewater:* CHLORIDE, 16th Ed., A. E. Greenberg, R. R. Trussell, L. S. Clesceri and Mary Ann H. Franson (eds.), pp. 286–288, Washington, DC, American Public Health Association, 1985.

FREE AND TOTAL CHLORINE

Chlorine is not found in nature. It is highly reactive and is manufactured by the electrolysis of sodium chloride. Chlorine is a powerful oxidizer, an excellent biocide and is used to treat municipal water supplies, municipal wastes, and is invariably used in swimming pools. It is also used to control odors in wastewater treatment plants and industrial applications. Today, chlorine is an essential raw material used in the manufacture of many chemical products such as insecticides, solvents, cleaning preparations, and plastics.

A chlorine concentration of 0.1 to 0.4 ppm is maintained in municipal water supplies to keep the water safe for human consumption. The chlorine can be added to the water supply in several forms: liquid sodium hypochlorite, chlorine gas, granular calcium hypochlorite, or as a chlorinated organic compound. The addition of a chlorine compound to a water supply is done to kill disease-causing micro-organisms. The undesirable side effect is to adversely intensify the odor and taste characteristics of phenols and certain other organic compounds, which are found in water supplies. There is also a possibility that chlorine can form potentially chloro-carcinogenic compounds under the right reaction conditions. Also, chlorine can combine with ammonia and nitrogen compounds to form a chloroamine-type compound which can adversely affect some aquatic organisms and human life. To avoid these potentially adverse effects, the water must be tested frequently and properly. If a water sample shows a presence of 0.5 ppm of chlorine, then a possibility of industrial effluent contamination or runoff from a process that uses extremely high concentrations of chlorine is very likely.

The use of a portable chlorine water testing kit enables the determination of both free chlorine and combined chlorine. Free available chlorine exists in water as hypochlorus acid and the hypochlorite ion. Free chlorine uses the chemical reagent, DPD (N-N-diethyl-p-phenylenediamine), which develops a pink color proportional to the amount of free chlorine in the sample. The presence of "combined chlorine," that chlorine which has reacted with ammonia or nitrogen compounds as amines, does not interfere with the analysis of chlorine as free chlorine, if the readings are taken after one minute.

Total chlorine is the sum of the free and combined chlorine. Combined chlorine is determined by adding an activator solution (potassium iodide) to the sample. Both the free chorine and the combined chlorine will react with the iodide to liberate iodine. The iodine then reacts with the DPD reagent to provide an accurate measurement of the total chlorine in the sample, expressed as ppm (mg/L) chlorine as Cl. The test kit will determine chlorine concentrations ranging from 0.1 to 2 ppm. The tests are applicable to water, wastewater, and seawater. The combined chlorine can be determined by subtracting the value for free chlorine from the total chlorine value when the terms are expressed in the same units.

The DPD Methods for free chlorine and total chlorine can be performed by several types of methods including titration, colorimetry, and spectrophotometry. Colorimetric methods include measurements by comparators and colorimeters, which are sufficiently accurate for most measurements. Spectrophotometric methods of analyses maintain the highest level of accuracy.

Free and total chlorine can also be determined by test strips and comparators. These inexpensive methods are appropriate for "approximate" values for chlorine in the range of 0 to 10 ppm.

FREE CHLORINE

Test Procedure*

1. Fill the sample cup to the 25 ml mark with the water sample.

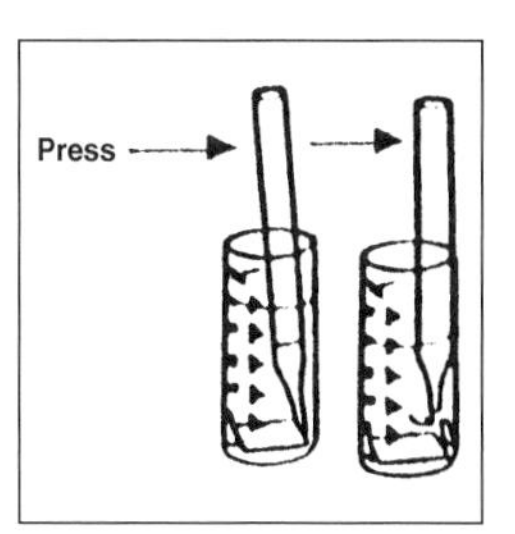

2. Place the tapered tip of the test ampoule into one of the four depressions in the bottom of the sample cup. Snap the tip by squeezing the test ampoule toward the side of the cup. The sample will fill the ampoule and begin to mix with the reagent inside.
   Note: A small gas bubble will remain in the ampoule to facilitate mixing.

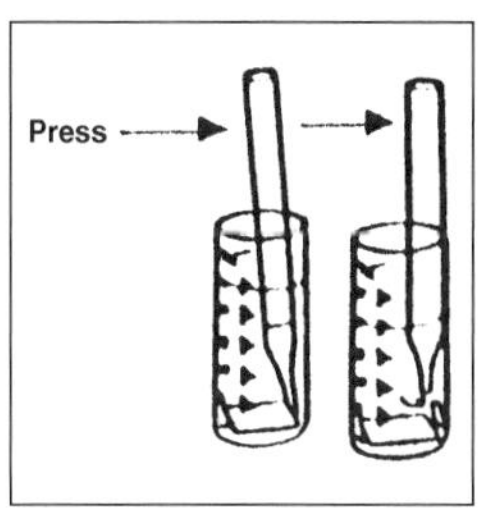

3. Remove the fluid filled ampoule from the sample cup. Mix the contents of the ampoule by inverting it several times, allowing the bubble to travel from end to end inside.
4. Wipe all liquid from the exterior of the ampoule and wait **ONE** minute.
5. Use the color comparator to determine the level of the free chlorine in the sample.

*Compliments of WARD'S Natural Science Establishment, Inc.

TOTAL CHLORINE—TEST PROCEDURE*

1. Fill the sample cup to the 25-ml mark with the water sample.

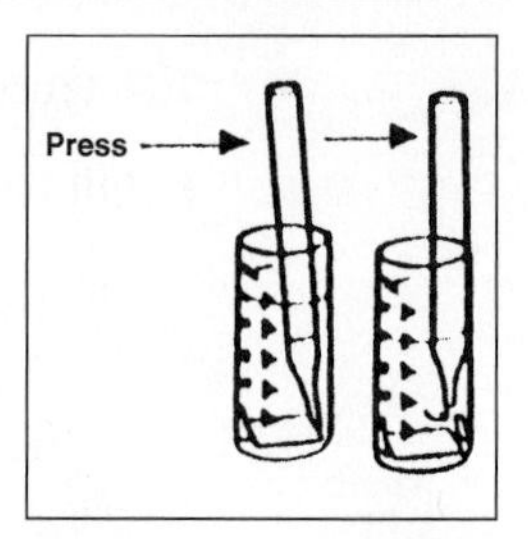

2. Add 5 drops of the activator solution. Stir briefly.

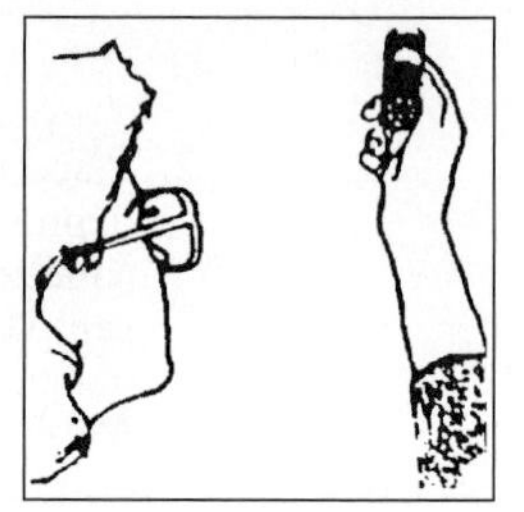

3. Using the pretreated sample, proceed with the **Free Chlorine** test procedure above.

COMPARATOR USE

Store the comparator in the **dark** when not in use. When testing, place the test ampoule in the center tube of the comparator, flat end downward. Direct the top of the cylinder toward a source of bright white light, while viewing from the bottom. Hold the comparator in a nearly horizontal position and rotate it until the color standard below the ampoule shows the closest match.

CAUTION AND SAFETY INFORMATION

1. Read the MSDS before performing this test procedure.
2. When the test kit is used as directed, no chemical reagent will come in contact with the user. Should contact occur, flush skin and eyes with water. Call a physician, should any reagent bc swallowed.
3. The person performing the tests should always wear safety glasses.
4. Do not break the tip of the ampoule unless it is completely immersed in the sample. Accidental breakage of the tip may produce a ''jackhammer effect'' and cause the ampoule to shatter.
5. Properly dispose of used ampoules as instructed by the MSDS.

SOURCES OF ERROR

Various free halogens (such as bromides and iodides) and halogenating agents produce color with the reagent.

CHLORINE, FREE (0.0 to 2.00 mg/L)

Method 8021 for water, wastewater, and seawater*
DPD Method using Powder Pillows**
USEPA Accepted for reporting drinking water analyses†

1. Enter the stored program number for free and total chlorine (Cl_2) powder pillows. Press: **8 0 ENTER.** The display will show: **Dial nm to 530.**
Note: The Pour-Thru Cell can be used with 25-mL reagents only.

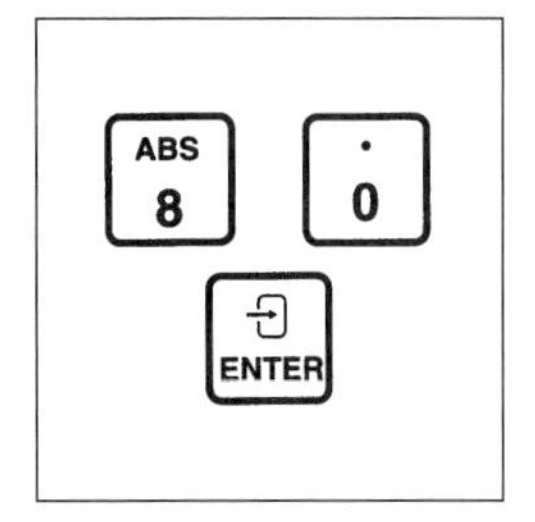

2. Rotate the wavelength dial until the small display shows: **530 nm.** When the correct wavelength is dialed in, the display will quickly show: **Zero Sample,** then: **mg/L CL_2**.

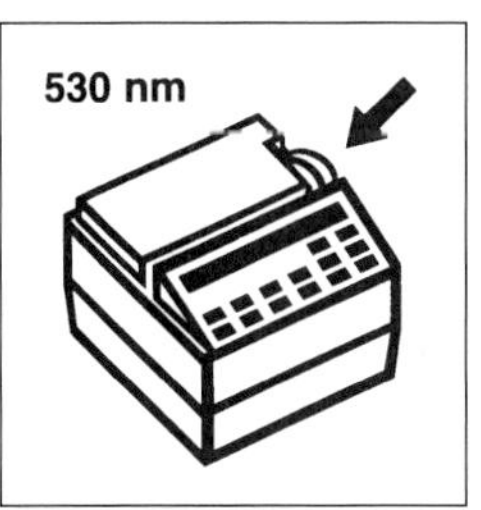

3. Insert the 10-mL Cell Riser into the sample compartment.

4. Fill a sample cell with 10 mL of sample (the blank). Place it into the cell holder. Close the light shield.
Note: Samples must be analyzed immediately and cannot be preserved for later analysis.

*Compliments of HACH Company. Analysis may be performed by DR/820, DR/850, and DR/890 Colorimeters and DR/2010 Spectrophotometer.
**Adapted from *Standard Methods for the Examination of Water and Wastewater, 408 E.*
†Procedure is equivalent to USEPA Method 330.5 for wastewater and Standard Method 4500-Cl G for drinking water.

5. Press: **ZERO.** The display will show: **Zeroing ...** then: **0.00 mg/L Cl_2.**

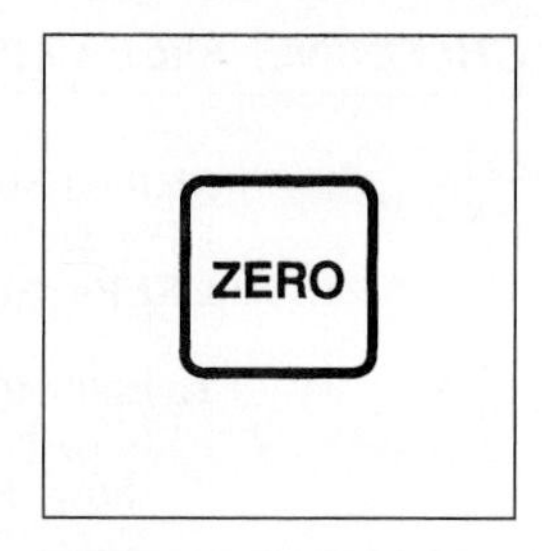

6. Fill another cell with 10 mL of sample.

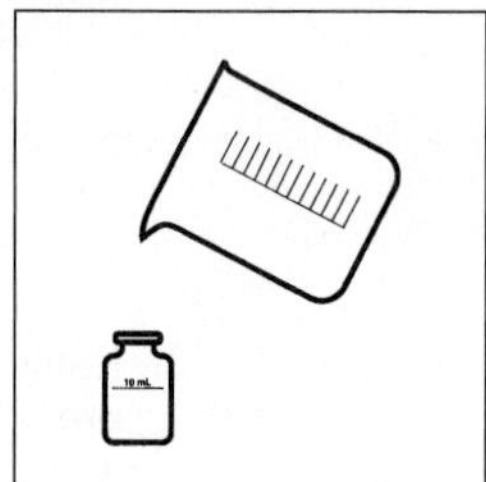

7. Add the contents of one DPD Free Chlorine Powder Pillow to the sample cell (the prepared sample). Stopper the cell and shake for 20 seconds.
Note: A pink color will develop if free chlorine is present
Note: Shaking dissipates bubbles which may form in samples containing dissolved gases.

8. Immediately (within one minute of reagent addition) remove stopper and place the prepared sample into the cell holder. Close the light shield.
Note: Proceed immediately to Step 9.

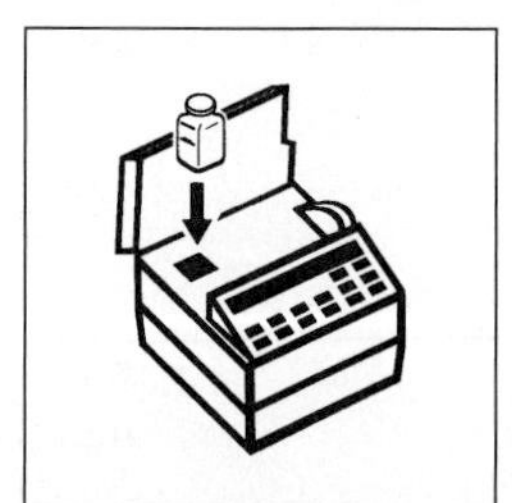

9. Press: **READ.** The display will show: **Reading ...** then the result in mg/L chlorine (Cl_2)will be displayed.
Note: If the sample temporarily turns yellow after reagent addition, or shows **OVER-RANGE,** dilute a fresh sample and repeat the test. A slight loss of chlorine may occur during dilution. Multiply the result by the appropriate dilution factor; see "Sample Dilution Techniques" in Section 1.

SAMPLING AND STORAGE

Analyze samples for chlorine **immediately** after collection. Free chlorine is a strong oxidizing agent, and it is unstable in natural waters. It reacts rapidly with various inorganic compounds and more slowly oxidizes organic compounds. Many factors, including reactant concentrations, sunlight, pH, temperature, and salinity influence decomposition of chlorine in water.

Avoid plastic containers since these may have a large chlorine demand. **Pretreat glass** sample containers to remove any chlorine demand by soaking in a dilute bleach solution (1 mL commercial bleach to 1 liter of deionized water) for at least 1 hour. Rinse thoroughly with deionized or distilled water. If sample containers are rinsed thoroughly with deionized or distilled water after use, only occasional pre-treatment is necessary.

Do not use the same sample containers for free and total chlorine. If trace iodide from the total chlorine reagent is carried over into the free chlorine determination, monochloramine will interfere. It is best to separate, dedicated sample containers for free and total chlorine determinations.

A common error in testing for chlorine is introduced when a representative sample is not obtained. If sampling from a tap, let the water flow for at least 5 minutes to ensure a representative sample. Let the container overflow with the sample several times, then cap the sample container so there is no headspace (air) above the sample. If sampling with a sample cell, rinse the cell several times with the sample, then fill to the 10-mL mark. Perform the chlorine analysis immediately.

ACCURACY CHECK

See p. 339 of the HACH Water Analysis Handbook, 3rd Ed., 1997.

INTERFERENCES

Samples containing more than 250 mg/L alkalinity or 150 mg/L acidity as $CaCO_3$ may not develop the full amount of color, or the color may instantly fade. Neutralize these samples to a pH of 6 to 7 with 1 N Sulfuric Acid, or 1 N Sodium Hydroxide. Determine the amount required on a separate 25 mL sample; then add the same amount to the sample to be tested. Samples containing monochloramine will cause a gradual drift to higher chlorine readings. When read within one minute of reagent addition, 3.0 mg/L monochloramine will cause an increase of less than 0.1 mg/L in the free chlorine reading.

Bromine, iodine, ozone, oxidized manganese, and chromium also may react and show as chlorine. To compensate for the effects of oxidized manganese or chromium adjust pH to 6 to 7 as described above, then add 3 drops of potassium iodide, 30 g/L, to 25 mL of sample, mix and wait 1 minute. Add 3 drops of sodium arsenite, 5 g/L, and mix. Analyze 10 mL of this sample as described above. (If chromium is present, allow exactly the same reaction period with the DPD for oxidized manganese and chromium). Subtract the result of this test from the original analysis to obtain the correct chlorine result.

DPD Free Chlorine Reagent Powder Pillows contain a buffer formulation which will withstand high (at least 1000 mg/L) levels of hardness without interference.

CHLORINE, FREE AND TOTAL (0 to 3.00 mg/L as Cl_2)

Method 8210*
Digital Titrator DPD-FEAS Method

1. Insert a clean delivery tube into a 0.00564 N Ferrous Ethylenediammonium Sulfate (FEAS) Titration Cartridge. Attach the cartridge to the titrator body.

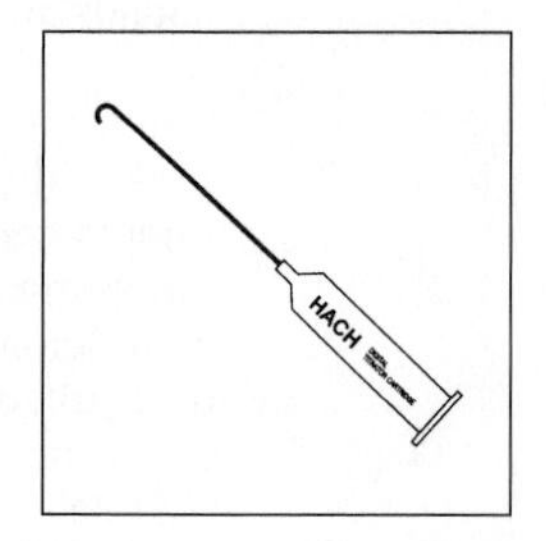

2. Turn the delivery knob to eject a few drops of titrant. Reset the counter to zero and wipe the tip.
   Note: For added convenience use the TitraStir stirring apparatus.

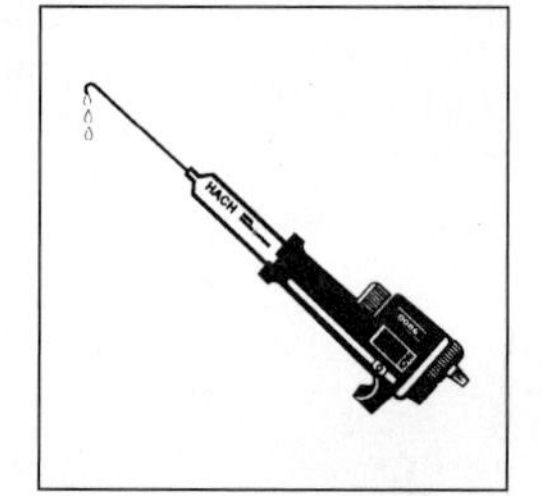

3. Pipet 25.0 mL of sample and transfer into a 50-mL erlenmeyer flask.

4. Add the contents of one DPD Free Chlorine Powder Pillow to the sample and swirl to mix.
   Note: Accuracy is unaffected if a small portion is undissolved.
   Note: See Sampling and Storage following these steps.

* Compliments of HACH Company.

5. Place the delivery tube tip into the solution and swirl the flask while immediately titrating with FEAS to a colorless end point. Record the number of digits required.
   Note: Complete the titration rapidly.

6. Calculate:
   Digits Required × 0.01 = mg/L Free Chlorine

7. If total residual chlorine is desired, return to Step 3 and substitute a DPD Total Chlorine Powder Pillow in Step 4. Wait 3 minutes before titrating. Continue with Step 5. The results will be expressed as mg/L total chlorine.
   mg/L Total Chlorine − mg/L Free Chlorine = mg/L Combined Chlorine

mg/L Total Chlorine
- mg/L Free Chlorine
mg/L Combined Chlorine

SAMPLING AND STORAGE

Chlorine in water is easily lost. Therefore, start chlorine determinations immediately after sampling, avoiding excessive light and agitation. Do not store samples.

ACCURACY CHECK

See p. 355 of the HACH Water Analysis Handbook, 3rd Ed., 1997.

INTERFERENCES

Higher room temperatures tend to lead to higher free chlorine residual due to reaction of chloramines. Higher room temperatures also result in increased color fading. If the sample contains more than 250 mg/L alkalinity or 150 mg/L acidity as $CaCO_3$, the sample may not develop the full amount of color or it may instantly fade. To overcome this interference, adjust the pH of a separate 25 mL sample to a 6 to 7 pH by adding 1 N Sulfuric Acid Standard Solution or 1 N Sodium Hydroxide Standard Solution in small increments and using a pH meter. Record the amount of acid or base required. Add this amount of acid or base to the sample to be tested and proceed with Step 4.

Bromine, iodine, ozone and oxidized forms of manganese and chromium will also react and read as chlorine. To compensate for the effects of manganese, Mn^{4+}, or chromium, Cr^{6+}, add 3 drops of Potassium Iodide, 30 g/L, to 25 mL of sample. Mix and wait 1 minute. Add 3 drops of Sodium Arsenite, 5 g/L, and mix. Analyze this solution as described above. (If chromium is present, allow exactly the same reaction period in Step 7 with the DPD for both analyses.) Subtract the result from the original analysis to correct for the interference.

CHLORINE, TOTAL (0.0 to 2.00 mg/L)

Method 8167*
DPD Method using Powder Pillows**
USEPA Accepted for reporting water and wastewater†

1. Enter the stored program number for free and total chlorine (Cl_2) powder pillows.
   Press: **8 0 ENTER.** The display will show: **Dial nm to 530.**
   Note: The Pour-Thru Cell can be used with 25-mL reagents only.

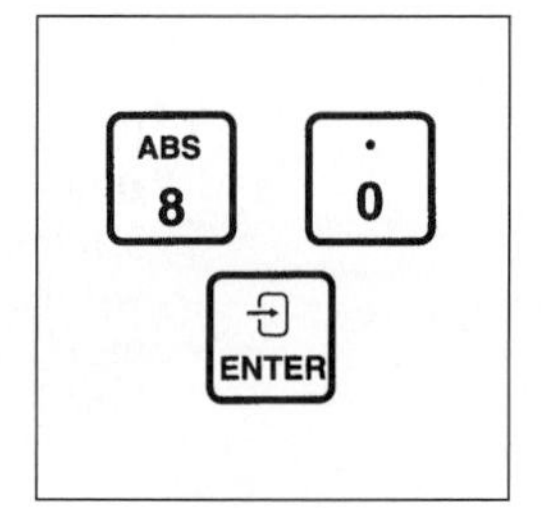

2. Rotate the wavelength dial until the small display shows: **530 nm.** When the correct wavelength is dialed in, the display will quickly show: **Zero Sample,** then: **mg/L CL_2.**

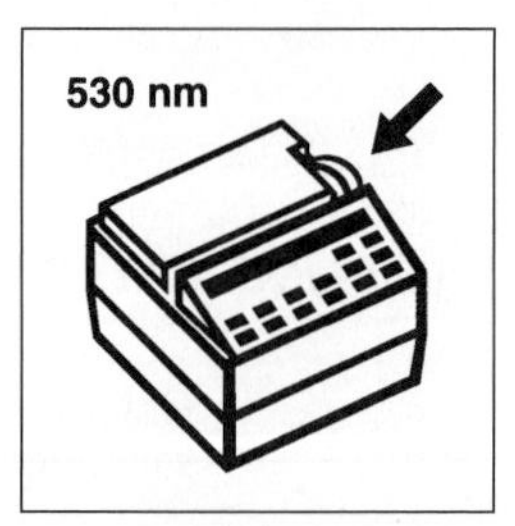

3. Insert the 10-mL Cell Riser into the sample compartment.

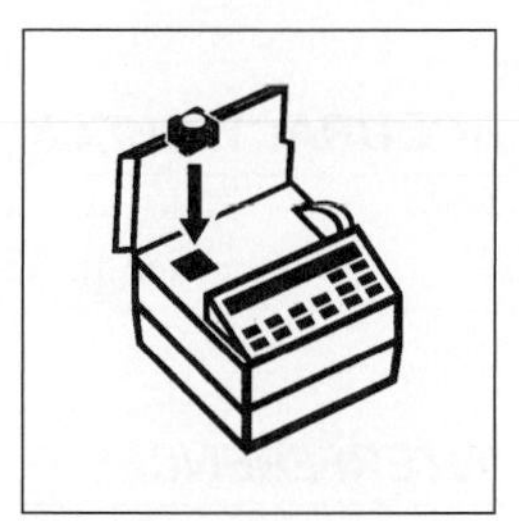

*Compliments of HACH Company. Analysis may be performed by DR/820, DR/850, and DR/890 Colorimeters and DR/2010 Spectrophotometer.

**Adapted from *Standard Methods for the Examination of Water and Wastewater, 408 E.*

† Procedure is equivalent to USEPA Method 330.5 for wastewater and Standard Method 4500-Cl G for drinking water.

4. Fill a 10-mL sample cell with 10 mL of sample.
Note: Samples must be analyzed immediately and cannot be preserved for later analysis.

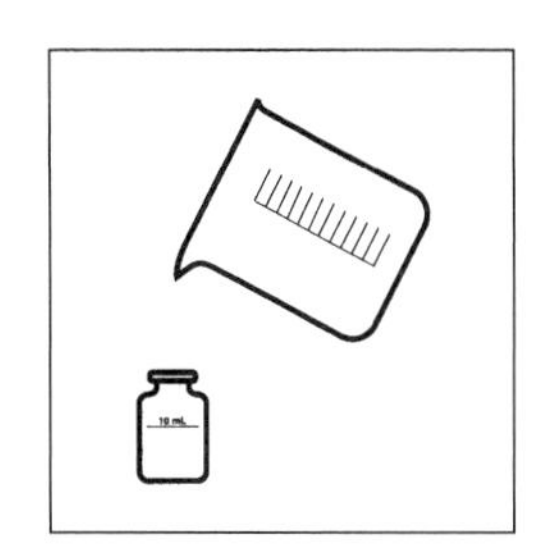

5. Add the contents of one DPD Total Chlorine Powder Pillow to the sample cell (the prepared sample). Stopper the sample cell and shake for 20 seconds. Remove the stopper.
Note: Shaking dissipates bubbles which may form.

6. Press: **SHIFT TIMER.** A 3-minute reaction period will begin.
Note: A pink color will develop if chlorine is present.

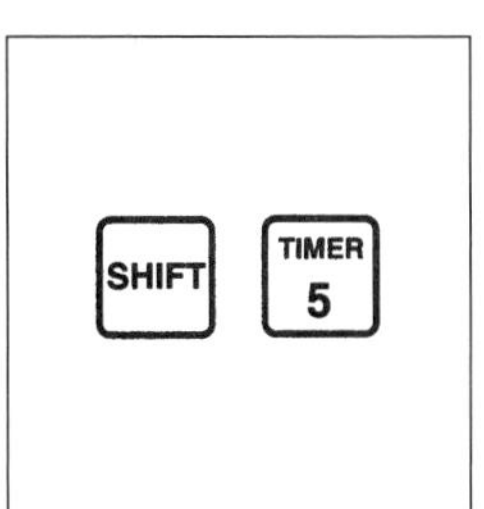

7. When the timer beeps, the display will show: **mg/L Cl_2.** Fill another sample cell (the blank) with 10 mL of sample. Place it into the cell holder. Close the light shield.

8. Press: **ZERO.** The display will show: **Zeroing ...** then: **0.00 mg/L Cl_2.**

9. Within 3 minutes after the timer beeps, place the prepared sample into the cell holder. Close the light shield.

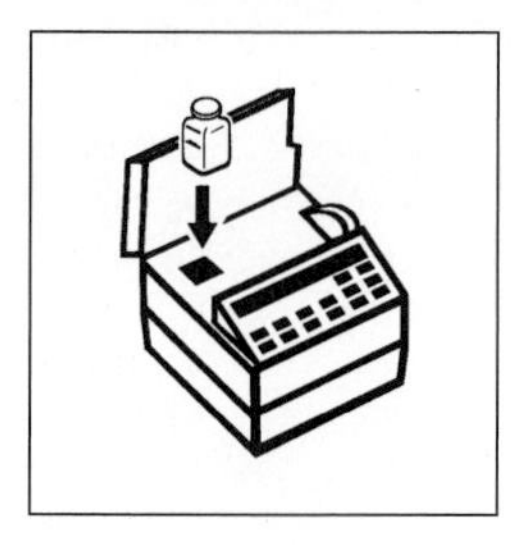

10. Press: **READ.** The display will show: **Reading ...** then the result in mg/L chlorine (Cl_2) will be displayed.
Note: If the sample temorarily turns yellow after sample addition, or shows ***OVER-RANGE,*** dilute a fresh sample and repeat the test. A slight loss of chlorine may occur during dilution. Multiply the result by the appropriate dilution factor; see Sample Dilution Techniques in Section 1.

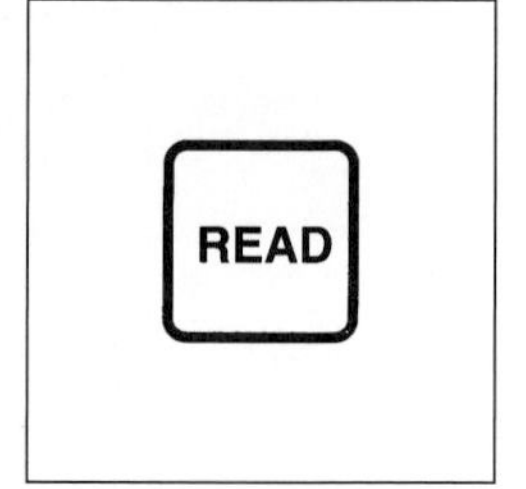

SAMPLING AND STORAGE

Analyze samples for chlorine **immediately** after collection. Chlorine is a strong oxidizing agent, and its is unstable in natural waters. It reacts rapidly with various inorganic compounds and more slowly oxidixes organic compounds. Many factors, including reactant concentrations, sunlight, pH, temperature, and salinity influence decomposition of chlorine in water.

Avoid plastic containers since these may have a large chlorine demand. **Pretreat glass** sample containers to remove any chlorine demand by soaking in a dilute bleach solution (1 mL commercial bleach to 1 liter of deionized water) for at least 1 hour. Rince thoroughly with dionized or distilled water. If sample containers are rinsed thoroughly with deionize dor distilled water after use, only occasional pre-treatment is necessary.

Do not use the same sample containers for free and total chlorine. IF trace iodide from the total chlorine reagent is carried over into the free chlorine determination, monochloramine will interfere. It is best to use separate, dedicated sample containers for free and total chlorine determinations.

A common error in testing for chlorine is introduced when a reprsentative sample is not obtained. If sampling from a tap, let water flow for at least 5 minutes to ensure a representative sample. Let the container overflow with the sample several times, then cap the sample containers so there is no headspace (air) above the sample. If sampling with a sample cell, rinse the cell several times with the sample, then carefully fill to the 10-mL mark. Perform the chlorine analysis immediately.

ACCURACY CHECK

See HACH water analysis Handbook, 3rd, Ed., p. 383.

INTERFERENCES

Samples containing more than 300 mg/L alkalinity or 150 mg/L acidity as $CaCO_3$ may not develop the full amount of color, or it may instantly fade. Neutralize these samples to a pH of 6 to 7 with 1 N Sulfuric Acid or 1 N Sodium Hydroxide. Determine the amount required on a separate 10 mL sample. Add the same amount to the sample to be tested. Correct for volume additions; see "Corrections for Volume Additions" in Section 1.

Bromine, iodine, ozone, oxidized manganeses, and chromium also may react and read as chlorine. To compensate for the effects of oxidized manganese or chromium, adjust the pH to between 6 and 7 as described above; then add 3 drops of potassium iodide, 30 g/L, to 25 mL of sample, mix, and wait one minute. Add 3 drops of sodium arsenite, 5 g/L, and mix. Analyze 10 mL of this sample as described above. (If chromium is present, allow exactly the same reaction period with DPD for both analyses.) Subtract the result of this test from the original analysis to obtain the correct chlorine result.

DPD Total Chlorine Reagent Powder Pillows contain a buffer formulation which will withstand high levels of hardness (at least 1000 mg/L) without interference.

CHLORINE, TOTAL (1 to 400 mg/L)

Method 8209 Using Sodium Thiosulfate*
Digital Titrator Iodomctric Burct

1. Select the sample volume and Sodium Thiosulfate Titration Cartridge corresponding to the expected chlorine concentration from Table 2.15.

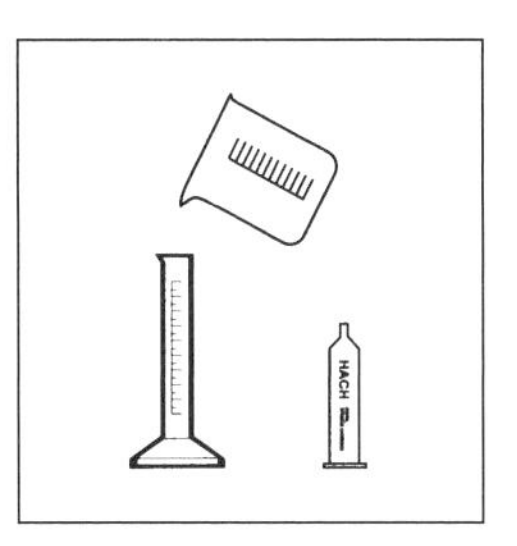

2. Insert a clean delivery tube into the titration cartridge. Attach the cartridge to the titrator body.

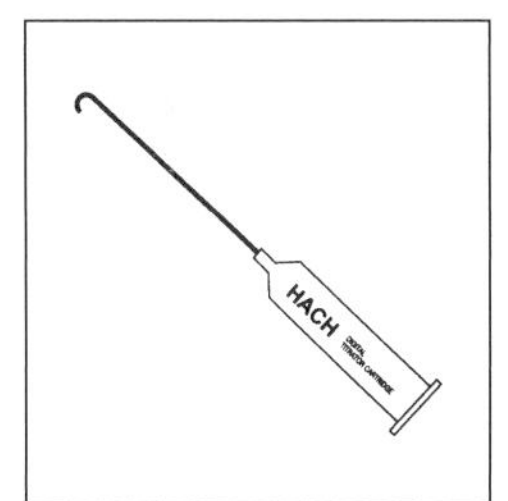

TABLE 2.15

Range (mg/L Cl_2)	Sample volume (mL)	Standard titrant Solution (N of $Na_2S_2O_3$)	HACH Catalog Number	Multiplier
1–4	100	0.02256	24091-01	0.01
2–8	50	0.02256	24091-01	0.02
5–20	20	0.02256	24091-01	0.05
100–400	1	0.02256	24091-01	1.00

*Compliments of HACH Company.

3. Flush the delivery tube by turning the delivery knob to eject a few drops of titrant. Reset the counter to zero and wipe the tip.
Note: For added convenience use the TitraStir stirring apparatus.

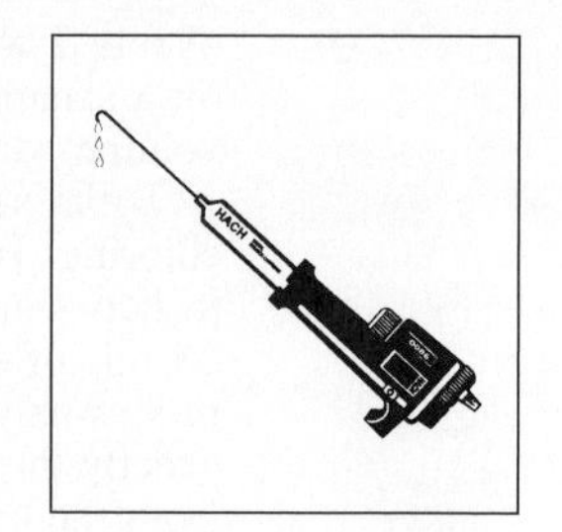

4. Use a clean graduated cylinder to take a water sample. Pour sample into a clean 125- or 250-mL erlenmeyer flask. Dilute to about the 100-mL mark with deionized water.
Note: See Sampling and Storage following these steps.

5. Add 2 droppers (2 mL) Acetate Buffer Solution, pH 4 and swirl to mix.

6. Clip open the end of 1 Potassium Iodide Powder Pillow. Add the contents to the flask. Swirl to mix.

7. Place the delivery tube tip into the solution and swirl the flask while titrating with sodium thiosulfate until the solution is a pale yellow.

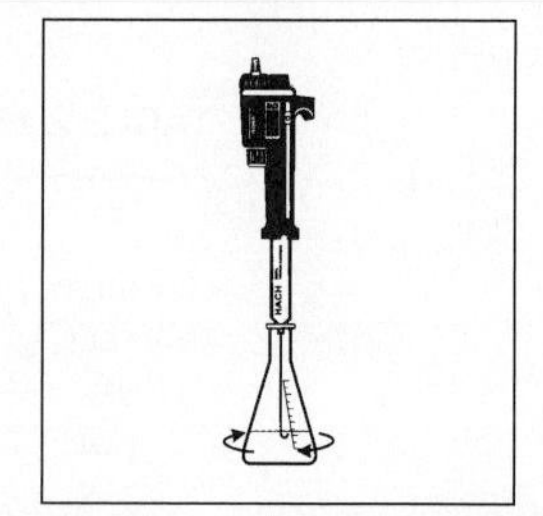

8. Add 1 dropper of starch indicator solution and swirl to mix. A dark blue color will develop.

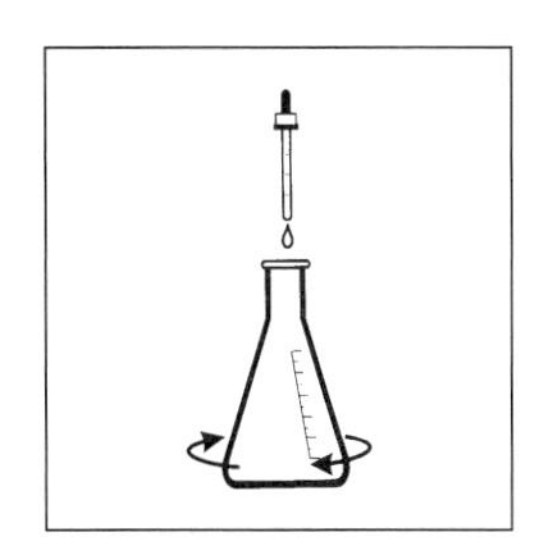

9. Continue the titration until the solution changes from dark blue to colorless. Record the number of digits required.

10. Calculate:
Digits Required × Digit Multiplier = mg/L Total Chlorine (Cl_2)
Note: These procedures can be used to check iodine and bromine concentrations if chlorine is not present. Multiply the test result (in mg/L chlorine) by 3.58 or 2.25, respectively, to accurately express the iodine or bromine content of your sample.

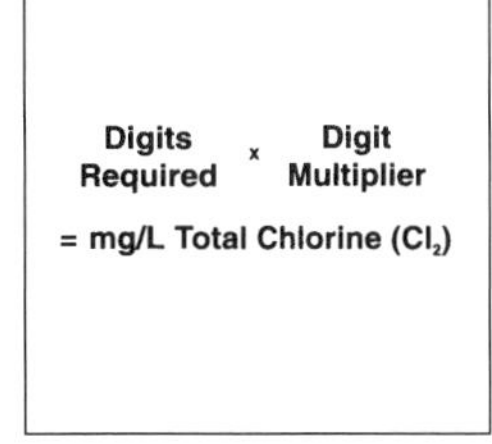

IODOMETRIC METHOD FOR CHLORINE, TOTAL (20–70,000 mg/L USING SODIUM THIOSULFATE)

1. Select the sample volume and Sodium Thiosulfate Titration Cartridge corresponding to the expected chlorine concentration from Table 2.16.

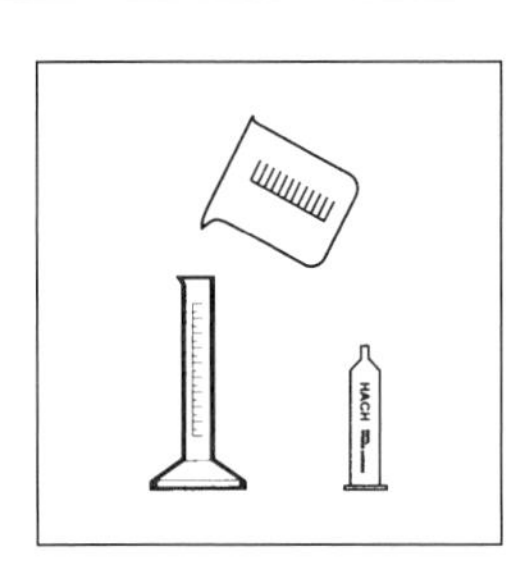

2. Insert a clean delivery tube into the titration cartridge. Attach the cartridge to the titrator body.

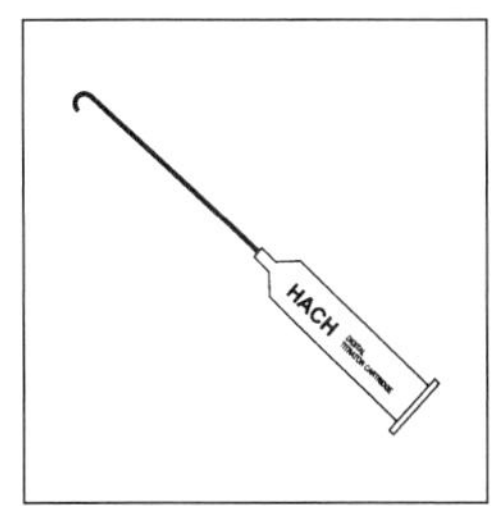

TABLE 2.16

Range (mg/L Cl_2)	Sample volume (mL)	Titration cartridge (N $Na_2S_2O_3$)	Catalog number	Digit multiplier
20–80	25	0.113	22673-01	0.2
50–200	10	0.113	22673-01	0.5
100–400	5	0.113	22673-01	1
250–1000	2	0.113	22673-01	2.5
500–2000	1	0.113	22673-01	5
2000–9000 (0.2–0.9%)	4	2.00	14401-01	22.2
5000–18,000 (0.5–1.8%)	2	2.00	14401-01	44.3
10,000–35,000 (1.0–3.5%)	1	2.00	14401-01	88.7
20,000–70,000 (2.0–7.0%)	0.5	2.00	14401-01	177

3. Flush the delivery tube by turning the delivery knob to eject a few drops of titrant. Reset the counter to zero and wipe the tip.
Note: For added convenience use the TitraStir stirring apparatus.

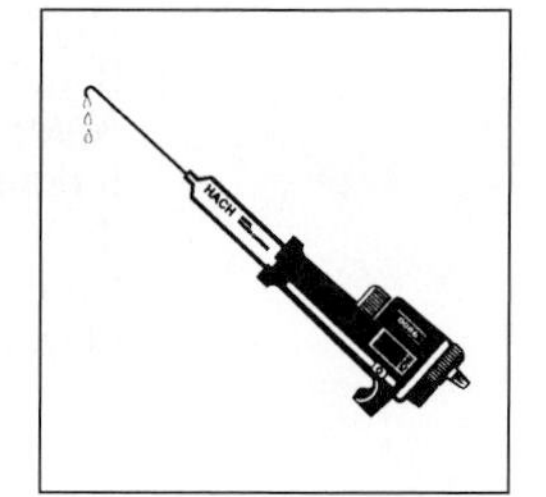

4. Use a pipet or graduated cylinder to measure the sample volume from Table 2.16. Transfer the sample into a 125-mL erlenmeyer flask and dilute to about the 50-mL mark with deionized water.

5. Add the contents of 1 Dissolved Oxygen 3 Powder Pillow.
Note: Normally the addition of the powder pillow will lower the pH to 4 or less. If the sample size is large and highly alkaline, verify the solution pH is 4 or less with a pH meter or pH paper before proceeding.

6. If you are using the 2.00 N Titration Cartridge, add the contents of one Potassium Iodide Powder Pillow, Cat. No. 20599-96, to the flask and swirl to mix.
Note: If you are using the 0.113 N Titration Cartridge, add the contents of 1 Potassium Iodide Powder Pillow, Cat. No. 1077-99, to the flask and swirl to mix.

7. Place the delivery tube tip into the solution and swirl the flask while titrating with sodium thiosulfate until the solution is a pale yellow.
Note: For added convenience use the TitraStir stirring apparatus.

8. Add 1 dropperful of starch indicator solution and swirl to mix. A dark blue color will develop.

9. Continue the titration until the solution changes from dark blue to colorless. Record the number of digits required.

10. Calculate:
Digits Required × Digit Multiplier = mg/L Total Chlorine (Cl_2).
To convert the above results to the equivalent percent chlorine (Cl_2), divide by 10,000.

Digits Required x Digit Multiplier
= mg/L Total Chlorine (Cl_2)

SAMPLING AND STORAGE

Collect at least 200 mL of sample in a clean glass or polyethylene container. Analyze on site or as soon as possible after collection.

ACCURACY CHECK

See p. 405 of the HACH Water Analysis Handbook, 3rd Ed., 1997.

QUESTIONS AND ANSWERS

1. Why is chlorination of a water supply essential?
Chlorine acts as a biocide and also helps to alleviate the adverse affects of iron, manganese and sulfide that may be present in the water supply.

2. What chlorine concentration is needed in drinking water to ensure the safety of consumption?
A chlorine concentration of 0.1 to 0.4 ppm is needed to ensure the safety of water for consumption.

3. What are the adverse effects that the chlorination of water supplies may produce?
Chlorination of water supplies may intensify the taste and odor characteristics of phenols and other organic compounds present in the water supply. The formation of potentially carcinogenic chloro-organic compounds is also possible as well as the chlorination of ammonia or nitrogenous compounds, forming combined chlorine (chloramine-type compounds), which adversely affects some aquatic organisms.

4. When can it be assumed (with respect to chlorine concentration) that drinking water is free from contaminating organisms?
Water is considered safe from organism contamination if chlorine is detectable in any quantity. For drinking purposes and for human health, chlorine should not exceed 0.4 ppm.

REFERENCES

1. *WARD's Snap Test,* Instant Water Quality Test Kits, WARD's Natural Science Establishment, Inc., Rochester, N.Y.
2. *HACH Water Analysis Handbook,* 3rd Ed., (Loveland, CO.: HACH Company, 1997).
3. *Standard Methods for the Examination of Water and Wastewater:* CHLORINE (RESIDUAL), 16th Ed., Method 408, A. E. Greenberg, R. R. Trussell, L. S. Clesceri and Mary Ann H. Franson (eds.), pp. 294–300, 316–317, Washington, DC, American Public Health Association, 1985.
4. *Hawley's Condensed Chemical Dictionary,* 11th Ed., rev'd by N. Irving Sax and Richard J. Lewis (New York, N.Y.: Van Nostrand Reinhold, 1987).

CHLORINE DIOXIDE

Chlorine dioxide is a deep yellow gas which is toxic and has an unpleasant odor. The volatile gas is a strong oxidizing agent and can react explosively under certain conditions. The gas should be handled with care in well vented areas.

Chlorine dioxide, ClO_2, is a strong bleaching agent used primarily in the paper and pulp industry. The compound is used to treat water supplies to combat taste and odor problems due to chemical contamination and specific types of algae. Chlorine dioxide is a strong disinfectant. Unlike chlorine, the compound does not form toxic trihalomethanes during the reaction with methane in water. In addition, chlorine dioxide oxidizes soluble forms of iron and manganese so that the metals can be removed from water more easily.

Samples should be analyzed immediately after collection and cannot be stored for later analysis. When testing for chlorine dioxide, samples should have limited exposure to sunlight. Collected samples of water and wastewater should not be agitated or stirred excessively. Samples and reference samples should be at the same temperature before measurements.

Oxidizing agents such as ClO^-, ClO_2^-, ClO_3^-, CrO_4^{2-}, ozone and iron (+3) can be significant interferences in the experiment. Hardness and turbidity can also interfere at relatively high levels.

Chlorine dioxide combines with chlorophenol red at pH 5.2 to form a colorless complex. The spectrometer measures the concentration (mg/L) of the chlorine dioxide at 575 nm. The ClO_2 method is unreactive to other moderate oxidizing agents such as chlorine.

CHLORINE DIOXIDE, LR (0 to 1.00 mg/L)

Method 8065 for water and wastewater*
Chlorophenol Red Method**

1. Enter the stored program number for chlorine dioxide (ClO_2), low range.
   Press: **7 2 ENTER.** The display will show: **Dial nm to 575.**
   Note: The Pour-Thru Cell can be used with this procedure. Rinse with deionized water after each analysis.

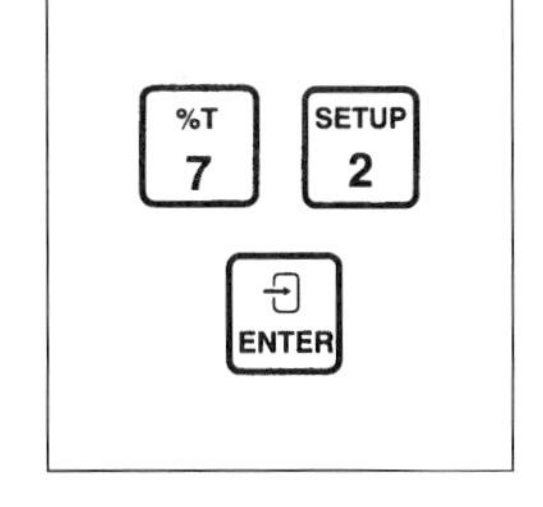

2. Rotate the wavelength dial until the small display shows: **575 nm.** When the correct wavelength is dialed in the display will quickly show: **Zero sample,** then: **mg/L ClO_2, LR.** (for Low Range)

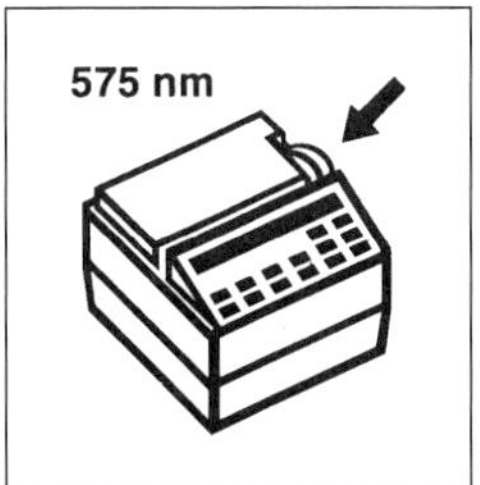

*Compliments of HACH Company.
**Adapted from Harp, Klein, and Schoonover, *Jour. Amer. Water Works Assn.*, (73): 387–388 (1981).

3. Fill two 50-mL graduated mixing cylinders with sample to the 50-mL mark.
Note: Analyze samples immediately after collection.
Note: For most accurate results, analyze each portion at the same sample temperature.

4. Using a TenSette Pipet or Class A pipet, add 1.00 mL Chlorine Dioxide Reagent 1 to each cylinder. Stopper. Invert several times to mix.

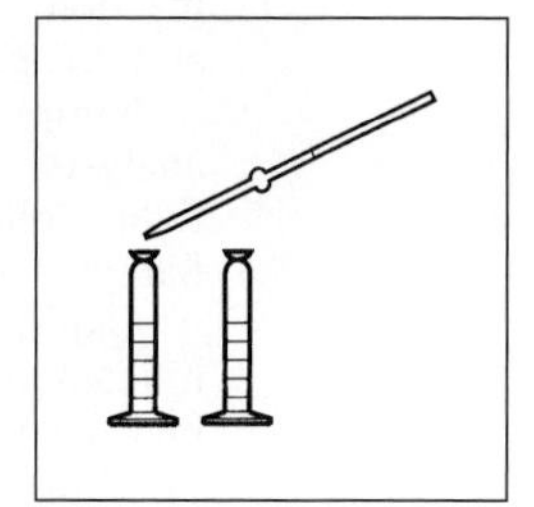

5. Add the contents of one Dechlorinating Reagent Powder Pillow to one cylinder. Invert several times until dissolved. This solution is the blank. The other solution is the prepared sample.

6. Using a Class A pipet, add exactly 1.00 mL Chlorine Dioxide Reagent 2 to each cylinder. Stopper. Invert several times to mix.

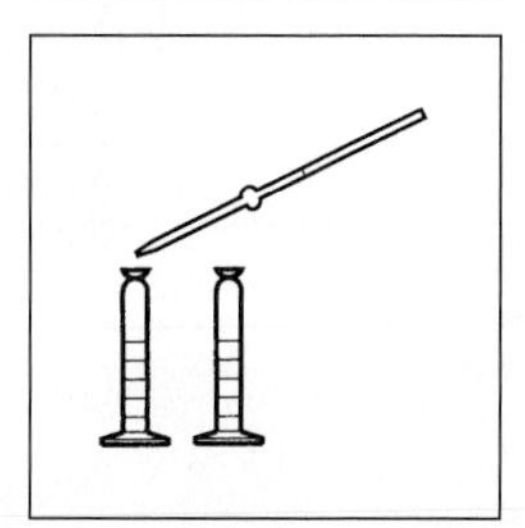

7. Using a Class A or Ten Sette pipet, add 1.00 mL of Chlorine Dioxide Reagent 3 to each cylinder. Stopper. Invert several times to mix.

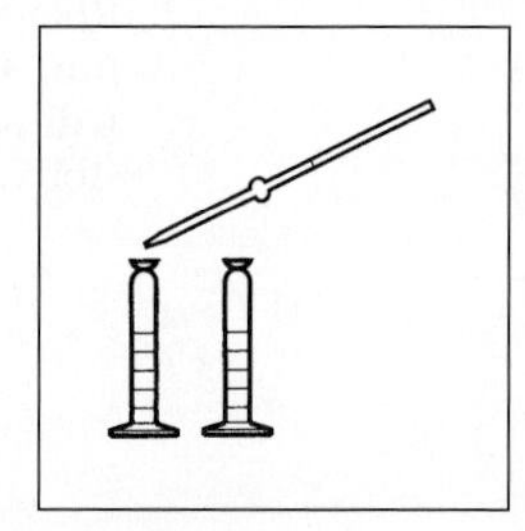

8. Pour 25 mL from each cylinder into respective sample cells.

9. Place the blank into the cell holder. Close the light shield.

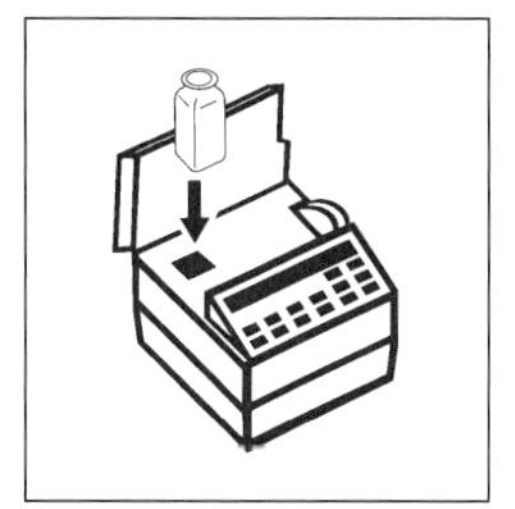

10. Press: **ZERO.** The display will show: **Zeroing ...** then: **0.00 mg/L ClO_2 LR.**

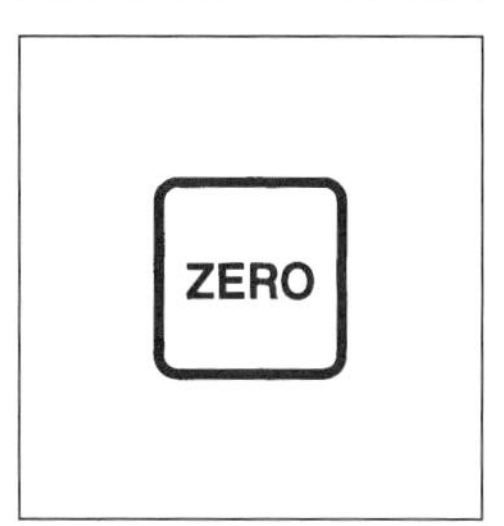

11. Place the prepared sample into the cell holder. Press: **READ.** The display will show: **Reading ...** then the result in mg/L ClO_2 will be displayed.

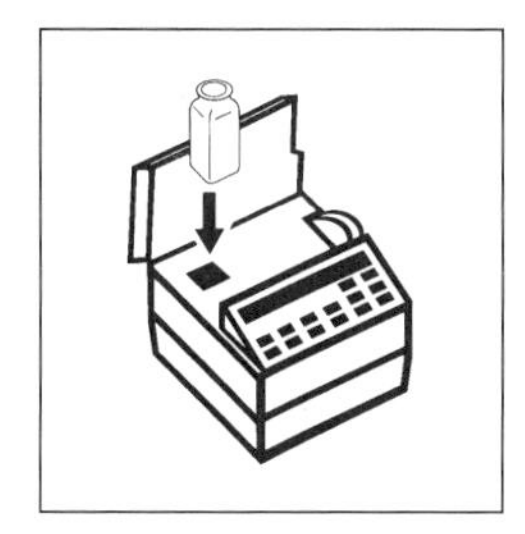

SAMPLING AND STORAGE

Collect samples in clean plastic or glass bottles. Fill completely and cap tightly. Avoid excessive agitation and exposure to light, especially sunlight. Samples must be analyzed immediately upon collection and cannot be preserved or stored for later analysis.

INTERFERENCES

- For highly acidic or alkaline water, 2.0 mL each of Chlorine Dioxide Reagent 1 and Chlorine Reagent 3 may be required instead of 1.0 mL.

- Ozone interferes at 0.5 mg/L.
- The following do not interfere at or below these concentrations:

ClO^-	5.5 mg/L	Fe^{3+}	5 mg/L
ClO_2^-	6 mg/L	Hardness	1000 mg/L
ClO_3^-	6 mg/L	Turbidity	1000 NTU
CrO_4^{2-}	3.6 mg/L		

QUESTIONS AND ANSWERS

1. What are some advantages for using chlorine dioxide instead of chlorine for water treatment?
Chlorine dioxide does not form toxic forms of halomethanes and haloethanes whereas chlorine does.

2. What are some disadvantages of using chlorine dioxide?
Compared to chlorine, chlorine dioxide is more explosive, more volatile, and more of a danger when handling.

3. Why should collected water samples not be stirred or agitated excessively?
The stirring or agitation of collected samples of water may decrease the amount of chlorine dioxide in the sample before tests.

4. What safety precautions should a person handling chlorine dioxide use?
Handle the containers with care and store the tanks in a proper place out of the sunlight and away from heat. Also ensure that all valves are properly working and closed when not in use. Also ensure that there are no leaks in the lines and around the meter. All areas should be well vented if chlorine dioxide is present.

REFERENCES

1. *HACH Water Analysis Handbook,* 3rd Ed. (Loveland, CO.: HACH Company, 1997).
2. *Standard Methods for the Examination of Water and Wastewater:* CHLORINE DIOXIDE, 16th Ed., Method 410, By A. E. Greenberg, R. R. Trussell, L. S. Clesceri and Mary Ann H. Franson (eds.), pp. 319–324, Washington, DC, American Public Health Association, 1985.

CHROMATE

Chromates are used primarily as oxidizing agents in industry. Primary applications are in the areas of metal oxidation, sea water analysis and the determination of radioactive strontium. Chromate compounds are frequently added to cooling water for corrosion control. Chromates are often used in indicator solutions, color standards, paint pigments, metallurgy, colored ceramics, wood preservatives, seed treatment, fungicides, dyeing mordant, corrosion inhibitor, battery depolarizer, metal coatings, organic analysis, rubber and plastics, inks, leather tanning, and preparation of other chromium compounds. Some transition metal chromates are toxic by inhalation or ingestion, while others exhibit carcinogenic characteristics.

Some industries have implemented a closed loop series of steps to minimize the release of chromates to the environment. Industries often recover wastes for additional applications, whereas others may manifest the waste for recovery or burial purposes. In the environment, chromate wastes are normally reduced due to reactions with other materials. Landfills contain a considerable amount of chromate and reduced-chromate compounds.

Interferences from substances capable of oxidizing iodide to iodine under acidic conditions sometimes produce high results for chromate. Oxidizing agents and ions such as ferric and copper can interfere with the analysis of chromate. The effects of iron and copper on chromate analysis can be minimized by a sample pretreatment procedure.

Chromate reacts with iodide under acidic conditions to form iodine as triiodide. Starch indicator is added to the solution to form a blue complex with the iodine. Then, the solution is titrated with sodium thiosulfate to a colorless end point. The volume of sodium thiosulfate used to titrate the solution is directly proportional to the chromate concentration.

CHROMATE (20 to > 400 mg/L as CrO_4^{-2})

Method 8211*
Digital Titrator Method using Sodium Thiosulfate

1. Insert a clean delivery tube into the Sodium Thiosulfate Titration Cartridge. Attach the cartridge to the titrator body.

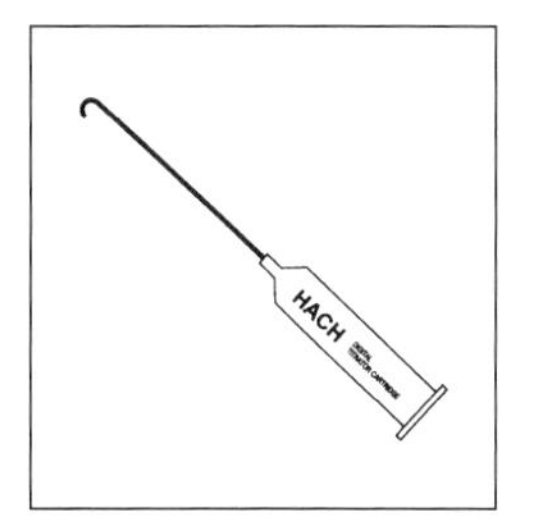

2. Turn the delivery knob to eject a few drops of titrant. Reset the counter to zero and wipe the tip.
Note: For added convenience use the TitraStir stirring apparatus.

3. Select a sample volume corresponding to the expected chromate (CrO_4^{-2}) concentration from Table 2.17.
Note: See Sampling and Storage following these steps.

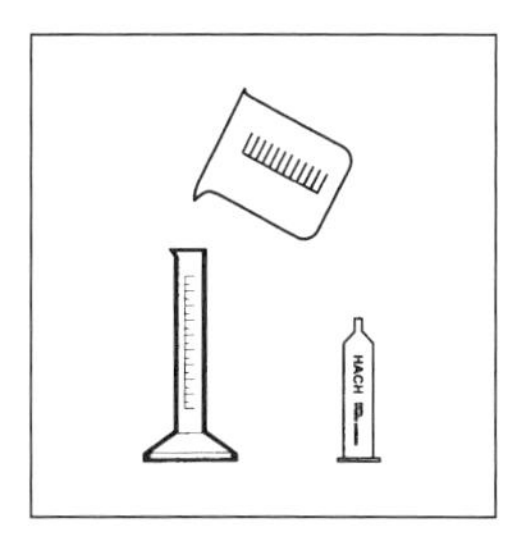

*Compliments of HACH Company.

TABLE 2.17

Range (mg/L as CrO_4^{2-})	Sample volume (mL)	Titration cartridge (N $Na_2S_2O_3$)	Digit multiplier
20–80	50	0.2068 N	0.2
50–200	20	0.2068 N	0.5
100–400	10	0.2068 N	1.0
>400	5	0.2068 N	2.0

4. Use a graduated cylinder or pipet to measure the sample volume from Table 2.17. Transfer the sample into a clean 125-mL erlenmeyer flask. Dilute to about the 50-mL mark with deionized water.

5. Add the contents of 1 Potassium Iodide Powder Pillow and swirl to mix.

6. Add the contents of 1 Dissolved Oxygen 3 Reagent Powder Pillow and swirl to mix. Wait at least 3 minutes, but not more than 10 minutes before completing Steps 7 to 9.
Note: A yellow or brown color indicates the presence of chromate.

7. Place the delivery tube tip into the solution and swirl the flask while titrating with sodium thiosulfate to a straw-yellow color.

8. Add 1 dropper of Starch Indicator Solution and swirl to mix.
Note: A blue color will form.

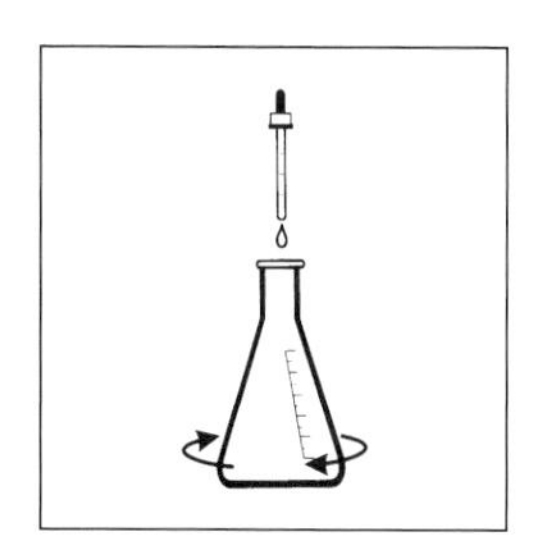

9. Continue titrating until the solution turns from blue to colorless. Record the number of digits required.

10. Calculate:
Total Digits Required × Digit Multiplier = mg/L chromate (CrO_4^{-2}).
Note: Results may be expressed as mg/L sodium chromate (Na_2CrO_4) or chromium (Cr) by multiplying the mg/L chromate by 1.4 or 0.448, respectively.

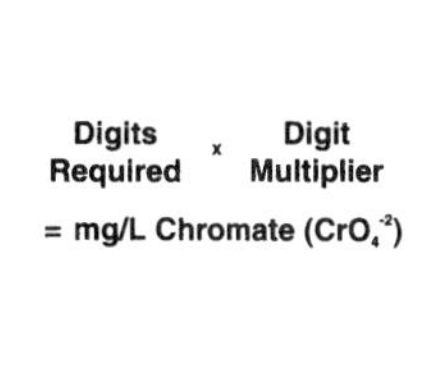

SAMPLING AND STORAGE

Collect 200 to 300 mL of sample in an acid-washed glass or polyethylene container. If the sample cannot be analyzed immediately add 1 mL concentrated sulfuric acid and swirl to mix.

ACCURACY CHECK

See p. 431 of the HACH Water Analysis Handbook, 3rd Ed., 1997.

INTERFERENCES

Substances capable of oxidizing iodide to iodine under acidic conditions (such as ferric iron and copper) will interfere to give high results. The effects of iron and copper may be masked by dissolving a Magnesium CDTA Powder Pillow, followed by two 1.0 g measuring spoons of sodium acetate in the sample between Steps 6 and 7.

QUESTIONS AND ANSWERS

1. What are the primary uses of chromate?
Chromates are used as indicator solutions in the determination of chloride and of salinity of seawater. Chromate compounds are used for corrosion control in cooling water of industrial processes and steam plants. Potassium and sodium dichromate are often used as an industrial and laboratory oxidant and for the determination of radioactive strontium. The compound is also used as permanent color solution during the determination of silica by the molybdosilicate method.

2. What indicator is used to indicate color change during titration?
Starch solution as indicator solution forms a blue colored complex in solution before titration.

3. What are some of the interferences during chromate determinations?
Substances (such as Fe^{+3} and Cu^{+2}) *capable of oxidizing iodide to iodine under acidic conditions will interfere unless pretreatment of the sample occurs. Additional oxidizing agents can easily provide high and misleading results.*

REFERENCES

1. HACH Water Analysis Handbook, 3rd Ed. (Loveland, CO.: HACH Company, 1997).
2. *Standard Methods for the Examination of Water and Wastewater:* 16th Ed., A. E. Greenberg, R. R. Trussell, L. S. Clesceri and Mary Ann H. Franson (eds.), pp. 109, 458, 287, and 648, Washington, DC, American Public Health Association, 1985).
3. *Hawley's Condensed Chemical Dictionary,* 11th Ed., rev'd by N. Irving Sax and Richard J. Lewis (New York, N.Y.: Van Nostrand Reinhold, 1987).

CHROMIUM

Chromium compounds are used primarily in industrial processes and may enter the environment and water supply through the discharge of wastes. Chromium exists primarily in the hexavalent state in water supplies and in industrial cooling towers, however some trivalent chromium has been occasionally observed.

Chromium compounds have varying degrees of toxicity. A number of specific chromium compounds are strong oxidizing agents, dangerous chemicals, and toxic by ingestion and inhalation. Hexavalent chromium compounds are carcinogenic (OSHA) and corrosive on tissue. Chromium in lower oxidation states tend to be less of a problem to humans and animals. The concentration of chromium in U.S. drinking waters appears to average less than 3.3 micrograms per liter. Reports of 40 micrograms of chromium per liter have been published.

Chromium compounds have several industrial and practical uses including the following: oxidizing agents, catalysts, inks, etching material, engraving, plating baths, medical applications, paints, textile dye, corrosion inhibitor, dyeing and printing, nuclear and high temperature research, and several additional applications. A number of compounds have limited solubility in water.

Companies that make toxic chromium compounds must control the level of effluent water contamination. Some industries have implemented loop systems to address the recovery and reuse issues of chromium in water. Additional industries are manifesting the chromium waste stream for burial purposes. Slag and sludge from metal manufacturing processes often contain significant amounts of toxic forms of chromium. Slag and sludge is sometimes deposited near water resources such as springs and flowing streams. A significant amount of the chromium waste generated in small industries and in society has been poorly characterized and improperly treated.

Vanadium, iron, and mercury (mercurous and mercuric ions) may interfere with the test for chromium. During the testing of most water, these metals at 1 ppm (mg/L) will not be a problem during the test for chromium. Molybdenum at much higher concentrations may also interfere.

Hexavalent chromium in water is directly determined by a spectrometric method at 540 nm. The hexavalent chromium can be determined colorimetrically by reaction with a 1,5-diphenylcarbohydrazide complex in an acidic buffer. In order to determine total chromium in water, all chromium must be converted to the hexavalent state by treatment with an alkaline hypobromite oxidation method or by digestion of the sample with a sulfuric-nitric acid mixture followed by oxidation with potassium permanganate. The amount of trivalent (and divalent) chromium in the sample can be determined by subtracting the results of a hexavalent chromium test from the results of the total chromium test.

CHROMIUM, HEXAVALENT (0 to 0.60 mg/L Cr^{6+})

Method 8023 for water and wastewater*
1,5-Diphenylcarbohydrazide Method** using Powder Pillows
USEPA accepted for wastewater analysis†

1. Enter the stored program number for hexavalent chromium (Cr^{6+}). Press: **90 ENTER.** The display will show: **Dial nm to 540.**
   Note: The Pour-Thru Cell can be used with 25-mL reagents only.

2. Rotate the wavelength dial until the small display shows: **540 nm.** When the correct wavelength is dialed in, the display will quickly show: **Zero sample,** then: **mg/L Cr^{6+}.**

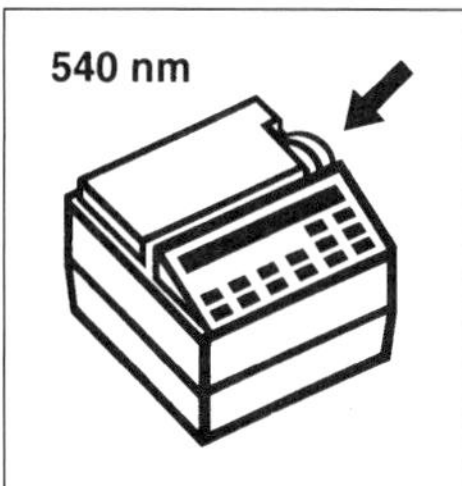

*Compliments of HACH Company. Analysis may be performed by DR/890 Colorimeter and DR/2010 Field Spectrophotometer.

**Adapted from *Standard Methods for the Examination of Water and Wastewater, 312 B.*

†Procedure is equivalent to USGS method I-1230-85 for wastewater.

3. Insert the 10-mL Cell Riser into the cell compartment.

4. Fill a sample cell with 10 mL of sample.

5. Add the contents of one ChromaVer 3 Reagent Powder Pillow to the cell (the prepared sample). Swirl to mix.
Note: A purple color will form if Cr^{6+} is present.
Note: At high chromium levels a precipitate will form. Dilute sample according to Sample Dilution Techniques *in Section 1.*

6. Press: **SHIFT TIMER.** A 5-minute reaction period will begin.

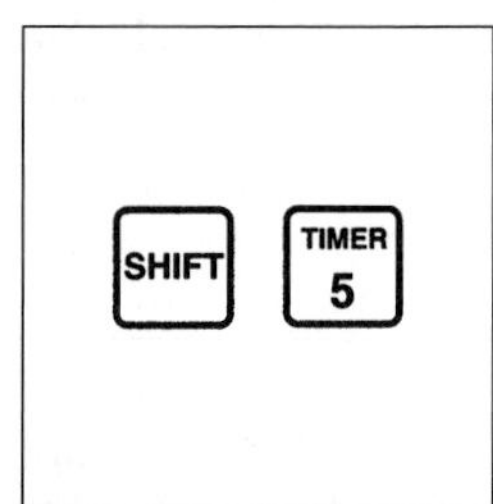

7. Fill another sample cell with 10 mL of sample (the blank).
Note: For turbid samples, add the contents of 1 Acid Reagent Powder Pillow. This ensures turbidity dissolved by the acid in the ChromaVer 3 Chromium Reagent is also dissolved in the blank.

8. When the timer beeps, the display will show: **mg/L Cr^{6+}.** Place the blank into the cell holder. Close the light shield.

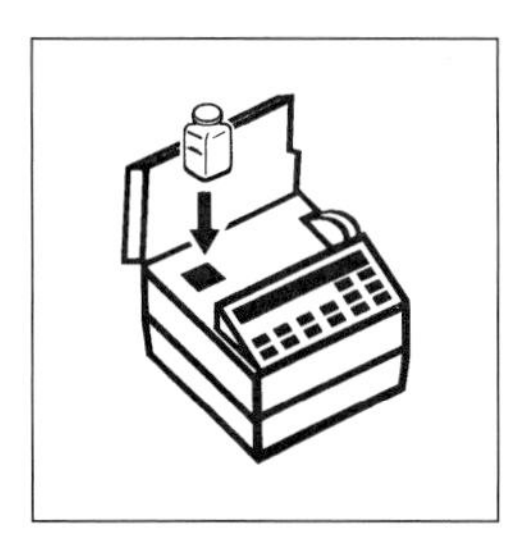

9. Press: **ZERO.** The display will show: **Zeroing ...** then: **0.00 mg/L Cr^{6+}.**

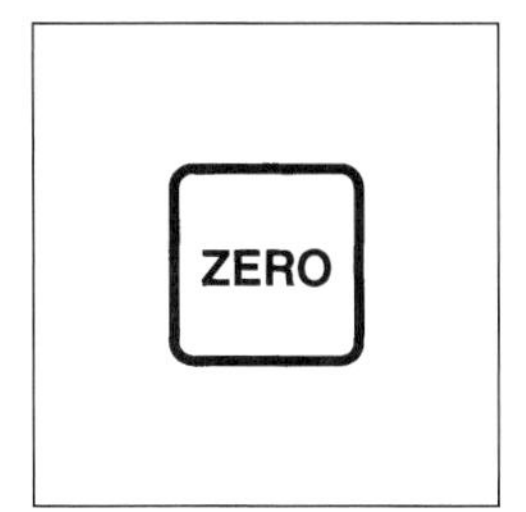

10. Place the prepared sample into the cell holder. Close the light shield.

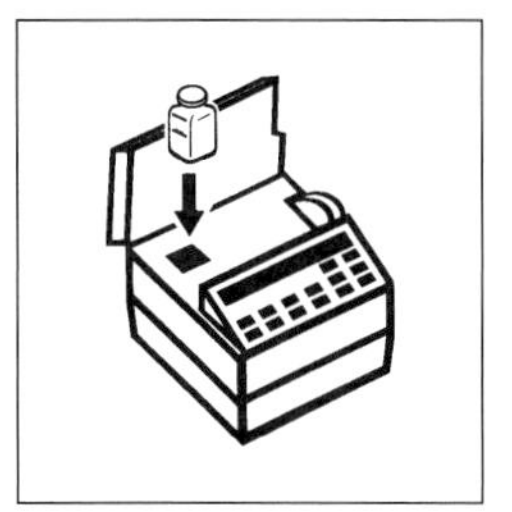

11. Press: **READ.** The display will show: **Reading ...** then the results in mg/L hexavalent chromium will be displayed.

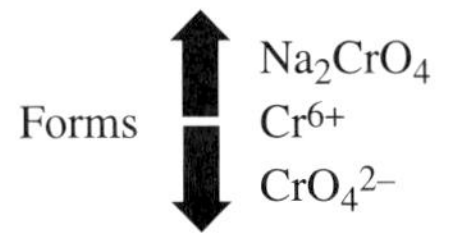

SAMPLING AND STORAGE

Collect samples in a cleaned glass or plastic container. Store at 4°C (39°F) up to 24 hours. Samples must be analyzed within 24 hours.

ACCURACY CHECK

See p. 436 of the HACH Water Analysis Handbook, 3rd Ed., 1997.

INTERFERENCES

The following substances do not interfere in the test, up to the following concentration:

Substance	Concentration
Mercurous & Mercuric Ions	Interferes slightly
Iron	1 mg/L
Vanadium	1 mg/L

Vanadium interference can be overcome by waiting 10 minutes before reading.
Highly buffered samples or extreme sample pH may exceed the buffering capacity of the reagents and require sample pretreatment; see "pH Interference" in Section 1.

CHROMIUM, Total (0 to 0.60 mg/L Cr)

Method 8024 for water and wastewater*
Alkaline Hypobromite Oxidation Method**†

1. Enter the stored program number for total chromium (Cr). Press: **1 0 0 ENTER.** The display will show: **Dial nm to 540.**
   Note: The Pour-Thru Cell can be used.

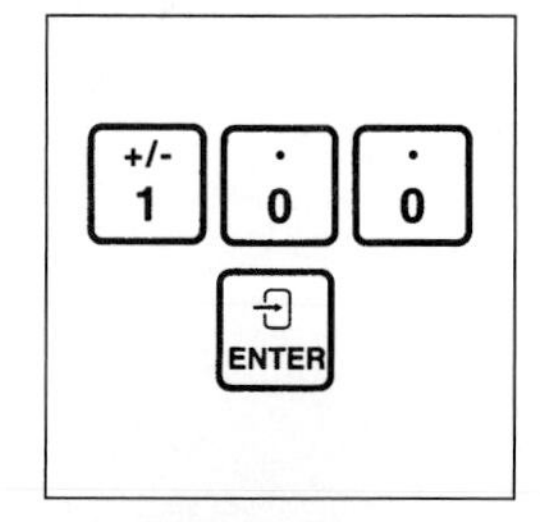

2. Rotate the wavelength dial until the small display shows: **540 nm.** When the correct wavelength is dialed in the display will quickly show: **Zero sample,** then: **mg/L Cr.**

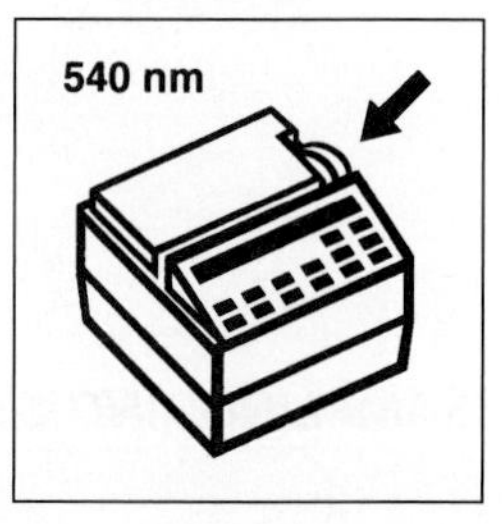

*Compliments of HACH Company.
**Adapted from *Standard Methods for the Examination of Water and Wastewater.*
†Procedure is equivalent to Standard Method 3500-Cr D for wastewater.

3. Fill a clean sample cell with 25 mL of sample.

4. Add the contents of one Chromium 1 Reagent Powder Pillow (the prepared sample). Swirl to mix.

5. Place the prepared sample into a boiling water bath.

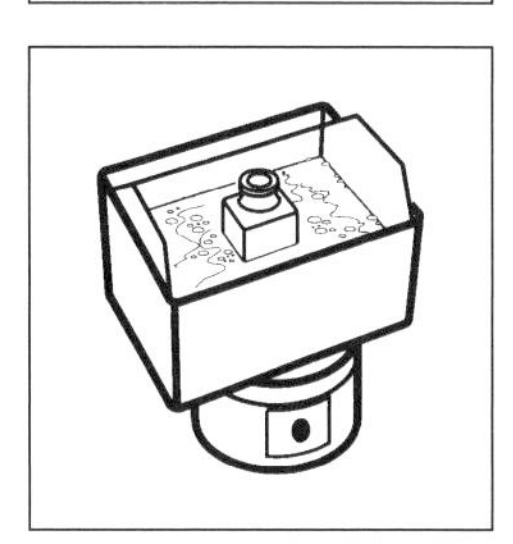

6. Press: **SHIFT TIMER.** A 5-minute reaction period will begin.

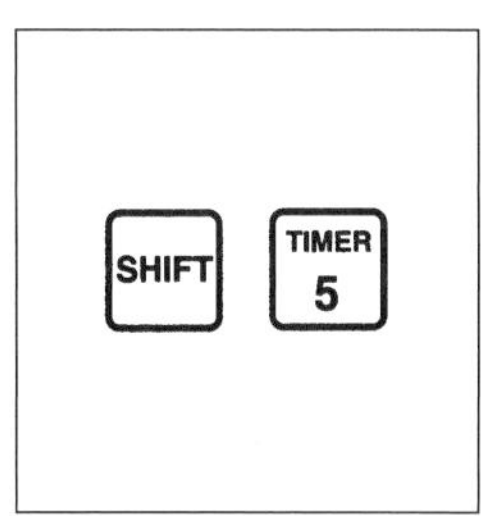

7. When the beeper sounds, remove the prepared sample. Using running tap water, cool the cell to 25°C.
Note: Use finger cots to handle the hot sample cell.

8. Add the contents of one Chromium 2 Reagent Powder Pillow. Swirl to mix.

9. Add the contents of one Acid Reagent Powder Pillow. Swirl to mix.

10. Add the contents of one ChromaVer 3 Chromium Reagent Powder Pillow. Swirl to mix.
Note: A purple color will form if chromium is present.
Note: ChromaVer 3 is white to tan in color. Replace brown or green powder. Undissolved powder does not affect accuracy.

11. Press: **SHIFT TIMER.** A 5-minute reaction period will begin.

12. When the timer beeps, fill another sample cell with 25 mL of sample (the blank). Place it into the cell holder. Close the light shield.
Note: For turbid samples, treat the blank as described in Steps 4 through 10.

13. Press: **ZERO.** The display will show: **Zeroing ...** then: **0.00 mg/L Cr.**

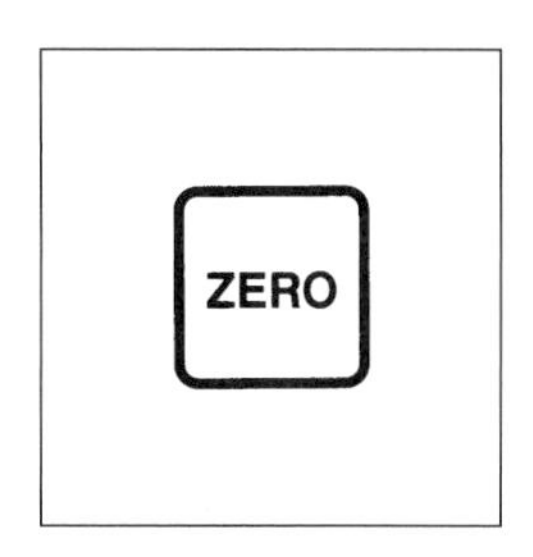

14. Place the prepared sample into the cell holder. Close the light shield.

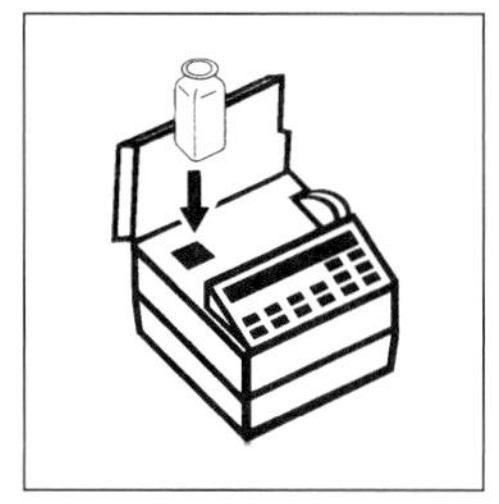

15. Press: **READ.** The display will show: **Reading ...** then the result in mg/L chromium will be displayed.
Note: Determine a reagent blank for each new lot of ChromaVer 3 using deionized water as the sample. Subtract this value from each result obtained.

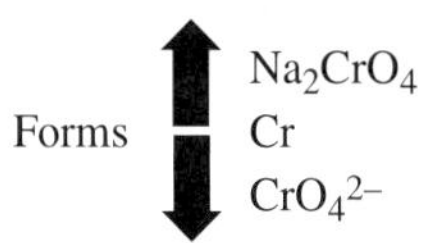

SAMPLING AND STORAGE

Collect samples in acid-washed glass or plastic containers. To preserve samples, adjust the pH to 2 or lower with nitric acid (about 2 mL per liter). Store preserved samples at room temperature up to six months. Adjust the pH to about 4 with 5.0 N sodium hydroxide before analysis. Correct the test results for volume additions.

ACCURACY CHECK

See p. 442 of the HACH Water Analysis Handbook, 3rd Ed., 1997.

INTERFERENCES

Large amounts of organic material may inhibit complete oxidation of trivalent chromium. If high levels of organic material are present, see "Digestion Tables" in Section 1 for instruction on sample digestion. Perform the analysis as described on the digested sample from the table and the procedure in Section 1 for liquids.

Note that iron does not interfere.

Highly buffered samples or extreme sample pH may exceed the buffering capacity of the reagents and require sample pretreatment; see "pH Interference" in Section 1.

QUESTIONS AND ANSWERS

1. What form does chromium usually assume in drinking water?
Chromium (VI) is usually observed in water supplies.
2. What substances interfere with the test for chromium (Hexavalent)?
Both vanadium and iron interfere at 1 mg/L and above. Vanadium interference can be overcome by waiting 10 minutes before reading. Mercurous and mercuric ions interfere slightly. Molybdenum at high concentrations can interfere with tests for chromium.
3. What oxidation states are expected for chromium?
Chromium compounds are usually found as Cr (+2), Cr (+3), Cr (+6) as ions and Cr (0) as metal.
4. Why is oxidation required to determine total chromium?
A purple compound develops with chromium (+6) and can be determined spectrophotometrically. Oxidation of chromium metal (Cr) and chromium ions (+2, +3) to Chromium (+6) and subsequent reaction with 1,5-diphenylcarbohydrazide in acidic solution allows the formation of the purple compound.

REFERENCES

1. *HACH Water Analysis Handbook,* 3rd Ed. (Loveland, CO.: HACH Company, 1997).
2. *Standard Methods for the Examination of Water and Wastewater*: CHROMIUM, 16th Ed., Method 312B, A. E. Greenberg, R. R. Trussell, L. S. Clesceri and Mary Ann H. Franson (eds.) pp. 201–202, Washington, DC, American Public Health Association, 1985.
3. *Hawley's Condensed Chemical Dictionary,* 11th Ed., rev'd by N. Irving Sax and Richard J. Lewis (New York, N.Y.: Van Nostrand Reinhold, 1987).

COBALT

Cobalt compounds and the metal have many practical uses. Cobalt compounds have variable toxicity levels and solubilities in water. The compounds decompose at many different temperatures and under a variety of conditions, while other cobalt compounds are very stable. Cobalt compounds have been used in pigments, cosmetics, whitener, paint and varnish drier, catalyst, ceramics, coloring enamels and glass, glazing pottery, oxygen stripping agent, semiconductors, mineral supplement, foam stabilizer, fertilizer additive, manufacture of vitamin B_{12}, storage batteries, hair dyes, hygrometers, vitamin preparation, nonionic surfactants (CTAS), bonding rubber to metals, electronic research, semiconductor, and medicine. Cobalt, as an additive, increases the corrosion resistance and the strength of alloys.

Cobalt consists of three useful radioactive substances with half lives of 72 days to 5.3 years. Radioactive cobalt has applications in biological and medical research, radiation therapy for cancer, radiographic testing of welds and castings, gas discharge tubes, portable radiation units, gamma radiation for wheat and potatoes, quality of marketable products, wool dyeing, oil consumption in internal combustion engines, and for locating buried telephone and electrical conduits.

Some of the cobalt waste is recovered and recycled to a variety of sources for use in many applications. Some facilities manifest the waste for either recovery or burial purposes, while other wastes are poured down the drain. In general, cobalt wastes are not very dangerous to humans and animals.

Cobalt easily forms coordination complexes with ammonia, water, and several other species. These colored compounds display none of the ordinary properties of cobalt. Also, cobalt forms a blue to green pigment named "cobalt blue" or "Thenard's Blue." The cobalt blue pigment is a very durable blue pigment resistant to both chemicals and weathering. This pigment is used in oil and water and as a cosmetic for eye shadows and grease paints.

A number of cations and anions interfere with the 1-(2-pyridylazo)-2-naphthol (PAN) test for cobalt. Cations of iron, copper, cadmium, chromium, manganese, lead, zinc, and aluminum, and the fluoride anion at concentrations below 40 mg/L (ppm) interfere with cobalt. After color development, EDTA is added to destroy all metal PAN complexes except cobalt and nickel. The PAN method is capable of detecting 0.1 mg/L of cobalt in the sample. Both nickel and cobalt can be determined on the same sample. The wavelength to determine the cobalt-PAN complex is 620 nm, whereas the wavelength to determine nickel in the sample is 560 nm. The analyses of both metals are noted in Method 8078 (Cobalt) and Method 8150 (Nickel).

COBALT (0 to 2.00 mg/L)

Method 8078 for water and wastewater*
1-(2-Pyridylazo)-2-Naphthol (PAN) Method**

1. Enter the stored program number for cobalt (Co).
   Press: **1 1 0 ENTER.** The display will show: **Dial nm to 620.**
   Note: Adjust the pH of stored samples before analysis.
   Note: The Pour-Thru Cell can be used with 25-mL reagents only. Rinse well with water after use.

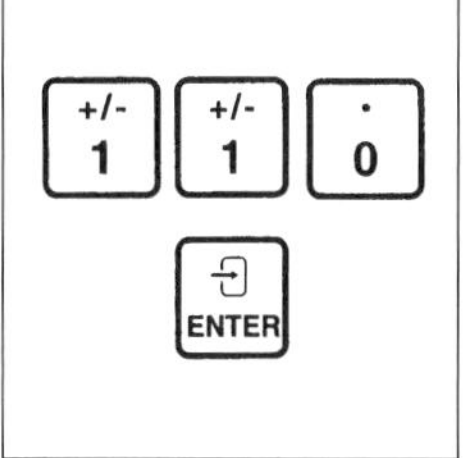

2. Rotate the wavelength dial until the small display shows: **620 nm.** When the correct wavelength is dialed in, the display will quickly show: **Zero sample,** then: **mg/L Co.**

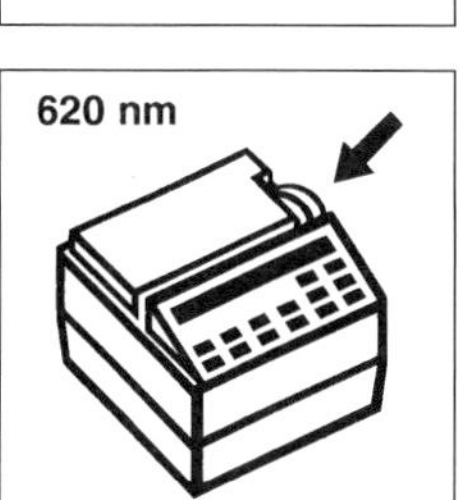

3. Insert the 10-mL Cell Riser into the sample compartment.
   Note: Total recoverable cobalt needs a prior digestion; use one of the procedures given in Digestion (Chapter 3) and the Sample and Analysis Volume Table for aqueous Liquids. If EDTA is present, use the vigorous digestion.

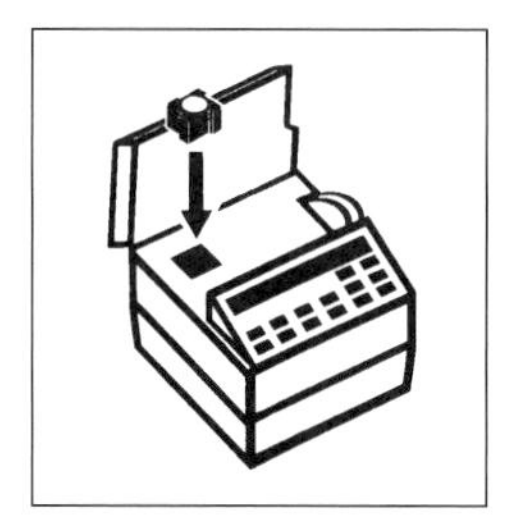

*Compliments of HACH Company.
**Adapted from H. Watanbe, Talanta, (21) 295 (1974).

4. Fill a 10-mL sample cell with 10 mL of sample (the prepared sample).
Note: If sample is less than 10°C (50°F), warm to room temperature prior to analysis.

5. Fill a second 10-mL sample cell with 10 mL of deionized water (the blank).

6. Add the contents of one Phthalate-Phosphate Reagent Powder Pillow to each sample cell. Stopper each cell. Immediately shake to dissolve.
Note: If sample contains iron (Fe^{3+}), all of the powder must dissolve completely before continuing with Step 7.

7. Add 0.5 mL of 0.3% PAN Indicator Solution to each sample cell. Stopper. Invert several times to mix.
Note: Use plastic dropper provided.

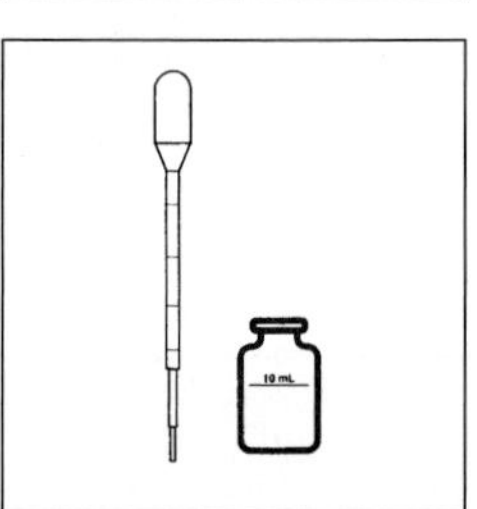

8. Press: **SHIFT TIMER.** A 3-minute reaction period will begin.
Note: During color development, the sample solution color may vary from green to dark red, depending on the chemical makeup of the sample. The deionized water blank should be yellow.

9. When the timer beeps, the display will show: **mg/L Co.** Add the contents of one EDTA Reagent Powder Pillow to each sample cell. Stopper. Shake to dissolve.

10. Remove both stoppers. Place the blank into the cell holder. Close the light shield.

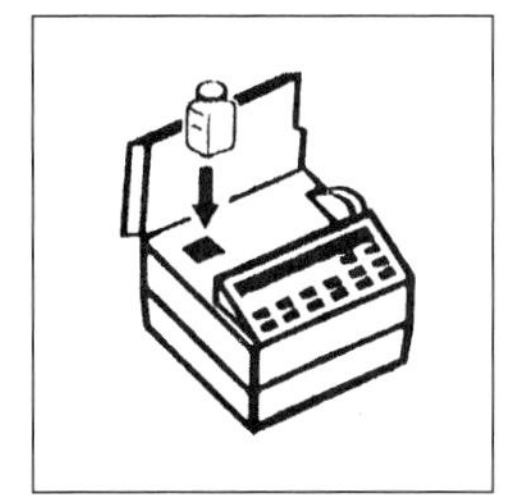

11. Press: **ZERO.** The display will show: **Zeroing ...** then: **0.00 mg/L Co.**

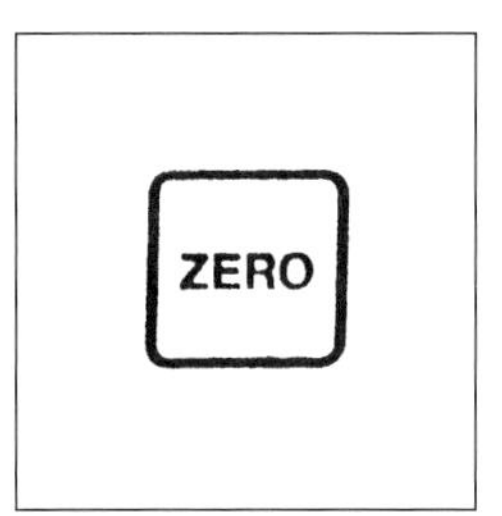

12. Place the prepared sample into the cell holder. Close the light shield. Press: **READ.** The display will show: **Reading ...** then the result in mg/L Co will be displayed.

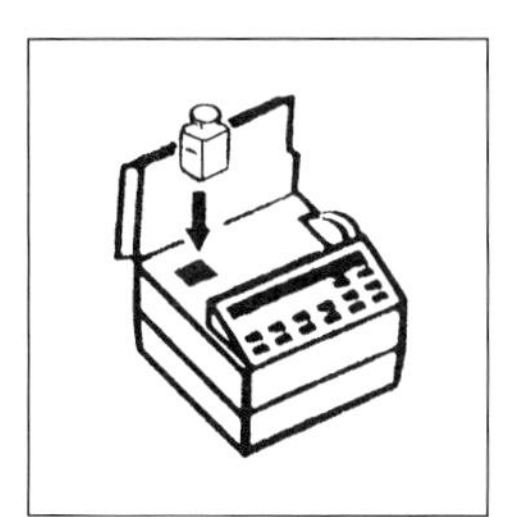

SAMPLING AND STORAGE

Collect samples in acid-washed plastic bottles. Adjust the sample pH to 2 or less with nitric acid (about 5 mL per liter). Preserved samples can be stored up to six months at room temperature. Adjust the sample pH between 3 and 8 with 5.0 Sodium Hydroxide Standard Solution just before analysis. Do not exceed pH 8 as this may cause some loss of cobalt as a precipitate. Correct test results for volume additions; see "Correcting for Volume Additions" in Section 1.

ACCURACY CHECK

See p. 446 of the HACH Water Analysis Handbook, 3rd Ed., 1997.

INTERFERENCES

The following may interfere in concentrations exceeding those listed below.

Al^{3+}	32 mg/L
Ca^{2+}	1000 mg/L as ($CaCO_3$)
Cd^{2+}	20 mg/L
Cl^-	8000 mg/L
Cr^{3+}	20 mg/L
Cr^{6+}	40 mg/L
Cu^{2+}	15 mg/L
F^-	20 mg/L
Fe^{2+}	Interferes directly; must not be present
Fe^{3+}	10 mg/L
K^+	500 mg/L
Mg^{2+}	400 mg/L
Mn^{2+}	25 mg/L
Mo^{6+}	60 mg/L
Na^+	5000 mg/L
Pb^{2+}	20 mg/L
Zn^{2+}	30 mg/L

Highly buffered samples or extreme sample pH may exceed the buffering capacity of the reagents and require sample pretreatment; see "pH Interference" in Section 1.

QUESTIONS AND ANSWERS

1. What is "cobalt blue?"
Cobalt blue is a blue pigment which is extremely durable and non-toxic. The pigment resists weathering and chemicals and is used for eye shadowing and painting of the body for display in shows.
2. How are most of the interferences avoided during a test for cobalt by the PAN method?
Interferences are avoided by adding EDTA to destroy the metal-PAN complexes in solution except those of cobalt and nickel.
3. How can nickel and cobalt be determined on the same sample?
Since EDTA does not destroy the Nickel-PAN complex, analysis for nickel can be performed at 560 nm (Method 8150) and cobalt can be determined at 620 nm (Method 8078).
4. Discuss some of the uses of radioactive cobalt.
Radioactive cobalt exists as cobalt-57 (Half-life 267 days), cobalt-58 (Half-life 72 days), and Cobalt-60 (Half-life 5.3 years). Cobalt-57 and cobalt-58 are used primarily for bi-

ological and medical research. Cobalt-60 is used in radiation therapy (cancer), radiographic testing of welds and castings, gas-discharge devices, liquid-level gauges, portable radiation units, gamma radiation for wheat and potatoes, wearing quality of floor wax, wool dyeing, oil consumption in internal combustion engines, the study of oil flow through porous media, and for locating buried telephone and electrical conduits.

REFERENCES

1. *HACH water analysis Handbook,* 3rd Ed. (Loveland, CO.: HACH Company, 1997).
2. *Hawley's Condensed Chemical Dictionary,* 11th Ed., rev'd by N. Irving Sax and Richard J. Lewis (New York, N.Y.: Van Nostrand Reinhold, 1987).
3. *Standard Methods for the Examination of Water and Wastewater:* Nonionic Surfactants as CTAS. 16th Ed., A. E. Greenberg, R. R. Russell, L. S. Clesceri and Mary Ann H. Franson (eds.), pp. 585–586, Washington, DC, American Public Health Association, 1985).

COLIFORM, GENERAL

For potable water, nonpotable water, and wastewater*

COLIFORM BACTERIA

Many of the microogranisms that cause serious disease, such as typhoid fever and dysentery, can be traced directly to polluted water. These disease-producing organisms, or pathogens, are discharged along with fecal wastes and are difficult to detect in water supplies. People may contact these pathogens in swimming pools, on bathing beaches, in rivers and streams, or from drinking contaminated water.

Testing for bacterial pathogens in water is impractical because it is hazardous, and the procedures are lengthy and involved. Most microbiological testing methods for water measure indicator organisms, not pathogens. Indicator organisms are bacteria which may not be pathogenic but usually are present when pathogens are present, and are more resistant to environmental stresses than pathogens. No organism or group of organisms satisfies all of the criteria for an indicator; however, one group, the coliforms, satisfies most of the requirements.

Coliform bacteria live in the intestines of man and other animals and are almost always present, even in healthy persons. The presence of coliforms in water is a warning signal that more dangerous bacteria may be present. Total coliform tests are used for potable water supplies. Fecal coliform tests usually are performed on untreated (nonpotable) water, wastewater, bathing water, and swimming water.

A number of methods are used to detect coliforms. For simultaneous detection of total coliforms and *Escherichia coli* (*E. coli*), a type of fecal coliform, a series of tests for m-ColiBlue24 Broth, Presence/Absence Bromcresol Purple Broth with MUG, and Lauryl Tryptose with MUG Broth can be implemented. A simple and accurate method for detecting both total coliforms and *Escherichia coli* is the coli-Mug procedure, which is a modification of the Most Probable Number (MPN, fermentation tube) technique. The m-ColiBlue24 Broth

*Compliments of HACH Company.

(Method 10029) membrane filtration technique can simultaneously determine total coliform bacteria and *E. coli* within 24 hours. The m-ColiBlue24 Broth method can be used to analyze all types of water including process waters for industrial and pharmaceutical applications, bottled water, beverages, wastewater, recreational waters, surface waters, ground water, and well water.

MEMBRANE FILTRATION METHOD

The Membrane Filtration (MF) method is a fast, simple way to estimate bacterial populations in water. The MF method is especially useful when evaluating large sample volumes or performing many coliform tests daily.

In the initial setup, an appropriate sample volume is passed through a membrane filter with a pore size small enough (0.45 microns) to retain the bacteria present. The filter is placed on an absorbent pad (in a petri dish) saturated with a culture medium that is selective for coliform growth. The petri dish containing the filter and pad is incubated, upside down, for 24 hours at the appropriate temperature. After incubation, the colonies that have grown are identified and counted using a low-power microscope.

MOST PROBABLE NUMBER METHOD

The Most Probable Number (MPN) Method (also referred to as the Multiple Tube Fermentation Technique) uses a specified number of test tubes to statistically predict the number of organisms present (based on the expected population of organisms in the sample). The MPN Method is ideal for wastewater samples and nonpotable samples, because the analyst can accommodate highly turbid samples by diluting prior to analysis. No filtering is necessary.

The MPN method is performed by using screw-capped tubes which contain sterile broth medium. Some tubes contained inverted inner vials (durham tubes) for gas collection. Simply add a sample to a tube of prepared medium and incubate. If coliforms are present, gas is produced and is trapped in the inner vial. Cloudiness, turbidity and/or fluorescence also can indicate the presence of coliforms.

PRESENCE/ABSENCE METHOD

The Presence/Absence (P/A) Method is a qualitative test that indicates only the presence or absence of organisms, not the number of organisms. The P/A Method is fast and suited to spot-checking applications. Only a minimal amount of analytical experience is required to perform the test.

Simply combine sample with medium, incubate for 24 to 48 hours, and check for a reaction indicating the presence of either total coliforms or *E. coli*.

U.S. Environmental Protection Agency drinking water regulations, effective December 31, 1990, require reporting only the presence or absence of coliforms. The World Health Organization recommends using the P/A Method for drinking water to ensure zero total coliforms and zero fecal coliforms or *E. coli*. The maximum contaminant goal of zero total coliforms eliminates the need to enumerate coliforms.

TECHNIQUE IS IMPORTANT

Good laboratory technique is essential when accuracy is important, particularly in microbiological laboratory procedures. Care in sample collection and preservation, a clean laboratory or work surface, proper sterilization and inoculation practices, and close temperature control help assure reliable results.

Using high-quality laboratory equipment and ready-to-use media can also save time and minimize errors. HACH's prepared media for MPN, MF and P/A testing helps eliminate contamination due to individual technician technique.

PREPARING SAMPLE CONTAINERS

Care must be taken to prevent contamination when conducting bacterial tests. All materials used for containing or transferring samples must be sterile! Presterilized plastic bags and bottles, or autoclavable glass and plastic bottles may be used to collect samples.

Presterilized Plastic Bags and Bottles

Plastic bags and bottles are available with or without dechlorinating agent.

Autoclavable Glass or Plastic Bottles

Glass or plastic bottles (at least 125 mL) may be used instead of plastic bags. These containers should be prepared as follows:

1. Wash in hot water and detergent.
2. Thoroughly rinse with hot tap water, followed by deionized water rinse to make sure that all detergent is removed.
3. If dechlorinating agent is needed (for chlorinated, potable water), add one Dechlorinating Reagent Powder Pillow to each 125-mL sample container. Add two powder pillows to a 250-mL sample container.
4. Steam sterilize glass and autoclavable plastic containers at 121°C for 15 minutes. Glass sample containers may be sterilized by hot air at 170°C for one hour.
5. Store sterile containers tightly capped in a clean environment until needed.
   Note: Dechlorinating agent should be used with potable or chlorinated water samples. It is not necessary for unchlorinated or nonpotable water samples. However, dechlorinating agent will not interfere with unchlorinated samples so, for simplicity, plastic bags containing dechlorinating agent may be used for all samples.

COLLECTING AND PRESERVING SAMPLES

Sampling should be properly carried out to ensure that seasonal variances are detected and that results are representative of that sample source.

Collect a sufficient volume of sample for analysis (usually a minimum of 100 mL of sample). World Health Organization guidelines prescribe 200 mL per sample while *Standard*

Methods for Examination of Water and Wastewater guidelines prescribe 100 mL of sample. Sample contamination during collection must be avoided.

No dechlorination is necessary if the sample is added directly to the medium on site. Otherwise, samples should be treated to destroy chlorine residual and transported for analysis immediately after collection. Sodium thiosulfate, which has been sterilized within the collection vessel, is generally used to destroy chlorine residual.

Analyze as soon as possible after collection. The maximum time between collection and examination of samples should be 8 hours (maximum transit time 6 hours, maximum processing time 2 hours). *If the time between collection and analysis will exceed 8 hours, maintain the sample at/or below 4°C, but do not freeze.* Maximum time between collection and analysis should not exceed 24 hours. Failure to properly collect and transport samples will cause inaccurate results.

Collect at least 100 mL of sample in presterilized plastic bags or bottles, or in sterile glass or plastic bottles. Do not fill sample containers completely. Maintain at least 2.5 cm (approximately one inch) of air space to allow adequate space for mixing the sample prior to analysis.

Faucets, Spigots, Hydrants or Pumps

Collect representative samples by allowing the water to run from a faucet, hydrant or pump at a moderate rate, without splashing, for two to three minutes before sampling. Do not adjust the rate of flow while the sample is being collected. Valves, spigots and faucets that swivel or leak, or those with attachments such as aerators and screens, should be avoided, or the attachments removed prior to sample collection.

Handle the sample containers carefully. Open them carefully just prior to collection, and close immediately following collection. Do not lay the lid or cap down and avoid contact near the mouths of the containers. Do not touch the inside of the containers. Do not rinse the containers. Properly label the sample containers immediately, and analyze the samples as soon as possible after collection.

Rivers, Lakes, and Reservoirs

When sampling a river, lake, or reservoir, fill the sample container below the water surface. Do not sample near the edge or bank. Remove the cap, grasp the sample container near the bottom and plunge the container, mouth down, into the water. (This technique excludes any surface scum.) Fill the container by positioning the mouth into the current or, in nonflowing water, by tilting the bottle slightly and allowing it to fill slowly. Do not rinse. Properly label the sample containers immediately, and analyze the samples as soon as possible after collection.

SAMPLE SIZE INFORMATION

Potable Water

Potable water should contain no coliforms per 100 mL, so testing should be done on undiluted samples. Use the MF method, the 5-tube or 10-tube MPN method, or the P/A method for analyzing potable water samples.

Nonpotable Water

Nonpotable water testing generally requires dilution of the original sample, based on probable coliform concentration. MPN and MF Methods require sample dilutions for nonpotable water. See the specific procedures for dilution instructions.

Disposing of Completed Tests

Active bacterial cultures grown during incubation must be disposed of safely. This may be accomplished in one of two ways:

Bleach. Used test containers may be sterilized by using a 10% bleach solution. Add approximately 12 mL of bleach to each test container. Allow 10 to 15 minutes contact time with the bleach. Pour the liquid down the drain, then dispose of the test containers in the normal waste.

Autoclave. Place used test containers in a contaminated-items bag or a biohazard bag and seal tightly. Test containers must be placed in a bag before autoclaving to prevent leakage into the autoclave. Autoclave used test containers in a bag at 121°C for 15 minutes at 15 pounds pressure. Once the test containers are sterile, they may be disposed of with the normal garbage. Place the bag of test containers in a separate garbage bag and tie tightly.

MEMBRANE FILTRATION METHOD (M-COLIBLUE24 BROTH)

Preparing Materials

To save time, start the incubator while preparing other materials. Set the incubator for the proper temperature setting described in the procedure (usually total coliforms are incubated at 35 ± 0.5°C and fecal coliforms are incubated at 45 ± 0.2°C).

Disinfect the work bench with a germicidal cloth, dilute bleach solutions, bactericidal spray or dilute iodine solution. Wash hands thoroughly with soap and water.

Mark each sample container with the sample number, dilution, date, and other necessary information. Take care not to contaminate the inside of the sample container in any way.

See the previous section for information about "Preparing Sample Containers" and "Collecting and Preserving Samples."

Convenient Packaging

Hach's PourRite Ampules contain prepared medium, which eliminates measuring, mixing and autoclaving steps necessary for preparing dehydrated medium. The ampules are designed with a large, unrestrictive opening that allows medium to pour out easily. Simply break off the top of the ampule and pour the medium onto an absorbent pad in a petri dish.

Each ampule contains enough selective medium for one test. Medium packaged in PourRite Ampules has a shelf-life of one year. Ampules are shipped with a Certificate of Analysis and have an expiration date printed on the label.

Using Presterilized Equipment and Media

You will need sterile materials, a disinfected work area and proper handling techniques, or contamination may give false results. To simplify technique and minimize the possibility of contamination, use presterilized equipment and media. HACH offers presterilized and disposable membrane filters, pipets, petri dishes, absorbent pads, inoculating loops, buffered dilution water in 99-mL bottles, sampling bags, and prepared growth media. MEL Portable Laboratories include presterilized consumables and a field filtration assembly.

Using Field Filtration Apparatus

1. Flame sterilize the top surface of the stainless steel Field Vacuum Support.
2. Attach the syringe tip to the vacuum support tubing.
3. Using sterile forceps, place a membrane filter, grid side up, onto the center of the vacuum support.
   Note: To sterilize forceps, dip forceps in alcohol and flame in an Alcohol or Bunsen Burner. Let forceps cool before use.
4. Open a package of funnels (start at the bottom of the package). Remove a funnel (base first) from the package.
5. Place the funnel onto the vacuum support. Do not touch the inside of the funnel. Push evenly on the funnel's upper rim to snap it onto the vacuum support.
6. Pour the sample into the funnel.
   Note: See specific procedures for the sample volume required.
7. Pull on the syringe plunger to draw the sample through the filter apparatus.
8. Remove the funnel.
9. Press the lever on the vacuum support stem to lift the membrane filter from the vacuum support surface.
10. Use sterile forceps to remove the membrane filter.
11. Place the membrane filter into a prepared petri dish and incubate according to the appropriate procedure.
12. Disconnect the syringe tip from the vacuum support tubing. Dispose of the liquid in the syringe.
13. Follow Steps 1 to 12 to filter remaining samples.

Using Autoclavable Apparatus

When numerous samples must be run on a routine basis, you may prefer to use an autoclave for nondisposable materials.

1. Wash sample bottles, pipets, petri dishes, filter holder with stopper and graduated cylinder (if needed) with hot water and detergent.
2. Rinse several times with tap water and then with deionized water. Dry thoroughly.
3. Prepare all equipment for autoclaving.
   - Loosely thread caps on sample bottles and cover caps and bottle necks with metal foil or paper.
   - Cover the openings of graduated cylinders with metal foil or paper.
   - Insert the filter funnel base into an autoclavable rubber stopper that will fit the filter flask.
   - Wrap the two parts of the filter funnel assembly separately in heavy wrapping paper and seal with masking tape.
   - Wrap petri dishes (borosilicate glass) in paper or place in aluminum or stainless cans.
4. Sterilize equipment in an autoclave at 121°C for 15 minutes. Borosilicate glass items may be sterilized with dry heat at 170°C for a minimum of 1 hour.

Preparing Autoclavable Filter Assembly

Disinfect the work bench or work area with a germicidal cloth, dilute bleach solution or dilute iodine solution. Wash hands thoroughly with soap and water.

1. After sterilization, remove the filter funnel assembly from the wrapping paper.
2. Do not contaminate the funnel by touching the inner surfaces that will be exposed to the sample.
3. Insert the funnel with rubber stopper into the filtering flask or filter funnel manifold and connect to the water trap and aspirator with rubber tubing.
4. Using sterile forceps, place a sterile membrane filter on the filter base and attach the filter tunnel top.
5. Filter a small quantity of sterile Buffered Dilution Water through the funnel to assure a good seal on the filter and connections before filtering the sample.

Sample Size

Sample size is governed by bacterial density as well as turbidity. Select a maximum sample size to give 20 to 200 colony-forming unity (CFU) per filter.

To accomplish this ideal situation, three different volumes should be filtered for samples where the bacterial count is uncertain. Table 2.18 lists recommended volumes for various types of samples.

When the sample is less than 20 mL (diluted or undiluted), 10 mL of sterile dilution water should be added to the filter funnel before vacuum is applied. This aids in the uniform distribution of the bacteria over the entire membrane filter.

Diluting Samples

As indicated by Table 2.18, very small sample volumes may be required for testing water samples high in turbidity or coliform number. Because it is almost impossible to measure these small volumes accurately, a series of dilutions should be made. The following procedure describes one method of preparing a series of dilutions.

TABLE 2.18 Suggested Sample Volumes for MF Total Coliform Test*

	Volume to be filtered (mL)							
Water Source	100	50	10	1	0.1	0.01	0.001	0.0001
Drinking water	X							
Swimming pools	X							
Wells, springs	X	X	X					
Lakes, reservoirs	X	X	X					
Water supply intake			X	X	X			
Bathing beaches			X	X	X			
River water				X	X	X	X	
Chlorinated sewage				X	X	X		
Raw sewage					X	X	X	X

**Standard Methods for the Examination of Water and Wastewater*, 19th Ed., p. 9–56.

Dilution Technique

1. Wash hands.
2. Open a bottle of sterile Buffered Dilution Water.
3. Shake the sample collection container vigorously, approximately 25 times.
4. Use a sterile transfer pipet to pipet the required amount of sample into the sterile Buffered Dilution Water.
5. Recap the buffered dilution water bottle and shake vigorously 25 times.
6. If more dilutions are needed, repeat Steps 3 to 5 using clean, sterile pipets and additional bottles of sterile Buffered Dilution Water.

Dilution Series

A. If 10-mL of sample is required:
Transfer 11 mL of sample into 99 mL of sterile Buffered Dilution Water. Filter 100 mL of this dilution to obtain the 10-mL sample.

B. If 1-mL sample is required:
Transfer 11 mL of the 10-mL dilution from A into 99 mL of sterile Buffered Dilution Water. Filter 100 mL of this dilution to obtain the 1-mL sample.

C. If 0.1-mL sample is required:
Transfer 11 mL of the 1-mL dilution from B into 99 mL of sterile Buffered Dilution Water. Filter 100 mL of this dilution to obtain the 0.1-mL sample.

D. If 0.01-mL sample is required:
Transfer 11 mL of the 0.1-mL dilution from C into 99 mL of sterile Buffered Dilution Water. Filter 100 mL of this dilution to obtain the 0.01-mL sample.

E. If 0.001-mL sample is required:
Transfer 11 mL of the 0.01-mL dilution from D into 99 mL of sterile Buffered Dilution Water. Filter 100 mL of this dilution to obtain the 0.001-mL sample.

F. If 0.0001-mL sample is required:
Transfer 11 mL of the 0.001-mL dilution from E into 99 mL of sterile Buffered Dilution Water. Filter 100 mL of this dilution to obtain the 0.0001-mL sample.

Sources of Laboratory Procedures

For additional information on various types of water test procedures by HACH, see the following:

1. *Portable Laboratory Manual* for potable water (MEL/850)
2. *Microbiology Environmental Laboratories—MEL Instrument and Procedures Manual for MEL P/A Safe Drinking Water Laboratory* (1993) includes several procedures for P/A Testing, MF Testing and MPN Testing.
3. The Coli-MUG Procedure Tube Method is included in a brochure.
4. Several microbiological test methods are discussed in the HACH water analysis Handbook.

TABLE 2.19 Reactions Using P/A Broth

Reaction	Comments	Report as
Color change from reddish purple to yellow or yellow-brown		Positive for total coliforms
No color change after 24 hours	Incubate for another 24 hours and re-check the sample for color change	
No color change after 48 ± 3 hours		Negative for total coliforms
Fluorescence under long-wave UV light (if using P/A Broth with MUG)		Positive for *E. coli*

Interpreting P/A Results

Confirming Positive Samples

Inoculum from incubated samples can be used to confirm the presence of bacteria. See Table 2.20 below. The media listed for fecal coliforms, total coliforms, and *E. coli* are USEPA-accepted for reporting purposes.

TABLE 2.20 Confirmation Media

Bacteria	Confirmation media	Incubation	Positive result
Total Coliforms (USEPA)	Brilliant Green Bile Broth	24–48 hours at 35 ± 0.5°C	Gas
Fecal Coliforms (USEPA)	EC Medium	24 hours at 44.5 ± 0.5°C	Gas
E. coli (USEPA)	EC Medium with MUG	24 hours at 44.5 ± 0.5°C	Gas = Positive for Fecal Coliforms. Fluorescence = Positive for *E. coli*

COLIFORM, TOTAL AND E. COLI

Using m- ColiBlue24 Broth PourRite Ampules*

1. Place a sterile absorbent pad in a sterile petri dish (use sterile forceps). Replace petri dish lid.
Note: Do not touch the pad or the inside of the dish.
Note: To sterlize forceps, dip forceps in alcohol and flame in an alcohol or Bunsen burner. Let forceps cool before use.

*Compliments of HACH Company.

2. Invert ampules 2 to 3 times to mix broth. Break open an ampule of m-ColiBlue24 Broth. Pour the contents evenly over the absorbent pad. Replace petri dish lid.

3. Set up the Membrane Filter Assembly. With sterile forceps, place a membrane filter with grid side up into the assembly. ***Note:*** Start the incubator while preparing other materials. Adjust the incubator temperature to 35°C. Use sterile materials and proper handling techniques in a disinfected area to avoid contamination.

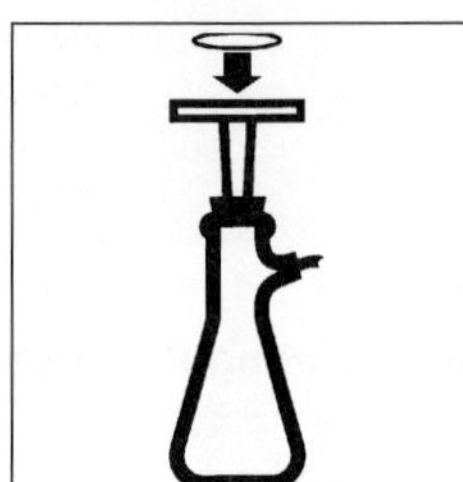

4. Shake the sample vigorously to mix. Pour 100 mL of sample or diluted sample into the funnel. Apply vacuum and filter the sample. Rinse the funnel walls 3 times with 20 to 30 mL of sterile buffered dilution water.

5. Turn off the vacuum and lift off the funnel top. Using sterile forceps, transfer the filter to the previously prepared petri dish.

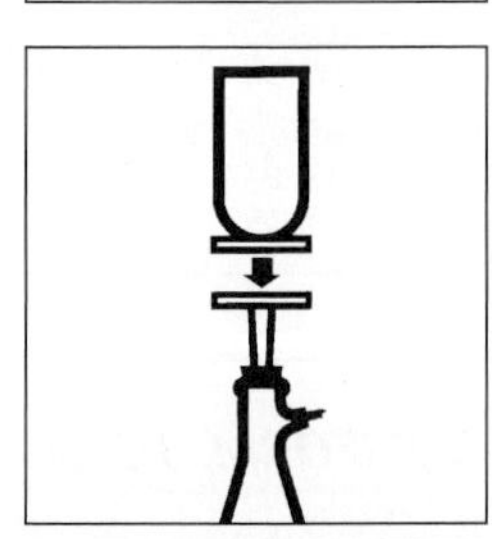

6. With a slight rolling motion, place the filter, grid side up, on the absorbent pad. Check for trapped air under the filter and make sure the filter touches the entire pad. Replace petri dish lid.

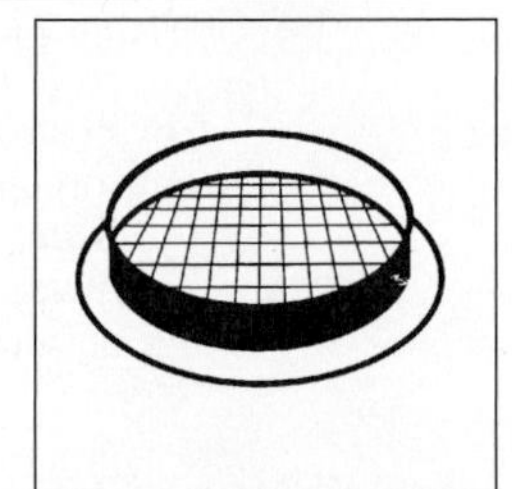

7. Invert the petri dish and incubate at 35 ± 0.5°C for 24 ± 4 hours.

8. Remove the petri dish from the incubator and count the colonies using a 10 to 15X stereoscopic microscope. Red and blue colonies indicate total coliforms and blue colonies indicate *E. coli.*
Note: Red colonies may vary in color intensity. Count all the red and blue colonies as total coliforms. Blue colonies may appear blue to purple. Count all blue to purple colonies as E. coli.

INTERPRETING RESULTS

Coliform density is reported as number of colonies per 100 mL of sample. Use samples that produce about 50 coliform colonies and not more than 200 colonies of all types per membrane to compute coliform density. Drinking water should produce very few colonies.

Equation A is used to calculate coliform density. Note that "mL sample" refers to actual sample volume, and not volume of dilution.

Equation A—Coliform density on a single membrane filter

$$\text{Coliform colonies per 100 mL} = \frac{\text{Coliform Colonies counted}}{\text{mL sample filtered}} \times 100$$

A) If growth covers the entire filtration area of the membrane or a portion of it, and colonies are not discrete, report results as "Confluent Growth With or Without Coliforms."

B) If the total colonies (coliforms plus non-coliforms) exceeds 200 per membrane or the colonies are too indistinct for accurate counting, report the results as "Too Numerous To Count" (TNTC).

In either case, a new sample must be run using a dilution that will give about 50 coliform colonies and not more than 200 colonies of all types.

When testing nonpotable water, if no filter meets the desired minimum colony count, the average coliform density can be calculated with Equation B below.

Equation B—Average coliform density for: 1) duplicates; 2) multiple dilutions; or 3) more than one filter/sample

$$\text{Coliform colonies per 100 mL} = \frac{\text{sum of all Colonies in all samples}}{\text{sum of volumes (in mL) of all samples}} \times 100$$

ADDITIONAL TESTING PROCEDURES

Total coliforms grow as red colonies on m-ColiBlue24 Broth. The percentage of red colonies that are actually non-coliforms (false positives) is comparable to the percentage of sheen colonies grown on m-Endo Broth which are non-coliforms (false positives). An oxidase test can confirm which red colonies are total coliforms. For further information on the oxidase test, see the *Standard Methods for the Examination of Water and Wastewater,* 18th Ed., 1992 or the HACH Analytical Procedure Bulletin on Membrane Filtration Technique for simultaneous total coliform and *E. coli* screening.

SUMMARY OF METHOD 8319

Presence/Absence (P/A) Bromcresol Purple Broth is ideal for screening drinking water samples for total coliforms. The method, a modification of the multiple-tube method, uses lactose and lauryl tryptose broths with bromcresol purple, which detects acidity formed during lactose fermentation by the bacteria.

Using sterile containers and proper handling conditions, 100 mL of sample is combined with P/A broth, the sample is incubated for 24 hours, and color change is checked. A yellow or yellow-brown color indicates the presence of total coliforms.

SUMMARY OF METHOD 8364

P/A Bromcresol Purple Broth with MUG allows simultaneous detection of total coliform bacteria and *E. coli.* In addition to the lactose and lauryl tryptose broths with bromcresol purple, this medium contains MUG reagent (4-methylumbelliferyl-D-glucuronide). MUG reagent produces a fluorogenic product when hydrolyzed by an enzyme (glucuronidase) specific to *E. coli.* MUG detects non-gas producing (anaerogenic) strains of *E. coli* and works well when competitive organisms are present.

Using sterile containers and proper handling conditions, a 100 mL sample is combined with P/A Broth with MUG, and the sample is incubated for 24 hours. After incubation, check for a color change and fluorescence. A yellow or yellow-brown color indicates the presence of total coliforms. To detect *E. coli,* examine samples under a long-wave ultraviolet (UV) light. Fluorescence indicates the presence of *E. coli.*

P/A BROTH IN DISPOSABLE BOTTLES

1. Collect 100 mL of sample in a sterile container. Do not contaminate the sample or sample container.
Note: Remove screens and other aeration devices from faucets and let water run for 2 or 3 minutes before collecting the stream. At times, the end of the spigot or faucet is caked with materials or rust and the spigot or faucet needs to be flamed with a match or cigarette lighter to avoid bacteria contamination of the collected water.

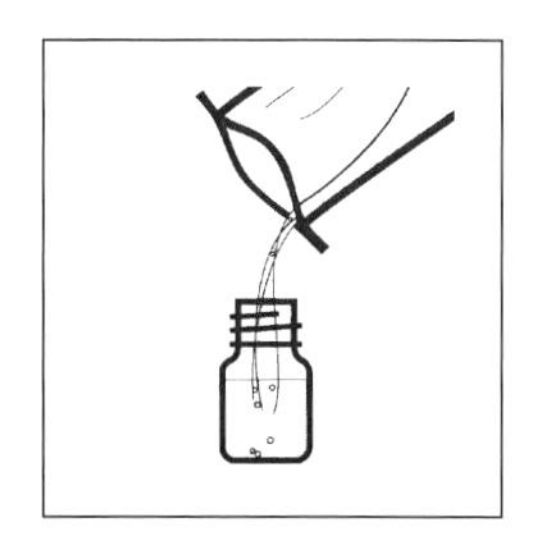

2. Add sample to the fill line on the P/A Broth Sample Bottle. Sample may be added from a sterile container, or directly from a faucet or spigot. Do not touch the inside or the bottom edges of the cap. Put the cap on.

3. Incubate the samples at 35 ± 0.5°C for 24 to 48 hours.

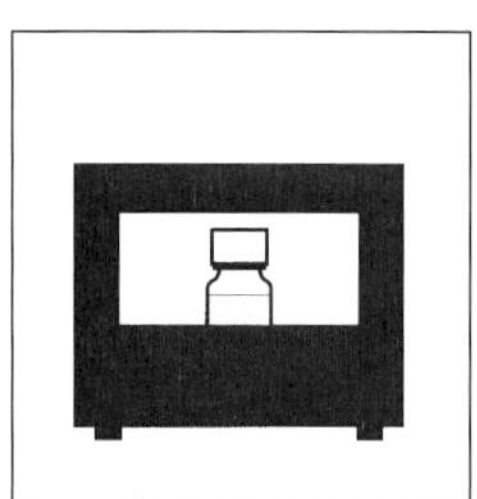

4. Note the reaction after 24 hours of incubation. If sample is negative, continue incubating for another 24 hours. See Table 2.21 below.

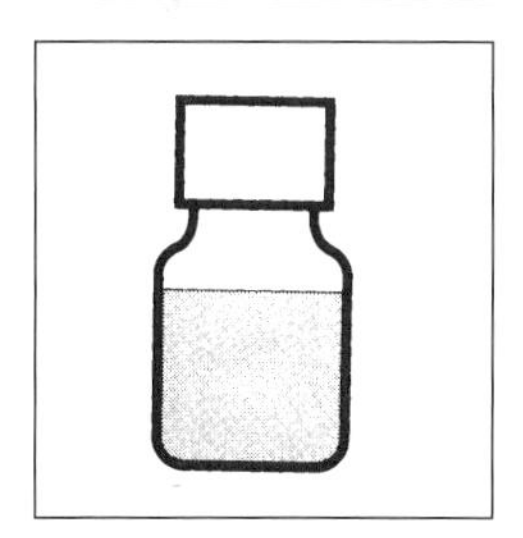

5. Confirm positive samples by inoculating the appropriate media from positive P/A Broth samples. See Table 2.22 below.

6. Dispose of completed tests appropriately. See "Disposing of Completed Tests: Bleach and Autoclave" noted previously.

Dispose of all completed tests

TABLE 2.21 Reactions Using P/A Broth

Reaction	Comments	Report as:
Color change from reddish purple to yellow or yellow brown		Positive for total coliforms
No color change after 24 hours	Incubate for another 24 hours and re-check the sample for color change	
No color change after 48 + 3 hours		Negative for total coliforms
Fluorescence under long-wave UV light (if using P/A Broth with MUG)		Positive for *E. coli*

Interpreting P/A Results*

Confirming Positive Samples*

Inoculum from incubated samples can be used to confirm the presence of bacteria. See Table 2.22 below.

The media listed for fecal coliforms, total coliforms, and *E. coli* are USEPA-accepted for reporting purposes.

TABLE 2.22 Confirmation Media

Bacteria	Confirmation media	Incubation	Positive result
Total coliforms (USEPA)	Brilliant green bile broth	24–48 hours at 35 + 0.5°C	Gas
Fecal coliforms (USEPA)	EC medium	24 hours at 44.5 + 0.5°C	Gas
E. coli (USEPA)	EC medium with MUG	24 hours at 44.5 + 0.5°C	Gas = positive for fecal coliforms. Fluorescence = positive for *E. coli*

*Compliments of HACH Company.

QUESTIONS AND ANSWERS

1. What do microbiological testing methods measure in water?
Most testing methods for water measure indicator organisms, not pathogens. Coliforms satisfy most of the criteria for an indicator organism.
2. What coliform test is used for potable water supplies?
Total coliform test is used for potable water supplies.
3. What are the advantages of the Membrane Filtration Method?
The method is fast and easy to perform. The MF Method can be used on large sample volumes. Many coliform tests can be performed daily. After incubation, the colonies that have grown can be identified and counted under a microscope.
4. What are the advantages of the MPN (Most Probable Number) Method?
Highly turbid samples of wastewater and nonpotable water can be diluted before analysis. Filtering is not necessary. The method statistically predicts the number of organisms present and requires less time to perform the task.
5. What are the advantages of the P/A (Presence/Absence) Method?
This method is fast and suited for spot-checking applications. A minimal amount of analytical experience is required to perform the test. The method reports only the presence or absence of total coliforms and E. coli. International agencies and the USEPA usually require reporting only the presence or absence of coliforms and E. coli.

REFERENCES

1. *HACH Water Analysis Handbook,* 3rd Ed. (Loveland, CO.: HACH Company, 1997).
2. *Standard Methods for the Examination of Water and Wastewater,* 16th Ed., A. E. Greenberg, R. R. Trussell, L. S. Clesceri and Mary Ann H. Franson (eds.), pp. 853–916, Washington, DC, American Public Health Association, 1985.
3. *Portable Laboratory Manual,* MEL/850 Portable Water Laboratory (Loveland, CO.: HACH Company, 1988).

COPPER

Copper is an essential element for the human body. It can be found on many labels of food such as cereals. The adult daily requirement is approximately 2.0 mg. Copper compounds are used in water supply systems to minimize biological growths (such as algae) in water reservoirs, lakes, ponds, and distribution pipes. Measurable amounts of copper may be introduced into drinking water by corrosion of copper-containing alloys in pipe fittings. Copper may occur in natural waters, wastewaters, and industrial waters as soluble salts and precipitated copper compounds.

Copper metal and copper compounds have numerous uses. Copper metal can be used in electrical devices, plumbing, heating, roofing and building construction, alloys, cooking utensils, coatings on a variety of metals, and many other applications. Copper compounds are found in fungicides, insecticides, pesticides, catalysts, pigments, photography, mildew and smut preventives, antifouling paints, wood preservative, dental cement, rodenticide, organic synthesis, batteries and electrodes, feed additive, metallurgy, electroplating, mordant in dyeing and printing fabrics, seed treatment, osmotic membranes, ceramics, enamels, mouth deodorant, light sensitive papers, reagents, pharmaceutical preparations, emulsifying agent, lube oil antioxidant, welding fluxes, brazing preparations, photocells, phosphors, dehydrating agent, leather, printing, treating fiber products, and nuclear reactors. In general, copper compounds generally lack the toxicity levels of other complexes of the same type.

In the experiment, a number of interferences can occur. A precipitate or sample turbidity may form if the solution is extremely acidic. Turbidity can be dissolved with the addition of a base such as KOH solution. Additional interferences include silver, cyanide, and high levels of iron, hardness, or aluminum found in seawater and industrial effluents. Each of these interferences can be minimized or eliminated by the addition of specific chemicals.

Copper ion tends to adsorb on the surface of sample containers. Therefore, samples should be analyzed as soon as possible after collection. If storage of a water sample is necessary, addition of a dilute solution of HCL will prevent adsorption. Copper exists in solution as cuprous (Cu^{+}) ion and cupric (Cu^{+2}) ion. Cuprous ion (Cu^{+}) is sufficiently stable in solution to react with 1, 10-phenanthroline-type compounds to form colored complexes which can be analyzed spectrophotometrically. In this experiment, copper in the sample reacts with a salt of bicinchoninic acid to form a purple colored complex in proportion to the copper concentration. The copper concentration can be determined by a spectrophotometer at 560 nm.

The Bicinchoninate Method is USEPA approved for reporting wastewater analysis for copper. The analysis for copper (0-5.00 ppm) can be applied to water, wastewater, seawater, and industrial waters. Seawater samples generally require pretreatment due to hardness and other specific interferences.

Strip tests have been developed to obtain quick and reliable approximations of total copper ion in solution. The strip tests have a range of measurement of 0 to 3 ppm and can be applied to many areas of water testing.

COPPER (0 to 5.00 mg/L)

Method 8506 for water, wastewater and seawater*
Bicinchoninate Method** using Powder Pillows
USEPA Approved for reporting wastewater analysis (digestion needed)†

1. Enter the stored program number for copper (Cu), bicinchoninate powder pillows. Press: **1 3 5 ENTER.** The display will show: **Dial nm to 560.**
   Note: The Pour-Thru Cell can be used for 25-mL reagents only.

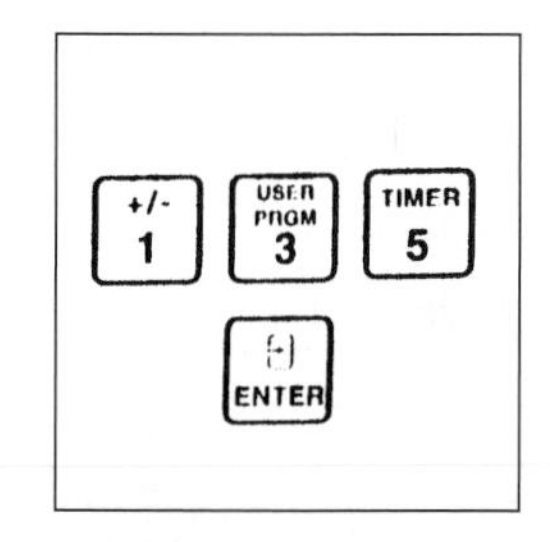

2. Rotate the wavelength dial until the small display shows: **560 nm.** When the correct wavelength is dialed in, the display will quickly show: **Zero sample,**then: **mg/L Cu Bicn.**
   Note: Determination of total copper needs a prior digestion. Digestion procedures and Sample and Analysis Volume Tables for Aqueous Liquids are listed in Section 1.

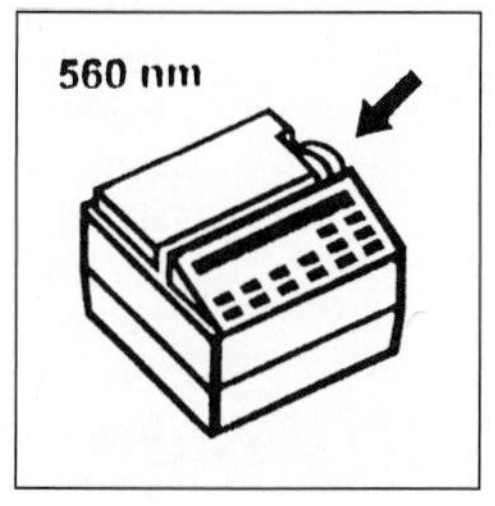

* Pretreatment required; see "Interferences (Using Powder Pillows)" in Section 1.
** Adapted from S. Nakano, *Yakugaku Zasshi*, 82 486–491 (1962) [*Chemical Abstracts*, 58 3390e (1963)].
† Powder Pillows only: Federal Register, 45 (105) 36166 (May 29, 1980).

3. Insert the 10-mL Cell Riser into the cell compartment.

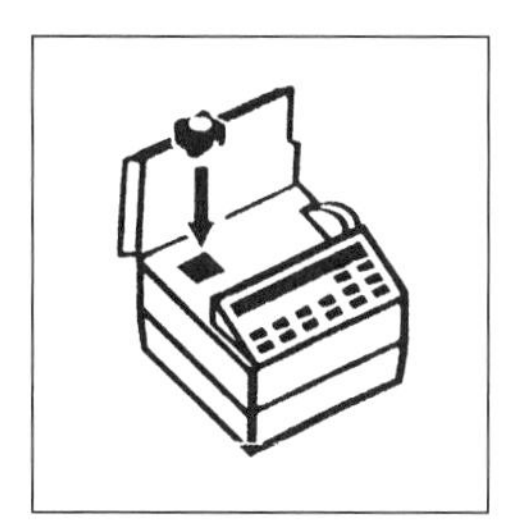

4. Fill a 10-mL sample cell with 10 mL of sample.
Note: Determine a reagent blank for each new lot of reagent. Use deionized water in place of the sample in the procedure. Subtract this value from each result obtained.
Note: Adjust pH of stored samples to between 4 and 6 before analysis.

5. Add the contents of one CuVer 1 Copper Reagent Powder Pillow to the sample cell (the prepared sample). Swirl to mix.
Note: A purple color will develop if copper is present.

6. Press: **SHIFT TIMER.** A two-minute reaction period will begin.
Note: Accuracy is not affected by undissolved powder.

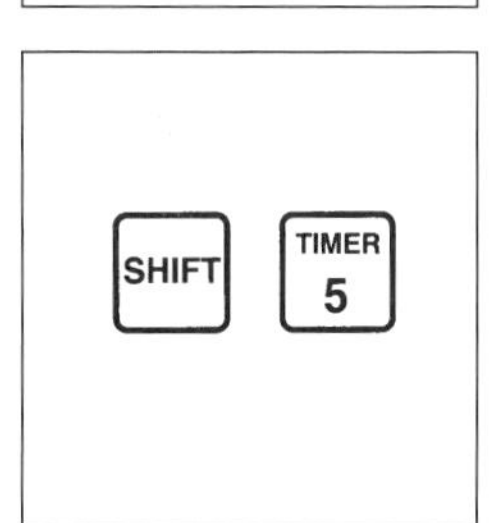

7. When the timer beeps, the display will show: **mg/L Cu Bicn.** Fill a second 10-mL sample cell (the blank) with 10 mL of sample.

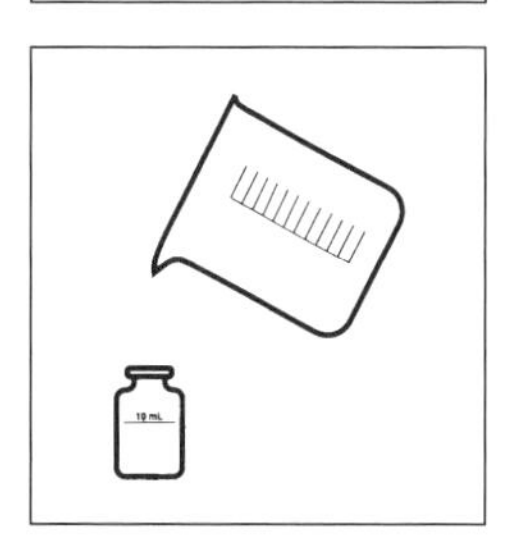

8. Place the blank into the cell holder. Close the light shield.

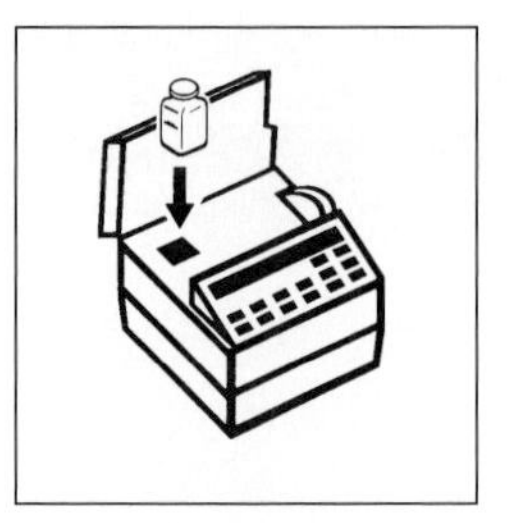

9. Press: **ZERO.** The display will show: **Zeroing ...** then: **0.00 mg/L Cu Bicn.**

10. Within 30 minutes after the timer beeps, place the prepared sample into the cell holder. Close the light shield.

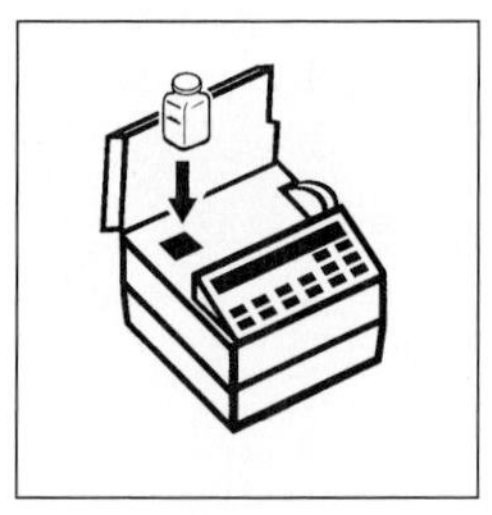

11. Press: **READ.** The display will show: **Reading ...** then the results in mg/L copper will be displayed.

SAMPLING AND STORAGE

Collect samples in acid-cleaned glass or plastic containers. Adjust the pH to 2 or less with nitric acid (about 2 mL per liter). Store preserved samples up to six months at room temperature. Before analysis, adjust the pH to 4 to 6 with 8 N potassium hydroxide. Do not exceed pH 6, as copper may precipitate. Correct the test result for volume additions; see "Correcting for Volume Additions" in Section 1 for more information. If only dissolved copper is to be determined, filter the sample before acid addition.

ACCURACY CHECK

See p. 533 of the HACH Water Analysis Handbook, 3rd Ed., 1997.

INTERFERENCES (USING POWDER PILLOWS)

If the sample is extremely acidic (pH 2 or less) a precipitate may form. Add 8 N Potassium Hydroxide Standard Solution drop-wise while swirling to dissolve the turbidity. Read the mg/L Cu. If the turbidity remains and turns black, silver interference is likely. Eliminate silver interference by adding 10 drops of saturated Potassium Chloride Solution to 75 mL of sample, followed by filtering through a fine or highly retentive filter. Use the filtered sample in the procedure.

Cyanide interferences prevent sufficient color development but can be overcome by adding 0.2 mL of formaldehyde to the 10-mL sample. Wait four minutes before taking the reading. Multiply the test results by 1.02 to correct for sample dilution by the formaldehyde.

To test samples such as seawater containing high levels of hardness, iron, or aluminum, analyze a 25-mL sample volume with a CuVer 2 Copper Reagent Powder Pillow instead of a CuVer 1 Pillow and the 10-mL sample volume (the taller 25-mL sample cells are required). Results obtained will include total dissolved copper (free and complexed).

To differentiate free copper from that complexed to EDTA or other complexing agents, analyze a 25-mL sample volume with a Free Copper Reagent Powder Pillow instead of the CuVer 1 pillow and the 10-mL sample volume (the taller 25-mL sample cells are required). Final results will be free copper only. Add a Hydrosulfite Reagent Powder Pillow to the developed sample and re-read the result. This result will include the total dissolved copper (free and complexed).

QUESTIONS AND ANSWERS

1. **How much copper is recommended for human consumption?**
   The adult daily requirement is 2.0 mg.
2. **Where is copper found in foods?**
   Copper is found in cereals, nuts, vegetables, animal tissues, and drinking water.
3. **Why is copper sulfate used in lakes, ponds, and other water bodies?**
   Copper sulfate is used in water supplies and bodies of water to minimize algae growth and biological growths.
4. **What interferences can occur in the experiment?**
   If the sample is acidic (pH 2 or less), a precipitate or turbidity may form. Silver and cyanide interferences must be addressed. High levels of iron, aluminum, and water hardness in seawater can be a problem.

REFERENCES

1. *HACH Water Analysis Handbook,* 3rd Ed. (Loveland, CO.: HACH Company, 1997).
2. *Hawley's Condensed Chemical Dictionary,* 11th Ed., rev'd by N. Irving Sax and Richard J. Lewis (New York, N.Y.: Van Nostrand Reinhold, 1987).
3. *Standard Methods for the Examination of Water and Wastewater:* COPPER, 16th Ed., Method 313, A. E. Greenberg, R. R. Trussell, L. S. Clesceri and Mary Ann H. Franson (eds.), pp. 204–208, Washington, DC, American Public Health Association, 1985.

CYANIDE

Cyanide compounds are extremely toxic and are observed primarily in industrial air emissions and water effluents. They can be present in fumigant processes, insecticide applications, gas scrubbing, chelating processes, coke ovens, metal-cleaning baths, electroplating processes, and ore flotation and extraction procedures. Some industries which use cyanides in the manufacturing process are those that produce metals, dyes, pigments, and nylon. Cyanide is sometimes found as a soluble or insoluble contaminate in solid waste and leachate produced by industry.

Cyanide complexes exist as inorganic and organic compounds. Inorganic compounds such as sodium cyanide are relatively stable in water that is not polluted with acid, oxidizing agents, and specific types of organic chemicals. Inorganic cyanides react with organic compounds containing carbonyl groups in water to form less toxic cyanohydrins and carboxylate complexes. Cyanides are generally stable and exist for a significant length of time in unpolluted water that is not moving.

Iron-cyanide complex ions are relatively stable and not acutely toxic in water. Complex cyanide ions in water tend to undergo photodecomposition on exposure to ultraviolet light to form hydrogen cyanide. During decomposition of iron-cyanide complexes in flowing river water, hydrogen cyanide evaporates into the air while bacterial and chemical destruction minimizes harmful concentrations of the free cyanide in water. Some of the transition metal cyanide complexes dissociate either partially or completely in water which can result in acute toxicity to fish at neutral pH.

Industries using cyanides can implement a closed loop system to recover and reuse the waste. Some facilities form much less toxic cyanates by oxidizing an alkaline solution of cyanide. Alternatively, sulfides react readily with cyanides to form biodegradable thiocyanates. Some industries manifest cyanide wastes for burial or modification purposes.

Cyanide compounds are not found in natural waters. Ionic forms of cyanide compounds tend to remain in basic or neutral solutions as simple cyanide ions. Hydrogen cyanide can be easily formed from both simple ionic compounds and organic complexes under acid conditions and volatilized into the air.

Chlorine in alkaline aqueous solution easily converts cyanide to intermediate complexes used for analysis purposes and ultimately to more stable cyanates. For USEPA reporting, samples containing cyanide should be distilled with acid. Sample distillation is also required to determine cyanide from transition metal cyanide complexes and heavy metal cyanide compounds. In this experiment, the pyridine-pyrazolone method gives an intense blue color with free cyanide and can be measured by the DR/850 and DR/890 Colorimeters and the DR/2010 Spectrophotometer in the field.

CYANIDE (0 to 0.200 mg/L)

Method 8027 for water, wastewater and seawater*
Pyridine-Pyrazalone Method**

1. Enter the stored program number for cyanide (CN^-). Press: **1 6 0 ENTER.** The display will show: **Dial nm to 612.**
Note: The Pour-Thru Cell can be used with 25-mL reagents only.

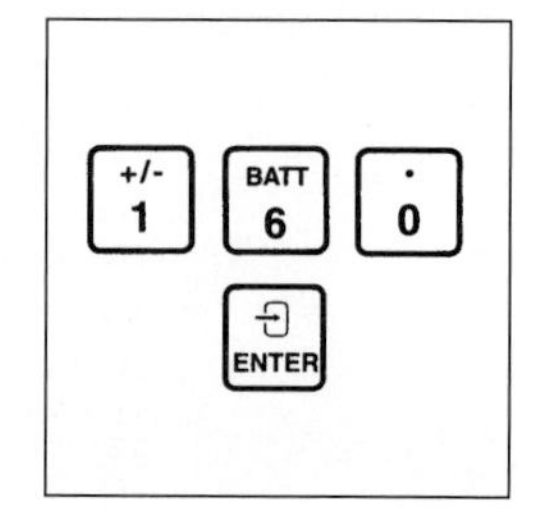

*Compliments of HACH Company. Analysis may be performed by DR/850 and Dr/890 Colorimeters and DR/2010 Spectrophotometer.
**Adapted from Joseph, Epstein, *Anal. Chem.,* 19(4), 272 (1947).

2. Rotate the wavelength dial until the small display shows: **612 nm.** When the correct wavelength is dialed in, the display will quickly show: **Zero sample,** then: **mg/L CN^-.**

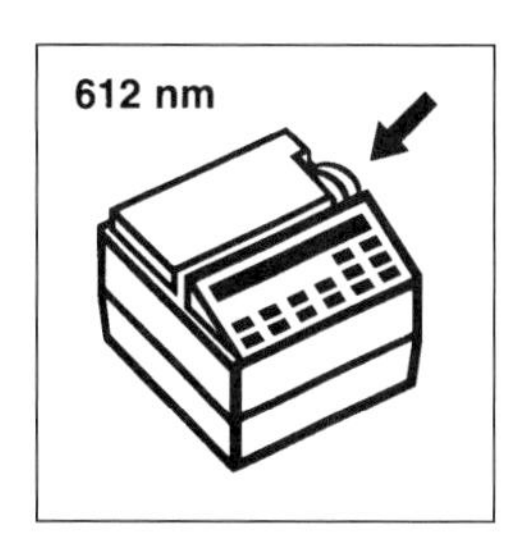

3. Insert a 10-mL Cell Riser into the cell compartment.

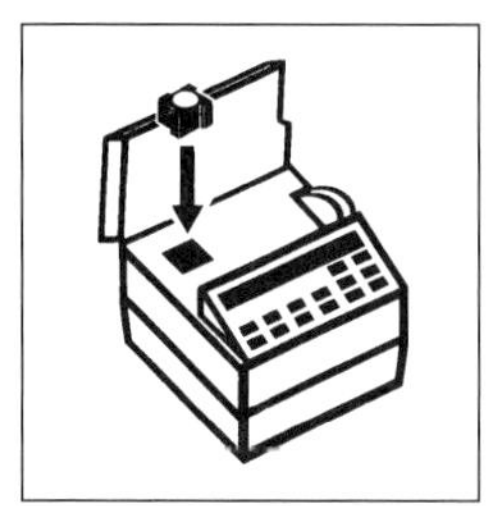

4. Using a graduated cylinder, pour 10 mL of sample into a 10-mL sample cell.

5. Add the contents of one CyaniVer 3 Cyanide Reagent Powder Pillow. Stopper the sample cell.

6. Shake the sample cell for 30 seconds.

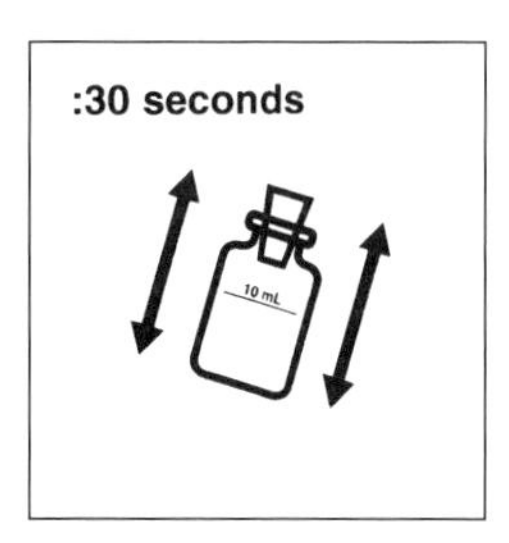

7. Wait an additional 30 seconds while the sample cell is undisturbed.

8. Add the contents of one CyaniVer 4 Cyanide Reagent Powder Pillow. Stopper the sample cell.

9. Shake the sample cell for 10 seconds. Immediately proceed with Step 10.
Note: Accuracy is not affected by undissolved CyaniVer 4 Cyanide Reagent Powder.

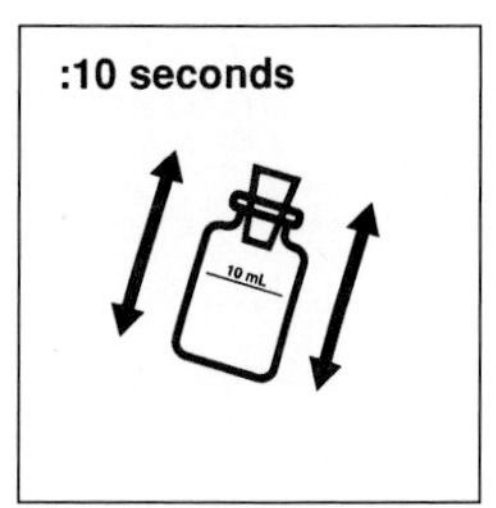

10. Add the contents of one CyaniVer 5 Cyanide Reagent Powder Pillow. Stopper the cell.
Note: Delaying the addition of the CyaniVer 5 Cyanide Reagent Powder for more than 30 seconds after the addition of the CyaniVer 4 Cyanide Reagent Powder will give lower test results.

11. Shake vigorously to completely dissolve the CyaniVer 5 Cyanide Reagent Powder (the prepared sample).
Note: If cyanide is present, a pink color will develop which then turns blue after a few minutes.

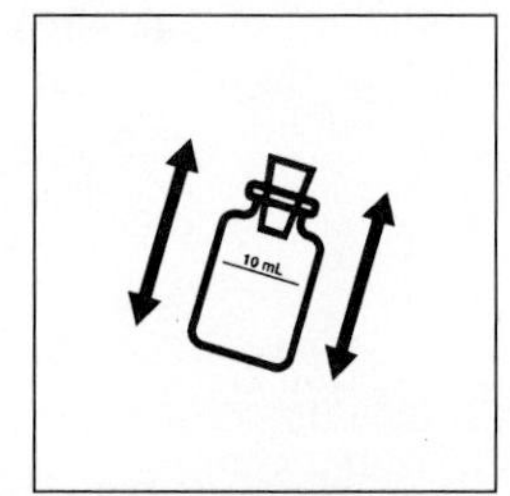

12. Press: **SHIFT TIMER.** A 30-minute reaction will begin. ***Note:*** Samples at less than 25°C require longer reaction time and samples at greater than 25°C give low test results.

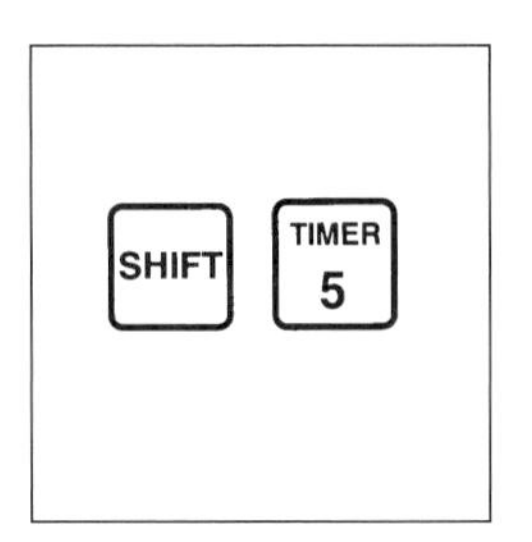

13. When the timer beeps, the display will show: **mg/L CN^-.** Fill another 10-mL sample cell (the blank) with 10 mL of sample.

14. Place the blank into the cell holder. Close the light shield.

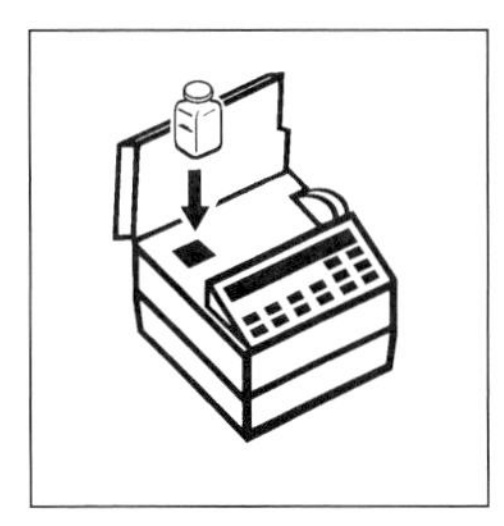

15. Press: **ZERO.** The display will show: **Zeroing ...** then: **0.000 mg/L CN^-.**

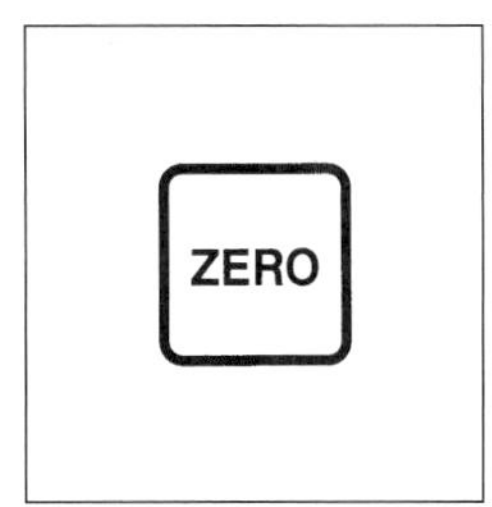

16. Remove the stopper. Place the prepared sample into the cell holder. Close the light shield. Press: **READ.** The display will show: **Reading ...** then the result in mg/L cyanide will be displayed.

SAMPLING AND STORAGE

Samples collected in glass or plastic bottles should be analyzed as quickly as possible. The presence of oxidizing agents, sulfides and fatty acids can cause cyanide loss during sample storage. Samples containing these substances must be pretreated as described in the following procedures before preservation with sodium hydroxide. If the sample contains sulfide and is not pretreated, it must be analyzed within 24 hours.

Preserve the sample by adding 4.0 mL of 5.0 N Sodium Hydroxide Standard Solution to each liter (or quart) of sample, using a glass serological pipet and pipet filler. Check the sample pH. Four mL of sodium hydroxide are usually enough to raise the pH of most water and wastewater samples to 12. Add more 5.0 N sodium hydroxide if necessary. Store the samples at 4°C (39°F) or less. Samples preserved in this manner can be stored for 14 days.

Before testing, samples preserved with 5.0 N sodium hydroxide or samples that are highly alkaline due to chlorination treatment processes or sample distillation procedures should be adjusted to approximately pH 7 with 2.5 N Hydrochloric Acid Standard Solution. Where significant amounts of preservative are used, a volume correction should be made; see "*Correcting for Volume Additions*" in Section 1 for more information.

OXIDIZING AGENTS

Oxidizing agents such as chlorine decompose cyanides during storage. To test for their presence and eliminate their effect, pretreat the sample as follows:

a) Take a 25-mL portion of the sample and add one drop of m-Nitrophenol Indicator Solution, 10 g/L. Swirl to mix.

b) Add 2.5 N Hydrochloric Acid Standard Solution drop-wise until the color changes from yellow to colorless. Swirl the sample thoroughly after the addition of each drop.

c) Add two drops of Potassium Iodide Solution, 30 g/L, and two drops of Starch Indicator Solution, to the sample. Swirl to mix. The solution will turn blue if oxidizing agents are present.

d) If Step c suggests the presence of oxidizing agents, add two level 1-g measuring spoonfuls of ascorbic acid per liter of sample.

e) Withdraw a 25-mL portion of sample treated with ascorbic acid and repeat Steps a to c. If the sample turns blue, repeat Steps d and e.

f) If the 25-mL sample remains colorless, adjust the remaining sample to pH 12 for storage with 5 N Sodium Hydroxide Standard Solution (usually 4 mL/L).

g) Perform the procedure given under Reducing Agents to eliminate the effect of excess ascorbic acid, before following the cyanide procedure.

SULFIDES

Sulfides will quickly convert to thiocyanate (SCN). To test for the presence of sulfide and eliminate its effect, pretreat the sample as follows:

a) Place a drop of sample on a disc of hydrogen sulfide test paper that has been wetted with pH 4 Buffer Solution.

b) If the test paper darkens, add a 1-g measuring spoon of lead acetate to the sample. Repeat Step a.

c) If the test paper continues to turn dark, keep adding lead acetate until the sample tests negative for sulfide.

d) Filter the lead sulfide precipitate through filter paper and a funnel. Preserve the sample for storage with 5 N Sodium Hydroxide Standard Solution or neutralize to a pH of 7 for analysis.

FATTY ACIDS

Caution: perform this operation in a hood as quickly as possible.

When distilled, fatty acids will pass over with cyanide and form soaps under the alkaline conditions of the absorber. If the presence of fatty acid is suspected, do not preserve samples with sodium hydroxide until the following pretreatment is performed. The effect of fatty acids can be minimized as follows:

a) Acidify 500 mL of sample to pH 6 or 7 with Acetic Acid Solution.

b) Pour the sample into a 1000-mL separatory funnel and add 50 mL of hexane.

c) Stopper the funnel and shake for one minute. Allow the layers to separate.

d) Drain off the sample (lower) layer into a 600-mL beaker. If the sample is to be stored, add 5 N Sodium Hydroxide Standard Solution to raise the pH to above 12.

ACCURACY CHECK

See p. 548 of the HACH Water Analysis Handbook, 3rd Ed., 1997.

INTERFERENCES (TURBIDITY)

Large amounts of turbidity will interfere and cause high readings. If the water sample is highly turbid, it should first be filtered before use in Steps 4 and 13. The test results should then be recorded as soluble cyanide.

OXIDIZING AND REDUCING AGENTS

Large amounts of chlorine in the sample will cause a milky white precipitate after the addition of the CyaniVer 5 Reagent. If chlorine or other oxidizing agents are known to be present, or if reducing agents (such as sulfide or sulfur dioxide) are known to be present, pretreat the sample before testing as follows using adequate ventilation:

Oxidizing Agents

a) Adjust a 25-mL portion of the alkaline sample to between pH 7 and 9 with 2.5 N Hydrochloric Acid Standard Solution. Count the number of drops of acid added.

b) Add 2 drops of Potassium Iodide Solution and 2 drops of Starch Indicator Solution to the sample. Swirl to mix. The sample will turn blue if oxidizing agents are present.

c) Add Sodium Arsenite Solution drop-wise until the sample turns colorless. Swirl the sample thoroughly after each drop. Count the number of drops.

d) Take another 25-mL sample and add the total number of drops of Hydrochloric Acid Standard Solution counted in Step a.

e) Subtract one drop from the amount of Sodium Arsenite Solution added in Step c. Add this amount to the sample and mix thoroughly.

f) Using 10 mL of this sample, continue with Step 4 of the cyanide procedure.

Reducing Agents

a) Adjust a 25-mL portion of the alkaline sample to between pH 7 and 9 with 2.5 N Hydrochloric Acid Standard Solution. Count the number of drops added.

b) Add 4 drops of Potassium Iodide Solution and 4 drops of Starch Indicator Solution to the sample. Swirl to mix. The sample should be colorless.

c) Add Bromine Water drop-wise until a blue color appears. Count the number of drops, and swirl the sample after addition of each drop.

d) Take another 25 mL sample and add the total number of drops of Hydrochloric Acid Standard Solution counted in Step a.

e) Add the total number of drops of Bromine Water counted in Step c to the sample and mix thoroughly.

f) Using 10 mL of this sample, continue with Step 4 of the cyanide procedure.

METALS

Nickel or cobalt in concentrations up to 1 mg/L do not interfere. Eliminate the interference from up to 20 mg/L copper and 5 mg/L iron by adding the contents of one HexaVer Chelating Reagent Powder Pillow to the sample and then mixing before adding the CyaniVer 3 Cyanide Reagent Powder Pillow in Step 4. Prepare a reagent blank of deionized water and reagents to zero the instrument in Step 15.

ACID DISTILLATION

For USEPA reporting purposes, samples must be distilled.

All samples to be analyzed for cyanide should be treated by acid distillation except when experience has shown that there is no difference in results obtained with or without distillation. With most compounds, a 1-hour reflux is adequate.

If thiocyanate is present in the original sample, a distillation step is absolutely necessary as thiocyanate causes a positive interference. High concentrations of thiocyanate can yield a substantial quantity of sulfide in the distillate. The "rotten egg" smell of hydrogen sulfide will accompany the distillate when sulfide is present. The sulfide must be removed from the distillate prior to testing.

If cyanide is not present, the amount of thiocyanate can be determined. The sample is not distilled and the final reading is multiplied by 2.2. The result is mg/L thiocyanate.

The distillate can be tested and treated for sulfide after the last step of the distillation procedure by using the following lead acetate treatment procedure.

a) Place a drop of the distillate (already diluted to 250 mL) on a disc of hydrogen sulfide test paper that has been wetted with pH 4.0 Buffer Solution.

b) If the test paper darkens, add 2.5 N Hydrochloric Acid Standard Solution drop-wise to the distillate until a neutral pH is obtained.

c) Add a 1-g measuring spoon of lead acetate to the distillate and mix. Repeat Step a.

d) If the test paper continues to turn dark, keep adding lead acetate until the distillate tests negative for sulfide.

e) Filter the blank lead sulfide precipitate through filter paper and funnel. This sample should now be neutralized to pH 7 and analyzed for cyanide without delay.

QUESTIONS AND ANSWERS

1. What are the forms of cyanide complexes and how stable are the cyanide compounds?
Some transition metal cyanide complexes exist as stable compounds where the cyanide anion can be removed by UV light frequencies, while other cyanide compounds of transition metals are less stable and tend to decompose more easily. Some organic cyanides are fairly stable whereas additional organic cyanides lack stability in the environment. Most ionic cyanides, such as sodium cyanide, are fairly stable in the powder form and in basic solution. However, hydrogen cyanide volatilizes rapidly and is an extremely dangerous chemical as a gas or as an acidified aqueous solution. The toxicity level of hydrogen cyanide gas is 50 ppm (NIOSH, 1997).

2. What are some of the interferences when testing for cyanide?
Oxidizing agents such as chlorine decompose cyanides during storage and in water. Sulfides react quickly with cyanides to form thiocyanates. Fatty acids form soaps in alkaline conditions with cyanides. Some organic compounds react easily with cyanides to form cyanohydrins and carboxylate complexes. Water turbidity will cause high readings.

3. When can cyanides be safely handled?
Complex cyanides such as molybdenum octacyanomolybdate and ferricyanides can be handled safely without any problems. Less stable transition metal cyanides can be normally handled as long as they do not come in contact with acid. Most cyanides, such as sodium cyanide, present greater problems but are relatively safe to handle in basic solution. Hydrogen cyanide and some of the lower molecular weight organic cyanides should be handled by someone who is trained in this area.

REFERENCES

1. *HACH Water Analysis Handbook,* 3rd Ed. (Loveland, C): HACH Company, 1997).
2. *Standard Methods for the Examination of Water and Wastewater:* CYANIDE, 16th Ed., A. E. Greenberg, R. R. Trussell, L. S. Clesceri and Mary Ann H. Franson (eds.), pp. 327–338, Washington, DC, American Public Health Association, 1985.
3. Hawley's Condensed Chemical Dictionary, 11th Ed., rev'd by N. Irving Sax and Richard J. Lewis (New York, N.Y.: Van Nostrand Reinhold, 1987).

FLUORIDE

Fluoride may occur naturally in water or it may be added in controlled amounts. The recommended limit for adults and school children of 1.0 mg/L (1.0 ppm) in drinking water effectively reduces tooth decay without harmful effects on health. For a child below six years old, the recommended level of fluoride by the American Dental Association is 0.3 to 0.6 ppm.* Too much fluoride in water can cause mottled enamel or "Dental Fluorosis," an objectionable discoloration of enamel. In a few instances, natural waters may contain greater levels of fluoride than the recommended limit. Fluoride tests should always be performed on drinking water sources before adding fluoride, especially in areas with a high pH or rich in limestone.

Industries manufacturing fluoride compounds must avoid the dumping of too much fluoride in effluents. Fluoride compounds are used in fluoridation of drinking water, toothpastes, phosphors, ceramics and ceramic bonded coatings, catalyst, wood preservation, electroplating, organic fluorination, degassing steel, fungicide, rodenticide, chemical cleaning, glass manufacture, high energy batteries, enamels, adhesive preservatives, disinfectant, dental prophylaxis, cryolite manufacture, lenses, spectroscopy, lasers, electronics, optics, high temperature dry film lubricants, and single windows in UV and infrared radiation detection systems. Some of the fluoride compounds are quite toxic while other compounds have little toxicity at low levels of concentration.

Seawater and wastewater samples require distillation to separate fluoride from interferences such as aluminum, ferric ion, orthophosphate, and hexametaphosphate at relatively low concentrations. Additional interferences can be eliminated by adding specific chemicals to the sample. The distillate will be free of substances that may interfere with colorimetric analysis for fluoride.

The SPADNS Method for fluoride determination involves the reaction of fluoride and a zirconium-dye solution. The fluoride combines with a portion of the zirconium to form a colorless complex anion (ZrF_6^{2-}), thus bleaching the red color in an amount proportional to the fluoride concentration. Fluoride can be determined in the field by the DR/850 and DR/890 Colorimeters and the DR/2010 Spectrophotometer. The spectrophotometric method is approved by the EPA for NPDES and NPDWR reporting purposes when the samples have been distilled.

FLUORIDE (0 to 2.00 mg/L F^-)

Method 8029 for water, wastewater and seawater**
SPADNS Method† (Reagent Solution)
USEPA accepted for reporting wastewater and drinking water analysis (distillation required)‡

1. Enter the stored program number for fluoride (F^-). Press: **1 9 0 ENTER.** The display will show: **Dial nm to 580.**
   Note: The Pour-Thru Cell cannot be used for this procedure.

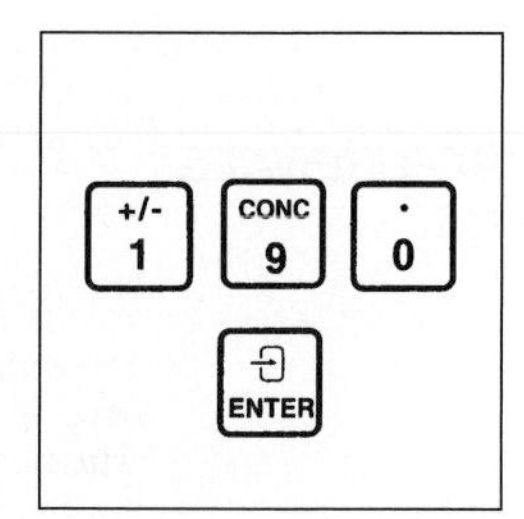

*Compliments of Ronald J. DeAngelis, DMD.

**Compliments of HACH Company. Analysis may be performed by DR/850 and DR/890 colorimeters and DR/2010 Spectrophotometer.

†Adapted from *Standard Methods for the Examination of Water and Wastewater, 413 C.*

‡Procedure is equivalent to Standard Methods 4500F-B,D for drinking water and wastewater.

2. Rotate the wavelength dial until the small display shows: **580 nm.** When the correct wavelength is dialed in, the display will quickly show: **Zero sample**, then: **mg/L F^-**.
Note: Approach the wavelength setting from higher to lower values.

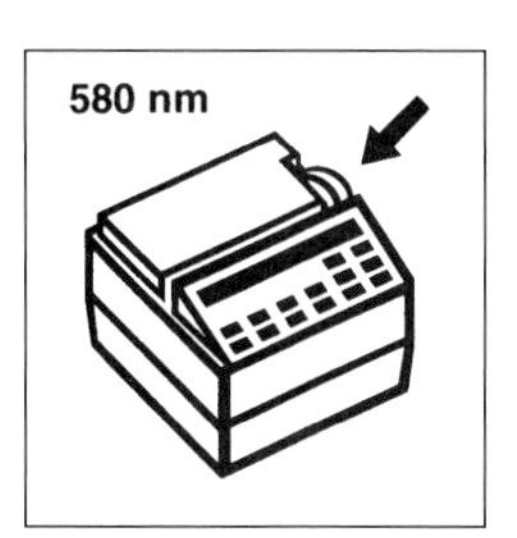

3. Insert a 10-mL Cell Riser into the cell compartment.

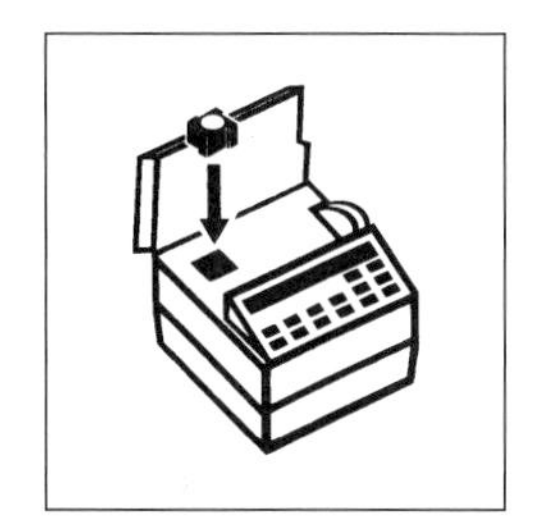

4. Measure 10.0 mL of sample into a dry 10-mL sample cell (the prepared sample).
Note: Use a graduated cylinder or pipet.
Note: For proof of accuracy, use a 1 mg/L Fluoride Standard in place of sample.

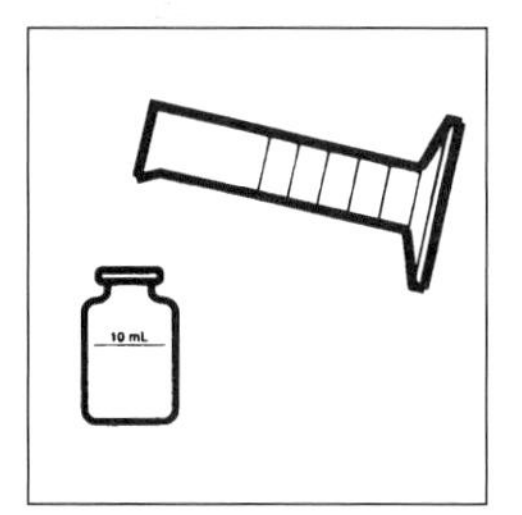

5. Measure 10.0 mL of deionized water into a second dry sample cell (the blank).
Note: Use a graduated cylinder or pipet.
Note: The sample and blank should be at the same temperature (± 1°C). Temperature adjustments may be made before or after reagent addition.

6. Pipet 2.00 mL of SPADNS Reagent into each cell. Swirl to mix.
Note: SPADNS Reagent is toxic and corrosive; use care while measuring.
Note: The SPADNS Reagent must be measured accurately.

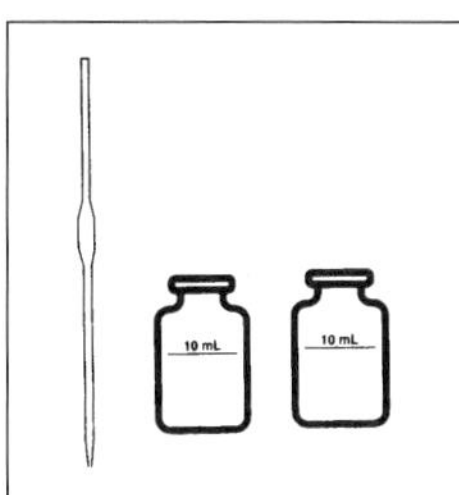

7. Press: **SHIFT TIMER.** A 1 minute reaction period will begin.

8. When the timer beeps, the display will show: **mg/L F⁻. Place the blank into the cell holder. Close the light shield.**

9. Press: **ZERO.** The display will show: **Zeroing ...** then: **0.00 mg/L F⁻.**

10. Place the prepared sample into the cell holder. Close the light shield.

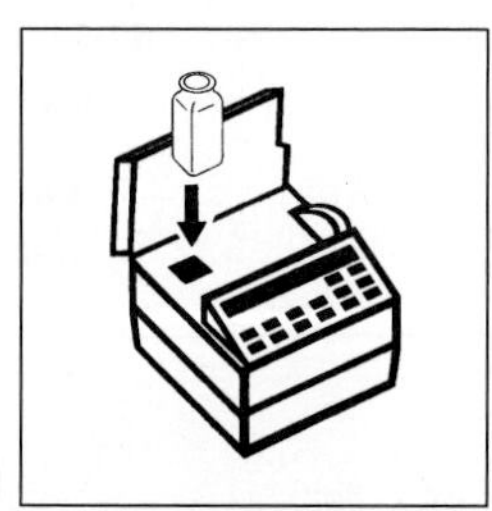

11. Press: **READ.** The display will show: **Reading ...** then the result in mg/L F⁻ will be displayed.

SAMPLING AND STORAGE

Collect samples in plastic bottles. Samples may be stored up to 28 days.

ACCURACY CHECK

See p. 567 of the HACH Water Analysis Handbook, 3rd Ed., 1997.

INTERFERENCES

This test is sensitive to small amounts of interference. Glassware must be very clean. Repeating the test with the same glassware is recommended to ensure that results are accurate.

The following substances interfere to the extent shown:

Substance	Concentration	Error
Alkalinity (as $CaCo_3$)	5000 mg/L	−0.1 mg/L F^-
Aluminum	0.1 mg/L	−0.1 mg/L F^-
Chloride	7000 mg/L	+0.1 mg/L F^-
Iron, ferric	10 mg/L	−0.1 mg/L F^-
Phosphate, ortho	16 mg/L	+0.1 mg/L F^-
Sodium Hexametaphosphate	1.0 mg/L	+0.1 mg/L F^-
Sulfate	200 mg/L	+0.1 mg/L F^-

SPADNS Reagent contains enough arsenite to eliminate interference up to 5 mg/L chlorine. For higher chlorine levels, add one drop of Sodium Arsenite Solution to 25 mL of sample for each 2 mg/L of chlorine.

To check for interferences from aluminum, read the concentration one minute after reagent addition, then again after 15 minutes. An appreciable increase in concentration suggests aluminum interference. Waiting two hours before making the final reading will eliminate the effect of up to 3.0 mg/L aluminum.

Most interferences can be eliminated by distilling the sample from an acid solution as described below:

a) Set up the distillation apparatus for the general purpose distillation. Turn on the water and make certain it is flowing through the condenser.

b) Measure 100 mL of sample into the distillation flask. Add a magnetic stirring bar and turn on the heater power switch. Turn the stir control to 5.

c) Cautiously measure 150 mL of StillVer Distillation Solution (2:1 sulfuric acid) into the flask. If high levels of chloride are present, add 5 mg silver sulfate for each mg/L chloride present.

d) Turn the heat control to setting 10, with the thermometer in place. The yellow pilot lamp shows when the heater is on.

e) When the temperature reaches 180°C (about one hour), turn the still off.

f) Dilute the collected distillate to 100 mL, if necessary. Analyze the distillate by the above method.

QUESTIONS AND ANSWERS

1. When does EPA accept this procedure of analysis?
EPA accepts the data from this method of analysis when the sample has been distilled.
2. Why do seawater and wastewater require distillation?
Seawater and wastewater require distillation to eliminate interferences from low concentrations of aluminum, ferric ion, orthophosphate, and hexametaphosphate and higher levels of chloride, sulfate and alkalinity. Interference from chlorine can be addressed by the addition of sodium arsenite solution.
3. What causes fluorosis?
Fluorosis or mottling of the teeth may occur when the fluoride level in drinking water is too high for extended periods of time.
4. What level of fluoride is permitted or recommended for drinking water in the USA?
A fluoride level of 1.0 mg/L is recommended for drinking water.

REFERENCES

1. *HACH Water Analysis Handbook,* 3rd Ed. (Loveland, CO.: HACH Company, 1997).
2. *Hawley's Condensed Chemical Dictionary,* 11th Ed., rev'd by N. Irving Sax and Richard J. Lewis (New York, N.Y.: Van Nostrand Reinhold, 1987).
3. *Standard Methods for the Examination of Water and Wastewater:* FLUORIDE, 16th Ed., Method 413, A. E. Greenberg, R. R. Trussell, L. S. Clesceri and MaryAnn H. Franson (eds.), pp. 352–360, Washington, DC, American Public Health Association, 1985.

FORMALDEHYDE

Formaldehyde diluted with water is used as a preservative for samples of plankton, fish, and other biological tissues and species used for examination and dissection in laboratories. The organic solvent has a variety of applications in society including, but not limited to, resin preparation, disinfectant, embalming fluids, corrosion inhibitor, industrial sterilant, cleaning solvent, treatment of grain smut, foam insulation, particle board and plywood, biocide, reducing agent for recovery of gold and silver, durable-press treatment of textile fabrics, chemical intermediate, metal plating baths, and several additional applications.

In most cases, formaldehyde solutions have been and continue to be dumped down the drain in many hospitals, research facilities, educational institutions, and some industries. In limited cases, formaldehyde is collected and purified by distillation or destroyed by incineration. Industries often distill and reuse the formaldehyde or manifest the waste for recovery or burial purposes. In some cases, small amounts of waste formaldehyde are allowed to evaporate in hoods in the laboratory.

The MBTH Method requires glassware to be washed with chromic acid to remove trace contaminates. Samples should be analyzed immediately after collection and a temperature controlled water bath is recommended to control the temperature of the sample. Several organic chemicals such as aldehydes, amines, and aniline interfere at relatively low levels while glycine, glucose, phenol and urea interfere near 1000 mg/L (ppm). Inorganic species including nitrite, ferric ion, copper, and ammonium (as N) also interfere at low concentrations. Additional cations and anions interfere at higher levels.

Formaldehyde reacts with MBTH (3-methyl-2-benzothiazoline hydrazone) in a developing solution to form a blue color which is proportional to the formaldehyde concentration. The MBTH Method is a sensitive colorimetric test for low range aldehyde measurements. The spectrometer measures the colored solution (micrograms per liter) at a wavelength of 630 nm.

FORMALDEHYDE (0 to 350 μg/L)

Method 8110 for water*
MBTH Method**

1. Enter the stored program number for formaldehyde (CH_2O). Press: **2 0 0 ENTER.** The display will show: **Dial nm to 630.**
Note: Samples should be analyzed immediately after collection.
Note: The Pour-Thru Cell cannot be used.

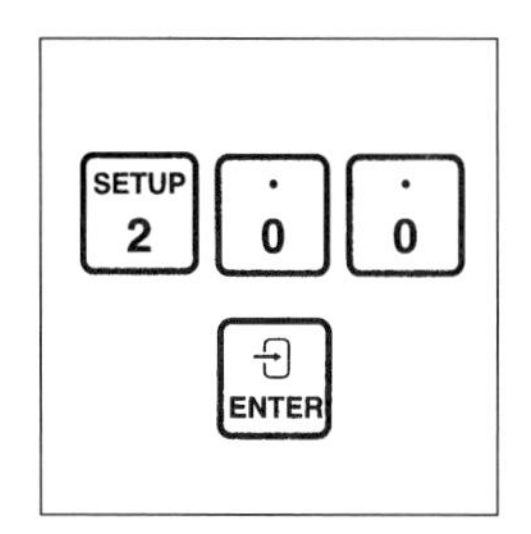

2. Rotate the wavelength dial until the small display shows: **630 nm.** When the correct wavelength is dialed in, the display will quickly show: **Zero sample**, then: **μg/L CH_2O.**

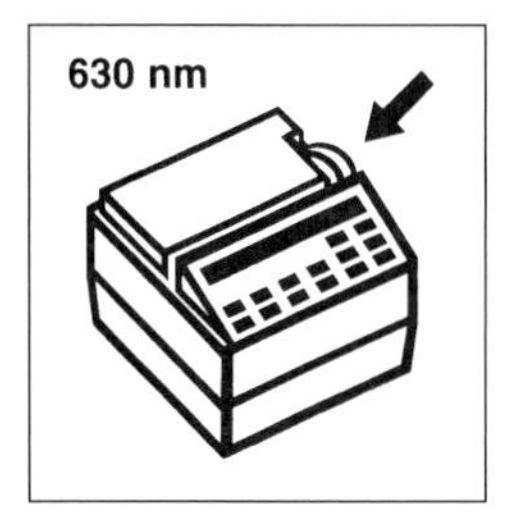

3. Accurately measure 25 mL of sample in a 50-mL mixing cylinder (the prepared sample).
Note: Wash glassware with chromic acid cleaning solution to remove trace contaminants.
Note: The sample should be at 25 ± 1°C, and the timing must be followed precisely. A temperature controlled water bath is recommended.

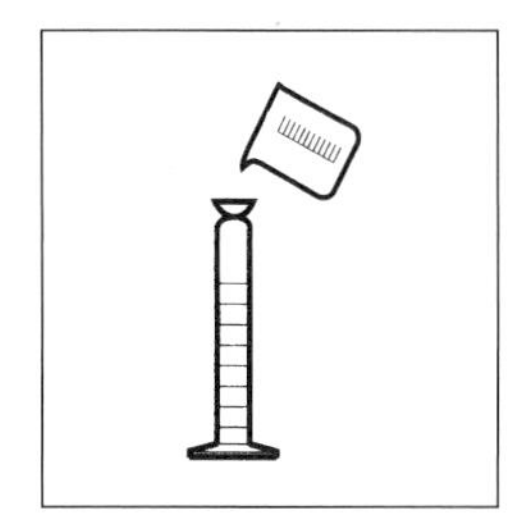

4. Accurately measure 25 mL of formaldehyde-free water in a second 50-mL mixing cylinder (the blank).
Note: Obtain formaldehyde-free water by distilling water from alkaline permanganate (4 g sodium hydroxide, 2 g potassium permanganate per 500 mL water). Discard the first 50 to 100 mL of distillate.

---

* Compliments of HACH Company.

** Adapted from T. G. Matthews, and T. C. Howell, *Journal of the Air Pollution Control Association,* 31(11), 1181–1184 (1981).

5. Add the contents of one MBTH Powder Pillow to the blank. Stopper the cylinder.

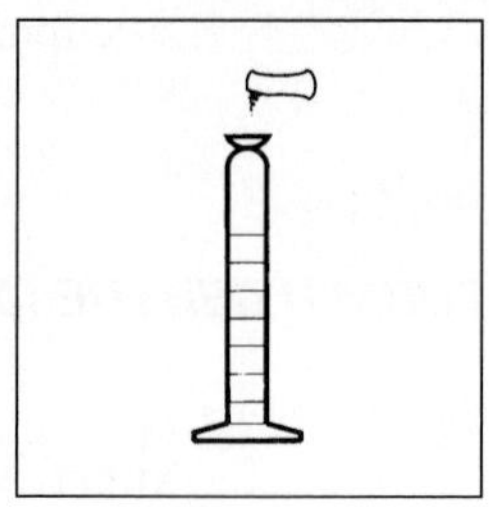

6. Immediately press: **SHIFT TIMER.** A 17-minute reaction period will begin.

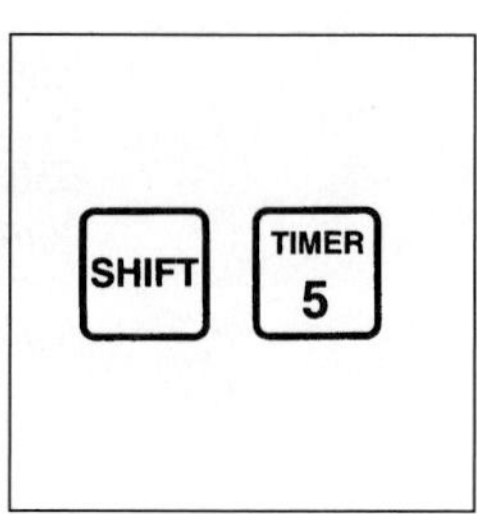

7. Immediately after the reaction period starts, shake the cylinder vigorously for 20 seconds.
Note: Do not wait for the timer to beep.

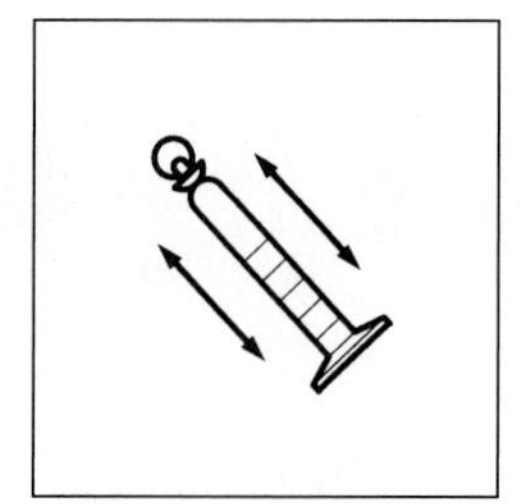

8. Add the contents of one MBTH Powder Pillow to the prepared sample when the timer shows 15:00. Stopper the cylinder.

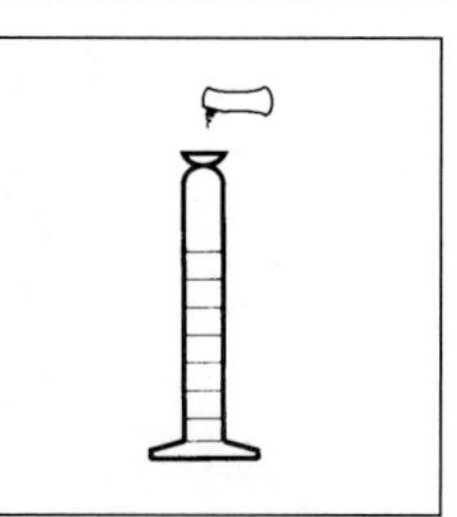

9. Shake the cylinder vigorously for 20 seconds.

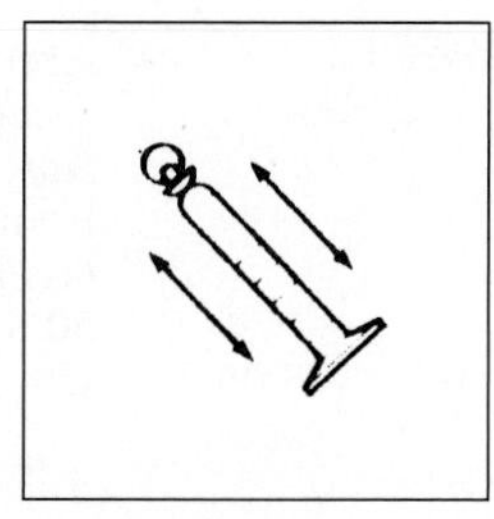

10. Add 2.5 mL of Developing Solution For Low Range Formaldehyde to the blank when the timer shows 12:00. Stopper. Invert to mix.

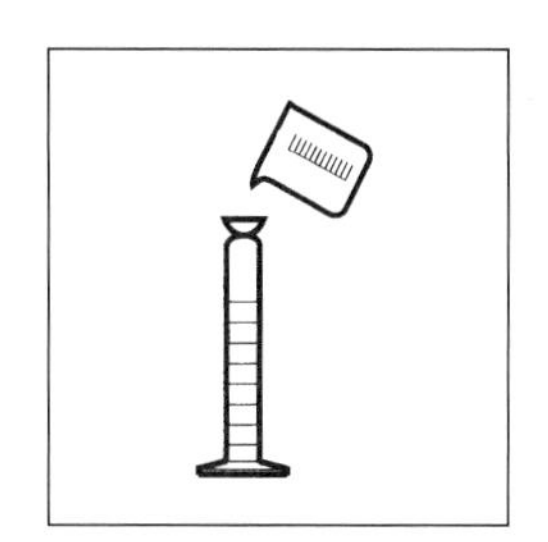

11. Add 2.5 mL of Developing Solution For Low Range Formaldehyde to the prepared sample when the timer shows 10:00. Stopper. Invert to mix.

12. Slowly pour the blank into the sample cell just before the timer shows 2:00. Place the blank into the cell holder. Close the light shield.
Note: If bubbles form on the cell walls, swirl the cell to dislodge them.

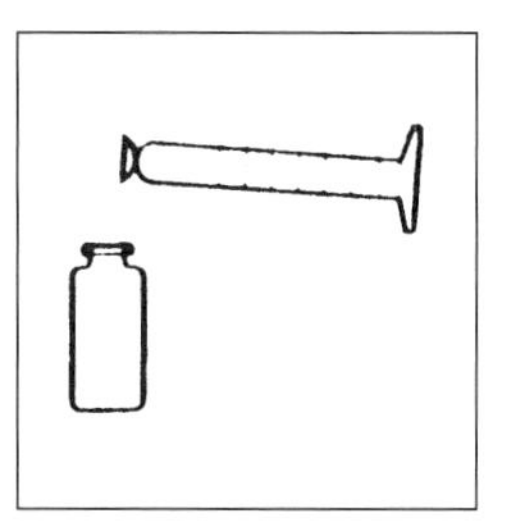

13. When the timer shows 2:00, press: **ZERO.** The display will show: **Zeroing ...** then: **0.μg/L CH_2O.**
Note: If desired, return the display to the timer mode by pressing SHIFT TIMER.

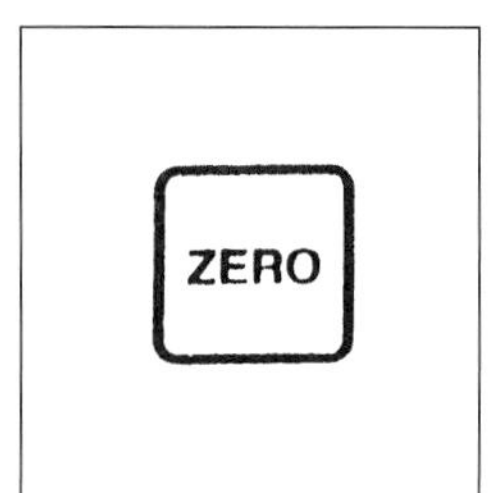

14. Pour the prepared sample into a sample cell. Place it into the cell holder. Close the light shield.

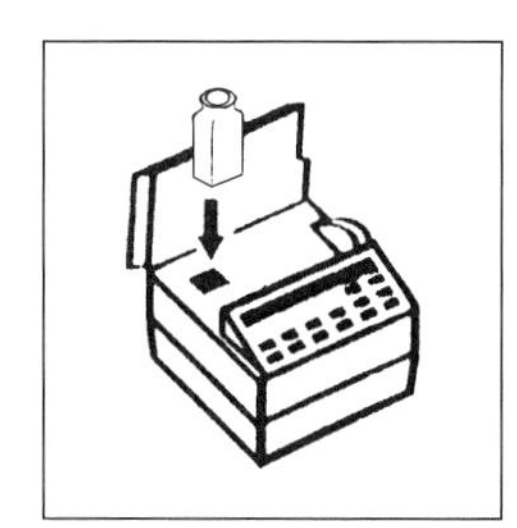

15. When the timer beeps, the display will show: **μg/L CH_2O.** Immediately press: **READ.** The display will show: **Reading ...** then the result in μg/L formaldehyde (CH_2O) will be displayed.

ACCURACY CHECK

See p. 573 of the HACH Water Analysis Handbook, 3rd Ed., 1997.

INTERFERENCES

The following may interfere when present in concentrations exceeding levels listed below.

Acetate	1000 mg/L	Iron (Fe^{3+})	12 mg/L
Ammonium (as N)	10 mg/L	Lead	100 mg/L
Aniline	10 mg/L	Manganese	500 mg/L
Bicarbonate	1000 mg/L	Mercury	70 mg/L
Calcium	3500 mg/L	Morpholine	0.36 mg/L
Carbonate	500 mg/L	Nitrate	1000 mg/L
Chloride	5000 mg/L	Nitrite	8 mg/L
Copper	1.6 mg/L	Phenol	1050 mg/L
Cyclohexylamine	250 mg/L	Phosphate	200 mg/L
Ethanolamine	33 mg/L	Silica	40 mg/L
Ethylenediamine	1.5 mg/L	Sulfate	10000 mg/L
Glucose	1000 mg/L	Urea	1000 mg/L
Glycine	1000 mg/L	Zinc	1000 mg/L
Iron (Fe^{3+})	12 mg/L		

QUESTIONS AND ANSWERS

1. Why is it necessary to wash glassware with chromic acid solution?
Glassware needs to be washed in chromic acid solution to remove trace contaminates.
2. How is formaldehyde-free water prepared?
Distill water with alkaline permanganate (4 g NaOH, 2g $KMNO_4$ per 500 mL water). Collect and use the 25 ml sample of water after distilling off and discarding the first 50 to 100 mL of water.
3. Does distilled water and bottled water have formaldehyde in it?
Some bottled water contains a low level of formaldehyde. Water and formaldehyde are

difficult to separate due primarily to hydrogen bonding. Even distilled water may contain a very low level of formaldehyde.

4. What types of compounds and ions interfere with formaldehyde analysis?
*At low concentration levels, some interferences are ammonium (as N), aniline, copper, ethanolamine, ethylenediamine, iron (**Fe^{+3}**), morpholine, nitrite, and silica. Several other organic and inorganic compounds and ions at higher concentrations can interfere with the analysis of formaldehyde.*

REFERENCES

1. *HACH Water Analysis Handbook*, 3rd Ed., (Loveland, CO.: HACH Company, 1997).
2. *Standard Methods for the Examination of Water and Wastewater:* 16th Ed., A. E. Greenberg, R. R. Trussell, L. S. Clesceri and Mary Ann H. Franson (eds.), pp. 1048 and 1137, Washington, DC, American Public Health Association, 1985.
3. *Hawley's Condensed Chemical Dictionary*, 11th Ed., rev'd by N. Irving Sax and Richard J. Lewis (New York, N.Y.: Van Nostrand Reinhold, 1987).

TOTAL HARDNESS

Originally, water hardness was described as the capacity of water to precipitate soap out of solution. Soap was precipitated primarily by the magnesium and calcium cations present. Today, total hardness is defined as the sum of the calcium (Ca) ions and magnesium (Mg) ions present in solution, both expressed as milligrams of calcium carbonate per liter. Additional cations, such as barium and strontium, also may precipitate soap. However, these cations normally contribute a minimal amount of hardness to the sample.

The hardness of water can, at times, be a cause for kidney stones. Kidney stones are formed from minerals, such as calcium and magnesium, that are filtered out of human waste. In the aquatic environment, there has been a limited amount of evidence that fish and other sea creatures may be somewhat susceptible to metals in hard water.

A "hard water" stream has an abundance of minerals like calcium, magnesium, carbonate, and other ions. "Hard water" streams usually have a large amount of plant growth and life due to the amount of nutrients that can be used for nourishment and growth. These streams support plants which produce their own food and is a process called "primary production." A major by-product of this process is oxygen. "Soft water" streams have a very low mineral content (nutrient poor) and have little primary production.

The measurement of total hardness is extremely important to many industries. Paper production, photo finishing, and pharmaceutical companies require "soft" water to prepare the finished product. Some companies that produce food and beverages must limit water hardness to maintain product quality. Any industrial plant or apartment complex that uses boilers and steam condensation units for water processing and treatment has to avoid the scale build up on equipment due to water hardness.

This test uses Ethylenediaminetetraacetic Acid (EDTA) to titrate a water sample for hardness. EDTA forms a soluble chelated complex with calcium and magnesium ions. Calmagite, an indicator, is added to the solution to produce a color change at the equivalence point during the titration. During the titration, the reaction occurs in a basic solution (pH of 8 to 14). The experiment detects total hardness of water in concentrations of 20 to 200 ppm (mg/L) as calcium carbonate ($CaCO_3$). Water hardness can be determined by titration methods and colorimetric measurements by spectrophotometers or colorimeters in the field. In addition, total hardness can be determined by test strips in the 0 to 25 gpg (grains per gallon) or the 0 to 425 ppm range. The test strip method is reliable but lacks the accuracy of more sophisticated methods.

A general guideline for total hardness of water is provided below:

mg/L as $CaCO_3$	Water classification
0 to 60	Soft
61 to 120	Moderately Hard
121 to 180	Hard
181 to 200	Very Hard

TOTAL HARDNESS

Test Procedure*

1. Fill the sample cup to the 25 mL mark with the sample.
2. Push the valve assembly onto the ampoule tip so that it fits snugly.
   Note: The valve assembly should reach the reference line on the neck of the ampoule.

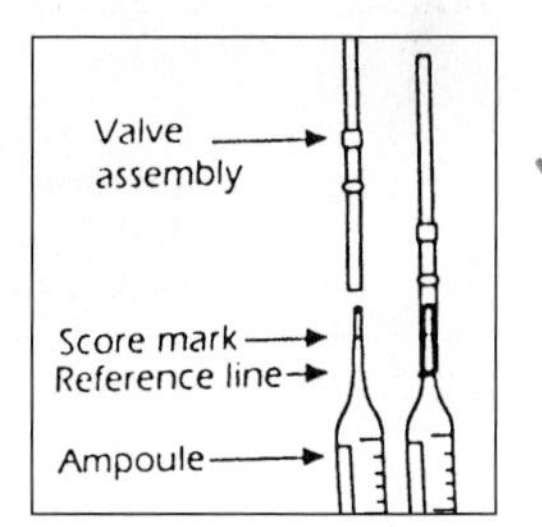

3. Gently snap the tip of the ampoule at the score mark.

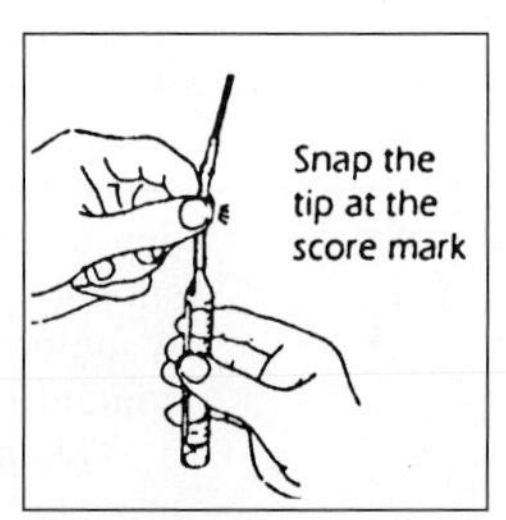

4. Lift the control bar and insert the test ampoule into the body of the titrating apparatus.

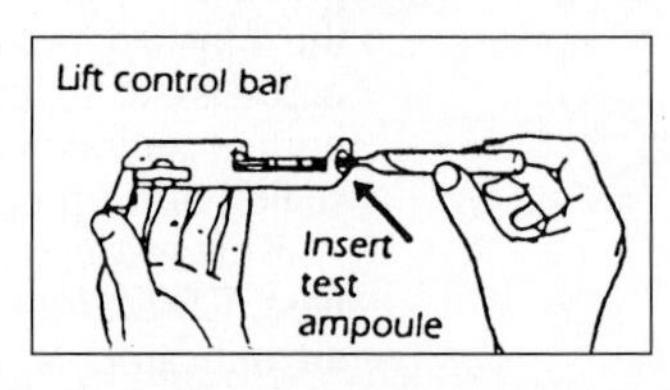

*Compliments of Ward's Natural Science Establishment, Inc.

5. Hold the titrating apparatus with the sample pipe immersed in the sample. Press the control bar firmly but briefly to pull in a small amount of the sample.
   Note: The contents will turn a blue color.

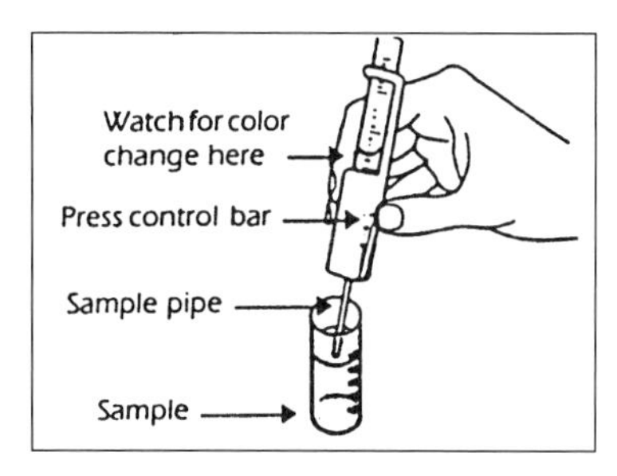

6. Press the control bar again briefly to allow another small amount of sample to be drawn into the test ampoule.
   Note: Do not press the control bar unless the sample pipe is immersed in the liquid.
7. After each addition, rock the entire assembly to mix the contents of the test ampoule. Watch for a color change from blue to pink.
8. Repeat Steps 6 and 7 until a permanent color change occurs.
9. When the color of the liquid in the test ampoule changes to pink, remove the ampoule from the titrating apparatus. Hold the test ampoule in a vertical position and carefully read the scale opposite the liquid level to obtain the test results in ppm (mg/L) as calcium carbonate.
   Note: Read the bottom of the meniscus of the titration apparatus. To convert to grains per gallon, divide ppm (mg/L) by 17.16.

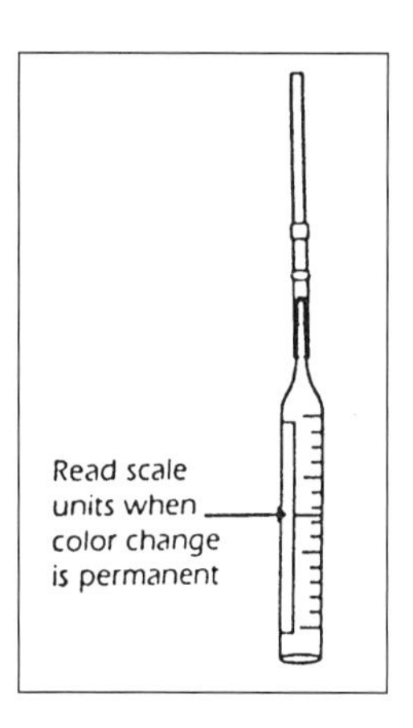

CAUTION AND SAFETY INFORMATION

1. Read the MSDS before performing this procedure.
2. Wash your hands after contact with any chemical. If contact with the chemical occurs, flush skin or eyes with water. If swallowed, call a physician.
3. Wear safety glasses when performing the chemical test.
4. Squeeze the control bar briefly during the test. If the control bar is pressed while the sample pipe is not immersed, air may enter the ampoule creating an insufficient vacuum to complete the test.
5. Mix the contents of the ampoule by rotating the ampoule.
6. If the ampoule fills completely without a color change (remains blue), the test result is less than the lowest scale reading. If the contents of the ampoule turn pink right away (see Step 5), the test result is greater than the highest scale reading. In this case, the

investigator may desire to dilute the sample with an equal volume distilled water and perform the test again to obtain an approximation value.

7. Dispose of used test ampoules as instructed in the MSDS.

QUESTIONS AND ANSWERS

1. What human disorder is believed to be influenced by water hardness?
   Water hardness is attributed to excess amounts of calcium and magnesium in the water. Kidney stones are caused by excessive water hardness.
2. What is meant by "hard water" and "soft water" in a stream?
   Hard water in a stream contains an abundance of calcium and magnesium. Plant growth will flourish in a hard water stream. In contrast, a soft water stream will contain few nutrients and plants will not grow in abundance.
3. Would the water be classified as hard or soft if it tested at 160 mg/L of $CaCO_3$?
   The water would be "hard."
4. Why is water hardness a problem in some industries?
   Water hardness creates a scale or crust on equipment such as steam condensation units and hot water boilers used in manufacturing and water treatment processes. The scale build-up on equipment can reduce equipment efficiency and cause breakdowns. Pharmaceutical companies and food and beverage industries must limit water hardness because it can affect product quality. Paper production and photo finishing laboratories require soft water for successful results. Hard water also impacts laundry facilities.

REFERENCES

1. *WARD's Snap Test,* Instant Water Quality Test Kits, WARD's Natural Science Establishment, Inc., Rochester, N.Y.
2. *Standard Methods for the Examination of Water and Wastewater:* HARDNESS, 16th Ed., Method 314B, A. E. Greenberg, R. R. Trussell, L. S. Clesceri and Mary Ann H. Franson (eds.) pp. 209–214, Washington, DC, American Public Health Association, 1985.
3. Eldon D. Enger and Bradley F. Smith, *Environmental Science,* 5th Ed. (Dubuque, IA: William C. Brown Publishing Company, 1992).

IODINE

Iodine has been used to disinfect potable waters and swimming pool waters. In water, active iodine exists in the forms of elemental I_2 and hypoiodous acid (HOI) or hypoiodite anion. Iodine is a strong oxidizing agent and the compound is unstable in natural water. The compound rapidly oxidizes many inorganic metals and anions and slowly oxidizes a number of organic compounds. The reaction and decomposition of iodine in water is influenced by reactant types and concentrations, sunlight, salinity, pH and temperature.

Iodine is used to manufacture dyes, iodides and iodates, antiseptics, germicides, medicinal soaps, stabilizers, and pharmaceuticals. The compound is used in catalysts, x-ray contrast media, food and feed additives, photographic film, water treatment, and as a unsaturation indicator. Iodine also has a radioactive isotope which is utilized for internal radiation therapy and to treat several thyroid related problems. Radioactive iodine can detect leaks in water lines and can be used to study thermal stabilities of food. The radioactive compound acts as a tracer in chemical reactions.

A representative sample of water from a tap or spring must be obtained for accurate results. Water should not be collected in plastic containers since these items have a large iodine demand. To remove iodine demand, all glass sample containers should be pretreated by soaking in bleach followed by a thorough rinse with distilled or deionized water.

Samples suspected of containing iodine should be analyzed immediately after collection to avoid changes of iodine to iodide. In the experiment, iodine reacts with DPD (N, N-diethyl-p-phenylenediamine) to form a magenta color which is proportional to the total iodine concentration.

IODINE (0 to 7.00 mg/L)

Method 8031 for water, wastewater, and seawater*
DPD Method** using Powder Pillows

1. Enter the stored program number for iodine (I_2) powder pillows. Press**: 2 4 0 ENTER.** The display will show: **Dial nm to 530.**
Note: The Pour-Thru Cell can be used with 25-mL reagents only.

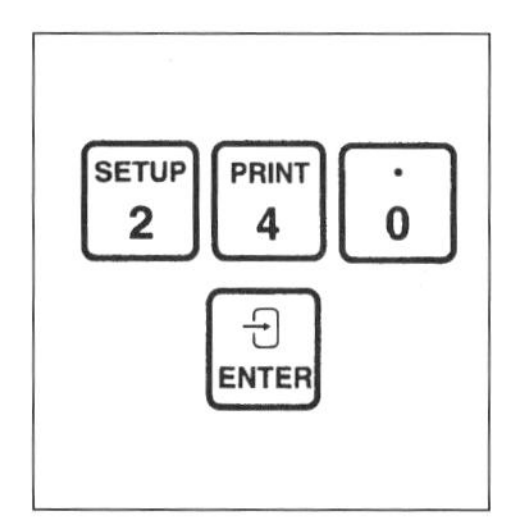

2. Rotate the wavelength dial until the small display shows: **530 nm.** When the correct wavelength is dialed in, the display will quickly show: **Zero sample**, then: **mg/L I_2.**

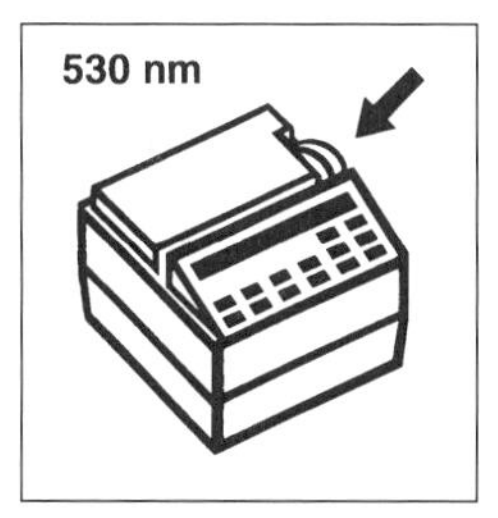

3. Insert the 10-mL Cell Riser into the cell compartment.

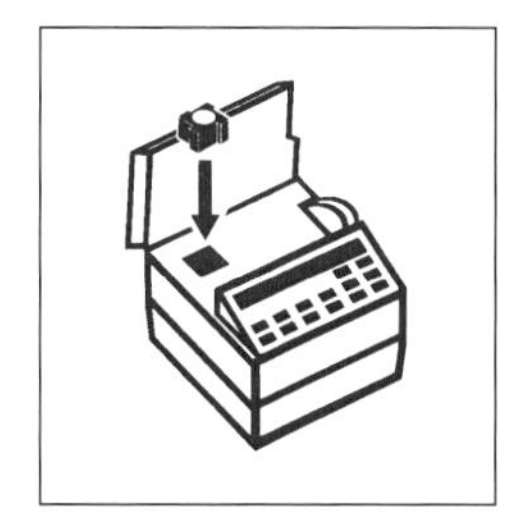

* Compliments of HACH Company.
** Adapted from A. T. Palin, *Inst. Water Eng.*, 21 (6) 537–547 (1967).

4. Fill a cell with 10 mL of sample.
Note: Samples must be analyzed immediately and cannot be preserved for later analysis.

5. Add the contents of one DPD Total Chlorine Powder Pillow to the sample cell (the prepared sample). Swirl to mix.
Note: A pink color will develop if iodine is present.

6. Press: **SHIFT TIMER.** A 3-minute reaction period will begin.

7. When the timer beeps the display will show: **mg/L I_2.** Fill a second sample cell with 10 mL of sample (the blank). Place it into the cell holder.

8. Press: **ZERO.** The display will show: **Zeroing ...** then: **0.00 mg/L I_2.**

9. Within 3 minutes after the timer beeps, place the prepared sample into the cell holder. Close the light shield.

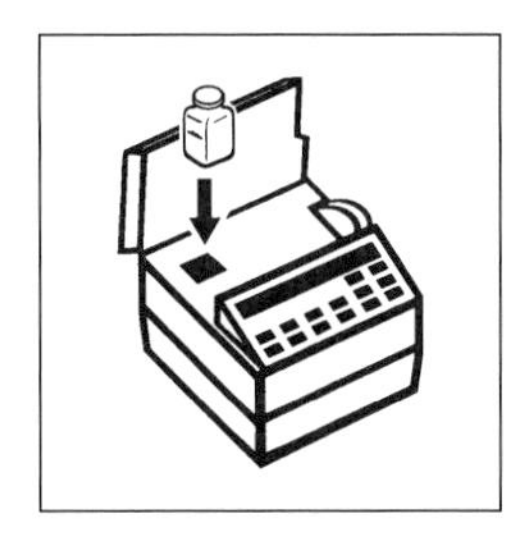

10. Press: **READ.** The display will show: **Reading ...** then the result in mg/L I_2 will be displayed.
Note: If the sample temporarily turns yellow after reagent addition, or reads **OVER-RANGE,** *dilute a fresh sample. Repeat the test. A slight loss of iodine may occur during dilution. Apply the appropriate dilution factor; see "Sample Dilution Techniques" in Section 1.*

SAMPLING AND STORAGE

Analyze samples for iodine **immediately** after collection. Free iodine is a strong oxidizing agent, and it is unstable in natural waters. It reacts rapidly with various inorganic compounds and more slowly oxidizes organic compounds. Many factors, including reactant concentrations, sunlight, pH, temperature and salinity influence decomposition of iodine in water.

Avoid plastic containers since these may have a large iodine demand. **Pretreat glass** sample containers to remove any iodine demand by soaking in a dilute bleach solution (1 mL commercial bleach to 1 liter of deionized water) for at least 1 hour. Rinse thoroughly with deionized or distilled water. If sample containers are rinsed thoroughly with deionized or distilled water after use, only occasional pre-treatment is necessary.

A common error in testing for iodine is introduced when a representative sample is not obtained. If sampling from a tap, let the water flow for at least 5 minutes to ensure a representative sample. Let the container overflow with the sample several times, then cap the sample containers so there is no headspace (air) above the sample. If sampling with a sample cell, rinse the cell several times with the sample, then carefully fill to the 10-mL mark. Perform the analysis immediately.

ACCURACY CHECK

See p. 637 of the HACH Water Analysis Handbook, 3rd Ed., 1997.

INTERFERENCES

Samples containing more than 300 mg/L alkalinity or 150 mg/L acidity as $CaCO_3$ may not develop the full amount of color, or it may instantly fade. Neutralize these samples to a pH of 6 to 7 with 1 N Sulfuric Acid or 1 N Sodium Hydroxide. Determine the amount required on a separate 25-mL sample. Add the same amount to the sample to be tested. Correct for volume additions.

Bromine, chlorine, ozone and oxidized forms of manganese and chromium also may react and read as iodine. To compensate for the effects of oxidized manganese or chromium adjust pH to 6 to 7 as described above. Add three drops of potassium iodide, 30 g/L, to 10 mL of sample, mix and wait one minute. Add three drops of sodium arsenite, 5 g/L, and mix. Analyze 10 mL of this sample as described above. (If chromium is present, allow the same reaction period with the DPD for oxidized manganese and chromium.) Subtract the result of this test from the original analysis to obtain the correct iodine result.

DPD Reagent Powder Pillows are formulated with a buffer which will withstand high levels (1000 mg/L) of hardness without interference.

QUESTIONS AND ANSWERS

1. Why are plastic containers avoided in this experiment?
   Plastic containers may have a large iodine demand. The oxidizing agent, iodine, may react with the plastic.
2. What are the interferences in the test?
   The color of the sample may fade or not develop if the alkalinity or acidity (as $CaCO_3$) is too high. Bromine, chlorine, ozone and oxidized forms of manganese and chromium may increase the iodine result.
3. Why should a sample suspected of containing iodine be tested immediately?
   Iodine is easily reduced to iodide in aqueous solution containing contaminates. In water, iodine, as an oxidizing agent, reacts relatively easily with metal cations, anions, and many organic complexes.

REFERENCES

1. *HACH Water Analysis Handbook,* 3rd ed. (Loveland, CO.: HACH Company, 1997).
2. *Hawley's Condensed Chemical Dictionary,* 11th Ed. rev'd by N. Irving Sax and Richard J. Lewis (New York, N.Y.: Van Nostrand Reinhold, 1987).
3. *Standard Methods for the Examination of Water and Wastewater:* IODINE., 16th Ed., A. E. Greenberg, R. R. Trussell, L. S. Clesceri and Mary Ann H. Franson (eds.) p. 369, Washington, DC, American Public Health Association, 1985.

IRON

Iron in water exists as ferrous (+2) ion and ferric (+3) ion. Ferric and ferrous compounds are numerous and have many practical uses. Iron is important to a healthy body, and deficiency in the body causes anemia. Iron in water can cause staining of laundry and porcelain. The iron concentration in surface water seldom reaches 1 mg/L.

Iron exists in the ferrous state under reducing conditions. Ferrous ion exposed to air or to oxidants is easily oxidized to ferric ion. In water, iron as ferrous and ferric ions may remain in solution, exist in a colloidal state, form organic and inorganic complexes, or form coarse suspended particles.

Silt and clay in suspension often contain significant amounts of iron. Flaking of rust (iron oxide) from water pipes and dissolution of iron from metal caps on storage or sampling bottles can produce results which are not attributed directly to water. Iron can also enter water systems from natural deposits, industrial waste effluents, and acidic mine drainage.

Iron has a tendency to form a colloidal suspension in solution. Adsorption of iron on various types of particles and deposition of iron on the sample container walls are common occurrences when iron has not been totally dissolved. The addition of acid to the suspension or colloidal solution will sometimes dissolve the iron.

Because ferrous iron is easily oxidized to the ferric form, the determination of the ferrous ion requires special precautions and should be determined immediately in the field by titration methods or with a Colorimeter or a Spectrophotometer. By using 1,10-phenanthroline, water samples of the ferrous ion should be limited to light exposure.

In the phenanthroline test for ferrous iron (Fe^{+2}), three molecules of phenanthroline chelate a single ferrous ion to form an orange-red complex. Ferric iron (Fe^{+3}) does not react with the phenanthroline. The ferric iron concentration can be determined by subtracting the ferrous iron concentration from the results of the total iron test.

Total iron includes both the soluble and insoluble forms of iron in the sample. The FerroVer Iron Reagent reacts with these different forms of iron to produce soluble ferrous ion. Then, the ferrous ion reacts with the 1,10-phenanthroline indicator to form an orange color solution in proportion to the "total" iron concentration, which can be easily measured by a Colorimeter or a Spectrophotometer. In many cases, digestion of the sample frees all of the iron for a total iron test by the FerroVer Method (Method 8008). Digestion of the sample before determination of total iron is required for reporting wastewater analysis to USEPA.

IRON, FERROUS (0 to 3.00 mg/L)

Method 8146 for water, wastewater, and seawater*
1,10-Phenanthroline Method** using Powder Pillows

1. Enter the stored program number for ferrous iron ($(Fe^{2+})^-$ powder pillows. Press: **2 5 5 ENTER.** The display will show: **Dial nm to 510.**
Note: The Pour-Thru Cell can be used with this procedure.

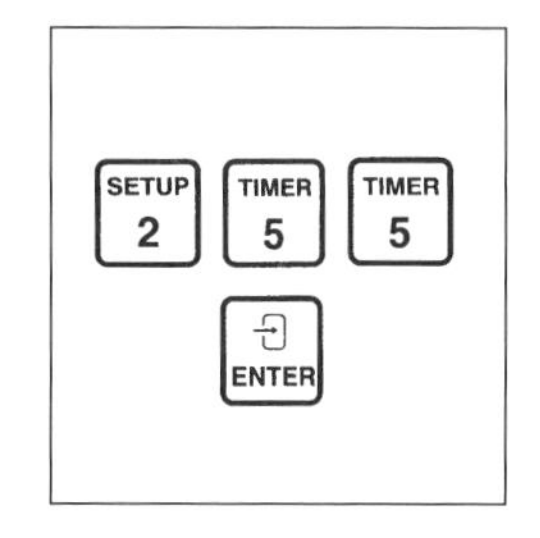

2. Rotate the wavelength dial until the small display shows: **510 nm.** When the correct wavelength is dialed in, the display will quickly show: **Zero sample,** then: **mg/L Fe^{2+}.**

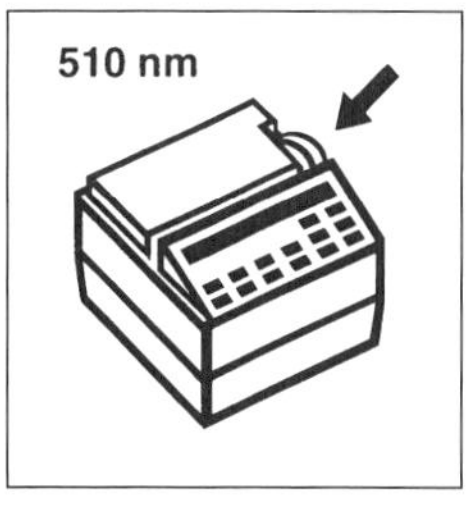

*Compliments of HACH Company. Analysis may be performed by DR/820, DR/850, and DR/890 Colorimeters and DR/2010 Spectrophotometer.

**Adapted from *Standard Methods for the Examination of Water and Wastewater, 315 B.*

3. Fill a sample cell with 25 mL of sample.
Note: Analyze samples as soon as possible to prevent oxidation of ferrous iron to ferric, which is not determined.

4. Add the contents of one Ferrous Iron Reagent Powder Pillow to the sample cell (the prepared sample). Swirl to mix.
Note: An orange color will form if ferrous iron is present.
Note: Undissolved powder does not affect accuracy.

5. Press: **SHIFT TIMER.** A 3-minute reaction period will begin.

6. When the timer beeps, the display will show: **mg/L Fe^{+2}.** Fill a second sample cell with 25 mL of sample (the blank).

7. Place the blank into the cell holder. Close the light shield.

8. Press: **ZERO.** The display will show: **Zeroing ...** then: **0.00 mg/L Fe^{2+}.**

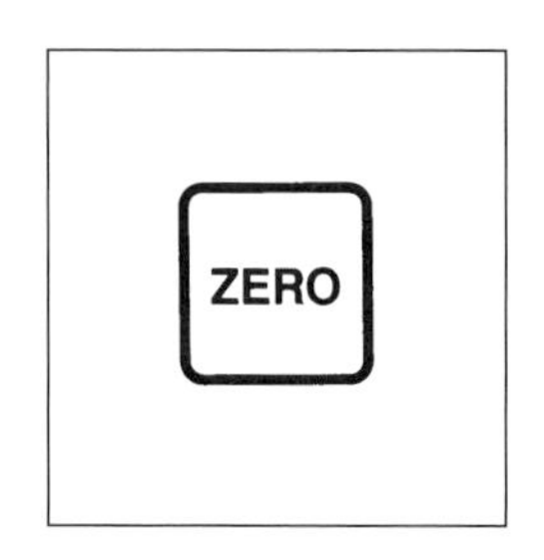

9. Place the prepared sample into the cell holder. Close the light shield.

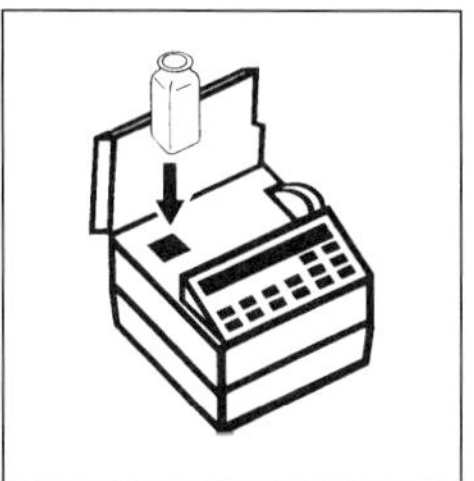

10. Press: **READ.** The display will show: **Reading ...** then the result in mg/L Fe^{2+} will be displayed.

ACCURACY CHECK

See p. 655 of the HACH Water Analysis Handbook, 3rd ed., 1997.

IRON, TOTAL (0 to 3.00 mg/L)

Method 8008 for water, wastewater, and seawater*
FerroVer Method** using Powder Pillows
USEPA Accepted for reporting wastewater analysis (Digestion is required)†

1. Enter the stored program number for iron (Fe) FerroVer, powder pillows. Press: **2 6 5 ENTER.** The display will show: **Dial nm to 510.**
Note: Adjust pH of stored samples before analysis.
Note: The Pour-Thru Cell may be used with 25-mL reagents only.

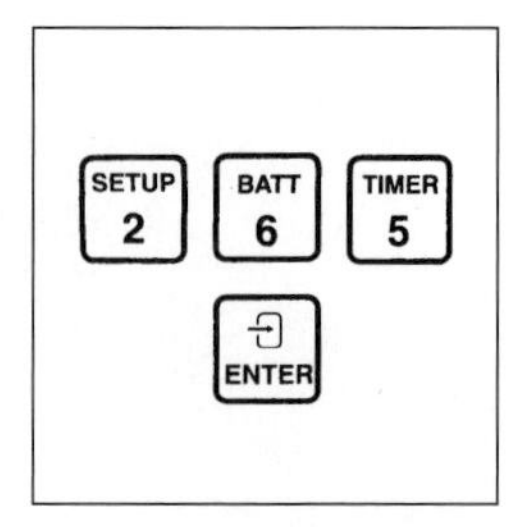

2. Rotate the wavelength dial until the small display shows: **510 nm.** When the correct wavelength is dialed in, the display will quickly show: **Zero sample,** then: **mg/L Fe FV.**

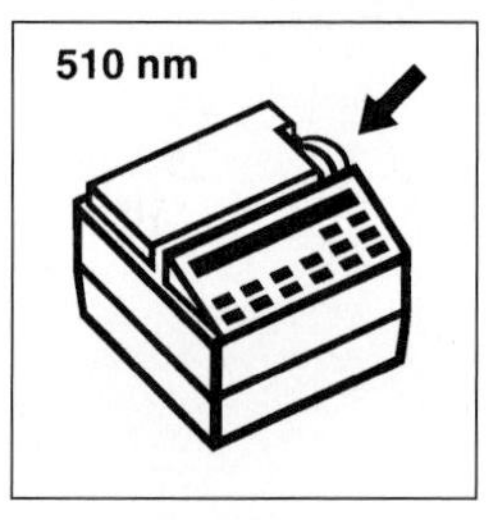

3. Insert the Cell Riser for 10-mL sample cells.

4. Fill a clean sample cell with 10 mL of sample.
Note: Determination of total iron needs a prior digestion; use the mild, vigorous or Digesdahl disgestion in Chapter 3.

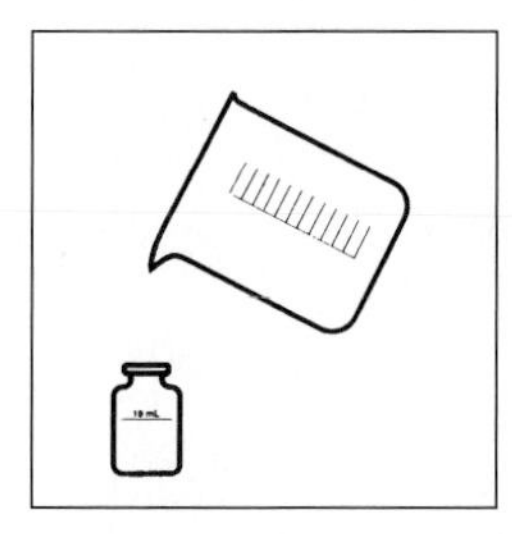

---

*Compliments of HACH Company. Analysis may be performed by DR/820, DR/850, and DR/890 Colorimeters and DR/2010 Spectrophotometer

**Adapted from *Standard Methods for the Examination of Water and Wastewater.*

†Federal Register, 45 (126) 43459 (June 27, 1980).

5. Add the contents of one FerroVer Iron Reagent Powder Pillow to the sample cell (the prepared sample). Swirl to mix.
Note: An orange color will form if iron is present.
Note: Accuracy is not affected by undissolved powder.

6. Press: **SHIFT TIMER.** A 3 minute reaction period will begin.
Note: Samples containing visible rust should be allowed to react at least 5 minutes.

7. When the timer beeps, the display will show: **mg/L Fe FV.** Another sample cell with 10 mL of sample (the blank).

8. Place the blank into the cell holder. Close the light shield.
Note: For turbid samples, treat the blank with one 0.1-g scoop of RoVer Rust Remover (use 0.2 for 25-mL samples). Swirl to mix.

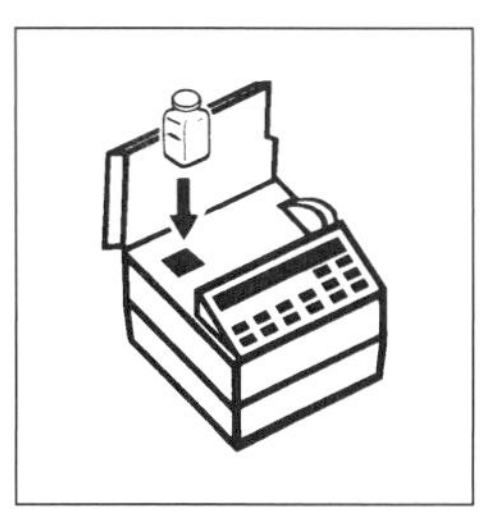

9. Press: **ZERO.** The display will show: **Zeroing ...** then: **0.00 mg/L Fe FV.**

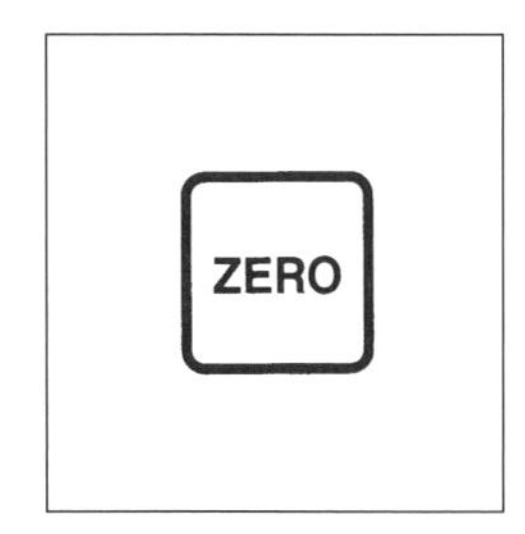

10. Within 30 minutes after the timer beeps, place the prepared sample into the cell holder. Close the light shield.

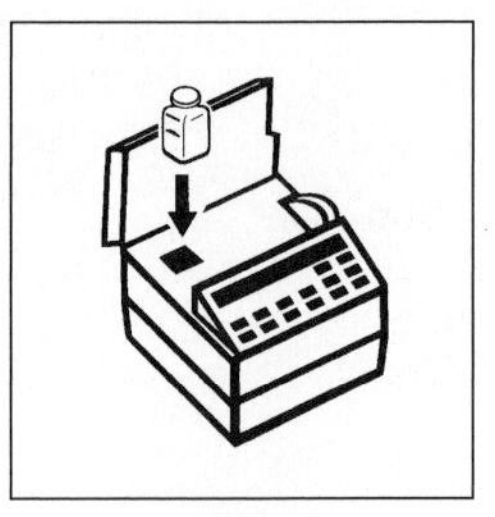

11. Press: **READ.** The display will show: **Reading ...** then the result in mg/L iron will be displayed.

SAMPLING AND STORAGE

Collect samples in acid-cleaned glass or plastic containers. No acid addition is necessary if analyzing the sample immediately. To preserve samples, adjust the pH to 2 or less with nitric acid (about 2 mL per liter). Preserved samples may be stored up to six months at room temperature. Adjust the pH to between 3 and 5 with 5.0 N Sodium Hydroxide Standard Solution before analysis. Correct the test result for volume additions; see "Correcting for Volume Additions" in Section 1 for more information.

If only dissolved iron is to be determined, filter the sample before acid addition.

ACCURACY CHECK

See p. 671 of the HACH Water Analysis Handbook, 3rd ed., 1997.

INTERFERENCES

The following will not interfere below the levels shown:

Chloride	185,000 mg/L
Calcium	10,000 mg/L as $CaCO_3$
Magnesium	100,000 mg/L as $CaCO_3$
Molybdate Molybdenum	50 mg/L as Mo

A large excess of iron will inhibit color development. A diluted sample should be tested if there is any doubt about the validity of a result.

FerroVer Iron Reagent Powder Pillows contain a masking agent which eliminates potential interferences from copper.

Samples containing some forms of iron oxide require the digestion. After digestion adjust the pH to between 2.5 and 5 with ammonium hydroxide.

Samples containing large amounts of sulfide should be treated as follows in a fume hood or well ventilated area: Add 5 mL of hydrochloric acid to 100 mL of sample and boil for 20 minutes. Adjust the pH to between 2.5 and 5 with 5 N sodium hydroxide and readjust the volume to 100 mL with deionized water. Analyze as described above.

Highly buffered samples or extreme sample pH may exceed the buffering capacity of the reagents and require sample pretreatment; see "pH Interference" in Section 1.

IRON (10-1000 mg/L as Fe)

Method 8214 Using the TitraVer Titration Cartridge*

1. Select a sample volume and a TitraVer Titration Cartridge corresponding to the expected iron (Fe) concentration from Table 2.23.

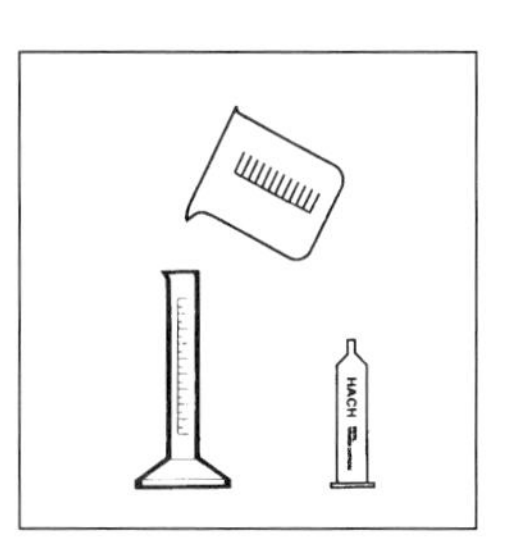

2. Insert a clean delivery tube into the titration cartridge. Attach the cartridge to the titrator body.

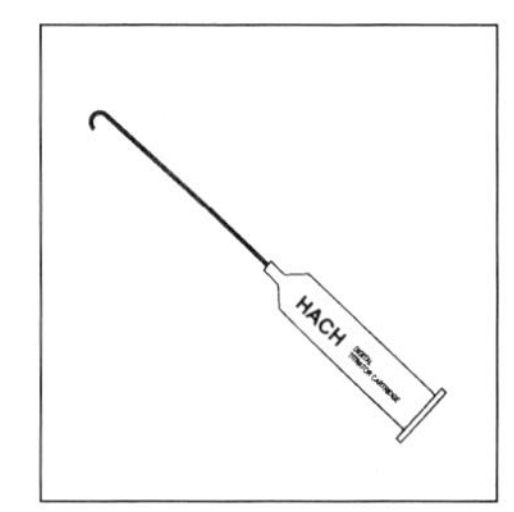

3. Turn the delivery knob to eject a few drops of titrant. Reset the counter to zero and wipe the tip.
Note: For added convenience use the TitraStir stirring apparatus.

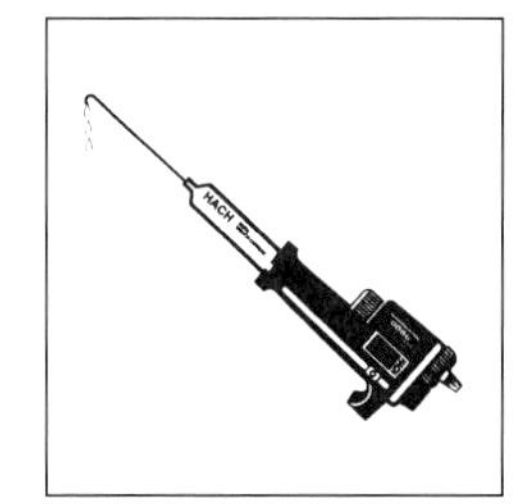

*Compliments of HACH Company.

4. Use a graduated cylinder to measure the sample volume from Table 2.23. Transfer the sample into a clean 125-mL erlenmeyer flask. Dilute to about the 50-mL mark with deionized water, if necessary.

5. Add the contents of one Citrate Buffer Powder Pillow and swirl to mix.

6. Add the contents of one Sodium Periodate Powder Pillow and swirl to mix.
   Note: A yellow color indicates the presence of iron.

7. Add the contents of one Sulfosalicylic Acid Powder Pillow and swirl to mix.
   Note: A red color will develop if iron is present.

8. Place the delivery tube tip into the solution and swirl the flask while titrating thc sample until the color changes from red to the original yellow. Record the number of digits required.

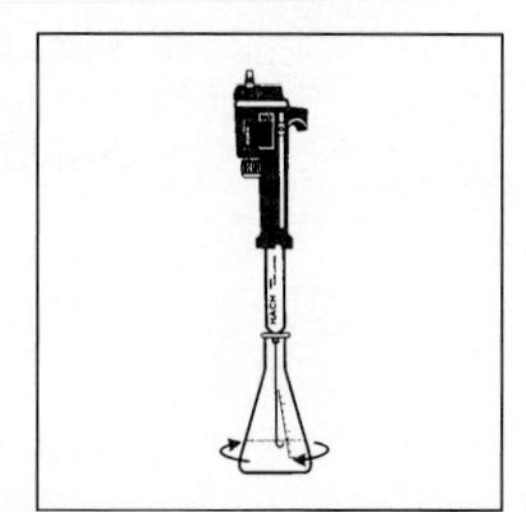

9. Calculate: Digits Required × Digit Multiplier = mg/L Iron (Fe)

Digits Required x Digit Multiplier
= mg/L Iron (Fe)

ACCURACY CHECK

See p. 650 of the HACH Water Analysis Handbook, 3rd ed., 1997.

QUESTIONS AND ANSWERS

1. What are the primary differences between ferrous ion and ferric ion?
Ferric compounds in solution have a more intense and bright color. Ferrous ions (Fe^{+2}) tend to oxidize to ferric (Fe^{+3}) ion in air or in the presence of oxidizing agents, whereas the ferric ion tends to be more stable. The ferrous cation complexes with 1,10-phenanthroline whereas ferric ion does not.
2. What are the problems with iron?
Rust and corrosion of metal pipes and containers are major problems. The dissolving of iron with a relatively small change in pH can impact items stored in containers with iron in the metal. Iron in water can cause staining of laundry and porcelain.
3. Why is total iron difficult to detect?
Iron exists as ferrous ion and ferric ion in solution, in a colloidal state, as suspended particles, and as organic and inorganic complexes. Iron compounds are found in silt, clay, and other media. Sample digestion can free iron before total iron analysis.
4. If the measured total iron concentration is 0.9 mg/L and the ferrous iron concentration is 0.2 mg/L, what is the concentration of the ferric ion?
Ferric ion concentration is 0.7 mg/L.

TABLE 2.23

Range (mg/L as Fe)	Sample Volume (mL)	Titration Cartridge (M TitraVer)	Catalog Number	Digit Multipler
10–40	50	0.0716	20817-01	0.1
25–100	20	0.0716	20817-01	0.25
100–400	50	0.716	20818-01	1.0
250–1000	20	0.716	20818-01	2.5

REFERENCES

1. *HACH Water Analysis Handbook*, 3rd ed., (Loveland, CO.: HACH Company, 1997)
2. *Hawley's Condensed Chemical Dictionary*, 11th Ed., rev'd by N. Irving Sax and Richard J. Lewis (New York, N.Y.: Van Nostrand Reinhold, 1987).
3. *Standard Methods for the Examination of Water and Wastewater:* IRON, 16th Ed., A. E. Greenberg, R. R. Trussell, L. S. Clesceri and Mary Ann H. Franson (eds.) pp. 214–219, Washington, DC, American Public Health Association, 1985.

LEAD

Lead accumulates in the body and is a poison. Some lead compounds are toxic by inhalation. A number of lead compounds are absorbed through the skin while others require ingestion. Lead and some lead compounds tend to accumulate in the bone structure when ingested in excessive amounts. Accumulation of lead in the body may cause permanent brain damage, convulsions and death. As a result of the potential for lead poisoning, a national program to reduce the concentration of lead in consumer products has been implemented. Humans have a tolerance level (Pb) of 0.15 mg per cubic meter of air.

Natural waters usually contain less than 20 micrograms per liter, although values higher than 300 micrograms per liter have been reported. A number of lead compounds are somewhat more soluble in water as the temperature of the water rises above 0°C. Sources of lead are industrial mines, smelter discharges, or the dissolution of old plumbing containing lead. Lead service pipes carrying soft, acid, and untreated water may dissolve some of the lead in the pipe.

Lead compounds are used in pigments, ceramic glazes and coatings, electrodes, battery components, electronic and optical applications. lasers, ceramic-bonded coatings, cloud seeding, bronzing, and printing. Additional uses of lead include stabilizers, organic preparations, insecticide, staining, dyeing, printing, tanning, high pressure lubricants, plastic preparation, stabilizers, metallurgy, matches, paints, radiation detectors, and photoconductors. In the past, some of the interior walls of older buildings were painted with lead-based paint.

Lead compounds that are toxic should not be released in effluents during preparation and manufacturing processes. Several industries have implemented a closed loop system to recover and reuse lead contaminated wastes. Some facilities manifest wastes for burial or reuse by other industries. A few industries utilize wastes containing small amounts of lead to produce other products. Organic wastes containing low level lead are sometimes burned in emission-controlled incinerators to recover energy. Wastes contaminated with lead are often disposed in landfills and by inappropriate procedures.

Metals such as bismuth, copper, mercury, silver and tin can interfere with the analysis for lead. These interferences can be minimized by an extraction procedure with chloroform. In the experiment, chloroform (D002) and cyanide (D003) solutions are regulated as hazardous wastes by the Federal RCRA. These wastes should be collected for disposal with laboratory solvents.

A Scanning Analyzer HAS-1000 (Hach Company) can provide low part per billion (ppb) lead readings for on-site testing. The instrument provides 2 ppb detection limit with a 1 ppb resolution. The lead analyzer is not subject to interferences typical in drinking water.

Lead ions in basic solution react with dithizone to form a pink to red complex which can be extracted with chloroform. The pink lead-dithizone complex is somewhat light sensitive and should be kept out of direct sunlight. The spectrophotometric measurement for lead at 515 nm is read in micrograms per liter. Digestion of wastewater samples is required for USEPA reporting.

LEAD (0 to 160 μg/L)

Method 8033 for water and wastewater*
Dithizone Method**
USEPA Accepted for reporting wastewater analysis (Digestion is required; see Section 1 Sample and Analysis Volume Table for Aqueous Liquids)†

1. Enter the stored program number for lead (Pb). Press: **2 8 0 ENTER.** The display will show: **Dial nm to 515.**
   Note: The Pour-Thru Cell cannot be used with this procedure.

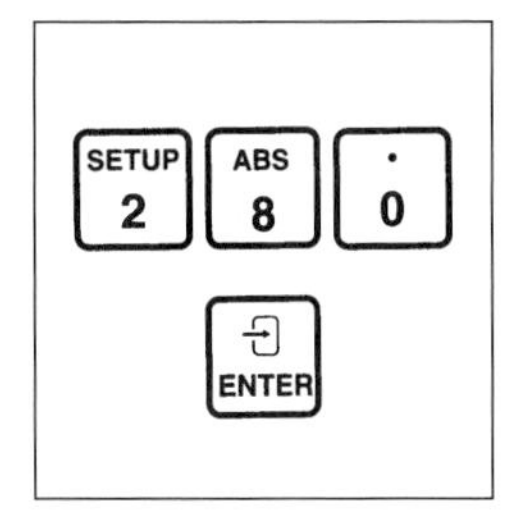

2. Rotate the wavelength dial until the small display shows: **515 nm.** When the correct wavelength is dialed in, the display will quickly show: **Zero sample,** then: **μg/L Pb.**

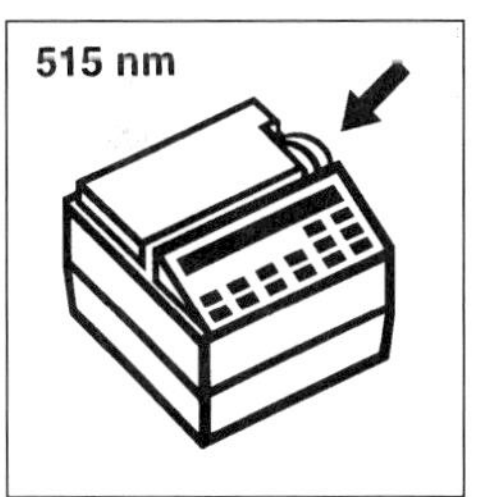

3. Fill a 250-mL graduated cylinder to the 250-mL mark with sample.
   Note: Clean all glassware with a 1:1 Nitric Acid Solution. Rinse with deionized water.
   Note: Cloudy and turbid samples may require filtering. Report results as μg/L soluble lead. Use a glass membrane filter to avoid loss of lead by adsorption on filter paper.

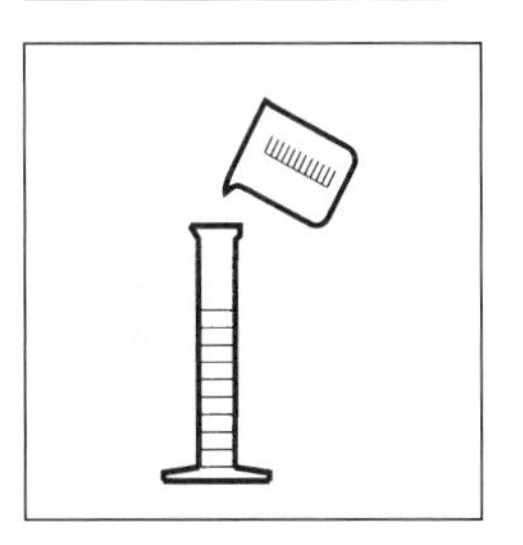

4. Transfer the sample into 500-mL separatory funnel.
   Note: Perform the procedure with proper ventilation or in a fume hood.

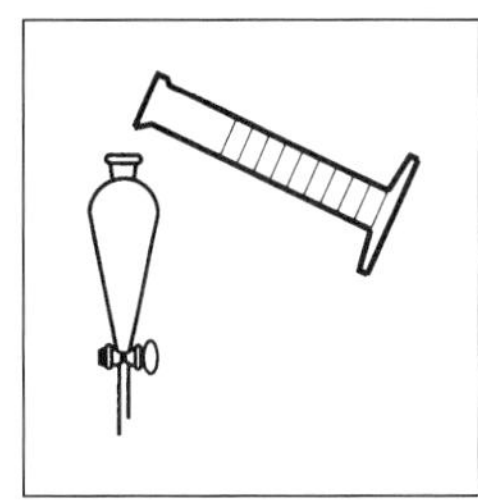

*Compliments of HACH Company.
**Adapted from L. J. Snyder, *Analytical Chemistry,* 19 684 (1947).
†Procedure is equivalent to Standard Method 3500-Pb D for wastewater.

5. Add the contents of one Buffer Powder Pillow, citrate type for heavy metals. Stopper the funnel. Shake to dissolve.
Note: Spilled reagent will affect test accuracy and is hazardous.

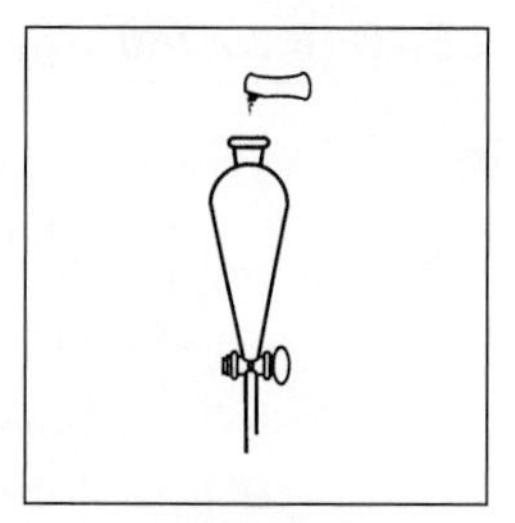

6. Add 50 mL of chloroform to a 50-mL graduated cylinder. Add the contents of one DithiVer Metals Reagent Powder Pillow. Stopper. Invert repeatedly to mix (DithiVer solution). Pour 30 mL of the DithiVer Solution into a second 50-mL graduated cylinder.
Note: Use adequate ventilation. The DithiVer Powder will not all dissolve. See DithiVer Solution Preparation below.

7. Add the 30 mL of DithiVer Solution from the cylinder to the separatory funnel. Stopper. Invert. Open stopcock to vent.

8. Add 5 mL of 5.0 N Sodium Hydroxide Standard Solution. Stopper. Invert. Open stopcock to vent. Shake the funnel once or twice and vent again.
Note: Add a few drops of 5.25 N Sulfuric Acid Standard Solution if the solution turns orange on shaking. The blue-green color will reappear. To avoid higher blanks, repeat procedure on new sample and use less sodium hydroxide.

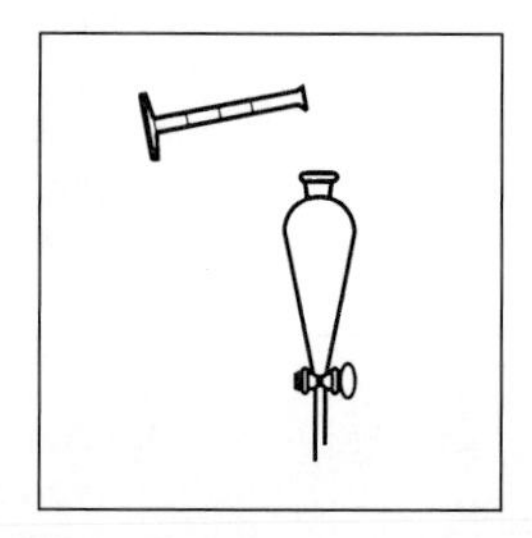

9. Continue adding 5.0 N Sodium Hydroxide Standard Solution dropwise until the color of the solution being shaken changes from blue-green to orange. Then add 5 more drops of 5.0 N Sodium Hydroxide Standard Solution.
Note: For most accurate results, adjust the sample to pH 11.0 to 11.5 using a pH meter, omitting the 5 additional drops of Sodium Hydroxide Standard Solution.

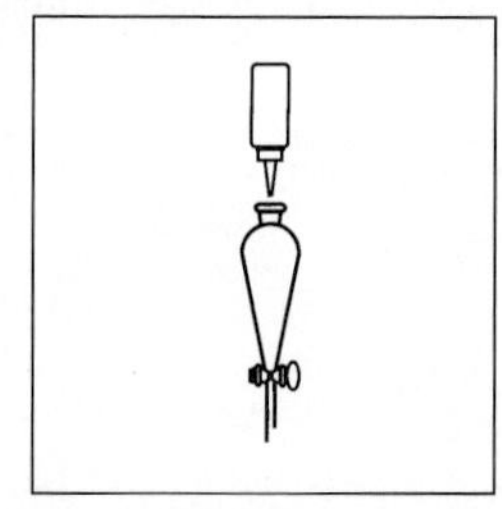

10. Add two heaping 1.0-g scoops of potassium cyanide to the funnel. Stopper. Shake vigorously until the potassium cyanide is all dissolved (about 15 seconds).
Note: Wait one minute for the layer to separate. The bottom (chloroform) layer will be pink if lead is present.

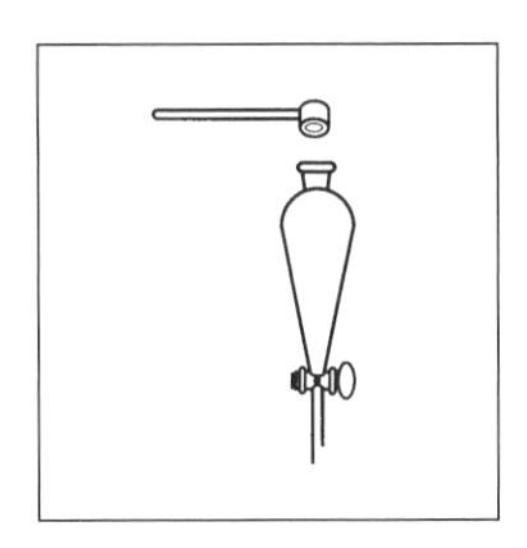

11. Insert a cotton plug the size of a pea into the delivery tube of the funnel and slowly drain the bottom (chloroform) layer into a dry 25-mL sample cell. Stopper. This is the prepared sample.
Note: The lead-dithizone complex is stable for hours if the sample cell is kept tightly capped and out of direct sunlight.

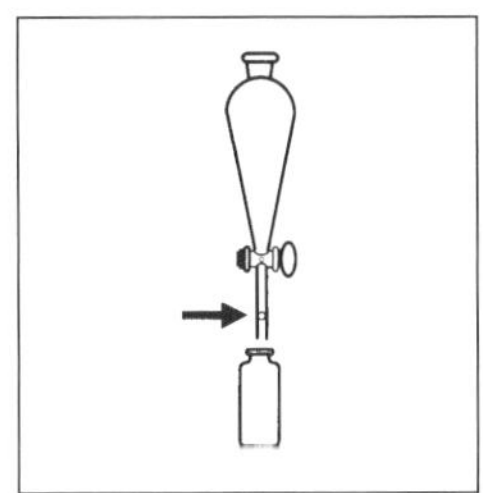

12. Fill a 25-mL sample cell (the blank) with chloroform. Stopper.

13. Place the blank into the cell holder. Close the light shield.

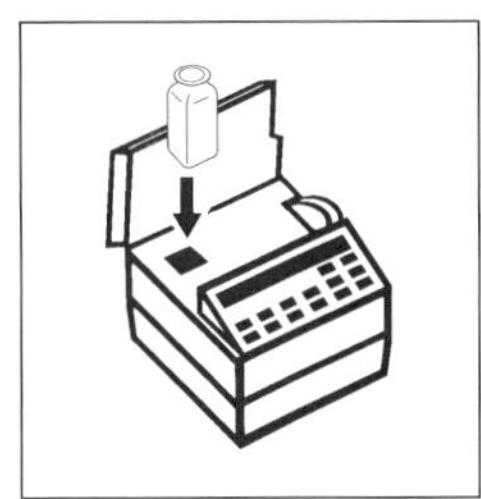

14. Press: **ZERO.** The display will show: **Zeroing ...** then: **0. μg/L Pb.**

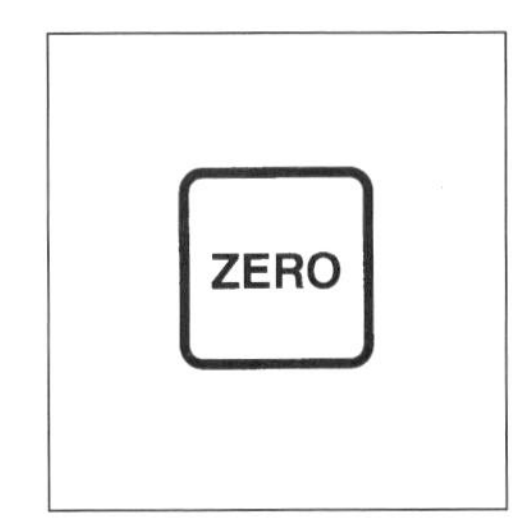

15. Place the prepared sample into the cell holder. Close the light shield.

16. Press: **READ.** The display will show: **Reading ...** then the result in $\mu g/L$ lead will be displayed.

DITHIVER SOLUTION PREPARATION, STORAGE AND BLANK

Store DithiVer Powder Pillows away from light and heat. A convenient way to prepare this solution is to add the contents of 10 DithiVer Metals Reagent Powder Pillows to a pint bottle of chloroform and invert several times until well mixed (carrier powder may not dissolve). Store dithizone solution in an amber glass bottle. This solution is stable for 24 hours.

A reagent blank on deionized water should be carried out through the entire method to obtain the most accurate results. The amount of reagent blank determined on each lot of DithiVer Metals Reagent Powder Pillow is then subtracted from each reading obtained in Step 16.

SAMPLING AND STORAGE

Collect sample in acid cleaned glass or plastic containers. Adjust the pH to 2 or less with nitric acid (about 2 mL per liter). Preserved samples can be stored up to six months at room temperature. Adjust the pH to 2.5 to 4.5 with 5.0 N sodium hydroxide before analysis. Correct the test result for volume additions; see "Correction for Volume Additions" in Section 1.

ACCURACY CHECK

See p. 685 of the HACH Water Analysis Handbook, 3rd Ed., 1997.

INTERFERENCES

The following do not interfere:

TABLE 2.24

Aluminum	Calcium	Magnesium
Antimony	Chromium	Manganese
Arsenic	Cobalt	Nickel
Cadmium	Iron	Zinc

The following interfere:

TABLE 2.25

Bismuth	Mercury	Tin
Copper	Silver	

Eliminate interference from these metals by the following treatment, beginning after procedure Step 6.

a) Measure about 5 mL of the prepared dithizone solution into the separatory funnel. Stopper the funnel, invert and open the stopcock to vent. Close the stopcock and shake the solution vigorously for 15 seconds.

b) Allow the funnel to stand undisturbed until the layers separate (about 30 seconds). A yellow, red, or bronze color in the bottom (chloroform) layer confirms the presence of interfering metals.

c) Draw off and discard the bottom (chloroform) layer.

d) Repeat extraction with fresh 5 mL portion of prepared dithizone solution (discarding the bottom layer each time) until the bottom layer shows a pure dark green color for three successive extracts. Extractions can be repeated a number of times without appreciably affecting the amount of lead in the sample.

e) Extract the solution with several 2 or 3 mL portions of pure chloroform to remove any remaining dithizone, again discarding the bottom layer each time.

f) Continue the procedure, substituting 28.5 mL of prepared dithizone solution for the 30 mL in Step 7.

Large amounts of zinc cause the color transition at the end point to be indistinct.

Highly buffered samples or extreme sample pH may exceed the buffering capacity of the reagents and require sample pretreatment; see "pH Interferences" in Section 1.

WASTE DISPOSAL

Both chloroform (D002) and cyanide (D003) solutions are regulated as hazardous wastes by the Federal RCRA. Do not pour these solutions down the drain. Chloroform solutions and the cotton plug used in the delivery tubes of the separatory funnel should be collected for disposal with laboratory solvent waste. Be sure that cyanide solutions are stored in a caustic solution with a pH > 11 to prevent potential release of hydrogen cyanide gas.

QUESTIONS AND ANSWERS

1. **Why should the lead-dithizone complex be tightly capped and out of direct sunlight?** *The complex is somewhat light sensitive. Capping of the container avoids evaporation of the chloroform from the sample.*
2. **How do you avoid problems with cyanide in the experiment?** *Before cyanide is added to the mixture, 5.0 N sodium hydroxide is added to the solution until the pH is a* ***minimum*** *of 11.0. Adding 5 extra drops of 5.0 N sodium hydroxide solution provides an extra sense of security.*
3. **Why should filter paper be avoided when filtering a cloudy or turbid sample?** *Filter paper adsorbs lead. Instead, use a glass membrane filter or a sintered glass filter.*
4. **Why should this experiment be performed in a hood or with proper ventilation?** *This experiment uses chloroform and potassium cyanide which require protection and open air. Both chemicals are regulated as hazardous wastes by the Federal RCRA. Do not pour these chemicals down the drain. The chemicals should be collected for disposal with laboratory solvent waste.*

REFERENCES

1. *HACH Water Analysis Handbook,* 3rd ed. (Loveland, CO.: HACH Company, 1997).
2. *Hawley's Condensed Chemical Dictionary,* 11th Ed., rev'd by N. Irving Sax and Richard J. Lewis (New York, N.Y.: Van Nostrand Reinhold, 1987).
3. *Standard Methods for the Examination of Water and Wastewater:* LEAD, 16th Ed., A. E. Greenberg, R. R. Trussell, L. S. Clesceri and Mary Ann H. Franson (eds.), pp. 221–223, Washington, DC, American Public Health Association, 1985.

MANGANESE

Manganese can exist in several different valence states. The heptavalent (+7) state of manganese represented by permanganate ion is a strong oxidizing agent capable of oxidizing most metals and organic matter. In water, divalent and trivalent manganese is often found as a stable soluble complex, and quadrivalent manganese is usually in a suspension. Manganese will exist as a soluble form in neutral water, but it has a tendency to oxidize to a higher oxidation state and precipitates or becomes adsorbed to the collection container walls. Manganese should be determined immediately after collection of sample. The determination of total manganese usually requires a digestion.

Manganese is normally present in water at levels below 1 mg/L, but the recommended allowable manganese level in public water supplies is 0.05 mg/L. Taste and staining problems can occur at 0.1 ppm Mn or above. The substance badly stains laundry and plumbing fixtures. A number of manganese compounds are increasingly soluble in water as the temperature of the water increases above 0°C. Manganese often requires special means of removal such as aeration, pH adjustment, chemical precipitation, and ion-exchange processes. Manganese can be found in receiving streams, domestic wastewater, and industrial effluents.

Manganese compounds are used in varnish and oil driers, deoxidizers, paints, fluorinating agents, thermocouples and electrical instruments, dyeing processes, catalysts, pharmaceutical preparations, fertilizer, feed additives and dietary supplements, colored glass, bleaching tallow, coating of metals, textiles, medicines, fungicides, antiknock agents and oxidizing agents. Manganese metal is primarily used as a ferroalloy (steel manufacture) and a nonferrous alloy (improves hardness and corrosion resistance). The metal is a purifying and scavenging agent in metal production.

Manganese compounds are generally not very toxic. Waste streams of manganese can be easily modified to a less toxic form by oxidation or reduction of the metal. Some forms of manganese are recovered and are recycled waste products. Solutions containing other toxic substances and manganese are sometimes buried or recycled.

The PAN Method is most appropriate as a sensitive, rapid procedure for determining low levels of manganese at 0 to 0.700 mg/L. An Ascorbic Acid Reagent reduces all oxidized forms of manganese to Mn^{+2}. An Alkaline-Cyanide Reagent masks interferences and the PAN Indicator forms an orange-colored complex with Mn^{+2}. For levels higher than 0.700 mg/L manganese, the sample will have to be diluted appropriately or the periodate oxidation method will have to be implemented on the sample. The amount of manganese in the samples and the goals of the measurements will dictate which field instrument or method is most appropriate. For many measurements, the DR/2010 Spectrophotometer and the DR/890 Colorimeter will be required.

MANGANESE, LR (0 to 0.700 mg/L)

Method 8149 for water and wastewater*
PAN Method**

1. Enter the stored program number for manganese (Mn). Press: **2 9 0 ENTER.** The display will show: **Dial nm to 560.** *Note:* The Pour-Thru Cell can be used with 25-mL reagents only.

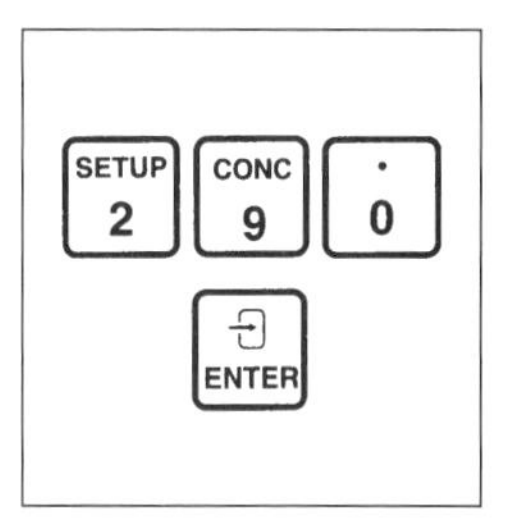

2. Rotate the wavelength dial until the small display shows: **560 nm.** When the correct wavelength is dialed in, the display will quickly show: **Zero sample,** then: **mg/L Mn LR (Low Range).**

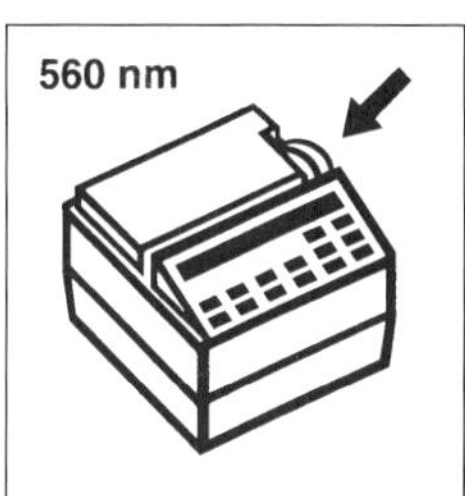

3. Insert the 10-mL Cell Riser into the cell compartment. *Note:* Total manganese determination requires a prior digestion; see "Digestion and Sample and Analysis Volume Table for Aqueous Liquids" in Section 1.

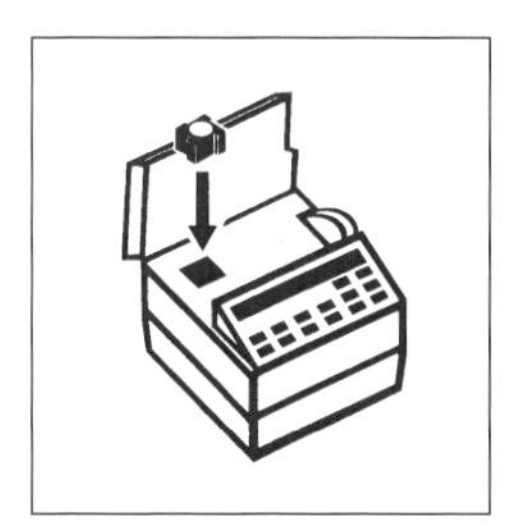

*Compliments of HACH Company. Analysis may be performed by DR/890 Colorimeter and DR/2010 Spectrophotometer.

**Adapted from K. Goto, et al., *Talanta,* 24, 752–753 (1977).

4. Fill a 10-mL sample cell with 10 mL of deionized water (this will be the blank).
Note: Rinse all glassware with 1:1 Nitric Acid Solution. Rinse again with deionized water.

5. Fill another 10-mL sample cell with 10 mL of sample (this will be the prepared sample).

6. Add the contents of one Ascorbic Acid Powder Pillow to each cell. Swirl to mix.
Note: For samples containing hardness greater than 300 mg/L $CaCO_3$, add four drops of Rochelle Salt Solution to the sample after addition of the Ascorbic Acid Powder Pillow.

7. Add 15 drops of Alkaline-Cyanide Reagent Solution to each cell. Swirl to mix.
Note: A cloudy or turbid solution may form in some samples after addition of the Alkaline-Cyanide Reagent Solution. The turbidity should dissipate after Step 8.

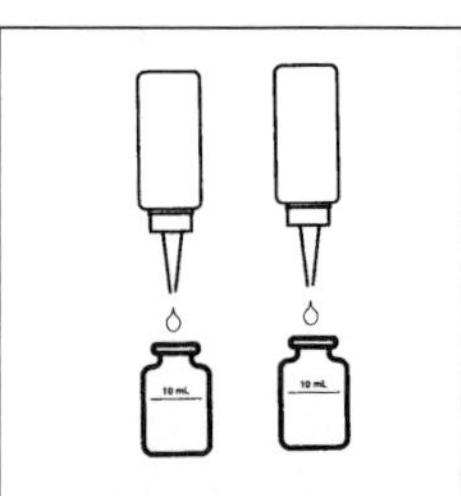

8. Add 21 drops of PAN Indicator Solution, 0.1%, to each sample cell. Swirl to mix.
Note: An orange color will develop in the sample if manganese if present.
Note: *For 25-mL reagents, use 1 mL of each liquid reagent in Steps 7 and 8.*

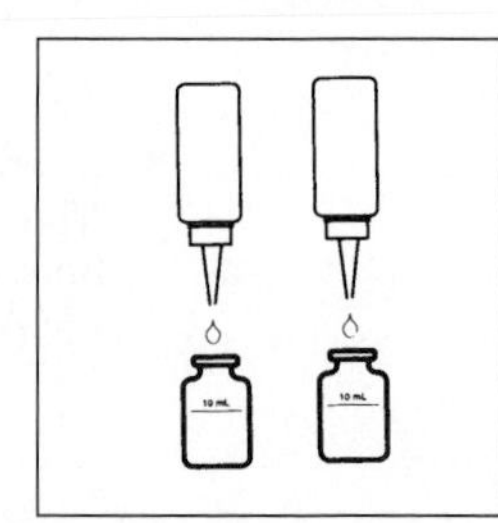

9. Press: **SHIFT TIMER.** A two-minute reaction period will begin.
Note: If the sample contains more than 5 mg/L iron, allow ten minutes for complete color development. To set the timer for 10 minutes, press **1000 SHIFT TIMER.**

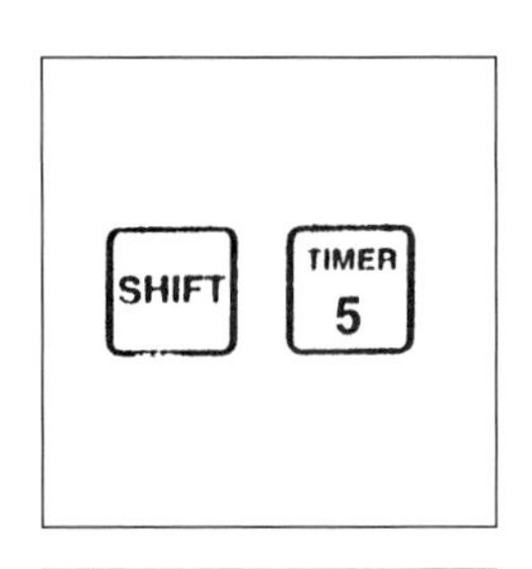

10. When the timer beeps, the display will show: **mg/L Mn LR.** Place the blank into the cell holder. Close the light shield.

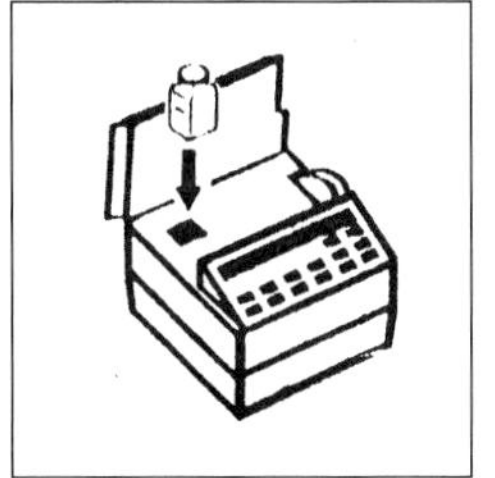

11. Press: **ZERO.** The display will show: **Zeroing ...** then: **0.00 mg/L Mn LR.**

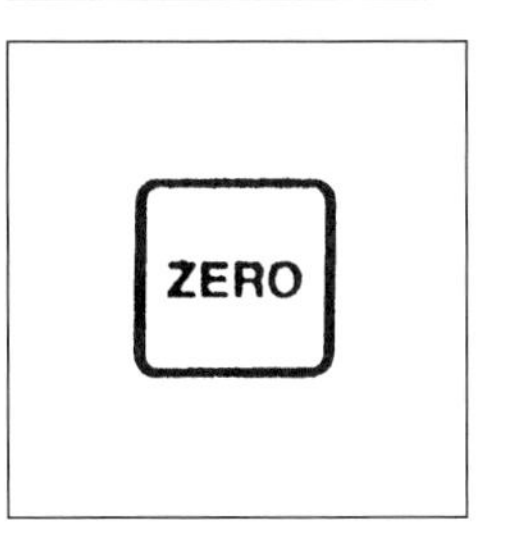

12. Place the prepared sample into the cell holder. Close the light shield. Press: **READ.** The display will show: **Reading ...** then the result in mg/L manganese will be displayed.
Note: See Waste Management below for proper disposal of cyanide containing wastes.

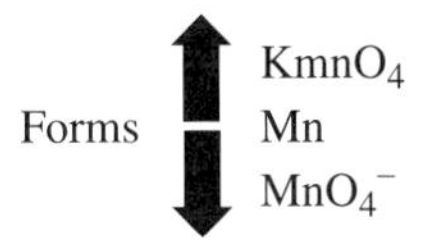

SAMPLING AND STORAGE

Collect samples in a clean glass or plastic container. Adjust the pH to 2 or less with nitric acid (about 2 mL per liter). Preserved samples can be stored up to six months at room temperature. Adjust the pH to 4.0 to 5.0 with 5.0 N sodium hydroxide before analysis. Correct the test result for volume additions; see "Correction for Volume Additions" in Section 1.

ACCURACY CHECK

See p. 715 of the HACH Water Analysis Handbook, 3rd ed., 1997.

INTERFERENCES

The following do not interfere up to the indicated concentrations:

Aluminum	20 mg/L
Cadmium	10 mg/L
Calcium	1000 mg/L as $CaCO_3$
Cobalt	20 mg/L
Copper	50 mg/L
Iron	25 mg/L
Lead	0.5 mg/L
Magnesium	300 mg/L as $CaCO_3$
Nickel	40 mg/L
Zinc	15 mg/L

WASTE MANAGEMENT

The alkaline cyanide solution contains cyanide. Cyanide solutions should be collected for disposal as reactive (D001) waste. Be sure cyanide solution is stored as a caustic solution with pH > 11 to prevent release of hydrogen cyanide gas.

QUESTIONS AND ANSWERS

1. **What are the most stable oxidation states for manganese ions?**
   Most manganese ions exist in the oxidation states including +2, +3, +4, and +7. Some of the manganese compounds are not very stable in water or in air.
2. **Why is a digestion required for total manganese?**
   In solution, manganese can form a stable soluble complex, a partially soluble suspended material, and a precipitate. Also, some manganese compounds can adsorb to the walls of a container.
3. **Why is too much manganese in water a problem?**
   Water containing too much manganese will badly stain laundry and plumbing fixtures.
4. **How is most manganese removed from water?**
   In many cases, manganese is not removed from water, but is changed to a different oxidation state where the manganese will not be an impact. The most common techniques in addressing problems with manganese are water aeration, pH adjustment, chemical precipitation, dilution, and ion exchange procedures.

REFERENCES

1. HACH Water Analysis Handbook, 3rd Ed. (Loveland, CO.: HACH Company, 1997).
2. *Hawley's Condensed Chemical Dictionary,* 11th Ed., rev'd by N. Irving Sax and Richard J. Lewis (New York, N.Y.: Van Nostrand Reinhold, 1987).
3. *Standard Methods for the Examination of Water and Wastewater:* MANGANESE, 16th ed., A. E. Greenberg, R. R. Trussell, L. S. Clesceri and Mary Ann H. Franson (eds.) pp. 228–229, Washington, D.C., American Public Health Association, 1985.

MERCURY

Mercury exists in a variety of chemical forms such as liquid mercury metal, simple inorganic and organic compounds, methylated mercury, and complexes of mercury. A number of mercury compounds have varying solubilities in different media and are sufficiently mobile to create problems. Mercury metal and many mercury compounds are highly toxic by ingestion, inhalation, and absorption. Mercury causes neurological damage, mutations and, at high levels, death of humans. Most organic and inorganic mercury compounds are very toxic and should be monitored closely in the environment. Tolerance levels (as Hg) are often at 0.05 mg per cubic meter of air for several important mercury compounds.

Mercury compounds and mercury metal are used in the following areas: antiseptic, silvering glass, analytical and organic chemistry, thermoscopy, germicide, batteries and battery electrolyte, pigments, antifouling paints, waterproofing, antisyphilitic, disinfectant, tanning, catalyst, sterilant, fungicide, insecticide and wood preservative, embalming fluids, photography, process engraving and lithography, germicidal soaps, vapor lamps, arc lamps, light fixtures, instruments, pharmaceuticals, chemical manufacturing, polishing compounds, pyrotechnics, electrodes, ceramics, anti-fungi coating on seeds, and many additional applications. Devices measuring pressure, temperature, flow rates and a variety of parameters also use liquid mercury.

Mercury metal and mercury compounds from a variety of pollution sources generally settle on the bottom of rivers and other water bodies. However, the metal is slowly transported down the river and concentrates at the river bottom in locations of low flow rates. From 0°C to 20°C, the solubility of mercury metal in water, even though small, increases by 6.5 times, whereas a 15-fold increase in water solubility is observed for mercury metal in the temperture range of 0°C to 30°C. In addition, a number of mercury compounds that have limited solubility in water become somewhat more soluble as the temperature of the water increases above 0°C.

Over a period of time, methylated mercury is formed from mercury metal combining with methane produced by decaying matter. Mercury compounds tend to solubilize by reacting with ammonia, water and other substances to form mercury-amine and mercury-hydroxyl complexes. Methylated mercury and other newly formed mercury compounds are more soluble in water and more easily transported in water than mercury metal and many mercury compounds. Mercury compounds and mercury metal tend to bioaccumulate in the food chain and eventually reach humans. Mercury has been identified in the organs of fish, turtles and other sea life.

Mercury is emitted when coal fired plants generate electricity and other furnaces lack proper emission controls. However, industry is not, by any means, the only source of mercury contamination. Many people do not know that mercury is present in many disposable items. Each person preparing or using mercury metal and many toxic mercury compounds must carefully isolate and recycle the products to avoid contamination of water, ground, and garbage. Unfortunately, it is nearly impossible to recover and recycle mercury metal and toxic forms of mercury compounds at this time.

In industry, mercury metal and several toxic mercury compounds require a closed loop system for recovery and reuse. A number of mercury wastes have been buried in the ground and recycled. Some wastes containing mercury have been dumped in undesirable places in the past. However, the system of isolating mercury wastes is a problem because mercury metal and mercury compounds are generally very stable and toxic. Once spilled on the ground or in river water, mercury metal will remain in the same area until it is "forced to move" or until it reacts and becomes more soluble. Environmental monitoring must be extensive and necessary because mercury is so toxic and does not decompose or change easily.

Inhalation of mercury vapor is hazardous to animals and people. Mercury metal should be used in areas of adequate ventilation. Exposed mercury metal in a container should be covered with a layer of water to inhibit mercury vaporization. If a mercury spill occurs, a mercury spill kit containing a powder, such as "CINNASORB," can be mixed with water to form a paste, which forms an amalgam with mercury droplets. Also, a powder called "RESISORB" can be sprinkled into cracks and crevices to minimize mercury vapors.

Mercury is difficult to analyze, and analysis procedures are expensive compared to other metals. Due to the sensitivity of the procedure, dedicated digestion glassware and sample cells are suggested for the determination of mercury. Glassware must be cleansed thoroughly and rinsed with 1:1 HCL solution followed by several deionized water rinses. In this experiment, extra care must be taken to avoid chemical exposure. Proper protection of personnel and appropriate handling procedures must be implemented at all times. Digestion and extraction procedures must be performed in a properly vented laboratory hood. Also, waste disposal of the mercury waste can be a major problem. Any solution containing above 0.2 mg/L mercury must be destroyed as a hazardous waste as defined by RCRA.

Sample digestion converts mercury compounds to mercuric (+2) ions. The mercuric ions in the digested sample are changed to mercury vapor, which is converted to mercuric chloride by a chemically active absorber column. After elution of the mercuric chloride from a specific column, a sensitive indicator is added to the solution to form a mercury-indicator complex. Later in the experiment, this complex is broken apart to form the indicator and the metal. The calorimeter is zeroed using the absorbance peak (412 nm) of the unreacted indicator. The measurement of the solution where the indicator and the metal has been broken apart provides evidence of the presence of mercury. The increase in solution absorbance is directly attributed to the concentration of mercury in the solution.

The MAS-50D Mercury Analyzer utilizes the EPA-approved Hatch and Ott Cold Vapor Method and features a sensitivity of 0.01 micrograms mercury and reproducibility of $\pm$ 0.05 μg at 1 μg levels.* In this determination, mercury in the sample is oxidized to mercuric ion with permanganate and persulfate in a nitric acid/sulfuric acid medium. Hydroxylamine hydrochloride is added to remove excess oxidant and stannous chloride reduces the mercury to metallic form. After the sample is connected to the analyzer, air is bubbled through the sample to evaporate the mercury and carry it through the absorption cell. The 253.7 nm mercury line emitted by the mercury lamp is absorbed by the vapor in proportion to the mercury concentration.

MERCURY (0.1 to 2.5 μg/L)

Method 10065 for water and wastewater**
Cold Vapor Mercury Concentration Method†

---

*Compliments of Bacharach, Inc.

**Compliments of HACH Company.

†Patents pending.

PHASE 1 SAMPLE DIGESTION: MUST BE DONE IN A HOOD! TOXIC GASES MAY BE PRODUCED.

1. Transfer one liter of the sample to a 2000-mL erlenmeyer flask. Add a 50-mm magnetic stir bar to the sample. Place the flask on a magnetic stirring hot plate and begin stirring.
Note: This procedure must be done in a fume hood. Toxic chlorine or other gases may be produced!
Note: HACH recommends using dedicated digestion glassware and sample cells for this procedure.

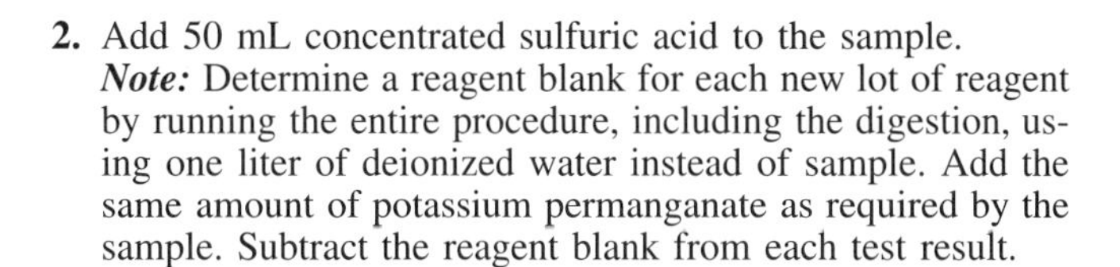

2. Add 50 mL concentrated sulfuric acid to the sample.
Note: Determine a reagent blank for each new lot of reagent by running the entire procedure, including the digestion, using one liter of deionized water instead of sample. Add the same amount of potassium permanganate as required by the sample. Subtract the reagent blank from each test result.

3. Add 25 mL concentrated nitric acid to the sample.

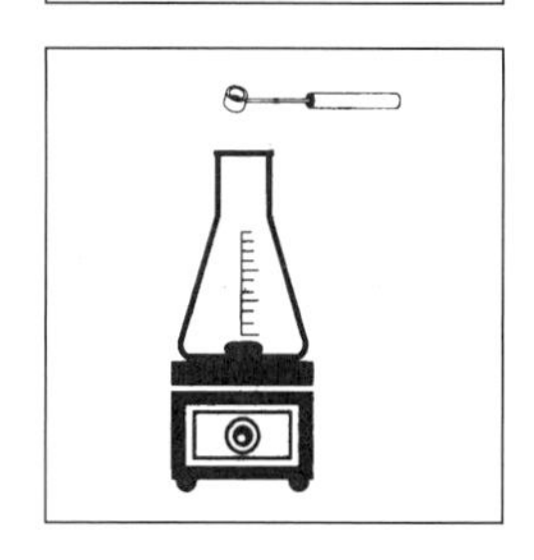

4. Add 4.0 g of potassium persulfate to the sample. Stir until dissolved.
Note: Alternatively, add one 5-g measuring scoop of potassium persulfate to the sample.

5. Add 7.5 g of potassium permanganate to the sample. Stir until dissolved.
Note: Alternatively, add a 10-g measuring scoop of potassium permanganate to the sample.

6. Cover the flask with a watch glass. Begin heating the sample to a temperature of 90°C *after* the reagents have dissolved. ***Note:*** For a mercury standard or reagent blank in distilled water, the heat step is not necessary.

7. Continue to stir and heat the sample at 90°C for two hours. ***Note:*** A dark purple color must persist throughout the two-hour digestion. Some samples (sea waters, industrial effluents or samples high in organic matter or chloride concentration) require additional permanganate. It may be difficult to see a dark purple color if the sample contains a black-brown manganese dioxide precipitate. You may add more potassium permanganate if the solution is not dark purple.

8. Cool the digested sample to room temperature. A brown-black precipitate of manganese dioxide may settle during cooling. If the digested sample does not have a purple color, the digestion may be incomplete. Add more potassium permanganate. Return the sample to the magnetic stirring hot plate and continue digestion until a purple color persists.

9. Return the cool, digested sample to the cool, magnetic stirring hot plate. Turn the stirrer on.

10. Using a 0.5-g measuring spoon, add 0.5 g-additions of hydroxylamine hydrochloride until the purple color disappears. Wait 30 seconds after each addition to see if the purple disappears. Add hydroxylamine hydrochloride until all the manganese dioxide is dissolved.

11. Remove the stir bar.

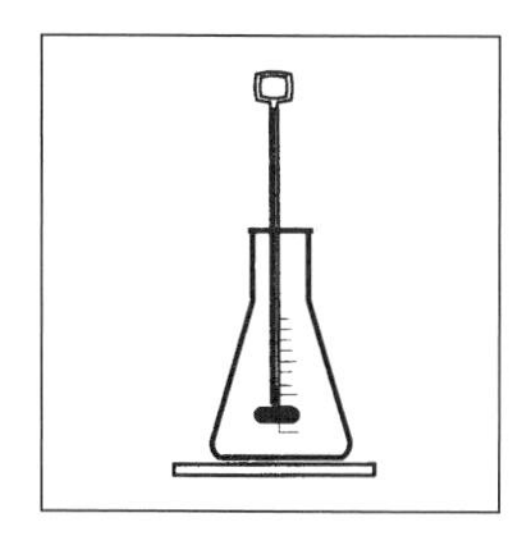

12. The digested sample is now ready for processing by cold vapor separation and preconcentration. Proceed to Phase 2.

PHASE 2 COLD VAPOR SEPARATION AND PRECONCENTRATION OF MERCURY

1. Enter the stored user program number for Cold Vapor Mercury. Press: **312 ENTER.** The display will show: **Dial nm to 412.**
Note: For a DR/2010 without this stored program, see Instrument Setup following these steps.

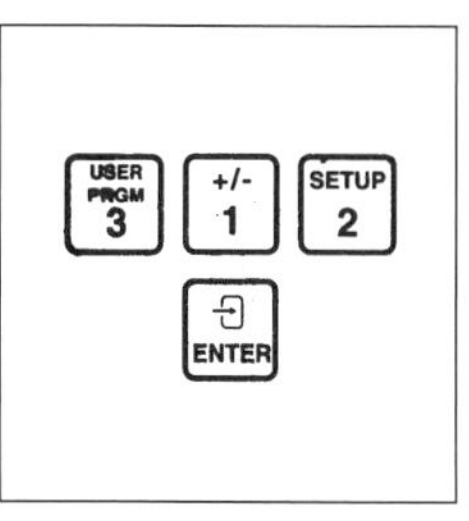

2. Rotate the wavelength dial until the small display shows: **412 nm.** When the correct wavelength is dialed in, the display will quickly show: **Zero Sample** then: **μg/L Hg CV.**

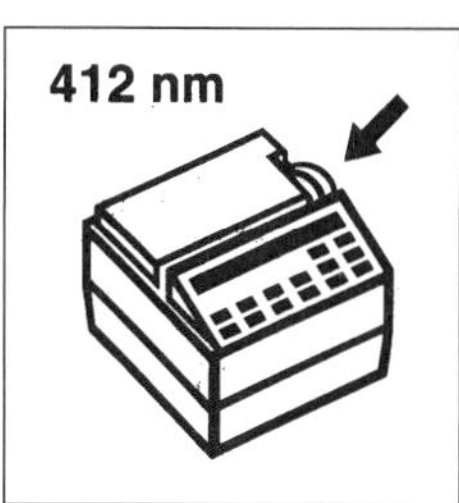

3. Transfer the digested sample to the Cold Vapor Gas Washing Bottle.
Note: The volume of digested sample should contain 0.1 to 2.5 μg Hg.

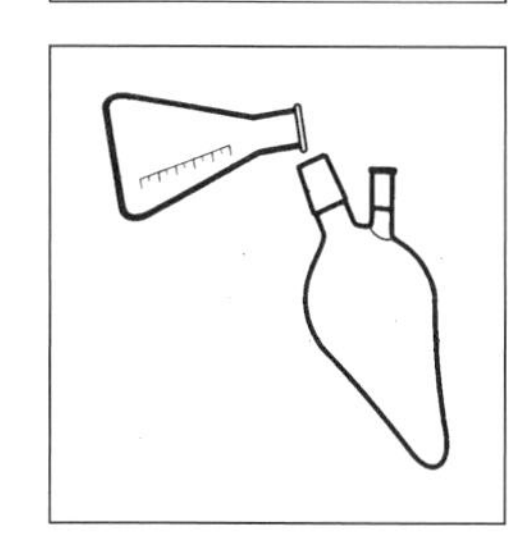

4. Set the Gas Washing Bottle in the support ring. Place the top on the Gas Washing Bottle. Wait until Step 11 to connect the Mercury Absorber Column to the Gas Washing Bottle.

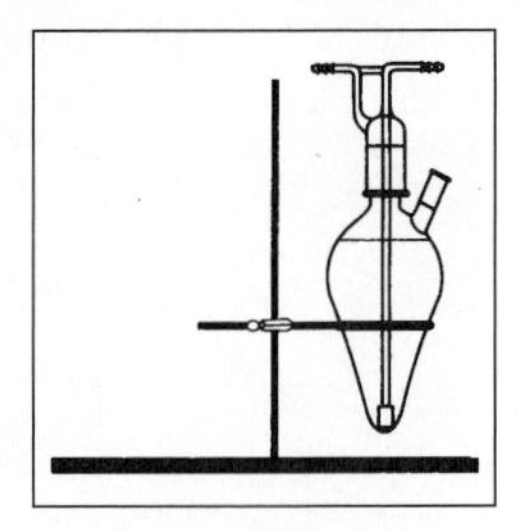

5. Connect the 100-mL erlenmeyer flask to the Mercury Absorber Column.

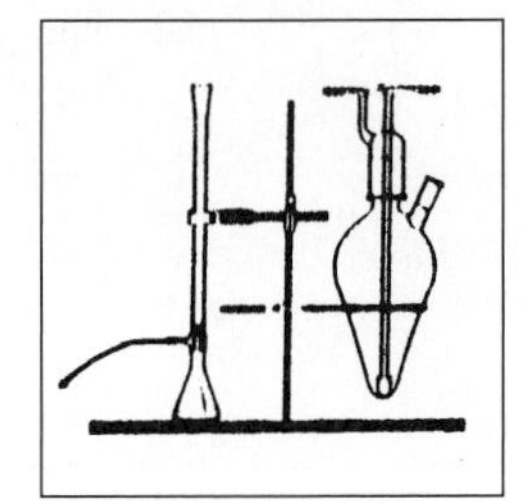

6. Pipet 8 mL of HgEx Reagent B into the Mercury Absorber Column.

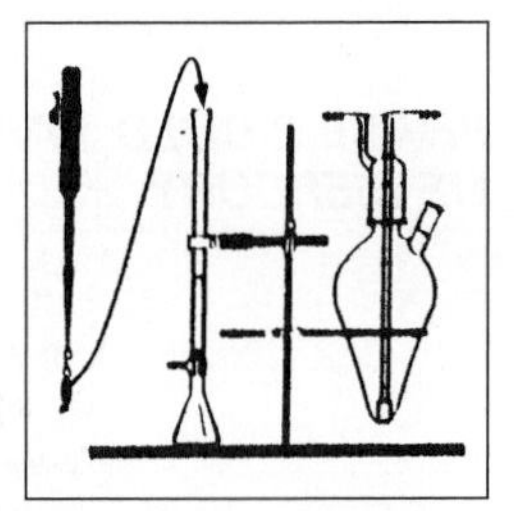

7. Connect the power to the vacuum pump and apply vacuum to the Mercury Absorber Column. Draw most of the HgEx Reagent B into the 100-mL erlenmeyer flask.

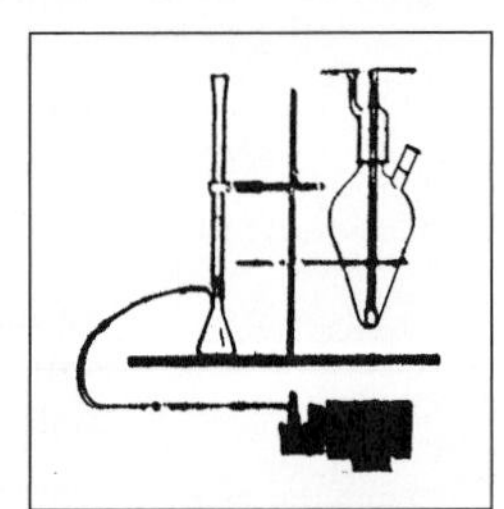

8. Disconnect the vacuum using the quick disconnect when HgEx Reagent B begins to drip from the inner delivery tube on the Mercury Absorber Column (about 10 seconds after starting the vacuum). Do not draw enough air through the column to begin drying the packing.

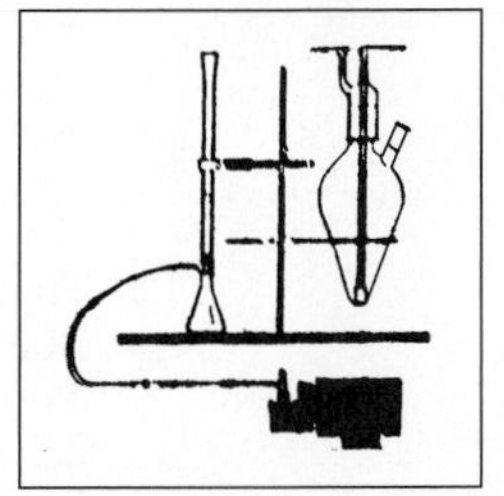

9. Remove the 100-mL erlenmeyer flask from the Mercury Absorber Column. Replace it with the 10-mL Distilling Receiver.

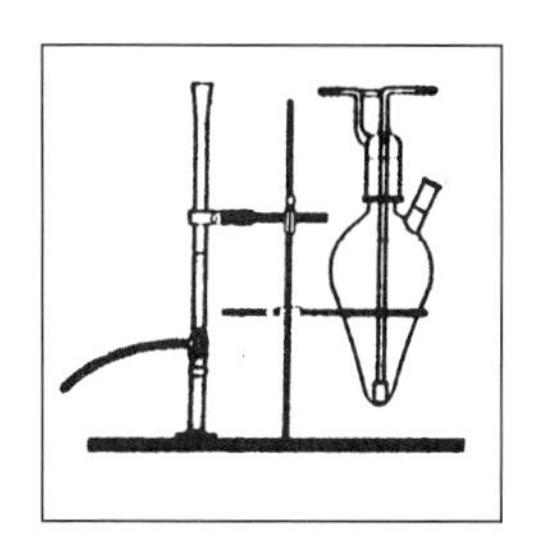

10. Pipet 2 mL of HgEx Reagent C into the Mercury Absorber Column.

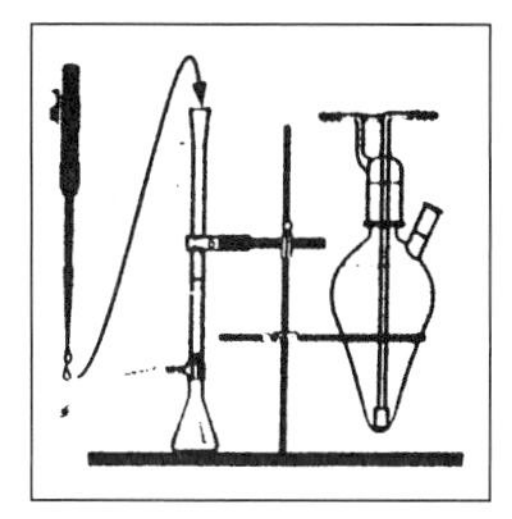

11. Connect the Mercury Absorber Column to the Gas Washing Bottle using the glass elbow.

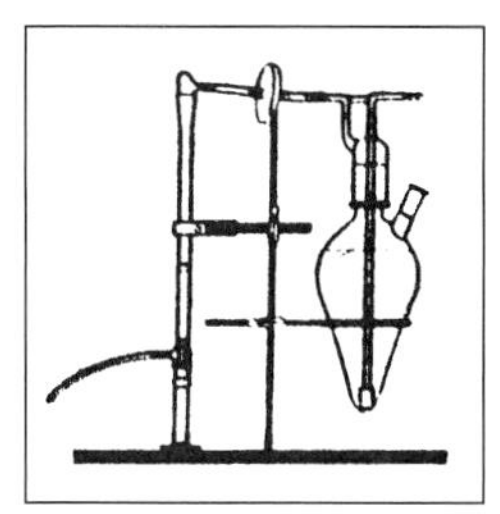

12. Shake an ampule of HgEx Reagent A to suspend the undissolved reagent. Open the ampule and gently shake the contents into the Gas Washing Bottle through the side neck. ***Note:*** Shaking the ampule is not necessary if there is no undissolved reagent in the ampule.

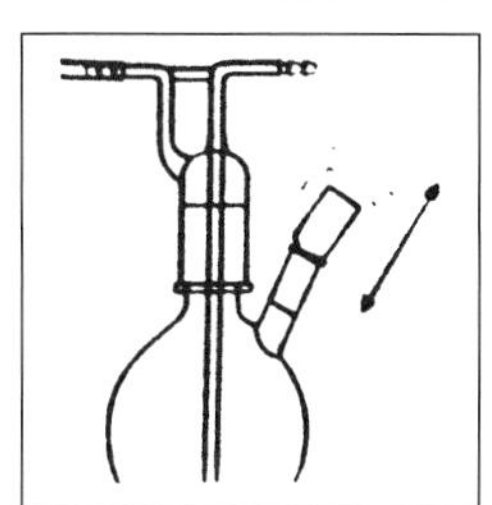

13. Stopper the side neck on the Gas Washing Bottle.

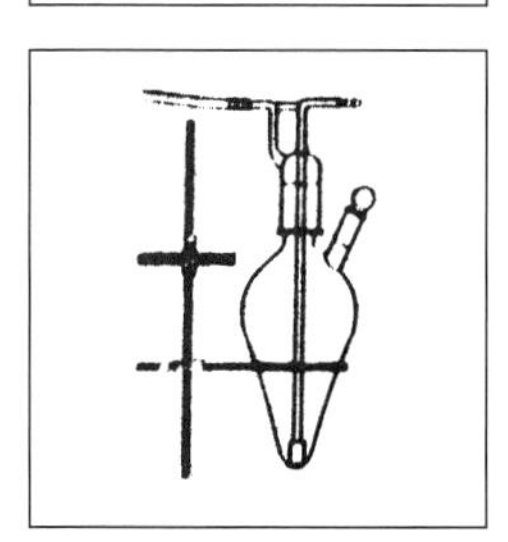

14. Reconnect the vacuum to the Mercury Absorber Column using the quick disconnect. The vacuum will pull HgEx Reagent C through the Mercury Absorber Column packing into the 10-mL receiver. Air bubbles should be produced at the gas dispersion tube in the Gas Washing Bottle.

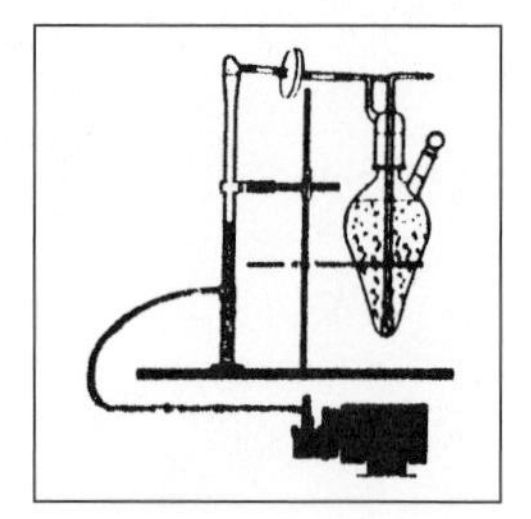

15. Press: **SHIFT TIMER.** A five-minute reaction period will begin. Let the solution bubble for this period.
Note: Air flow rate through the Gas Washing Bottle should be 1 to 5 liters per minute. Allow more time for lower air flow rates (i.e., if air flow rate is 1 liter/minute, let the solution bubble for 10 minutes).

16. After the timer beeps, remove the glass elbow from the top of the Mercury Absorber Column. Keep the vacuum pump on.

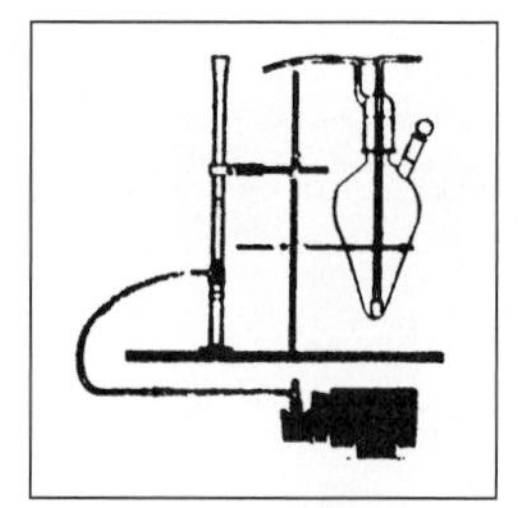

17. Pipet 8 mL of HgEx Reagent B into the Mercury Absorber Column to elute the captured mercury. Continue to apply vacuum to pull the HgEx Reagent B into the Distilling Receiver.

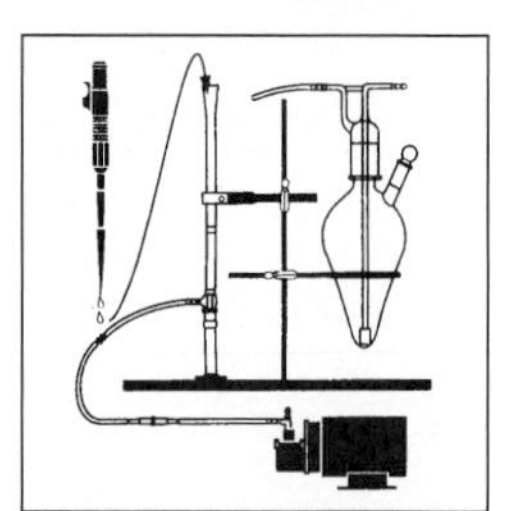

18. Turn off or disconnect power to the vacuum pump when the volume in the Distilling Receiver reaches the 10-mL mark.
Note: If necessary, bring the volume in the Distilling Receiver up to 10 mL with HgEx Reagent B. To avoid low volumes, disconnect the vacuum a little sooner in Step 6. This leaves more HgEx Reagent B in the packing of the Mercury Absorber Column.

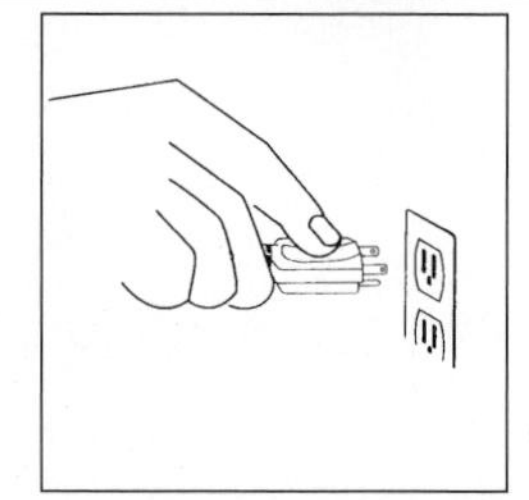

19. Remove the Distilling Receiver from the Mercury Absorber Column. Reconnect the 100-mL erlenmeyer flask to the column.

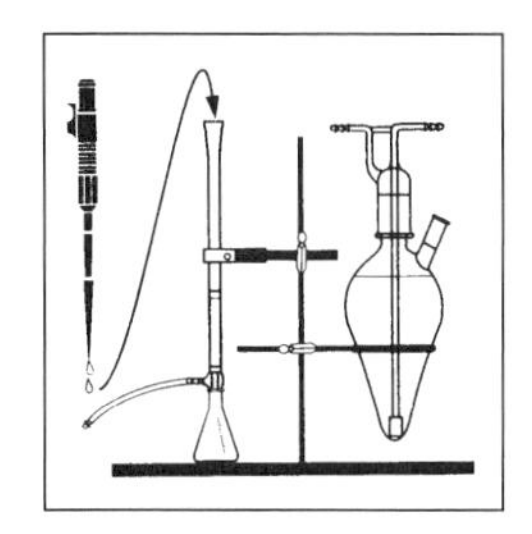

20. Pipet 3 mL of HgEx Reagent B into the Mercury Absorber Column (do not apply vacuum). This keeps the absorber packing wet between tests. The Mercury Absorber Column eluate in the Distilling Receiver is ready for analysis. Proceed to Phase 3.

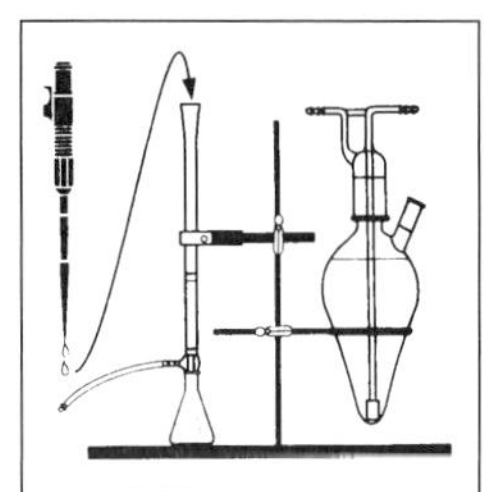

PHASE 3 COLORIMETRIC ANALYSIS

1. Insert the 10-mL Cell Riser into the sample cell compartment.

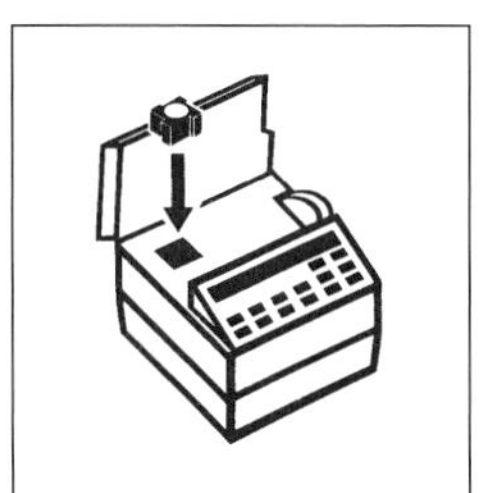

2. Using the funnel provided, add the contents of one HgEx Reagent 3 foil pillow to the eluate in the Distilling Receiver. Stopper the receiver. Invert the receiver to dissolve the reagent.

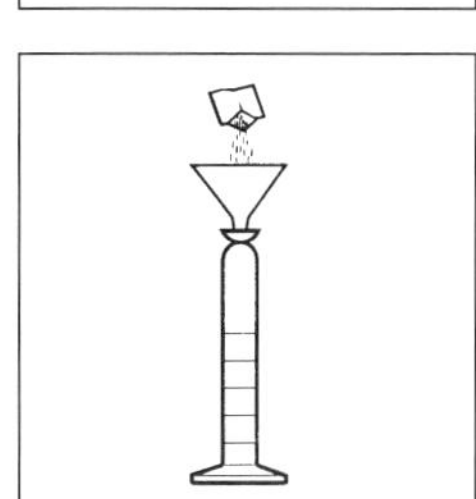

3. Add the contents of one HgEx Reagent 4 foil pillow to the Distilling Receiver using the funnel provided. Stopper the receiver. Invert the receiver to dissolve the reagent.

4. Add 8 drops of HgEx Reagent 5 to the Distilling Receiver. Stopper the receiver. Invert to mix.

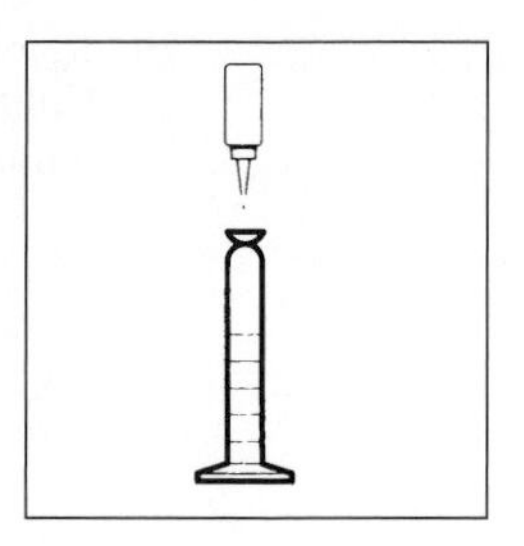

5. Press: **SHIFT TIMER.** A 2-minute reaction period will begin.

6. During the reaction period, transfer the solution to a 10-mL sample cell. Wipe the sample cell sides with a clean tissue. ***Note:*** The Pour-Thru Cell cannot be used with this procedure.

7. After the timer beeps, place the prepared sample into the cell holder and close the light shield.

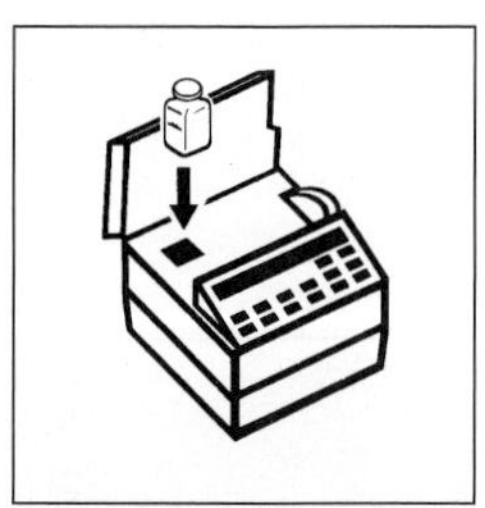

8. Press: **ZERO.** The display will show: **Zeroing ...** then: **0.1 μg/L Hg CV.**
Note: This program uses a non-zero intercept.

9. Remove the cell from the cell holder. Add the contents of one HgEx Reagent 6 foil pillow to the solution. Swirl the cell until the reagent is completely dissolved. Immediately go to Step 10.
Note: Do not use the funnel to add HgEx Reagent 6 to the sample cell. Any HgEx Reagent 6 in the funnel will make mercury undetectable in subsequent tests.

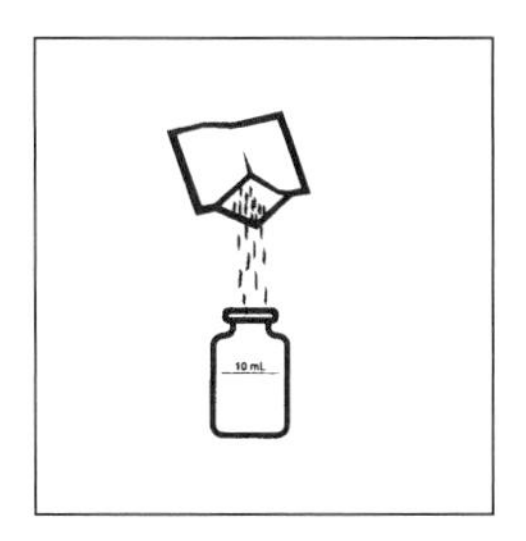

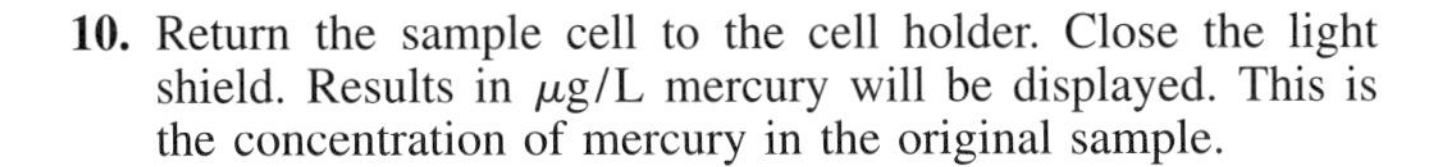

10. Return the sample cell to the cell holder. Close the light shield. Results in μg/L mercury will be displayed. This is the concentration of mercury in the original sample.

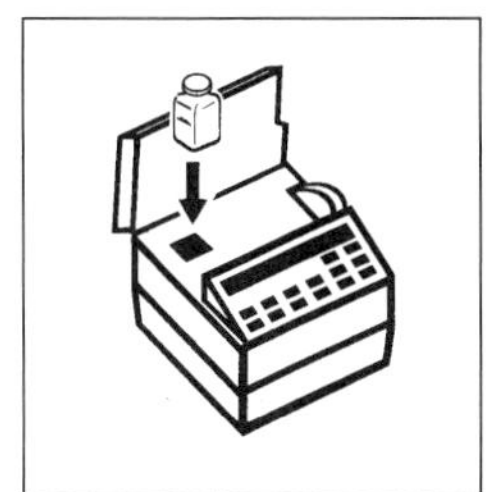

SAMPLING AND STORAGE

Collect 1000 mL of sample in an analytically clean, glass or polyethylene terephthalate (PET) container. Add 10 mL of concentrated hydrochloric acid to preserve the sample before sample collection. Fill the container completely to minimize air space when closed. Close a glass container with a ground glass stopper. Close a PET container with a PET cap or a polypropylene cap (no liner).

Store aqueous samples at 2 to 6°C. Acid-preserved samples are stable for at least 6 months.

ACCURACY CHECK: STANDARD ADDITIONS METHOD

See HACH Water Analysis Handbook, 3rd edition, 1997, page 742.

STANDARD SOLUTION METHOD

a) Transfer 800 mL of deionized water into the Gas Washing Bottle.

b) Add 50 mL of concentrated sulfuric acid and 25 mL of concentrated nitric acid to the water. Swirl to mix.

c) Prepare a 0.1-mg/L mercury standard solution by serially diluting a 1000-mg/L Mercury Standard Solution:

- To make a 10.0-mg/L standard, add 1.0 mL of concentrated nitric acid to a 500-mL volumetric flask. Dilute 5.00 mL of a 1000 mg/L standard to 500 mL with deionized water. Mix well.

- To make a 1.0-mg/L standard solution, add 0.2 mL of concentrated nitric acid to a 100-mL volumetric flask. Dilute 10.0 mL of the 10.0-mg/L standard to 100 mL with deionized water. Mix well.
- To make a 0.1-mg/L standard solution, add 0.2 mL of concentrated nitric acid to a 100-mL volumetric flask. Dilute 10.00 mL of the 1.0-mg/L standard to 100 mL with deionized water. Mix well.

d) Pipet 10.0 mL of the 0.1-mg/L mercury standard solution into the Gas Washing Bottle. Swirl to mix.

e) Begin at Step 2 of Phase 2. Follow the procedure steps.

f) Test the eluate as described in Phase 3. The displayed concentration should be between 0.9 to 1.1 μg/L Hg.

SYSTEM START UP

HACH recommends that the analyst perform a few examinations on mercury standards and blanks for system equilibration before beginning sample testing. This allows the system to stabilize before processing samples.

STARTUP STANDARD

Test a mercury standard solution by following the procedure under Accuracy Check using the Standard Solution Method. Continue with Step g (below) if the value is not within specified limits.

g) Pipet 10.0 mL of the 0.1-mg/L mercury standard solution into the purged solution in the Gas Washing Bottle. Immediately stopper the Gas Washing Bottle.

h) Begin at Step 5 of Phase 2. Follow the procedure steps.

i) Test the eluate as described in Phase 3. The displayed concentration should be between 0.9-1.1 μg/L Hg. Repeat Steps g to i if the value is not within these limits.

STARTUP BLANK

Run a system blank by using the purged solution in the Gas Washing Bottle after a satisfactory test of the Startup Standard has been completed.

j) Leave the purged solution in the Gas Washing Bottle. Do not add an aliquot of mercury standard.

k) Begin at Step 5 of Phase 2. Follow the procedure steps.

l) Test the eluate as described in Phase 3. The displayed concentration should be <0.2 μg/L Hg. Repeat the Startup Blank procedure until a reproducible value is obtained.

INTERFERENCES

Standards were used to prepare a single test solution with the following matrix. A second test solution containing only mercury at the same concentration was prepared as the control.

The two solutions were digested then analyzed concurrently. There was no interference from the matrix of the test solution at the concentrations listed below in Table 2.26.

In addition, no interference occurred with a test solution containing 1000 mg/L Na^+, 1000 mg/L K^+, 1000 mg/L Mg^{2+}, and 400 mg/L Ca^{2+}.

TABLE 2.26 Interference Test Solution Matrix

Ion or substance	Concentration
Ag^{+2}	7 mg/L Ag^{+2}
Al^{+3}	10 mg/L Al^{+3}
Au^{3+}	500μg/L Au^{3+}
Cd^{+2}	10 mg/L Cd^{+2}
Co^{+2}	10 mg/L Co^{+2}
Cr^{+6}	10 mg/L Cr^{+6}
Cu^{+2}	10 mg/L Cu^{+2}
F^-	1.0 mg/L F^-
Fe^{+2}	100 mg/L Fe^{+2}
Hg^{+2}	1 μg/L Hg^{+2}
Mo^{+6}	10 mg/L Mo^{+6}
Ni^{+2}	10 mg/L Ni^{+2}
NO_3^--N	50 mg/L NO_3^--N
Pb^{2+}	10 mg/L Pb^{2+}
SiO_2	100 mg/L SiO_2
Zn^{+2}	10 mg/L Zn^{+2}

STORAGE AND MAINTENANCE OF THE COLD VAPOR MERCURY APPARATUS

Storage

Store the apparatus as follows for fastest system stabilization and greatest sensitivity:

- Store the Gas Washing Bottle filled with deionized water containing 15 mL of concentrated sulfuric acid. Seal the bottle with the Gas Washing Bottle stopper and top.
- Store the Mercury Absorber Column with the packing wetted with HgEx Reagent B. The erlenmeyer flask should be kept attached underneath the column. The top of the Mercury Absorber Column should be attached to the Gas Washing Bottle with the glass elbow as in the procedure.

Glassware Care

HACH recommends using dedicated glassware and sample cells because of the sensitivity of this procedure. Thoroughly clean the glassware and sample cells between tests. After washing, rinse with 1:1 hydrochloric acid solution, then rinse several times with deionized water.

Maintaining the System

- With proper care and storage, the Mercury Absorber Column may be used an unlimited number of times.
- Replace the Mercury Scrubber in the air trap housing at least once for every reagent set used.
- Moisture build-up on the Gas Washing Bottle side of the Acro® 50 Vent Filter will reduce the purging air flow rate. If this occurs replace the filter or dry it in an oven at 110°C.

SAFETY

Wear personal protective equipment such as safety glasses with side shields, or a face shield to protect your eyes. Use other protective equipment as necessary (such as a fume hood) to avoid chemical exposure. Perform all steps exactly as prescribed in the procedure.

WASTE DISPOSAL

Proper management and disposal of waste is the responsibility of the waste generator. Waste disposal information is provided as a guideline only. It is up to the generator to arrange for proper disposal and comply with applicable local, state, and federal regulations governing waste disposal. HACH Company and the authors of this text make no guarantees or warranties, express or implied, for the waste disposal information represented in this procedure.

1. Dispose of the solution in the Gas Washing Bottle by neutralizing the solution to a pH of 6 to 9 and flushing to the sanitary sewer with water for several minutes.
2. The mercury contained in one liter of sample is concentrated by a factor of 100 by the Mercury Absorber Column. Mercury analysis within the range of the test may produce a solution in the sample cell that is above the RCRA Toxicity Characteristic limit of 0.20 mg/L Hg. The sample cell will contain 0.25 mg/L mercury if the original sample was at 2.5 μg/L mercury (the upper limit of the test range). Dispose of the solution in the sample cell as a hazardous waste if the test result was greater than 2 μg/L mercury in the original sample. Otherwise, pour the solution into the sanitary sewer and flush with water for several minutes.
3. The mercury scrubber will capture mercury vapor if the Mercury Absorber Column is not properly activated using HgEx Reagent B and HgEx Reagent C. In addition, mercury is also captured if the capacity of the Absorber Column is exceeded. If the Mercury Scrubber has captured mercury vapor, it must be disposed of according to applicable regulations.

QUESTIONS AND ANSWERS

1. Where do we find liquid mercury?
Liquid mercury can be found in thermometers, manometers and several other devices to measure pressure, temperature, flow rates and environmental conditions. Medical facilities and different types of laboratories (chemistry, physics, oceanography, meteorology, metrology, geology, etc.) often use mercury containing devices to measure specific characteristics.

2. How toxic are mercury compounds?
   All inorganic compounds of mercury are considered to be highly toxic by ingestion, inhalation, and skin absorption. Many organic compounds of mercury are highly toxic. Tolerance (as Hg) for alkyl compounds is 0.01 mg per cubic meter of air, whereas tolerance (as Hg) for many less toxic organic compounds and inorganic compounds is 0.05 mg per cubic meter of air.
3. Why is liquid mercury considered to be a problem at the bottom of a body of water such as a river, stream, pond or ocean?
   Inorganic mercury can be converted to methylmercury by bacteria in water. The soluble form of methylmercury can be easily transported in drinking water. Methylmercury concentrates in the bodies of fish, turtles, and other edible species.
4. Why is mercury a major problem?
   Liquid mercury and mercury compounds have many practical uses for which there are limited substitutes. Mercury metal is stable and has a history of being stored in bottles in homes, small business, and factories. People played with liquid mercury and used it for all types of practical measurements around the home and in the medical field and the laboratory. Mercury tends to accumulate in the organs of animals, sea life, and man.

REFERENCES

1. *HACH Water Analysis Handbook,* 3rd Ed. (Loveland, CO.: HACH Company, 1997).
2. *Hawley's Condensed Chemical Dictionary,* 11th Ed., rev'd by N. Irving Sax and Richard J. Lewis (New York, N.Y.: Van Nostrand Reinhold, 1987).
3. *Standard Methods for the Examination of Water and Wastewater:* MERCURY, 16th Ed., A. E. Greenberg, R. R. Trussell, L. S. Clesceri and Mary Ann H. Franson (eds.), p. 232, Washington, DC, American Public Health Association, 1985.

NICKEL

Nickel metal and a number of nickel compounds are flammable. Nickel metal is relatively nontoxic to humans and most animals. The toxicity of nickel to aquatic life varies widely whereas nickel salts at concentration levels near 1.0 mg/L appear to cause damage to specific types of plants. A number of nickel compounds are increasingly soluble in water as the temperature of the water increases above 0°C. Some nickel compounds are toxic and known carcinogens (OSHA).

Nickel metal displays some level of carcinogenic characteristics as a dust in air and is primarily used in metal alloys, coatings, battery, fuel cells and catalytic processes. Nickel compounds are used in dyes, ceramics, electroplating, glazes, antioxidants, paints, cosmetics, storage batteries, fuel cell electrodes, capacitors, catalysts, mordant in dyeing and printing, electroplating, metallurgy, reagents, and porcelain painting.

Nickel wastes that are toxic should not be released into the air or into the water resources. Nickel wastes can be recovered and reused in a semi-closed or closed loop system. Some industries manifest the waste stream especially if the nickel waste is mixed with other types of wastes that can not be easily separated. Manifested wastes containing nickel compounds are often recovered for applications in other industries or buried in containers. Nickel wastes are often found as discards or in landfills.

Nickel has a tendency to form complexes with chelating agents, such as EDTA. Vigorous digestion on the sample will eliminate these interferences. Common interferences in concentrations less than 50 mg/L (ppm) are ions of aluminum, cadmium, chromium, copper, fluoride, iron (+2, +3), manganese (+2), lead and zinc. Most of the metals react with 1-(2-pyridylazo-2-naphthol (PAN) to form colored complexes. Iron (+3) is masked by adding pyrophosphate. EDTA is added to destroy all metal-PAN complexes except those of nickel and cobalt.

This method can be used to determine nickel at relatively low concentration levels (0 to 1.000 mg/L) in water, wastewater, and water from industrial operations. Samples with high concentrations of nickel may have to be diluted appropriately or be tested by the Heptoxime procedure (HACH Method 8037).

The wavelength to determine the nickel-PAN complex is 560 nm, whereas the wavelength to determine cobalt is 620 nm. A correction for cobalt interference can be applied to the spectrophotometer readings. The sensitive PAN procedure for detecting nickel and cobalt at concentrations near 1 mg/L can be made on the same sample by adjusting the wavelength and zeroing the instrument before both measurements.

NICKEL (0 to 1.000 mg/L)

Method 8150 for water and wastewater*
1-(2 Pyridylazo)-2-Naphthol (PAN) Method**

1. Enter the stored program number for nickel (Ni), PAN. Press: **3 4 0 ENTER.** The display will show: **Dial nm to 560.** *Note:* The Pour-Thru Cell cannot be used.

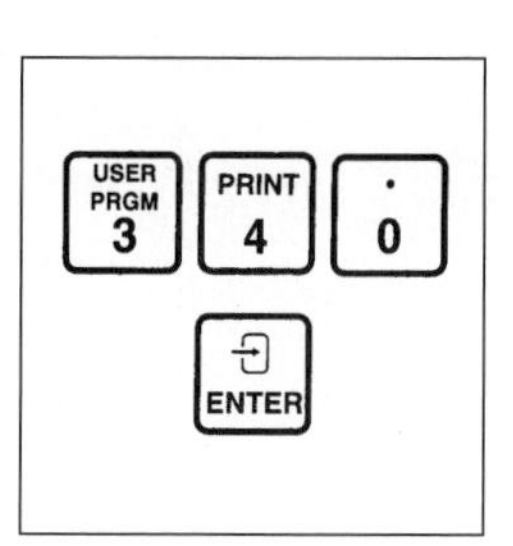

2. Rotate the wavelength dial until the small display shows: **560 nm.** When the correct wavelength is dialed in, the display will quickly show: **Zero sample,** then: **mg/L Ni PAN.**

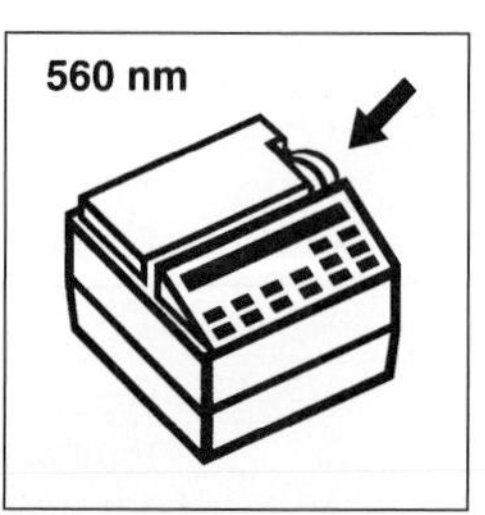

3. Insert the 10-mL Cell Riser into the cell compartment.

*Compliments of HACH Company. Analysis may be performed by DR/890 Colorimeters and DR/2010 Spectrophotometer.

**Adapted from H. Watanabe, Talanta, 21, 295 (1974).

4. Fill a 10-mL sample cell with 10 mL of deionized water (the prepared sample).
Note: If sample is less than 10°C (50°F), warm to room temperature before analysis. Adjust the pH of stored samples.

5. Fill a second 10-mL sample cell with 10 mL of deionized water (the blank).

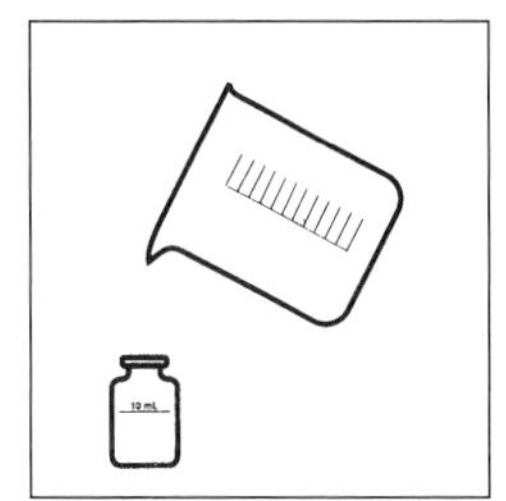

6. Add the contents of one Phthalate-Phosphate Reagent Powder Pillow to each cell. Stopper. Shake immediately to dissolve.
Note: If sample contains iron (Fe^{3+}), all the powder must be dissolved completely before continuing with Step 7.
Note: Two #16 HDPE stoppers are required.

7. Add 0.5 mL of 0.3% PAN Indicator Solution to each cell. Stopper. Invert several times to mix.
Note: Use the plastic dropper provided.

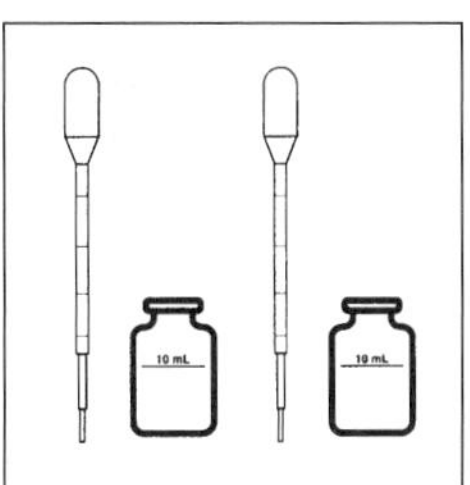

8. Press: **SHIFT TIMER.** A 15-minute reaction period will begin.
Note: During color development, the sample solution color may vary from yellowish orange to dark red. The blank should be yellow.

9. When the timer beeps, the display will show: **mg/L Ni PAN.** Add the contents of one EDTA Reagent Powder Pillow to each cell. Stopper. Shake to dissolve.

10. Place the blank into the cell holder. Close the light shield.

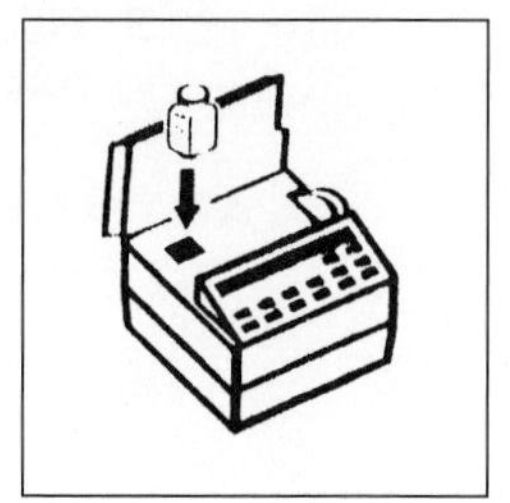

11. Press: **ZERO.** The display will show: **Zeroing ...** then: **0.00 mg/L Ni PAN.**

12. Place the prepared sample into the cell holder. Close the light shield.

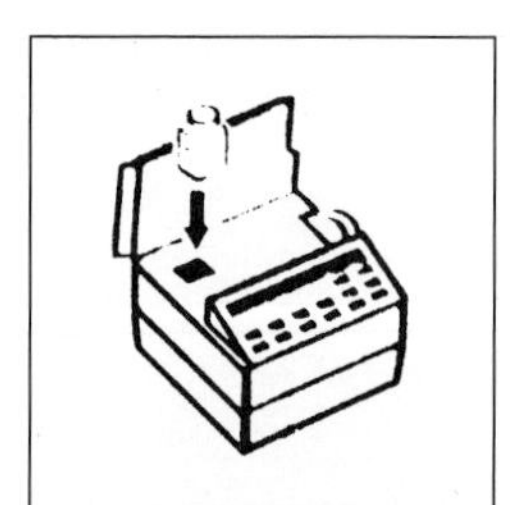

13. Press: **READ.** The display will show: **Reading ...** then the result in mg/L nickel will be displayed.
Note: If the sample contains cobalt, continue with Steps 14 to 19.

14. To correct for the presence of cobalt, rotate the wavelength dial to change the wavelength display to show: **620 nm.** Press: **ENTER.**

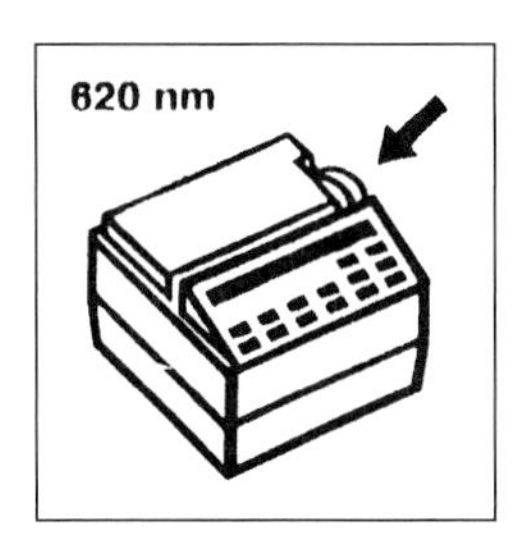

15. Place the blank into the cell holder. Close the light shield.

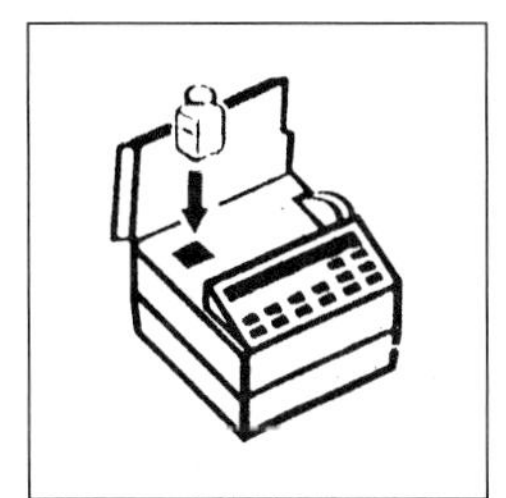

16. Press: **ZERO.** The display will show: **Zeroing ...** then: **0.000 mg/L Ni PAN.**
Note: The wavelength setting will continue to flash.

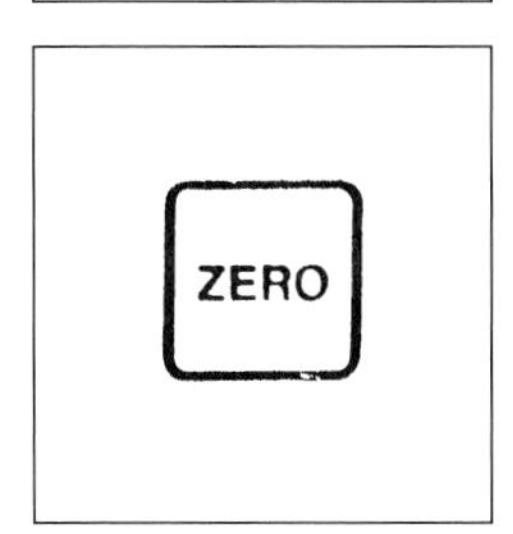

17. Place the prepared sample into the cell holder. Close the light shield.

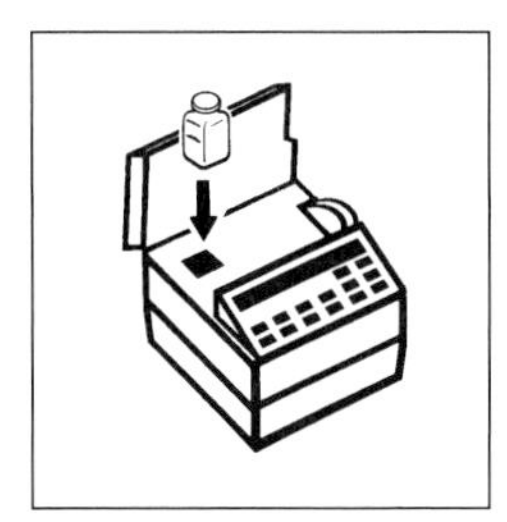

18. Press: **READ.** The display will show: **Reading ...** then the apparent nickel concentration in mg/L nickel due to the cobalt present will be displayed.

19. Use the equation below to correct for cobalt interference:

$$\text{mg/L Ni (Step 13)} - (0.89) \times \text{mg/L Ni (Step 18)} = \text{mg/L Ni}$$

Note: You may measure the cobalt concentration with the same prepared sample by using the Cobalt Stored Program No. 110.
Note: For Ni in chapter 3, a procedure of analysis is not required because the procedure is provided for extraction in the text for Ni in solids and then analysis is performed by procedures in chapter 2 as indicated in the text.

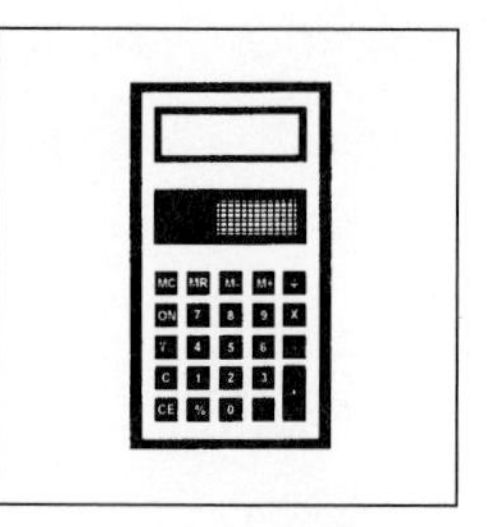

SAMPLING AND STORAGE

Collect samples in acid-washed plastic bottles. Adjust the pH to 2 or less with nitric acid (about 5 mL per liter). Preserved samples can be stored up to six months at room temperature. Adjust the sample pH to between 3 and 8 with 5.0 N Sodium Hydroxide Standard Solution just before analysis. Do not exceed pH 8 as this may cause some loss of nickel as a precipitate. Correct the test results for volume additions; see "Correction for Volume Additions" in Section 1 for more information.

ACCURACY CHECK

See p. 769 of the HACH Water Analysis Handbook, 3rd Ed., 1997.

INTERFERENCES

Highly buffered samples or extreme sample pH may exceed the buffering capacity of the reagents and require sample pretreatment; see "Correcting for Volume Additions" in Section 1.

Chelating agents, such as EDTA, interfere. Use either the Digesdahl or vigorous digestion to eliminate this interference.

The following may interfere when present in concentration exceeding those listed below in Table 2.27a.

TABLE 2.27a

Al^{3+}	32 mg/L
Ca^{2+}	1000 mg/L as ($CaCO_3$)
Cd^{2+}	20 mg/L
Cl^-	8000 mg/L
Cr^{3+}	20 mg/L
Cr^{6+}	40 mg/L
Cu^{2+}	15 mg/L
F^-	20 mg/L
Fe^{3+}	10 mg/L
Fe^{2+}	Interferes directly and must not be present
K^+	500 mg/L
Mg^{2+}	400 mg/L
Mn^{2+}	25 mg/L
Mo^{6+}	60 mg/L
Na^+	5000 mg/L
Pb^{2+}	20 mg/L
Zn^{2+}	30 mg/L

QUESTIONS AND ANSWERS

1. How are interferences by metals avoided during the test for nickel?
   Iron is masked by adding pyrophosphate. Metals, except nickel and cobalt, that form the metal-PAN complexes are destroyed by adding EDTA to the solution.
2. Why is it important to maintain the pH of the solution below pH = 8?
   When the pH exceeds 8, nickel will begin to precipitate.
3. A sample analysis reads 10.0 mg/L nickel at 560 nm. At 620 nm, after zeroing the instrument, the same sample reads 2.0 mg/L. How much nickel is in the sample?
   To correct for cobalt interference,

$$mg/L\ Ni = 10.0 - (0.89) \times 2.0 \text{ or essentially } 8.2\ mg/L \text{ of Nickel}$$

4. How are nickel compounds different than most metals and metal compounds?
   Nickel and nickel compounds are generally more toxic than most people realize. The type of nickel compound dictates the level of toxicity and carcinogenic character.

REFERENCES

1. *HACH Water Analysis Handbook,* 3rd Ed. (Loveland, CO.: HACH Company, 1997).
2. *Hawley's Condensed Chemical Dictionary,* 11th Ed., rev'd by N. Irving Sax and Richard J. Lewis (New York, N.Y.: Van Nostrand Reinhold, 1987).
3. *Standard Methods for the Examination of Water and Wastewater:* NICKEL, 16th Ed., A. E. Greenberg, R. R. Trussell, L. S. Clesceri and Mary Ann H. Franson (eds.) p. 234, Washington, DC, American Public Health Association, 1985.

MONOCHLORAMINE NITROGEN AND FREE AMMONIA NITROGEN

Free ammonia (as NH_3 and NH_4^+) and monochloramine (NH_2CL) can be found in drinking water. Ammonia and chlorine are found primarily in human waste and cleaning solutions. Under the appropriate conditions, ammonia and chlorine react to form monochloramines and dichloramines. The chloramine compounds are considered to be very toxic.

Dilution water used in the experiment must be free of ammonia, chlorine and chlorine demand. Ammonia contamination from air can yield high test results. Ammonia from the air can accumulate in solutions and deionized water as well as on the sides of glassware used in the experiment. Extensive rinsing of glassware with excess sample is necessary before sample analysis.

The important interferences in the experiment are pH (less than 7) and chlorine demand (non-ammonia, greater than 2 mg/L as CL_2). Additional interferences are high levels of calcium (3000 mg/L), magnesium (1600 mg/L), and sulfate (> 900 mg/L).

In the experiment, hypochlorite combines with free ammonia to form monochloramine. Monochloramine reacts with salicylate to form an aminosalicylate compound which produces a green solution in the presence of a cyanoferrate catalyst. Free ammonia is determined by comparing the color intensities of samples with and without hypochlorite.

Monochloramine nitrogen and free ammonia analyses can be performed by the salicylate method on the DR/850 and the DR/890 Colorimeters and the DR/2010 Spectrophotometer in the field. The powder pillow and the AccuVac ampul methods can determine levels in the range of 0.0 to 0.5 mg/L NH_2CL-N.

MONOCHLORAMINE AND FREE AMMONIA NITROGEN (0.00 to 0.50 mg/L NH_2Cl-N)

Method 10045 for drinking water*
Salicylate Method** Using Powder Pillows

1. Enter the stored program number for monochloramine nitrogen (NH_2Cl-N), salicylate method. Press: **3 8 6 ENTER.** The display will show: **Dial nm to 655.**

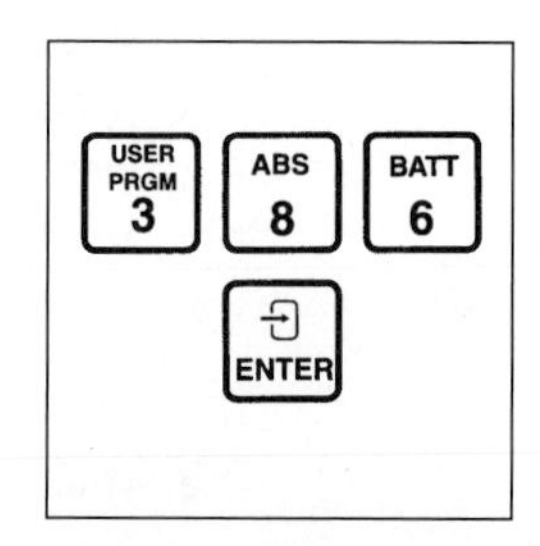

2. Rotate the wavelength dial until the small display shows: **655 nm.** When the correct wavelength is dialed in the display will quickly show: **Zero sample,** then: **mg/L NH_2Cl-N.**

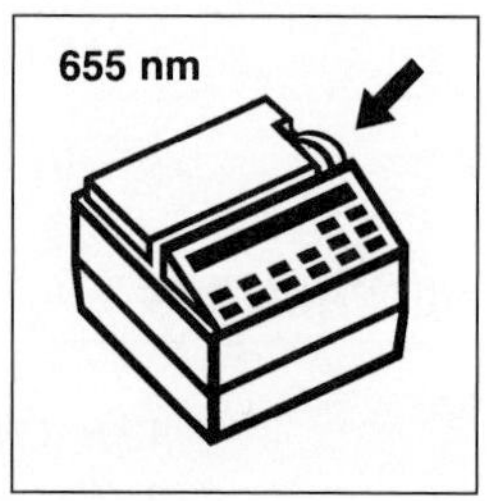

*Compliments of HACH Company. Analysis may be performed by DR/850 and DR/890 Colorimeters and DR/2010 Field Spectrophotometer.

**Adapted from *Clin. Chim. Acta.*, 14 403 (1966).

3. Fill three 10-mL round sample cells to the 10-mL line with sample.

4. Label one cell "blank," one cell "free ammonia" and one cell "monochloramine."

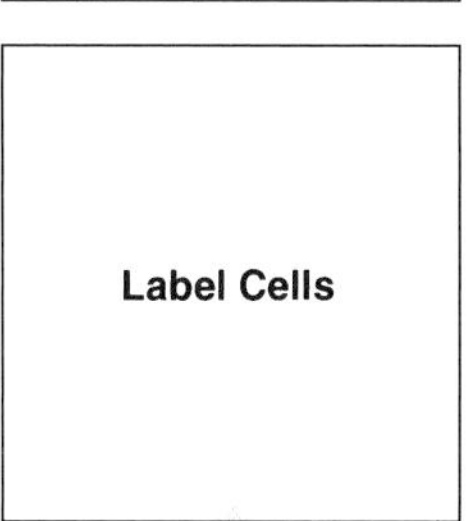

5. Cap the cell labeled "blank." Nothing will be added to it.

6. Add one drop of Hypochlorite Solution to the cell labeled "free ammonia." Cap the cell and mix.
Note: Occasionally shake the Hypochlorite Solution bottle to ensure proper dispensation.

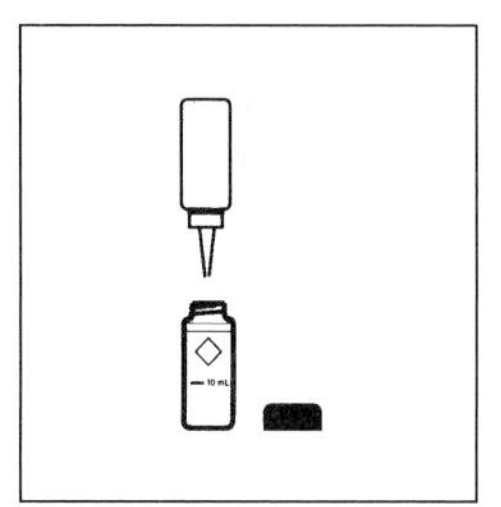

7. Promptly add the contents of one Monochloramine Reagent pillow to the cells labeled "free ammonia" and "monochloramine." Cap and shake both cells to dissolve.
Note: Free ammonia is the ammonia (NH_3) and ammonium NH_4^+) present in the sample, corrected for the monochloramine present.

8. Press: **SHIFT TIMER.** A 15-minute reaction period will begin. Wipe fingerprints, liquid, etc. off of the cells.

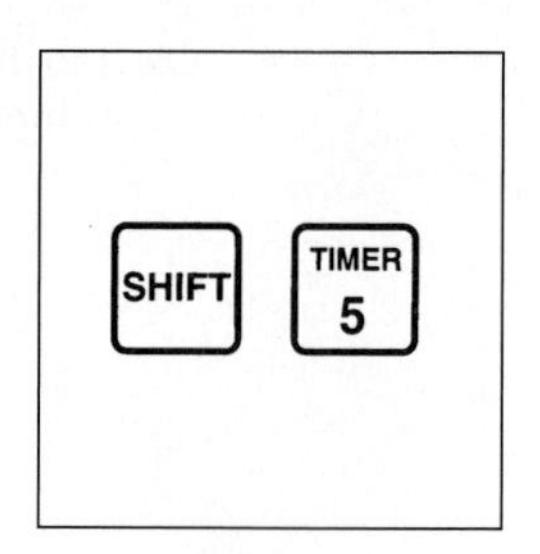

9. Place an AccuVac Vial Adapter into the cell holder. ***Note:*** Place the grip tab at the rear of the cell holder.

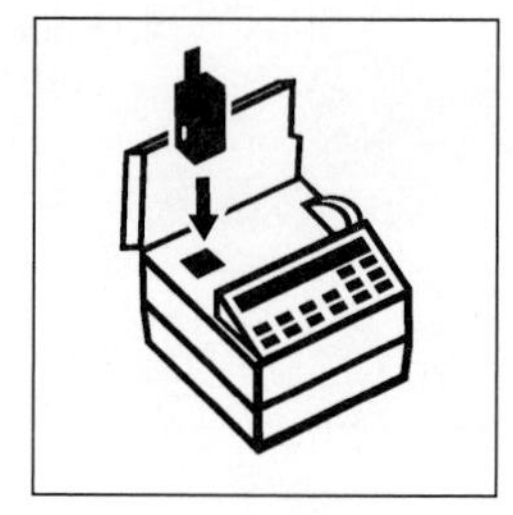

10. When the timer beeps, place the blank into the cell holder. Close the light shield.

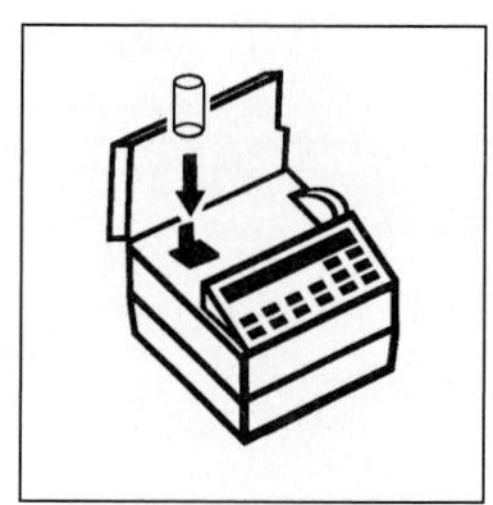

11. Press: **ZERO.** The display will show: **Zeroing ...** then: **0.00 mg/L NH_2Cl-N.**

12. Place the "monochloramine" cell into the cell holder. Close the light shield.

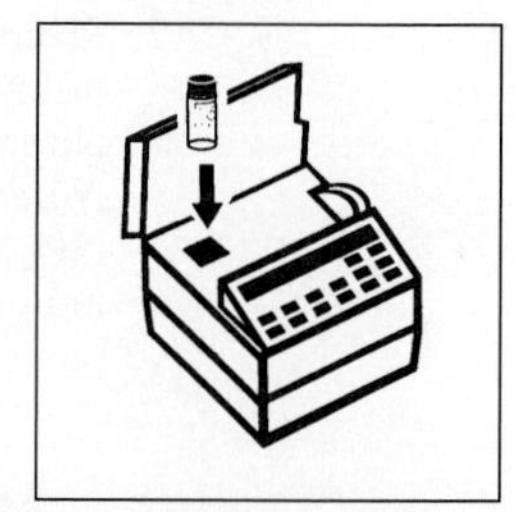

13. Press: **READ.** The display will show: **Reading ...** then the monochloramine result in mg/L as nitrogen will be displayed. Store or record this value.
Note: Do not remove the sample cell.

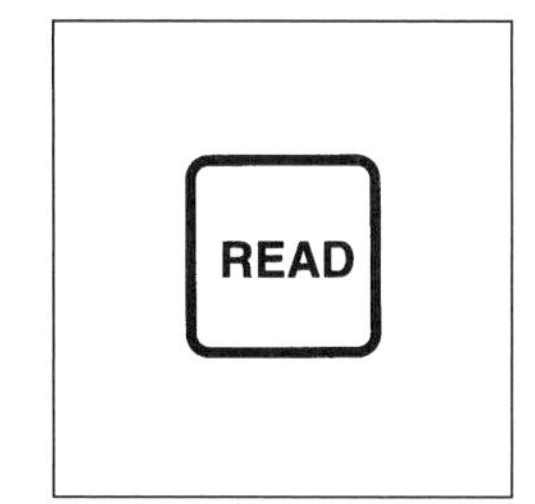

Forms NH_2Cl
NH_2Cl-N
Cl_2

14. Press: **SELECT PRGM.** Enter the stored program number for free ammonia (NH_3-N). Press: **3 8 7 READ.** The display will show: **Zero sample** then: **mg/L NH_3-N Free.**

15. Press: **ZERO.** The display will show: **Zeroing ...** then: **0.00 mg/L NH_3-N Free.**

16. Place the "free ammonia" cell into the cell holder. Close the light shield.

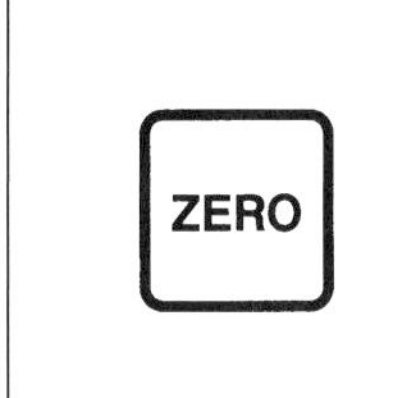

17. Press: **READ.** The display will show: **Reading ...** then the free ammonia result in mg/L as nitrogen will be displayed. ***Note:*** If the free ammonia and monochloramine results total over 0.5 mg/L as nitrogen, repeat the test with a diluted sample for best results. See "Accuracy Check."

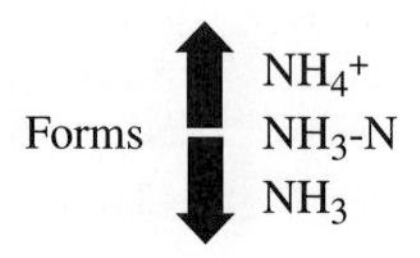

SAMPLING AND STORAGE

Collect samples in clean glass bottles. Most reliable results are obtained when samples are analyzed as soon as possible after collection.

ACCURACY CHECK

See p. 851 of the HACH Water Analysis Handbook, 3rd Ed., 1997.

INTERFERENCES

Ammonia contamination from air is a common cause of high test results. Beakers and other containers may require rinsing with excess sample just before use. Samples, solutions and deionized water will accumulate ammonia from the air.

The following ions may interfere when present at the concentrations listed in Table 2.28 below:

Mixtures of small amounts of differing substances may produce test interference. To assess the influence of interferences in your sample, perform the "Accuracy Check, Standard Additions Method" in HACH Water Analysis Handbook, 3rd Ed., pp. 851–852, 1997.

TABLE 2.28

Calcium	3000 mg/L as $CaCO_3$
Chlorine demand, non-ammonia	greater than 2 mg/L as Cl_2
Magnesium	1600 mg/L as $CaCO_3$
PH	Less than 7
Sulfate	greater than 900 mg/L as SO_4^{2-}

QUESTIONS AND ANSWERS

1. Why is the measure of monochloramine and free ammonia important?
The measure of free ammonia and monochloramine indicates polluted drinking water from sewage, wash wastewater or some other source. Also, monochloramine is very toxic.
2. Why should glassware and containers used to collect and measure samples be rinsed with excess sample before analysis?
Glassware and containers used in the experiment must be rinsed with excess sample to minimize the amount of contamination from ammonia.
3. Should tap water be used in the experiment?
All water including dilution water must be free of chlorine, chlorine demand and ammonia before the final test is completed.

REFERENCE

1. *HACH Water Analysis Handbook,* 3rd Ed. (Loveland, CO.: HACH Company, 1997).

NITRATE

Nitrates are found in many different materials. Nitrate-forming bacteria convert nitrites to nitrates under aerobic conditions. Lightening converts atmospheric nitrogen to nitrates. Nitrates have many practical uses such as catalyst, reagent intermediate, latex coagulant, mordant, oxidizing agent, fertilizer, glass manufacture, curing foods, matches, pyrotechnics and rocket propellants, dynamite, black powders, dyes, pharmaceuticals, preservative, enamel, modifying burning properties of tobacco, hair dying, antiseptic, indelible inks, plating, ceramics, silvering mirrors, insecticide, electroplating, corrosion inhibitor, steel and metal processing, explosives, tanning, acids, and coatings. Nitrates are often found in water effluents from areas of relatively high population density and from effluents from a variety of industries and small business operations.

Recently, the USEPA has established a maximum concentration level in drinking water in accordance with the Safe Drinking Water Act. A number of drinking water sources have exceeded the drinking water standard for nitrate at 10.0 ppm. These wells are located in agriculture areas of the west and midwest and the urban areas of the northeast in the USA.* Nitrate levels in streams have increased and have been attributed to industrial effluents and fertilizer applications on nearby land. Nitrate levels in water promotes plant growth and appears to trigger algae blooms. In excessive amounts, nitrate contributes to an infant illness known as methemoglobinemia known as "blue babies." Health officials believe that high levels of nitrate in water can result in suffocation of infants six months old and younger, although it has not been proven in 1999. Water system owners are required to warn customers when nitrate levels exceed 10 mg/L. Traditionally, nitrates in water have not been treated as a major problem.

A tubular membrane diffusion cell analyzer sold by Timberline Instruments can detect levels of inorganic nitrogen in aqueous solutions at the part per billion (ppb) level. The system is appropriate for analysis of nitrate by the zinc reduction method. This instrument is more sensitive to nitrates in liquids than most methods and instruments.

*U.S. Geological Survey, 1985.

Nitrate (NO_3^-) procedures are applied in limited concentration ranges and often are impacted by interfering substances. In this procedure, cadmium is added to the solution to reduce nitrate to nitrite. The nitrite ion reacts with sulfanilic acid in an acidic solution to form an intermediate diazonium salt. The salt combines with an acid to form an amber-colored product that is measured colorimetrically by the DR/2010 Spectrophotometer or the DR/890 Colorimeter. The method determines nitrates and nitrite nitrogen levels in a sample. At higher levels of nitrate in water, the sample may have to be diluted with nitrate free water for analysis purposes. Alternatively, either high range or low range nitrates in wastewater, seawater, and water can be determined by a cadmium reduction method on the DR/820, DR/850, and DR/890 Colorimeters and the DR/2010 Spectrophotometer. In addition, color developing test strips provide reliable results in the 0 to 50 ppm range for industrial, research and municipal settings.

Some interferences in the experiment are strong oxidizing agents, reducing agents, ferric iron, and high levels of chloride. Nitrite interferences can be diminished by addition of bromine water and phenol solution to the sample.

NITRATE (0 to 4.5 mg/L NO_3^--N)

Method 8171* for water, wastewater and seawater**
Cadmium Reduction Method (Using Powder Pillows)

1. Enter the stored program number for medium range nitrate nitrogen (NO_3^--N). Press: **3 5 3 ENTER.** The display will show: **Dial nm to 400.**
Note: The Pour-Thru Cell can be used if rinsed well with deionized water after use.

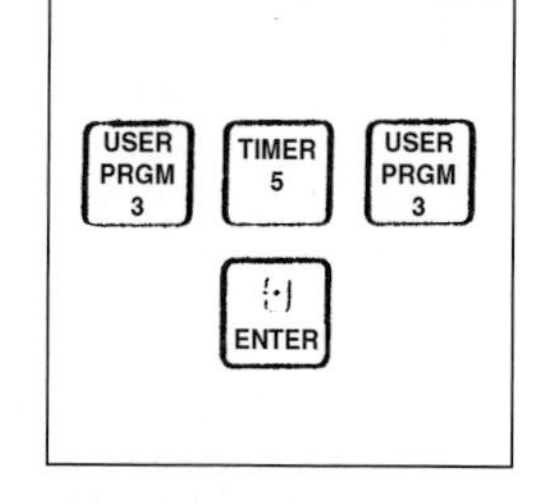

2. Rotate the wavelength dial until the small display shows: **400 nm.** When the correct wavelength is dialed in, the display will quickly show: **Zero sample,** then: **mg/L NO_3^-–N MR (Mid Range).**

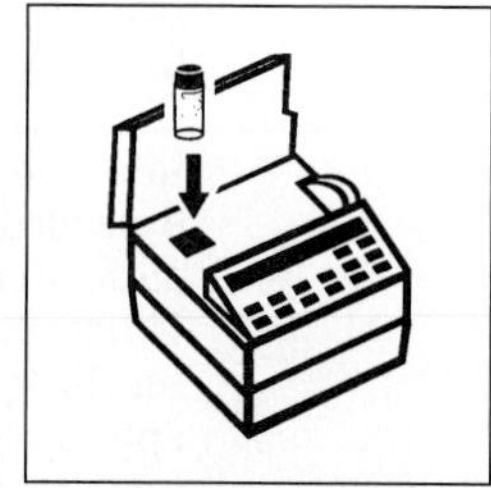

*Compliments of HACH Company. Analysis may be performed by DR/890 Colorimeters and DR/2010 Spectrophotometer.

**Seawater requires a manual calibration; see "Interferences" Section 1.

3. Fill a sample cell with 25 mL of sample (the prepared sample).

4. Fill another cell with 25 mL of deionized water (the blank).

5. Add the contents of one NitraVer 5 Nitrate Reagent Powder Pillow to each cell. Stopper.

6. Press: **SHIFT TIMER.** A one-minute reaction period will begin. Shake until the timer beeps.
Note: Shaking time and technique influence color development. For most accurate results, do successive tests on a standard solution and adjust the shaking time to obtain the correct result.

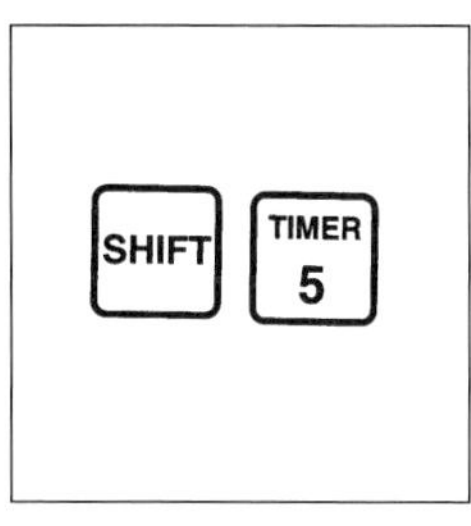

7. When the timer beeps, press: **SHIFT TIMER.** A five-minute reaction period will begin.
Note: A cadmium deposit will remain after the NitraVer 5 Nitrate Reagent Powder dissolves and will not affect test results.
Note: An amber color will develop if nitrate nitrogen is present.

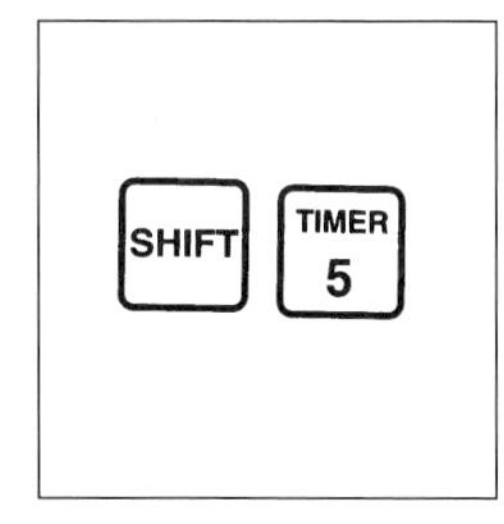

8. When the timer beeps, the display will show: **mg/L NO_3^--N MR.** Remove the stopper. Place the blank into the cell holder. Close the light shield.

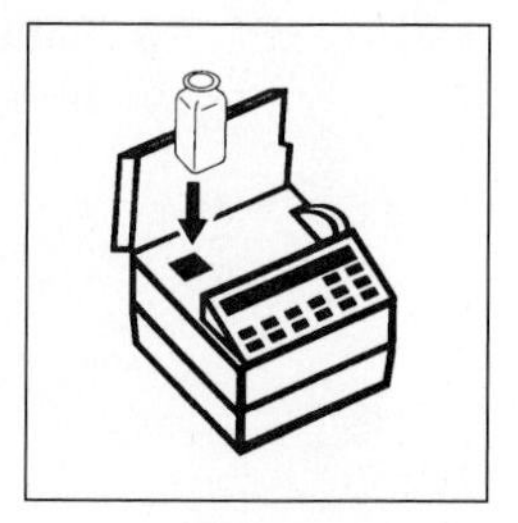

9. Press: **ZERO.** The display will show: **Zeroing ...** then: **0.0 mg/L NO_3^--N MR.**

10. Place the prepared sample into the cell holder. Close the light shield.

11. Press: **READ.** The display will show: **Reading ...** then the result in mg/L nitrate expressed as nitrogen (NO_3^--N) will be displayed.

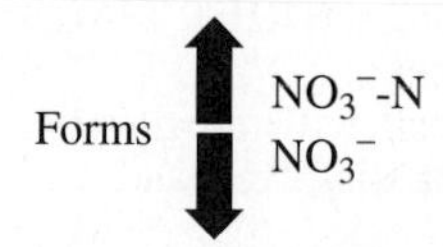

SAMPLING AND STORAGE

Collect samples in clean plastic or glass bottles. Store at 4°C (39°F) or lower if the sample is to be analyzed within 24 to 48 hours. Warm to room temperature before running the test. For longer storage periods, adjust sample pH to 2 or less with sulfuric acid, ACS (about 2 mL per liter). Sample refrigeration is still required.

Before testing the stored sample, warm to room temperature and neutralize with 5.0 N Sodium Hydroxide Standard Solution.

Do not use mercury compounds as preservatives.

Correct the test results for volume additions; see "Correcting for Volume Additions" in Section 1 for more information.

ACCURACY CHECK

See p. 789 of the HACH Water Analysis Handbook, 3rd Ed., 1997.

INTERFERENCES

Compensate for nitrite interference as follows:

a) Add Bromine Water dropwise to the sample in Step 3 until a yellow color remains.

b) Add one drop of Phenol Solution to destroy the color.

c) Proceed with Step 3. Report results as total nitrate and nitrite.

Strong oxidizing and reducing substances will interfere.

Ferric iron causes high results and must be absent.

Chloride concentrations above 100 mg/L will cause low results. The test may be used at high chloride levels and in seawater, but a calibration must be performed using standards spiked to the same chloride concentration. See "User Entered Methods" in the *DR/2010 Instrument Manual* for more information.

Highly buffered samples or extreme sample pH may exceed the buffering capacity of the reagents and require sample pretreatment; see "pH Interference" in Section 1.

QUESTIONS AND ANSWERS

1. What are the interferences in the experiment?
   Interferences are ferric ion, strong oxidizing and reducing agents, and high levels of chloride. Nitrite interferences can be diminished by treatment procedures.
2. How are nitrate procedures limited?
   Nitrate analysis procedures are limited by the range of analysis. Sensitivity is changed for each concentration range of measurements. As the concentration range changes, the wavelength for the measurement must be adjusted.
3. Why is shaking and mixing the sample important after adding the contents of one NitraVer 5 Reagent Powder Pillow?
   Shaking time and technique influence the color development. For good color development, use right to left wrist action to invert the sample and to mix the sample thoroughly.
4. What is the acceptable level of nitrate in drinking water?
   The acceptable drinking water level for nitrates is 10 mg/L. However, a number of water sources have levels that exceed 30 mg/L.

REFERENCES

1. *HACH Water Analysis Handbook,* 3rd Ed. (Loveland, CO.: HACH Company, 1997).
2. *Hawley's Condensed Chemical Dictionary,* 11th Ed., rev'd by N. Irving Sax and Richard J. Lewis (New York, N.Y.: Van Nostrand Reinhold, 1987).
3. *Standard Methods for the Examination of Water and Wastewater:* NITROGEN (NITRATE), 16th Ed., A. E. Greenberg, R. R. Trussell, L. S. Clesceri and Mary Ann H. Franson (eds.), p. 391, 394–396, Washington, DC, American Public Health Association, 1985.

NITRITE

Nitrite anion along with nitrate anion and ammonia is one of the primary nitrogen sources available to plants. Specific types of bacteria convert ammonia under aerobic conditions to nitrites. In surface water exposed to air, nitrites tend to oxidize to nitrates. In waters that are oxygen deficient, the nitrite anion survives until oxygen and the appropriate reaction conditions are present. The presence of large quantities of nitrites indicates partially decomposed organic wastes in the water. Concentrations of nitrite in drinking water normally are below 0.1 mg/L.

Most nitrites are used as corrosion inhibitors, food curing agents, and analysis reagents for other compounds. Additional applications include rubber accelerators, color fixative, pharmaceuticals, photographic reagent, dye manufacture, and antidote for cyanide poisoning.

Strong oxidizing and reducing agents interfere with nitrite in solution. Ferrous and cupric ions cause low results during measurements of nitrite. Several metal ions can interfere by causing precipitation, however addition of a special reagent hinders precipitation of common interfering ions.

A solution with a high nitrate concentration can easily contain small quantities of nitrite in the same solution. Both nitrite and nitrate should be measured nearly at the same time to avoid oxidation-reduction processes. The nitrite analysis using the diazotization method is USEPA approved for reporting wastewater and drinking water analysis. A number of the methods determine nitrite appropriate for water, wastewater and seawater using the DR/2010 Spectrophotometer and the DR/820, DR/850 and DR/890 Colorimeters. A method of test strip testing at 0 to 3 ppm nitrite is readily available for testing of water from different sources.

For nitrite in the range of 0 to 150 mg/L NO_2^-, the Ferrous Sulfate Method (Method 8153) is recommended for water and wastewater. Nitrite is reduced to nitrous oxide by ferrous sulfate in an acidic solution. Nitrous oxide combines with the ferrous ions to form a greenish brown complex in direct proportion to the nitrite present in the solution.

When using the ceric standard solution, nitrite at much higher levels (100 to 2500 mg/L) can be determined on samples obtained from the market place and in industry. In this procedure, sodium nitrite can be titrated with tetravalent cerium ion in the presence of ferroin indicator. The cerium ion, a strong oxidant, causes a color change of the solution from orange to pale blue. The concentration of sodium nitrite is proportional to the amount of titrant used.

NITRITE, Low Range (0 to 0.300 mg/L NO_2^--N)

Method 8507 for water, wastewater and seawater*
Diazotization Method (Using Powder Pillows)
USEPA approved for reporting wastewater and drinking water analysis**

1. Enter the stored program number for low range nitrite nitrogen (NO_2^--N) powder pillows. Press: **3 7 1 ENTER.** The display will show: **Dial nm to 507.**
Note: The Pour-Thru Cell can be used with 25-mL reagents only.

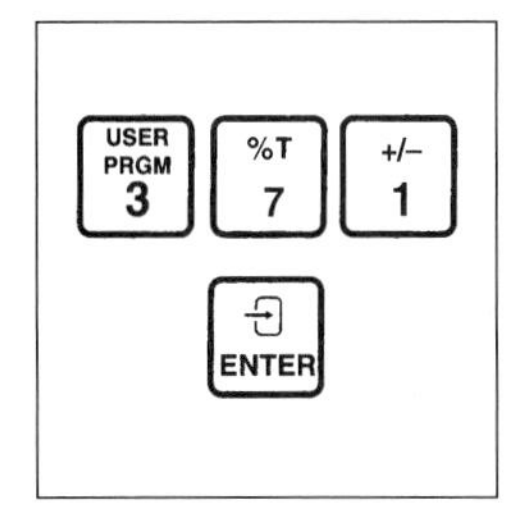

2. Rotate the wavelength dial until the small display shows: **507 nm.** When the correct wavelength is dialed in, the display will quickly show: **Zero sample,** then: **mg/L NO_2^--N LR (Low Range).**

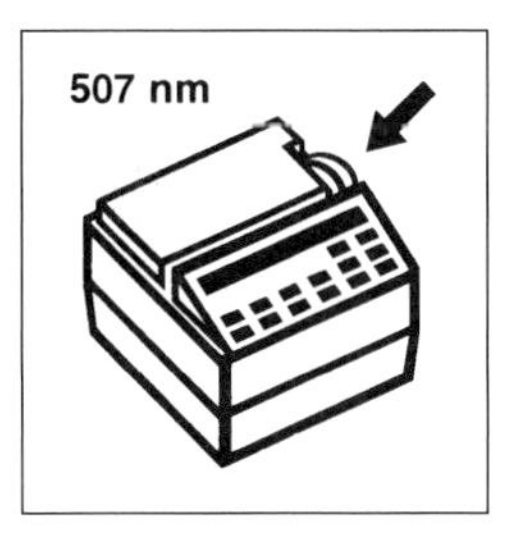

3. Insert the 10-mL Cell Riser into the cell compartment.

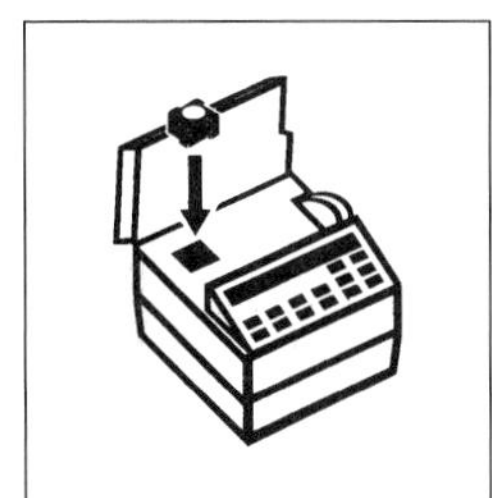

4. Fill a 10-mL sample cell with 10 mL of sample.

*Compliments of HACH Company. Analysis may be performed by DR/820, DR/850, and DR/890 Colorimeters and DR-2010 Spectrophotometer.

**Federal Register, 44(85) 25505 (May 1, 1979).

5. Add the contents of one NitriVer 3 Nitrite Reagent Powder Pillow (the prepared sample). Stopper. Shake the cell to dissolve the powder.
Note: A pink color will develop if nitrite nitrogen is present.

6. Press: **SHIFT TIMER.** A 20-minute reaction period will begin.

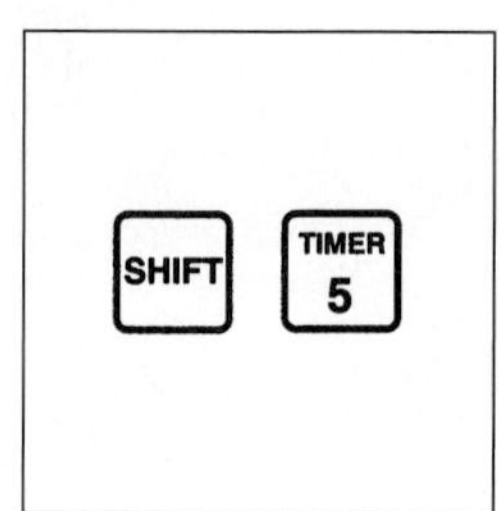

7. When the timer beeps, the display will show: **mg/L NO_2^--N LR.** Fill a second 10-mL sample cell with 10 mL of sample (the blank).

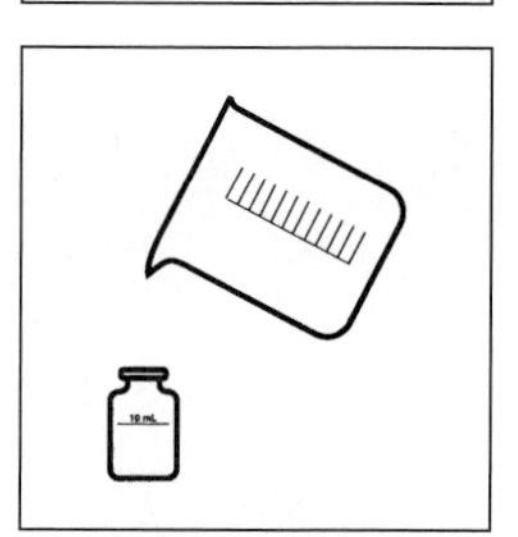

8. Place the blank into the cell holder. Close the light shield.

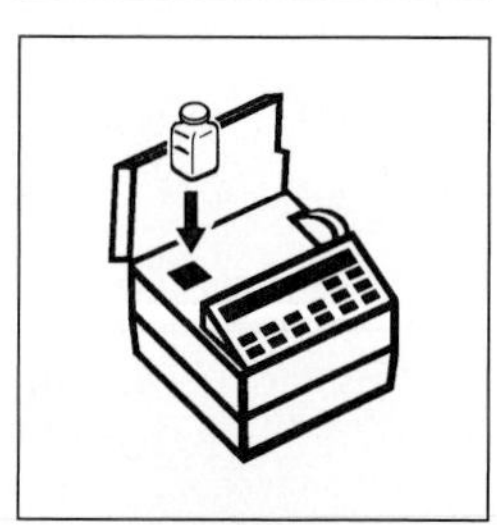

9. Press: **ZERO.** The display will show: **Zeroing ...** then: **0.000 mg/L NO_2^-N LR.**

10. Remove the stopper from the prepared sample. Place the cell into the cell holder. Close the light shield.

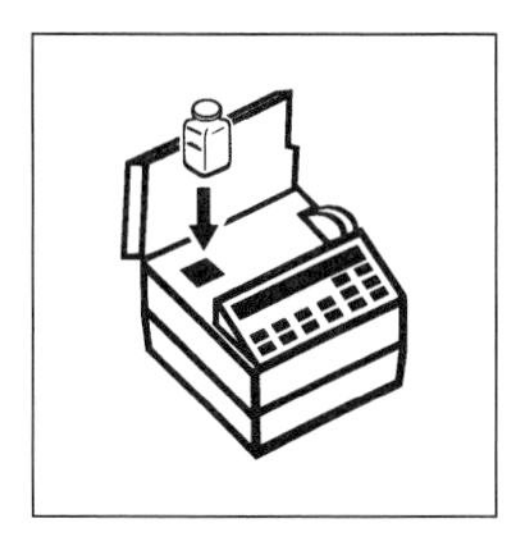

11. Press: **READ.** The display will show: **Reading ...** then the result in mg/L nitrite expressed as nitrogen (NO_2^--N) will be displayed.

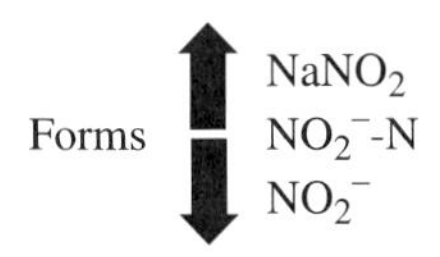

SAMPLING AND STORAGE

Collect samples in clean plastic or glass bottles. Store at 4°C (39°F) or lower if the sample is to be analyzed within 48 hours. Warm to room temperature before running the test.

ACCURACY CHECK

See p. 810 of the HACH Water Analysis Handbook, 3rd Ed., 1997.

INTERFERENCES

- Strong oxidizing and reducing substances interfere.
- Cupric and ferric ions cause low results
- Ferric, mercurous, silver, bismuth, antimonous, lead, auric, chloroplatinate and metavanadate ions interfere by causing precipitation.
- Very high levels of nitrate (100 mg/L nitrate as N or more) appear to undergo a slight amount of reduction to nitrite, either spontaneously or during the course of the test. A small amount of nitrite will be found at these levels.

NITRITE, (Medium to High Range) (0 to 150 mg/L NO_2^-)

Method 8153 for water and wastewater*
Ferrous Sulfate Method**

1. Enter the stored program number for high range nitrite (NO_2^-). Press: **3 7 3 ENTER.** The display will show: **Dial nm to 585.**
Note: The Pour-Thru Cell cannot be used.

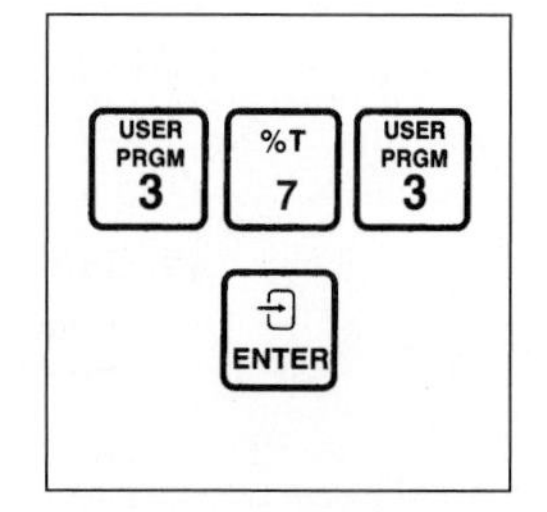

2. Rotate the wavelength dial until display shows: **585 nm**. When the correct wavelength is dialed in, the display will quickly show: **Zero Sample,** then: **mg/L NO_2^- HR.**

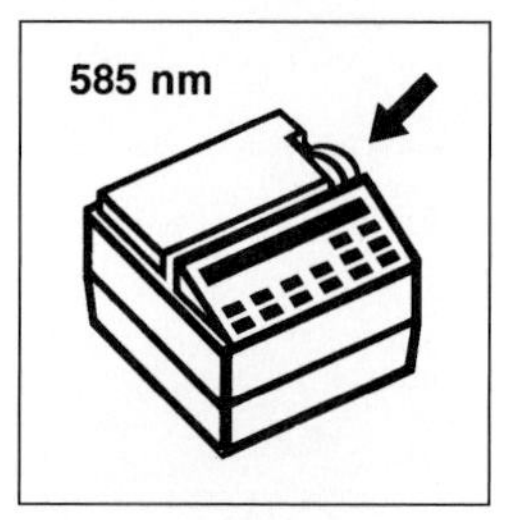

3. Insert the 10-mL Cell Riser into the cell compartment.

4. Fill a sample cell with 10 mL of sample.

*Compliments of HACH Company. Analysis may be performed by DR/890 Colorimeters and DR-2010 Spectrophotometer.

**Adapted from R. McAlpine, and B. Soule, *Qualitative Chemical Analysis,* New York, 476, 575 (1933).

5. Add the contents of one NitriVer 2 Nitrite Reagent Powder Pillow, stopper and shake to dissolve (the prepared sample).
Note: A greenish brown color will develop if nitrite is present.

6. Press: **SHIFT TIMER.** A 10 minute reaction period will begin. It is critical to leave the sample undisturbed on a flat surface for the reaction period or low results may occur.

7. Fill another sample cell with 10 mL of sample (the blank). Place it into the cell holder.

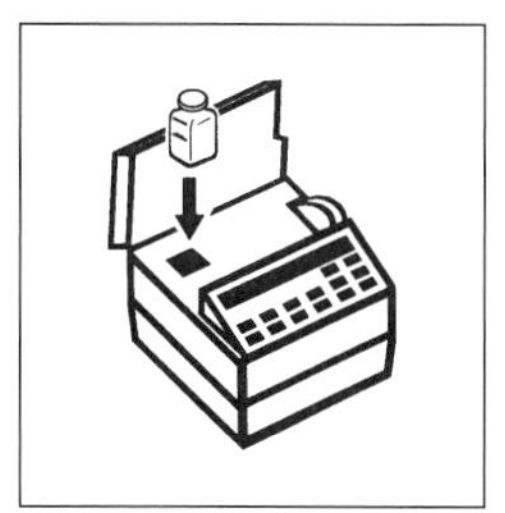

8. When the timer beeps the display will show: **mg/L NO_2^- HR.** Press: **ZERO.** The display will show: **Zeroing ...** then: **0. mg/L NO_2^- HR.**

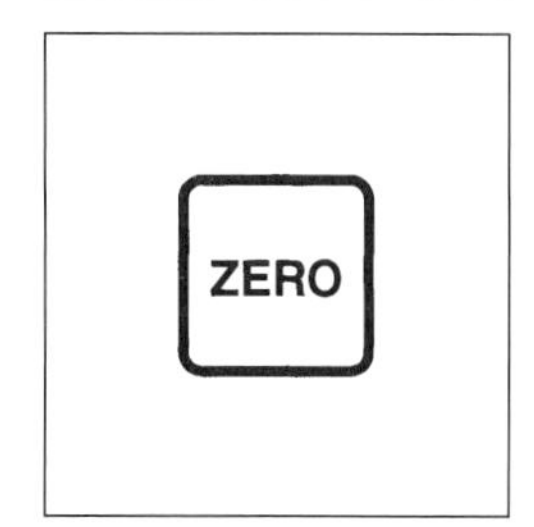

9. Gently invert the prepared sample twice. Remove the stopper. Place the prepared sample into the cell holder. Close the light shield.
Note: Avoid excessive mixing or low results may occur.

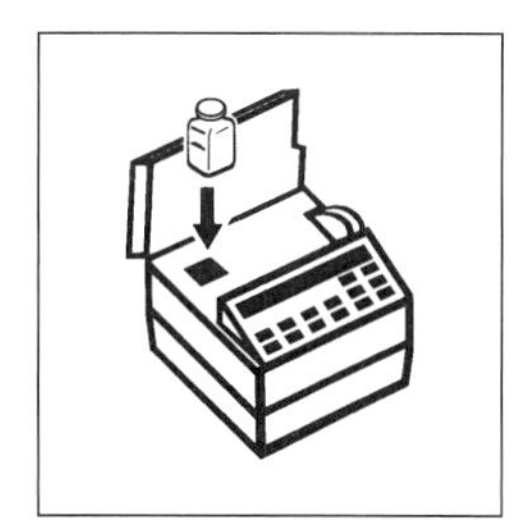

10. Press: **READ**. The display will show: **Reading ...** then the result in mg/L NO_2^- will be displayed.

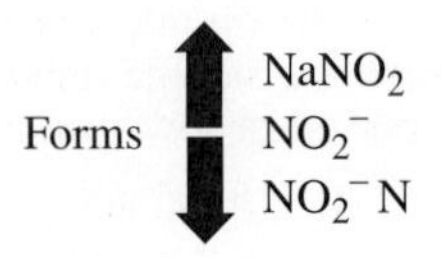

SAMPLING AND STORAGE

Collect samples in clean plastic or glass bottles.

The following storage instructions are necessary only when prompt analysis is impossible. Store at 4°C (39°F) or lower if the sample is to be analyzed within 48 hours. Warm to room temperature before analysis.

ACCURACY CHECK

See HACH Water Analysis Handbook, 3rd Ed., 1997, p. 805.

INTERFERENCES

This test does not measure nitrates nor is it applicable to glycol based samples. Dilute glycol based samples and follow the Nitrite, Low Range Procedure (Stored Program No. 371).

NITRITE (100-2500 mg/L $NaNO_2$)

Method 8351 Using Ceric Standard Solution*

1. Select the sample volume from Table 2.28a which corresponds to the expected sample sodium nitrite concentration (as $NaNO_2$).

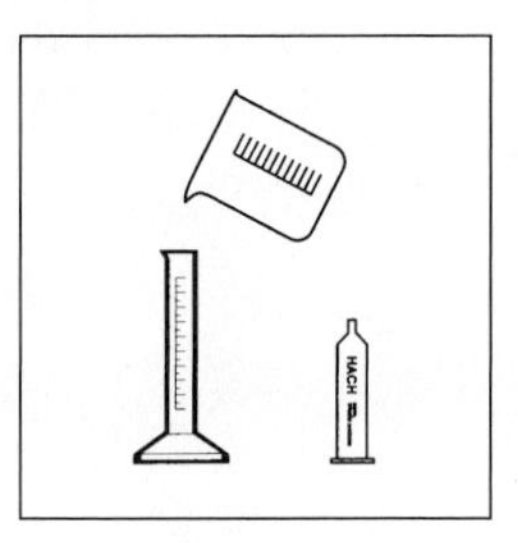

*Compliments of HACH Company.

TABLE 2.28a

Expected Sample Concentration (as $NaNO_2$)	Sample Volume (mL)	Digit Multiplier
100–400	25	0.86
400–800	10	2.15
800–1500	5	4.31
1500–2500	2	10.78

2. Insert a clean delivery tube into the Ceric Standard Solution Titration Cartridge. Attach the cartridge to the titrator body.

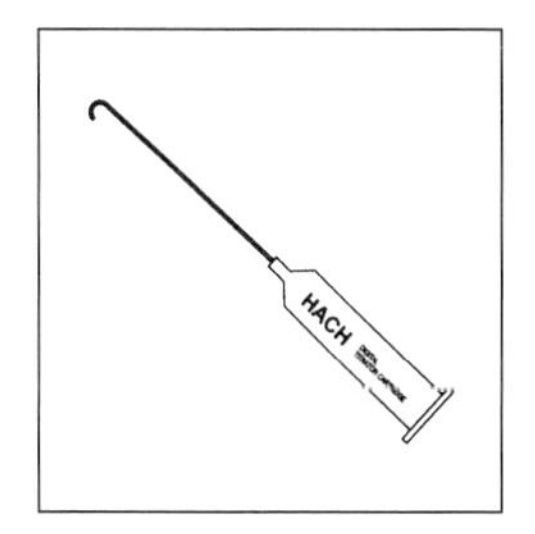

3. Hold the Digital Titrator with the cartridge tip pointing up. Turn the delivery knob until a few drops of titrant are expelled. Reset the counter to zero and wipe the tip.
Note: For added convenience use the TitraStir stirring apparatus.

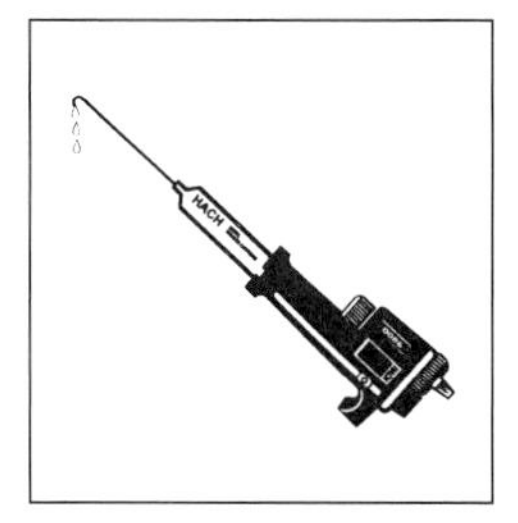

4. Use a graduated cylinder or pipet to measure the sample volume from Table 2.26. Transfer the sample to a clean 125-mL erlenmeyer flask. Add deionized water to about the 75-mL mark, if necessary.
Note: A pipet is recommended for sample volumes less than 10 mL.

5. Add 5 drops of 5.25 N Sulfuric Acid Standard Solution to the flask. Swirl to mix.

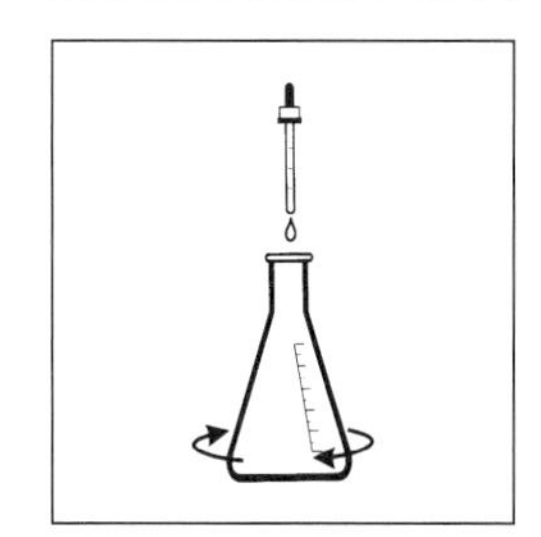

6. Add 1 drop of Ferroin Indicator Solution to the flask. Swirl to mix.

7. Place the delivery tip into the solution. While titrating with Ceric Standard Solution, swirl the flask until the solution color changes from orange to pale blue. Record the number of digits required.

8. Calculate: Digits Required × Digit Multiplier = mg/L Sodium Nitrite ($NaNO_2$)
Note: See Standardization of Ceric Solution to verify the normality.

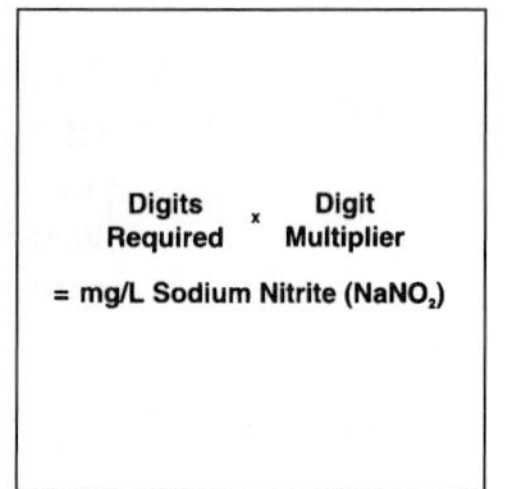

STANDARDIZATION OF CERIC SOLUTION

The normality of the Ceric Solution will sometimes decrease over time. Before use, verify the normality with the following procedure. This standardization should be done monthly.

1. Use a graduated cylinder or pipet to measure 50 mL of deionized water into a 125-mL erlenmeyer flask.
2. Add 5 mL of 19.2 N Sulfuric Acid Standard Solution. Swirl to mix.
3. Insert a clean delivery tube into a Ceric Standard Titration Cartridge.
4. Hold the Digital Titrator with the cartridge tip pointing up. Turn the delivery knob until a few drops of titrant are expelled. Reset the counter to zero and wipe the tip.
5. Place the delivery tube tip into the solution. While swirling the flask, add 200 digits of Ceric Standard.
6. Insert a clean delivery tube into a 0.200 N Sodium Thiosulfate Titration Cartridge.
7. Hold the Digital Titrator with the cartridge tip pointing up. Turn the delivery knob until a few drops of titrant are expelled. Reset the counter to zero and wipe the tip.
8. Place the delivery tube tip into the solution. While swirling the flask, titrate with the sodium thiosulfate from an intense yellow color to a faint yellow color. Record the number of digits required. This step should require about 400 to 450 digits of titrant.

9. Add one drop of Ferroin Indicator Solution. Swirl to mix. The solution will turn a faint blue.
10. Continue titrating with the Ceric Standard Solution from a faint blue to orange color. Record the number of digits required.
11. Calculate the correction factor:

$$\text{Correction factor} = \frac{\text{Digits Required}}{500}$$

12. Multiply the mg/L sodium nitrite from Step 8 of the nitrite titration procedure by the correction factor to obtain the correct sodium nitrite concentration.

SAMPLING AND STORAGE

Collect samples in clean plastic or glass bottles. Prompt analysis is recommended. If prompt analysis is impossible, store samples for 24 to 48 hours at 4°C (39°F) or lower. Warm to room temperature before analysis. Do not use acid preservatives.

ACCURACY CHECK

See p. 823 of the HACH Water Analysis Handbook, 3rd Ed., 1997.

QUESTIONS AND ANSWERS

1. How stable is nitrite compared to nitrate?
   Nitrites tend to oxidize to nitrates in air and in surface waters.
2. Which metal ions interfere during the test for nitrite?
   Precipitation interferes with the test and is caused by the following ions: ferric, mercurous, silver, bismuth, antimonous, lead, auric, chloroplatinate, and metavanadate. Ferrous and cupric ions can cause low results.
3. Why should nitrites be measured at the same time as nitrates?
   Nitrites and nitrates should be determined close together to avoid oxidation of nitrites to nitrates in water.
4. How are the three methods of nitrite applied?
   The Diazotization Method (Method 8507) is used primarily for measurements of a low level of nitrite (0 to 0.300mg/L) in water, wastewater and seawater. This method is USEPA approved for reporting wastewater analysis. The Ferrous Sulfate Method analyzes samples in the middle range (0 to 150mg/L) of nitrite and is primarily used for water and wastewater analysis. The Ceric Standard Solution Method (Method 8351) analyzes higher concentrations of nitrite (100 to 2500 mg/L $NaNO_2$) found in commercial applications and waste streams.

REFERENCES

1. *HACH Water Analysis Handbook,* 3rd Ed. (Loveland, CO.: HACH Company, 1997).
2. *Hawley's Condensed Chemical Dictionary,* 11th Ed., rev'd by N. Irving Sax and Richard J. Lewis (New York, N.Y.: Van Nostrand Reinhold, 1987).

3. *Standard Methods for the Examination of Water and Wastewater:* NITROGEN (NITRITE), 16th Ed., A. E. Greenberg, R. R. Trussell, L. S. Clesceri and Mary Ann H. Franson (eds.) p. 404, Washington, DC, American Public Health Association, 1985).

NITROGEN (AMMONIA)

Ammonia is produced by microbiological decay of animal and plant protein. In groundwater, it is normally attributed to microbiological processes. Ammonia and ammonium compounds in fertilizer and the applications of cleansing agents and useful solvents are sources of ammonia nitrogen. Domestic pollution in surface water usually indicates the presence of ammonia nitrogen.

In surface water, ammonia is present at a low concentration because it adsorbs to clays and soil particles and is not leached from soils easily. Ammonia concentrations in natural surface water and ground water is usually less than 10 micrograms of ammonia nitrogen per liter. The concentration of ammonia in wastewater is often greater than 30 milligrams per liter (mg/L).

In wastewater treatment operations, ammonia and chlorine react to form mono- and dichloramines which are quite toxic. These chemicals are usually diluted by the waters of the receiving stream or river. If chlorine is present in the sample, a small amount of 0.1 N sodium thiosulfate or sodium arsenite will eliminate the interference.

The Nessler Method is the most common procedure of determining ammonia nitrogen in wastewater, seawater and many water samples. The method requires distillation of wastewater and seawater samples to remove substances that cause excessive hardness, color and turbidity. Less common interferences of organic chemicals may cause greenish colors and turbidity. In this experiment, it is necessary to use ammonia free water which can be prepared by ion-exchange and distillation methods. Normally, deionized water is appropriate after being checked for nitrogen (ammonia). Deionized water will have to be distilled if it contains nitrogen (ammonia).

The Mineral Stabilizer complexes hardness attributed to calcium and magnesium in the sample. The Polyvinyl Alcohol Dispersing Agent aids the development of the color during the reaction of Nessler Reagent with ammonium ions. The yellow color of the solution at 425 nm is proportional to the ammonia concentration. The Nessler Method is performed on the DR/2010 Spectrophotometer.

The Salicylate Method for determination of ammonia nitrogen in water, wastewater and seawater is appropriate for field studies, however the methods are not USEPA approved. The DR/850 and DR/890 Colorimeters and the DR/2010 Spectrophotometer can be used to determine both low range and high range ammonia nitrogen. For reliable and quick results, test strips can provide the NH_3-N results in the 0 to 6.0 ppm range.

AMMONIA NITROGEN (0 to 2.50 mg/L NH_3-N)

Method 8038* for water, wastewater** and seawater**
Nessler Method**
USEPA Accepted for wastewater analysis. Distillation is required.

1. Enter the stored program number for ammonia nitrogen (NH_3-N).
Press: **3 8 0 ENTER.** The display will show: **Dial nm to 425.**
Note: Adjust the pH of stored samples before analysis.
Note: The Pour-Thru Cell can be used with this procedure. Clean the cell by pouring a few sodium thiosulfate pentahydrate crystals into the cell funnel. Flush it through the funnel and cell with enough deionized water to dissolve. Rinse out the crystals.

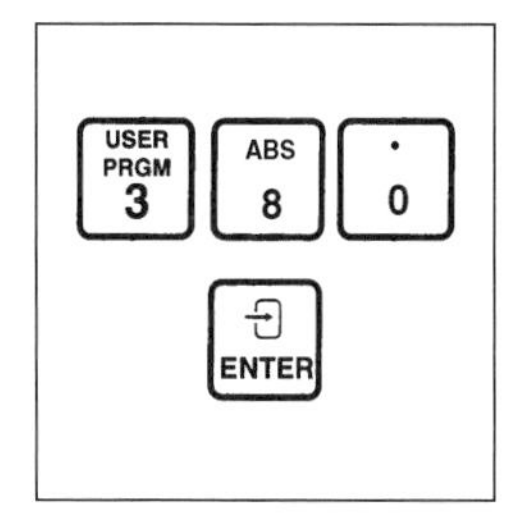

2. Rotate the wavelength dial until the small display shows: **425 nm.** When the correct wavelength is dialed in the display will quickly show: **Zero Sample,** then: **mg/L NH_3-N Ness.**
Note: This test is sensitive to the wavelength setting. To assure accuracy, run the test using a 1.0 mg/L standard solution and deionized water blank. Repeat Steps 10 to 12 at slightly different wavelengths, setting the dial from higher to lower values, until the correct result is obtained. The wavelength should be 425 ± 2 nm. Always set this wavelength by approaching from high to low values.

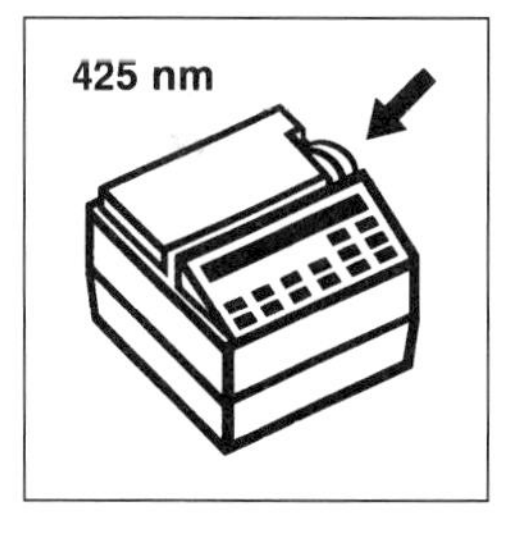

3. Fill a 25-mL mixing graduated cylinder (the prepared sample) to the 25-mL mark with sample.

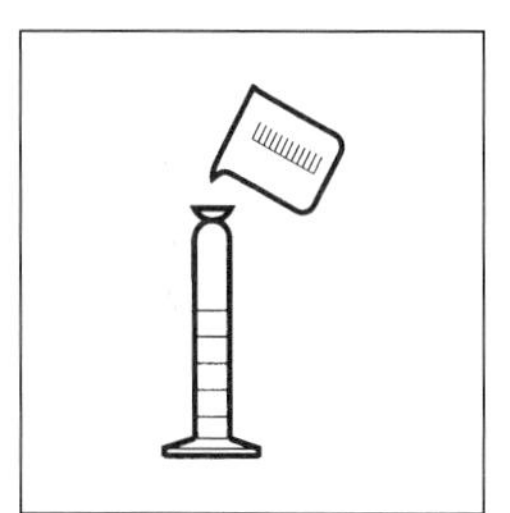

4. Fill another 25-mL mixing graduated cylinder (the blank) with deionized water.

*Compliments of HACH Company.

**Requires distillation: Adapted from *Standard Methods for the Examination of Water and Wastewater. 417 A and 417 B.*

5. Add 3 drops of Mineral Stabilizer to each cylinder. Invert several times to mix. Add 3 drops of Polyvinyl Alcohol Dispersing Agent to each cylinder. Invert several times to mix.

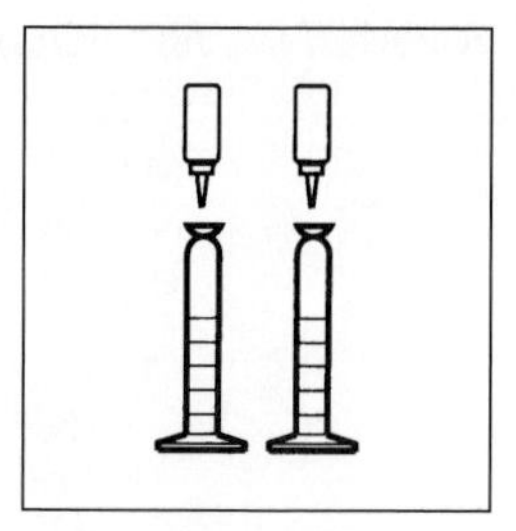

6. Pipet 1.0 mL of Nessler Reagent into each cylinder. Stopper. Invert several times to mix.
Note: A yellow color will develop if ammonia is present. (The reagent will cause a faint yellow color in the blank.)

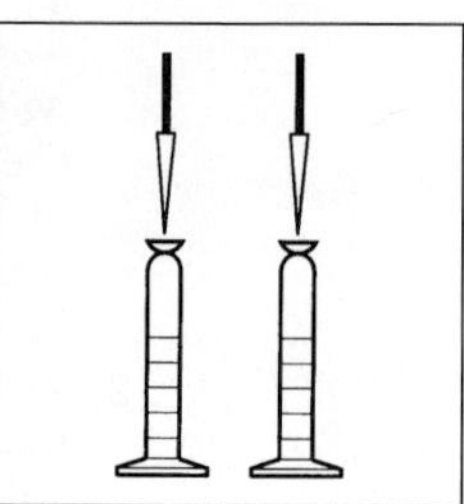

7. Press: **SHIFT TIMER.** A one-minute reaction period will begin.
Note: Continue with Step 8 while timer is running.

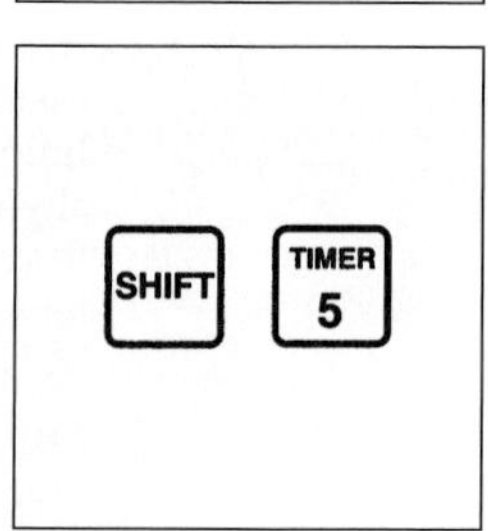

8. Pour each solution into a sample cell.

9. When the timer beeps, the display will show: **mg/L NH_3-N Ness.** Place the blank into the cell holder. Close the light shield.

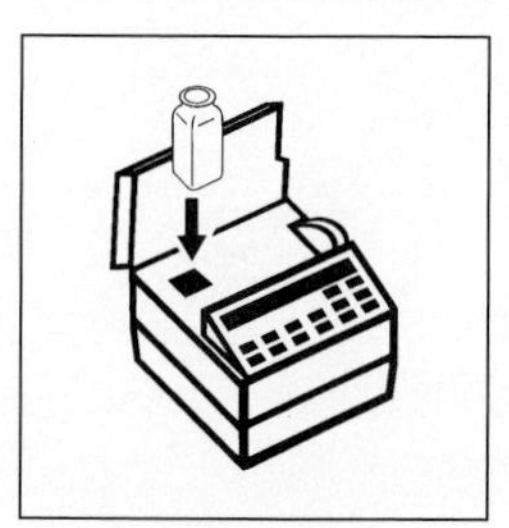

10. Press: **ZERO.** The display will show: **Zeroing ...** then: **0.00 mg/L NH_3-N Ness.**

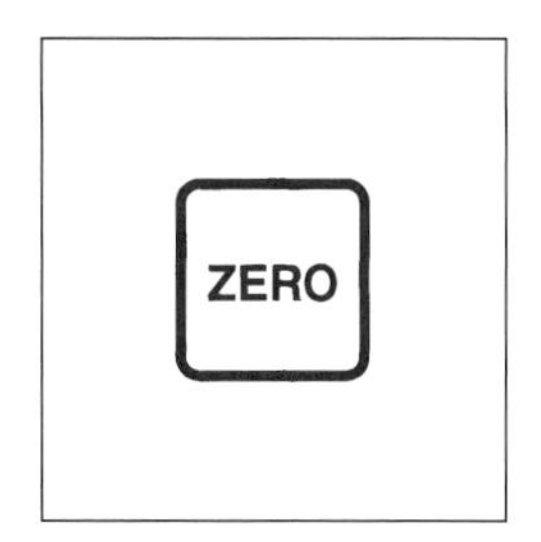

11. Place the prepared sample into the cell holder. Close the light shield.
Note: Do not wait more than five minutes after reagent addition (Step 6) before performing Step 12.

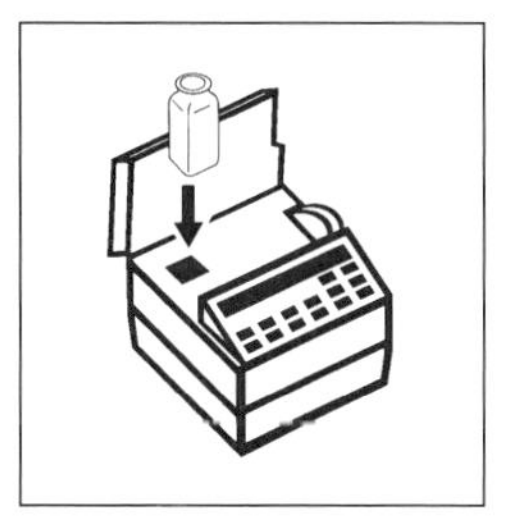

12. Press: **READ.** The display will show: **Reading ...** then the result in mg/L ammonia expressed as nitrogen (NH_3–N) will be displayed.

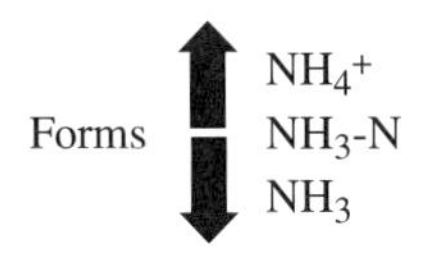

SAMPLING AND STORAGE

Collect samples in clean glass or plastic bottles.

If chlorine is present, add one drop of 0.1 N sodium thiosulfate for each 0.3 mg/L Cl_2 in a 1-liter sample. Preserve the sample by reducing the pH to 2 or less with sulfuric acid (at least 2 mL). Store at 4°C (39°F) or less. Preserved samples may be stored up to 28 days. Before analysis, warm samples to room temperature and neutralize with 5 N sodium hydroxide. Correct the test result for volume additions: see *"Correcting for Volume Additions"* in Section 1.

ACCURACY CHECK

See page 827 of the HACH Water Analysis Handbook, 3rd Ed., 1997.

INTERFERENCES

A solution containing a mixture of 500 mg/L as $CaCO_3$ and 500 mg/L Mg as $CaCO_3$ does not interfere. If the hardness concentration exceeds these concentrations, extra Mineral Stabilizer should be added.

Iron and sulfide interfere by causing a turbidity with Nessler Reagent.

Residual chlorine must be removed by addition of sodium arsenite solution. Use two drops to remove each mg/L Cl from a 250-mL sample. Sodium thiosultate can be used in place of sodium arsenite. See "Sample Collection, Preservation and Storage" in Section 1.

Less common interferences, such as glycine, various aliphatic and aromatic amines, organic chloramines, acetone, aldehydes and alcohols may cause greenish or other off colors or turbidity. It may be necessary to distill the sample if these compounds are present.

Seawater samples may be analyzed by addition of 1.0 mL (27 drops) of Mineral Stabilizer to the sample before analysis. This will complex the high magnesium concentrations found in sea water, but the sensitivity of the test will be reduced by 30% due to the high chloride concentration. For best results, perform a calibration, using standards spiked to the equivalent chloride concentration, or distill the sample as described below.

DISTILLATION

a) Measure 250 mL of sample into a 250-mL graduated cylinder and pour into a 400-mL beaker. If necessary, destroy chlorine by adding 2 drops of sodium arsenite solution per mg/L Cl_2.

b) Add 25 mL of borate buffer solution and mix. Adjust the pH to about 9.5 with 1 N sodium hydroxide solution. Use a pH meter.

c) Set up the general purpose distillation apparatus. Pour the solution into the distillation flask. Add a stir bar.

d) Use a graduated cylinder to measure 25 mL of deionized water into a 250-mL erlenmeyer flask. Add the contents of one boric acid powder pillow. Mix thoroughly. Place the flask under the still drip tube. Elevate so the end of the tube is immersed in the solution.

e) Turn on the heater power switch. Set the stir control to 5 and the heat control to 10. Turn on the water and adjust to maintain a constant flow through the condenser.

f) Turn off the heater after collecting 150 mL of distillate. Immediately remove the collection flask to avoid sucking solution into the still. Measure the distillate to ensure 150 mL was collected (total volume 175 mL).

g) Adjust the pH of the distillate to about 7 with 1 N sodium hydroxide. Pour the distillate into a 250-mL volumetric flask; rinse the erlenmeyer flask with deionized water. Add the rinsings to the volumetric flask. Dilute to the mark. Stopper. Mix thoroughly. Analyze as described above.

QUESTIONS AND ANSWERS

1. Why is distillation required for wastewater and seawater?
Wastewater and seawater contain substances such as chloride, magnesium and calcium that interfere with the test. Water samples contaminated with organic chemicals will develop off-colors (non-yellow) and turbidity.

2. How is residual chlorine removed from the water sample?
Residual chlorine can be removed by the addition of sodium thiosulate or sodium arsenite.

3. If ammonia is present in the sample, what color develops in the presence of the Nessler Reagent?
A yellow color develops when the sample is inverted several times and ammonia is present.

4. Can deionized water be used as ammonia-free water?
Properly deionized water should not contain ammonia. Ammonia-free water will depend on the type of unit making the deionized water. Distillation or ion exchange methods may have to be implemented.

REFERENCES

1. *HACH Water Analysis Handbook,* 3rd Ed. (Loveland, CO.: HACH Company, 1997).
2. *Hawley's Condensed Chemical Dictionary,* 11th Ed., rev'd by N. Irving Sax and Richard J. Lewis (New York, N.Y.: Van Nostrand Reinhold, 1987).
3. *Standard Methods for the Examination of Water and Wastewater:* NITROGEN (AMMONIA), 16th Ed., A. E. Greenberg, R. R. Trussell, L. S. Clesceri and Mary Ann H. Franson (eds.), pp. 374–382, Washington, DC, American Public Health Association (1985).

NITROGEN, TOTAL KJELDAHL

The Total Kjeldahl Method determines the amount of ammonia and organic nitrogen in a sample. The ammonia and organic nitrogen is attributed to decomposition of organic matter. This method determines the amount of nitrogen in the trinegative state. The method fails to account for nitrogen in the form of nitrate, nitrite, nitrile, nitro, nitroso, azide, azine, azo, hydrazone, oxime, and semi-carbazone.

An analysis for ammonia and organic nitrogen (Total Kjeldahl Nitrogen) requires a digestion procedure. The digestion process oxidizes carbon compounds to carbon dioxide and converts organic forms of nitrogen (amino acids, proteins, peptides, etc.) to ammonia. The digestion procedure can be performed on sludge, oils, fats, and dry samples such as soil, plants, nuts and other products containing organic nitrogen. Samples of dried materials often require chopping or grinding into small pieces before the digestion process.

A known volume of acid is added to a specific weight (usually 0.100 to 0.500 grams) of a solid, gel, sludge, or semisolid sample in a digestion flask. For determination of organic nitrogen and ammonia in oils, 0.10 to 0.25 grams of material are usually required for dilution to a specific volume. After boiling for four minutes, 10 mL of 50% hydrogen peroxide is added to the charred sample and the sample is heated for 1 minute. The digested mixture is allowed to cool and is diluted to 100 mL with deionized water. After selecting the appropriate sample size and diluting the sample to 25.0 mL or a volume satisfactory for analysis, the clear sample can be analyzed for Total Kjeldahl Nitrogen (Nessler Method) by the procedure (HACH Method 8075) provided in Chapter 2. A pH adjustment on the sample will be required before analysis for total nitrogen.

Organically-bound nitrogen reacting with hydrogen peroxide and sulfuric acid is converted into ammonium salts. A Mineral Stabilizer is added to the sample to complex calcium and magnesium. Polyvinyl Alcohol Dispersing Agent helps to develop the color in the reaction of Nessler Reagent with ammonium ion. In the Modified Nessler Method Test, a yellow color proportional to the ammonia concentration in solution is measured at 460 nm by a spectrometer.

NITROGEN, TOTAL KJELDAHL (0 to 150 mg/L)

Method 8075 for water, wastewater, and sludge*
Nessler Method** Digestion required

1. A User-Entered Calibration is necessary to obtain the most accurate results. See the User Calibration section at the back of this procedure. Program 399 can be used directly for process control or applications where a high degree of accuracy is needed.
Note: Sensitivity to wavelength setting and reagent lot variation necessitate user calibration for best results.

Choose Desired Program Accuracy

2. Enter the stored program number for total Kjeldahl nitrogen. Press: **3 9 9 ENTER** for the factory stored program or **9 ? ? ENTER** for the user stored program (see "User Calibration"). The display will show: **Dial nm to 460.**
Note: The Pour-Thru Cell can be used. Periodically clean the cell by pouring a few sodium thiosulfate pentahydrate crystals into the cell funnel. Flush it through the funnel and cell with enough deionized water to dissolve. Rinse out the crystals.

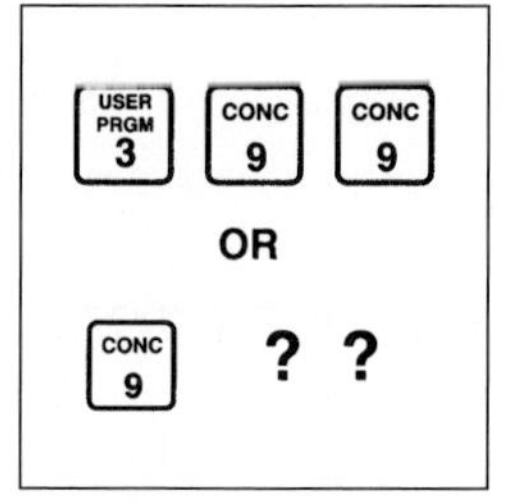

3. Rotate the wavelength dial until the small display shows: 460 nm. When the correct wavelength is dialed in, the display will quickly show: **Zero Sample,** then: **mg/L TKN.**
Note: Always set the wavelength by approaching from high to low values. For even greater accuracy, run the accuracy check and a deionized water blank. If the correct result is not obtained, repeat the accuracy check at slightly different wavelengths, again setting the dial from higher to lower values. The wavelength should be 460 ± 2 nm.

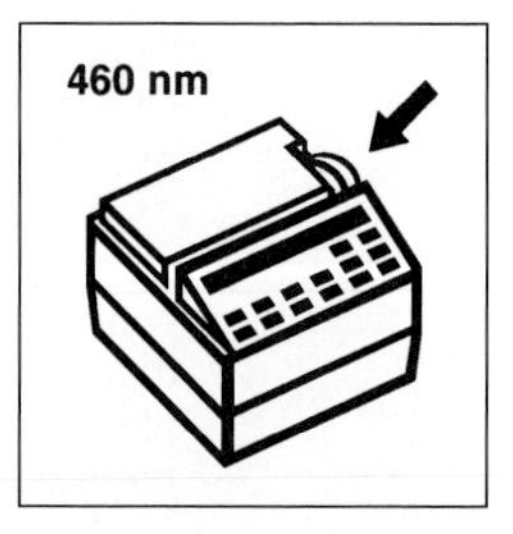

*Compliments of HACH Company. Analysis may be performed by DR/890 Colorimeters or DR/2010 Spectrophotometer.

**Adapted from: Hach et al., *Journal of Association of Official Analytical Chemists,* 70 (5) 783–787 (1987); Hach et al., *Journal of Agricultural and Food Chemistry,* 33 (6) 1117–1123 (1985); *Standard Methods for the Examination of Water and Wastewater, 417 A and 417 B.*

4. Digest the sample as described in this chapter, or in the Digesdahl Digestion Apparatus Instruction Manual. Digest an equal amount of deionized water as the blank.

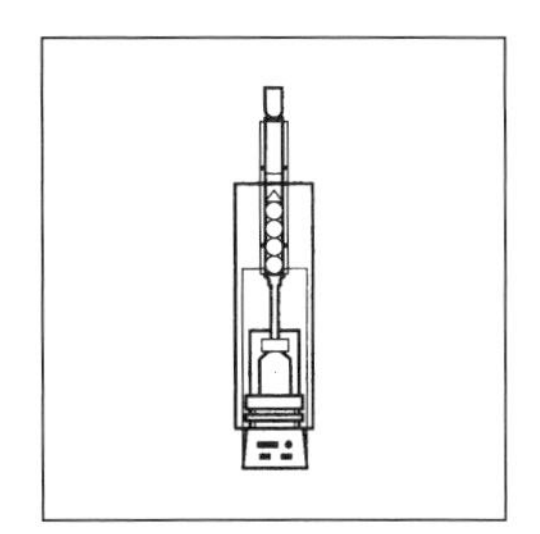

5. Select the appropriate *analysis* volume of the digested sample given in Table 2.29 Pipet the analysis volume from the sample and the digested blank into separate 25-mL mixing graduated cylinders.

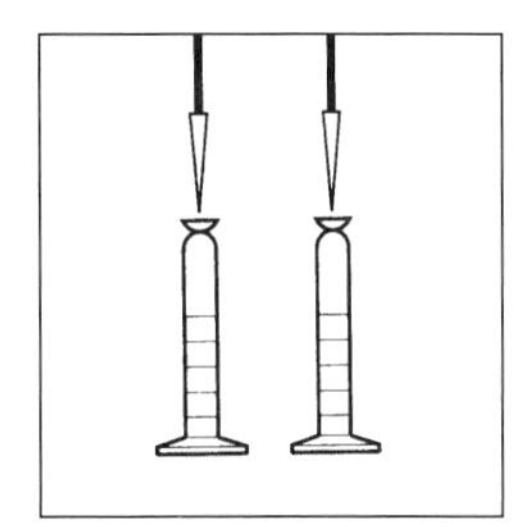

6. Add one drop TKN Indicator to each cylinder. Add 8.0 N KOH dropwise to each cylinder, mixing after each addition. Continue until the first apparent blue color is visible.

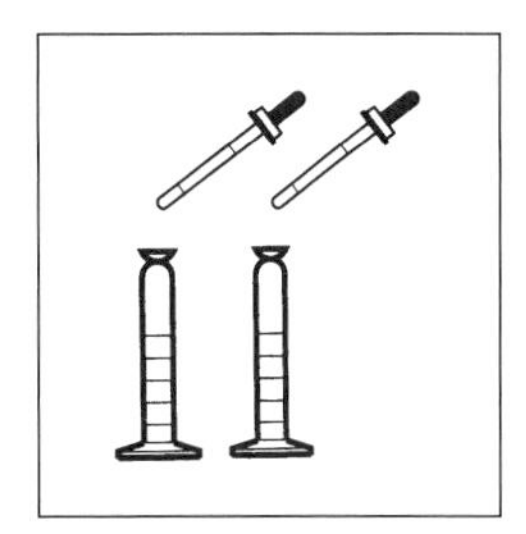

7. Add 1.0 N KOH one drop at a time, mixing after each addition, until the first permanent blue color appears.

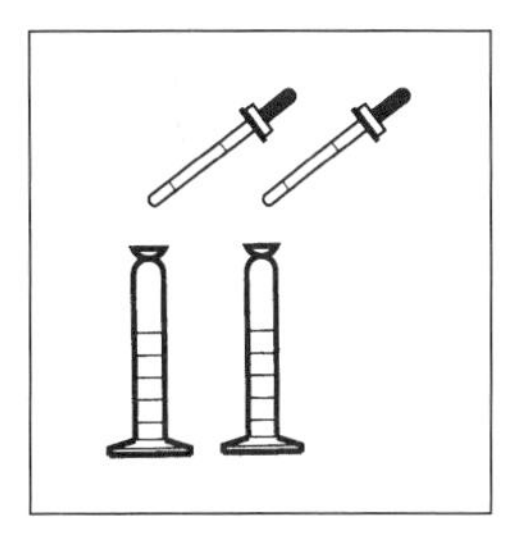

TABLE 2.29 AQUEOUS SAMPLES (Solutions of suspensions in water-less than 1% solids)

Expected nitrogen concentration (mg/L)	Analysis volume (mL)
0.5–35	20
2–140	10.0
11–700	5.00
45–2800	1.00
425–28000	0.500

8. Fill both mixing cylinders to the 20-mL mark with deionized water. Add 3 drops of Mineral Stabilizer to each cylinder. Invert several times to mix. Add 3 drops of Polyvinyl Alcohol Dispersing Agent to each cylinder. Invert several times to mix.
Note: Hold the dropping bottles upright while dispensing.

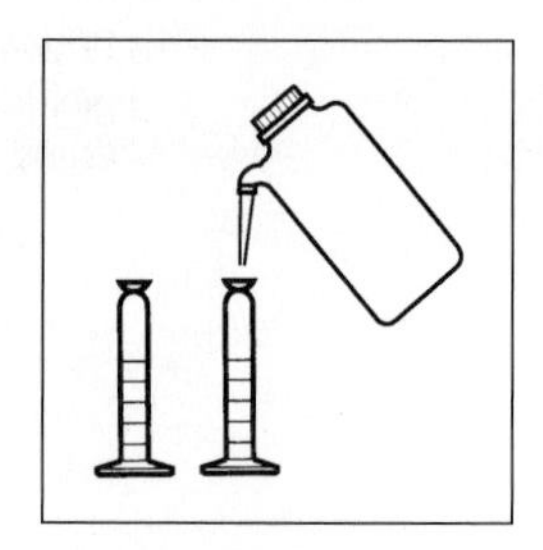

9. Fill both cylinders to the 25-mL mark with deionized water.

10. Pipet 1 mL of Nesslers Reagent to each cylinder. Stopper, invert repeatedly. The solution should not be hazy.
Note: Any haze (or turbidity) will cause incorrect results.

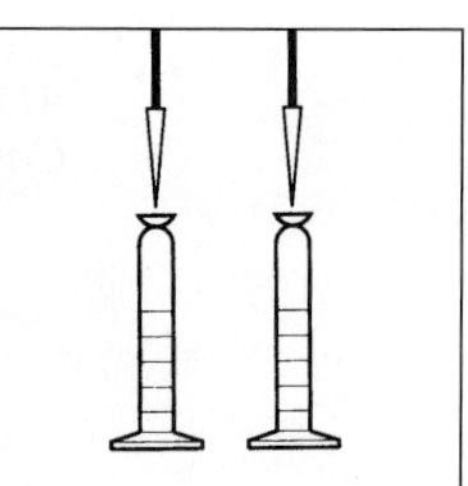

11. Press: **SHIFT TIMER.** A two-minute reaction period will begin.

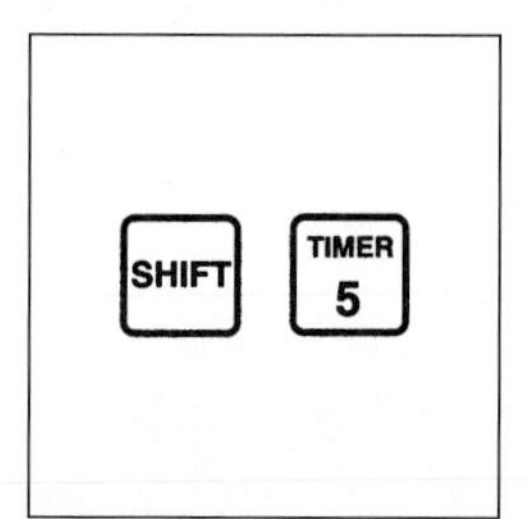

12. When the timer beeps, the display will show: **mg/L TKN.** Pour the contents of each cylinder into a 25-mL sample cell.

13. Place the blank into the cell holder. Close the light shield.

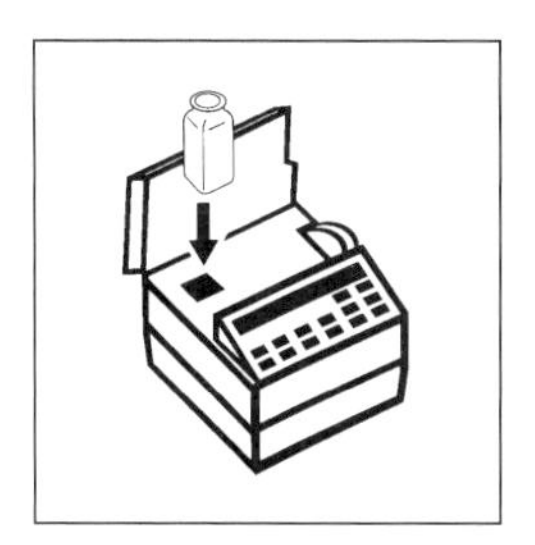

14. Press: **ZERO.** The display will show: **Zeroing ...** then: **0.mg/L TKN.**

15. Place the prepared sample into the cell holder. Close the light shield.

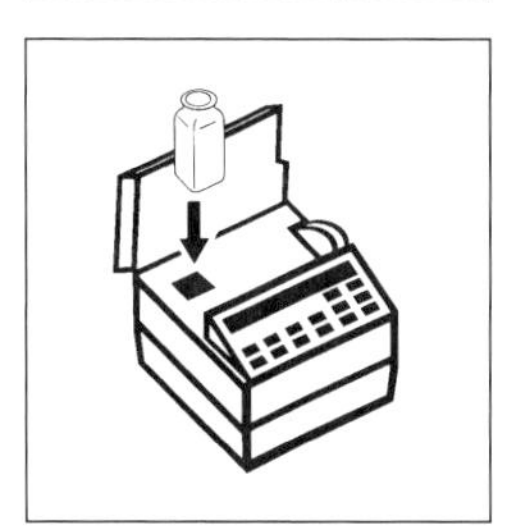

16. Press: **READ.** The display will show: **Reading ...** then the result in mg/L total kjeldahl nitrogen as N will be displayed. ***Note:*** The readout is the actual concentration of total kjeldahl nitrogen when the sample amount is 25 mL and the analysis volume is 3 mL.

17. Use the formula shown to calculate the final TKN (Total Kjeldahl Nitrogen) value:

$$\text{ppm TKN} = \frac{75 \times A}{B \times C}$$

where: A = mg/L read from the display
B = g (or mL of water) of sample taken for digest
C = mL analysis volume of digested sample (Step 5)

Note: For water samples ppm TKN = mg/L TKN.

SAMPLING AND STORAGE

Collect samples in a cleaned glass or plastic container. Adjust the pH to 2 or less with sulfuric acid (about 2 mL per liter) and cool to 4°C. Preserved samples can be stored up to 28 days.

ACCURACY CHECK

See p. 865 of the HACH Water Analysis Handbook, 3rd Ed., 1997.

KJELDAHL NITROGEN STANDARD METHOD

This checks the efficiency of the digestion. There are standards available for doing this test. For a complete procedure, see "Digestion Accuracy Check" in the "Digesdahl Digestion Apparatus" Instruction Manual.

USER CALIBRATION

For most accurate results, the use of a user-calibrated program is highly recommended. The HACH stored Program 399 is intended for process control samples or other applications where a high degree of accuracy is not necessary.

A one-time setup of a program for TKN is recommended for each new lot of reagents. A new calibration may be performed for each lot of Nessler Reagent as follows.

STANDARD PREPARATION

Prepare standards representing concentrations of 20, 60, 80, 100, 140 and 160 mg/L NH_3-N as follows:

a) Using volumetric pipets, transfer 5.0, 15.0, 20.0, 25.0, 35.0, and 40.0 mL of 100 mg/L NH_3-N standard solution into six separate 100-mL volumetric flasks. Dilute to volume with deionized water, stopper, and invert to mix.

b) Begin at Step 4 of the procedure using a 3-mL aliquot for the sample volume. Also prepare a blank solution by substituting a 3 mL aliquot of deionized water for sample in Step 4.
Note: Standard solutions are prepared as if a 25-mL volume was used for the digestion. Actual concentrations prepared in Step 1 are 5, 15, 20, 25, 35, and 40 mg/L NH_3-*N. These represent original concentrations of 20, 60, 80, 100, 140, and 160 mg/L* NH_3*-N, based on the 25 to 100 mL dilution in the digestion.*

INITIAL SETUP OF TKN PROGRAM

a) Press **SHIFT USER PRGM.** Use the **UP** arrows to scroll to **Copy Program.** Press **ENTER.**

b) Scroll to or enter the template number for TKN (904). Press **ENTER.**

c) Scroll to or enter the desired user program number for TKN (>950). Press **ENTER.** Record the program number for reference.

d) The display will show **Program Copied.** Press **EXIT.** The program is now ready to be calibrated.
Note: The templates within the User Program cannot be used directly. They must be copied into a usable program number (*greater than 950) as in Steps c and d. For process control, the copied program can be used directly. For best results, however, the program should be calibrated with each new lot of reagents.*

USER CALIBRATION OF TKN PROGRAM

a) Use the test procedure to develop color in the standards just before recording the absorbance values for the calibration.

b) Press **SHIFT USER PRGM.** Use the **UP** arrow key to scroll to **Edit Program.** Press **ENTER.**

c) Scroll to or enter the program number for TKN (from Step c in Initial Setup). Press **ENTER.**

d) Use the **DOWN** arrow to scroll down to **Calib Table:X** (X = denotes a number which indicates the number of data points in the table). Press **ENTER.**

e) The instrument will prompt **Zero Sample.** Place the blank solution in the cell holder. Close the light shield. Press **ZERO.** The instrument will prompt you to adjust to the proper wavelength if necessary.

f) The first concentration point will be displayed. Press **ENTER** to display the stored absorbance value of the first concentration point.

g) Place the first developed standard solution (same concentration as the value displayed) in the cell holder. Close the light shield. Press **READ** to display the measured absorbance of the standard. Press **ENTER** to accept the displayed absorbance value.

h) The second concentration point will be displayed. Press **ENTER** to display the stored absorbance value of the second concentration. Place the second developed standard solution in the cell holder. Close the light shield. Press **READ** to display the measured absorbance value of the standard.

i) Press **ENTER** to accept the absorbance reading. The next concentration point will then be displayed.

j) Repeat Steps h and i as necessary for the remaining standards.

k) When finished reading absorbance values, press **EXIT.** Scroll down to **Force Zero.** Press **ENTER.** Change the Force Zero setting to **ON** by using an arrow key. Press **ENTER** again.

l) Scroll down to **Calib. Formula.** Press **ENTER** twice or until only the **0** in **F(0)** is flashing. Press the DOWN arrow to select **F1** (linear calibration). Press **ENTER** to select **F1.**
Note: Other calibration fits may be used, if appropriate.

m) Press **EXIT** twice. The display will show **Store Changes?** Press **ENTER** to store the changes.

n) Press **EXIT.** The program is now calibrated and ready for use. Start on Step 2 of the procedure.

QUESTIONS AND ANSWERS

1. Why is a digestion process required in this experiment?
   Digestion of samples is required to determine the amount of bound nitrogen called organic nitrogen. The bound nitrogen would not be released for analysis if digestion of the sample was not implemented.
2. What types of samples require digestion before analysis?
   Aqueous samples, oils and fats, soil samples, hay and grass samples, specific types of fertilizers, and anything that contains organic nitrogen would require a digestion process.
3. What is the purpose of the dispersing agent?
   A dispersing agent is added to the solution to aid in the development of color for analysis purposes.
4. Will this experiment measure all of the nitrogen in the sample?
   This experiment does not determine the all of the nitrogen in the ionic form and the organic form. Nitrogen as nitrate, nitrite, nitrile, nitroso, nitro, azide, azine, azo, hydrazone, oxime, and semi-carbazone are not included in the determination.

REFERENCES

1. *HACH Water Analysis Handbook,* 3rd Ed. (Loveland, CO.: HACH Company, 1997).
2. *Standard Methods for the Examination of Water and Wastewater:* Macro-Kjeldahl Method, 16th Ed., A. E. Greenberg, R. R. Trussell, L. S. Clesceri and Mary Ann H. Franson (eds.) pp. 408–409, Washington, DC, American Public Health Association, 1985.

ORGANICS IN WATER AND WASTEWATER

A number of industries must monitor effluents for low level chlorinated hydrocarbons and petroleum hydrocarbons such as gasoline and kerosene. Industry, dry cleaning operations, garages, or individuals contaminate water bodies and soil by dumping cleaning agents and gasoline products. Gasoline tank leaks, vehicle accidents, and gasoline spills can add sufficient amount of gasoline to streams and ground waters to be a problem.

In specific areas of the USA, chemical plants, dry cleaners, sewage treatment facilities, and public waterworks are required to monitor effluents for chlorinated organic solvents. Petroleum hydrocarbons, perchloroethylene, 1,1,1-trichloroethane and trichloroethylene can be identified and detected in the effluents of local business and industry. The solvents are used in either liquid or vapor form to clean and degrease equipment. Additional applications of the solvents include the following: drying agent for metals and certain solids, vermifuge, heat transfer medium, manufacture of fluorocarbons, and dry-cleaning solvent.

A measured wastewater sample or industrial effluent is placed in a sample bottle of specific dimensions. A detector tube attached to a gas detector piston pump measures the concentration of the target substance in the air in the bottle's head space.* The air concentration is correlated to a water concentration by using correction graphs or correction tables and by applying a temperature correction factor. Depending on the sample size, each detector tube measures a toxic chlorinated organic chemical or petroleum hydrocarbons within a specific measuring range (ppm) at 20°C. In addition, hazardous spill response teams can also use this detector tube technology to assess the environmental impact of a gasoline spill or a chlorinated hydrocarbon.

A test for volatile organic halides in water has been developed to detect trichloroethylene (TCE), perchloroethylene (PCE), carbon tetrachloride (CCL_4), trihalomethanes (THMs) and other toxic solvents in the parts per billion (ppb) range. The Quick Test Volatile Organic Halides Water Test Kit can be used for site characterization, site mapping, ground water mapping, selecting samples for laboratory analysis, and monitoring processes.** The test is based on a photochemical reaction which forms a colored solution proportional to the concentration of the water contaminate. After extracting the analyte from the water with a solvent and Teflon Tape, the analyte is forced from the tape and forms a complex with a reagent. The reagent-analyte complex is exposed to ultraviolet light and the absorbance of the complex is measured by an "Envirometer." The recorded absorbance is compared to an standard curve stored in the instrument and the concentration of the analyte (total organic halides) is displayed in $\mu g/L$ or parts per billion (ppb). The assay range is 5 to 2000 parts per billion with sample dilution required above 200 ppb. The Quick Test Volatile Organic Halides Water Test Kit curve can be calibrated with a number of chlorinated organic complexes.

ORGANICS IN WATER AND WASTEWATER: CHLORINATED SOLVENTS AND PETROLEUM HYDROCARBONS IN WATER

Sampling the Wastewater or Effluent

1. Prior to sampling the effluent or wastewater, carefully wash the plastic beaker and the sampling bottle to remove any oily substances. Rinse the inside of the beaker and the bottle using distilled water.
2. Wash the glass bottle 2 to 3 times in the wastewater or effluent to be measured, then fill the bottle with the wastewater or effluent. Wait one minute to condition the bottle.
3. Dispose of the wastewater or effluent in the bottle and plug the sampling bottle.
4. Using a clean plastic beaker, collect a 200-mL wastewater or effluent sample.

MEASUREMENT PROCEDURE

1. Unplug the sampling bottle. Pour the 200 mL of sample from the plastic beaker into the sampling bottle.
2. Plug the sampling bottle and shake it well for one minute.
3. Break the tips off a fresh detector tube by bending each tube end in the tube tip breaker of the pump (supplied with the kit). Using the Sensidyne/Gas Model 800 Gas Sampling pump, insert the tube securely into the rubber inlet of the pump. Make certain the arrow on the tube is pointing **toward** the pump.
   Note: If Sensidyne Detector Tube 135L for 1,1,1-Trichloroethane is used, break the tips off a fresh primary tube and an analyzer tube. Connect both tubes (*at the ends marked with a c) using the rubber tubing supplied with the kit.*
4. Make certain that the pump handle is all the way in. Align the guide marks on the shaft and pump body. Position the tube and the pump so that the tube inlet is just ***above*** the water level inside the sample bottle.

---

*Compliments of Sensidyne.

**Compliments of Strategic Diagnostics, Inc., 111 Pencader Drive, Newark, DE.

5. Pull the handle until it locks on the 1/2 pump stroke (50 mL). Wait until staining stops on the detector tube.
6. Read the concentration level indicated on the tube at the maximum level of the stain. If the stain in the tube is uneven due to channeling, use an average of the highest and lowest points of the maximum stain.
   Note: The concentration level reading is located at the point where the stained and unstained portions of the tube meet.
7. If the discoloration ends before the first calibration mark (50 mL sampling), take another 50 mL sampling. This is accomplished by pulling the handle on the Model 800 pump to the 100 mL (full stroke) position.
8. After obtaining the tube reading, measure the temperature of the sample (wastewater or effluent). If the temperature of the sample is higher or lower than 20°C, adjust the reading using the **multiplying factor** found in the Temperature Correction Table for the appropriate substance under testing. The Temperature Correction Table for each substance is listed below.
9. Once the tube reading has been adjusted for the temperature of the sample, proceed to the Correction Procedure section below. The experiment may be repeated by collecting a fresh 200 mL water sample and by following **Steps 1 to 9 above.**

Tube Measurements

Substance	Tube no. & color change	Chart	Measuring range (ppm)	Sampling size
Tetrachloroethylene or	Tube 133M	Figure 2.3	0.3–6.0	50 mL
Perchloroethylene	Yellow to Purple		0.1–2.6	100 mL
	Tube 133L	Figure 2.4	0.08–1.9	50 mL
	Yellow to Purple		0.03–0.65	100 mL
Trichloroethylene	Tube 132M	Figure 2.5	0.35–7.0	50 mL
	Yellow to Purple		0.15–2.8	100 mL
	Tube 132L	Figure 2.6	0.07–1.75	50 mL
	Yellow to Purple		0.03–0.7	100 mL
1,1,1-Trichloroethane or	Tube 135L	Figure 2.7	1.4–14.0	50 mL
Methyl Chloroform	White to Orange		0.4–4.0	100 mL
Petroleum Hydrocarbons	Tube 101L	Figure 2.8	11.4–312	50 mL
(Gasoline)	Brown/Yellow		2–55	100 mL
	to Dark		0.9–24	200 mL
	Green		0.4–12	300 mL

CORRECTION PROCEDURE

A correction table or a correction graph may be used to obtain the true concentration level of the hydrocarbon or the chlorinated hydrocarbon in the sample of wastewater or effluent. The Sensidyne Company can provide appropriate Correction Graphs for each tube of contaminate.

The Correction Factor can be obtained without the aid of the graphs by using the appropriate Correction Factor Table. Locate the sample size used (usually 50 mL or 100 mL).

Multiply the adjusted tube reading by this factor to obtain the true concentration (ppm) of contaminate in the sample.

Temperature Correction Table (Tube 133M) for Perchloroethylene

Water Temperature (°C)	0	5	10	15	20	25	30
Correction Factor	2.8	1.9	1.4	1.2	1.0	0.9	0.8

Correction Table (Tube 133M)

Sampling volume	Correction factor
50 mL (1/2 stroke)	0.063
100 mL (Full stroke)	0.026

FIGURE 2.3 Correction Procedure: Perchloroethylene (Tetrachloroethylene)—Tube 133M

Temperature Correction Table (Tube 133L) for Perchloroethylene

Water Temperature (°C)	0	5	10	15	20	25	30
Correction Factor	3.0	2.2	1.7	1.25	1.0	0.8	0.7

Correction Table (Tube 133L)

Sampling volume	Correction factor
50 mL (1/2 stroke)	0.077
100 mL (Full stroke)	0.026

FIGURE 2.4 Correction Procedure: Perchloroethylene (Tetrachloroethylene)—Tube 133L

Temperature Correction Table (Tube 132M) for Trichloroethylene

Water Temperature (°C)	0	5	10	15	20	25	30
Correction Factor	2.8	2.0	1.5	1.2	1.0	0.9	0.8

Correction Table (Tube 132M)

Sampling volume	Correction factor
50 mL (1/2 stroke)	0.07
100 mL (Full stroke)	0.028

FIGURE 2.5 Correction Procedure: Trichloroethylene—Tube 132M

Temperature Correction Table (Tube 132L) for Trichloroethylene

Water Temperature (°C)	0	5	10	15	20	25	30
Correction Factor	2.8	2.0	1.5	1.2	1.0	0.9	0.8

Correction Table (Tube 132L)

Sampling volume	Correction factor
50 mL (1/2 stroke)	0.07
100 mL (Full stroke)	0.028

FIGURE 2.6 Correction Procedure: Trichloroethylene—Tube 132L

Temperature Correction Table (Tube 135L) for 1,1,1-Trichloroethane (Methyl Chloroform)

Water Temperature (°C)	0	5	10	15	20	25	30
Correction Factor	3.0	2.0	1.5	1.2	1.0	0.85	0.75

Correction Table (Tube 135L)

Sampling volume	Correction factor
50 mL (1/2 stroke)	0.07
100 mL (Full stroke)	0.02

FIGURE 2.7 Correction Procedure: 1,1,1-Trichloroethane (Methyl Chloroform)—Tube 135L

A non-linear graph of true petroleum hydrocarbon concentration (ppm) versus Tube 101L reading at various temperatures provides the correction data for gasoline and other petroleum hydrocarbons. The correction data may be obtained from a graph with the equipment from Sensidyne or from the tables below. The table below provides estimates or approximations from the graph. The source of gasoline and the interpretation of the graph will have an impact on the accuracy of the table below.

Tube 101L Readings—Gasoline (Petroleum Hydrocarbons)

	Temperature	10°C	15°C	20°C	25°C	30°C
True	0	0	0	0	0	0
Petroleum	10	105	150	200	230	302
Hydrocarbon	20	190	260	340	410	510
Concentration	30	225	340	460	580	700
(ppm)	40	275	405	570	700	870
	50	295	470	650	820	>1000
	55	300	490	705	900	—

Correction Table (Tube 101L)

Sampling volume	Correction factor
50 mL (1/2 stroke)	5.68
100 mL (1 Stroke)	1.00
200 mL (2 strokes)	0.44
300 mL (3 strokes)	0.22

FIGURE 2.8 Correction Procedure: Petroleum Hydrocarbons (Gasoline)—Tube 101L

QUESTIONS AND ANSWERS

1. **Why are the chlorinated hydrocarbons used today?**
   Chlorinated hydrocarbons will remove grease and oil stains from clothing, equipment and machinery parts when other solvents do not perform adequately. The compounds are good drying agents for metals and specific solids. The substances are also used as heat transfer medium. The substances are still used because other materials are more expensive and do not perform as well.
2. **Why are perchloroethylene,1,1,1-trichloroethane and trichloroethane a handling problem?**
   The chlorinated compounds are carcinogenic. With a high density, the compounds tend to cling onto solids and other materials and cannot be removed easily. The compounds are very slightly soluble in water and very soluble in organic compounds.
3. **A 133M Tube to test perchloroethylene in industrial effluent is inserted into a Model 800 Gas Pump and measures the air over a properly prepared 200 mL of industrial effluent in a glass bottle. After the handle has been locked at 50 mL, a purple stain develops at a maximum level of 6.2 ppm. Using a thermometer, the temperature of the water sample is 25°C. Determine a) the concentration value with temperature correction, and b) the true concentration of perchloroethylene in the sample.**
   a) 6.2 ppm $\times$ 0.9 = 5.58 ppm with temperature correction
   b) 5.58 ppm $\times$ 0.063 = 0.35 ppm as the true concentration of perchloroethylene in the industrial effluent.

REFERENCES

1. *Wastewater Test Kit for Chlorinated and Petroleum Hydrocarbons,* Sensidyne, Inc., 2nd Ed., July 1992.
2. *Hawley's Condensed Chemical Dictionary,* 11th Ed., rev'd by N. Irving Sax and Richard J. Lewis (New York, N.Y.: Van Nostrand Reinhold, 1987).
3. Joe Dautlick and Kia Wyatt, Strategic Diagnostics Inc., 111 Pencader Drive, Newark, DE.
4. Alan Seidman, Zero Waste Co., La Jolla, CA.

BIOCHEMICAL OXYGEN DEMAND (BOD)

The test determines the amount of oxygen required for the biochemical degradation of organic material (carbonaceous demand) and the oxygen used to oxidize inorganic materials such as ferrous iron, sulfides and nitrites. In limited cases, the test may also measure the oxygen required to oxidize the reduced forms of nitrogen (nitrogenous demand). A Nitrification Inhibitor is added to the solution to prevent ammonia oxidation. If ammonia oxidizes in the sample, errors would result because oxygen use in the sample would not be due exclusively to pollutants in the sample.

The determination of BOD is accomplished by diluting appropriate portions of the sample with water saturated with oxygen and measuring the dissolved oxygen in the mixture both immediately and after a period of incubation for usually five days. The BOD is computed by a specific formula using a graphic method solution.

Dilution of sample before incubation is necessary to bring the oxygen demand and supply into appropriate balance. Specific nutrients and a buffer are added to the dilution water to support bacterial growth in a sample with appropriate pH and a controlled incubation temperature at 20°C. The standard incubation period is 5 days in the dark.

Many industrial effluents and chlorinated effluents will require modifications to the test procedure. In most cases, the effluent can be collected before the chlorination process and tested by the standard procedures. Toxins will affect the microorganisms in a sample and lower the BOD of the sample. The effect of toxins can sometimes be eliminated by diluting the sample with high quality distilled water. Alternatively, seed used in the dilution water may be acclimatized to tolerate toxins, i.e. mixture of 90% sewage and 10% toxic waste.

BIOCHEMICAL OXYGEN DEMAND

Method 8043 for water and wastewater*
Dilution Method**
USEPA Accepted

1. Prepare sample dilution water using a BOD Nutrient Buffer Pillow; see "Dilution Water Preparation" following this procedure.

2. Determine the range of sample volumes required for your sample; see "Choosing Sample Size" following this procedure.
Note: If the minimum sample volume is 3 mL or more, determine the dissolved oxygen in the undiluted sample. This can be omitted when analyzing sewage and settled effluents known to have a dissolved oxygen content near 0 mg/L.

3. With a serological pipet, measure a graduated series of at least four, but preferably five or six, portions of well-mixed sample and transfer to separate 300-mL glass-stoppered BOD bottles. Stir the sample with the pipet before pipetting each portion.
Note: See "Interferences" following this procedure when analyzing chlorinated or industrial effluents.
Note: Do not add sample to one BOD bottle. This will be the dilution water blank. For additional proof of accuracy, see "Accuracy Check" following these steps.

*Compliments of HACH Company.

**Adapted from *Standard Methods for the Examination of Water and Wastewater* and from R. L. Klein and C. Gibbs, *Journal Water Pollution Control Federation*, 51(9), 2257, 1979.

4. Add two shots of Nitrification Inhibitor (approx. 0.16 g) to each bottle, if desired.
Note: Use of Nitrification Inhibitor will inhibit the oxidation of nitrogen compounds if only carbonaceous oxygen demand is desired. It is especially recommended for samples with low BODs.

5. Fill each bottle to just below the lip with seeded or unseeded dilution water. When adding the water, allow it to flow slowly down the sides of the bottle to prevent formation of bubbles.

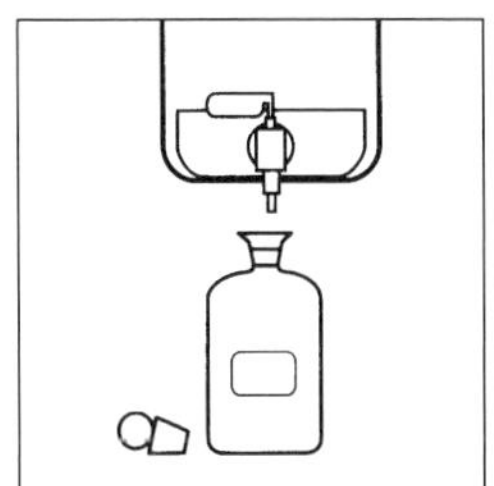

6. Stopper the bottle, being careful not to trap any air bubbles. Press on the stopper of the bottle with your finger; then invert the bottle several times to mix.
Note: Determine initial DO's if following the Standard Methods Procedure. This is not necessary if using the graphical method.

7. Add enough dilution water to the lip of the BOD bottle to make a water seal.

8. Place a plastic overcap over the lip of each bottle and place bottles in an incubator at 20 + 1°C. Incubate in the dark for five days.

9. When the five-day incubation period is complete, determine the dissolved oxygen content (mg/L DO remaining) in each bottle as described in the Dissolved Oxygen Procedure or potentiometrically by using a dissolved oxygen probe.
Note: This procedure has been EPA approved, but the graphical method outlined in Step 10 has not. See "Calculating Results: Standard Methods" (following these steps) for the EPA approved calculation.

Determine DO after five days

10. Determine the BOD using the graphical method as follows; see "Calculating Results: Graphical Method" for more information.
a) Plot the mg/L DO remaining in each diluted sample versus the mL sample taken; then draw the best straight line through the plotted points.
Note: An erroneous point is visually evident at this time and can be disregarded. However, at least three points should be on the line or very close to it. For unseeded dilution water, the line should cross the "mg/L oxygen remaining" scale near or below the oxygen saturation value for the altitude of the laboratory as discussed in "Dilution Water Preparation."
b) To calculate the BOD, use the following equation which is mathematically equivalent to the BOD equation in Standard Methods.

Calculate BOD

$$\text{mg/L BOD} = (A \times 300) - B + C$$

where:
A = the slope. The slope of the line is equal to the mg/L DO consumed per mL of sample taken. Take any point on the line and subtract the mg/L DO remaining at that point from the mg/L DO where the line crosses the DO scale (Y intercept, mg/L DO remaining). Divide the difference by the mL of sample at the point chosen. 300 = the volume of the BOD bottle.
B = the Y intercept. This is the DO value where the line crosses the "DO remaining" scale. (This should be very close to the actual dilution water blank value).
C = the sample DO. This is the DO of the undiluted sample.
Another way to write this equation is:

$$\text{mg/L BOD} = (\text{Slope} \times 300) - (\text{Y intercept} + \text{Sample DO})$$

Note: If the best straight line is obtained by linear regression through use of a calculator, the sign (−) of the slope must be changed (+) before multiplying by 300.

DILUTION WATER PREPARATION

Note: The DO uptake in 5 days at 20°C should not exceed 0.2 mg/L.
Note: The bubbling apparatus should be cleaned before and after use.

The BOD test requires very high quality water be used for diluting samples. Water must be free of all toxic substances, such as small amounts of chlorine, copper and mercury, as well as free of organic matter. If organic matter is present in dilution water, it will create an oxygen demand.

The most practical way to produce water of low organic content on a consistent basis is by distillation from alkaline permanganate (sodium hydroxide pellets and potassium permanganate). Commercial stills which automatically produce high quality distilled water are available.

Direct use of deionized water from ion exchange columns is not recommended because of erratic release of organic materials from the cartridges, especially new ones. These organic materials will not be detected with conductivity measurements but may show up in the final results as an oxygen demand. Bacterial growth also may be present on the column.

Distilled water as it is produced from a still usually is warm in temperature and not saturated with oxygen. The temperature of the BOD dilution water must by 20°C at the time of use and near or at saturation with oxygen. It is recommended that distilled water be stored in a BOD incubator until it reaches 20°C and dilution water be prepared immediately before use. The distilled water can be placed in one-gallon jugs by filling each of them with three liters or by filling two-gallon jugs with six liters. The jugs should be capped and placed in the incubator for storage. After 24 hours or more, the temperature will be 20°C and the water will be saturated or nearly saturated with oxygen furnished by the air above the water in the jugs. If five-gallon containers are used, the distilled water should be saturated with oxygen by bubbling in filtered air from a hose connected to an aquarium pump or air compressor.

It is not necessary to use seeded dilution water when analyzing sewage, sewage plant effluent (unless it has been chlorinated) or river water. However, there are certain samples such as industrial or trade wastes or chlorinated sewage which do not contain sufficient bacteria to oxidize the organic matter that may be present.

To test such samples, some bacterial seed must be added to the samples. This is done by adding a small, measured volume of water known to contain a good bacterial population to the dilution water. Raw sewage is recommended as a source of seed. This material should be stored at 20°C for 24 to 36 hours before use. When using domestic sewage as seed, it should be allowed to stand undisturbed until most solids settle. Pipet from the upper portion of the bottle of seed material. It has been found that the addition of 3.0 mL of raw domestic sewage seed to each liter of dilution water is ample. Seed that has a BOD of 200 mg/L (a typical range for domestic sewage) when added at the rate of 3 mL per liter of dilution water will deplete 0.6 mg/L DO.

USING BOD NUTRIENT BUFFER PILLOWS

To prepare dilution water, select the BOD Nutrient Buffer Pillow for the amount of dilution water you wish to prepare; see Table 2.30. Shake the pillow, cut it open, and add the contents to a jug containing the proper amount of 20°C distilled water. Choose a container which will be only partially filled by the solution. Cap the jug and shake vigorously for one minute to dissolve the slurry and to saturate the water with oxygen.

FOLLOWING CONVENTIONAL METHOD

To prepare dilution water by the conventional method, pipet 1 mL of each of the following solutions per liter of distilled water at 20°C: Calcium Chloride Solution, Ferric Chloride

TABLE 2.30

Description	Catalog Number (HACH)
BOD Nutrient Buffer Pillows:	
for preparing 300 mL of dilution water	14160-66
for preparing 3 L of dilution water	14861-98
for preparing 6 L of dilution water	14862-98
for preparing 19 L of dilution water	14863-98

Solution, Magnesium Sulfate Solution, and Phosphate Buffer Solution. Cap the bottle and shake vigorously for one minute. The Phosphate Buffer Solution should be refrigerated to decrease the rate of biological growth. Use care with all solution to avoid contamination.

CHOOSING SAMPLE SIZE

The range of sample volumes to be diluted depends on two factors: type of sample and the laboratory's elevation.

If the sample contains high levels of organic material, such as raw sewage, its BOD will be high and small portions must be diluted in the test. On the other hand, if a sample has a low BOD, such as polluted river water, larger portions will be necessary; see Table 2.31.

The laboratory's elevation influences the amount of oxygen that can be dissolved in the dilution water. At sea level and normal barometric pressure, water can be saturated with up to 9.2 mg/L DO at 20°C. At higher elevations, the amount of oxygen that can dissolve in water decreases, so less oxygen is available to microorganisms. Refer to Table 2.32 and Table 2.33. Smaller portions of sample must be taken so there will be dissolved oxygen

TABLE 2.31 Determining Minimum Sample Volume

Sample type	Estimated BOD mg/L	mL of sample*
Strong Trade Waste	600	1
Raw and Settled Sewage	300	2
	200	3
	150	4
	120	5
	100	6
	75	8
	60	10
Oxidized Effluents	50	12
	40	15
	30	20
	20	30
	10	60
Polluted River Waters	6	100
	4	200
	2	300

*mL of sample taken and diluted to 300 mL in standard BOD bottle.

remaining in the BOD bottle after five days of incubation. For most accurate results, sample size should be chosen so that at least 2.0 mg/L of dissolved oxygen are consumed during the incubation period, but 1.0 mg/L DO is left in the BOD bottle. (Table 2.31 has done this for you).

Follow these steps to determine the range of sample volumes to use:

a. Estimate the BOD of your particular sample type (see Table 2.31). Sewage has approximately 300 mg/L BOD; oxidized effluents have about 50 mg/L or less.

b. Determine the minimum sample volume that can be used for the estimated BOD of your sample from Table 2.31. For example, if a sewage sample is estimated to contain 300 mg/l BOD, the smallest allowable sample volume is 2 mL. For sewage effluent with an estimated BOD of 40 mg/L, the volume is 15 mL.

c. Determine the laboratory's altitude.

d. Determine the maximum sample volume for the altitude of your laboratory from Table 2.32. At 1,000 feet an estimated BOD of 300 mg/L, the largest sample portion should be 8 mL. For a BOD of 40 mg/L, the maximum volume is 60 mL.

e. Choose three other sample volumes between the minimum and maximum volumes so the total number of portions is five or greater. In the two cases given in Step d above, a series of 2, 4, 5, 6 and 8 mL portions is suggested for an estimated BOD of 300 mg/L and a series of 15, 25, 35, 45 and 60 mL portions for a BOD of 40 mg/L.

INTERFERENCES

Many chlorinated and industrial effluents require special handling to ensure reliable BOD results. Usually, careful experimentation with the particular sample will indicate what modifications should be made to the test procedure.

Toxins in the sample will adversely affect any microorganisms present and result in lower BODs.

a. To eliminate small amounts of residual chlorine, allow the sample to stand for one to two hours at room temperature. For larger quantities, determine the amount of sodium thiosulfate to add to the sample as follows:

1) Measure 100 mL of sample into a 250-mL erlenmeyer flask. Using a 10-mL serological pipet and a pipet filler, add 10 mL of 0.020 N Sulfuric Acid Standard Solution and 10 mL of Potassium Iodide Solution, 100 g/L, to the flask.

2) Add three droppersful of Starch Indicator Solution and swirl to mix.

3) Fill a 25-mL buret with 0.025 N Sodium Thiosulfate Standard Solution and titrate the sample from dark blue to colorless.

4) Calculate the amount of 0.025 N Sodium Thiosulfate Standard Solution to add to the sample:

$$\text{mL 0.025N sodium thiosulfate required} = \frac{\text{mL titrant used} \times \text{volume of remaining sample}}{100}$$

5) Add the required amount of 0.025 Sodium Thiosulfate Standard Solution to the sample. Mix thoroughly. Wait 10 to 20 minutes before running the BOD test.

TABLE 2.32 Determining Maximum Sample Volume

Estimated BOD at:			
Sea level	1000 feet	5000 feet	mL of sample*
2460	2380	2032	1
1230	1189	1016	2
820	793	677	3
615	595	508	4
492	476	406	5
410	397	339	6
304	294	251	8
246	238	203	10
205	198	169	12
164	158	135	15
123	119	101	20
82	79	68	30
41	40	34	60
25	24	21	100
12	12	10	200
8	8	7	300

*mL of sample taken and diluted to 300 mL in standard BOD bottle.

b. To eliminate the effect of phenols, heavy metals or cyanide, dilute the sample with high quality distilled water. Alternately, the seed used in the dilution water may be acclimatized to tolerate such materials. Acclimatize seed as follows:

1) Fill a one-gallon stainless steel or plastic container with domestic sewage and aerate for 24 hours. Allow the heavier material to settle.
2) After settling for one hour, siphon off three quarts of material and discard.
3) Fill the container with a mixture of 90% sewage and 10% wastes containing the toxic material.
4) Aerate for 24 hours. Repeat Steps 2) and 3) with increasing amounts of waste until the container holds 100% toxic waste material.

c. Optimum pH for BOD test is between 6.5 and 7.5. Adjust samples to pH 7.2 with Phosphate Buffer Solution or 1 N (or more dilute) Sulfuric Acid or Sodium Hydroxide Standard Solution if the pH is not in this range.

d. Cold samples may be supersaturated with oxygen and will have low BOD results. Fill a one-quart bottle about halfway with cold sample and shake vigorously for two minutes. Allow sample temperature to reach 20°C before testing.

TABLE 2.33 Oxygen Saturation Values at Various Altitudes

Oxygen saturation value	Sea level	1000 ft.	2000 ft.	3000 ft.	4000 ft	5000 ft.	6000 ft.
(at 20°C)	9.2 mg/L	8.9 mg/L	8.6 mg/L	8.2 mg/L	7.9 mg/L	7.6 mg/L	7.4 mg/L

CALCULATING RESULTS (GRAPHICAL METHOD)

The mg/L DO remaining was determined for a series of four dilutions of domestic sewage after five days of incubation. Results were as follows:

mL of Sample taken	mg/L DO remaining
2.0	7.50
3.0	6.75
6.0	4.50
9.0	2.25

The DO values (mg/L DO remaining) were plotted versus the mL of sample taken and a straight line drawn as in Figure 1 on p. 922 of the HACH Water Analysis Handbook, 1997. If a set of BOD dilutions is run correctly with a homogeneous sample, a graph of the mg/L DO remaining versus the sample volume would result in a straight line. The value where the line intersects the y axis is equal to the DO content of the dilution water after incubation, although this is not actually measured. In this case, it was equal to 9.0 mg/L and the DO of the domestic sewage sample was assumed to be zero. If another type of sample is used, the DO of an undiluted sample should be measured either by the Winkler titration method or potentiometrically.

STANDARD METHODS

When dilution water is not seeded:

$$BOD_{5}, \text{ mg/L} = \frac{D_1 - D_2}{P}$$

When dilution water is seeded:

$$BOD_{5}, \text{ mg/L} = \frac{(D_1 - D_2) - (B_1 - B_2)f}{P}$$

where:

D_1 = DO of diluted sample immediately after preparation, mg/L,
D_2 = DO of diluted sample after 5 d incubation at 20°C, mg/L,
P = decimal volumetric fraction of sample used,
B_1 = DO of seed control before incubation, mg/L,
B_2 = DO of seed control after incubation, mg/L, and
f = ratio of seed in diluted sample to seed in seed control
= (% seed in diluted sample)/(% seed in seed control)

If seed material is added directly to sample or to seed control bottles:

f = (volume of seed in diluted sample)/(volume of seed in seed control)

Report results as $CBOD_5$ if nitrification is inhibited. If more than one sample dilution meets the criteria of a residual DO of at least 1 mg/L and a DO depletion of at least 2 mg/L and there is no evidence of toxicity at higher sample concentrations or the existence of an obvious anomaly, average results in the acceptable range.

ACCURACY CHECK

See p. 923 of the HACH Water Analysis Handbook, 3rd Ed., 1997.

QUESTIONS AND ANSWERS

1. Define BOD.
BOD is Biochemical Oxygen Demand. BOD is an empirical measurement of the oxygen requirements of municipal and industrial wastewaters and sewage. The test results determine the effect of waste discharges on the oxygen resources of the receiving water.
2. Why is the BOD test of limited value?
The actual oxygen demand is dependent on temperature change, biological population, water movement, sunlight, oxygen concentration, pollution type, and other environmental factors. As these conditions change, the actual oxygen demand can easily change. The BOD test has limited value because the controlled environment for the tests in the laboratory cannot take these changes into consideration.
3. Can all water be tested for BOD?
Toxins present in water will decrease the microorganism population which results in a lower BOD value. Chlorinated effluents require special procedures to ensure accurate BOD results. Many industrial effluents containing chemicals such as phenols, heavy metals or cyanide will have to be diluted with distilled water before BOD tests are attempted.

REFERENCES

1. HACH Water Analysis Handbook, 3rd Ed. (Loveland, CO.: HACH Company, 1997).
2. *Standard Method for the Examination of Water and Wastewater:* OXYGEN DEMAND (BIOCHEMICAL), 16th Ed., A. E. Greenberg, R. R. Trussell, L. S. Clesceri and Mary Ann H. Franson (eds.) pp. 525–527, Washington, DC, American Public Health Associaton, 1985.

CHEMICAL OXYGEN DEMAND (COD)

Chemical Oxygen Demand (COD) estimates the amount of organic matter in wastewater, industrial effluents and water from a variety of sources. Chemical oxygen demand (mg/L

COD) is defined as the amount of oxygen consumed per liter under the conditions of the experiment. Traditionally, chemical oxygen demand attempts to measure the oxygen equivalent of the organic matter content of a sample that is susceptible to oxidation by a strong chemical oxidant, such as potassium dichromate in a 50% sulfuric acid solution.* The COD test can provide a good estimate of BOD (Biochemical Oxygen Demand) results when the tested sample contains only readily available organic bacterial food and no toxic matter.

During the oxidation of organic materials by dichromate in sulfuric acid with a silver compound as a catalyst, carbon is oxidized to form carbon dioxide, and hydrogen in the organic compound is converted to water. A mercuric compound is normally added to reduce interference from the oxidation of chloride ions by the dichromate.*

Additional products of the reaction are chromium in various oxidation states and ions of mercury and silver, which will require appropriate methods of handling and disposal. The dichromate digestion method has superior oxidizing ability and can be applied to a wide variety of samples. Compounds such as straight chain aliphatic compounds, pyridine, and related compounds resist oxidation and do not interfere in the analysis.

The COD test of salt water requires addition of mercuric sulfate or dilution of the sample being tested. Sampling and storage bottles must be free of organic contamination. Water containing chemicals with a measurable COD must not be used to wash or rinse equipment before analysis begins.

In this experiment, the dichromate ion is reduced to green chromic ion (Cr^{+3}) by oxidizable organic compounds. The amount of yellow Cr^{+6} remaining in solution is determined when the 0 to 150 mg/L (low range) colorimetric measurement is implemented at 420 nm. The amount of green Cr^{+3} produced in solution is measured in the high range (0 to 1500 mg/L or the 0 to 15000 mg/L) of colorimetric measurement at 620 nm.

The determination of COD can be performed in the field or in the laboratory with various Colorimeters and the DR/2010 Spectrophotometer. However, most of the digestion experiments will be performed in the laboratory unless adequate arrangements can be made for the field. Many of the procedures in this experiment require caution and some laboratory experience.

After the oxidation step is completed, the amount of dichromate consumed is determined titrimetrically or colorimetrically. Either the amount of reduced chromium (chromic ion) or the amount of unreacted dichromate can be measured. If the latter method is chosen, the analyst must know the precise amount of dichromate added.*

CHEMICAL OXYGEN DEMAND

Method 8000**
DIGESTION

*HACH Water Analysis Handbook, 3rd Ed., 1997, pp. 1249–1250.

**Compliments of HACH Company. DR/820 (Manganese III Method only), DR/850, and DR/890 Colorimeters can be used to determine COD.

1. Homogenize 100 mL of sample for 30 seconds in a blender.
Note: Mix the sample prior to homogenization. To improve accuracy and reproducibility, pour the homogenized sample into a 250-mL beaker and gently stir with a magnetic stir plate. For samples containing large amounts of solids, increase the homogenization time.
Note: Some of the chemicals and apparatus used in this procedure may be hazardous to the health and safety of the user if inappropriately handled or accidentally misused. Please read all warnings and the safety section of this manual. Wear appropriate eye protection and clothing for adequate user protection. If contact occurs, flush the affected area with running water. Follow instructions carefully.

2. Turn on the COD Reactor. Preheat to 150°C. Place the plastic shield in front of the reactor.
Note: Ensure safety devices are in place to protect analyst from splattering should reagent leaking occur.

3. Remove the cap of a COD Digestion Reagent Vial for the appropriate range:

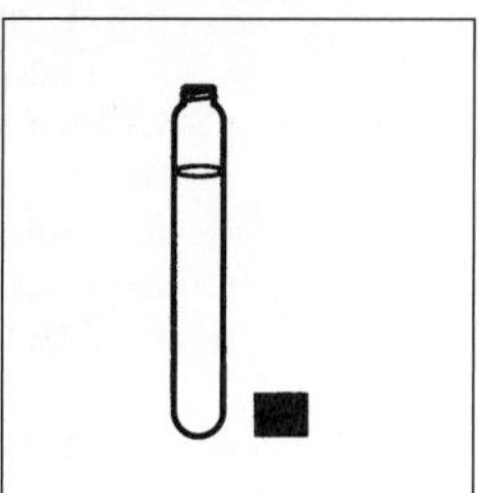

Sample concentration range (mg/L)	COD digestion reagent vial type
0 to 40	Ultra Low Range
0 to 50	Low Range
0 to 1500	High Range
0 to 15000	High Range Plus

Note: The reagent mixture is light-sensitive. Keep unused vials in the opaque shipping container, in a refrigerator if possible. The light striking the vials during the test will not affect results.

*The High Range Plus (0 to 15000 mg/L) COD Vials are not USEPA approved.

4. Hold the vial at a 45-degree angle. Pipet 2.00 mL (0.2 mL for the 0 to 15000 mg/L range) of sample into the vial.
Note: For the 0 to 15000 mg/L range,* pipet only 0.20 mL of sample, not 2.00 mL of sample, using a TenSette Pipet. For greater accuracy, a minimum of three replicates should be analyzed and the results average.
Note: Spilled reagent will affect test accuracy and is hazardous to skin and other materials. Do not run tests with vials which have been spilled. If spills occur, wash with running water.

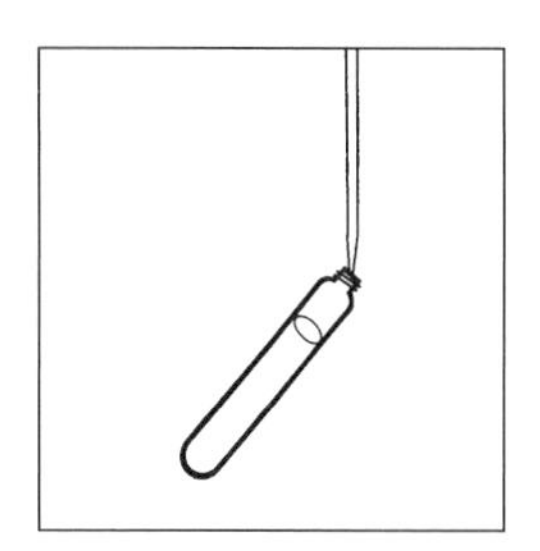

5. Replace the vial cap tightly. Rinse the outside of the COD vial with deionized water and wipe the vial clean with a paper towel.

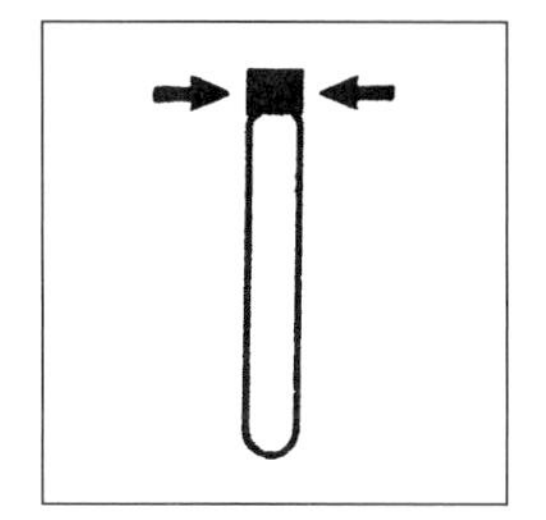

6. Hold the vial by the cap and over a sink. Invert gently several times to mix the contents. Place the vial in the preheated COD Reactor.
Note: The vial will become very hot during mixing.

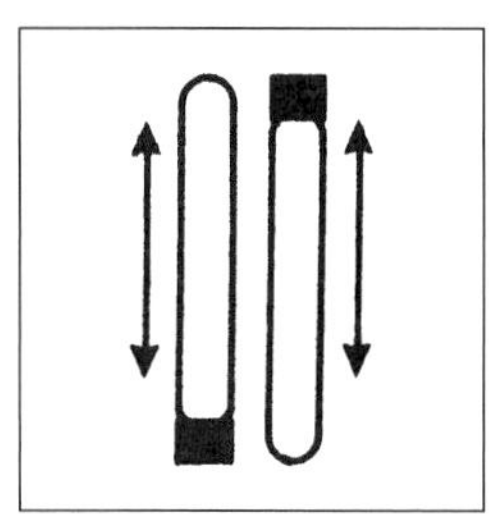

7. Prepare a blank by repeating Steps 3 to 6, substituting 2.00 mL (0.2 mL for the 0 to 15000 mg/L range) deionized water for the sample.
Note: Be sure the pipet is clean.
Note: One blank must be run with each set of samples. Run samples and blanks with the same lot of vials.

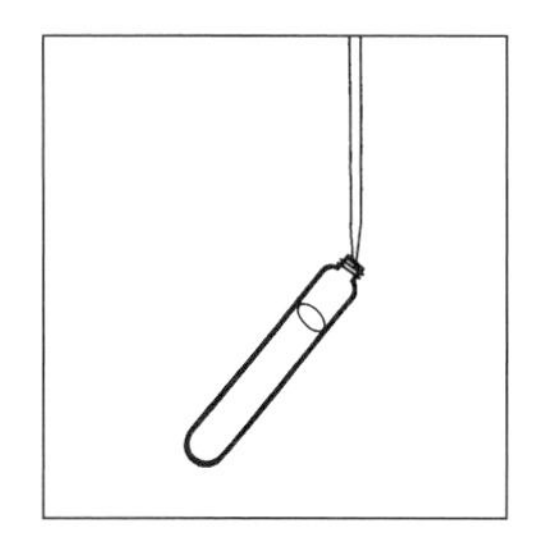

8. Heat the vials for two hours.
Note: Many samples are digested completely in less than two hours. If desired, measure the concentration (while still hot) at 15 minutes intervals until the reading remains unchanged. Cool the vials to room temperature for final measurement.

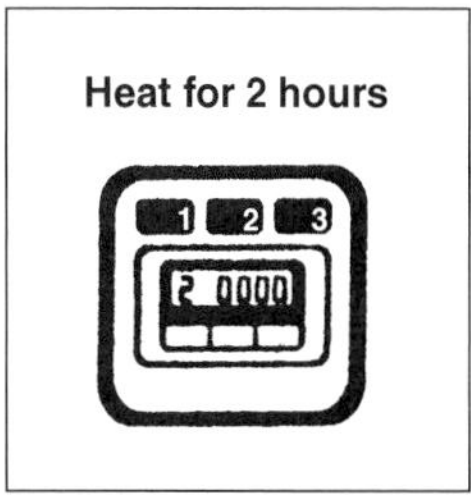

9. Turn the reactor off. Wait about 20 minutes for the vials to cool to 120°C or less.

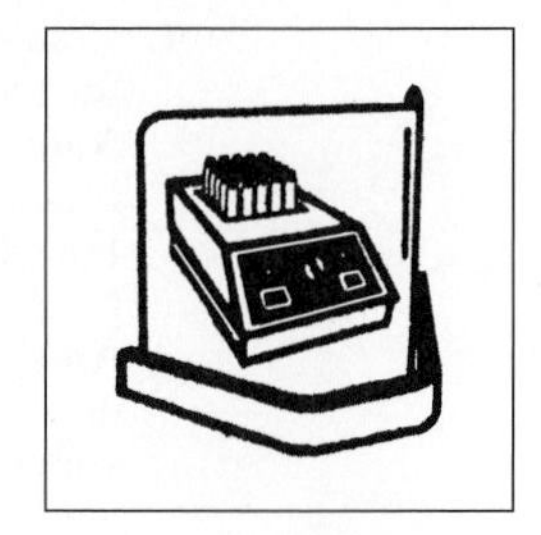

10. Invert each vial several times while still warm. Place the vials into a rack. Wait until the vials have cooled to room temperature.
Note: If a pure green color appears in the reacted sample, measure the COD and, if necessary, repeat the test with a diluted sample.

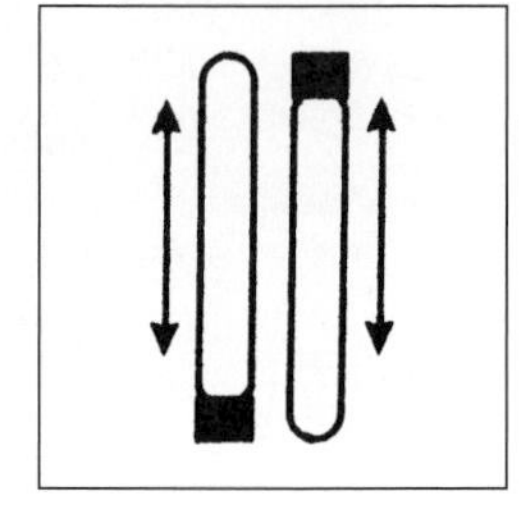

11. Use one of the following techniques to measure the COD:

- Colorimetric method, 0 to 150 mg/L COD
- Colorimetric method, 0 to 1500 mg/L COD
- Colorimetric method, 0 to 15000 mg/L COD

For the 0 to 40 mg/L Colorimetric method, see the HACH Water Analysis Handbook, 3rd Ed., 1997, pp. 945–946.

COLORIMETRIC MEASUREMENT, 0 TO 150 mg/L COD

1. Enter the stored program number for chemical oxygen demand (COD), low range. Press: **4 3 0 ENTER.** The display will show: **Dial nm to 420.**

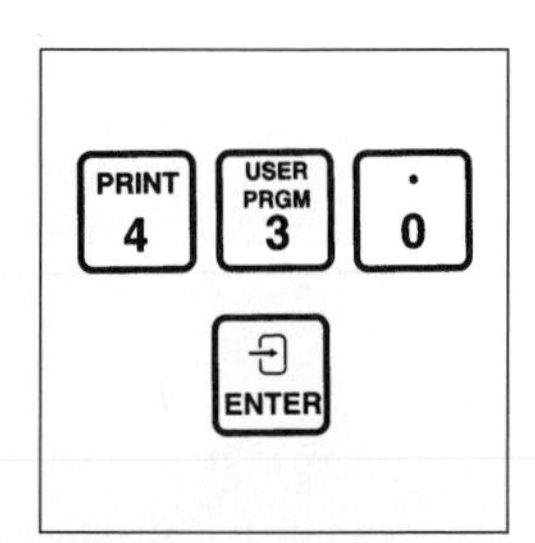

2. Rotate the wavelength dial until the small display shows: **420 nm.** When the correct wavelength is dialed in, the display will quickly show: **Zero sample,** then: **mg/L COD LR.**
Note: Approach the wavelength setting from the higher to lower values.

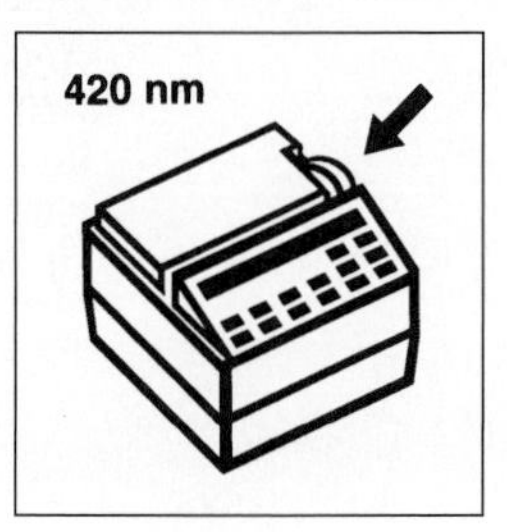

3. Place the COD Vial Adapter into the cell holder with the marker to the right.

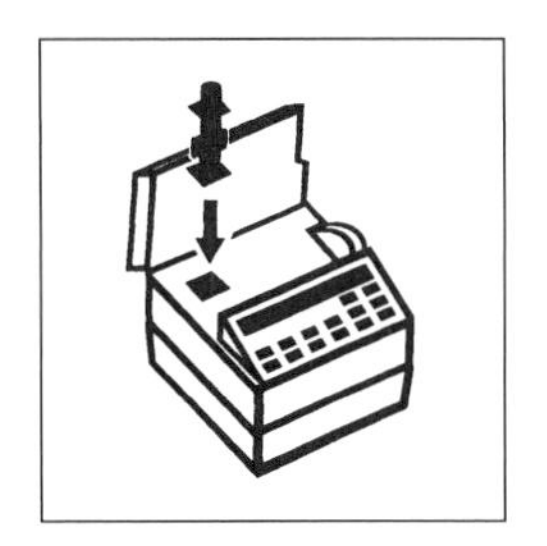

4. Clean the outside of the blank with a towel.
Note: Wiping with a damp towel, followed by a dry one, will remove fingerprints or other marks.

5. Place the blank into the adapter with the HACH logo facing the front of the instrument. Place the cover on the adapter.
Note: The blank is stable when stored in the dark; see "Blanks for Colorimetric Determination" following these procedures.

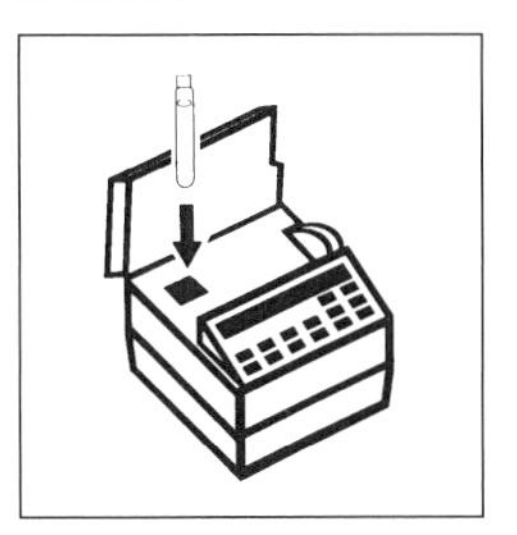

6. Press: **ZERO.** The display will show: **Zeroing ...** then: **0.mg/L COD LR.**

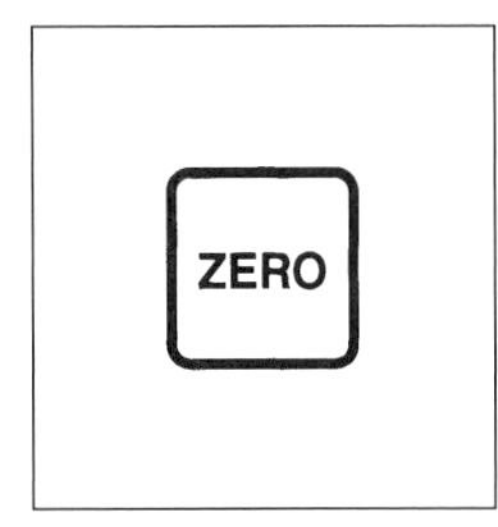

7. Clean the outside of the sample vial with a towel.

8. Place the sample vial into the adapter with the HACH logo facing the front of the instrument. Place the cover on the adapter.

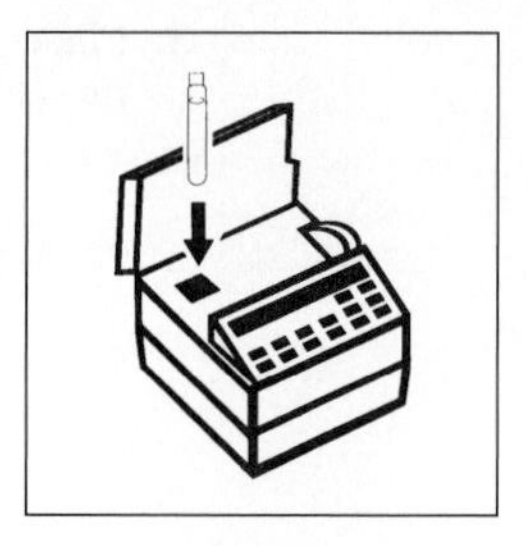

9. Press: **READ.** The display will show: **Reading ...** then the result in mg/L COD will be displayed.
Note: For most accurate results with samples near 150 mg/L COD, repeat the analysis with a diluted sample.

COLORIMETRIC MEASUREMENT, 0 TO 1500 AND 0 TO 15000 mg/L COD*

1. Enter the stored program number for chemical oxygen demand (COD), high range. Press: **4 3 5 ENTER.** The display will show: **Dial nm to 620.**

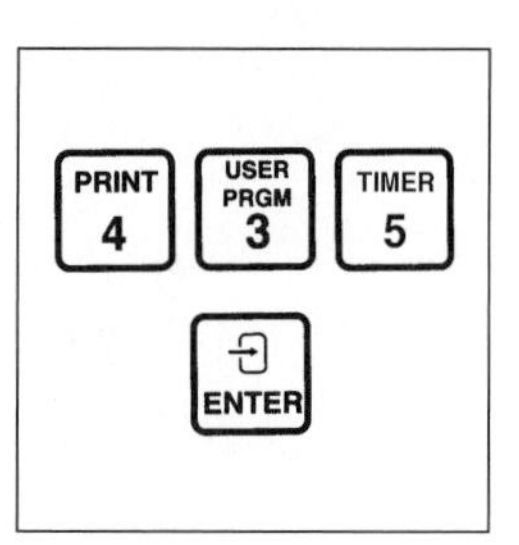

2. Rotate the wavelength dial until the small display shows: **620 nm.** When the correct wavelength is dialed in, the display will quickly show: **Zero sample,** then: **mg/L COD HR.**

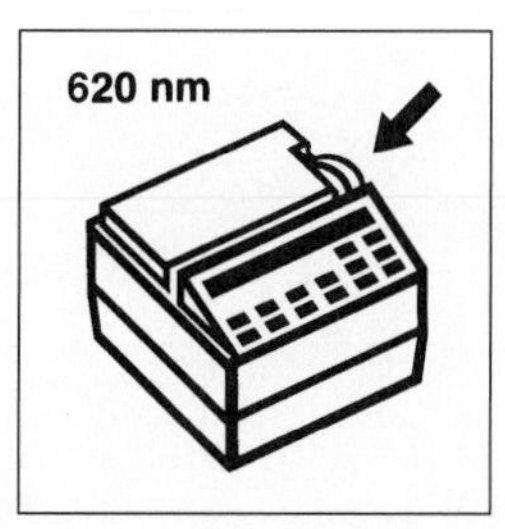

*The High Range Plus (0 to 15000 mg/L) COD Vials are not USEPA approved.

3. Place the COD Vial Adapter into the cell holder with the marker to the right.

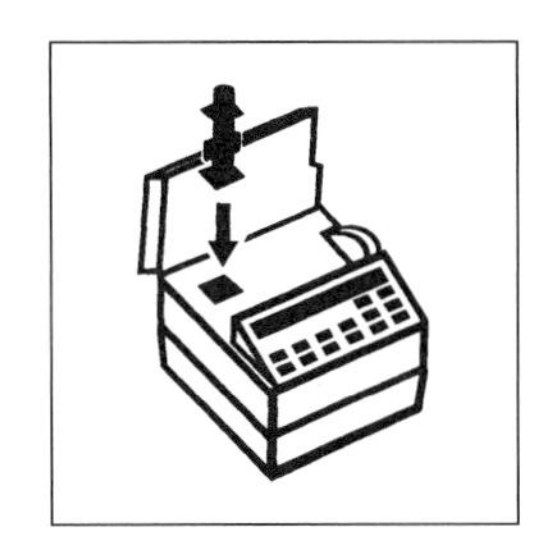

4. Clean the outside of the blank with a towel.
Note: Wiping with a damp towel, followed by a dry one, will remove fingerprints or other marks.

5. Place the blank into the adapter with the HACH logo facing the front of the instrument. Place the cover on the adapter.
Note: The blank is stable when stored in the dark; see "Blanks for Colorimetric Determination" following these procedures.

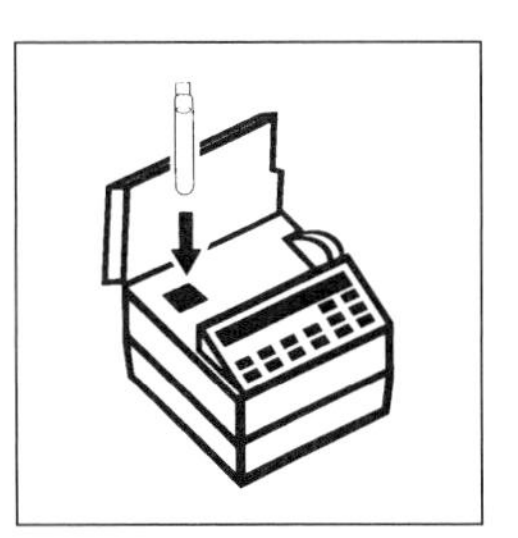

6. Press: **ZERO.** The display will show: **Zeroing ...** then: **0.mg/L COD HR.**

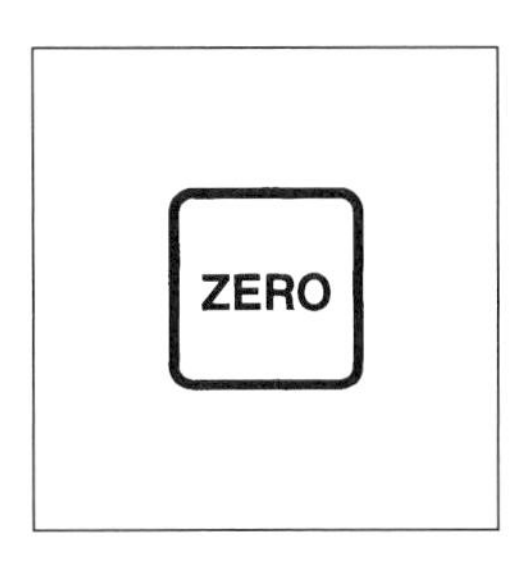

7. Clean the outside of the sample vial with a towel.

8. Place the sample vial into the adapter with the HACH logo facing the front of the instrument. Place the cover on the adapter.

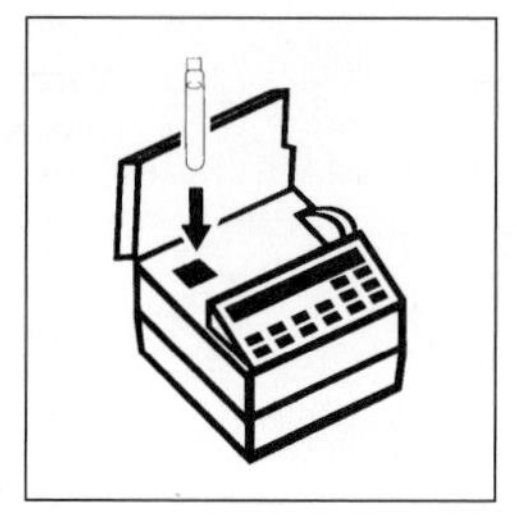

9. Press: **READ.** The display will show: **Reading ...** then the result in mg/L COD will be displayed.
Note: When using High Range Plus COD Digestion Reagent Vials multiply the reading by 10.
Note: For most accurate results with samples near 1500 or 15000 mg/L COD, repeat the analysis with a diluted sample.

SAMPLING AND STORAGE

Collect samples in glass bottles. Use plastic bottles only if they are known to be free of organic contamination. Test biologically active samples as soon as possible. Homogenize samples containing solids to assure representative samples. Samples treated with sulfuric acid to a pH of less than 2 (about 2 mL per liter) and refrigerated at 4°C can be stored up to 28 days. When significant amounts of preservative are used, a volume correction should be made for the extra acid by dividing the total volume (sample + acid) by the sample volume and multiplying this value by the final test reading.

ACCURACY CHECK

See p. 952 in the HACH Water Analysis Handbook, 3rd Ed., 1997.

PREPARING ORGANIC-FREE WATER

To prepare organic-free water with no measurable COD:

1. Pour 1.0 liter of deionized water with low COD in a 2-liter erlenmeyer flask.
2. Add the contents of one Potassium Persulfate Powder Pillow to the flask. Swirl to dissolve.
3. Suspend a UV lamp in the flask so the glass portion of the bulb is immersed and the black bakelite portion is above the solution. Follow the safety and operation instructions recommended in the UV lamp kit. Safety UV goggles should be worn for eye protection.
4. Irradiate the solution with UV light for at least two hours (overnight is fine).
5. Remove the lamp from the solution. Add one level 0.05-g scoop of nickel sulfate to the solution.
6. Heat the water to a boil. Remove the flask from the hot plate and cover it with a watch glass.
7. Let the flask cool to room temperature. The water will have zero oxygen demand. Seal the flask top with aluminum foil to prevent organic contamination. The water should stay free of oxygen demand for one week if properly sealed.

BLANKS FOR COLORIMETRIC MEASUREMENT

The blank may be used repeatedly for measurements using the same lot of vials. Store it in the dark. Monitor decomposition by measuring the absorbance at the appropriate wavelength (350, 420, or 620 nm). Zero the instrument in the absorbance mode, using the culture tube containing 5 mL of deionized water. Measure the absorbance of the blank and record the value. Prepare a blank when the absorbance has changed by approximately 0.010 absorbance units.

INTERFERENCES

Chloride is the primary interference when determining COD concentration. Each COD vial contains mercuric sulfate that will eliminate chloride interference up to the level specified in column 1 in the table below. Samples with higher chloride concentrations should be diluted. Dilute the sample enough to reduce the chloride concentration to the level given in column 2.

If sample dilution will cause the COD concentration to be too low for accurate determination, add 0.50 g of mercuric sulfate ($HgSO_4$) to each COD vial before the sample is added. The additional mercuric sulfate will raise the maximum chloride concentration allowable to the level given in column 3.

Vial type used	Maximum Cl^- concentration in sample (mg/L)	Suggested Cl^- concentration of diluted sample (mg/L)	Maximum Cl^- concentration in sample with 0.5 g $HgSO_4$ added (mg/L)
Ultra Low Range	2000	1000	NA
Low Range	2000	1000	8000
High Range	2000	1000	4000
Ultra High Range	20000	10000	40000

Bromide interference will not be controlled by mercuric sulfate.

QUESTIONS AND ANSWERS

1. Why would washing a vial with water and rinsing the vial with acetone or alcohol be a problem during the COD experiment?
   Acetone and alcohol would interfere in the COD test and produce a measurable COD.
2. What is the primary interference in the experiment?
   Chloride is the primary interference in this experiment. Mercuric sulfate will eliminate chloride interference up to a specific level as indicated in a table under "Interferences." In some cases, the sample may have to be diluted.
3. How important is organic-free water in this experiment?
   All water must be free of organic chemicals in order to have a zero oxygen demand. A procedure for preparing organic-free water is provided on p. 953 of the HACH Water Analysis Handbook, 3rd Ed., 1997.

REFERENCES

1. *HACH Water Analysis Handbook,* 3rd Ed. (Loveland, CO.: HACH Company, 1997).
2. *Standard Methods for the Examination of Water and Wastewater:* OXYGEN DEMAND (CHEMICAL). 16th Ed., E. Greenberg, R. R. Trussell, L. S. Clesceri and Mary Ann H. Franson (eds.), pp. 532–533, Washington, DC, American Public Health Association, 1985).

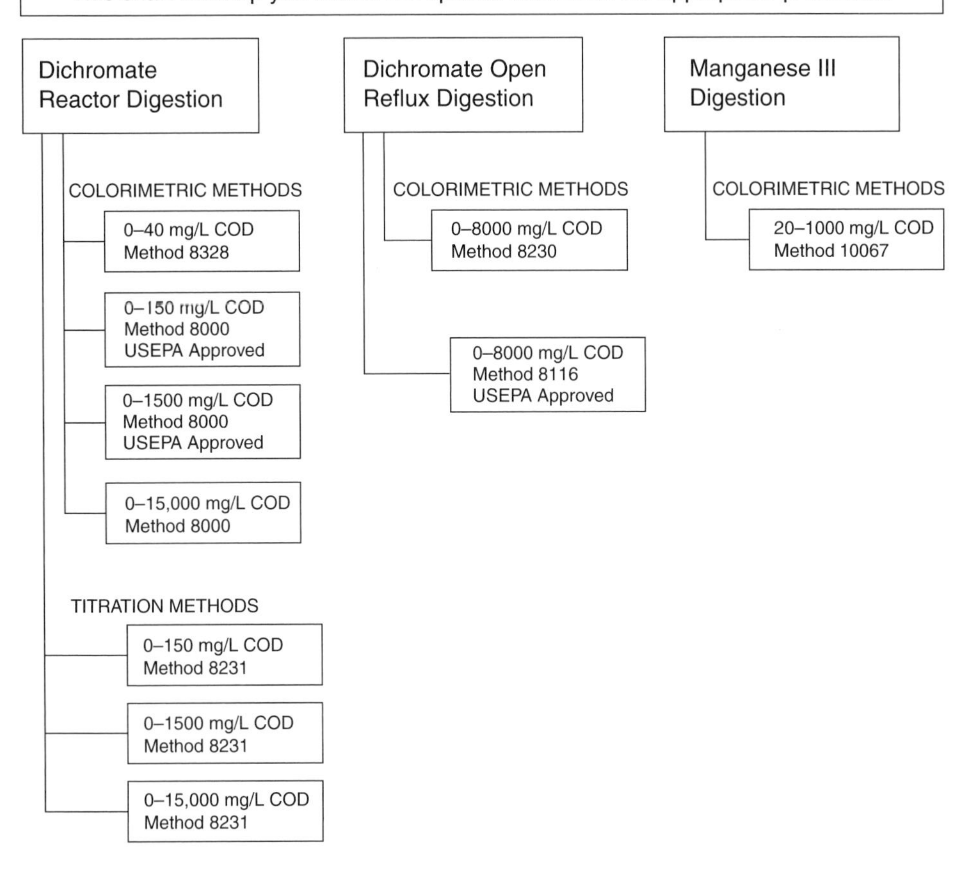

DISSOLVED OXYGEN

Dissolved oxygen (DO) is essential for the life of fish and other aquatic organisms. DO is an indicator of corrosivity of water, photosynthetic activity, septicity and "life of the water body." The distribution of dissolved oxygen in water depends upon depth, turbulence and mixing, temperature, light, sludge deposits, microbial action, respiration of oxygen consuming species, and photosynthetic activity (oxygen producers) of phytoplankton. Wind currents and water turbulence tend to increase the amount of dissolved oxygen in water. Oxygen levels in flowing water increase with disturbance caused by rocks, fallen trees and waterfalls. A significant variation in dissolved oxygen concentration can be observed along a stream or river.

The depletion of dissolved oxygen is primarily responsible for the ecological degradation of rivers and lakes. Bacteria use dissolved oxygen in the decaying process of plant and animal matter in water. A significant increase in Biochemical Oxygen Demand (BOD) in water causes a rapid decline in dissolved oxygen (DO). As water flows rapidly downstream away from organic wastes from a populated area, BOD will decrease and DO will ultimately increase as long as the river can be cleansed through natural processes.

Untreated sewage and rotting materials (such as garbage, leaves, sewage, and wood) are often responsible for oxygen depletion in water bodies. Wastes require a great amount of oxygen to decompose. For this reason, the dissolved oxygen levels in water are often monitored near dumps, sewage treatment plants, industrial sites, and areas of high population density.

Water with a dissolved oxygen content of about 4 mg/L is essential for the survival of most aquatic plants and animal life. Depending on the temperature of the water, the dissolved oxygen content in quality waters will be in the range of 8 to 15 mg/L for adequate fish population (pike, walleye, or bass). Waters containing organic suspended solids or considerable amounts of pollution may display some errors in measurements.

The activity of living organisms may cause rapid oxygen depletion. These effects can be minimized by immediate analysis at the site of water collection. The sample of water should not be stirred or agitated when collecting the sample. The sampler should be tilted slightly so that water can run into the sampler without bubbling. The stored samples and most dyes used for DO analysis should be protected from the sunlight.

When collecting water samples below surface, entrapping air bubbles in the water sample must be avoided. Special samplers and equipment are used to collect water samples from the depths of a lake, ocean, pond, or stream. The collected sample of water should be tested immediately.

A portable Dissolved Oxygen System is a microprocessor-based digital meter with an attached dissolved oxygen probe. The battery powered unit displays temperature in °C and dissolved oxygen in either mg/L (milligrams per liter) or % air saturation. The approximate salinity of the water and the approximate altitude of the region will be required to accurately calibrate the modern DO meter. All calibrations should be completed at a temperature which is as close as possible to the sample temperature. After turning off the instrument, recalibration of the instrument may be required before taking measurements. In general, dissolved oxygen readings by instruments are only as good as the calibration technique.

Dissolved oxygen in water can be determined by several types of procedures. Common test procedures include titration, Dissolved Oxygen (DO) meters and probes, colorimetric methods, and spectrophotometric procedures. Measurements are usually performed on water supplies, wastewater, aquaculture systems, boiler feedwater, industrial influents and effluents, water in research applications, and water habitats.

In this experiment, a water sample reacts with manganous sulfate and alkaline iodide-azide reagent to form an orange-brown precipitate. After acidification of the sample, the floc reacts with iodide to produce free iodine as triiodide in proportion to the oxygen concentration. The iodine is titrated with sodium thiosulfate to a colorless endpoint.

DISSOLVED OXYGEN (1 to greater than 10 mg/L DO)

Method 8215*
Azide Modification of Winkler Method using the Digital Titrator

*Compliments of HACH Company. The DR/820, DR/850 and DR/890 Colorimeters measure HRDO (High Range DO—Method 8166) and the DR/850 and DR/890 Colorimeters determine DO by the Indigo Carmine Method (Low range DO- Method 8316) during field studies.

Using a 300-mL BOD Bottle*

1. Collect a water sample in a clean 300-mL BOD Bottle.
Note: Allow the sample to overflow the bottle for 2 to 3 minutes to ensure air bubbles are not trapped.
Note: If samples cannot be analyzed immediately, see "Sampling and Storage" following these steps.

2. Add the contents of one Manganous Sulfate Powder Pillow and one Alkaline Iodide-Azide Reagent Powder Pillow.

3. Immediately insert the stopper so air is not trapped in the bottle. Invert several times to mix.
Note: A flocculent precipitate will form. It will be orange-brown if oxygen is present or white if oxygen is absent. The floc settles slowly in salt water and normally requires 5 additional minutes before proceeding to Step 5.

4. Wait until the floc in the solution has settled. Again invert the bottle several times and wait until the floc has settled.
Note: Waiting until floc has settled twice assures complete reaction of the sample and reagents.

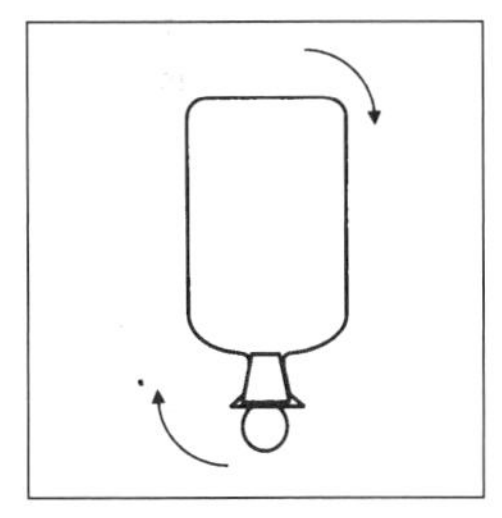

5. Remove the stopper and add the contents of one Sulfamic Acid Powder Pillow. Replace the stopper without trapping air in the bottle and invert the prepared sample several times to mix.
Note: The floc will dissolve and leave a yellow color if oxygen is present.

*If the procedure is modified, a 60-mL BOD bottle can be used to determine Dissolved Oxygen.

6. Select a sample volume and Sodium Thiosulfate Titration Cartridge corresponding to the expected dissolved oxygen (DO) concentration from Table 2.34.

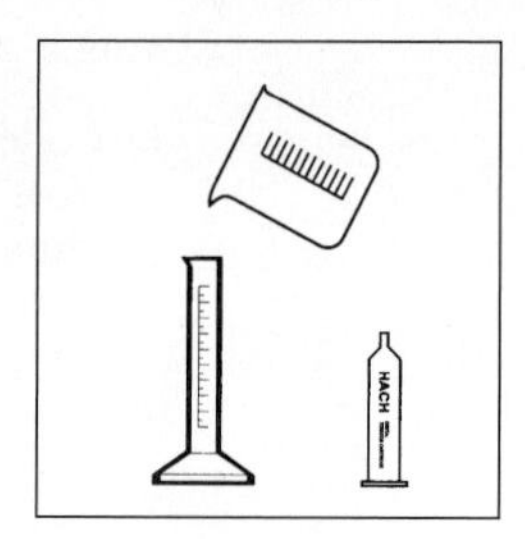

7. Insert a clean delivery tube into the titration cartridge. Attach the cartridge to the titrator body.

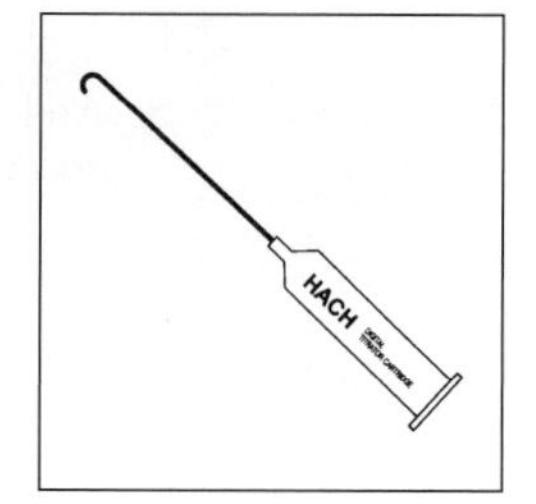

8. Turn the delivery knob to eject a few drops of titrant. Reset the counter to zero and wipe the tip.
Note: For added convenience use the TitraStir stirring apparatus.

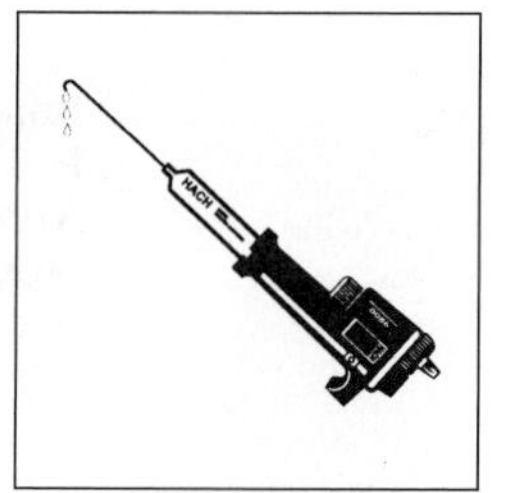

9. Use a graduated cylinder to measure the sample volume from Table 2.34. Transfer the sample into a 250-mL erlenmeyer flask.

TABLE 2.34

Range (mg/L DO)	Sample volume (mL)	Titration cartridge (N $Na_2S_2O_3$)	HACH Catalog number	Digit multiplier
1–5	200	0.200	22675-01	0.01
2–10	100	0.200	22675-01	0.02
> 10	200	2.000	14401-01	0.1

10. Place the delivery tube tip into the solution and swirl the flask while titrating with sodium thiosulfate to a pale yellow color.

11. Add two 1-mL droppers of Starch Indicator Solution and swirl to mix.
Note: A dark blue color will develop.

12. Continue the titration to a colorless end point. Record the number of digits required.
Note: See DO saturation table in Oxygen, Dissolved section of Appendix A of the HACH Water Analysis Handbook, 3rd Ed., 1997, p. 1252–1253.

13. Calculate: Digits Required × Digit Multiplier = mg/L Dissolved Oxygen

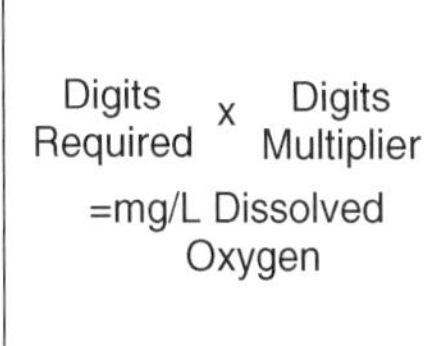

DISSOLVED OXYGEN (1 to greater than 10 mg/L DO)

Method 8332* using the Digital Titrator
Using a 60-mL BOD Bottle

*Compliments of HACH Company. The DR/820, DR/850 and DR/890 Colorimeters measure HRDO (High Range DO—Method 8166) and the DR/850 and DR/890 Colorimeters determine DO by the Indigo Carmine Method (Low range DO—Method 8316) during field studies.

1. Collect a water sample in a clean 60-mL glass-stoppered BOD Bottle.
Note: Allow the sample to overflow the bottle for 2 or 3 minutes to ensure air bubbles are not trapped.
Note: If samples cannot be analyzed immediately, see Sampling and Storage following these steps.
Note: Follow this procedure when using the 60 mL glass-stoppered BOD bottle supplied with DREL Portable Laboratories.

2. Add the contents of one Dissolved Oxygen 1 Reagent Powder Pillow and one Dissolved Oxygen 2 Reagent Powder Pillow.

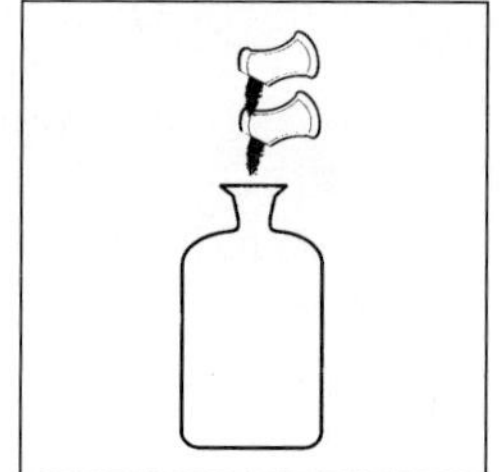

3. Immediately insert the stopper so air is not trapped in the bottle. Invert several times to mix.
Note: A flocculent precipitate will form. It will be orange-brown if oxygen is present or white if oxygen is absent. The floc settles slowly in salt water and normally requires 5 additional minutes before proceeding to Step 5.

4. Wait until the floc in the solution has settled and the top half of the solution is clear. Again invert the bottle several times and wait until the floc has settled.
Note: Results are not affected if the floc does not settle or if some of the reagent powder does not dissolve.

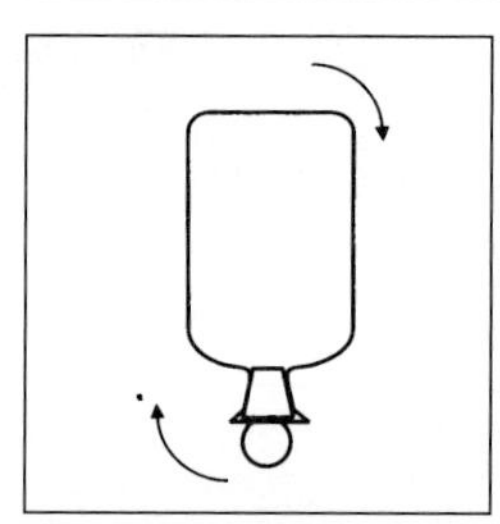

5. Remove the stopper and add the contents of one Dissolved Oxygen 3 Powder Pillow. Replace the stopper without trapping air in the bottle and invert several times to mix.
Note: The floc will dissolve and leave a yellow color if oxygen is present.

6. Accurately measure 20 mL of the prepared sample and transfer it to a 250-mL erlenmeyer flask.

7. Attach a clean straight-stem delivery tube to a 0.2000 N Sodium Thiosulfate Titration Cartridge. Twist the cartridge onto the titrator body.

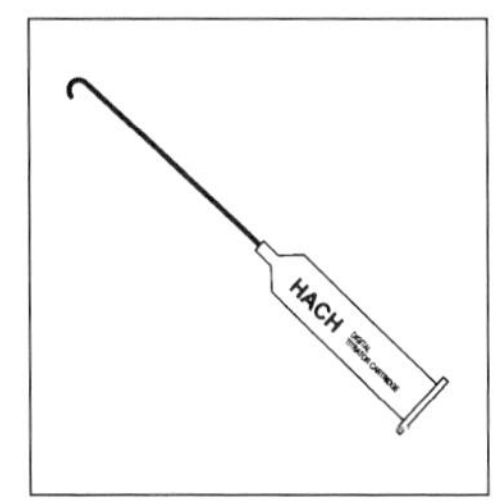

8. Flush the delivery tube by turning the delivery knob to eject a few drops of titrant. Reset the counter to zero and wipe the tip.
Note: For added convenience use the TitraStir stirring apparatus.

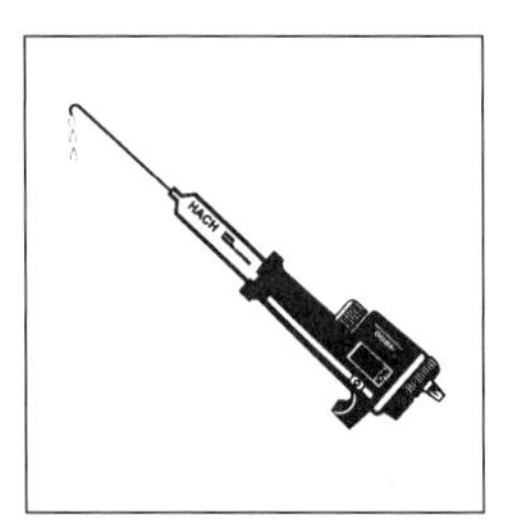

9. Titrate the prepared solution with 0.2000 N Sodium Thiosulfate until the sample changes from yellow to colorless. Record the number of digits.

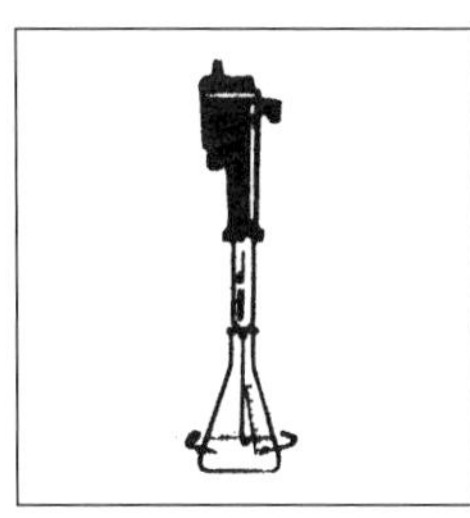

10. Calculate:
(Digits Required × 0.1 = mg/L Dissolved Oxygen)

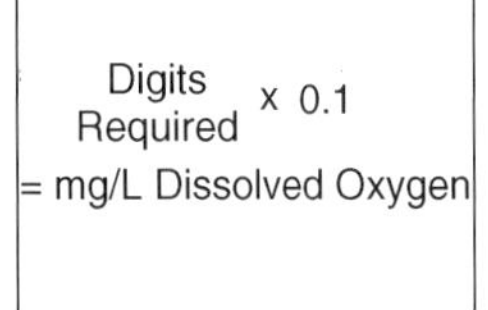

SAMPLING AND STORAGE

Sampling and sample handling are important in obtaining meaningful results. The dissolved oxygen content of the sample changes with depth, turbulence, temperature, sludge deposits, light, microbial action, mixing, travel time, and other factors. A single dissolved oxygen test rarely reflects the over-all condition of a body of water. Several samples taken at different times, locations and depths are recommended for most reliable results.

Collect samples in clean BOD Bottles (see Step 1). If storage is necessary, run Steps 1 to 4 of the procedure and store in the dark at 10 to 20°C. Seal the bottle with water by pouring a small volume of water into the flared lip area of a stopper bottle. Snap a BOD Bottle Cap over the flared lip. Samples preserved like this can be held for 4 to 8 hours. Begin with Step 6 when analyzing.

ACCURACY CHECK

See p. 895 of the HACH Water Analysis Handbook, 3rd Ed., 1997.

INTERFERENCES

Nitrite interference is eliminated by the azide in the reagents. Other reducing or oxidizing substances may interfere. If these are present, use an alternate method, such as the High Range Dissolved Oxygen Method (colorimetric) in the HACH Water Analysis Handbook, 3rd Ed., 1997, or a dissolved oxygen electrode.

QUESTIONS AND ANSWERS

1. What is the relationship between Biochemical Oxygen Demand (BOD) and Dissolved Oxygen (DO)?
   DO decreases due to the fact that BOD increases.
2. How do you prevent air from collecting in the BOD bottle?
   The bottle should be tilted slightly and submerged partially in the water. Allow the water to flow into the bottle ***without*** *bubbling. After the bottle is near full, submerse the bottle gently until it fills. Below the water surface, tilt the bottle upright and stopper the bottle. If any bubbles or air pockets are observed in the sample, pour out the contents of the bottle and perform the filling procedure again.*
3. What is the standard test for dissolved oxygen?
   The standard test for dissolved oxygen is the Winkler Method or the Azide Modification of the Winkler Method.
4. What color is the flocculent precipitate of a sample of water containing oxygen in a 300 mL BOD bottle?
   The flocculent precipitate will be orange-brown if the water contains dissolved oxygen. If the water does not contain oxygen, the floc will be white.
5. How does salt water impact the formation of floc during the DO test?
   The floc will settle very slowly in salt water. Floc settling usually will require an additional 5 to 8 minutes settling time depending on the amount of salt in the water (density).

6. What is the level of dissolved oxygen in water?
Approximately 4 mg/L DO is required to maintain most aquatic plants and animal life. Quality waters often have DO levels of 8mg/L to 14 mg/L. This value is temperature dependent.

REFERENCES

1. *HACH Water Analysis Handbook,* 3rd Ed. (Loveland, CO.: HACH Company, 1997).
2. *WARD's Snap Test,* Instant Water Quality Test Kits, WARD's Natural Science Establishment, Inc., Rochester, N.Y.
3. *Standard Methods for the Examination of Water and Wastewater:* OXYGEN (DISSOLVED), 16th Ed., A. E. Greenberg, R. R. Trussell, L. S. Clesceri and Mary Ann H. Franson (eds.), pp. 413–421, Washington, DC, American Public Health Association, 1985.

OZONE IN WATER

Ozone is a fire and explosion risk in contact with organic materials, but is relatively safe to handle in water. The substance is toxic by inhalation and is listed as a strong irritant. The EPA standard for ambient air is 0.12 ppm.

Ozone, O_3, is a strong oxidizing agent that is often used to treat water and wastewater in place of chlorine. Ozone is used to treat industrial wastes, deodorize air and sewage gases, remove chlorine from nitric acid, oxidize phenols and cyanides, oxidize materials in several chemical processes, bleach common items such as waxes and textiles, identify double bonds in organic chemistry, prepare peroxides, and act as a bactericide. Often, ozone is used in swimming pools, spas, bottled water and beverage industries.

Ozone provides microbial sterilization and disinfection, and can eliminate bad tastes and odor in water and destroy organic materials that cause coloring in water. It also oxidizes metals such as ferrous ion and reduced forms of manganese to form insoluble oxides in water. The insoluble metal oxides can be precipitated or filtered from water.

Ozone loss during sample collection and handling can be a major problem. Ozone loss can easily occur by warming the water sample and by disturbing or shaking the sample. The aqueous sample containing ozone should be collected and analyzed immediately without transfer from one container to another.

The indigo reagent is light sensitive and should be stored in the dark at all times. The indigo dye reacts quantitatively with ozone. The blue color of indigo is immediately oxidized in proportion to the amount of ozone in the sample and is measured at 600 nm with a colorimeter (DR/850 or DR/890) or a DR/2010 Spectrophotometer. Chlorine interference is avoided by reagents present in the ampul used for testing.

OZONE (0 to 0.25 mg/L O_3, 0 to 0.75 mg/L O_3 or 0 to 1.50 mg/L O_3)

Method 8311 for water*
Indigo Method using AccuVac Ampuls

*Compliments of HACH Company. Sample analysis may be performed with the DR/850 and DR 890 Colorimeters and the DR/2010 Field Spectrophotometer.

1. Enter the stored program number for ozone (O_3) AccuVac ampuls.
Press: **4 5 4 ENTER** for low range (0-0.25 mg/L) OR
Press: **4 5 5 ENTER** for mid range (0-0.75 mg/L) OR
Press: **4 5 6 ENTER** for high range (0-1.50 mg/L).
The display will show: **Dial nm to 600.**

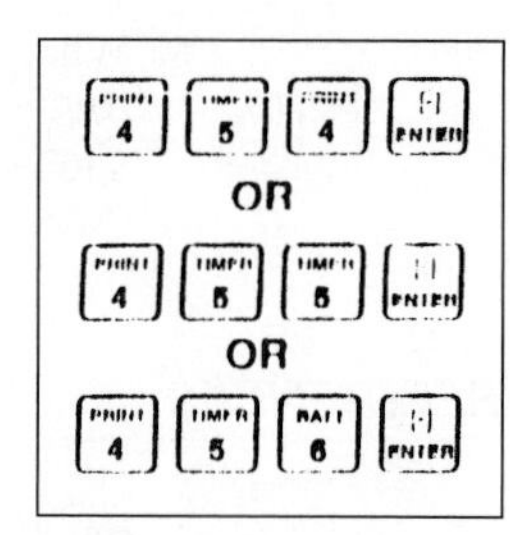

2. Rotate the wavelength dial until the small display shows: **600 nm.** When the correct wavelength is dialed in the display will quickly show: **Zero sample,** then:
for #454, **mg/L O_3 Indigo L;**
for #455, **mg/L O_3 Indigo M;**
for # 456, **mg/L O_3 Indigo H.**

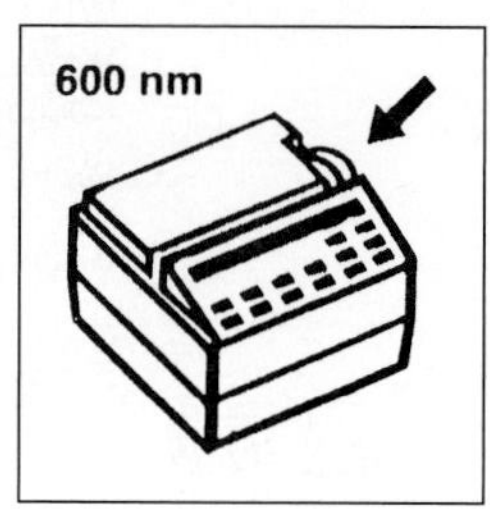

3. Gently collect at least 40 mL of sample in a 50-mL beaker.
Note: Samples must be analyzed immediately and cannot be preserved for later analysis.

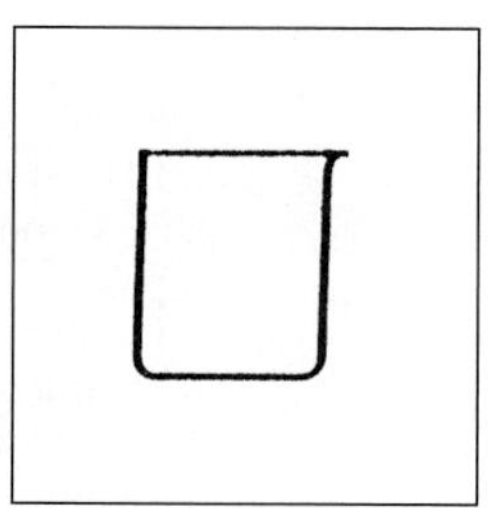

4. Collect at least 40 mL of ozone free water (blank) in another 50-mL beaker.
Note: Ozone-free water used for the blank may be deionized water or tap water.

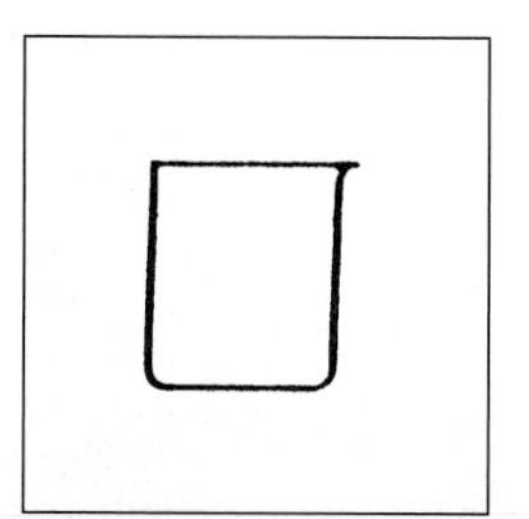

5. Fill one Indigo Ozone Reagent AccuVac Ampul with the sample and one ampul with the blank.
Note: Keep the tip immersed while the ampul fills.

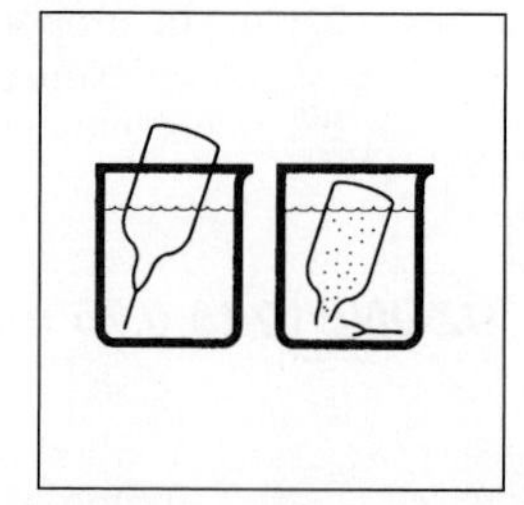

6. Quickly invert both ampuls several times to mix. Wipe off any liquid or fingerprints.
Note: Part of the blue color will be bleached if ozone is present. (The sample will be lighter than the blank).

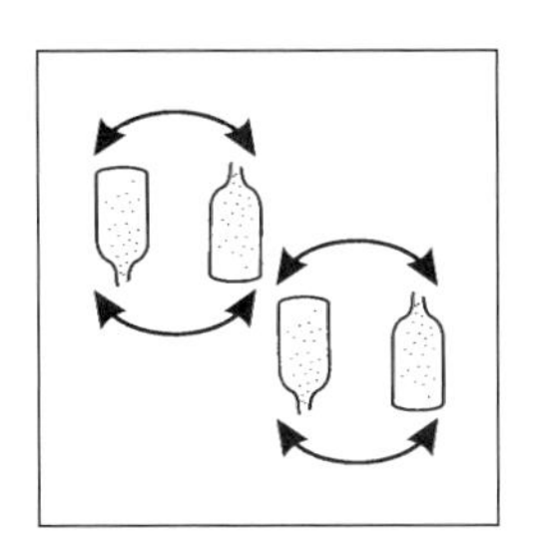

7. Place the AccuVac Vial Adapter into the cell holder.
Note: Place the grip tab at the rear of the cell holder.

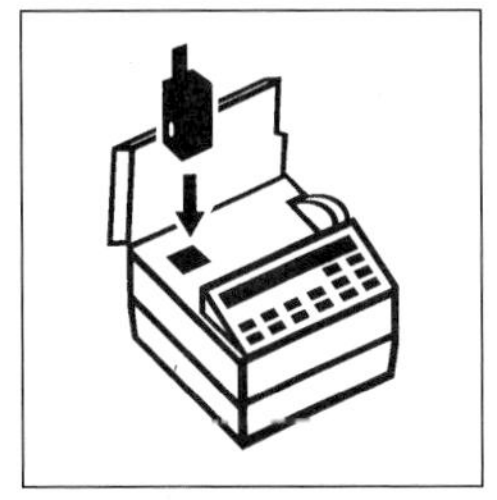

8. Place the **sample** AccuVac Ampul into the cell holder. Close the light shield.

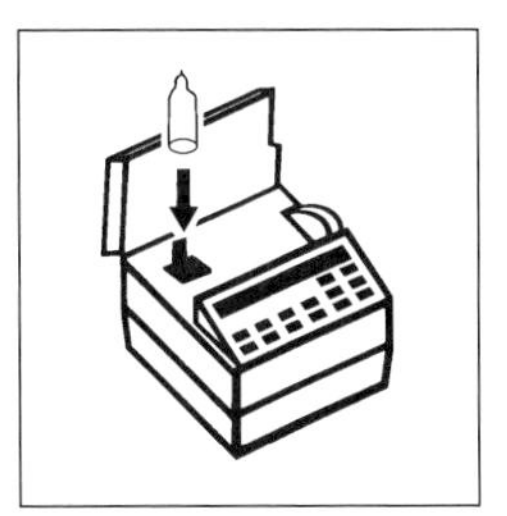

9. Press: **ZERO.** The display will show: **Zeroing ...** then: **0.00 mg/L O_3.**

10. Place the AccuVac Ampul containing the **blank** into the cell holder. Close the light shield.

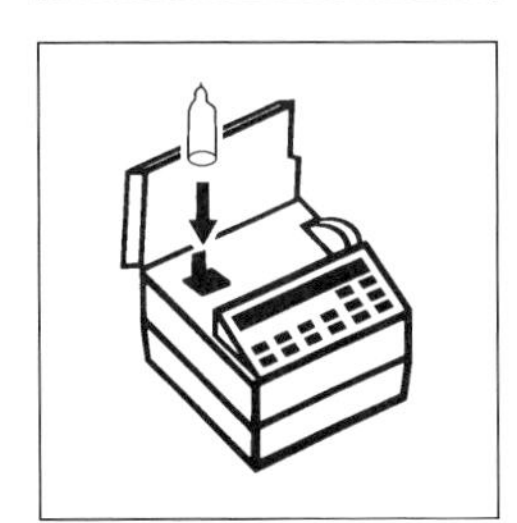

11. Press: **READ.** The display will show: **Reading ...** then the result in mg/L ozone (O_3) will be displayed.

SAMPLING

The chief consideration when collecting a sample is to prevent the escape of ozone from the sample. The sample should be collected gently and analyzed immediately. Warming the sample, or disturbing the sample by stirring or shaking will result in ozone loss. After collecting the sample, do not transfer it from one container to another unless absolutely necessary.

STABILITY OF INDIGO REAGENT

Indigo is light-sensitive. Therefore, the AccuVac Ampuls should be kept in the dark at all times. However, the indigo solution decomposes slowly under room light after filling with sample. The blank ampul can be used for multiple measurements during the same day.

QUESTIONS AND ANSWERS

1. Why should the indigo reagent be stored in the dark?
The indigo reagent is light sensitive and should be stored in the dark at all times.

2. Why should extra precautions be taken while handling the water sample containing ozone?
Ozone can easily escape from the water sample. Avoid warming, shaking, and stirring the sample. Transfer of the sample from one container to another can decrease the amount of ozone in the sample. Gently collect the sample and analyze it immediately.

3. Could this method be applied to analyze ozone in the air?
The method could be used for the determination of ozone in the air by absorption of the ozone in the indigo reagent for an appropriate period of time.

REFERENCES

1. *HACH Water Analysis Handbook,* 3rd Ed. (Loveland, CO.: Hach Company, 1997).
2. *Standard Methods for the Examination of Water and Wastewater:* OZONE (RESIDUAL), 16th Ed., A. E. Greenberg, R. R.Trussell, L. S. Clesceri and Mary Ann H. Franson (eds.), pp. 426–427, Washington, DC, American Public Health Association, 1985.
3. Hawley's Condensed Chemical Dictionary, 11th Ed., rev'd N. Irving Sax and Richard J. Lewis (New York, N.Y.: Van Nostrand Reinhold, 1987).

PALLADIUM

Palladium is a heterogeneous hydrogenation catalyst during organic alkene reactions with hydrogen to form alkanes. Palladium and hydrogen form a Lindlar Catalyst which easily converts alkynes to Syn-alkenes. Also, palladium is a catalyst in the hydrogenation of alkylbenzenes to form cycloalkane products. Palladium is a primary constituent during the reduction of aromatic nitro-organic compounds to form amine complexes. In addition, the metal is used in industry as a dehydrogenation catalyst.

Palladium is used in activator solutions for the autocatalytic plating of printed circuit boards. Additional uses of palladium include the coating of metals with electrolytic and electroless baths in industrial applications. The metal is often used to decrease reaction time of chemical reactions and provides coatings in a variety of applications in the plastic industry and the electronics industry. Palladium chloride is used to determine the presence of radioactive iodine during monitoring procedures by precipitation, ion-exchange or distillation methods.

The primary applications of tests for palladium are remediation sites, industries, and wastewater treatment facilities. The test for palladium involves a hypobromite oxidation procedure which destroys any reducing agent present in palladium-tin activator solutions. After excess hypobromite is destroyed by sodium metabisulfite, palladium reacts with N,N'-dimethyldithiooxamide under acid conditions to form a yellow solution proportional to the amount of palladium present. The presence of copper in solution is masked by 2,2'-bipyridine.

The method of analysis may be applied to palladium as long as the metal has been dissolved into solution. Palladium plating baths and solutions containing the dissolved metal may be analyzed by this method if the samples have been diluted to less than 250 mg/L palladium.

PALLADIUM (0 to 250 mg/L)

N,N'-Dimethyldithiooxamide Method
Method 8144 for palladium -tin activator baths

1. Enter the stored program number for palladium (Pd). Press: **460 ENTER** The display will show: **Dial nm to 420**
Note: The Pour-Thru Cell can be used if rinsed well with deionized water immediately after use.

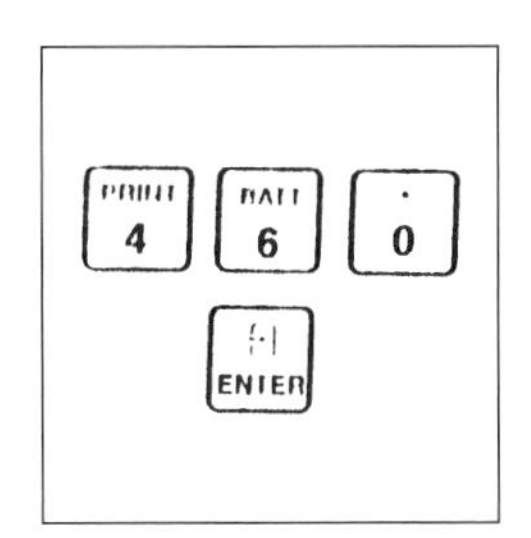

2. Rotate the wavelength dial until the small display shows: **420 nm.** When the correct wavelength is dialed in the display will quickly show: **Zero Sample** then: **mg/L Pd.**

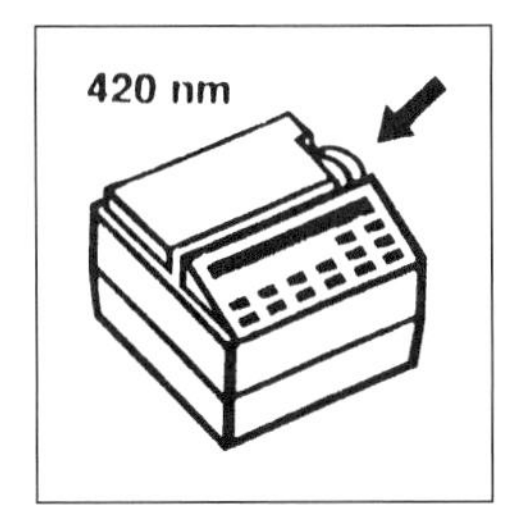

3. Fill two 25-mL mixing cylinders to the 20-mL mark with deionized water.

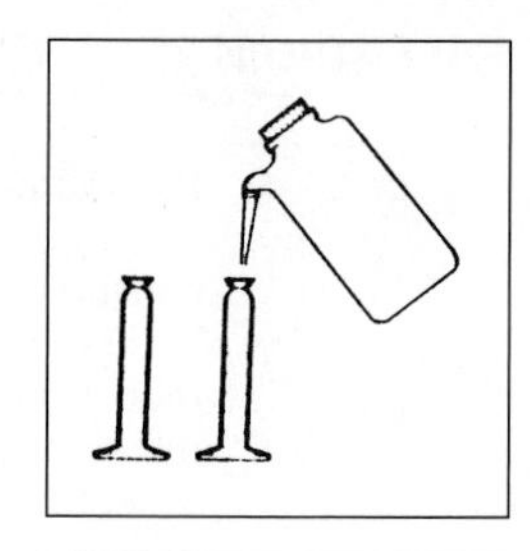

4. Add 5.0 mL of concentrated hydrochloric acid to each cylinder. Swirl to mix.
Note: Use a Mohr pipet and pipet filler.

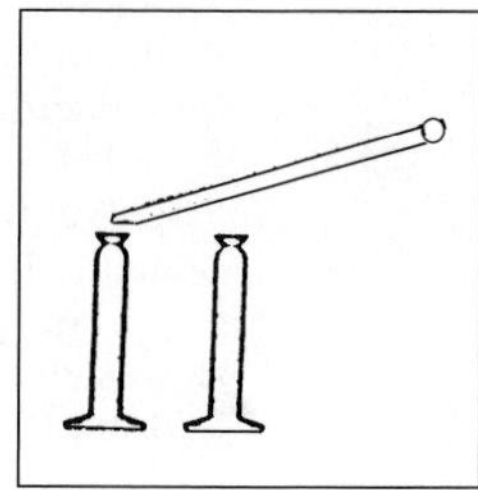

5. Add one 0.2-g scoop of 2,2'-bipyridine to each cylinder. Cap and invert to mix.
Note: Because of the 2,2'-bipyridine density, approximately 0.1 g fills a 0.2-g scoop.

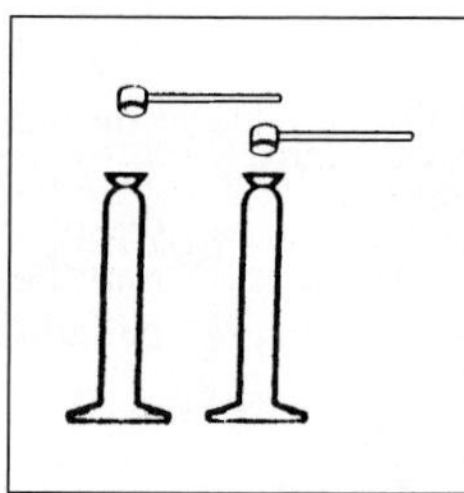

6. Add 0.5 mL of palladium activator sample to one of the mixing cylinders (the prepared sample). The other cylinder is blank. ***Note:*** Use a Tensette Pipet or 0.5 mL volumetric pipet.

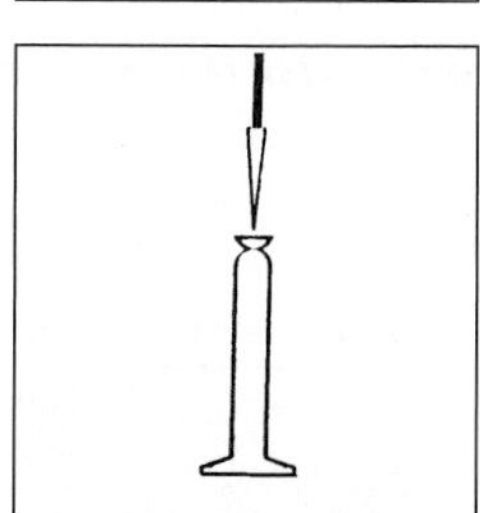

7. Add one Chromium 1 Reagent Powder Pillow to each cylinder. Stopper. Invert several times to mix.

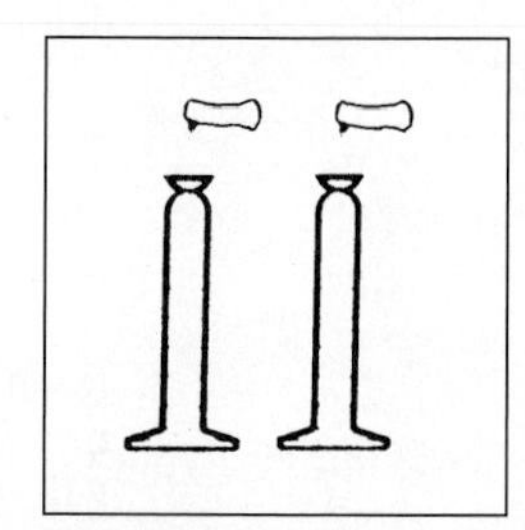

8. Press: **Shift Timer** A five-minute reaction will begin.

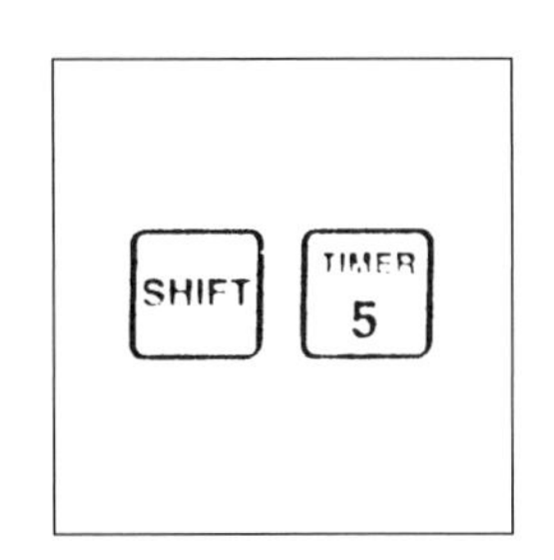

9. When the timer beeps, add one Sodium Meta Bisulfite Reagent Powder Pillow to each cylinder. Stopper. Invert several times to mix.

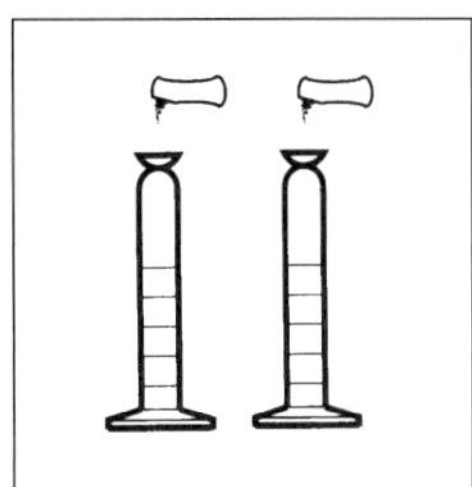

10. Add 10 mL of N,N'-Dimethyldithiooxamide Indicator Solution to each cylinder. Stopper. Invert several times to mix. ***Note:*** A pressure build up sometimes occurs when the indicator is added. Use a paper towel to remove the stopper if this occurs.

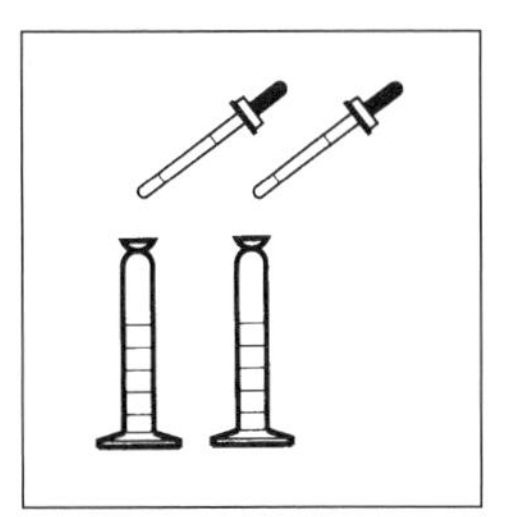

11. Press: SHIFT TIMER A two-minute reaction period will begin.

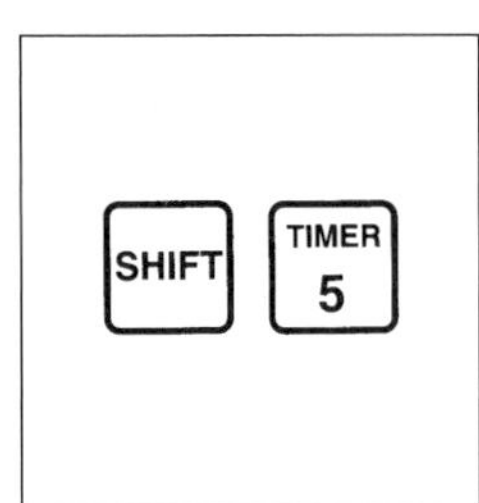

12. When the timer beeps, the display will show: **mg/L Pd** Pour the contents of each cylinder into a sample cell.

13. Place the blank in the cell holder. Close the light shield.

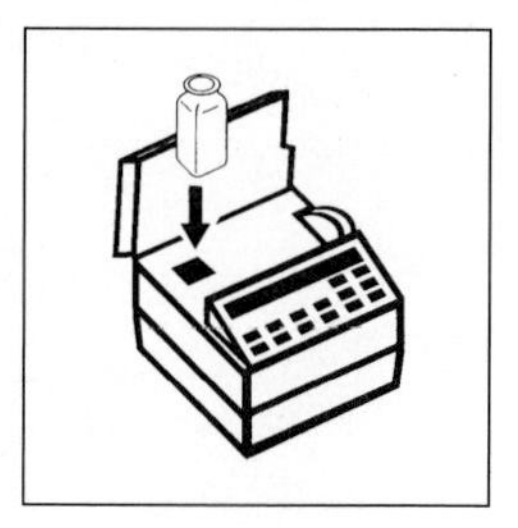

14. Press: **Zero** The display will show: **Zeroing ...** then: **0. mg/L Pd**

15. Place the prepared sample into the cell holder. Close the light shield. Press: **READ.** The display will show: **Reading ...** then the result in mg/L Pd will be displayed.

SAMPLING AND STORAGE

Samples must be collected in clean glass or plastic bottles. Activator bath samples should be analyzed as soon as possible after collection. Samples should be mixed well before pipetting into graduated cylinders.

INTERFERENCES

Copper and nickel do not interfere under the reaction conditions of the test. Gold does not interfere in concentrations up to 50 mg/L.

QUESTIONS AND ANSWERS

1. Where is palladium found?
The metal is often found in waste effluent streams from industries and as buried wastes in remediation sites. Palladium is primarily found in the electronics industry, the

chemical-producing industries, polymer industries, metals industry, and radioactivity detection laboratories.

2. What are some of the practical uses for palladium and compounds of palladium?
Palladium forms catalysts with hydrogen to reduce multiple bond compounds to alkanes and to alkenes. The metal is used in a number of chemical industries to form amines and dehydrogenation catalysts. Additional applications include autocatalytic plating baths for circuit boards, electrolytic and electroless plating baths for metal and electronic parts, and catalysts in polymer producing industries. Palladium chloride is used to detect radioactive iodine at low levels in drinking water.

3. What is the purpose of 2,2'-bipyridine in the experiment?
The chemical masks copper during the analysis procedure.

4. Can this method of analysis be applied to plating baths where the concentration of palladium is high?
This method of analysis can be applied to plating baths by diluting the sample from the bath to less than 250 mg/L palladium.

REFERENCES

1. *DR/2010 Spectrophotometer Procedures Manual,* 3rd Ed. (Loveland, CO.: HACH Company, 1998).
2. *Standard Methods for the Examination of Water and Wastewater:* RADIOACTIVE IODINE, 16th Ed., A. E. Greenberg, R. R. Trussell, L. S. Clesceri and Mary Ann H. Franson (eds.), pp. 674–679, Washington, DC, American Public Health Association, 1985.
3. *Hawley's Condensed Chemical Dictionary,* 11th Ed., rev'd by N. Irving Sax and Richard J. Lewis (New York, N.Y.: Van Nostrand Reinhold, 1987).

pH VALUE

A definition for pH is pH= $-$ log [H^+]. The term pH or hydrogen ion activity represents the intensity of the acidic or basic character of a solution. As pH increases pOH decreases correspondingly, because pOH + pH = 14. At pH=7.0, the activities of hydrogen ion and hydroxyl ion are equal, and as a result a neutral system exists at 25°C.

Natural waters usually have pH values in the range of 6 to 8.3, although pH values as low as 4 and as high as 9 have been observed. Most natural waters are slightly basic due to the presence of bicarbonates and carbonates of the alkali and alkaline earth metals.

pH may be measured by a pH strip (paper), pH indicator, or a pH meter. The pH meter can be a single or double electrode model. For routine analysis, a pH meter accurate and reproducible to 0.1 pH unit with a 0 to 14 pH range and a temperature adjustment system is appropriate for measurements of pH on water sources. The pH of water and wastewater during treatment processes depends on the degree of water softening, coagulation, precipitation, disinfection, corrosion control, and acid-base neutralization processes.

Buffer capacity is the amount of strong acid or base needed to change the pH value of a one-liter sample by one pH unit. Buffer solutions of pH 4, 7, and 10 are used to calibrate the meter before measurements. For reliable results, it is important to rinse the electrode with distilled water or a portion of the sample and gently blot the electrode dry with a kimwipe or another type of "soft tissue." For most samples of water, the calibration of the meter against the pH 7 buffer solution and rinsing and gently drying of the electrode(s) before measuring the sample pH is standard practice. When several samples are measured in a few hours, calibration of the pH meter before measuring each sample is not necessary.

Over a one-hour period, the calibration of the pH meter should be rechecked with buffer solution after the measurement of 10 samples of water. However, it is necessary to properly rinse the electrode(s) after and before each pH measurement.

A pH electrode should be stored in a buffer solution (pH= 7) when not in use. The glass sensing membrane needs proper care to function properly. Proper storage instructions are provided in the manual provided with the pH meter or the electrode.

A colorimetric pH determination using phenol red indicator is practical for field studies. This method uses the DR/820, DR/850 or DR/890 Colorimeters to determine the pH of water and wastewater samples in the pH range of 6.5 to 8.5. This method is appropriate for a number of environmental applications, but has limited applicability to industrial sites.

pH

Method 8156 for water and wastewater*
Hach One Combination pH Electrode Method with an EC10 Portable pH Meter
USEPA approved for reporting**

SAMPLE pH MEASUREMENT (CALIBRATION IS REQUIRED)

1. Press the Dispenser Button once (it will click).

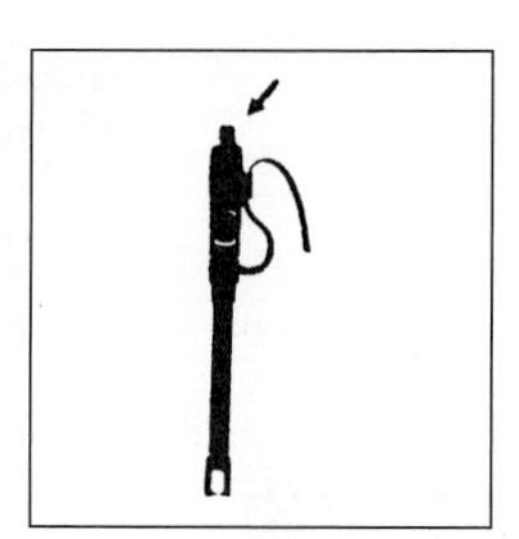

2. Inspect the end of the electrode for the presence of gel. If gel is not visibly oozing from the tubing, press the Dispenser Button again.

3. Place the electrode in the sample. Be sure the entire sensing end is submerged and that there are no air bubbles under the electrode.

*Compliments of HACH Company. pH is included in the DR/820, DR/850 and DR/890 Colorimetric procedures.
**Procedure is equivalent to USEPA method 150.1 and Standard 4500-H^+ for water and wastewater.

4. Record the pH value when the display is stable.

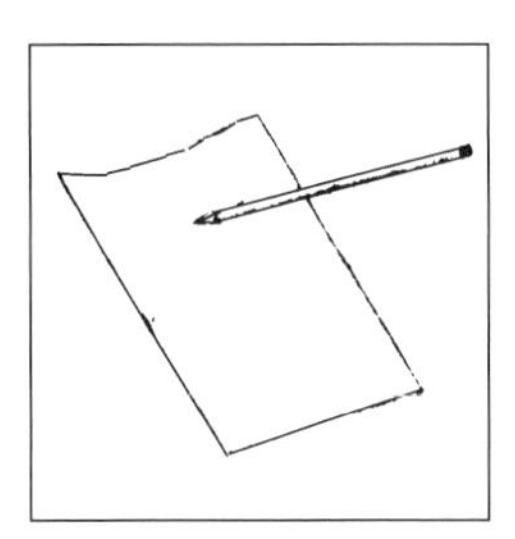

5. Rinse the electrode thoroughly with deionized water and blot dry.

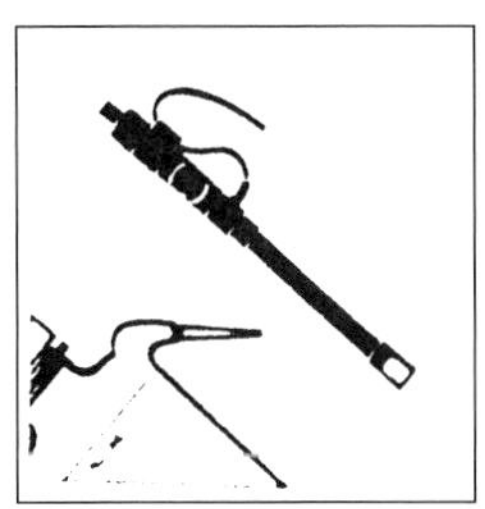

TWO-POINT CALIBRATION IN AUTOMATIC MODE WITH TEMPERATURE PROBE

1. Press the **POWER** key. The display will light.

2. Ensure the meter is in pH mode.
Note: Be sure no air bubbles are trapped inside the glass bulb of the electrode and that gel is present at the tubing end.

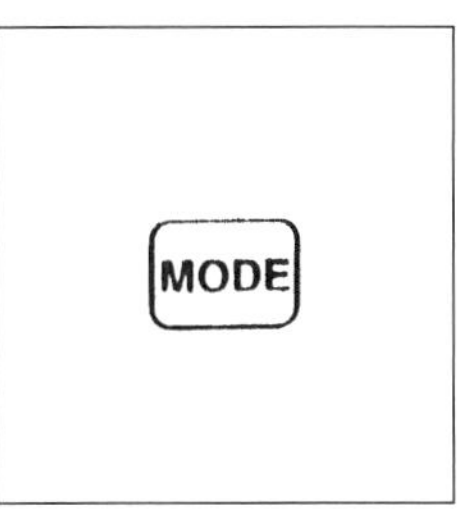

3. Press the **CAL** key. The slope data of the previous calibration or a default slope will appear. Then the meter will go into measure mode with a "P1" (Point 1) in the display.
Note: HACH buffers are available as powders or solutions. They are color-coded for added convenience.

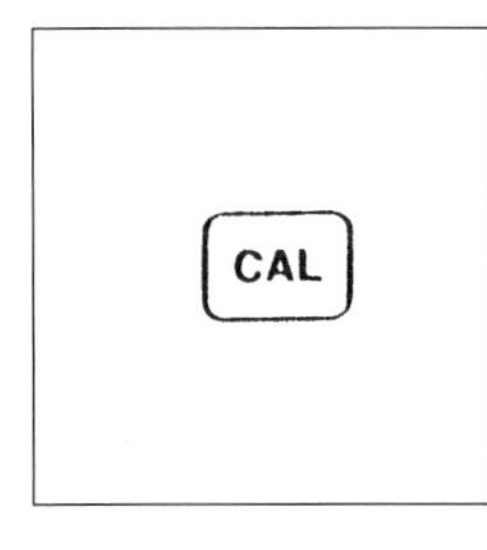

4. Place the electrode into a pH 4 buffer solution. The temperature will show the actual solution temperature and "ATC" (Automatic Temperature Compensation) will appear in the display if the temperature probe is connected.
Note: Buffer solution (pH 4, 7 and 10) may be used in any order.

5. When the display stabilizes, the meter will beep and show the temperature corrected value for the pH 4 buffer and a ready indicator. Press **YES** to accept this value for the first buffer.

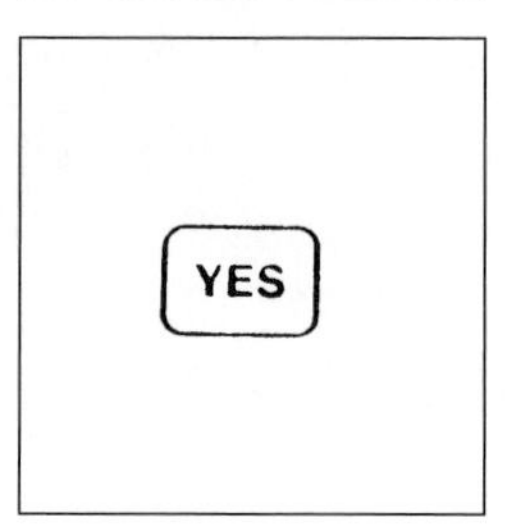

6. Rinse the electrode with deionized water and blot dry.

7. Place the electrode in the pH 7 buffer and press the dispenser button.

8. When the display stabilizes it will beep and show the temperature corrected value for the pH 7 buffer and "ready." Press the **YES** key to accept this value for the second buffer.
Note: Pressing any key other than the **YES** key at this point will cancel the calibration values entered and the meter will default to the previous slope and adjust the offset only.

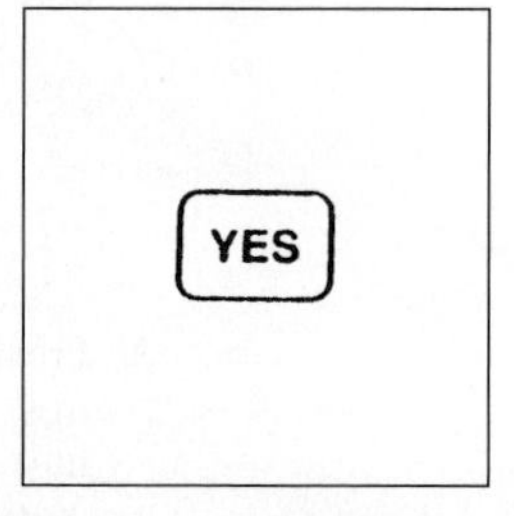

9. The slope will appear in the display. It should be 59.0 ± 3 (mV/decade). Press the **YES** key to accept this slope. Rinse the electrode with deionized water or a portion of the sample to be measured. Blot dry.

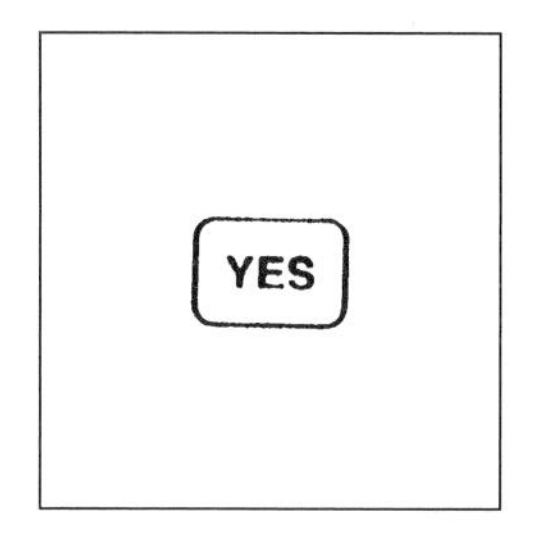

10. Place the electrode into the sample. Press the Dispenser Button. The meter will measure the sample pH.
Note: Repeat Step 10 for each sample. Check the calibration against 7.0 pH Buffer after every 10 samples. A value outside the range of 6.95 to 7.05 indicates the analyst should recalibrate.

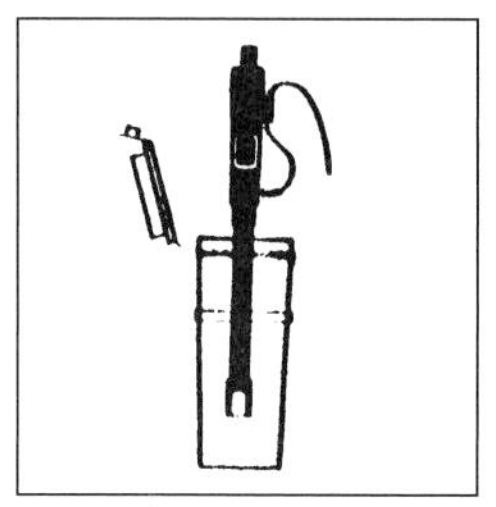

11. To review, press the **CAL** key to show the electrode slope in the display. The slope should be 59.0 ± 3 mV/decade. The meter will then attempt to re-calibrate. Press **MODE** again to return to measurement mode.
Note: For other calibrations or more complete operation instructions, refer to the instrument manual.

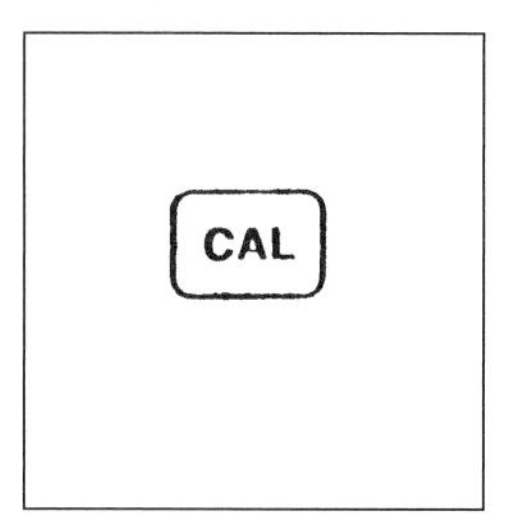

SAMPLING AND STORAGE

Collect samples in clean plastic or glass bottles. Fill completely and cap tightly. Cool to 4°C (39°F) and determine pH within six hours. If samples cannot be analyzed within six hours, report the actual holding time with the results.

INTERFERENCES

Acid error is negligible.

Sodium error, usually present in alkaline solutions, is low, but increases at pH values higher than pH 11.

QUESTIONS AND ANSWERS

1. What procedures should be standard practice when measuring the pH of several solutions?
The electrode of the pH meter must be rinsed with distilled water or sample water and gently blotted with a kimwipe or another soft tissue before each measurement. After each

measurement of pH, the same routine should be implemented. Calibration of the pH meter should be checked after every 10 samples measured over a one-hour period. If the pH meter is used intermittently, check the calibration of the pH meter for every sample if the buffer (pH = 7) solution has a measured pH value outside the range of 6.95 to 7.05.

2. What is the pH range of natural waters?
Most natural water has a pH range of 6 to 8.3. However, some natural waters have been observed with pH as low as 4 and as high as 9. These waters are usually associated with "special" springs, mines and dumps.

3. What is a buffer?
A buffer solution is added to water to control pH of the solution when other contaminates or species are added to the solution. Buffer capacity is defined as the amount of strong acid or strong base needed to change the pH value of a one liter sample by one pH unit. These solutions are used to standardize the pH meter.

REFERENCES

1. *HACH Water Analysis Handbook,* 3rd Ed. (Loveland, CO.: HACH Company, 1997).
2. *Standard Methods for the Examination of Water and Wastewater:* pH VALUE, 16th Ed., A. E. Greenberg, R. R. Trussell, L. S. Clesceri and Mary Ann H. Franson (eds.), pp. 429–432, Washington, DC, American Public Health Association, 1985).

PHENOLS

Phenols, or hydroxybenzene compounds, can be found in potable waters, domestic wastewater, industrial wastewaters, and natural waters. Natural waters normally contain less than one microgram of phenol per liter. During chlorination of drinking water, a small amount of phenol in water can impart an objectionable taste to the water. Chlorination of water can produce small amounts of chlorophenols under the appropriate conditions. During water treatment, a process of superchlorination, ozonation, activated carbon adsorption, chlorine dioxide addition, or chloramine treatment is implemented to remove phenol from drinking water.

Oil refineries, coke plants and several chemical manufacturing plants produce phenolic wastes and can contaminate waters with phenol and phenolic compounds. Phenolic compounds are used to make resins, salicylic acid, phenolphthalein indicator, chlorophenol compounds, germicidal paints, disinfectants, pharmaceuticals, dyes and indicators, fuel oil sludge inhibitors, solvents, rubber chemicals, and 2,4-D. The compound is also used as a selective solvent for refining lubricating oils and acts as a slimicide in various applications.

The 4-aminoantipyrine method is a colorimetric method that determines phenol and ortho- and meta- substituted phenols. This method does not determine most of the para-substituted phenols such as paracresol which may be present in industrial wastewaters and in polluted surface waters. The reactive phenols combine with 4-aminoantipyrine in the presence of potassium ferricyanide to form a colored dye. After extraction of the dye from aqueous solution with chloroform, the amount of the antipyrine dye is measured at 460 nm with a spectrometer.

Wastewater and water samples may require pretreatment due to interferences by reducing agents, oxidizing agents, sulfides, or suspended matter. The pH of the sample should be in the range of 3 to 11.5 for accurate results. The type of phenolic compound impacts the sensitivity of the method. Since various types of phenolic compounds may be present in the sample, the test results are expressed as the equivalent concentration of phenol. This method is USEPA accepted for reporting wastewater analysis when the sample has been distilled.

PHENOLS (0 to 0.200 mg/L)

Method 8047 for water, wastewater and seawater*
4-Aminoantipyrine Method**
USEPA Accepted for reporting wastewater analysis (distillation is required; see Section 1).†

1. Measure 300 mL of deionized water in a 500-mL graduated cylinder.
Note: Analyze samples within four hours to avoid oxidation; see "Sampling and Storage" following these steps.

2. Pour the measured deionized water into a 500-mL separatory funnel (the blank).

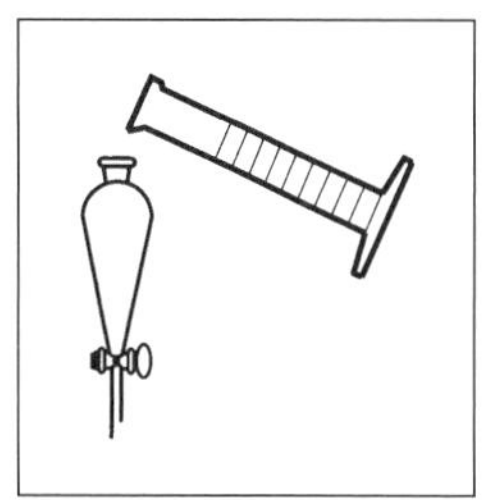

3. Measure 300 mL of sample in a 500-mL graduated cylinder.

4. Pour the measured sample into another 500-mL separatory funnel (the prepared sample).

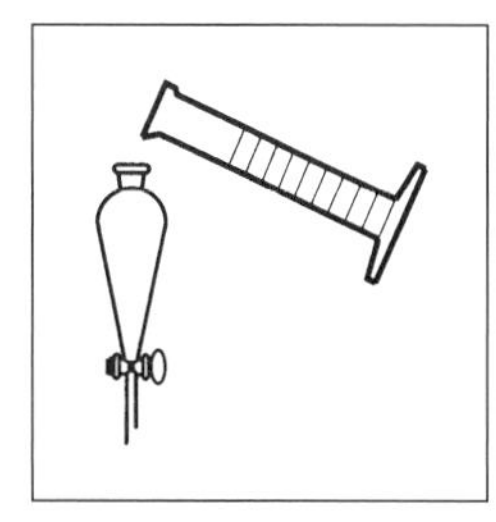

*Compliments of HACH Company.
**Adapted from *Standard Methods for the Examination of Water and Wastewater, 510 C.*
†Procedure is equivalent to USEPA Method 420.1 for wastewater.

5. Add 5 mL of Hardness 1 Buffer to each separatory funnel. Stopper. Shake to mix.

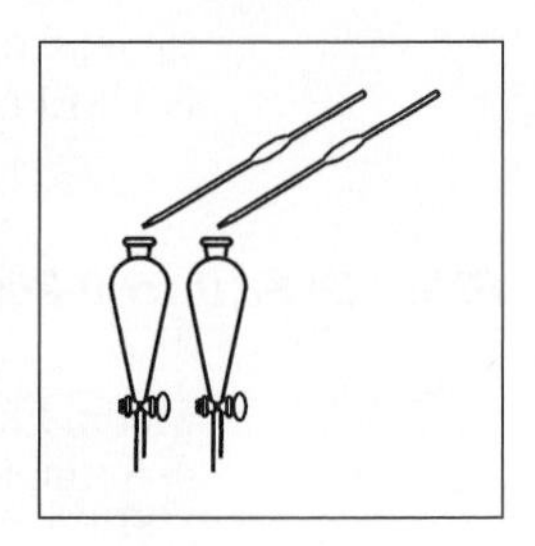

6. Add the contents of one Phenol Reagent Powder Pillow to each separatory funnel. Stopper. Shake to dissolve.
Note: Spilled reagent will affect test accuracy and is hazardous to skin and other materials.

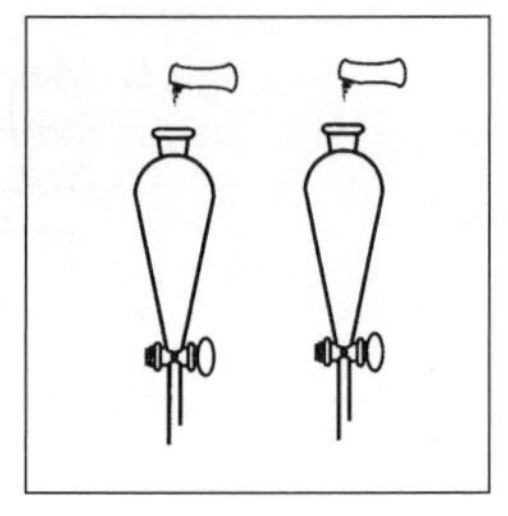

7. Add the contents of one Phenol 2 Reagent Powder Pillow to each separatory funnel. Stopper. Shake to dissolve.

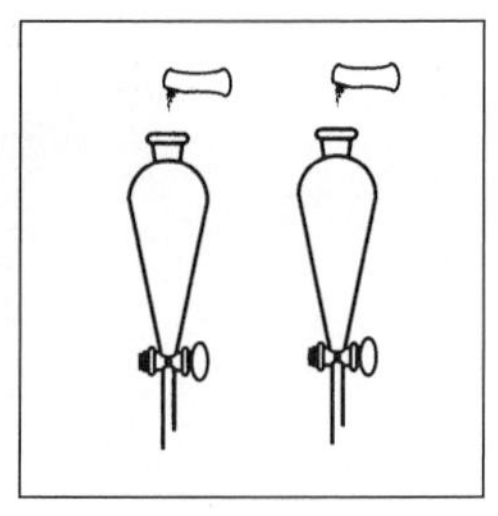

8. Add 30 mL of chloroform to each separatory funnel. Stopper each funnel.
Note: Use chloroform only with proper ventilation. A fume hood is ideal.

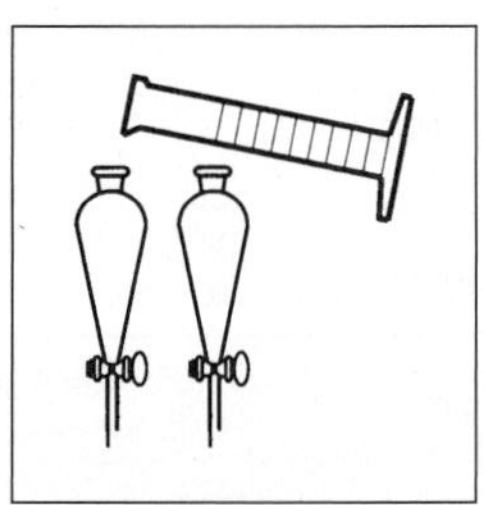

9. Invert each funnel and temporarily vent. Shake each funnel briefly and vent. Then vigorously shake each funnel for a total of 30 seconds.

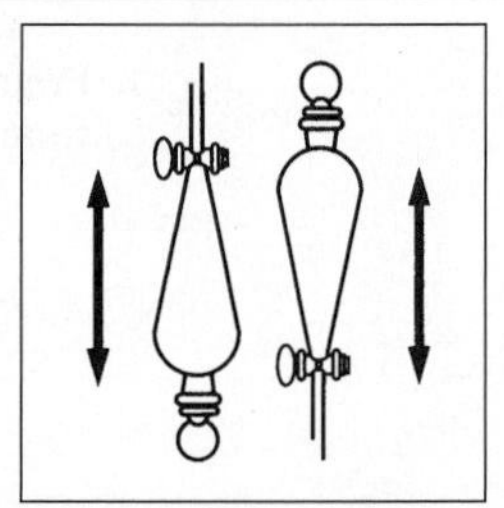

10. Remove the stoppers. Allow both funnels to stand until the chloroform settles to the bottom of the funnel.
Note: The chloroform will be yellow to amber if phenol is present.

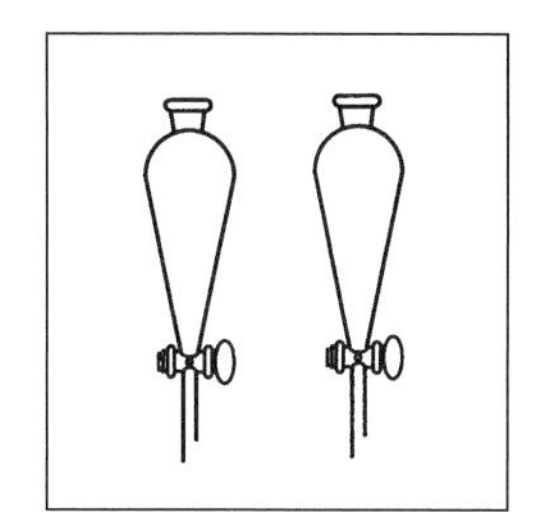

11. Insert a pea-sized cotton plug into the delivery tube of each funnel.

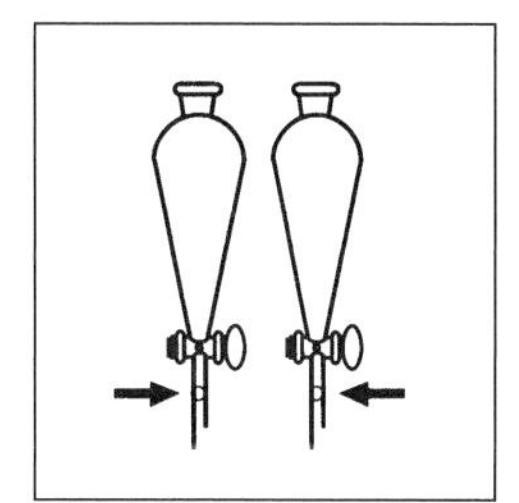

12. Drain the chloroform layer into separate 25-mL sample cells.
Note: The cotton removes any suspended water or particles from the chloroform.
Note: Proceed promptly through the rest of the procedure; chloroform will evaporate, causing high readings.

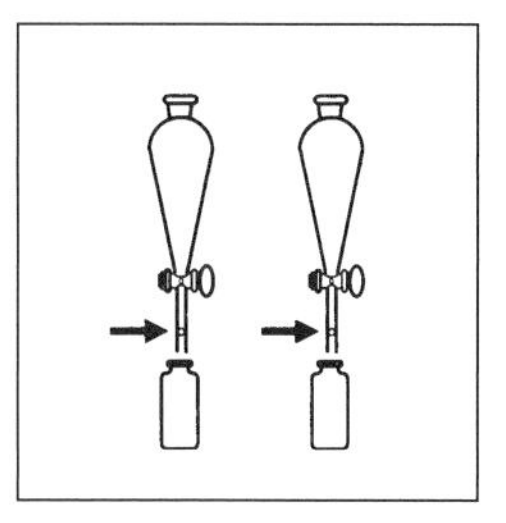

13. Enter the stored program number for phenols. Press: **4 7 0 ENTER.** The display will show: **Dial nm to 460.**
Note: The Pour-Thru Cell cannot be used with this procedure.

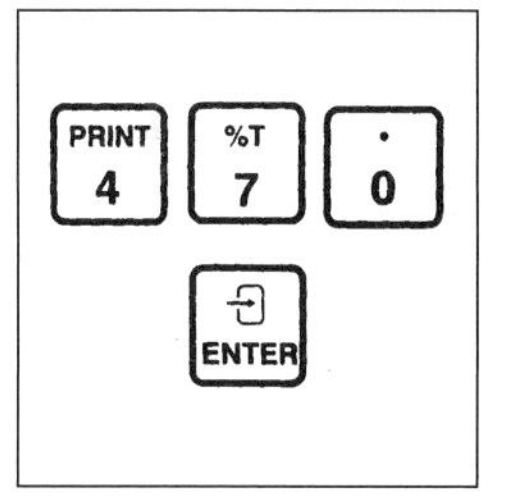

14. Rotate the wavelength dial until the small display shows: **460 nm.** When the correct wavelength is dialed in, the display will quickly show: **Zero Sample** then: **mg/L PHENOL.**

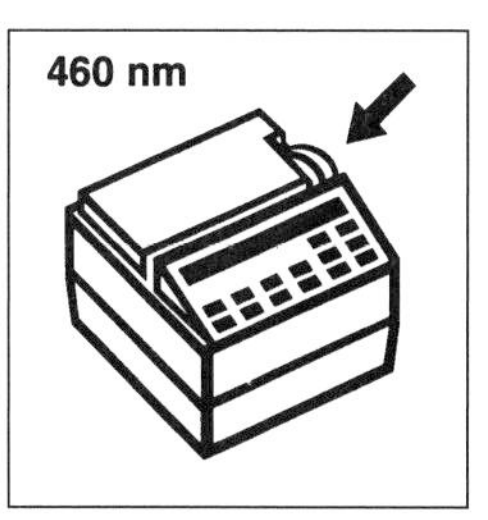

15. Place the blank into the cell holder. Close the light shield.

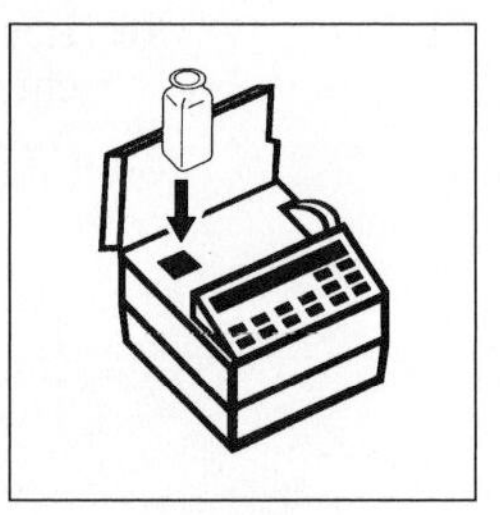

16. Press: **ZERO.** The display will show: **Zeroing ...** then: **0.000 mg/L PHENOL.**

17. Place the prepared sample into the cell holder. Close the light shield.

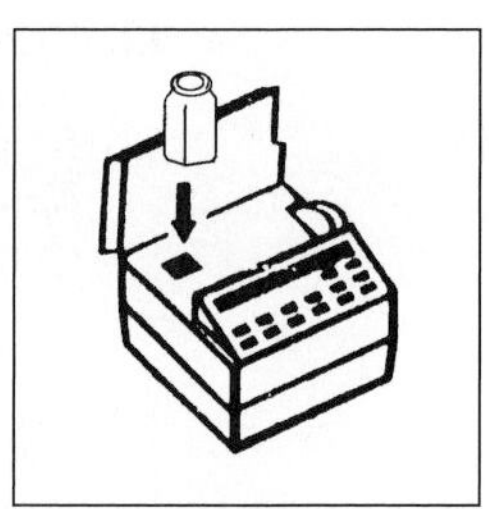

18. Press: **READ.** The display will show: **Reading ...** then the results in mg/L as phenol will be displayed.

SAMPLING AND STORAGE

The most reliable results are obtained when samples are analyzed within four hours after collection. The following storage instructions are necessary only when prompt analysis is impossible. Collect 500 mL of sample in clean glass containers and add the contents of two Copper Sulfate Powder Pillows. Adjust the sample pH to 4 or below with 10% Phosphoric Acid Solution. Store at 4°C (39°F) or lower and analyze within 24 hours.

ACCURACY CHECK

See p. 992 of the HACH Water Analysis Handbook, 3rd Ed., 1997.

INTERFERENCES

The sample pH must be between 3 and 11.5 for the best results. In the presence of sulfides or suspended matter, the following pretreatment will be necessary:

a) Fill a clean 500-mL graduated cylinder to the 350-mL mark with sample. Pour the sample into a clean 500-mL erlenmeyer flask.

b) Add the contents of one Sulfide Inhibitor Reagent Powder Pillow. Swirl to mix.

c) Filter 300 mL of the sample through a folded filter paper. Use this solution in Step 4.

Interference can be caused by reducing agents and oxidizing agents such as chlorine. Sample distillation as described in the following steps will eliminate interferences.

a) Set up the distillation apparatus for the test by assembling the general purpose distillation apparatus. Use the 500-mL erlenmeyer flask to collect the distillate. It may be necessary to elevate the flask with a laboratory jack. Place a stirring bar into the distillation flask.

b) Measure 300 mL of sample in a 500-mL graduated cylinder. Pour it into the distillation flask.

c) Using a serological pipet, add 1 mL of Methyl Orange Indicator Solution to the distillation flask. Turn on the heater power switch. Set the stir control to 5.

d) Add Phosphoric Acid Solution, 10%, drop-wise, until the indicator changes from yellow to orange.

e) Add the contents of one Copper Sulfate Powder Pillow and allow to dissolve (omit this step if copper sulfate was used to preserve the sample).

f) Turn on the water and adjust so a constant flow of water is maintained through the condenser. Set the heat control to setting 10.

g) Turn off the still after collecting 275 mL of distillate.

h) Fill a 25-mL graduated cylinder to the 25-mL mark with deionized water. Turn the still back on. Add the water to the flask. Resume heating until another 25 mL of distillate is collected.

QUESTIONS AND ANSWERS

1. Why should the samples be analyzed within 4 hours of sampling?
   The most reliable results are obtained if the sample is analyzed almost immediately after collection to avoid oxidation problems.
2. What are some of the interferences in the experiment?
   Sample pretreatment is required for sulfides and suspended matter in the sample. For bests results, the pH of the sample should be 3 to 11.5. Interferences caused by reducing agents and oxidizing agents can be avoided by distillation of the water sample in the presence of methyl orange indicator, phosphoric acid solution, and copper sulfate.
3. Why is sensitivity an important issue in this experiment?
   Method sensitivity varies with the type of phenolic compound present in the solution. The results are reported as the equivalent concentration of phenol.

REFERENCES

1. *HACH Water Analysis Handbook,* 3rd Ed. (Loveland, CO.: Hach Company, 1997).
2. *Standard Methods for the Examination of Water and Wastewater:* PHENOLS, 16th Ed., A. E. Greenberg, R. R. Trussell, L. S. Clesceri and Mary Ann H. Franson (eds.), pp. 556–560, Washington, DC, American Public Health Association, 1985.
3. *Hawley's Condensed Chemical Dictionary,* 11th Ed., rev'd by N. Irving Sax and Richard J. Lewis (New York, N.Y.: Van Nostrand Reinhold, 1987).

PHOSPHATE

High quantities of phosphate in water can contribute to eutrophication problems in rivers, ponds, and lakes. Excess phosphate in water promotes an algal bloom. The rapid growth of algae and phytoplankton can decrease the amount of dissolved oxygen in a body of water. Algae reproduce and die rapidly, and the decomposition of algae depletes dissolved oxygen in the water. Thus, the organism population can be seriously minimized.

Phosphates occur in many different forms in our environment and exhibit various degrees of solubility and physical properties. Phosphates are classified as orthophosphates, condensed phosphates (pyrophosphates, metaphosphates, and polyphosphates), and organically bound phosphates. These compounds can be found in water and wastewater treatment facilities, commercial detergents, fertilizers, body wastes, food additives and dietary supplements, textile processing, pharmaceuticals, fireproofing materials, medicine, research, emulsifing agents, photographic developers, antifreezes, gasoline purification processes, and during manufacturing of galvanoplastics, paper, ceramic glazes, and dyes. Most human and animal waste contain some phosphate material.

A few industries treat phosphates as a waste which can be recovered and recycled as an usuable material. Some phosphate wastes are manifested for reuse or burial. Many phosphate containing materials are dumped in water bodies due to water and sewage treatment processes and detergents. Additional phosphate materials enter our water sources through the use of fertilizer, sewage dumping, and garbage disposal in landfills. Sources of phosphate in water supplies above ground or below ground are found in many everyday materials including food.

Phosphates exist under slightly varying conditions of pH. Reactive phosphorus, acid-hydrolyzable phosphorus and organic phosphorus can occur in both the dissolved and suspended fractions in a sample, resulting in difficult analyses and possibly erroneous results. Therefore, phosphate must be in the appropriate form for accurate results during analysis.

During analysis of water, samples must be collected and stored in glass bottles because phosphates may be adsorbed onto the walls of plastic bottles. All glass containers must first be rinsed with dilute hydrochloric acid and then with distilled water. Commercial detergents containing phosphate must be avoided when cleaning glassware used for phosphate analysis.

Orthophosphate is a primary constituent of water pollution. The Molybdenum Blue Test Method measures orthophosphates in water, wastewater, and seawater at 1 to 10 ppm. In the presence of orthophosphate, an acidic solution of ammonium molybdate is reduced to a blue colored solution by the reducing agent stannous chloride. Phosphates in the form of soluble pyrophosphates, metaphosphates, and polyphosphates will not respond to the test unless 25 mL of the sample is first boiled for 30 minutes with 2.0 mL of 5.25 N Sulfuric Acid Solution.

Sample turbidity may yield incorrect phosphate results. Also, analysis of water exposed to an illegal "moonshine" still on a stream bank can provide very high and inconsistent results for orthophosphates, depending on how far measurements are taken from the source of the still.

If the analyses for various types of phosphates include HACH equipment, reactive phosphorus (orthophosphate) can be determined by either the Amino Acid Method or the PhosVer 3 (Ascorbic Acid) Method using the DR/2010 Spectrophotometer, the DR/850, or the DR/890 Colorimeters in the field. Test strips can provide fast and "approximate" results in the range of 0 to 50 ppm phosphate. Total phosphorus can be determined by the same methods or similar field methods after persulfate digestion of the bound phosphate to orthophosphate.

REACTIVE PHOSPHORUS (0.0 to 2.50 mg/L PO_4^{3-})

PhosVer 3 (Ascorbic Acid) Method using Powder Pillows for Orthophosphate Analysis*
USEPA Accepted for wastewater analysis reporting**
Method 8048 for water, wastewater, and seawater†

1. Enter the stored program number for reactive phosphorus, ascorbic acid method, powder pillows. Press: **4 9 0 ENTER.** The display will show: **Dial nm to 890.**
Note: The Pour-Thru Cell can be used with 25-mL reagents only.

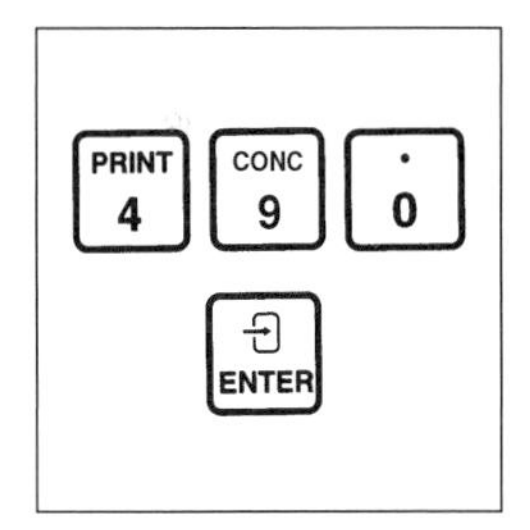

2. Rotate the wavelength dial until the small display shows: **890 nm.** When the correct wavelength is dialed in, the display will quickly show: **Zero Sample,** then: **mg/L PO_4^{3-} PV.**

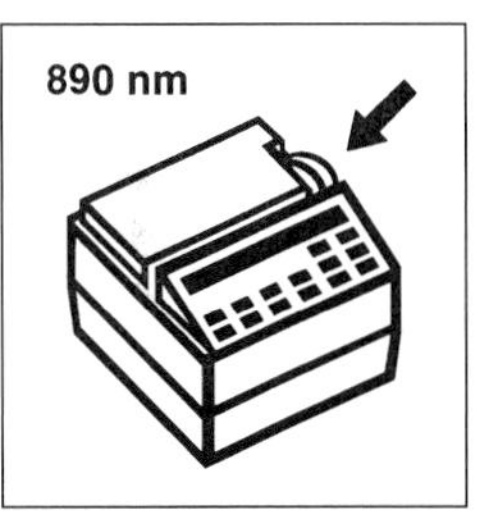

3. Insert a 10-mL Cell Riser into the cell compartment.

* Adapted from *Standard Methods for the Examination of Water and Wastewater.*
** Procedure is equivalent to USEPA method 365.2 and Standard Method 4500-P-E for wastewater.
† Compliments of HACH Company.

4. Fill a clean 10-mL sample cell with 10 mL of sample. (Clean and dry the outside of the vial with a soft tissue or Kimwipe).

5. Add the contents of one PhosVer 3 Phosphate Powder Pillow for 10-mL sample to the cell (the prepared sample). Swirl immediately to mix.
Note: A blue color will form if phosphate is present.

6. Press: **SHIFT TIMER.** A two-minute reaction period will begin.
Note: Use a 10 minute reaction period if determining total phosphorus following the acid-persulfate digestion.

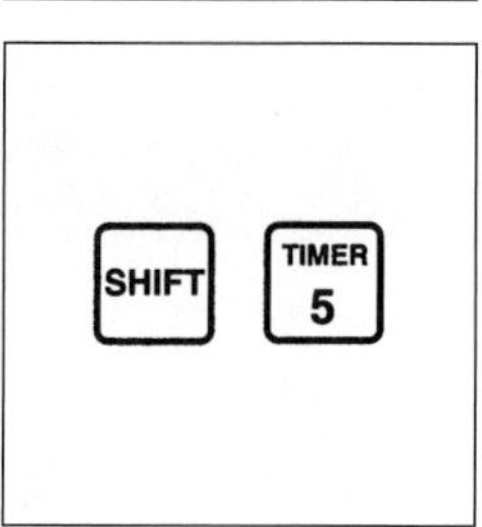

7. Fill a second 10-mL sample cell with 10 mL of sample (this is the blank).

8. When the timer beeps, the display will show: **mg/L PO_4^{3-} PV.** Place the blank into the cell holder. Close the light shield.

9. Press: **ZERO.** The display will show: **Zeroing ...** then: **0.00 mg/L PO_4^{3-} PV.**

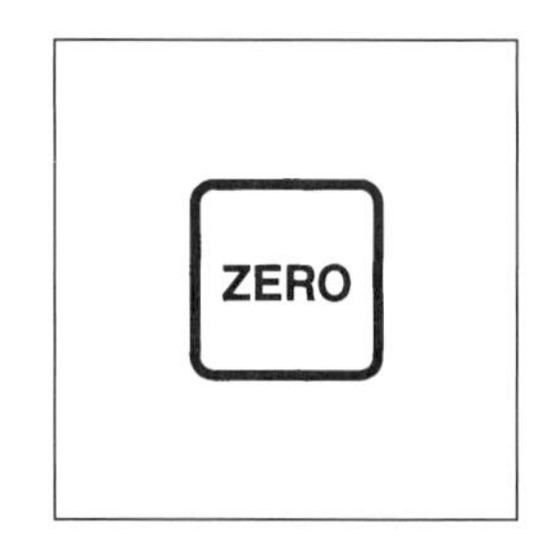

10. Place the prepared sample into the cell holder. Close the light shield.
Note: Run a reagent blank for this test. Use deionized water in place of the sample in Steps 4 and 7. Subtract this result from all test results run with this lot of PhosVer 3.

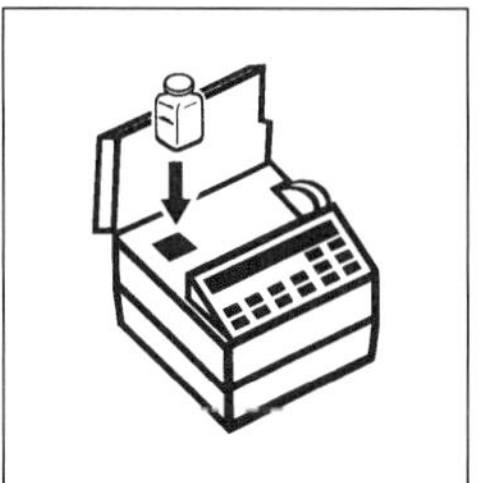

11. Press: **READ.** The display will show: **Reading ...** then the results in mg/L PO_4^{3-} will be displayed.

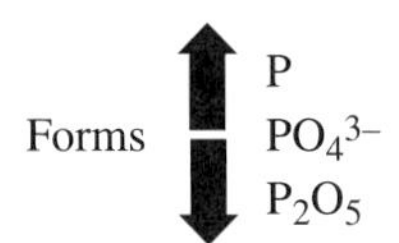

INTERFERENCES

Cleaning glassware with commercial detergents containing phosphates will increase the level of phosphate in the tested sample. Large amounts of turbidity may cause inconsistent results in the phosphate test. Highly buffered samples or extreme sample pH may exceed the buffering capacity of the reagents and require sample pretreatment. A few metal ions such as aluminun, chromium, iron, nickel, and zinc in solution may interfere with the test for phosphate when present in concentrations of 100 mg/L or more. Copper interferes at 10 mg/L. Silica interferes at 50 mg/L and silicates interfere above 10 mg/L. Refer to HACH Water Analysis Handbook, 3rd Ed., *Reactive Phosphorus,* p. 1015–1020.

TOTAL PHOSPHORUS

Method 8190 for Organic and Acid Hydrolyzable (meta-, pyro-, and poly-) phosphates*
Acid Persulfate Digestion Method**
USEPA Accepted for reporting wastewater analysis†

1. Measure 25 mL of sample into a 50-mL erlenmeyer flask using a graduated cylinder.
Note: Rinse all glassware with 1:1 Hydrochloric Acid Solution. Rinse again with deionized water.

2. Add the contents of one Potassium Persulfate Powder Pillow. Swirl to mix.

3. Add 2.0 mL of 5.25 N Sulfuric Acid Solution.
Note: Use the 1-mL calibrated dropper provided.

4. Place the flask on a hot plate. Boil gently for 30 minutes.
Note: Sample should be concentrated to less than 20 mL for best recovery. After concentration, maintain the volume near 20 mL by adding small amounts of deionized water. Do not exceed 20 mL.

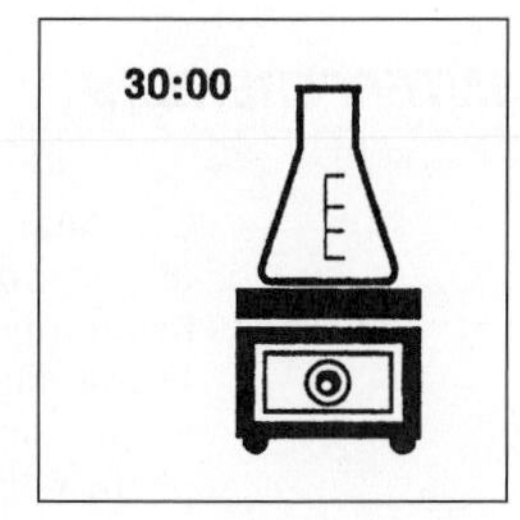

* Compliments of HACH Company.
** Adapted from *Standard Methods for the Examination of Water and Wastewater.*
† Procedure is equivalent to USEPA Method 365.2 and Standard Method 4500-P B,5 and P E.

5. Cool the sample to room temperature.

6. Add 2.0 mL of 5.0 N Sodium Hydroxide Solution. Swirl to mix.
Note: Use the 1-mL calibrated dropper provided. pH should be 8.2 pH after addition of sodium hydroxide.

7. Pour the sample into a 25 mL graduated cylinder. Return the volume to 25 mL. Proceed with the PhosVer 3 (Ascorbic Acid) Method using Powder Pillows for the determination of the total phosphorus content of the sample. The sample may have to be diluted before analysis by the PhosVer 3 (Ascorbic Acid Method)
Note: Use deionized water rinsings from the flask to adjust the volume. Results of the reactive phosphorus test at this point will include the organic phosphate plus the orthophosphate and the acid-hydrolyzable (*condensed forms of meta-, pyro-, and polyphosphates*) *phosphates. The organic phosphate concentration is determined by subtracting results of an acid hydrolyzable phosphorus test (See HACH Water Analysis HANDBOOK) from this result. Make sure that both results are in the same units before taking the difference.*

SAMPLING AND STORAGE

Collect samples in plastic or glass bottles that have been acid cleaned with 1:1 Hydrochloric Acid Solution and rinsed with deionized water. Do not use commercial detergents containing phosphate for cleaning glassware used in this test.

Analyze samples immediately after collection for best results. If prompt impossible, see the procedure listed under Total Phosphorus, HACH Water Analysis Handbook, 3rd Ed., p. 1040.

QUESTIONS AND ANSWERS

1. Which form of phosphate can be determined directly without digestion or pretreatment?
Orthophosphate is the only form which can be analyzed without pretreatment or digestion.

2. Why are phosphates added to water in municipal and industrial water treatment processes?
Phosphates, in polymer form, are added to water to control corrosion of pipes and equipment.

3. Why is it necessary to put restrictions on phosphate in laundry detergents?
People living on the waterfront often allow their wash water to drain directly into the lake, river, or stream. If the water contains phosphate, algal blooms will be prolific and be devastating to aquatic life. As huge quantities of algae die, their decaying biomass depletes the dissolved oxygen concentration in water and causes fish and other species to die.

4. What effects would a concentration of 20 ppm of orthophosphate have on a lake?
Anything above 10 ppm (mg/L) is considered to be too high. The resulting algal bloom would seriously deplete oxygen.

5. What is reactive phosphorus?
Reactive phosphorus is a measure of orthophosphate plus a small fraction of condensed phosphates (meta-, poly, and pyrophosphates) that may be hydrolyzed during the testing procedure.

REFERENCES

1. *WARD's Snap Test,* Instant Water Quality Test Kits, WARD's Natural Science Establishment, Inc., Rochester, N.Y.
2. *HACH Water Analysis Handbook,* 3rd Ed. (Loveland, CO.: HACH Company, 1997).
3. *Standard Methods for the Examination of Water and Wastewater:* PHOSPHORUS, 16th Ed., Method 424, A. E. Greenberg, R. R. Trussell, L. S. Clesceri and Mary Ann H. Franson (eds.), pp. 437–438, Washington, DC, American Public Health Association, 1985.
4. *Hawley's Condensed Chemical Dictionary,* 11th Ed., rev'd N. Irving Sax and Richard J. Lewis (New York, N.Y.: Van Nostrand Reinhold, 1987).

POLYACRYLIC ACID

Acrylic acid is an irritant and corrosive to the skin, whereas polyacrylic acid tends to be less corrosive and less toxic towards humans. Additional derivatives such as acrylamides are possible carcinogens in humans and can cause slurred speech and loss of balance.

Polyacrylic acid forms a number of polymer-type compounds and resins. Depending on the monomer in the formulation and the method of polymerization, acrylic resins can be hard, brittle solids or fibrous, elastomeric structures or viscous liquids. The hard, shatterproof, transparent or colored materials are known as lucite and plexiglas and have been used as glass substitutes, decorative illuminated signs, contact lenses, and furniture components. The materials are used in dentures and medical instruments and during specimen preservation. Suspension-polymerized beads and molding powders are used as headlight lenses, chromatography adsorbents, and ion-exchange resins. Aqueous emulsions are applied to adhesives, laminated structures, fabric coatings, and nonwoven fabrics. Acrylonitrile-derived acrylics are used in synthetic fibers and synthetic elastomers. Most of these final products are safe to use and do not decompose easily in the environment.

The low molecular weight polyacrylic acids, polyacrylates and associated copolymers have appeared in the environment and are often significantly soluble in water. Compounded prepolymers are used as exterior auto paints and other paint applications. Commercially available polyacrylic acid is available in a variety of strengths. Some of the copolymers have been used as commercial scale inhibitors. In the USA, most of these substances are generally controlled by regulations during applications and the hazardous waste manifest.

The primary use of polyacrylic acid tests is for acrylamide pollution in water. The acrylamides, derivatives of polyacrylic acids, are often water soluble polymers used as thickeners and flocculants in the paper and pulp industry and in the water and wastewater treatment facilities. The compounds are also used for soil conditioning agents, enhanced oil recovery, and for permanent press fabrics.

Polyacrylic acids (pAA) in the sample are selectively adsorbed onto a liquid chromatographic column using a technique developed by Rohm and Haas Company.* After separation from the sample, the polyacrylic acid is eluted off the column and the concentration is determined colorimetrically. Calibrations are based on Rohm and Haas Company Acumer (low molecular weight) polyacrylic products. Because commercially available polyacrylic acid can come in many strengths, preparation of the standards used for the calibrations has been based on a 100 % total solids basis and 100 % active polymer. For accurate work, diluted standards prepared from the product in use may be used to establish a calibration.

POLYACRYLIC ACID (0 to 20.0 mg/L as Acumer 1000 or 1100)

Method 8107** for Water and Brines
Absorption-Colorimetric Method†

SAMPLE PREPARATION

1. Remove the syringe plunger. Attach the prefilter to the syringe barrel. Twist to lock it on.
Note: The Pour-Thru Cell cannot be used.

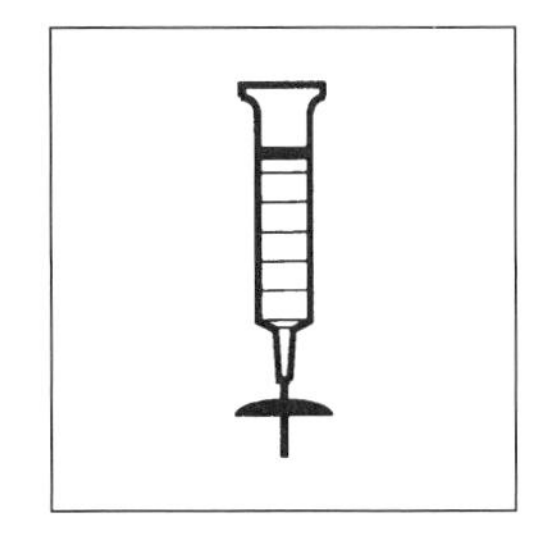

2. Rinse the syringe with the sample. Fill to the 30-cc mark.
Note: The syringe markings may wear off with continued use. They can be made more permanent by scoring with a knife.

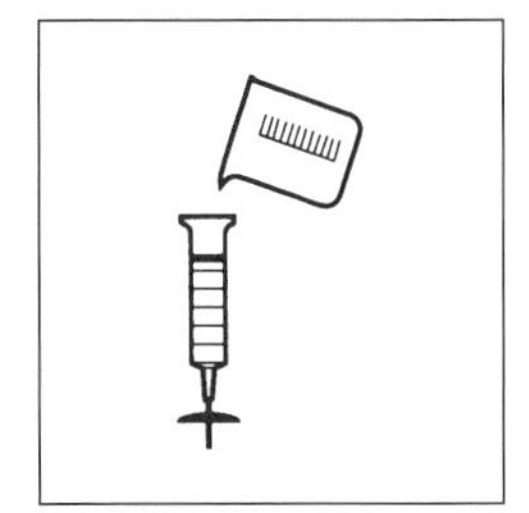

*DR/2010 Spectrophotometer Procedures Manual, 1998 Ed., p. 698.
**Compliments of HACH Company.
†Acumer is a trademark of Rohm and Haas Company.

3. Insert the plunger and force the sample through the filter into a 50-mL erlenmeyer flask labeled "sample."

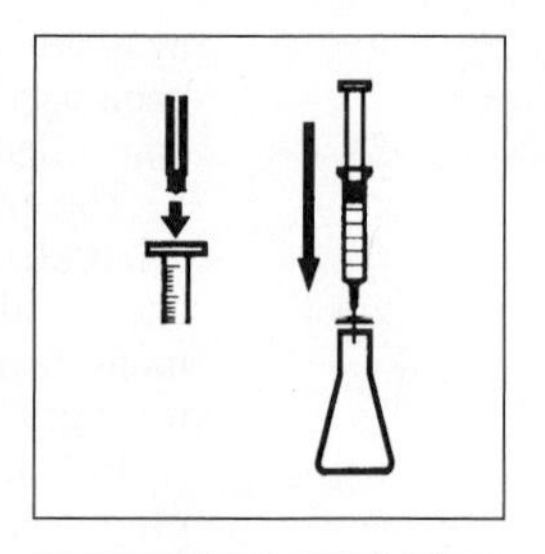

4. Fill another clean flask, labeled "reagent blank," with approximately 30 mL of deionized water.

5. Add 0.5 mL of Buffer Solution, pH 2.5, to each flask. Swirl to mix.
Note: Check the sample pH with pH indicator paper. If necessary, adjust to pH 2 or 3 with 1:1 Nitric Acid Solution.

6. Fill a third flask, labeled "eluant," with approximately 30 mL of Polyacrylic Acid Eluant Solution.

7. Assemble the syringe apparatus as shown in the kit directions. Place the long end of the LC cartridge on the male tip of the three-way valve. Turn the valve to the aspirate (down) position. Draw about 5 cc of reagent blank through the tubing into the syringe. Draw in air to the 30-cc mark.

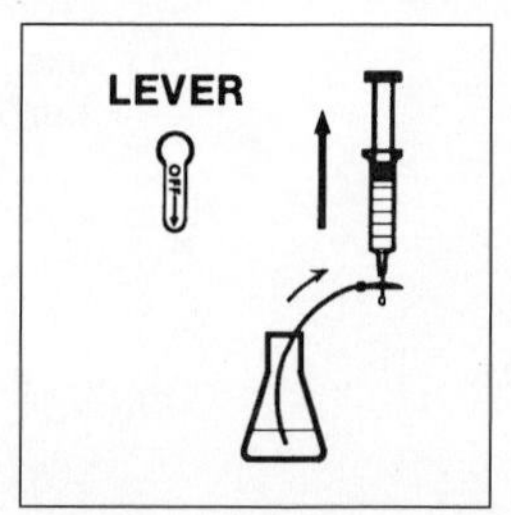

8. Rinse the syringe. Discard the solution through the tubing. *Note:* Move the plunger up and down several times to clear the tubing.

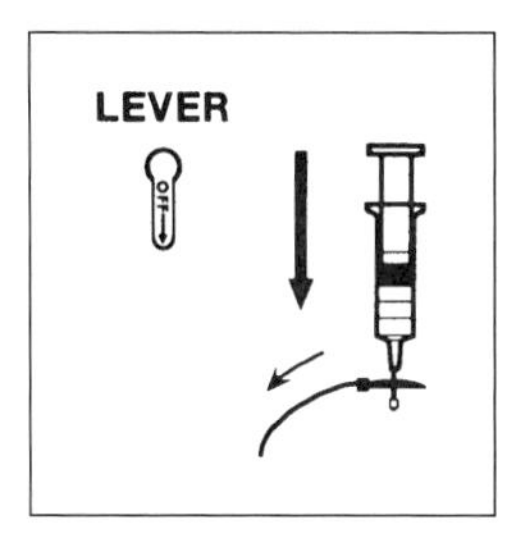

9. Draw the remaining reagent blank into the syringe through the tubing followed by a small volume of air, past the 30-cc mark.
Note: A small volume of air above the solution facilitates complete elution from the LC column.

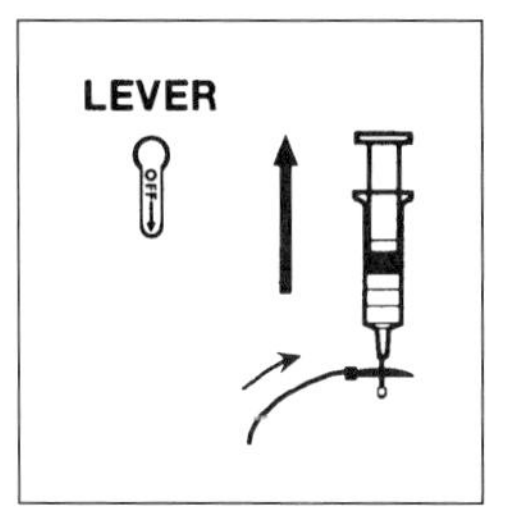

10. Push the plunger down to adjust the solution volume to exactly the 20-cc mark.

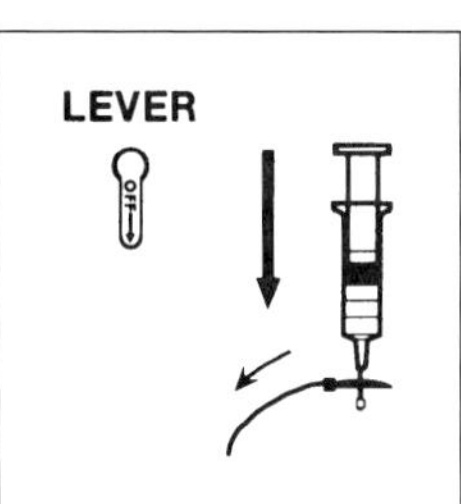

11. Rotate the valve lever to the pump (left) position and slowly force the solution through the LC column over a period of at least 15 seconds, discarding the solution. Follow the directions in the kit.

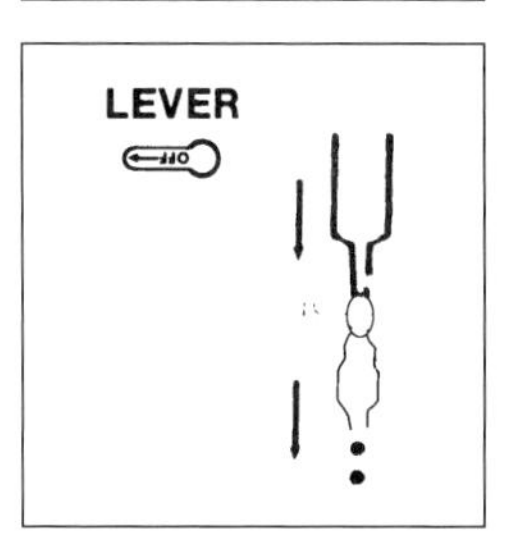

12. Again rotate the valve lever to aspirate (down) position. Draw about 5 cc of eluant into the syringe through the tubing followed by air past the 25-cc mark.

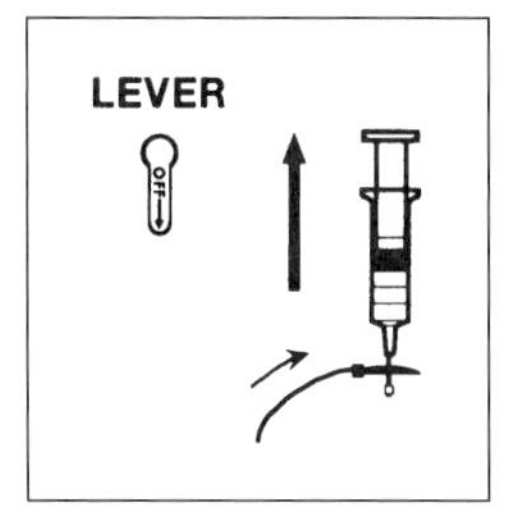

13. Rinse the syringe by shaking. Discard the eluant through the tubing.
Note: Move the plunger up and down several times to clear the tubing.

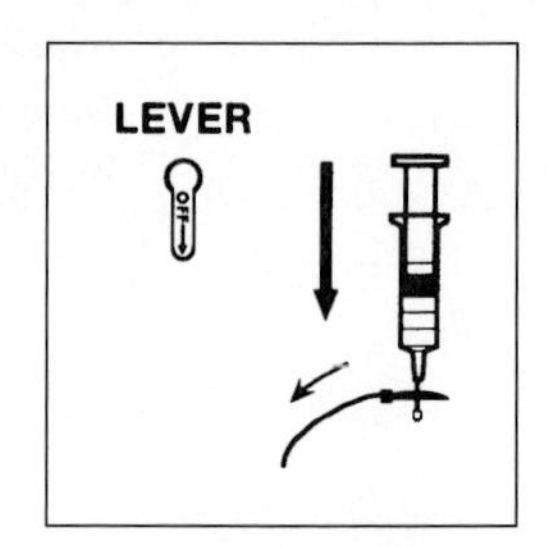

14. Draw at least 10 cc of eluant into the syringe through the tubing followed by air, past the 25-cc mark.

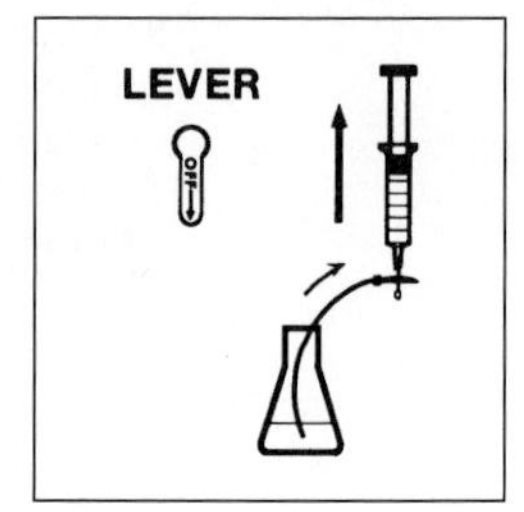

15. Push the plunger down to adjust the eluant volume to exactly the 10-cc mark.

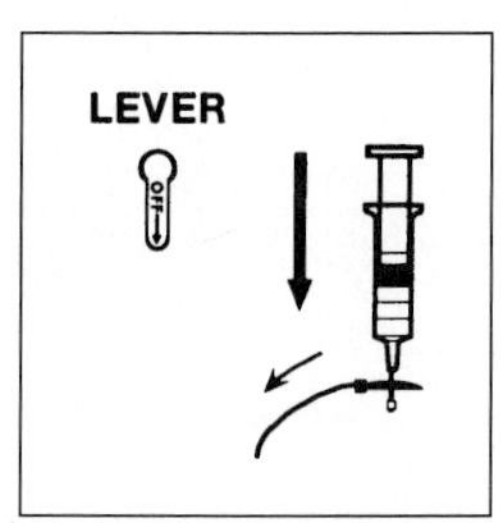

16. Rotate the valve lever to the pump (left) position. Over a period of 30 seconds, force the eluant through the cartridge. Collect the eluant in a 25-mL tall-form graduated cylinder.

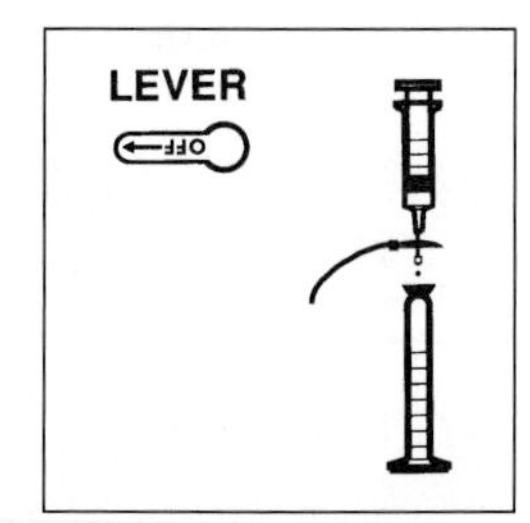

17. Fill the cylinder to exactly 25 mL with deionized water. Stopper. Invert to mix. Label this cylinder "reagent blank."
Note: Up to five samples can be run with one reagent blank.
Note: *Volumes are critical at this point. The cylinders can be matched by pipetting 25 mL of deionized water into each and marking them, if necessary, at the correct volume. Tall-form cylinders must be used.*

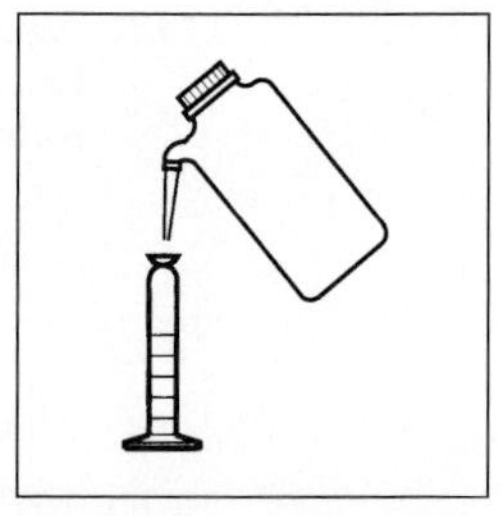

18. Clean the syringe and cartridge by drawing in 25 cc of deionized water with the valve lever in the aspirate (down) position. Discard the water back through the tubing.

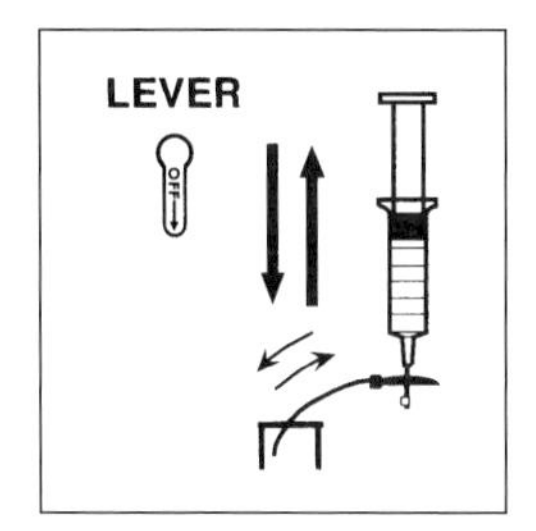

19. Repeat, discarding the water through the cartridge with the valve lever in the pump (left) position.
Note: The cartridge must be rinsed to remove any traces of eluant, which would affect adsorption of pAA from the next sample.

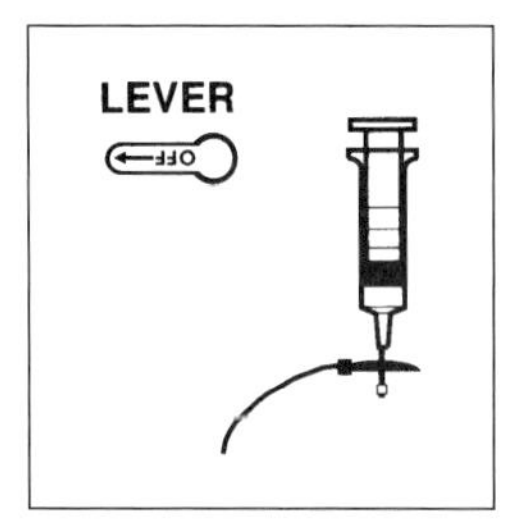

20. Return the valve to the aspirate (down) position and repeat Steps 8 through 20, using the buffered sample in place of the reagent blank. Label the glassware "sample."
Note: After use, rinse the LC cartridge with 2 cc of eluant solution, then deionized water. Store the cartridge in the vial supplied, with a few drops of eluant solution.

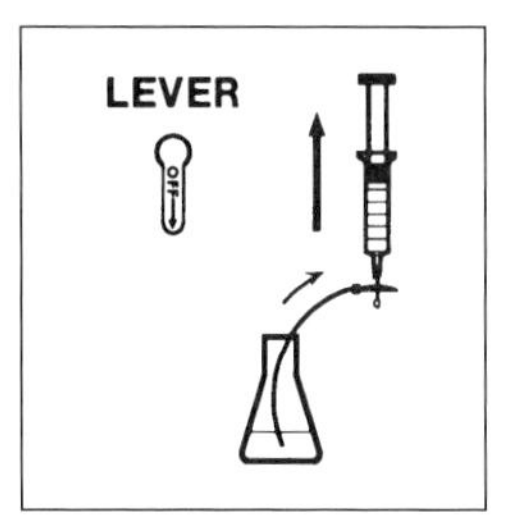

COLORIMETRIC ANALYSIS FOR POLYACRYLIC ACID

1. Enter the stored program number for polyacrylic acid (pAA). Press: **555 ENTER** for units in Acumer 1000 **OR 560 ENTER** for units in Acumer 1100. The display will show: **Dial nm to 482.**
Note: The Pour-Thru Cell cannot be used with this procedure.

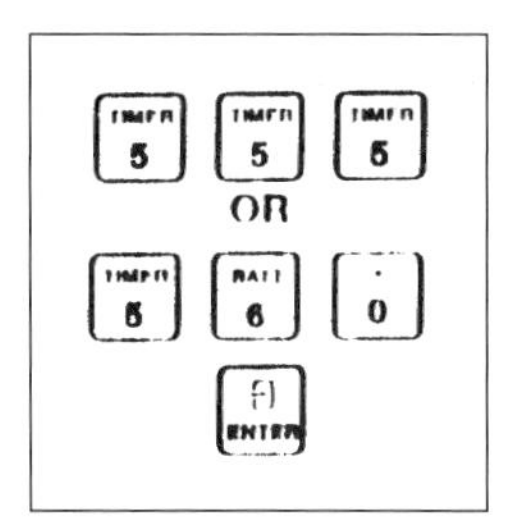

2. Rotate the wavelength dial until the small display shows: **482 nm.** When the correct wavelength is dialed in the display will show: **Zero Sample** then: **mg/L Acumer 1000 OR mg/L Acumer 1100.**

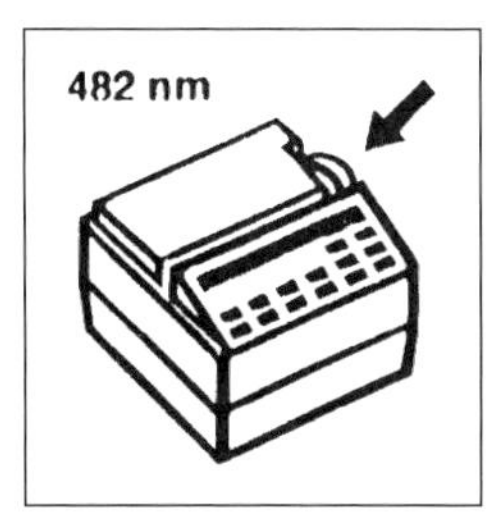

3. Pipet exactly 1.00 mL of Polyacrylic Acid 1 Reagent into each mixing cylinder. Stopper and mix. Proceed immediately to Step 4.
Note: Use a volumetric pipet or TenSette Pipet to measure this volume.

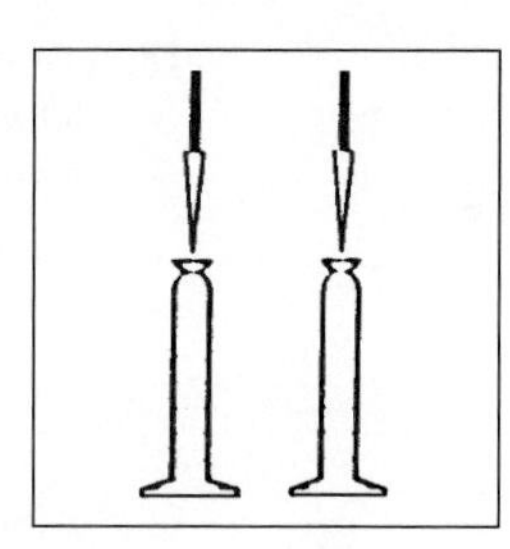

4. Press: **SHIFT TIMER.** A five-minute reaction period will begin. Place cylinders in the dark immediately.

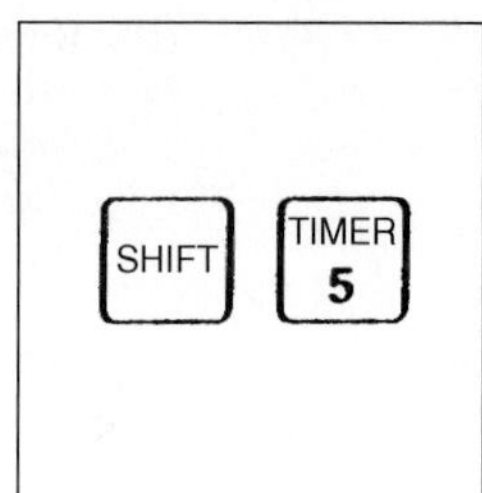

5. When the timer beeps, add exactly 1.00 mL of Polyacrylic Acid 2 Reagent to each cylinder. Stopper and mix. Proceed rapidly through Steps 6 and 7.
Note: Use a volumetric pipet or TenSette Pipet to measure this volume.

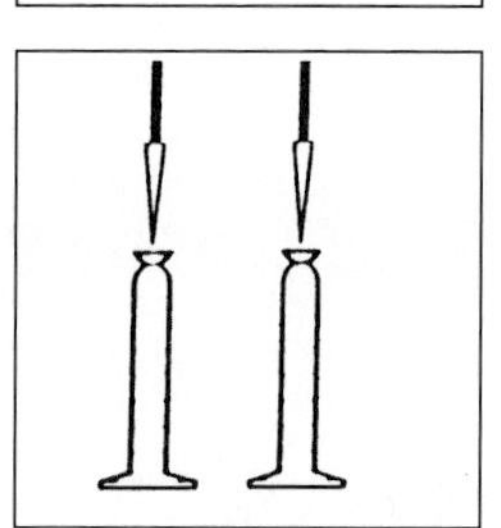

6. Press: **SHIFT TIMER.** A second five-minute reaction period will begin. Do Step 7 immediately.

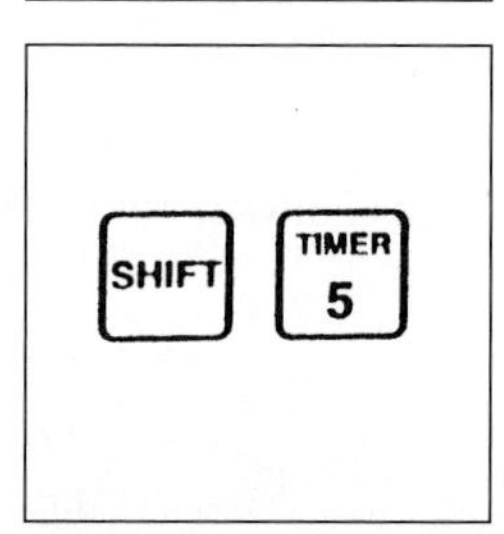

7. Transfer the solutions to two dry 25-mL sample cells, labeled "reagent blank" and "sample." Immediately place the sample cells in the dark.

8. When the timer beeps, the display will show the appropriate units. Place the cell labeled "sample" into the cell holder.
Note: The sample is used to zero the instrument.

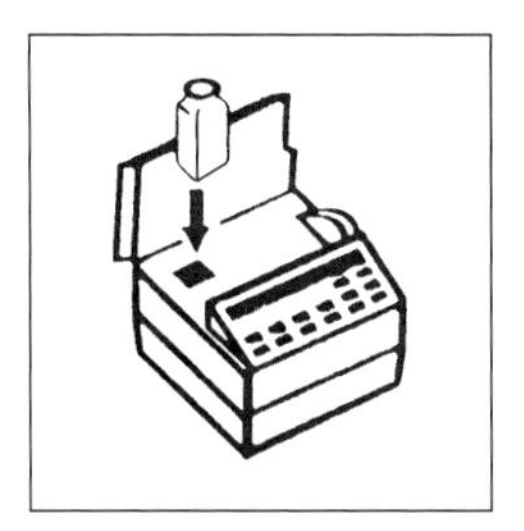

9. Press: **ZERO.** The display will show: **Zeroing ...** then: **0.0 mg/L** in the **appropriate** polyacrylic acid units.
Note: The reagent blank is stable in the dark for approximately 15 minutes if needed for determining up to five samples.

10. Immediately place the cell labeled "reagent blank" into the cell holder.

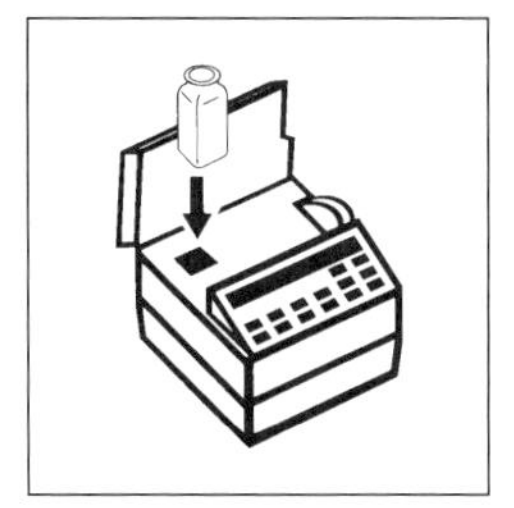

11. Press: **READ.** The display will show: **Reading ...** then result in mg/L polyacrylic acid as Acumer (total solids basis) will be displayed.
Note: If concentrations of less than 1 mg/L are being determined, see the "Interferences" section.
Note: For concentrations above the range of the test, dilute the sample by an appropriate factor and repeat.

INTERFERENCES

Filter turbid or oily samples through glass wool or a moderately rapid paper, such as S & S No. 560, before beginning the test.

The test works in brines having up to 75,000 mg/L total dissolved solids (TDS) and 20,000 mg/L chloride. Minimize the effect of most interferences by flushing the LC cartridge after the polyacrylic acid is adsorbed on the column. To flush, prepare 30 mL of deionized water buffered with 0.5 mL of Buffer Solution, pH 2.5. Repeat Steps 7 to 11 in the Sample Preparation section with this solution. Continue with Step 12.

Samples with concentrations of less than 1 mg/L should be repeated. Filter two 30-cc volumes of sample in Step 2. Use two 20-cc volumes (instead of one) of sample through the cartridge in Sample Preparations Steps 10 and 11. Continue with Step 12. Divide the resulting concentration by 2. Up to five 20-cc volumes of the sample can be run through the LC cartridge. It may be necessary to flush the LC cartridge as stated above.

Avoid use of facial tissue when drying glassware or apparatus, as it may contain interfering substances. Kimwipes or Kaydry wipers are recommended.

QUESTIONS AND ANSWERS

1. What are polyacrylic acid and polyacrylamides?
Polyacrylic acid is used to prepare a number of resins that have many practical and safe uses. Acrylamides or polyacrylamides are water soluble polymers that are used extensively in the wastewater and water treatment facilities and in the paper and pulp industries as cationic and ionic exchange media. The polymers have been used for soil conditioning, oil recovery, and permanent press fabrics in laundries.

2. Which forms of the polyacrylic compounds present the greatest environmental problem?
Most of the low molecular weight polyacrylic acid, polyacrylates, and associated copolymers used in paints, commercial scale inhibitors, and other applications do not appear to be as carcinogenic or as great a human health problem as the acrylamides or polyacrylamides.

3. Why are low molecular weight polyacrylic compounds an environmental problem?
Low molecular weight polyacrylic compounds can be irritants and corrosive to skin. Some of the more volatile compounds are somewhat toxic by inhalation. The tolerance level varies depending on the molecular weight and volatility of the compound. Some of the compounds have a relatively low flash point and are combustible. A significant number of the polyacrylic compounds are water soluble polymers, which can present a problem with control and isolation.

REFERENCES

1. DR/2010 Spectrophotometer Procedures Manual, 3rd Ed. (Loveland, CO.: HACH Company, 1997).
2. *Hawley's Condensed Chemical Dictionary,* 11th Ed., rev'd by N. Irving Sax and Richard J. Lewis (New York, N.Y.: Van Nostrand Reinhold. 1987).

SALINITY

Salinity is defined as the total solids in water after all carbonates have been converted to oxides, all bromide and iodide compounds have been replaced by chloride, and all organic matter has been oxidized. Salinity is an important issue in seawater and specific industrial wastes. The salinity of normal seawater is 35 parts salt per 1000 parts seawater.

Salinity is sometimes expressed in terms of grams per kilogram, while other times the term is expressed in terms of mg/L chloride or as mg/L sodium chloride or as ppt (parts per thousand). The units and description must always be recorded with the numerical salinity value.

The Mercuric Nitrate Method of chloride analysis produces a sharp yellow to pinkish purple end point due to the diphenylcarbazone in the presence of a buffer. The titration has been applied to water with salinity levels of 0-100 ppt (parts per thousand).

SALINITY

(0-100 ppt* Salinity)
Method 10073**
Digital Titrator with Mercuric Nitrate

1. Insert a clean, straight-stem delivery tube to a Mercuric Nitrate Titration Cartridge. Attach the cartridge to the titrator body.

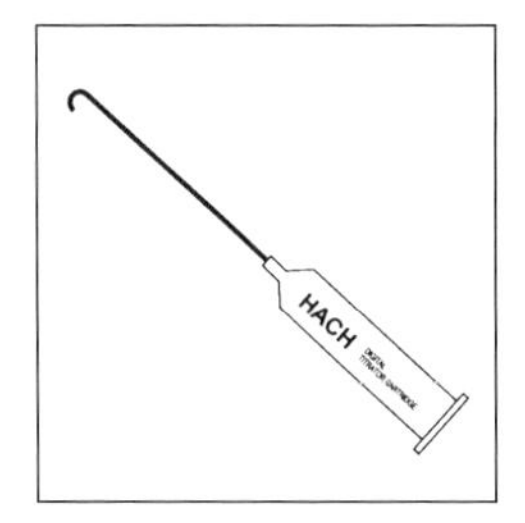

2. Flush out the delivery tube by turning the knob until titrant begins flowing from the end of the tube. Wipe the tip and reset the counter to zero.
Note: For added convenience use the TitraStir stirring apparatus.

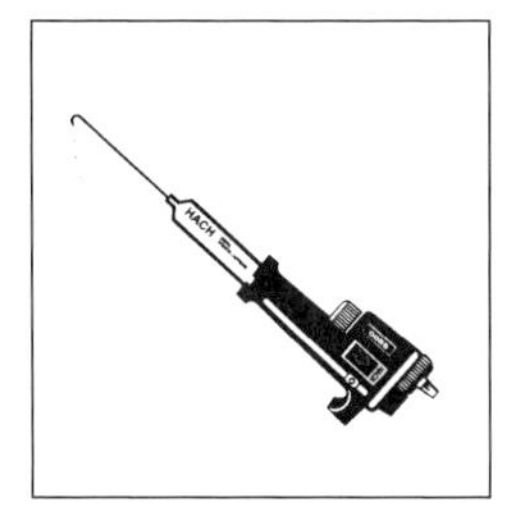

3. Using the 3 mL (3-cc) syringe, collect a 2.0-mL water sample. Add to the vial provided.

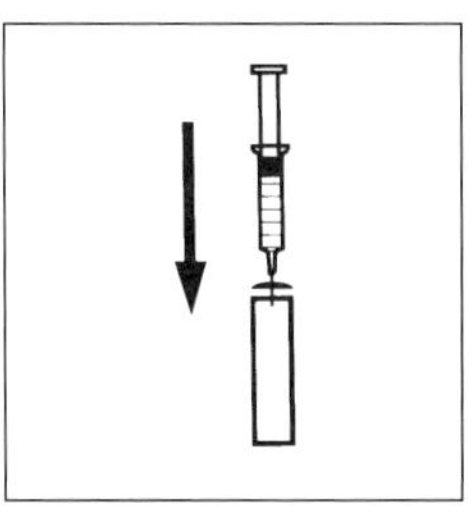

4. Fill the vial to the 10-mL mark with deionized water.

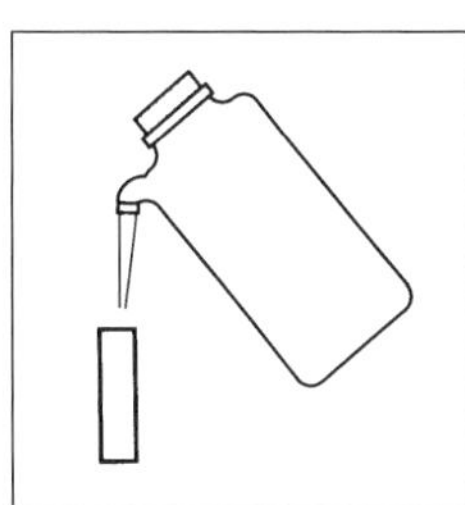

*ppt = parts per thousand.
**Compliments of HACH Company.

5. Add the contents of one Diphenylcarbazone Reagent Powder Pillow to the vial and mix.
Note: Results will not be affected if a small portion of the diphenylcarbazone reagent powder does not dissolve.

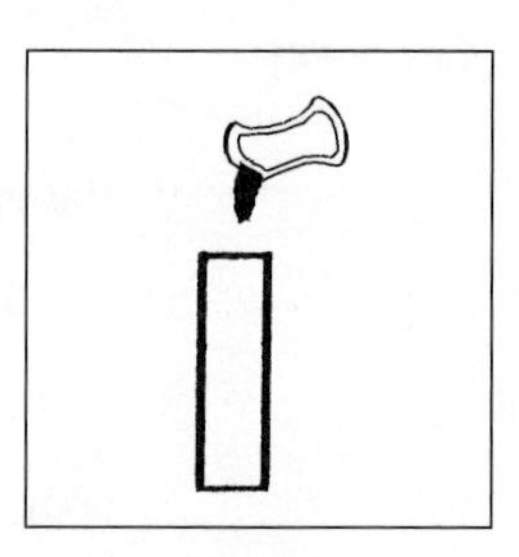

6. Titrate the sample with mercuric nitrate until the color changes from yellow to light pink. Record the number of digits.

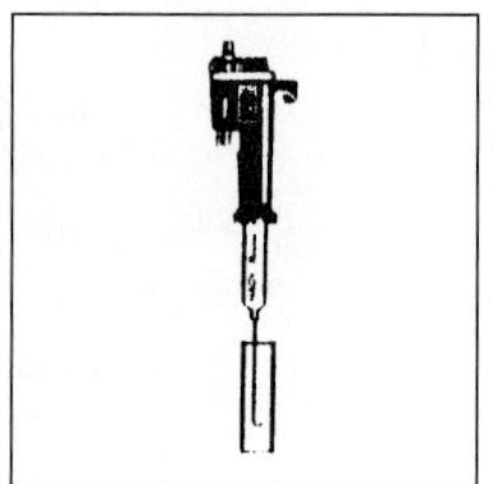

7. Determine the salinity of the water sample in parts per thousand (ppt) by multiplying the reading in the digital titrator counter window by 0.1.
Note: Results may be expressed as mg/L chloride (Cl^-) by multiplying the ppt salinity by 569. Results may be expressed as mg/L sodium chloride (NaCl) by multiplying the ppt salinity by 940.

QUESTIONS AND ANSWERS

1. Determine the salinity of a sample of water in parts per thousand (ppt) when the digital titrator reads 250.

$$250 \times 0.1 = 25 \text{ ppt Salinity}$$

2. Convert ppt (parts per thousand salinity) to mg/L chloride.

$$\text{mg/L chloride} = \text{\# parts per thousand Salinity} \times 569$$

3. Convert the ppt to mg/L of sodium chloride.

$$\textit{mg/L sodium chloride} = \textit{\# parts per thousand Salinity} \times 940$$

REFERENCES

1. *HACH Water Analysis Handbook,* 3rd Ed., (Loveland, CO.: HACH Company, 1997).
2. *Standard Methods for the Examination of Water and Wastewater:* SALINITY, 16th Ed., A. E. Greenberg, R. R. Trussell, L. S. Clesceri and Mary Ann H. Franson (eds.), p. 101, Washington, DC, American Public Health Association, 1985.

3. *Hawley's Condensed Chemical Dictionary,* 11th Ed., rev'd by N. Irving Sax and Richard J. Lewis (New York, N.Y.: Van Nostrand Reinhold, 1987).

SELENIUM

Selenium is toxic to animals and appears to be somewhat toxic to humans. Traces of selenium are essential to maintain normal body metabolism, but higher levels of selenium are suspected of causing dental caries and of being the cause of cancer. Supposedly, selenium helps to guard cells and tissues in the human body against radical damage. A few compounds of selenium are increasingly soluble in water as the temperature of the water increases above 0°C. This issue could lead to a potential health problem. Selenium, depending on its valence state, can be a significant pollutant with high toxicity or a material which has limited toxicity.

A maximum concentration level for selenium in drinking water has been established by the USEPA in accordance with the Safe Drinking Water Act. The selenium concentration in most drinking water is less than 10 micrograms per liter. On rare occasions, concentrations of selenium in water have been measured at 500 micrograms per liter near soils rich in selenium-containing materials.

Water can be contaminated with selenium from industrial sources or mining practices. Wastes with selenium in any of its four valence states can contaminate water. A number of organic compounds of selenium are fairly common. Selenium is a nonmetallic element used in electronics, xerographic plates, TV cameras, photocells, magnetic computer cores, solar batteries, ceramics, catalyst, feed additive, and trace supplements for humans. Selenium compounds are used in medicine, medicated shampoo, lubricating oil as an antioxidant, reagents, catalysts, and as vulcanization agents.

Industries manufacturing selenium metal and selenium compounds must protect the water bodies by eliminating and severely minimizing discharges of the toxic materials to the environment. Several industries have implemented a closed loop or semi-closed loop system to recover and reuse waste. Some industries manifest the waste stream for burial purposes or for reuse by other industries.

Strong oxidizing agents such as chlorine and bromine can react with the indicator to give low results. Interferences can be eliminated by performing distillation on the raw sample. The distillation procedure includes the addition of a strong acid and an oxidizer to the sample. Safety precautions must be observed and proper protective clothing must be worn during this experiment.

An EDTA masking agent is added to the sample to remove interferences during the test. After a buffer adjusts the pH, diaminobenzidine reacts with selenium (Se^{+4}) to form a yellow colored complex. After the complex is extracted, the color intensity is measured colorimetrically at 420 nm. The sample must be distilled in order to detect selenium in other oxidation states such as selenium (−2 and +6).

SELENIUM (0 to 1.00 mg/L)

Method 8194 for water and wastewater*
Diaminobenzidine Method**

*Compliments of HACH Company.

**Adapted from *Standard Methods for the Examination of Water and Wastewater, 323 B and 323 C.*

1. Enter the stored program number for selenium (Se). Press: **6 4 0 ENTER.** The display will show: **Dial nm to 420.**

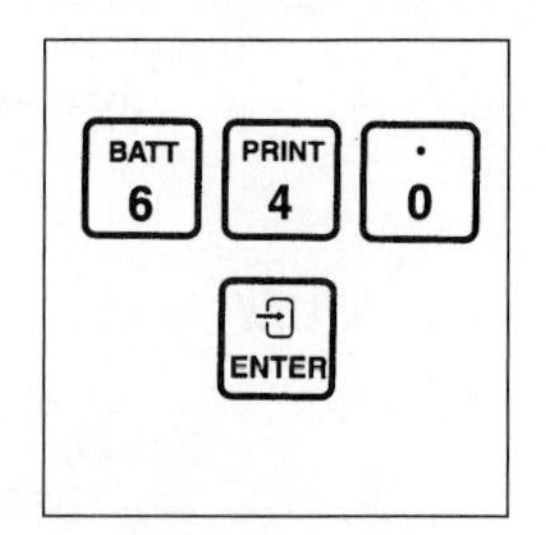

2. Rotate the wavelength dial until the small display shows: **420 nm.** When the correct wavelength is dialed in, the display will quickly show: **Zero sample,** then: **mg/L Se.**

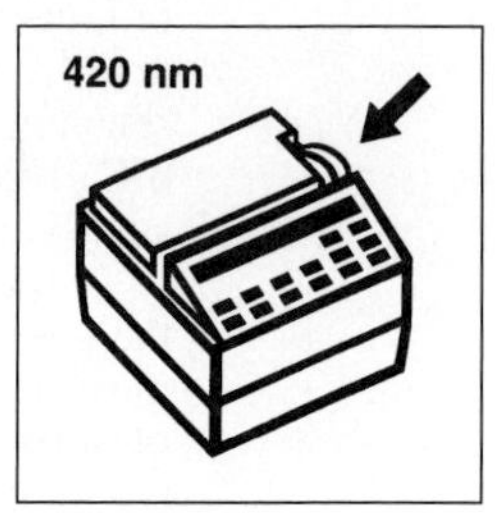

3. Measure 100 mL of deionized water into a 500-mL erlenmeyer flask (label the flask "blank").

4. Measure 100 mL of sample into a second 500-mL erlenmeyer flask (label the flask "sample").
Note: To determine a total selenium, perform a distillation (see "Distillation" following these steps). Use the distillate in Step 4.

5. Add a 0.2-g scoop of TitraVer Hardness Reagent to each flask. Swirl to mix.

6. Add a 0.05-g scoop of diaminobenzidine tetrahydrochloride to each flask. Swirl to mix.

7. Add 5.0 mL of Buffer Solution, sulfate type, pH 2.0, to each flask. Swirl to mix.
Note: If the sample has been distilled, omit the Buffer Solution. Adjust the pH of the sample distillate to 2.7 ($\pm$0.2 pH) using 5 N Sodium Hydroxide Standard Solution. Adjust the blank to the same pH value using 5.25 N Sulfuric Acid Standard Solution.

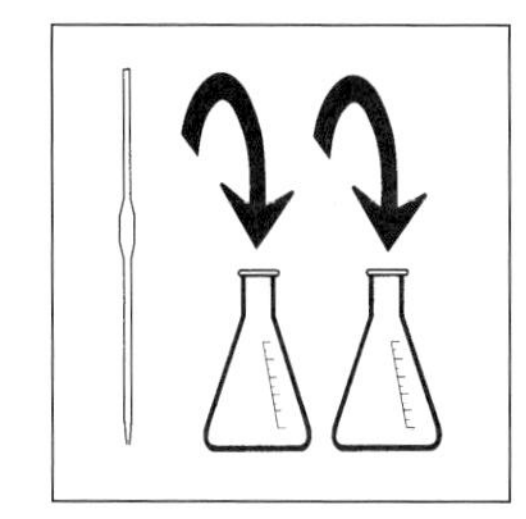

8. Heat each flask on a hot plate (no open flames), bringing the contents to a gentle boil.

9. Press: **SHIFT TIMER.** A five-minute reaction period will begin. Continue to boil the contents gently during this time period.
Note: A yellow color will develop if selenium is present.

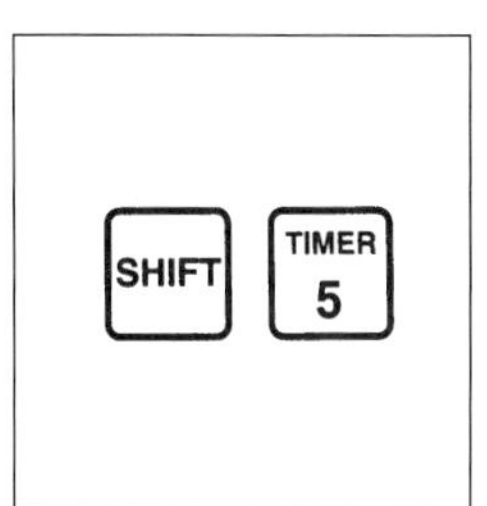

10. When the timer beeps, remove both flasks. Cool to room temperature using a water bath.
Note: Do not boil more than one minute after the timer beeps.

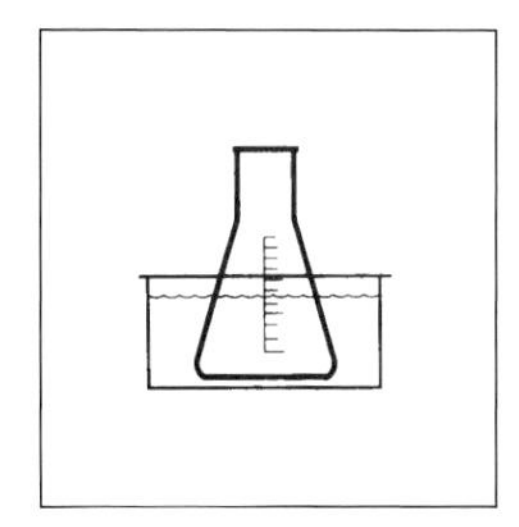

11. Transfer the contents of each flask to separate 250-mL separatory funnels. Label the funnels "blank" and "sample."

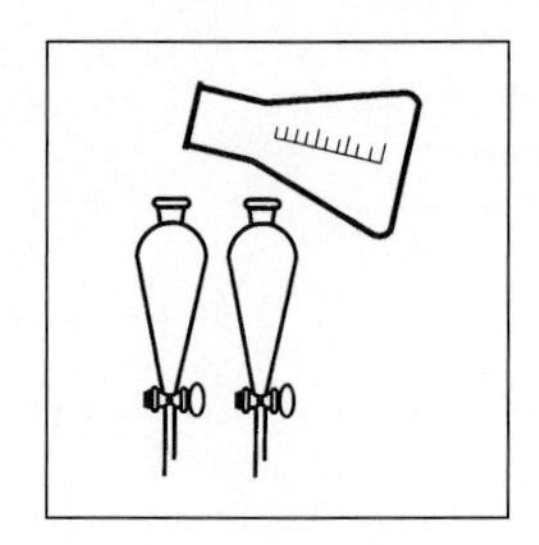

12. Add 2.0 mL of 12 N Potassium Hydroxide Standard Solution to each funnel using a calibrated 1.0-mL plastic dropper. Stopper. Shake each funnel to mix.

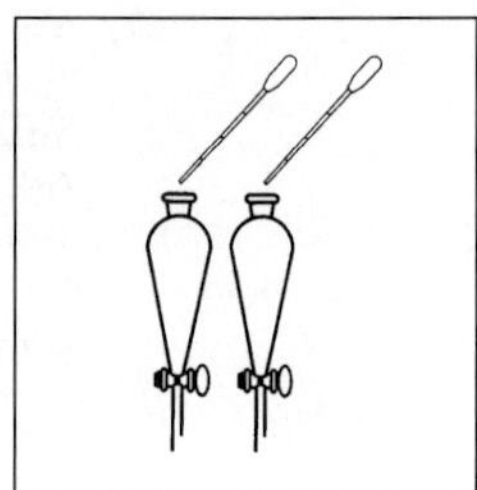

13. Add 30 mL of toluene to each funnel. Stopper. Shake each funnel vigorously for 30 seconds.
Note: Use toluene only with adequate ventilation.

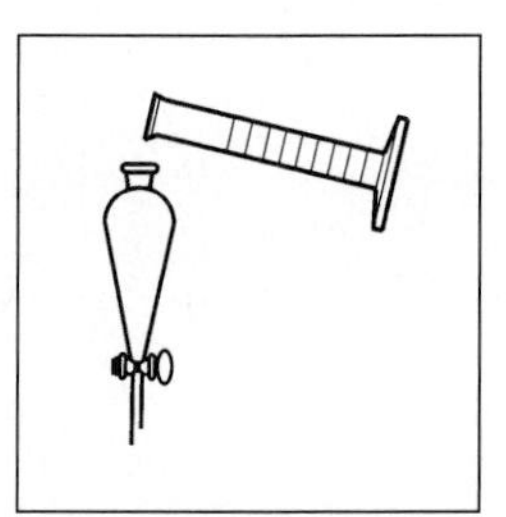

14. Press: **SHIFT TIMER.** A three-minute reaction period will begin.

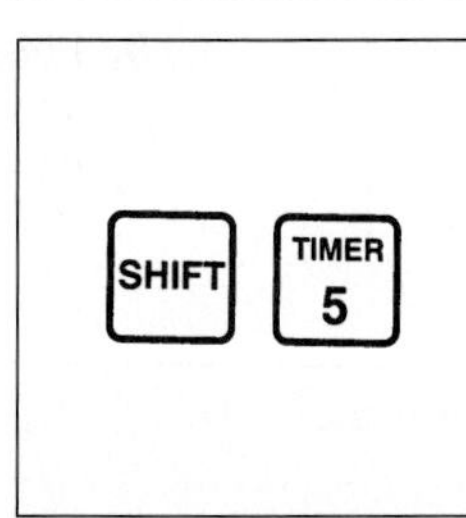

15. When the timer beeps, the display will show: **mg/L Se.** Drain the lower water layer from each funnel and discard.
Note: Do not wait more than five minutes after the timer beeps before competing Steps 16 through 20.

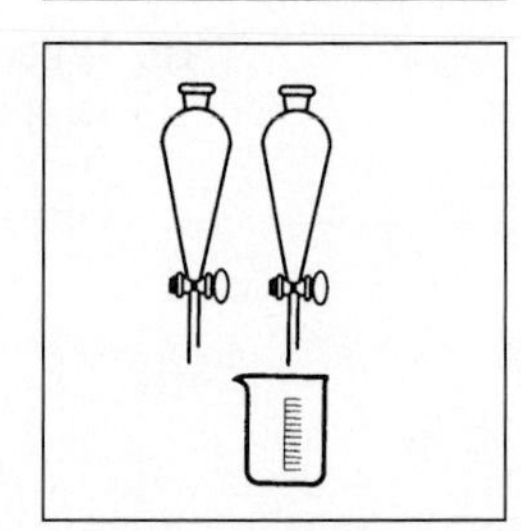

16. Insert a cotton plug into the delivery tube of each separatory funnel. Slowly drain the toluene into respective sample cells labeled "blank" and "sample." Stopper the sample cells.
Note: Filtering the toluene through dry absorbent cotton will remove any water or suspended particles.
Note: The developed color is stable but should be measured as soon as possible.

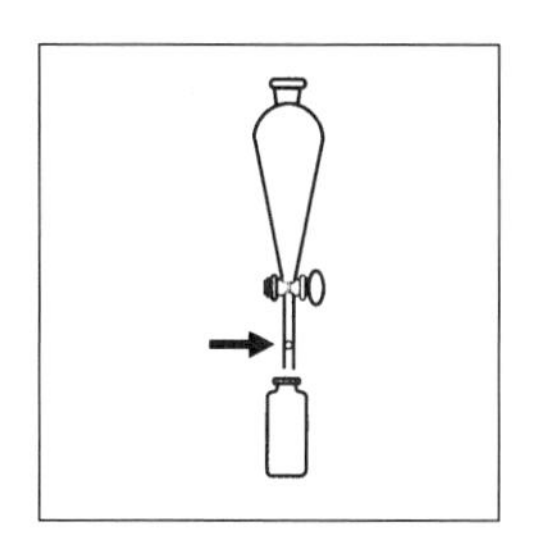

17. Place the blank into the cell holder. Close the light shield.

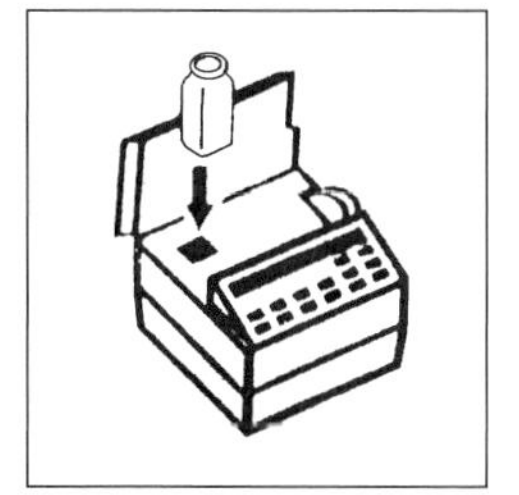

18. Press: **ZERO.** The display will show: **Zeroing ...** then: **0.00 mg/L Se.**

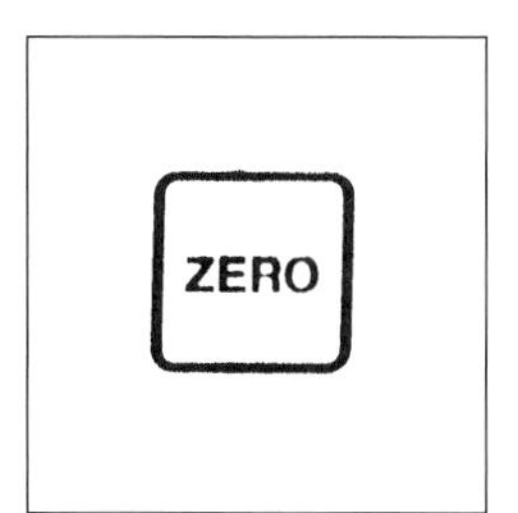

19. Place the prepared sample in the cell holder. Close the light shield.
Note: Acetone is a suitable solvent for removing toluene from glassware.

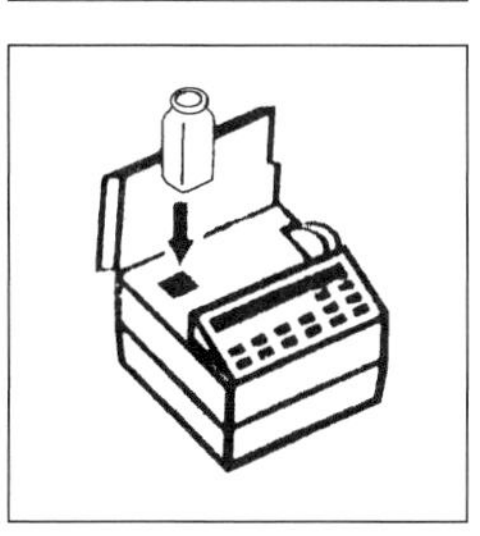

20. Press: **READ.** The display will show: **Reading ...** then the result in mg/L Se will be displayed.

SAMPLING AND STORAGE

Collect samples in clean glass or plastic containers. Adjust the pH to 2 or less with nitric acid (about 1.5 mL per liter). Preserved samples can be stored for up to six months at room temperature. Correct the test result for volume additions (see Section 1).

DISTILLATION

Always perform this procedure under a fume hood! This distillation involves the use of a strong acid and oxidizer at high temperatures. To avoid personal injury, observe all laboratory safety precautions when operating the distillation apparatus.

a) Measure 500 mL of sample into a 1000-mL beaker.

b) Add 1 mL of Methyl Orange Indicator Solution. Stir with a glass rod.

c) Using a dropper, add 0.1 N Hydrochloric Acid Standard Solution drop-wise until the solution becomes pink. Then add an additional 2 mL.

d) Pipet 5.00 mL Calcium Chloride Solution. Mix well.

e) Using a dropper, add 1-g/L Potassium Permanganate Standard Solution drop-wise until the solution is purple.

f) Place the beaker on a hot plate. Evaporate the solution to approximately 250 mL. Periodically add 1-g/L Potassium Permanganate Solution to keep the solution purple. *Note:* Any precipitate forms at this step is manganese dioxide and may be ignored.

g) Cool the solution. While cooling, set up the distillation apparatus for the general purpose distillation.

h) Pour the treated sample solution into the distillation flask. Add a stirring bar to the flask.

i) Pipet 5.0 mL of 0.1 N Sodium Hydroxide Standard Solution into the flask. Turn the stirrer power switch to ON. Set the stir control to 5.

j) Turn on the water and adjust so a constant flow is maintained through the condenser. Set the heat control to 10.

k) When only a few milliliters are left in the distillation flask, turn the power switch off. The distillate in the erlenmeyer flask may be discarded.

l) **Perform this step under a hood.** When the flask has cooled, add 50 mL of 19.2 N Sulfuric Acid Standard Solution to the flask. Add the contents of one Potassium Bromide Powder Pillow to the flask.

m) Fill a 250-mL beaker to the 75-mL mark with deionized water. Place it under the drip tube. Elevate the beaker with a laboratory jack so the tube extends below the level of the water.

n) Add 1.0 mL of 30% Hydrogen Peroxide Solution to the flask. Turn the stir control to 5 and the heat control to 10. Cap the distillation flask.

o) Heat the distillation flask until the yellow color is gone from the complete distillation apparatus, including the J-tube and condenser. Remove the beaker from under the drip tube.

p) Turn off the heater switch. When the J-tube and condenser have cooled, rinse them with deionized water. Add the washings to the 250-mL beaker. Total volume in the beaker should be approximately 100 mL.

q) Add the Phenol Solution drop-wise to the distilled sample to discharge the bromine color (a white precipitate of tribromophenol will form).

r) Allow the precipitate to settle. Using a dropper, collect about 5 mL of the clear, colorless distillate and transfer to a test tube.

s) Test the solution for completeness of precipitation by adding 2 drops of Phenol Solution. If the solution becomes cloudy or white precipitate forms, residual bromine is still present (proceed to next step). If no cloudiness occurs, the sample is ready for analysis.

t) Transfer the 5-mL aliquot back to the beaker and continue to add Phenol Solution until no turbidity is formed in subsequent 5-mL aliquots.

u) Transfer the sample into a 500-mL volumetric flask. Rinse the beaker with deionized water and add to the flask.

v) Dilute to volume with deionized water, stopper and mix well. The distillate is now ready for analysis.

ACCURACY CHECK

See p. 1082 of the HACH Water Analysis Handbook, 3rd Ed., 1997.

INTERFERENCES

There are no positive inorganic interferences with this method. Strong oxidizing agents such as iodine, bromine, or chlorine can react with the indicator to give low results. Manganese and up to 2.5 mg/L ferric iron will not interfere. Interferences will be eliminated by following the distillation process.

QUESTIONS AND ANSWERS

1. Why is a distillation procedure often required for selenium?
   Distillation is required to determine total selenium and to eliminate interferences. Without distillation, only selenium in the +4 state is measured.
2. What safety precautions are required in this experiment?
   The procedure must be performed under a fume hood. The distillation involves the use of a strong acid and oxidizer at high temperatures. Practice safe handling of chemicals and observe all laboratory safety precautions when operating the distillation apparatus.
3. How are interferences removed in the experiment?
   Interferences such as iron are removed from the sample by the addition of an EDTA masking agent.
4. How much selenium is safe for animals and humans?
   Above trace levels, ingested selenium is toxic to animals and appears to be somewhat toxic to humans. The concentration of selenium in most drinking water is less than 10 micrograms per liter.

REFERENCES

1. *HACH Water Analysis Handbook,* 3rd Ed., (Loveland, CO.: HACH Company, 1997).
2. *Hawley's Condensed Chemical Dictionary,* 11th Ed., rev'd by N. Irving Sax and Richard J. Lewis (New York, N.Y.: Van Nostrand Reinhold, 1987).

3. *Standard Methods for the Examination of Water and Wastewater:* SELENIUM, 16th Ed., A. E. Greenberg, R. R. Trussell, L. S. Clesceri and Mary Ann H. Franson (eds.), p. 238, Washington, DC, American Public Health Association, 1985).

SILVER

Silver can cause a blue-gray discoloration to the skin and eyes of a person. Concentrations as low as 0.17 micrograms per liter have displayed toxic effects on fresh water fish. The concentration of silver in the U.S. drinking water has been reported to be as high as 2 micrograms per liter.

Relatively small quantities of silver have limited use in the disinfection of swimming pools. Silver compounds have been used for photography, x-ray, detonators, laboratory agents, photographic film and plates, photochromic glass, organic synthesis, batteries, photometry, optics, silver plating, antiseptics, absorption cells, cloud seeding, catalysts, indelible inks, silvering mirrors, germicide, hair dying, and wound cauterizing. Silver compounds have extensive applications in the plating industry. Some silver compounds are shock sensitive and are strong irritants to skin and tissue. Silver metal is used to manufacture a number of silver compounds. The metal is used as a liner for chemical vats and equipment and as a coating on numerous items used everyday. Colloidal silver is a nucleating agent in photography and during its combination with protein in medicine.

Since silver has significant value, many industries recover silver for investment purposes and for industrial applications. The amount of silver dumped into water resources and buried in hazardous waste sites is usually quite small. Small quantities of silver in complex mixtures are normally recovered by various separation procedures.

When determining silver, there are a significant number of negative interferences by metals, chloride and ammonia at relatively high concentrations. A few metals, especially mercury at 2 mg/l, provided a positive interference. During the analysis for silver, digestion of samples containing organic matter, thiosulfate and cyanide is required to eliminate interferences.

Silver ions in basic solution react with an additive to form a colored complex. Sodium thiosulfate performs as a decolorizing agent in the blank while other additives act as buffer, indicator and masking agents. Organic extractions are avoided in this procedure. This procedure has fewer interferences than the traditional dithizone method and can be used for electroplating baths and silver strike solutions.

SILVER (0 to 0.60 mg/L)

Method 8120 for water and wastewater*
Colorimetric Method

*Compliments of HACH Company.

1. Enter the stored program number for silver (Ag). Press: **6 6 0 ENTER.** The display will show: **Dial nm to 560.**
   Note: If cyanide is present, digest the sample; see "Digestion," below.
   Note: The Pour-Thru cell cannot be used with this procedure.

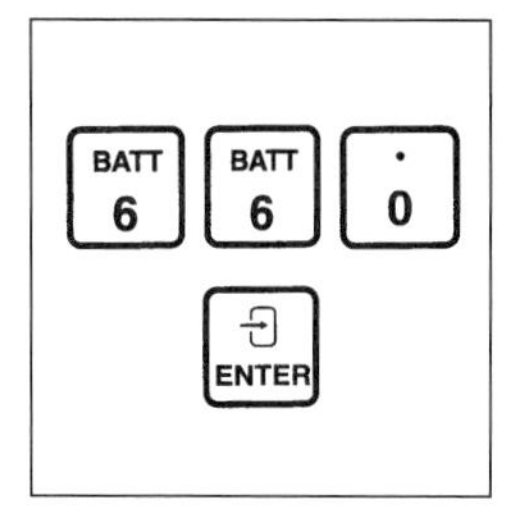

2. Rotate the wavelength dial until the small display shows: **560 nm.** When the correct wavelength is dialed in the display will quickly show: **Zero sample,** then: **mg/L Ag.**

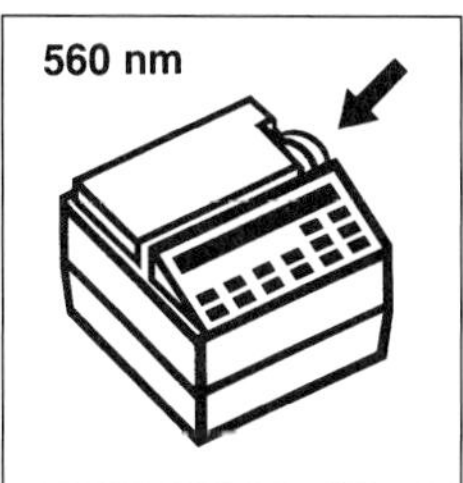

3. Add the contents of one Silver 1 Powder Pillow to a dry 50-mL graduated mixing cylinder.
   Note: If the Silver 1 Powder becomes wet at this point, the powder will not dissolve completely, which will inhibit color development.

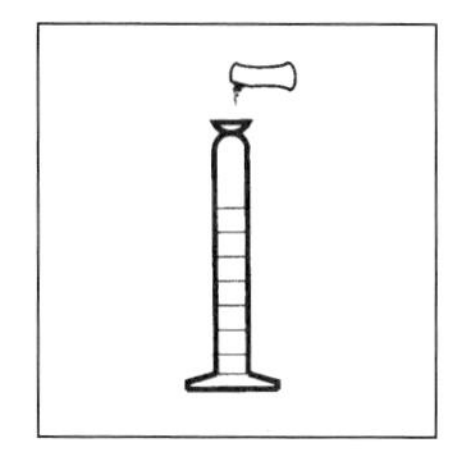

4. Add the contents of one Silver 2 Reagent Solution Pillow to the cylinder. Swirl to completely wet the powder.
   Note: If clumps of dry powder are present when the sample is poured in, the powder will not dissolve completely. This will inhibit color formation.

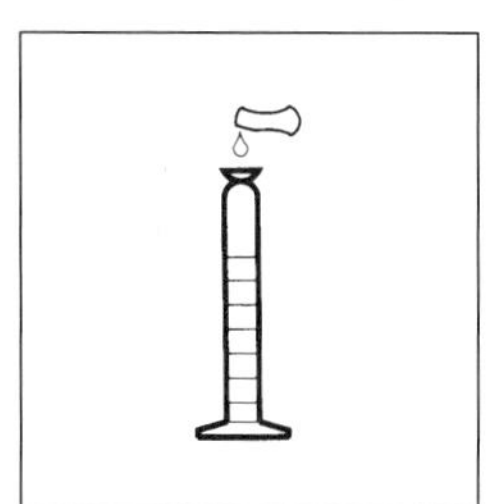

5. Using a 50-mL graduated cylinder, add 50 mL of sample to the cylinder. Stopper. Invert repeatedly for one minute.
   Note: Adjust the pH of stored samples before analysis.

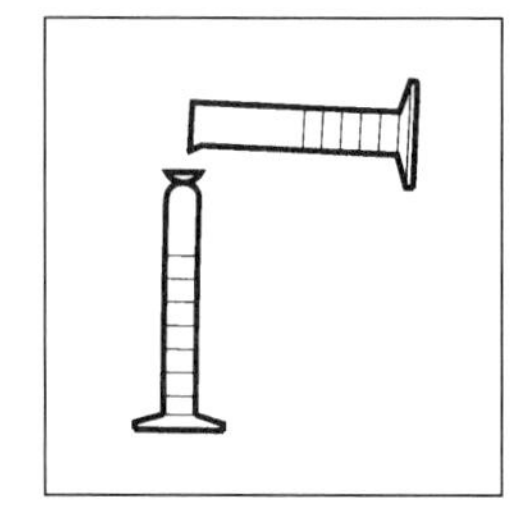

6. Pour 25 mL of the mixture into a sample cell (the blank). Add the contents of one Thiosulfate Powder Pillow to the sample cell. Swirl for 30 seconds to mix.
Note: It is important to generate a blank for each sample.

7. Press: **SHIFT TIMER.** A two-minute reaction period will begin.

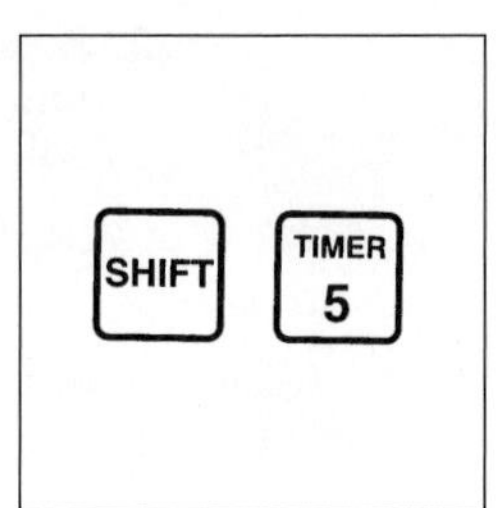

8. Pour the portion remaining in the cylinder into a second sample cell (the prepared sample).

9. When the timer beeps, the display will show: **mg/L Ag.** Place the blank in the cell holder. Close the light shield.

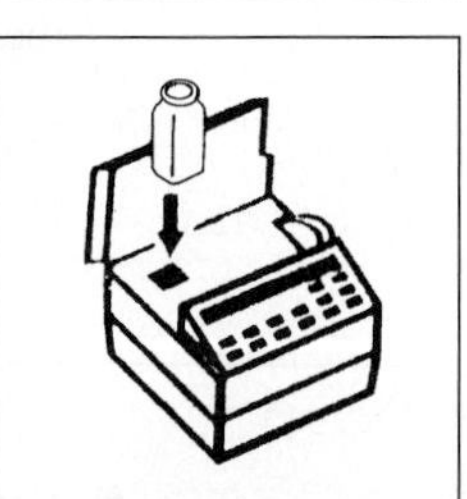

10. Press: **ZERO.** The display will show: **Zeroing ...** then: **0.00 mg/L Ag.**

11. Place the prepared sample in the cell holder. Close the light shield.

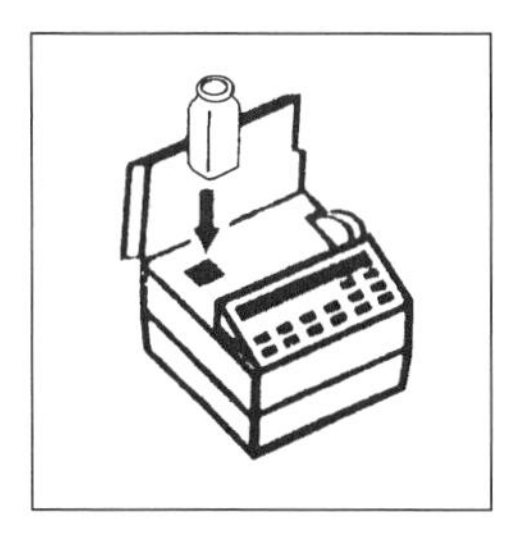

12. Press: **READ.** The display will show: **Reading ...** then the result in mg/L silver will be displayed.
Note: Rinse the cells carefully between samples to avoid development of a film on the cell walls.

SAMPLING AND STORAGE

Collect samples in acid-cleaned plastic or glass bottles. Using pH paper, adjust the pH to 2 or less with nitric acid (about 2 mL/liter). Store preserved samples at room temperature for up to 6 months. Before analysis, adjust the pH to 9 to 10 with 5.0 N sodium hydroxide. Do not use a pH meter because of silver contamination from the electrode. Correct for volume additions (see Section 1).

ACCURACY CHECK

See p. 1112 of the HACH Water Analysis Handbook, 3rd Ed., 1997.

INTERFERENCES

Interference studies were conducted by preparing a known silver solution (about 0.4 mg/L) and the potential interfering ion. The ion was said to interfere when the silver concentration changed by $\pm 10\%$. Interferences are indicated in Table 2.35.

DIGESTION

This digestion is for samples containing organic matter, thiosulfate or cyanide. Possible sources for these compounds are wastewater, silver electroplating baths and silver strike solutions. Digestion should be done with a Digesdahl Digestion Apparatus or with a regular digestion apparatus.
Note: Poisonous hydrogen cyanide gas is generated during this digestion; use a fume hood.

TABLE 2.35

Negative interference:	
Aluminum	30 mg/L
Ammonia	750 mg/L
Cadmium	15 mg/L
Chloride	19 mg/L
Chromium^{6+}	90 mg/L
Copper	7 mg/L
Iron	30 mg/L
Lead	13 mg/L
Manganese	19 mg/L
Nickel	19 mg/L
Zinc	70 mg/L
Positive interference:	
Calcium	600 mg/L
Magnesium	2000 mg/L
Mercury	2 mg/L

Note: Always wear safety glasses and use a safety shield, or operate the Digesdahl within a closed fume hood. Follow the additional safety precautions in the Digesdahl Apparatus Manual.

a) Add an appropriate size sample to the 100-mL volumetric flask of the Digesdahl. Add several boiling chips to prevent bumping.
Note: Appropriate sample size is determined experimentally. The final sample concentration (after dilution to 100 mL) should be between 0 and 0.5 mg/L. Larger dilutions may be necessary for electroplating baths and silver strike solutions. Do not exceed the maximum sample volume of 25 mL. Several 25-mL aliquots may be digested in succession to concentrate a very dilute sample.

b) Turn on the water aspirator and make sure there is suction in the fractionating head.

c) Add 3 mL of concentrated sulfuric acid to the sample in the volumetric flask. Immediately place the head on the volumetric flask. Never use less than 3 mL of acid.

d) Place the volumetric flask on the heater. Turn the temperature dial to 440°C (825°F).

e) After the sample begins to char or the sulfuric acid reflux line becomes visible, wait 3 to 5 minutes.

f) Visually confirm the presence of liquid in the flask before adding hydrogen peroxide.

g) Add 10 mL of 50% hydrogen peroxide to the sample via the capillary funnel in the fractionating head.

h) After the hydrogen peroxide has boiled off, heat the sample until heavy white sulfuric acid fumes are present. Continue heating and reduce the sample volume to near dryness. Do not let the sample go completely dry at any time.
Note: If the sample goes to dryness, turn the Digesdahl off and cool completely. Add water to flask before handling. Repeat digestion from the beginning.
Note: If only thiosulfate is present in the sample, proceed to Step l.

i) Add another 3 mL of sulfuric acid via the capillary funnel.

j) Add another 5 mL of hydrogen peroxide. Check the solution for digestion completion. If digestion is not complete, continue adding hydrogen peroxide in 5 to 10 mL portions.

Several portions may be necessary.
Note: Digestion is complete when the digestate is colorless or the color of the digestate does not change upon addition of hydrogen peroxide. Also, a completely digested sample will not foam.

k) After digestion is complete and all the hydrogen peroxide is boiled off, reduce the volume of the digestate to near dryness. Do not allow the sample to become completely dry. Remove the flask from the heater. Cool to room temperature.

l) Slowly add about 25 mL of deionized water to the cooled flask. Add 2 drops of 1-g/L Phenolphthalein Indicator Solution.

m) Add 2 drops of 1-g/L Thymolphthalein Indicator Solution.

n) Using sodium hydroxide, adjust the pH of the solution to between 9 and 10. The solution will be pink in this pH range.
Note: A purple color indicates a pH greater than 10. If this occurs, add a drop of sulfuric acid and 2 drops of each indicator; repeat pH adjustment. Initially, use 50% sodium hydroxide, then 1.0 N sodium hydroxide as the end point is approached.

o) Filter turbid digestates. Quantitatively transfer the filtrate (or unfiltered sample) to a clean 100-mL volumetric flask. Dilute to the mark with deionized water. The sample is ready for analysis.

QUESTIONS AND ANSWERS

1. What are the major interferences in the experiment?
Negative interferences below 20 mg/L are cadmium, copper, lead, manganese, nickel and chloride. A positive interference below 20 mg/L is mercury at 2 mg/L.

2. Why is this method easier than the traditional dithizone method?
This colorimetric method has fewer interferences than the dithizone method. Also, organic extractions are as necessary as in other methods.

3. When should digestion be performed on a sample?
Digestion should be performed on samples containing organic matter, thiosulfate, or cyanide. Sources of these compounds are wastewater, silver electroplating baths and silver strike solutions.

4. Why is silver seldom dumped in water bodies or buried in waste sites?
Silver has value. The cost of recovery can be minimized by recovery of silver.

REFERENCES

1. *HACH Water Analysis Handbook,* 3rd Ed. (Loveland, CO.: HACH Company, 1997).
2. *Standard Methods for the Examination of Water and Wastewater:* SILVER, 16th Ed., A. E. Greenberg, R. R. Trussell, L. S. Clesceri and Mary Ann H. Franson (eds.), p. 242, Washington, DC, American Public Health Association, 1985.
3. *Hawley's Condensed Chemical Dictionary,* 11th Ed., rev'd by N. Irving Sax and Richard J. Lewis (New York, N.Y.: Van Nostrand Reinhold. 1987).

SULFATE

Sulfate is often present in natural waters in concentrations up to several thousand milligrams per liter. High concentrations of sulfate in water (> 250 mg/L) tend to increase the amount of lead dissolved from pipes containing lead. Most sulfate compounds are stable in water and are not altered by normal environmental factors.

Mine drainage may contribute large amounts of sulfate to the environment. Insoluble metal sulfate cakes form in oil fields when two or more types of water are mixed. Bacterial reduction in industrial and wastewater treatment plants can convert sulfates to sulfides noticeable by a distinctive hydrogen sulfide odor.

Metal sulfates have many practical uses in our society. Some applications of sulfate compounds are in dietary supplements and animal feed as a zinc compound, preservative, pharmaceuticals, freezing mix, cosmetic lotions, mineral waters, filler in synthetic detergents, flour enrichment, medicine, and the brewing industry. Additional applications include analytical reagents, cement, alum manufacture, pigments, glass manufacture, chemical manufacturing, coatings, ceramics, paperboard manufacture, processing textile, fibers, tanning, fireproofing, catalyst, fertilizers, laboratory reagent, plating, metal priming paints, rust inhibitor in paints, lubricants, plastic and rubber products, rayon manufacturing, battery electrolytes, herbicide, sulfur and sulfuric acid sources, metallurgy, drying industrial gases, desiccant, wallboard, dyes, surgical casts, and many additional applications.

Non-toxic sulfate compounds are often poured down the drain or discarded in municipal waste. More toxic compounds are sometimes recycled or recovered for use in the industrial process. The extremely toxic and dangerous sulfate materials, such as sulfuric acid, are isolated from the environment or neutralized before being discarded.

This sulfate analysis is the SulfaVer 4 Method for water, wastewater and seawater. The method is USEPA accepted for reporting wastewater analysis. Sulfate ions in the sample react with barium to form an insoluble barium sulfate compound which produces turbidity in the sample. The amount of turbidity formed in the sample is proportional to the sulfate concentration in the sample. The sulfate analysis on water, wastewater and seawater can be performed in the field on the DR/2010 Spectrophotometer and the DR/820, DR/850, and DR/890 Colorimeters.

Color or suspended matter may interfere at 450 nm. Some suspended matter may be removed by filtration before the sample is added to a clean 25 mL sample cell. After measurements, the sample cell should be cleaned with soap and brush to avoid a dry crust on the inside of the cell.

SULFATE (0 to 70 mg/L)

Method 8051 for water, wastewater, and seawater*
SulfaVer 4 Method** using Powder Pillows
USEPA accepted for reporting wastewater analysis†

*Compliments of HACH Company. Analysis may be performed by the DR/820, DR/850, and DR/890 Colorimeters, and the DR/2010 Field Spectrophotometer.

**Adapted from *Standard Methods for the Examination of Water and Wastewater, 426 C.*

†Procedure is equivalent to USEPA Method 375.4 for wastewater.

1. A User-Entered Calibration is necessary to obtain the most accurate results. See the "User Calibration" section at the end of this procedure. Program 680 can be used directly for process control or applications where a high degree of accuracy is not needed.
Note: The nature of turbidimetric tests and reagent lot variation necessitate user calibration for best results.

Choose Desired Program Accuracy

2. Enter the appropriate stored program number for Sulfate (SO_4^{2-}) Powder Pillows.
Press: **6 8 0 ENTER** or **9 ? ? ENTER.** The display will show: **Dial nm to 450.**
Note: The Pour-Thru Cell cannot be used with this procedure.

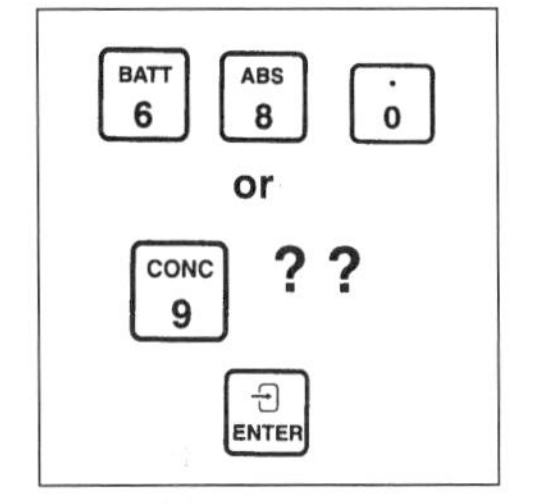

3. Rotate the wavelength dial until the small display shows: **450 nm.** When the correct wavelength is dialed in the display will quickly show: **Zero sample,** then: **mg/L SO_4^{2-}.**
Note: For greater accuracy, perform a user calibration for each new lot of SulfaVer 4 Sulfate Reagent Powder Pillows; see "User Calibration" following these steps.

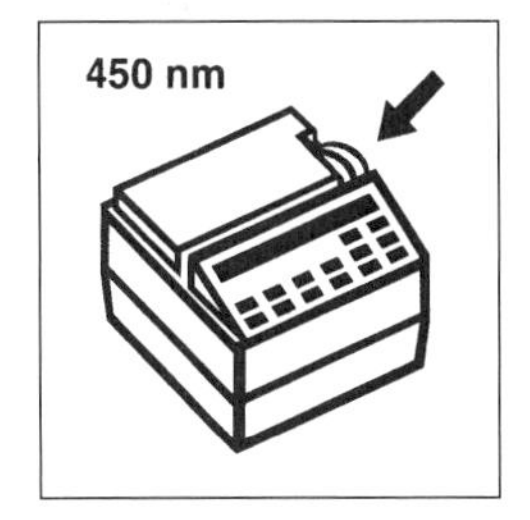

4. Fill a clean sample cell with 25 mL of sample.
Note: Filter highly turbid or colored samples. Use filtered sample in this step and Step 7.

5. Add the contents of one SulfaVer 4 Sulfate Reagent Powder Pillow to the sample cell (the prepared sample). Swirl to dissolve.
Note: A white turbidity will develop if sulfate is present.
Note: Accuracy is not affected by undissolved powder.

6. Press: **SHIFT TIMER.** A five-minute reaction period will begin.
Note: Allow the cell to stand undisturbed.

7. When the timer beeps, the display will show: **mg/L SO_4^{2-}.** Fill a second sample cell with 25 mL of sample (the blank).

8. Place the blank into the cell holder. Close the light shield.

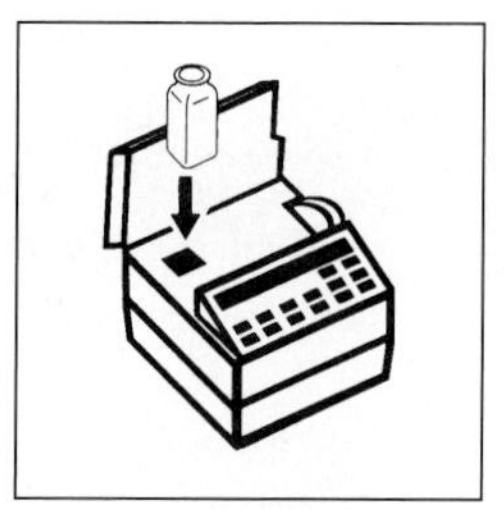

9. Press: **ZERO.** The display will show: **Zeroing ...** then: **0.mg/L SO_4^{2-}.**

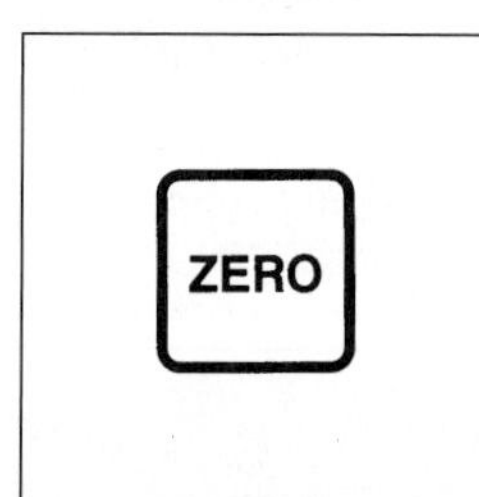

10. Within five minutes after the timer beeps, place the prepared sample into the cell holder. Close the light shield.

11. Press: **READ.** The display will show: **Reading ...** then the result in mg/L SO_4^{2-} will be displayed.
Note: Clean the sample cells with soap and a brush.

SAMPLING AND STORAGE

Collect samples in clean plastic or glass bottles. Samples may be stored up to 28 days by cooling to 4°C (39°F) or lower. Warm to room temperature before analysis.

ACCURACY CHECK

See p. 1141 of the HACH Water Analysis Handbook, 3rd Ed., 1997.

INTERFERENCES

The following interfere at levels above those concentrations listed:

Calcium	20,000 mg/L as $CaCO_3$
Chloride	40,000 mg/L as Cl^-
Magnesium	10,000 mg/L as $CaCO_3$
Silica	500 mg/L as $CaCO_3$

INITIAL SETUP OF SULFATE PROGRAM

The templates within User Program cannot be run directly. They must be copied into a usable program number (> 950) as in Steps c and d. Then, calibrate the program. The preprogrammed calibration (680 or 685) may be suitable for process control.

a) Press **SHIFT USER PRGM.** Use the **UP** arrow key to scroll to **Copy Program.** Press **ENTER.**

b) Scroll to or enter the template number for sulfate [906, 907 (AV)]. Press **ENTER.**

c) Scroll to or enter the desired user program number for sulfate (>950). Press **ENTER.** Record the program number for reference.

d) The display will show **Program Copied.** Press **EXIT.** The program is now ready to be calibrated.

USER CALIBRATION OF SULFATE PROGRAM

a) Use the test procedure to develop the turbidity in the standards just before recording the absorbance values for the calibration.

b) Press **SHIFT USER PRGM.** Use the **UP** arrow to scroll to **Edit Program.** Press **ENTER.**

c) Scroll to or enter the program number for sulfate (from Step c in Setup). Press **ENTER.**

d) Use the **DOWN** arrow to scroll down to **Calib Table:X** (X=denotes a number which indicates the number of data points in the table). Press **ENTER.**

e) The instrument will prompt **Zero Sample.** Place the blank solution in the cell holder. Close the light shield. Press **ZERO.** The instrument will prompt you to adjust to the proper wavelength if necessary.

f) The first concentration point will be displayed. Press **ENTER** to display the stored absorbance value of the first concentration point.

g) Place the first developed standard solution (same concentration as the value displayed) in the cell holder. Close the light shield. Press **READ** to display the measured absorbance of the standard. Press **ENTER** to accept the displayed absorbance value.

h) The second concentration point will be displayed. Press **ENTER** to display the stored absorbance value of the second concentration. Place the second developed standard solution in the cell holder. Close the light shield. Press **READ** to display the measured absorbance value of the standard.

i) Press **ENTER** to accept the absorbance reading. The next concentration point will then be displayed.

j) Repeat Steps h and i as necessary for the remaining standards.

k) When you are finished reading the absorbance values of the standards, press **EXIT.** Scroll down to **Force Zero.** Press **ENTER** to change the setting. Change to **ON** by pressing the arrow key, then press **ENTER.**

l) Scroll down to **Calib Formula.** Press **ENTER** twice or until only the **0** in **F(0)** is flashing. Press **DOWN** arrow to select **F3** (cubic calibration). Press **ENTER** to select **F3.**

m) Press **EXIT** twice. The display will show **Store Changes?** Press **ENTER** to confirm.

n) Press **EXIT.** The program is now calibrated and ready for use. Start on Step 2 of the iconed procedure.

QUESTIONS AND ANSWERS

1. How prevalent are sulfates in our society?
Compounds of sulfate have extensive applications in our society.

2. What are the interferences during sulfate measurements?
The primary interferences are highly turbid samples and colored samples.

3. Why should the sample cell be cleaned immediately after measurements?
If the sample cell is not cleaned soon after the measurement, the excess sulfate may cake onto the sides of the sample cell. Later, the difficulty in cleaning the sample cell may warrant the purchase of a new cell due to scratches and scrapes.

4. How can you be certain that the bottle filled with the blank and the sample bottle match spectrophotometrically?

Both the sample bottle and the bottle used for the blank should be filled with water and measured with the spectrometer at 450 nm. The reading for both the sample cell and the blank should be "zero." If the readings on pure water in clean sample cells do not match, either clean the cells and perform the experiment again or obtain a sample cell that does match spectrophotometrically.

REFERENCES

1. *HACH Water Analysis Handbook,* 3rd Ed. (Loveland, CO.: HACH Company, 1997).
2. *Hawley's Condensed Chemical Dictionary,* 11th Ed., rev'd by N. Irving Sax and Richard J. Lewis (New York, N.Y.: Van Nostrand Reinhold, 1987).
3. *Standard Methods for the Examination of Water and Wastewater:* SULFATE, 16th Ed., A. E. Greenberg, R. R. Trussell, L. S. Clesceri and Mary Ann H. Franson (eds.), pp. 464–468, Washington, DC, American Public Health Association, 1985.

SULFIDE

Sulfides are sometimes present in hot springs, oil field waters, and groundwater and are commonly found in sewage and industrial wastes. They are produced by the anaerobic decomposition of organic matter or by the bacterial reduction of sulfates.

Sulfides in waste streams produce hydrogen sulfide which has a rotten egg odor. Hydrogen sulfide is extremely toxic and has claimed the lives of numerous workers in sewers. The sulfide gas attacks metals directly and has caused serious damage to concrete sewers.

Sulfide compounds are used as catalysts, antifouling paints, pigments, source of lead, infrared radiation detector, semiconductor, luminous paint, lubricant additive, phosphor, depilatory, medicine, analytical reagent, sheep dips, and fungicide. They are used in ceramics, dyeing, engraving and lithography, paper pulp, isolation of metals from ores, and the preparation of rubber and plastics. The metal sulfides are used to produce hydrogen sulfide and other sulfides for practical uses.

Sulfide wastes have a variety of solubilities. Industries manufacturing a toxic sulfide waste stream must modify the waste product to reduce the toxicity level or protect the water bodies by eliminating discharges of the toxic materials to the environment. Sulfide wastes should be recovered and recycled for applications in the industrial process. However, some of the sulfide waste has been dumped into the environment and not handled appropriately. Some wastes containing sulfides and several other contaminates have been manifested for burial purposes.

Samples containing sulfides should be capped tightly and analyzed immediately to prevent loss of hydrogen sulfide. Substances such as sulfite, thiosulfate, and hydrosulfite interfere with sulfides by preventing the development of the blue colored solution. Hydrogen sulfide and acid-soluble metal sulfides react with N, N-dimethyl-p-phenylenediamine oxalate to form methylene blue in the presence of a mild oxidizing agent. The intensity of the blue color is proportional to the sulfide concentration in the sample and is measured at 665 nm. High levels of sulfides in water can be analyzed after proper dilution. The Methylene Blue Method is USEPA accepted for reporting wastewater analysis and can be performed on the DR/2010 Spectrophotometer and the DR/850 and the DR/890 Colorimeters in the field.

SULFIDE (0 to 0.600 mg/L S^{2-})

Method 8131 for water, wastewater, and seawater*
Methylene Blue Method**
USEPA Accepted for reporting wastewater analysis†

1. Enter the stored program number for sulfide (S^{2-}).
Press: **6 9 0 ENTER.** The display will show: **Dial nm to 665.**
Note: The Pour-Thru Cell can be used with this procedure.

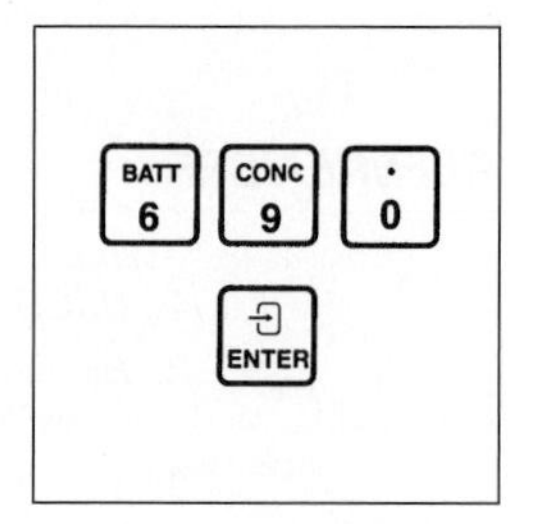

2. Rotate the wavelength dial until the small display shows: **665 nm.** When the correct wavelength is dialed in the display will quickly show: **Zero sample,** then: **mg/L S^{2-}.**

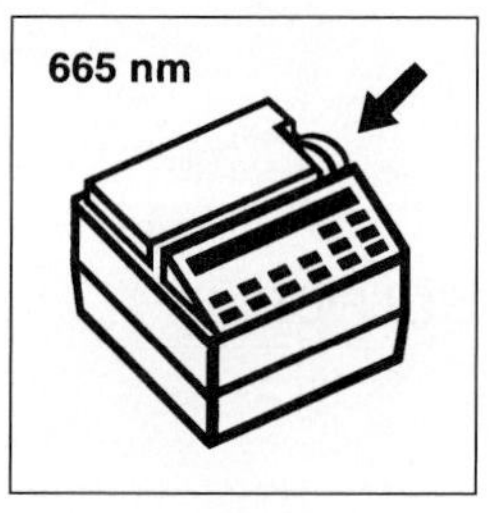

3. Pipet 25 mL of sample into a clean sample cell.
Note: Samples must be analyzed immediately and cannot be preserved for later analysis. Avoid excessive agitation as it causes loss of sulfide. For turbid samples, see "Interferences."

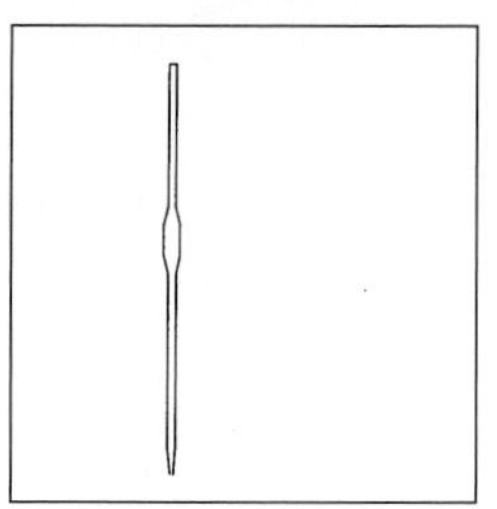

4. Fill a second sample cell with 25 mL of deionized water (the blank).

---

*Compliments of HACH Company. Analysis may be performed by the DR/850 and the DR/890 Colorimeters and the DR/2010 Field Spectrophotometer.

**Adapted from *Standard Methods for the Examination of Water and Wastewater, 427 C.*

†Procedure is equivalent to USEPA Method 376.2 or Standard Method 4500-S^{2-} D for wastewater.

5. Add 1.0 mL of Sulfide 1 Reagent to each cell. Swirl to mix.

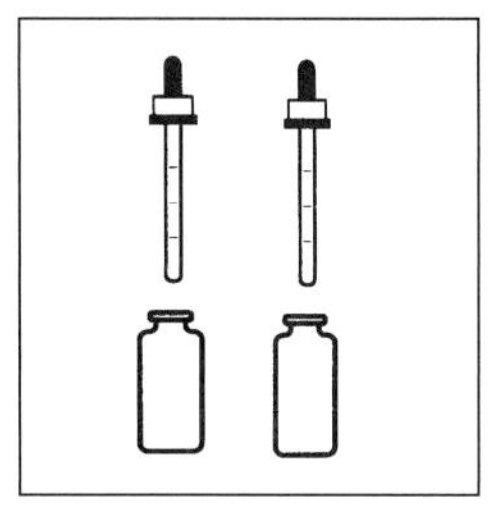

6. Add 1.0 mL of Sulfide 2 Reagent to each cell. Immediately swirl to mix.
Note: A pink color will develop, then the solution will turn blue if sulfide is present.

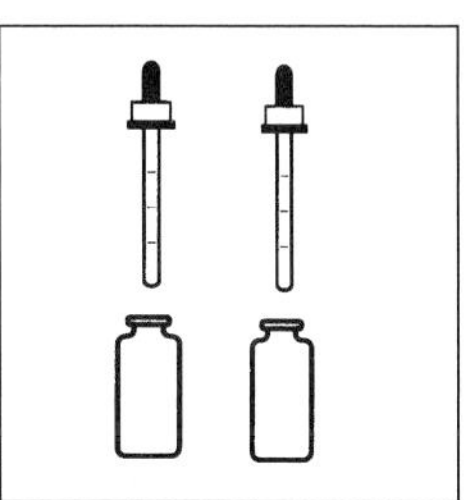

7. Press: **SHIFT TIMER.** A five-minute reaction period will begin. When the timer beeps, the display will show: **mg/L S^{2-}.**

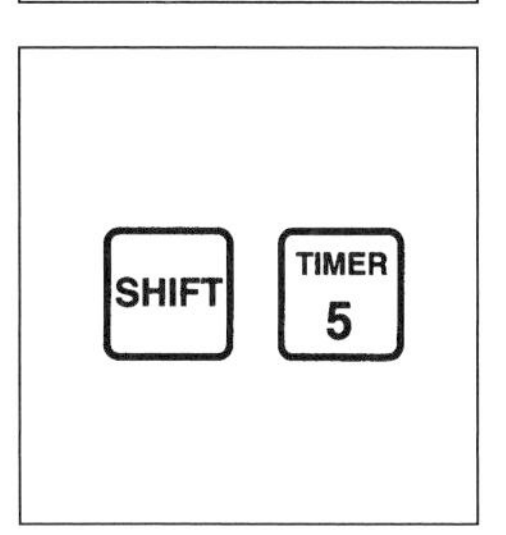

8. Place the blank into the cell holder. Close the light shield.

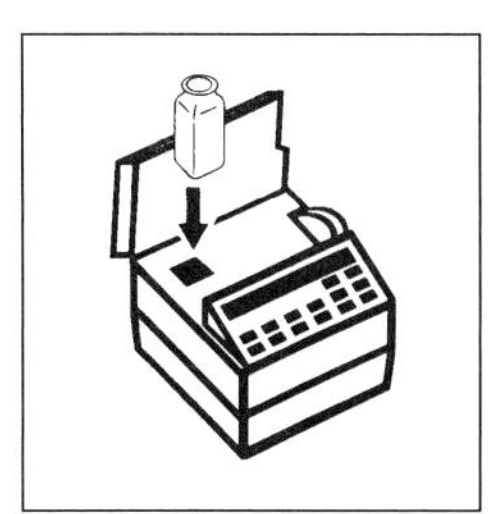

9. Press: **ZERO.** The display will show: **Zeroing ...** then: **0.000 mg/L S^{2-}.**

10. Immediately place the prepared sample into the cell holder. Close the light shield.

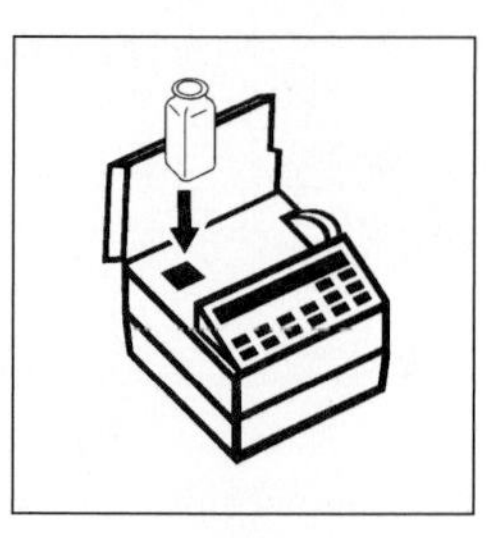

11. Press: **READ.** The display will show: **Reading ...** then the result in mg/L sulfide (S^{2-}) will be displayed.

SAMPLING AND STORAGE

Collect samples in clean plastic or glass bottles. Fill completely and cap tightly. Avoid excessive agitation or prolonged exposure to air. Analyze samples immediately.

ACCURACY CHECK

See p. 1147 of the HACH Water Analysis Handbook, 3rd Ed., 1997.

INTERFERENCES

For turbid samples, prepare a sulfide-free blank as follows. Use it in place of the deionized water blank in the procedure.

a) Measure 25 mL of sample into a 50-mL erlenmeyer flask.

b) Add Bromine Water dropwise with constant swirling until a permanent yellow color just appears.

c) Add Phenol Solution dropwise until the yellow color just disappears. Use this solution in Step 5 in place of deionized water.

Strong reducing substances such as sulfite, thiosulfate and hydrosulfite interfere by reducing the blue color or preventing its development. High concentrations of sulfide may inhibit full color development and require sample dilution. Some sulfide loss may occur when the sample is diluted.

SOLUBLE SULFIDES

Determine soluble sulfides by centrifuging the sample in completely filled, capped tubes and analyzing the supernatant. Insoluble sulfides are then estimated by subtracting the soluble sulfide concentration from the total sulfide result.

QUESTIONS AND ANSWERS

1. What is the threshold odor concentration of hydrogen sulfide in clean water?
   The threshold odor concentration in clean water is between 0.025 and 0.25 micrograms per liter.
2. What substances interfere with the sulfide determination method?
   Strong reducing agents such as sulfite, thiosulfate, and hydrosulfite interfere by preventing the full development of the blue color after the addition of the Sulfide 2 Reagent.
3. Why must samples be analyzed immediately after collection?
   Due to potential loss of hydrogen sulfide, the samples should be analyzed immediately without prolonged exposure to air and without excessive agitation.

REFERENCES

1. *HACH Water Analysis Handbook,* 3rd Ed., (Loveland, CO.: HACH Company, 1997).
2. *Hawley's Condensed Chemical Dictionary,* 11th Ed., rev'd by N. Irving Sax and Richard J. Lewis (New York, N.Y.: Van Nostrand Reinhold, 1987).
3. *Standard Methods for the Examination of Water and Wastewater:* SULFIDE, 16th Ed., A. E. Greenberg, R. R. Trussell, L. S. Clesceri and Mary Ann H. Franson (eds.), pp. 470–471, Washington, DC, American Public Health Association, 1985.

SULFITE

Sulfite readily oxidizes to sulfate in natural waters. Sulfite ions may occur in wastewaters and natural waters as a result of industrial pollution, boiler waters, and food preservatives. Sulfite ions in wastewater treatment appears to have a fairly rapid oxygen demand and have some toxicity toward fish and other aquatic life. Sulfites are also used for the treatment of boilers and boiler feedwaters to control dissolved oxygen. Excess sulfite in boiler water lowers the pH and promotes corrosion. At times, sulfite ions may occur in treatment plant effluents dechlorinated with sulfur dioxide.

Sulfite compounds have applications in the paper mill industry, dyes, photographic developer, food preservative and antioxidant, textile bleaching, water treatment, medicine, food and wine production, and the sugar industry as disinfectant. Additional uses include brewing, paper manufacture, biological cleansing, discoloration retarder of food, and preservative for anatomical specimens,

Sulfite wastes are sometimes recycled and reused in industry for a variety of purposes. If the sulfite waste contains other more toxic substances, then the waste stream is buried or modified to a less toxic form. In many cases, the sulfite wastes can easily be oxidized to sulfates which tend to be more stable and less of a problem environmentally.

Sulfite samples must not be agitated too much due to the possibility of oxidation. Metals such as copper catalyze the oxidation of sulfite to sulfate. Oxidizing agents and reducing agents can easily react with sulfite to form other more stable substances

Under acidic conditions, sulfite ion is titrated with a standard solution of potassium iodate-iodide containing starch indicator to form a solution with a blue starch end point. The blue color develops as the iodine reacts with the starch. The volume of titrant to reach the end point is proportional to the sulfite concentration.

SULFITE (4 to greater than 400 mg/L as SO_3^{2-})

Method 8216*
Iodate-Iodide Method**

1. Select a sample volume corresponding to the expected sulfite (SO_3^{2-}) concentration from Table 2.36..

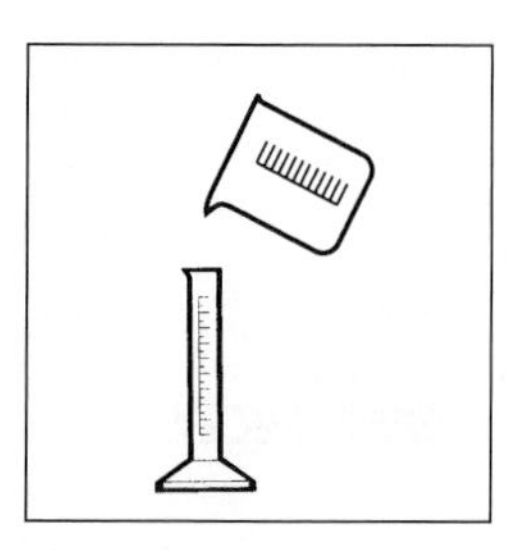

2. Insert a clean delivery tube into the Iodate-Iodide Titration Cartridge (KIO_3-KI). Attach the cartridge to the titrator body.

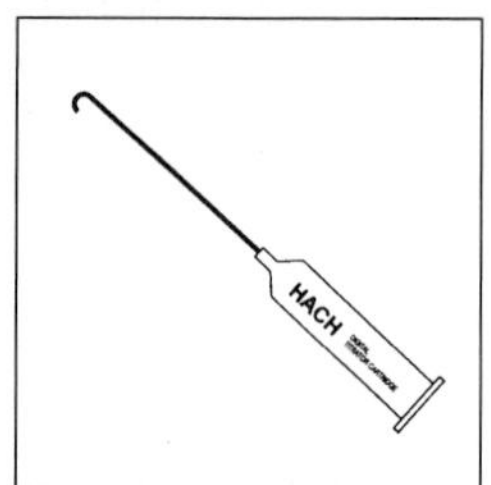

TABLE 2.36

Range (mg/L as SO_3^{2-})	Sample volume (mL)	Titration cartridge (N KIO_3-KI)	Catalog number	Digit multiplier
Up to 160	50	0.3998	14961-01	0.4
100–400	20	0.3998	14961-01	1.0
>400	5	0.3998	14961-01	4.0

* Compliments of HACH Company.
** Adapted from *Standard Methods for the Examination of Water and Wastewater, 428 A.*

3. Turn the delivery knob to eject a few drops of titrant. Reset the counter to zero and wipe the tip.
Note: For added convenience use the TitraStir stirring apparatus.

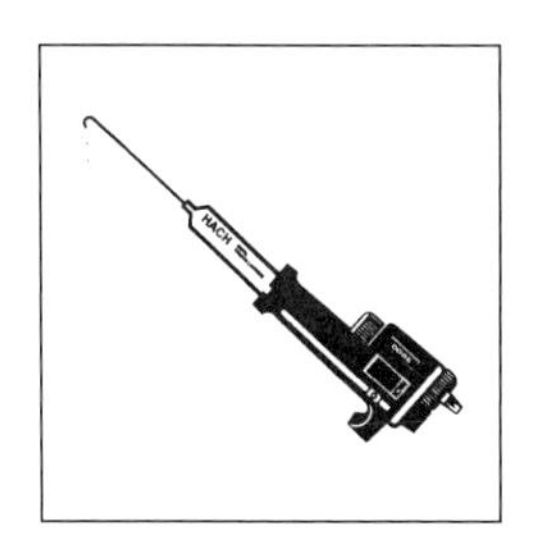

4. Use a graduated cylinder or pipet to measure the sample volume from Table 2.36. Transfer the sample into a clean 125-mL erlenmeyer flask. Dilute to about the 50-mL mark with deionized water.
Note: Avoid unnecessary agitation throughout the procedure.
Note: See "Sampling and Storage" after these steps.

5. Add the contents of one Dissolved Oxygen 3 Reagent Powder Pillow and swirl gently to mix.
Note: 0.5 mL of 19.2 N Sulfuric Acid Standard Solution may be substituted for the powder pillow.

6. Add one dropperful of Starch Indicator Solution and swirl to mix.

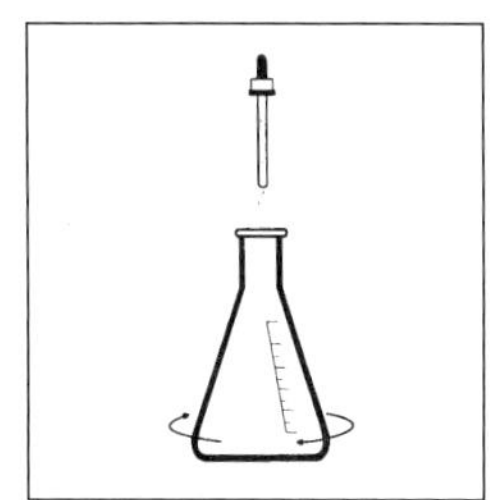

7. Place the delivery tube tip into the solution and swirl the flask while titrating with the iodate-iodide to a permanent blue end point. Record the number of digits required.

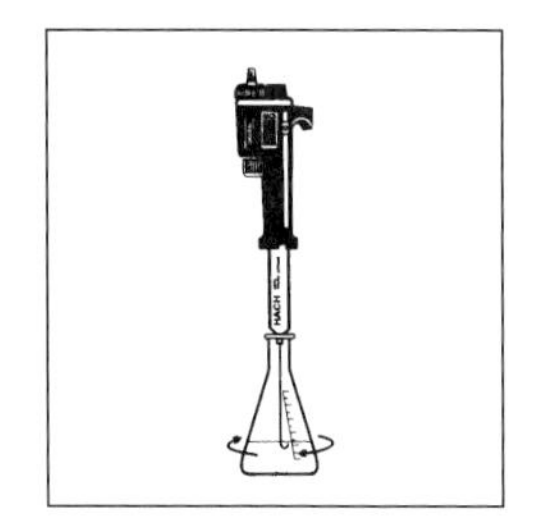

8. Calculate: Digits Required × Digit Multiplier = mg/L Sulfite (SO_3^{2-}).

Note: To obtain the concentration of other sulfite forms, multiply the mg/L SO_3^{2-} determined in Step 8 by the appropriate multiplier from Table 2.37.

Digits Required x Digits Multiplier

= mg/L Sulfite (SO_3^{2-})

TABLE 2.37

Form	Multiplier
Bisulfite, Hydrogen Sulfite (HSO_3^-)	1.01
Sodium Bisulfite, Sodium Hydrogen Sulfite ($NaHSO_3$)	1.30
Sodium Metabisulfite, Sodium Pyrosulfite ($Na_2S_2O_5$)	2.37
Sodium Sulfite (Na_2SO_3)	1.58

SAMPLING AND STORAGE

Samples must be analyzed immediately. Cool hot samples to 50°C or lower.

ACCURACY CHECK

See p. 1151 of the HACH Water Analysis Handbook, 3rd Ed., 1997.

INTERFERENCES

Sulfide, organic matter and other oxidizable substances will cause positive error in the titration. Nitrite will react with sulfite to cause low results. Some metals, especially copper, catalyze the oxidation of sulfite to sulfate. Addition of one Dissolved Oxygen 3 Powder Pillow per liter of sample immediately upon sampling will help eliminate the effects of nitrite and copper.

QUESTIONS AND ANSWERS

1. Why should the sample not be shaken violently?
   Excessive shaking of the sample will oxidize the sulfite to sulfate by atmospheric oxygen and low results will be observed.

2. Why should the sample be analyzed immediately?
Samples must be analyzed immediately to avoid oxidation of sulfite. In some cases, reduction of sulfite may result if the solution contains sulfide, organic matter, and other oxidizable contaminants.

3. What is the purpose of the starch solution?
The starch solution is the indicator used in the titration. A blue color will develop due to the excess iodine reacting with the starch.

REFERENCES

1. *HACH Water Analysis Handbook,* 3rd Ed. (Loveland, CO.: HACH Company, 1997).
2. *Hawley's Condensed Chemical Dictionary,* 11th Ed., rev'd by N. Irving Sax and Richard J. Lewis (New York, N.Y.: Van Nostrand Reinhold, 1987).
3. *Standard Methods for the Examination of Water and Wastewater:* SULFITE, 16th Ed., A. E. Greenberg, R. R. Trussell, L. S. Clesceri and Mary Ann H. Franson (eds.), p. 479, Washington, DC, American Public Health Association, 1985).

SURFACTANTS, ANIONIC (DETERGENTS)

High amounts of detergent in the environment can cause problems. They contain significant levels of phosphates or organo-phosphate products. Detergents are commonly used as cleansing agents in health facilities, industry, restaurants, resorts, schools, homes, and many other institutions.

Some areas of the country have severely restricted detergents containing phosphates. Less populated areas have generally not been concerned about this potential problem. The U.S. Public Health Service has determined that drinking water should not exceed 0.5 ppm in detergent concentration.

A small fraction of detergent is used in specific oils and gasoline. In oils, detergents consist of a long carbon chain attached to a sulphonic acid group or phosphate group. The carbon chain is nonpolar and will dissolve in nonpolar solutions such as oil. The phosphate group is polar and will dissolve in polar solutions such as water. These detergents are often used as cleansing solvents for engines and vehicle parts and sometimes discarded down the drain.

Phosphate is a significant part of specific types of laundry detergent. An abundance of phosphate in water can contribute to eutrophication, especially when large amounts of nitrogen are also present. Excess phosphate can cause an algal bloom. The rapid growth of algae and phytoplankton causes the depletion of dissolved oxygen in a body of water. Algae reproduce and die rapidly. The decomposition process of algae depletes an excessive amount of dissolved oxygen. As a result, the population of fish and organisms dependent on the oxygen supply can be impacted severely and die.

Detergents, ABS (Alkyl benzene sulfonate) and LAS (linear alkylate sulfonate) develop an ion-pair complex in the presence of crystal violet dye in benzene. The anionic surfactants are determined spectrophotometrically at 605 nm in samples obtained from wastewater, water and seawater sources. Interferences in the experiment are perchlorate and periodate ions. High amounts of chloride from salt brines and seawater will cause low results during detergent analysis.

SURFACTANTS, ANIONIC (0 to 0.275 mg/L)

Method 8028 for water, wastewater, and seawater*
Crystal Violet Method**

1. Enter the stored program number for anionic surfactants. Press: **7 1 0 ENTER.** The display will show: **Dial nm to 605.**
Note: The Pour-Cell cannot be used with this procedure.

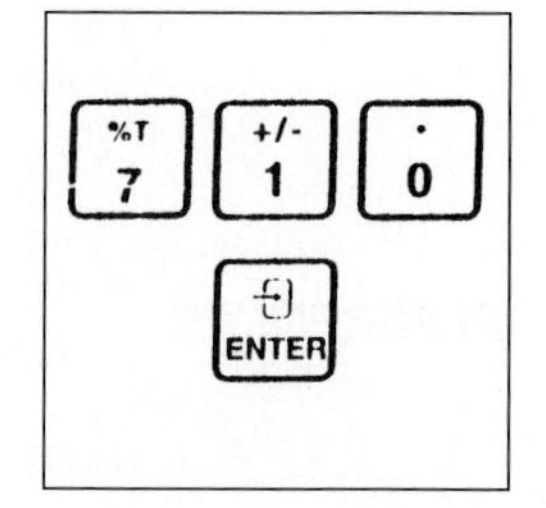

2. Rotate the wavelength dial until the small display shows: **605 nm.** When the correct wavelength is dialed in, the display will quickly show: **Zero sample** then: **mg/L SURF.ANION.**

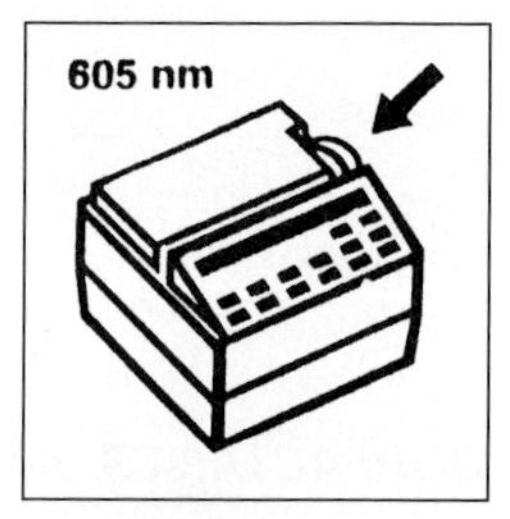

3. Fill a clean 500-mL graduated cylinder to the 300-mL mark with sample. Pour the sample into a clean 500-mL separatory funnel.

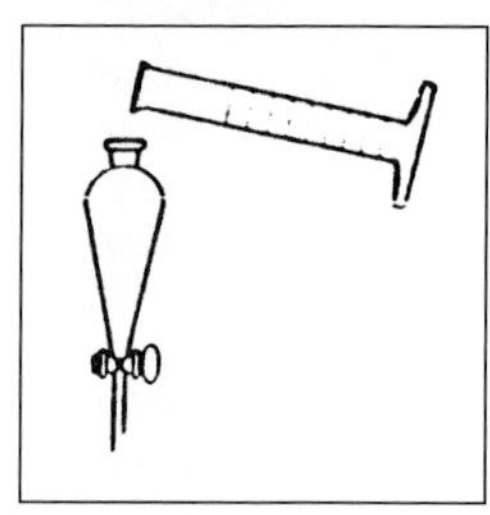

4. Add 10 ml of Sulfate Buffer Solution. Stopper the funnel. Shake the funnel for five seconds.

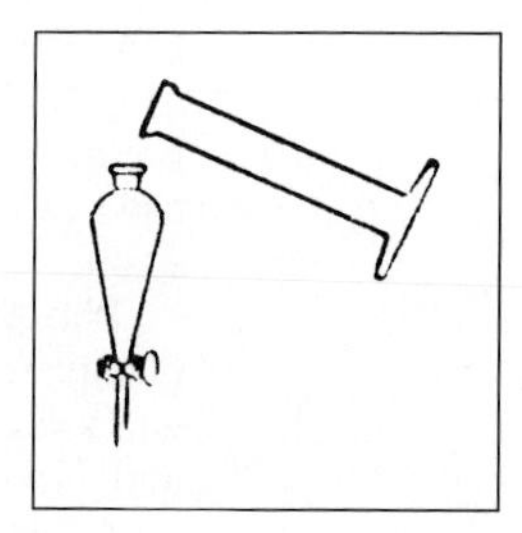

*Compliments of HACH Company.
***Analytical Chemistry,* 38, 791 (1966).

5. Add the contents of one Detergents Reagent Powder Pillow to the funnel. Stopper the funnel and shake to dissolve the powder.

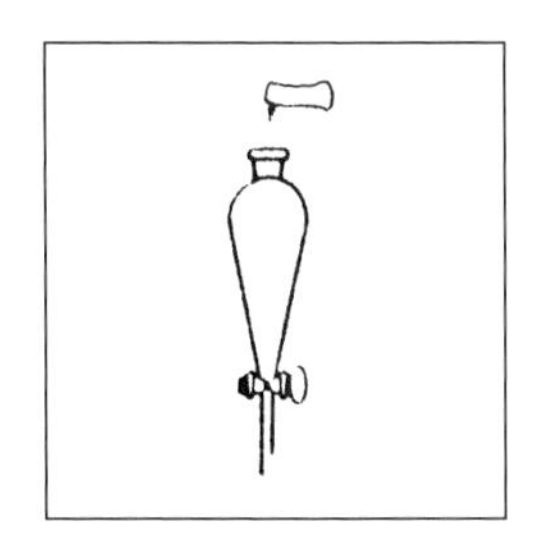

6. Add 30 mL of benzene to the funnel. Stopper the funnel and shake gently for one minute.
Note: Spilled reagent will affect test accuracy and is hazardous to the skin and other materials.
Note: Use benzene only in a well-ventilated area.

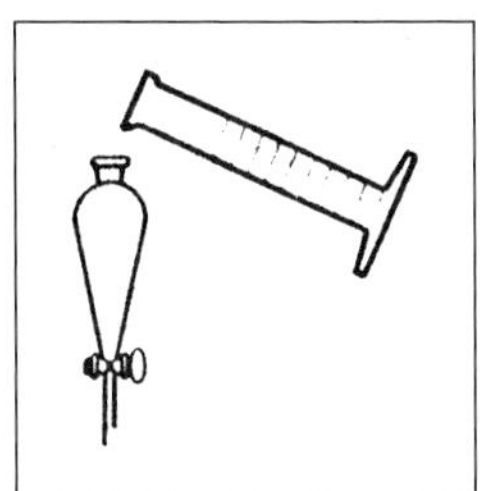

7. Place the separatory funnel in a support stand.

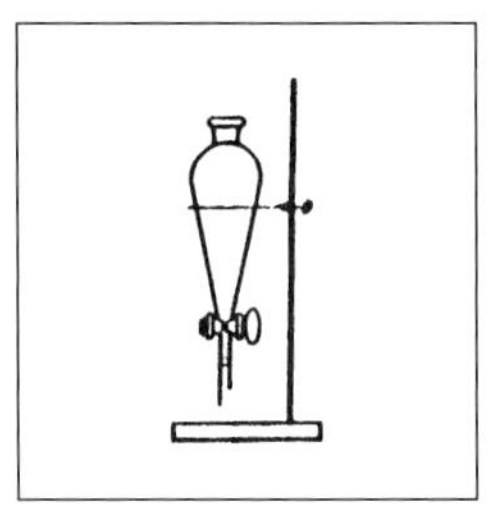

8. Press: **SHIFT TIMER.** A 30-minute reaction period will begin.
Note: Excessive agitation may cause an emulsion to form, requiring a longer time for phase separation. For these samples, remove most of the water layer, then gently agitate the funnel with a clean inert object in the funnel such as a Teflon-coated magnetic stirring bar.

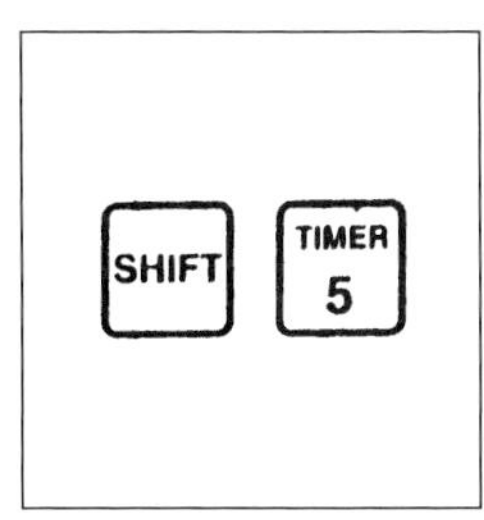

9. After the timer beeps, the display will show: **mg/L SURF.ANION.** Remove the stopper and drain the bottom water layer. Discard this layer.

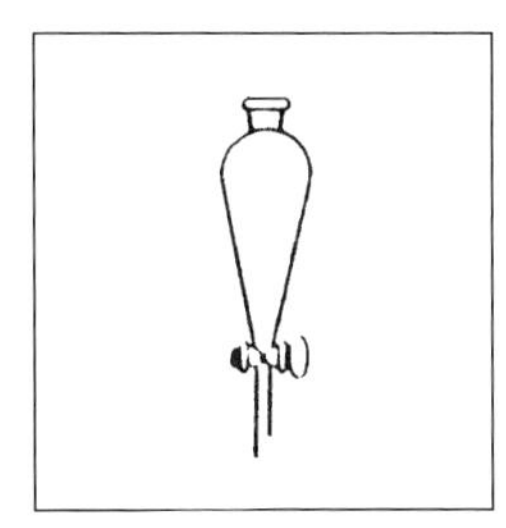

10. Drain the top benzene layer into a clean 25-mL sample cell (the prepared sample).
Note: The benzene layer cannot be filtered before color measurement. Filtration removes the blue color.

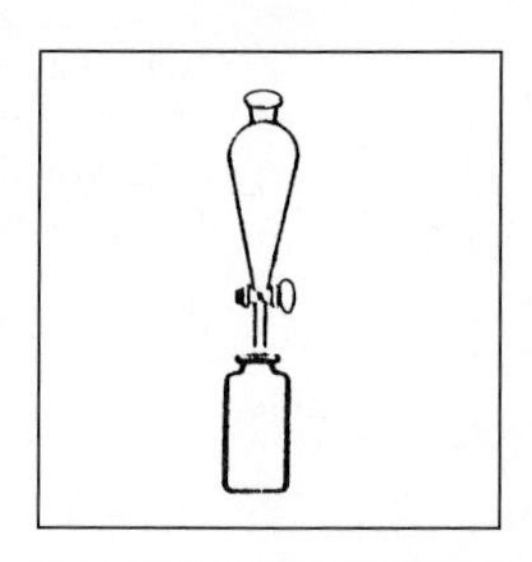

11. Fill another sample cell to the 25-mL mark with pure benzene (the blank).

12. Place the blank in the cell holder. Close the light shield.

13. Press: **ZERO.** The display will show: **Zeroing ...** then: **0.000 mg/L SURF.ANION.**

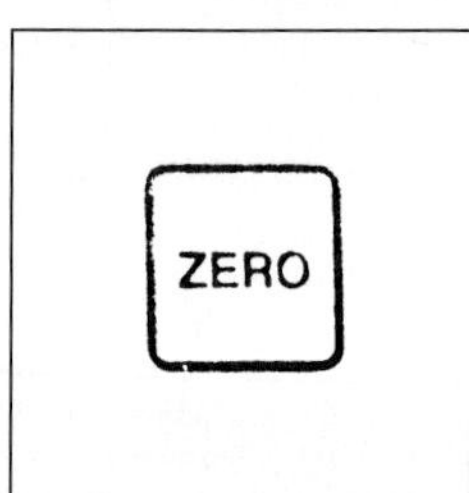

14. Place the prepared sample into the cell holder. Close the light shield.

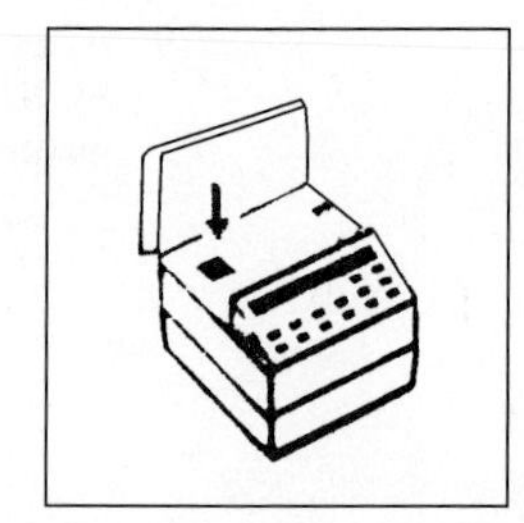

15. Press: **READ.** The display will show: **Reading ...** then the result in mg/L anionic surfactants will be displayed.
Note: Acetone may be used to clean benzene from glassware.
Note: The prepared sample and blank must be disposed of according to current Federal, State, and local regulations for benzene.

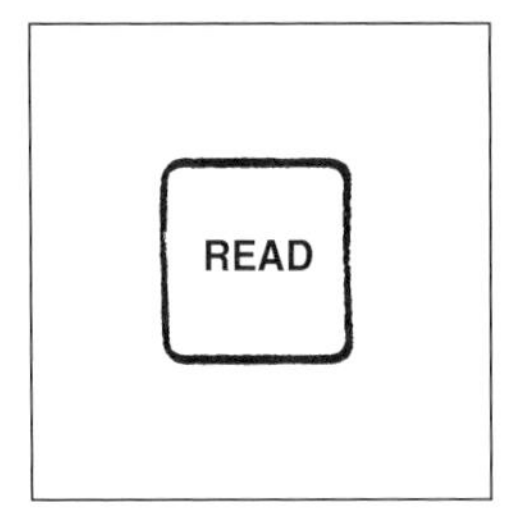

SAMPLING AND STORAGE

Collect samples in clean plastic or glass bottles. Analyze samples as soon as possible, but they may be stored at least 24 hours by cooling to 4°C (39°F). Warm to room temperature before testing.

ACCURACY CHECK

See p. 1160 in the HACH Water Analysis Handbook, 3rd Ed., 1997.

INTERFERENCES

Perchlorate and periodate ions will interfere. High amounts of chloride, such as those levels found in brines and seawater, will cause low results.

QUESTIONS AND ANSWERS

1. Why do detergents pose a serious problem for aquatic systems?
Wastewater from homes, car washes, hospitals and nursing homes, industries and other facilities often use detergents to clean floors, dishes and other equipment. Detergents often contain phosphorus which promotes the growth of algae and the depletion of dissolved oxygen in a body of water. Oxygen depletion can wipe out an entire population of organisms.
2. Would a detergent concentration of 1.0 ppm be considered safe for drinking water?
The U.S. Public Health Service has established an upper limit for detergent concentration in drinking water at 0.5 ppm. Therefore, 1.0 ppm of detergent would not be acceptable for drinking water.
3. What are the types of phosphate compounds in detergents?
Inorganic phosphate compounds used in some detergents consist of the PO_4^{3-} *unit. Additional detergents consist of organic phosphates which are more likely to remove organic substances such as grease and oil.*

REFERENCES

1. *HACH Water Analysis Handbook,* 3rd Ed., (Loveland, CO.: HACH Company, 1997).
2. *WARD's Snap Test,* Instant Water Quality Test Kits, WARD's Natural Science Establishment, Inc., Rochester, N.Y.
3. *Standard Methods for the Examination of Water and Wastewater:* PHOSPHORUS, 16th Ed., pp. 437–438, A. E. Greenberg, R. R. Trussell, L. S. Clesceri and Mary Ann H. Franson (eds.), Washington, DC, American Public Health Association, 1985).

TOTAL PETROLEUM HYDROCARBONS (TPH)

Total Petroleum Hydrocarbons (TPH) consists of compounds such as gasoline, kerosene, diesel fuel, Jet Fuel A, mineral spirits (a grade of naphtha), ethylbenzene, and m-xylene found in petroleum fuels. The TPH components are primarily substituted aromatic organic compounds and alkane-type compounds consisting of straight chain paraffins, branched chain paraffins and cycloparaffins. Jet Fuel A is a kerosine type fuel used primarily for short and medium flights.

The materials are flammable and a dangerous fire risk. The substances are moderately toxic by ingestion, inhalation, and skin absorption. Most materials are irritants to the skin and eyes. Human tolerance levels are limited but somewhat variable depending on the type of TPH.

Specific petroleum products are used for fuels, domestic heating, cleaning solvent, insecticidal sprays, petroleum chemicals, paint and varnish thinners, dry-cleaning fluid, asphalt and road tar solvents, and rubber cement solvents. Meta-xylene is used primarily as a solvent, aviation fuel, insecticide, and an intermediate for dyes and organic synthesis. Ethylbenzene is used as a solvent and as an intermediate in the production of styrene.

An analytical technique using a Mega-TPH Analyzer can quantify Total Petroleum Hydrocarbons at 3.4 μm as specified in EPA Method 418.1. The Mega-TPH Analyzer is a wavelength-specific, portable infrared instrument. By measuring the intensity of the characteristic infrared absorption bands, the concentration of each of the materials can be determined.

This test for TPH is a semi-quantitative test using an immunoassay procedure.* The TPH Stabilizing Agent stabilizes the substituted aromatic organic compounds in samples. Sample, standard and color development reagents are added to test tubes that are coated with an antibody specific for petroleum fuels. The TPH concentration in samples is determined by comparing the developed color intensity to the color of the TPH standard. The final TPH concentration is inversely proportional to the intensity of color development. Lighter color development indicates higher levels of TPH in the sample.

Various thresholds are obtained by sample dilution. Specific compounds may be tested using the sensitivity values as described in the sensitivity section. Surfactants in the sample may cause false positives.

TOTAL PETROLEUM HYDROCARBON (TPH) in WATER (Thresholds vary according to analyte)**

Method 10052† for Water
Immunoassay Method

*DR/2010 Spectrophotometer Procedures Manual, pp. 797–804, 1998.

**Test is semi-quantitative. Results are expressed as greater or less than the threshold value used.

†Compliments of HACH Company.

1. Enter the stored program for absorbance. Press: **0 ENTER.**
   Note: The Pour-Thru Cell cannot be used.

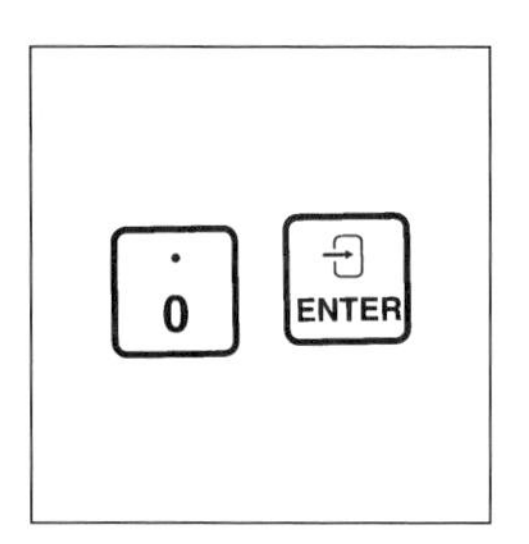

2. Rotate the wavelength dial until the small display shows: **450 nm**

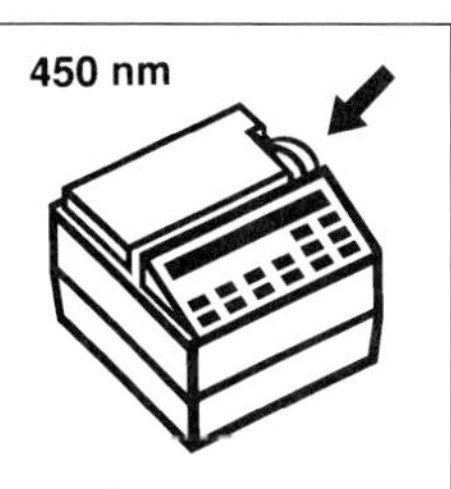

TPH PHASE 1: SAMPLE PREPARATION

1. Label a sample collection container. Collect at least 100 mL of sample. Before analysis, chill the samples.
   Note: Read "Measuring Hints" following these steps before testing.

2. Choose the desired volume from Table 2.38 below. Measure the volume into a 50-mL graduated mixing cylinder. Add deionized water to the 50-mL mark. Stopper and mix.

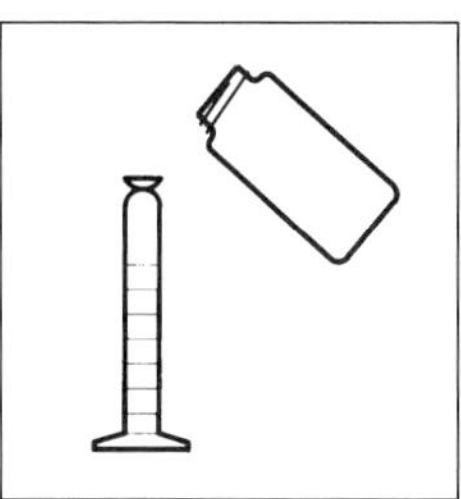

3. Add the contents of one TPH Stabilizing Agent Powder Pillow to an empty sample bottle. Label the bottle.
   Note: Discard or recycle bottles after one use to prevent cross contamination.

TABLE 2.38

Threshold	Sample volume	Threshold	Sample volume
220 ppb (m-xylene)	50 mL	5.5 ppm (m-xylene)	2 mL
550 ppb (m-xylene)	20 mL	11 ppm (m-xylene)	1 mL
1.1 ppm (m-xylene)	10 mL	22 ppm (m-xylene)	0.5 mL
2.2 ppm (m-xylene)	5 mL	55 ppm (m-xylene)	0.2 mL

Note: 1.1 ppm = 1100 ppb; 2.2 ppm = 2200 ppb

4. Pour the sample into the labeled sample bottle until the bottle is full. Cap. Mix until TPH Stabilizing Agent is dissolved. Proceed to Phase 2 promptly.

TPH PHASE 2: IMMUNOASSAY- STEPS IN THIS PHASE REQUIRE EXACT TIMING.

1. Label four TPH Enzyme Conjugate Tubes Standard 1, Standard 2, Sample 1 and Sample 2 for use in Step 7.
Note: The TPH Conjugate and Antibody Tubes are matched lots. Mixing with other lots will cause erroneous results.

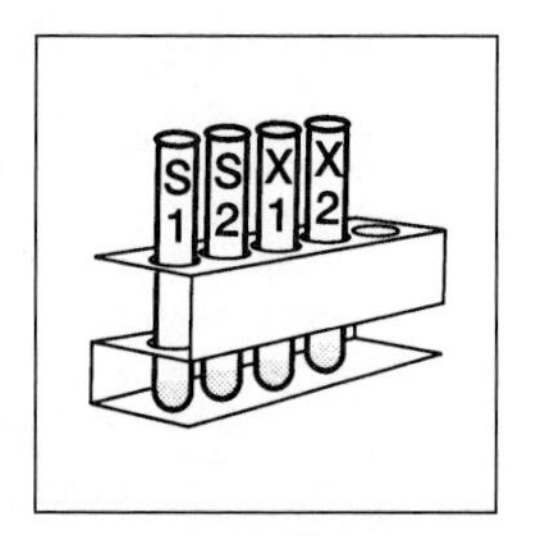

2. Label four TPH Antibody Tubes Standard 1, Standard 2, Sample 1 and Sample 2.
Note: Sample 1 and Sample 2 tubes can be duplicate samples, different dilutions of the same sample, or 2 different samples.

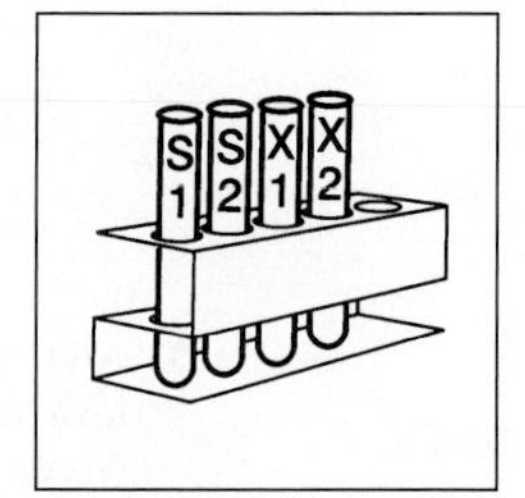

3. Using a TenSette Pipet, add 2.0 mL of prepared sample to each of the TPH Enzyme Conjugate Tubes labeled Sample 1 and Sample 2.

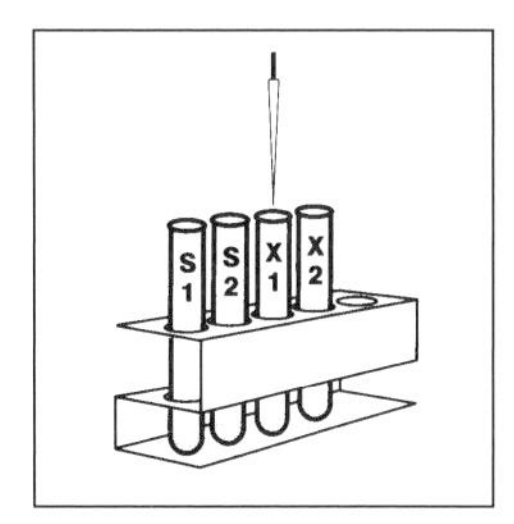

4. Add the contents of 2 TPH Standard Buffer Solution Pillows to each of the TPH Enzyme Conjugate Tubes labeled Standard 1 and Standard 2.
Note: Squeeze the top of the inverted pillow to empty. If necessary, cut a top corner of the inverted pillow with clippers.

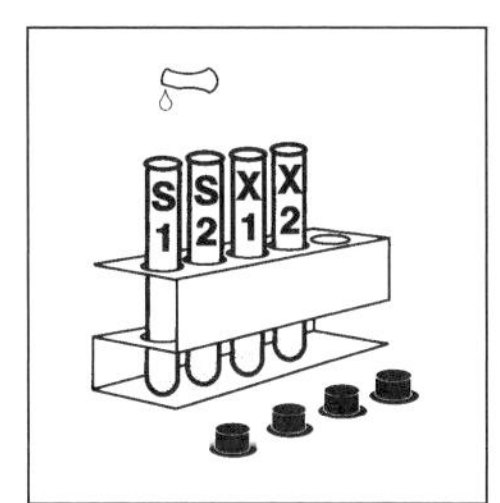

5. Snap open a TPH Standard Ampule.

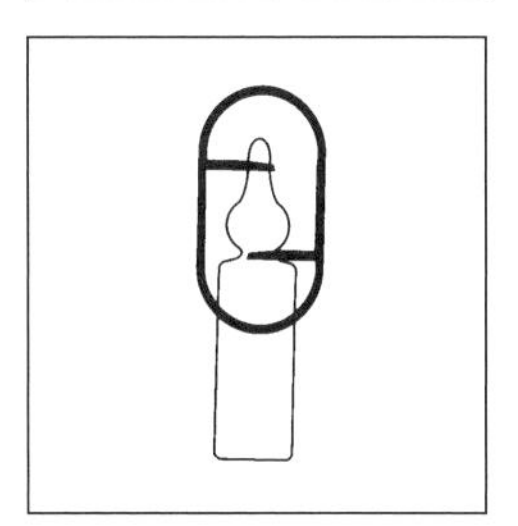

6. Using the WireTrol pipet, add 50 μL of TPH Standard to the Standard 1 and Standard 2 TPH Enzyme Conjugate Tubes. Swirl to mix thoroughly.
Note: Dispense standard below the level of the solution in the Enzyme Conjugate Tubes. Use a new tip for each tube.

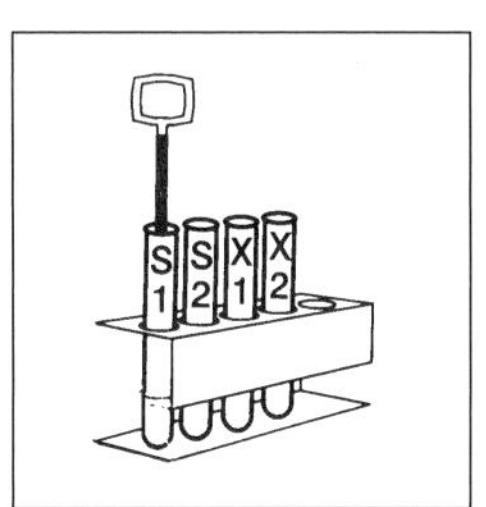

7. Pour the contents of the TPH Enzyme Conjugate Tubes into the correct TPH Antibody Tubes. Swirl to mix.

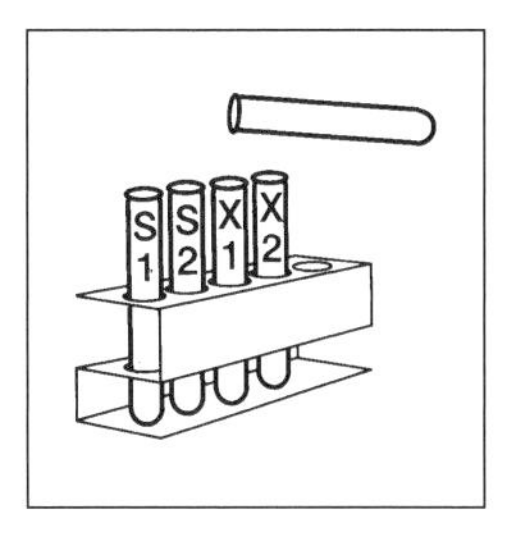

8. Begin a 10-minute reaction period.
Note: During this phase, m-xylene and other substituted aromatics in the sample compete with the enzyme conjugate for a limited number of binding sites on the inside of the antibody tubes.

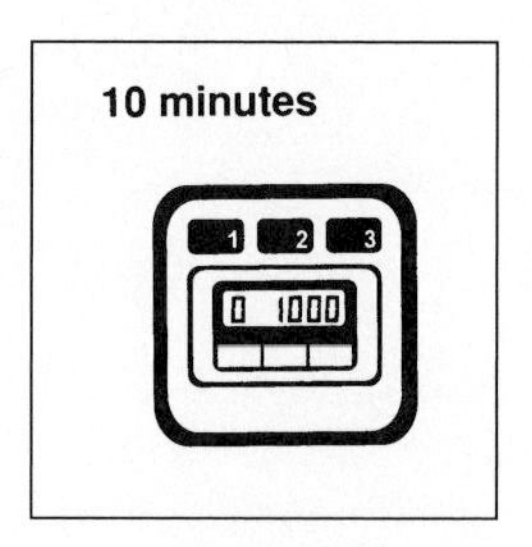

9. After the 10-minute period, discard the contents of the TPH Antibody Tubes into an appropriate container.

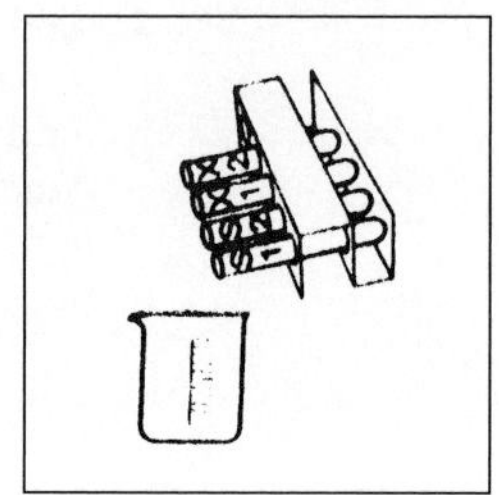

10. Wash each tube thoroughly and forcefully 4 times with Wash Solution. Empty the tubes into an appropriate container. Shake well to ensure most of the Wash Solution drains after each wash.
Note: Wash Solution is a harmless dilute detergent.

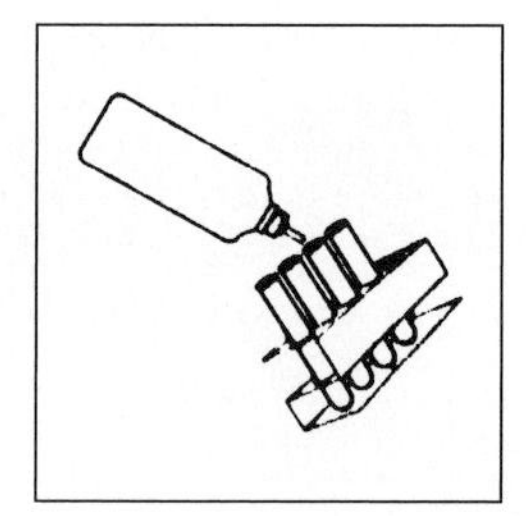

11. Continue to the next phase immediately.
Note: Ensure most of the Wash Solution is drained from the tubes. Turn the tubes upside down and gently tap on a paper towel to drain. Some foam may be left from the Wash Solution; this will not affect the results.

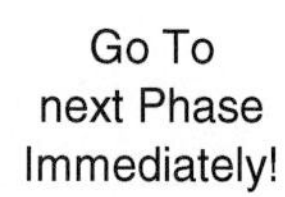

TPH PHASE 3: COLOR DEVELOPMENT

Check reagent labels carefully! Reagents must be added in proper order for valid test results.

1. Add 5 drops of Solution A to each tube. Replace the bottle cap.
   Note: Hold reagent bottles vertically for accurate delivery or erroneous results may occur.

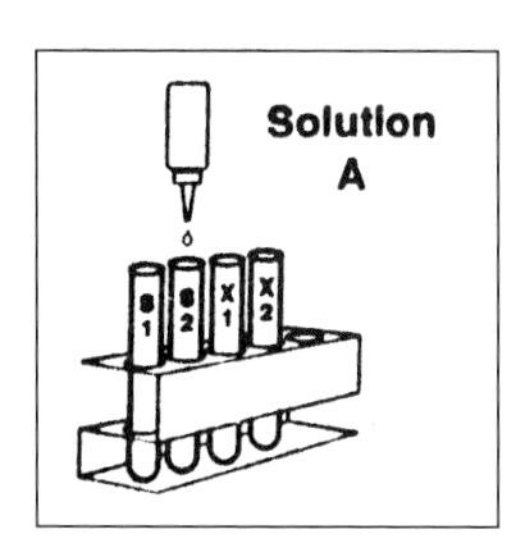

2. Begin a 2.5-minute period and immediately add 5 drops of Solution B to each tube. Swirl to mix. Replace the bottle cap.
   Note: Add drops to the tubes in the same order to ensure proper timing (i.e., left to right). Solution will turn blue in some or all of the tubes.

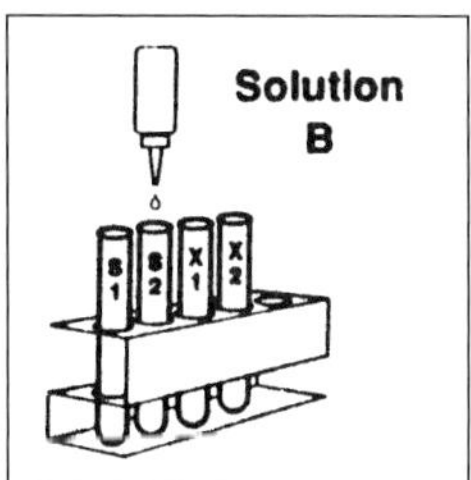

3. Let each tube react for exactly 2.5 minutes. Then add 5 drops of Immunoassay Stop Solution to each tube. Replace the bottle cap.
   Note: Blue solutions will turn yellow when Stop Solution is added. TPH concentration is inversely proportional to color development; a lighter color indicates higher levels of TPH.

4. Using the TenSette Pipet with a new tip, add 0.5 mL of deionized water to each tube. Swirl to mix.

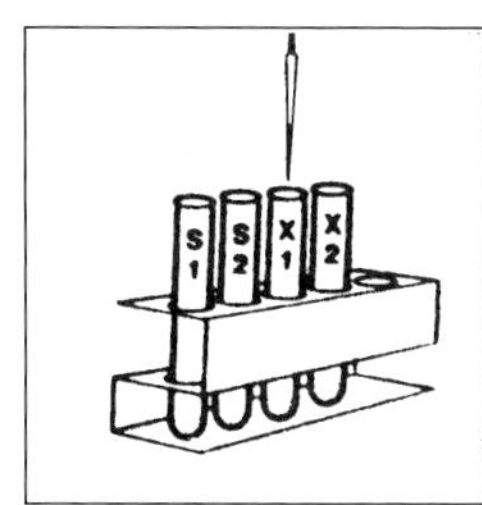

TPH PHASE 4: MEASURING THE COLOR

1. Fill a Zeroing Tube with deionized water (the blank). Wipe the outside of all tubes with a tissue to remove smudges and fingerprints.

2. Insert the Immunoassay Adapter into the sample cell compartment.
Note: Align the adapter so the light beam openings face the sides of the DR/2010. Press firmly on the adapter to seat it.

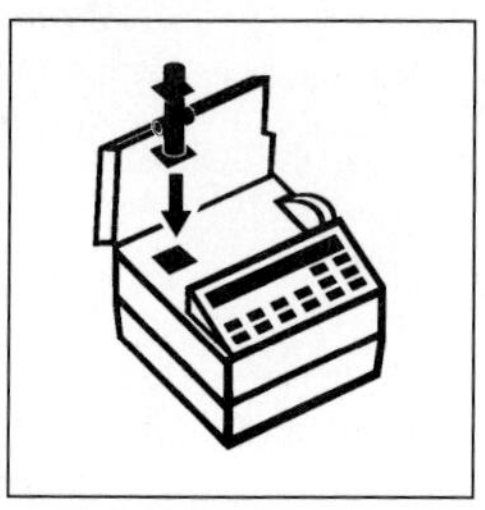

3. Place the blank in the cell holder. Place the cover on the adapter.

4. Press: **ZERO.** The display will show: **0.000 Abs.**

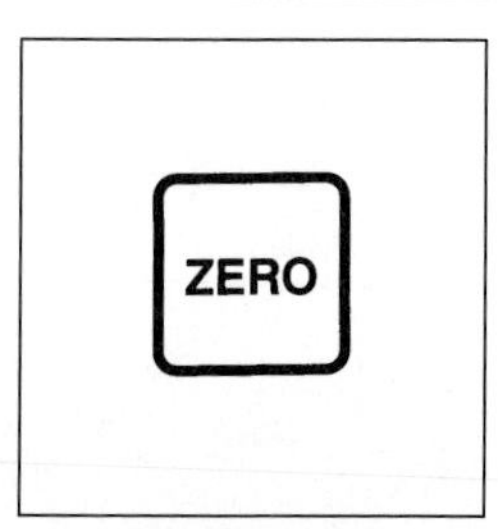

5. Insert Standard 1 tube into the cell holder. Place the cover on the adapter.

6. Record the absorbance reading.

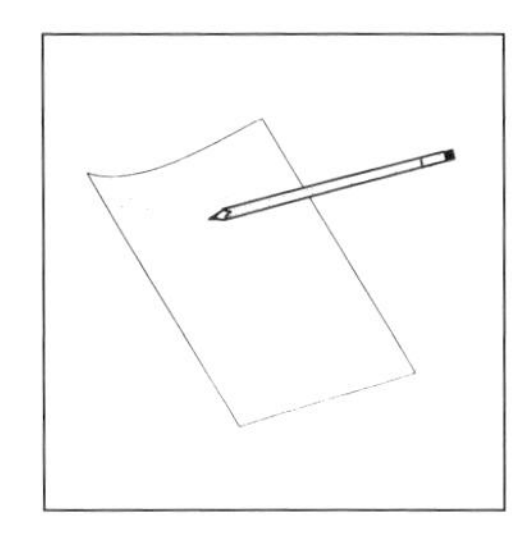

7. Repeat Steps 5 and 6 for the Standard 2 tube.
Note: If Standard 1 and Standard 2 are more than 0.350 absorbance units apart, repeat the test beginning at Phase 2, Immunoassay.

Repeat Steps 5 and 6 for Standard #2

8. Insert the Sample 1 tube into the cell holder. Place the cover on the adapter.
Note: TPH concentration is inversely proportional to the color intensity (or absorbance value). More color means less TPH in the sample.

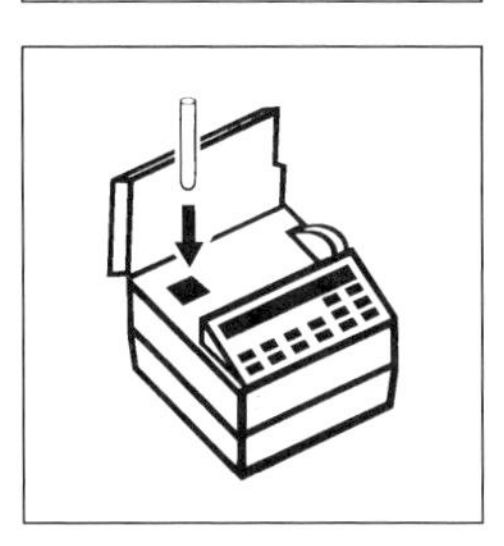

9. Record the absorbance reading.

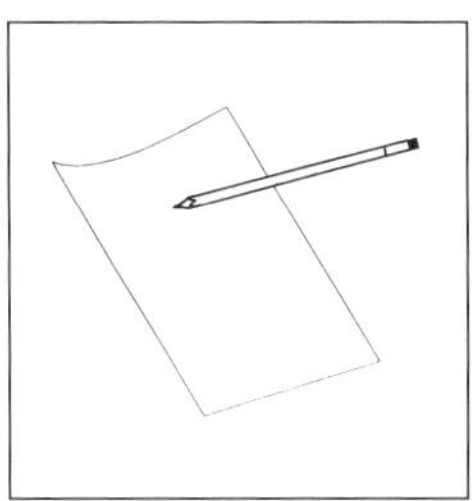

TABLE 2.39

If sample absorbance is ...	Sample TPH Concentration is ...
... less than the highest standard absorbance	... greater than the chosen threshold value
... greater than the highest standard absorbance	... less than the chosen threshold value

10. Repeat Steps 8 and 9 for the Sample 2 tube. See Table 2.39 on page 2.321 to interpret results.
Note: Results are read as m-xylene. For other compounds, see "Sensitivity" below.

Repeat Steps 8 and 9 for Sample #2

INTERPRETING THE RESULTS

Use Table 2.39 to interpret the results.

SENSITIVITY

The levels listed in Table 2.40 are the concentrations in an undiluted water sample necessary to give a positive response.

- Threshold values may vary depending on the composition of the source contamination in real-world samples.
- Surfactants in the sample may cause false positives.
- To convert results to one of the compounds listed in Table 2.40 use the following equation:

$$\text{Diluted threshold Value for compound} = (A/B) \times C$$

where:
A = Undiluted threshold value for that compound (Table 2.40)
B = 220(sensitivity to m-xylene)
C = Threshold chosen from Table 2.38

TABLE 2.40

Compound	ppb	Uncertainty (90% Cl)*
Gasoline	600	±100 ppb
Kerosene	920	±170 ppb
Diesel	940	±130 ppb
Jet Fuel A	490	±100 ppb
Mineral Spirits	1160	±320 ppb
Toluene	380	±50 ppb
Ethylbenzene	90	±25 ppb
m-Xylene	220	±40 ppb

*The CI, or Confidence Interval, at 90% is an expression of the uncertainty in the determined threshold value for each component listed above.

For an example of a calculation, see p. 802 of the *DR/2010 Spectrophotometer Procedures Manual,* 1998.

SAMPLING AND STORAGE

If prompt analysis is not possible, chill the sample in an ice bath, refrigerator or cooler to limit the loss of volatile compounds. Analyze the sample as soon as possible after collection. Storing sample longer than 24 hours after collection may compromise the integrity of the sample. When collecting samples, fill containers completely (no headspace) and cover the container with a tightly sealed lid immediately after collection.

MEASURING HINTS

- Timing is crucial; follow instructions carefully.
- Handle the Antibody Tubes carefully. Scratching the inside or outside may cause erroneous results. Wipe tubes with a paper towel to remove smudges or fingerprints before measuring.
- Hold all dropper bottles vertically and direct drops to the bottom of the tube.
- Antibody Tubes and Enzyme Conjugate Tubes are made in matched lots. Do not mix with other reagent lots.
- Paper towels, liquid waste containers and laboratory tissue are required.
- Chilling samples prior to analysis helps limit the loss of volatile compounds during Sample Preparation and Immunoassay Steps.

SAFETY AND STORING AND HANDLING OF REAGENTS

- Wear protective gloves and eye wear.
- Store reagents at room temperature and out of direct sunlight (less than 80°F or 27°C).
- Keep aluminized pouch that contains antibody-coated tubes sealed when not in use.
- If Stop Solution or liquid from the extraction jar comes in contact with eyes, wash thoroughly with cold water for 15 minutes. Seek immediate medical help.
- Operational temperature of the reagents is 40 to 90°F (5 to 32°C).

QUESTIONS AND ANSWERS

1. What is TPH?
TPH is defined as Total Petroleum Hydrocarbons and represents the type of compounds found in petroleum fuels of various types. The TPH components are straight chain paraffins, branched chain paraffins, cycloalkanes, naphtha, and substituted aromatic organic compounds.

2. Why should the analyst avoid scratching the Antibody Tubes?
Scratching the inside or outside of the Antibody Tubes may cause erroneous results.

3. During color development, why is it necessary to hold the delivery bottle vertically over each solution tested?

Hold the delivery bottle vertically to minimize inaccuracies in drop size and to reduce erroneous results.

4. What is the purpose of the TPH Stabilizing Agent?
The TPH Stabilizing Agent stabilizes the substituted aromatic organic compounds, such as ethylbenzene and m-xylene, in the samples.

REFERENCES

1. DR/2010 Spectrophotometer Procedures Manual, 3rd Ed. (Loveland, CO.: HACH Company, 1998).
2. *Hawley's Condensed Chemical Dictionary,* 11th Ed., rev'd by N. Irving Sax and Richard J. Lewis (New York, N.Y.: Van Nostrand Reinhold, 1987).

TOTAL TRIHALOMETHANES (TTHM)

Many drinking water sources contain small amounts of the trihalomethanes. Chloroform and other chlorinated trihalomethanes can be produced in trace quantities during disinfection of drinking water by chlorination. In addition, chloroform, bromoform, and chlorinated trihalomethanes are traceable to industrial effluents and laboratory solvents. The presence of trihalomethane solvents in groundwater can be attributed to their widespread usage, relative mobility in soils and groundwater, resistance to biodegradation, and the absence of removal processes such as volatilization and photolysis.[1] Most of the trihalomethane solvents are nonflammable and are slightly soluble in water.

Humans have a tolerance level of 10 ppm in air to chloroform, and 0.5 ppm in air to bromoform. Bromoform is toxic by ingestion, inhalation, and skin absorption. The U.S. Safe Drinking Water Act regulates trihalomethanes (sum of chloroform, bromodichloromethane, dibromochloromethane and tribromomethane) at 100 μg/L.[1]

The trihalomethane compounds can create a variety of health issues for humans. The primary effects of these compounds are on the central nervous system. Extensive human exposure to the compounds can cause liver and kidney damage, ventricular fibrillation, and cardiac problems. Prolonged inhalation or ingestion of these chemicals may be fatal. Some of the chemicals such as chloroform are toxic by inhalation and have a narcotic effect on people. Chloroform is a known carcinogen (OSHA) and has been prohibited by the FDA. The compound can no longer be used in drugs, cosmetics, food packaging, cough medicines, toothpastes, etc.

In the past, chlorofluorocarbons have been used as refrigerants and aerosol sprays. Most of these applications have been prohibited because of their depleting effect on stratospheric ozone. Chlorofluorocarbons have limited applications and are not of major importance today.

Methylene chloride, or dichloromethane, is similar to the trihalomethanes. Humans have a tolerance level of 100 ppm in air to methylene chloride. The compound is used in paint remover, solvent degreasing, plastics processing, blowing agents in foams, and during the preparation of chemicals. Dichloromethane has been used for extraction of substances from a variety of materials. In some countries, the solvent is still used for decaffeination of coffee even though the process is not permitted in the United States.

Chloroform , or trichloromethane, is used in fumigants, insecticides, fluorocarbon refrigerants and during plastic manufacture. Chloroform has many practical uses as a solvent and as an animal anesthetic in a variety of laboratories and research centers. Bromoform, known as tribromomethane, has applications in medicine (sedative), organic synthesis, geological assaying, and as a solvent for waxes, greases and oils. Dibromochloromethane has applications in organic synthesis. Iodoform exists as a solid and can be applied as an antiseptic for external use only.

Normally, water containing halomethanes is injected directly into a chromatograph to detect small levels of halomethanes. If foreign components are present, halomethanes can be extracted into a light petroleum solvent and the solution is injected directly into a chromatograph.

The total trihalomethane (TTHM) content of water can be detected by a THM Plus™ colorimetric method provided by the HACH Company. In the experiment, the THM Plus method reacts with the trihalogenated disinfection by-products formed as a result of the disinfection of drinking water with chlorine in the presence of naturally occurring organic materials.[4] The formation of the disinfection by-products (DBPs) is dependent on the temperature, pH, chlorine contact time, precursor concentration, bromide concentration, chlorine dose, and residuals.

THM compounds in the sample react with N,N-diethylnicotinamide under heated alkaline conditions to form a dialdehyde intermediate. After cooling and acidifying the sample to pH 2.5, the dialdehyde intermediate is reacted with 7-napthylamine-1,3-disulfonic acid to form a Schiff base which absorbs at 515 nm. The color of the solution is directly proportional to the total amount of THM compounds present in the sample. The DR/2010 spectrophotometer determines the concentration of trihalomethanes in drinking water at 515 nm in the range of 0 to 200 ppb as chloroform.

THM PLUS™: TRIHALOMETHANES (0–200 ppb as Chloroform)*

Method 10132
For screening THMs in drinking water

1. Enter the stored program number for Trihalomethane (THM) Plus. Press: **725 ENTER** The display will show: **Dial nm to 515**
Note: For the most precise results, use matched cells.

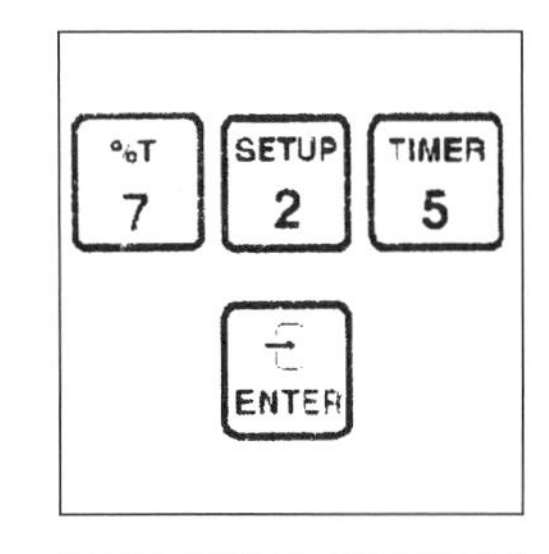

2. Rotate the wavelength dial until the small display shows: **515 nm** When the correct wavelength is dialed in, the display will quickly show: **Zero Sample** then: **ppb $CHCL_3$**

*Compliments of HACH Company.

3. Prepare a hot water bath by adding 500 mL of water to an evaporating dish. Put the dish on a hot plate and turn heater on high.
Note: If analyzing more than four samples, use 450 mL water.

4. Prepare a cooling bath by adding 500 mL of cold water (18 to 25°C) tap water to a second evaporating dish.
Note: Maintain the water temperature between 18 and 25°C. If analyzing more than four samples, use 450 mL water.

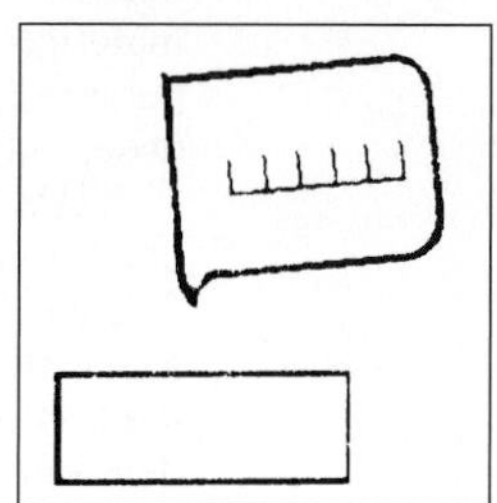

5. Fill two sample cells to the 10 mL mark with sample. Label one sample and the other blank.
Note: Perform steps 5 through 9 **rapidly** so as to not lose volatile THMs from the sample. If you are testing more than one sample, complete steps 5 through 9 for one sample before going on to the next step.
Note: If dispensing sample with a pipette, the pipette must dispense quickly without causing aeration or back pressure.

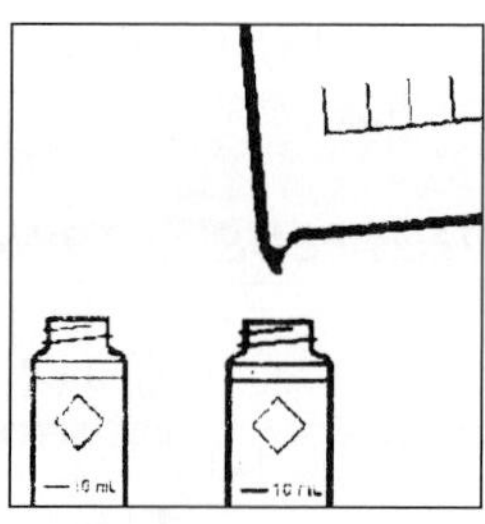

6. Add three drops of THM Plus Reagent 1 to each cell.

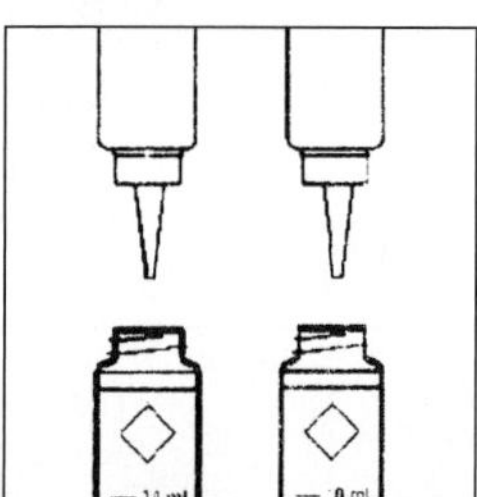

7. Cap tightly and mix gently by swirling each cell three times.
Note: Vigorous shaking can cause loss of THMs.

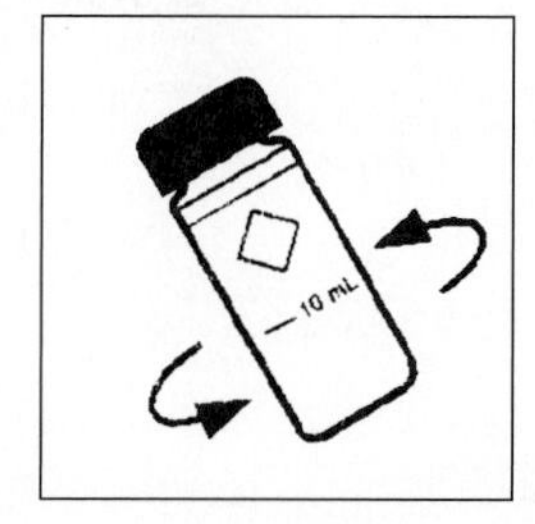

8. Use a TenSette pipette® to add 3 mL of THM Plus Reagent 2 to each cell.
Note: The liquid is viscous and a small amount may remain in the tip after dispensing. This will not affect the results. Also, the THM Plus reagent must be at room temperature before use.

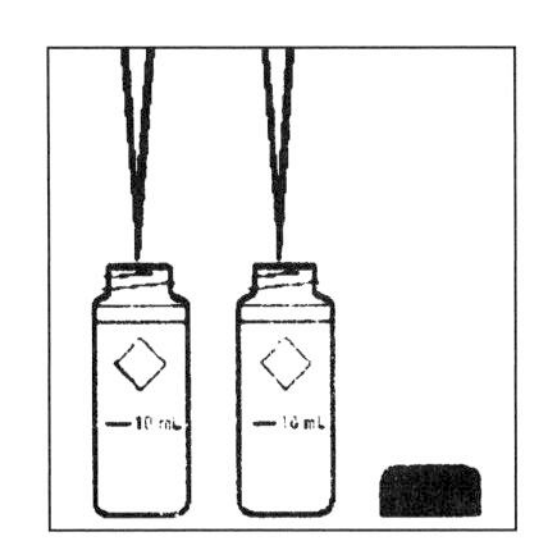

9. Cap tightly and mix by shaking ten times.
Note: Thorough mixing ensures that all of the THM goes into the liquid and does not accumulate in the head space.

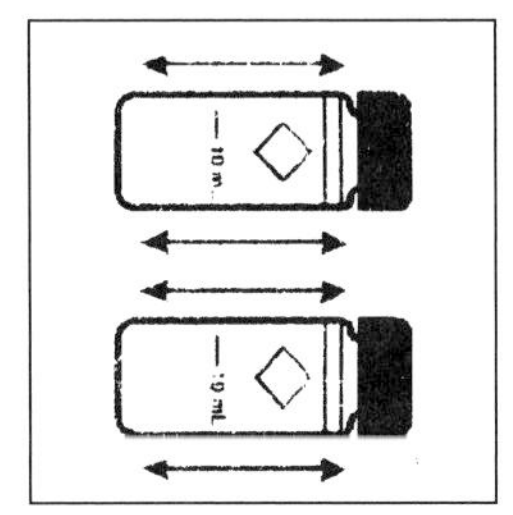

10. Place the sample cell in the cell holder assembly. Set the blank aside.

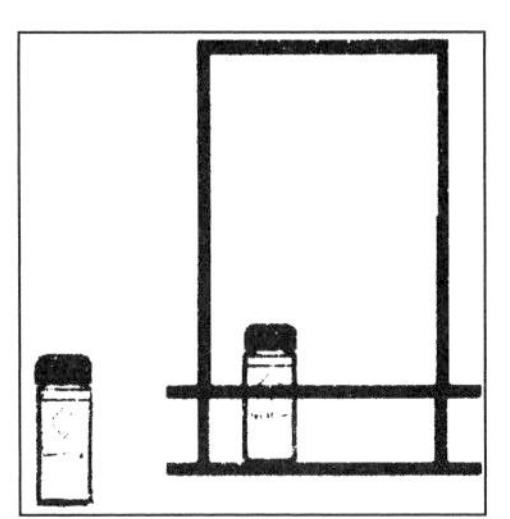

11. Place the assembly in the hot-water bath when the bath is boiling rapidly.
Note: Do not allow water to rise above the white line near the top of the sample cells.

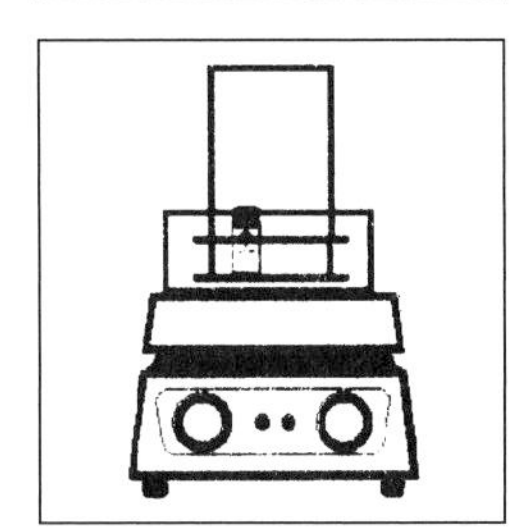

12. Press **SHIFT TIMER** to begin a five-minute reaction period.

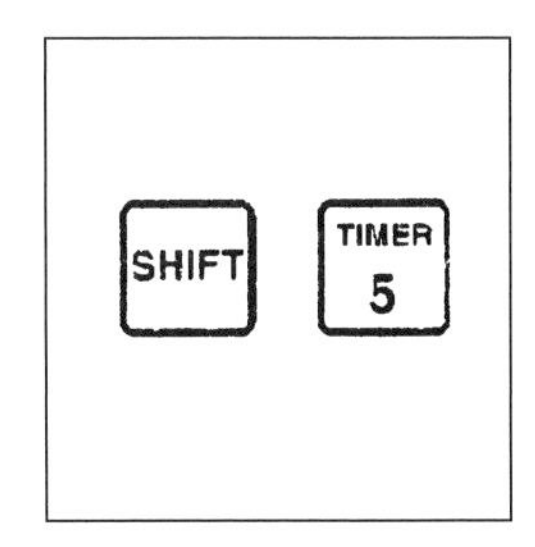

13. At the end of the reaction period, remove the assembly and sample cell from the hot-water bath and place in the cooling bath.

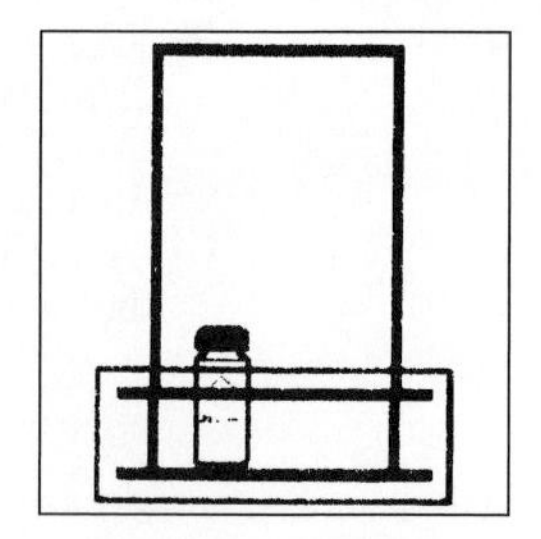

14. Press **SHIFT TIMER.** Cool for three minutes. At the end of the cooling period, remove the cell from the cooling bath.

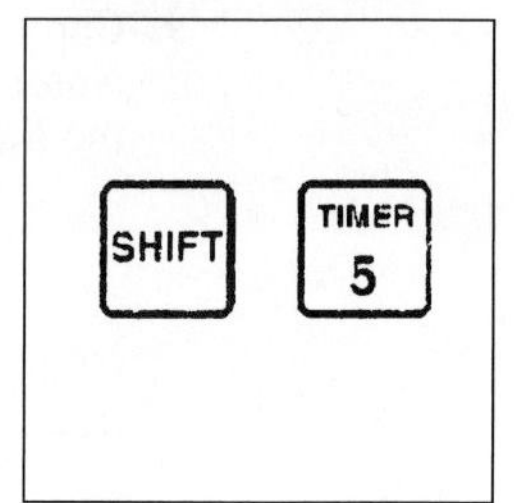

15. Use the Repipet Jr. to add 1 mL of THM Plus Reagent 3 to the sample cell and to the blank. Swirl to mix.
Note: The sample and blank will become warm.
Note: The liquid is viscous and may not be entirely dispensed if measured using any other pipetting method.

16. Replace the cooling water with fresh, cold tap water. Place the assembly containing the sample and blank cells into the cooling bath.

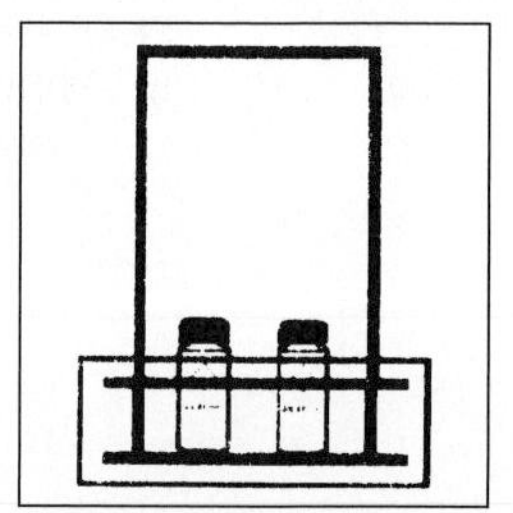

17. Press **SHIFT TIMER** to begin a three-minute cooling time. At the end of the cooling, remove the cells from the cooling bath.
Note: *At the end of the cooling time, the temperature of the sample should be between 15 and 25°C.*

18. Add one THM Plus Reagent 4 Powder Pillow to the sample cell and one to the blank.

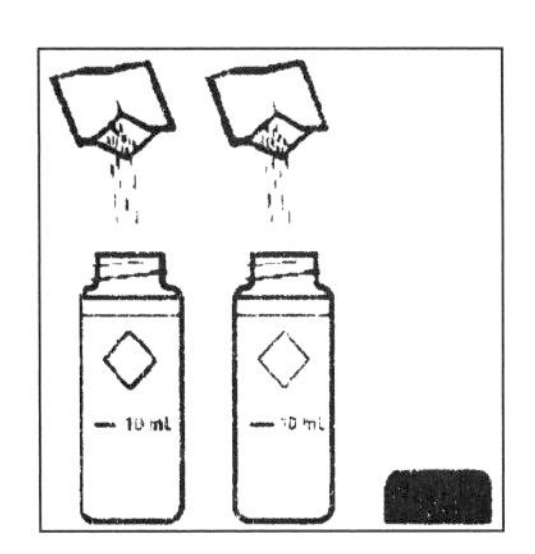

19. Cap each cell tightly and mix by shaking ten times. *Note:* All the powder should dissolve.

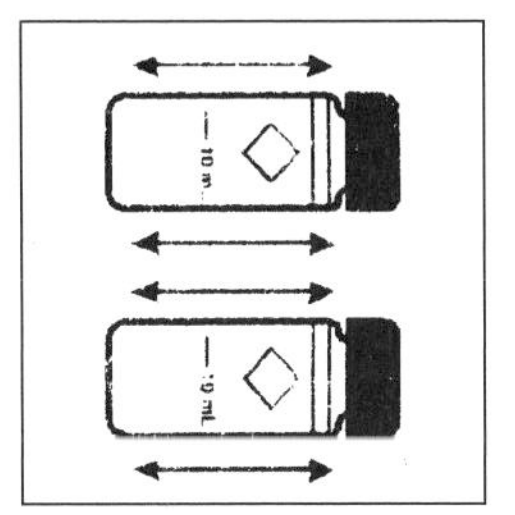

20. Press **SHIFT TIMER** to begin a 15-minute color development time.

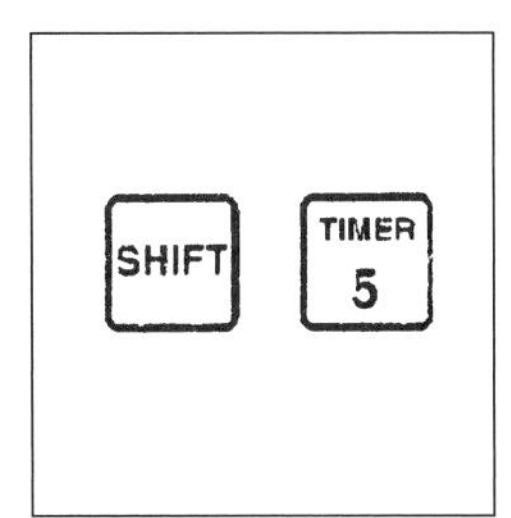

21. While the color is developing, insert the AccuVac® Ampul Adapter into the instrument.
Note: Place the grip tab at the rear of the cell holder.

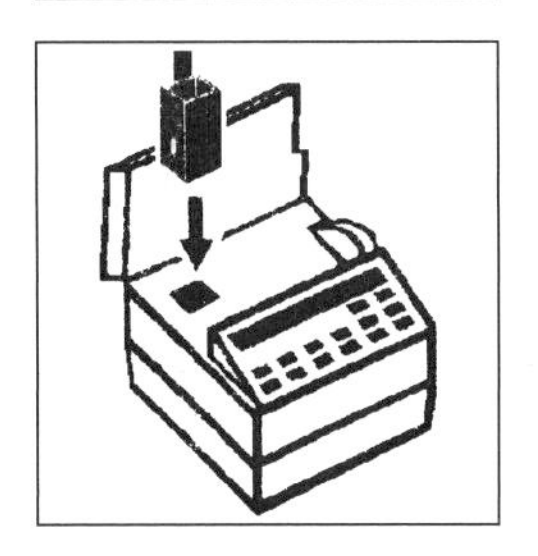

22. Wipe the reagent blank with a damp towel, followed by a dry one, to remove fingerprints or other marks.

23. When the timer beeps, place the blank into the adapter with the HACH logo facing the front of the instrument. Close the sample compartment lid.

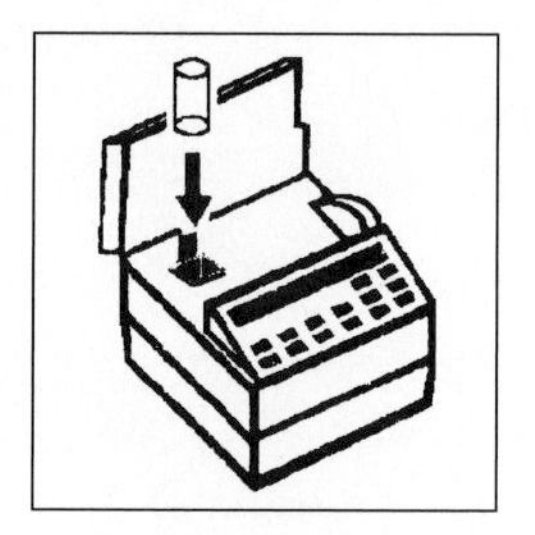

24. Press: **ZERO** The display will show **Zeroing ...** then: **ppb $CHCL_3$.**
Note: For multiple samples from the same source, zero only on the first sample.

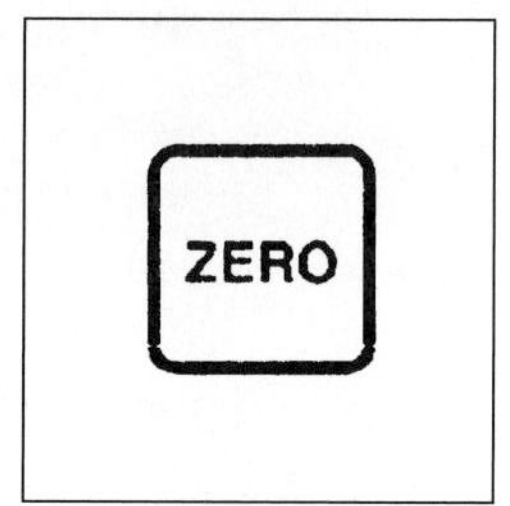

25. Wipe the sample cell with a damp towel, followed by a dry one, to remove fingerprints or other marks.

26. Place the prepared sample into the adapter with the HACH logo facing the front of the instrument. Close the sample compartment.

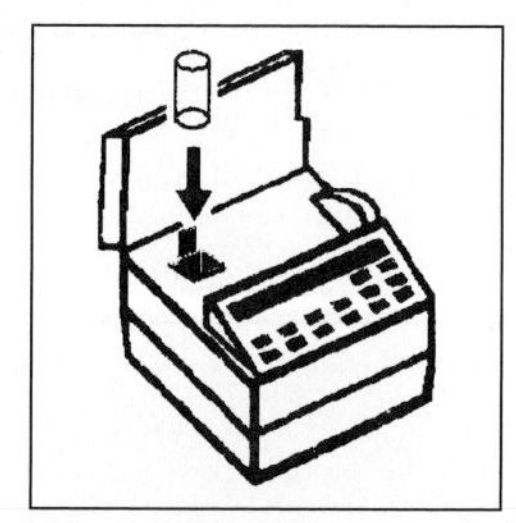

27. Press: **READ** The display will show: Reading ... then the display will show the results in **ppb $CHCL_3$.**
Note: In the Constant-On mode, pressing **Read/Enter** is not required. **Wait** will not appear. When the display stabilizes, read the result.

SAMPLING AND STORAGE

Collect samples in 40-mL glass bottles sealed with Teflon[R]-lined septa caps. Fill the bottle slowly to overflowing so that no air is included with the sample. Seal the bottles tightly and invert to check that no air has been trapped.

Because trihalomethanes compounds are extremely volatile, immediate analysis yields the greatest accuracy. If the samples cannot be analyzed immediately, add one drop of 1.0 N Sodium Thiosulfate to each bottle to prevent the formation of additional THM compounds in the chlorinated samples. Store the preserved samples at 4°C in an atmosphere free of organic vapors. Samples should not be stored for more than 14 days. Allow samples to equilibrate to room temperatures before analyzing.

Ascorbic acid *cannot* be used as a preservative with the THM Plus method.

The THM Plus Reagent 2 should be stored in a refrigerator to avoid reagent degradation.

ACCURACY CHECK: STANDARD ADDITIONS AND STANDARD SOLUTIONS METHOD

See HACH section titled "TMH Plus™: Trihalomethanes" on pages 833 and 834 of the *DR/2010 Spectrophotometer Handbook.*.

SAMPLE CELL MATCHING

The sample cells must be matched for accurate analysis purposes. In order to properly match sample cells, a procedure is provided on pages 834 and 835 of the *DR/2010 Spectrophotometer Handbook.*

INTERFERENCES

Interference	Maximum Level Tested
Chlorine	<10 ppm
Copper	<1000 ppm
Hardness, Ca	<1000 ppm as $CaCO_3$: May have some turbidity until Reagent 3 is added.
Hardness, Mg	<4000 ppm as $CaCO_3$: May have some turbidity until Reagent 3 is added.
Iron	<10 ppm
Lead	<2 ppm
Mercury	<10 ppm
Monochloramine	<20 ppm
Nickel	<10 ppm
Sodium Bisulfite	<100 ppm
EDTA	Interferes negatively at all levels

The list of disinfection by-products which react (interferes positively) during the test for trihalomethanes include the following: chloral hydrate; 1,1,1-trichloro-2-propanone; 1,1,1-trichloroacetonitrile; dibromochloroacetic acid, dichlorobromoacetic acid, tribromoacetic acid, trichloroacetic.

QUESTIONS AND ANSWERS

1. Why are the trihalomethanes found in water?
Trihalomethanes, such as chloroform, are produced in trace quantities during disinfection of drinking water by chlorination. In addition, trace amounts of the compounds are found in industrial and business effluents and can be attributed to laboratory solvents, process solvents, and chlorine disinfection of drinking water, process waters and waste waters.

2. Why are chlorofluorocarbons seldom used today?
Chlorofluorocarbons, primarily used as aerosol sprays and refrigerants, have a depleting effect on stratospheric ozone.

3. What health problems can trihalomethanes cause for humans?
These compounds attack the central nervous system of humans. Extensive human exposure to these compounds can cause liver damage, kidney damage, ventricular fibrillation, and cardiac problems. Prolonged inhalation or ingestion of these chemicals can be fatal. Chloroform is a known carcinogen (OSHA) and has a narcotic effect on people. At a low concentration, bromoform is toxic by ingestion, inhalation, and skin absorption.

4. What are the human tolerance levels and drinking water standards for the trihalomethanes?
Humans have a tolerance level of 10 ppm in air to chloroform, and 0.5 ppm in air to bromoform. The U. S. drinking water standard for trihalomethanes (sum of chloroform, bromodichloromethane, dibromochloromethane and tribromomethane) is 100 μg/L.

REFERENCES

1. *Fate of Pesticides and Chemicals in the Environment.* Jerald L. Schnoor (ed.) (New York, N.Y.: John Wiley & Sons, Inc., 1992).
2. Roger N. Reeve, *Environmental Analysis.* (ACOL), John D. Barnes (ed.) (New York, N.Y.: John Wiley & Sons, 1994).
3. *Hawley's Condensed Chemical Dictionary,* 11th Ed., rev'd by N. Irving Sax and Richard J. Lewis (New York, N.Y.: Van Nostrand Reinhold, 1987).
4. *DR/2010 Spectrophotometer Handbook,* THM Plus™: Total Trihalomethanes, HACH Company, 1999.

ToxTrak™ TOXICITY TEST

The ToxTrak™ Toxicity Test is a simple test for evaluating the inhibitory effect of wastes and chemicals in wastewater treatment processes.* The toxicity test can be used to screen influent, effluent and/or process waters from wastewater treatment plants, chemical process and production plants, power plants, metal plating processes, petroleum manufacturers, pulp

*Compliments of HACH Company.

and paper facilities, specific dye manufacturing plants, toxic metal manufacturing plants, and other industrial facilities.

The toxicity test is used to identify toxins in receiving waters and in the wastewater treatment plant biomass. Bacteria from the plant's biomass can be used to assess toxicity levels. The results represent the effect of toxins on the biomass and are often more realistic than many microbiological tests. Also, the test provides an evaluation of toxicity levels of chemicals in the plant and in the laboratory. Because of the relatively short time required for the testing procedures, corrective actions can be implemented relatively quickly.

A number of methods are used to monitor toxicity, such as respiration, growth, specific enzyme activity, and bioluminescence. These methods require lengthy incubation times, solvent extraction, centrifugation, colony counting procedures, and expensive equipment. Long-term bioassays, such as *Ceriodaphnia* (chronic toxicity test), can require several days to complete.

A redox-active dye, called resazurin, changes color from blue to pink when reduced by bacterial dehydrogenases.* Toxic substances can often inhibit the rate of dye (resazurin) reduction. A chemical accelerant shortens the reaction time. Biomass or bacteria from the sample modify the solution environment. In certain cases, an industrial effluent or solution may be studied by adding the redox-active dye, a ToxTrak Accelerator Solution and bacteria (*Escherichia coli*) to a sample contaminated with chemicals and other pollutants. In each instance, the solution absorbance is measured with a colorimeter, spectrophotometer, or a color disc comparator. When the absorbance of the negative control has decreased to a specific level, the absorbance of the sample is measured again. A simple calculation determines the toxic effects or the percent inhibition of a sample on bacteria. Finally, the effect of a toxin on a system can be predicted with significant accuracy.

ToxTrak TOXICITY TEST*

Method 10017** for Wastewater
Colorimetric Method†

*U.S. Patent Number 5,413,916.

**Compliments of HACH Company.

†D. Liu, *Bull. Environ. Contam. Toxicol.*, 26, 145–149 (1981).

INOCULUM DEVELOPMENT

1. Using one of the pipets provided, add 1.0 mL of source culture to a Tryptic Soy Broth Tube.
   Note: Use Indigenous Biomass

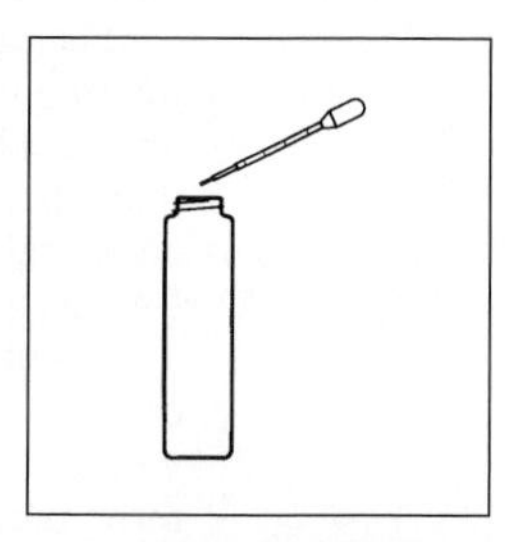

2. Incubate until the vial contents are visibly turbid (turbidity indicates bacterial growth).

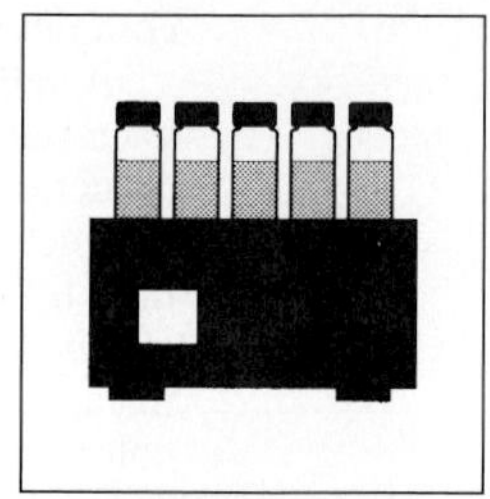

PREPARATION OF BACTROL DISKS

1. Flame sterilize forceps by dipping into alcohol and flame in an alcohol or Bunsen Burner. Let the forceps cool.

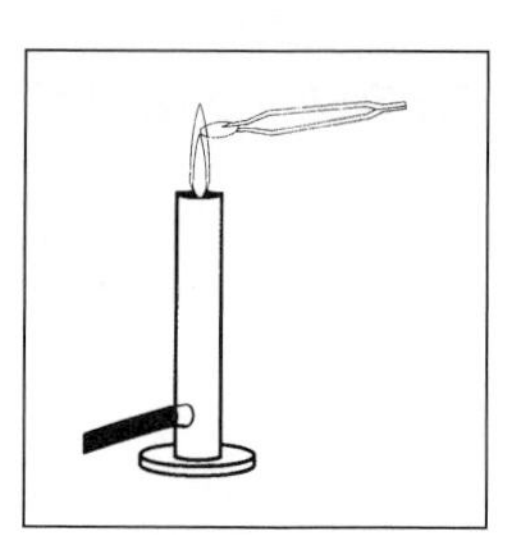

2. Remove the cap from the Bactrol Inoculum Bottle. Pick out one Bactrol Inoculum Disk with the sterilized forceps.

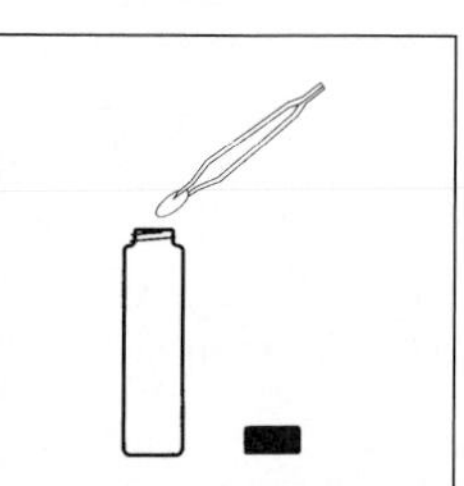

3. Remove the cap from a Lauryl Tryptose Broth Tube and drop in the Inoculum Disk using the sterilized forceps. Shake to dissolve the disk.

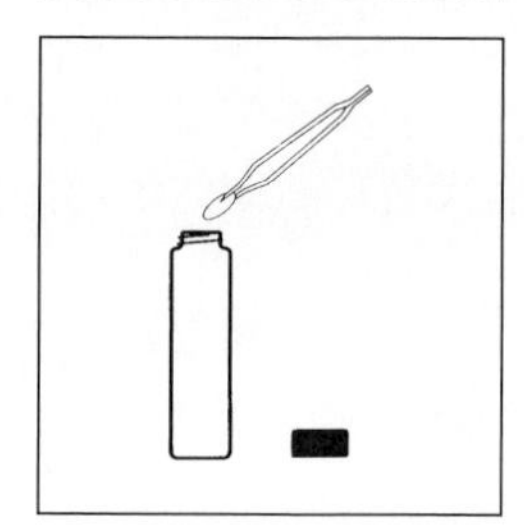

4. Incubate the Lauryl Tryptose Broth Tube until the medium is visibly turbid. Turbidity will develop much faster if incubation is done at 35°C instead of room temperature. At 35°C, 10 hours is usually sufficient.

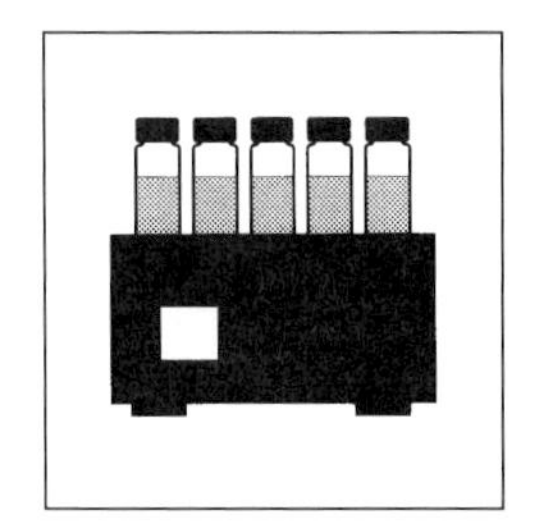

5. Inoculate a new Lauryl Tryptose Broth Tube by first inverting the tube in Step 4, and then switching the caps of the two tubes. Then invert the new tube. After incubation, use this new vial in subsequent tests.
Note: In this way, several medium vials may be inoculated from one Bactrol Disk.
Note: If testing on consecutive days, keep inoculum several days or at room temperature.

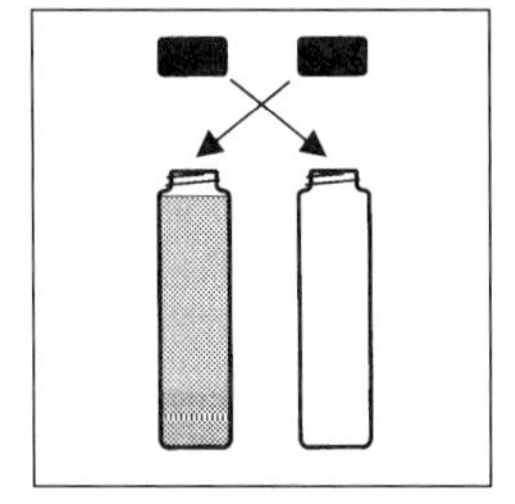

COLORIMETRIC REACTION

1. Turn on the DR/2010 and put the instrument in constant mode. Allow the instrument to warm up for at least 15 minutes.

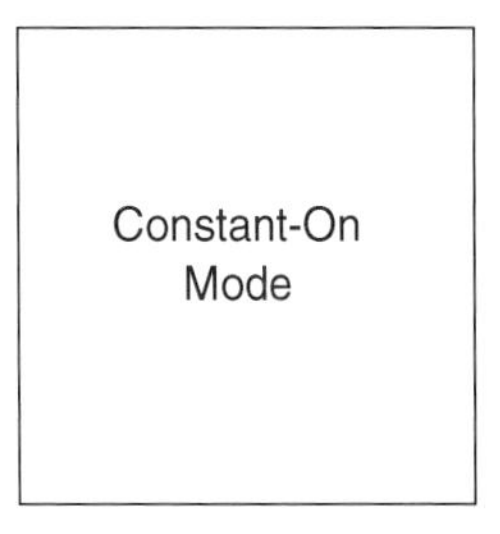

2. Place the 13-mm Test 'N Tube adapter into the cell holder with the marker to the right.

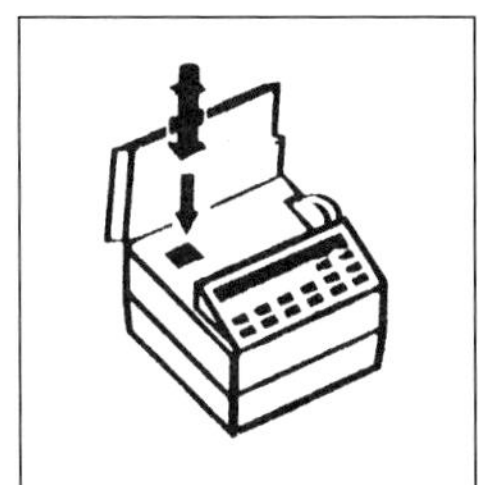

3. Enter the stored program number for absorbance. Press: **0 ENTER** The display will show: **Abs**

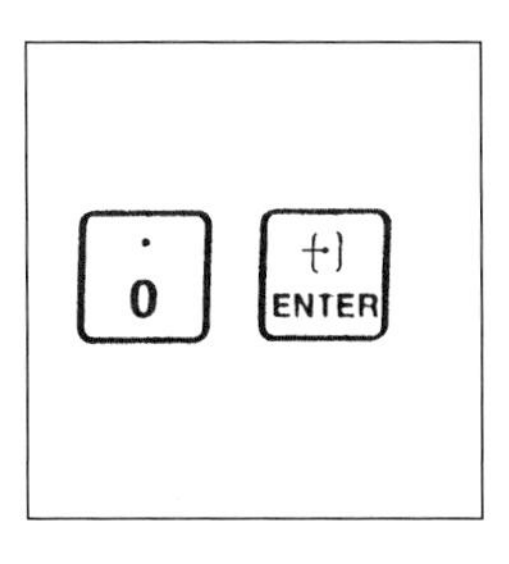

4. Rotate the wavelength dial until the small display shows: **603 nm**

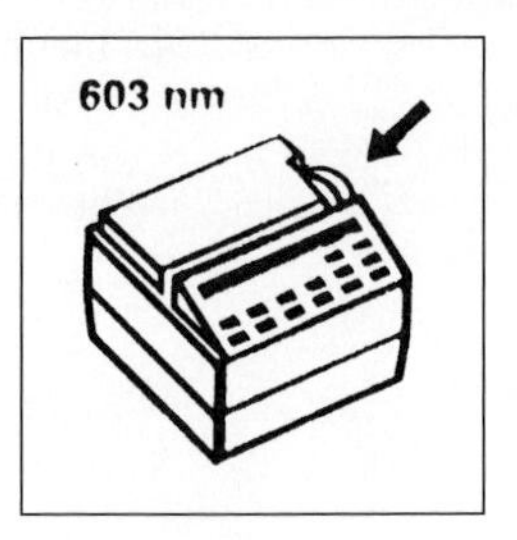

5. Fill a 13-mm reaction tube with deionized water. Label this tube as "blank." Place the tube in the Test 'N Tube adapter and place the lid on the adapter.

6. Press: **ZERO** The display will show: **Zeroing ....** then: **0.000 Abs**

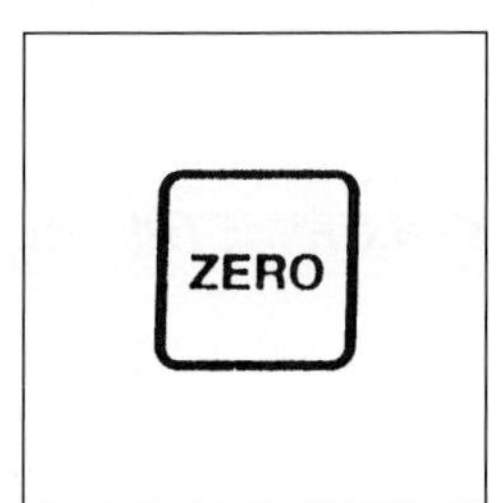

7. Label a tube "control." Open one ToxTrak Reagent Powder Pillow and add the contents to the empty reaction tube.

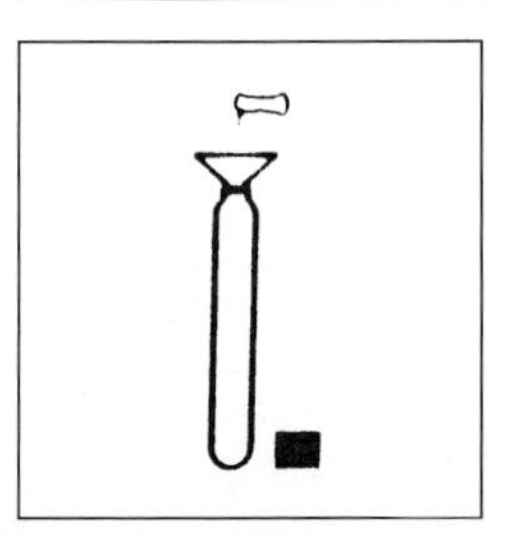

8. For each sample or dilution, repeat Step 7. Label each tube clearly.

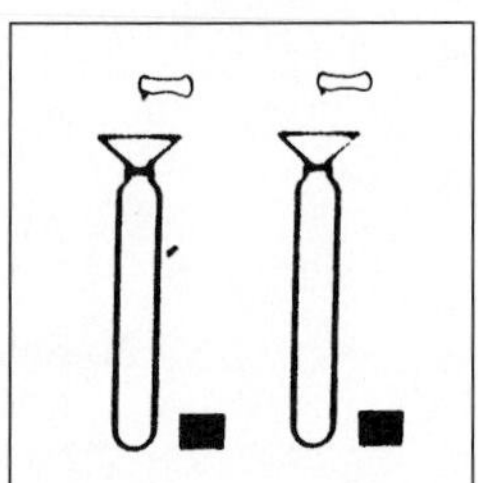

9. Add 5.0 ml of deionized water to the control tube.

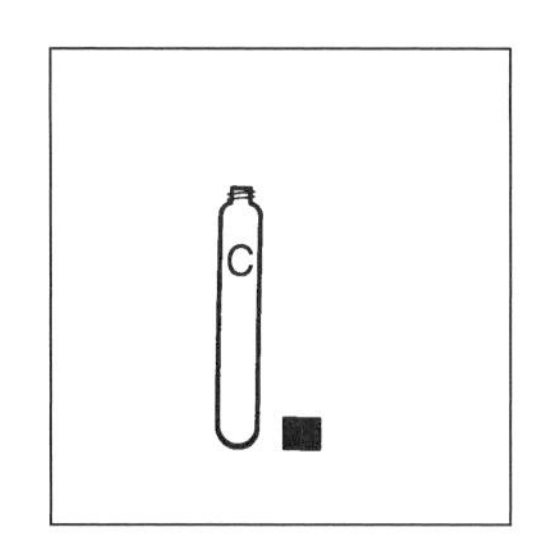

10. Add 5.0 mL of sample (or dilutions) to the sample tube.

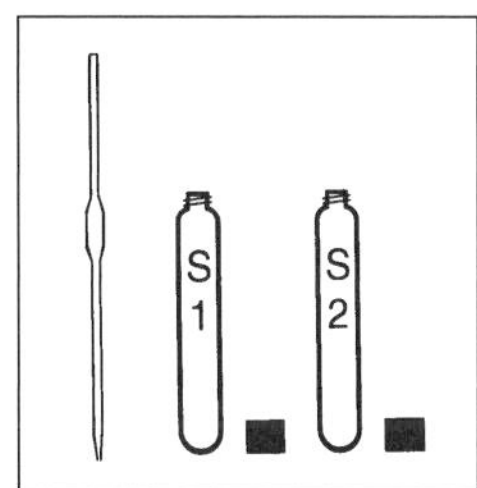

11. Add 2 drops of Accelerator Solution to each tube. Cap and invert to mix.

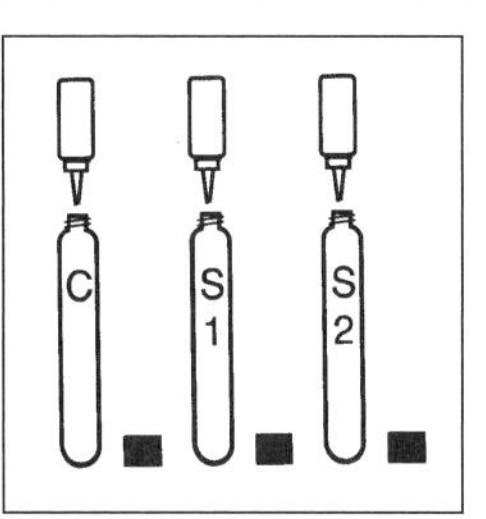

12. Add 0.5 mL of inoculum (previously prepared) to each tube. Cap and invert the tube.

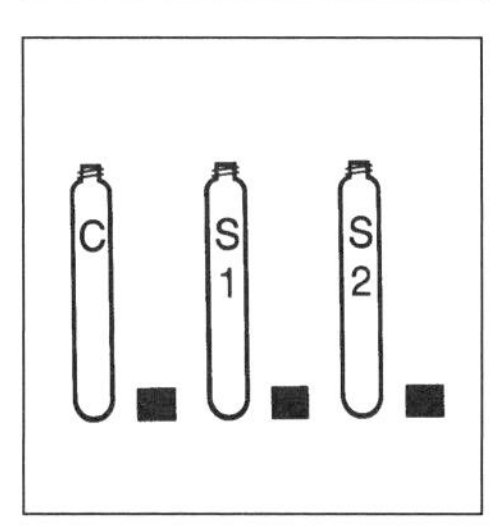

13. Place the control tube in the cell holder. Place the lid on the adapter. Record the absorbance.

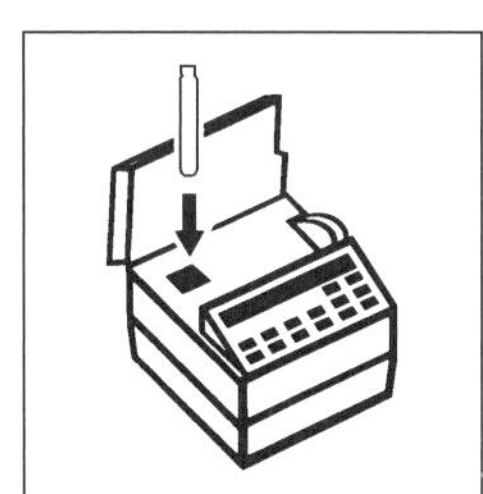

14. Repeat Step 13 for all samples and dilutions. Be sure to record each absorbance.

15. Allow the solutions in the tubes to react until the absorbance of the **control tube** decreases 0.60 ±0.10. This should take about 45 to 75 minutes.

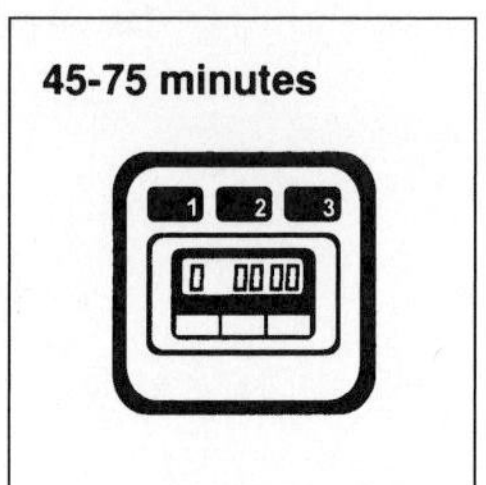

16. Place the blank tube in step 5 in the Test 'N Tube adapter. Place the lid on the adapter.

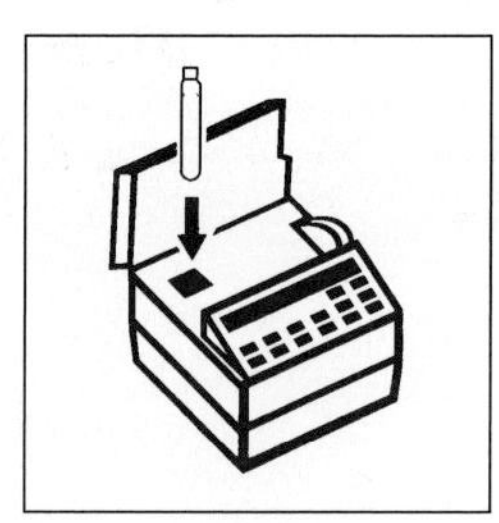

17. Press: **ZERO** The display will show: **0.000**

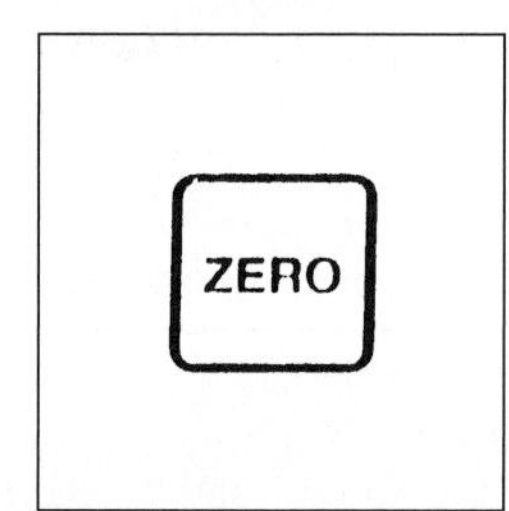

18. Place the control tube into the cell holder. Place the lid on the adapter. Record the absorbance value. Place each sample cell into the cell holder. Place the lid on the adapter. Record the absorbance value of each sample.

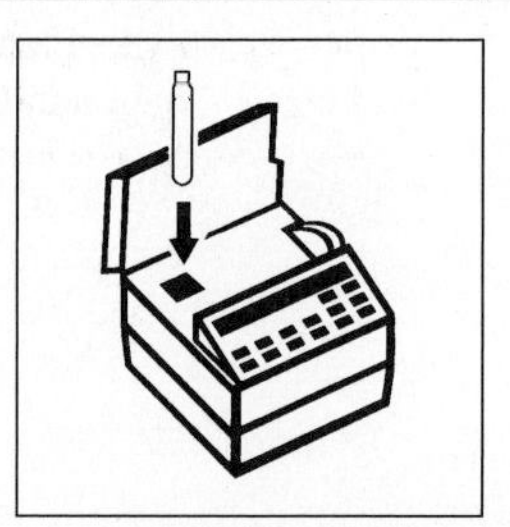

19. Calculate the % Inhibition as follows:

$$\% \ 1 = \left[1 - \left[\frac{\Delta\text{Abs sample}}{\Delta\text{Abs control}}\right]\right] \times 100$$

Note: Some toxins increase respiration and will give a negative % inhibition on all respiration-based toxicity tests. After repeated testing, samples which always give a % inhibition in Step 19 that is more negative than 10% should be considered toxic.

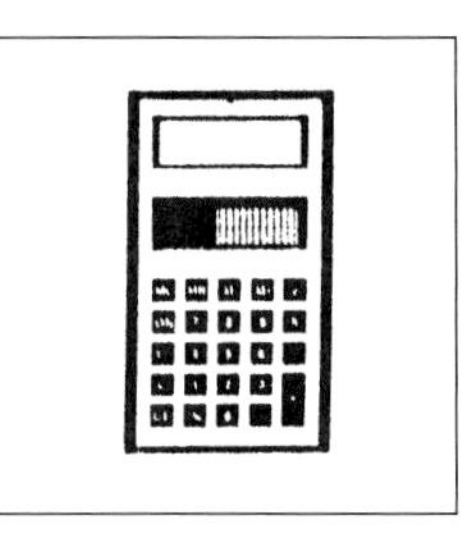

DISPOSAL OF TEST CULTURES

Dispose of active bacterial cultures by using one of these methods:

1. Autoclave used test containers at 121°C for 15 minutes at 15 pounds of pressure. Once the containers are sterile, pour the contents down the drain with running water. The reaction tubes may be washed and re-used.
2. Sterilize test containers by using 1:10 dilution of commercial laundry bleach. Pour the test container contents and test containers into the bleach solution. Allow 10 to 15 minutes of contact time with the bleach solution. Then pour the liquid down the drain and ash the reaction tubes for re-use.

QUESTIONS AND ANSWERS

1. Briefly explain the method of toxicity determination.
The method is based on the reduction of resazurin, a redox-active dye, by bacterial respiration. During the reduction due to bacterial respiration, resazurin changes color from blue to pink and is measured colorimetrically.
2. How are active bacterial cultures disposed of?
Sterilize test containers with 1:10 diluted commercial laundry bleach. Pour the contents of the test containers and the test containers into the bleach solution. After 15 minutes of contact time, the bleach solution can be poured down the drain. A second procedure involves autoclaving the tubes and their contents at 121°C for 15 minutes under pressure. After the containers are sterile, the contents can be poured down the drain. In both cases, the reaction tubes can be washed and re-used.
3. The control tube has an initial absorbance of 1.6 and decreases to 1.0 Abs. The sample tube has an initial absorbance of 1.7 and decreases to 1.3 Abs. Calculate the % Inhibition.

$$\Delta \text{ Abs. Sample} = 1.7 - 1.3 = 0.4 \qquad \textit{and}$$

$$\Delta \text{ Abs. Control} = 1.6 - 1.0 = 0.6$$

$$\%\text{I} = [1 - (0.4/0.6)] \times 100$$

$$\%\text{I} = 33.3$$

4. Briefly outline the procedure for handling the broth solutions and the equipment.
In each case, it is important to properly handle equipment (forceps, caps, etc.), broth solutions, dilution water, inoculum and inoculum disk without contaminating these items.

When removing the cap from an inoculum bottle, do not finger the inside of the cap or mishandle it. All sample bottles and pipets or dropper pipet should be sterilized in an autoclave. Some of these items maybe available in sterile packages.

REFERENCES

1. DR/2010 Spectrophotometer Procedures Manual, 3rd Ed., (Loveland, CO.: HACH Company, 1998).
2. ToxTrak™ Toxicity Test, HACH Company, 1998.

TURBIDITY

Turbidity in water occurs as a result of silt, suspended clay, finely divided organic matter, suspended inorganic matter, plankton, and other microorganisms. Turbidity measures an optical property of water that results from the scattering of light and absorbing of light by the particulate matter present in the sample. Smaller particles scatter shorter wavelengths of light more intensely, while larger particles scatter longer wavelengths of light more readily. Light scattering intensifies as the number of particles in solution increases. The amount of light scattered and absorbed by the particulate matter depends on size and shape of the particles, color, and the refractive properties of the particles.

Turbidity can be measured as "Nephelometric Units" (NTU) or "Jackson Turbidity Units" (JTU) or "Formazin Attenuation Units" (FAU) depending on which instrument is used for the measurement. Normally, there is limited relationship between the measurements on one instrument and those measured by a different type of measuring device.

Some instruments are calibrated using formazin turbidity standards. The readings are in terms of Formazin Attenuation Units (FAU) and can be used for in-plant monitoring. Based on a Formazin Standard, 1 NTU = 1 FAU. However, the optical method of measurement for FAU is very different than the NTU Method. The FAU test can not be used for USEPA reporting purposes, whereas the NTU is USEPA accepted.

Formazin is prepared by accurately weighing and dissolving 5.000 grams of hydrazine sulfate and 50.0 grams of hexamethylenetetramine in exactly 1.0 liter of distilled water. The formazin solution develops a white turbidity after standing at 25°C for 48 hours. This prepared solution measures 4000 NTU. The randomness of particles and particle sizes within formazin standards yield statistically reproducible scatter on various types of turbidimeters.

Turbidity can be determined on samples that are free of debris and rapidly settling coarse sediments. Error in measurements can occur from dirty glassware, the presence of air bubbles, particles that settle too fast, and the effects of vibrations that disturb the visibility of the water surface. Temperature extremes and sample color can modify the turbidity of the sample.

Turbidity can be measured on samples of water, wastewater, or seawater by the Absorptometric Method (0 to 1000 FAU) with the DR/820, DR/850 and DR/890 Colorimeters. Alternatively, turbidity of samples in the field or in the laboratory that require traceable numbers to standards must be determined by the Nephelometric Method (0 to 999 NTU). For monitoring purposes, the Attenuation Method (0-4400 FAU) with the DR/2010 Spectrophotometer and the Absorptometric Method with Colorimeters are adequate.

Alternatively, a battery operated portable microprocessor turbidity meter used to determine turbidity of water and wastewater generally covers a 0-1000 FTU range in two scales: 0 to 50 FTU and 50-1000 FTU. A single point calibraiton at 10 FTU with a specific stable primary standard is often used instead of the more toxic formazin standard. During measurements, the instrument passes a beam of infrared light through a vial containing the sample. A sensor, poisitioned at 90° with respect to the direction of light, detects the amount of light scattered by the undissolved particles in the sample. The microprocessor converts the readings into FTU values. These types of instruments have been designed to perform measurements according to the ISO 7027 International Standard.

TURBIDITY (0 to 999 NTU)

Method 8195 for water and wastewater*
Nephelometric Method**

1. Collect a representative sample in a clean container. Fill a sample cell to the line, taking care to handle the sample cell only by the top. Cap the cell.
   Note: Refer to your Turbidimeter Instruction Manual for information on collecting a representative sample.

2. Wipe the cell with a soft, lint-free cloth to remove water spots and fingerprints.

3. Apply a thin film of silicone oil. Wipe with a soft cloth to obtain an even film over the entire surface.

4. Turn the instrument on. Allow it to warm up per the manufacturer's instructions.

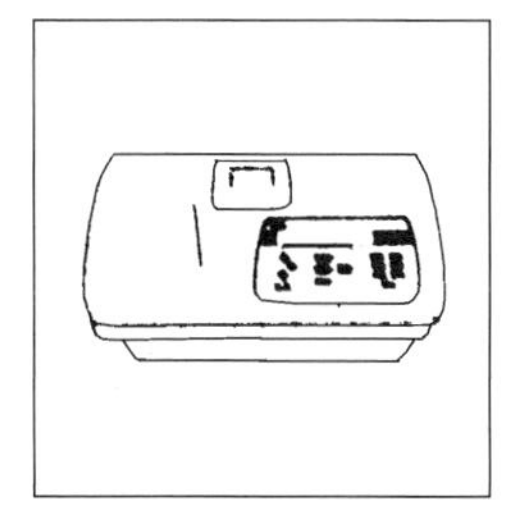

*Compliments of HACH Company. Method 8237 uses the DR/820, DR/850 and the DR/890 Colorimetric Absorptometric Method (0 to 1000 FAU) whereas this experiment uses the Nephelometric Method (0 to 999 NTU).

**Meets or exceeds the specification criteria outlined in EPA 180.1.

5. Gently invert the sample cell 2 to 3 times.
Note: Do not shake or invert rapidly. Entrained air bubbles and erroneous readings could result.

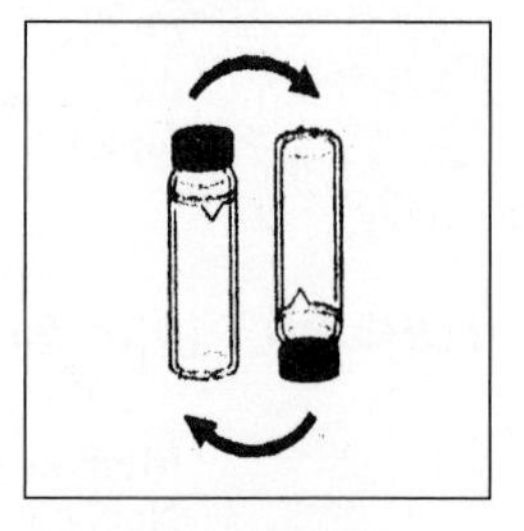

6. Set the appropriate range and signal average on the turbidimeter (if applicable).

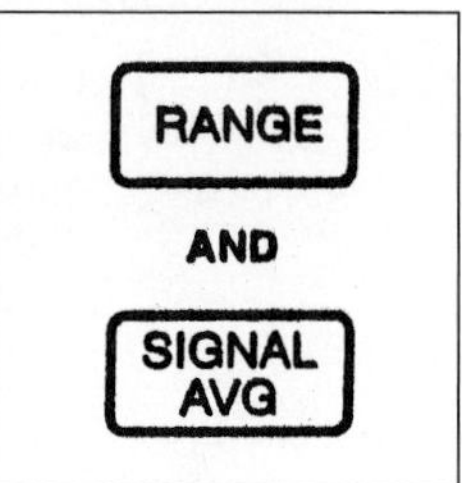

7. Place the sample cell in the instrument cell compartment. Close the lid.
Note: Make sure the sample cell index is consistent when performing multiple readings.

8. Record the stable reading (or, if the instrument in use has datalogging capabilities, store the reading in the instrument's memory).

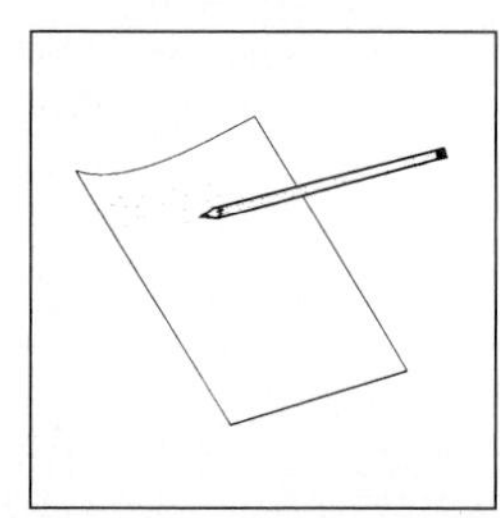

SAMPLING AND STORAGE

Collect samples in clean plastic or glass bottles. Analyze samples as soon as possible after collection. Samples may be stored for 7 days by cooling to 4°C (39°F) or lower. Measure samples at the same temperature as when collected.

INTERFERENCES

Interfering substance	Interference levels and treatments
Air Bubbles	Interfere at all levels. Degass the sample using the optional Degassing Kit or Ultrasonic Bath.
Color	Interferes, depending on color. Use ratio mode or 860 nm light.
Temperature Extremes	May interfere by changing the turbidity of the sample. Analyze samples as soon as possible after collection. Analyze at the same temperature as the original sample.

SAMPLING AND STORAGE

Collect samples in acid-washed plastic bottles. For storage, adjust the pH to 2 or less with nitric acid (about 2 mL per liter). The preserved samples can be stored up to six months at room temperature.

Adjust the pH to 4 to 5 with 5.0 N sodium hydroxide before analysis. Do not exceed pH 5, as zinc may be lost as a precipitate. Correct the test result for volume additions; see "Sample Collection, Preservation and Storage," and "Correcting for Volume Additions," in Section 1 for more information.

If only dissolved zinc is to be determined, filter the sample before the acid addition.

ACCURACY CHECK

See p. 1186 of the HACH Water Analysis Handbook, 3rd Ed., 1997.

QUESTIONS AND ANSWERS

1. **What is the relationship between NTU and JTU?**
   There is no direct relationship between NTU and JTU measurements.
2. **What causes turbidity?**
   Turbidity in a sample is caused by silt, suspended clay, suspended organic chemicals, suspended inorganic chemicals, plankton, microorganisms and other finely divided matter.
3. **What causes errors in the measurement?**
   Turbidity measurements by different instruments on the same sample provide results that are not consistent. Dirty glassware, the presence of air bubbles, the rate of particle fall, the type of particle, and the effects of vibrations, temperature, and color impact the results obtained from turbidity measurements.
4. **What is the relationship between NTU and FAU?**
   One FAU is equivalent to one NTU WHEN Formazin is the standard. However, the optical method of measurement for FAU is very different than the NTU method. Therefore, the FAU test cannot be used for USEPA reporting.

REFERENCES

1. *HACH Water Analysis Handbook,* 3rd Ed. (Loveland, CO.: HACH Company, 1997).
2. *Standard Methods for the Examination of Water and Wastewater:* TURBIDITY, 16th Ed., A. E. Greenberg, R. R. Trussell, L. S. Clesceri and Mary Ann H. Franson (eds.), pp. 133–139, Washington, DC, American Public Health Association, 1985.

VOLATILE ACIDS

The measurement of volatile acids can be used as a control test for anaerobic digestion. Volatile fatty acids are water-soluble fatty acids that can be removed from aqueous solution by distillation. These acids contain up to six carbon groups and have relatively high boiling points.

The distillation method gives incomplete and somewhat variable recovery. Heating rate and proportion of the sample recovered as distillate affect the experimental result. The determination of a recovery rate for each apparatus and establishing a standard set of operating conditions will minimize variability in the experiment. In addition, the removal of sludge solids from the sample reduces the possibility of hydrolysis of complex materials to volatile acids. Even with these problems, the process is suitable for control purposes and routine analysis.

Because the method is empirical, the still-heating rate, the presence of sludge solids and the final distillate volume affect recovery quality. Hydrogen sulfide (H_2S) and carbon dioxide are liberated during distillation and can provide a positive error during titration. The error can be minimized by discarding the first 15 mL of distillate and accounting for the distillate in the recovery factor.

To determine the recovery factor, f, for a given apparatus, dilute an appropriate volume of acetic acid stock solution to 250 mL in a volumetric flask to approximate the expected sample concentration. Distill this volume as a normal sample. Calculate the recovery factor, $f=a/b$, where: a=volatile acid concentration recovered in distillate, mg/L, and b=volatile acid concentration in standard solution used, mg/L.*

The field sample should be analyzed as soon as possible after collection to avoid volatilization and changes in the fatty acids in the solution. After a sample acidified with sulfuric acid is distilled, the distillate is titrated to the phenolphthalein end point with sodium hydroxide. The volume of titrant required is proportional to the volatile acid concentration. The results are expressed in terms of mg/L as acetic acid.

VOLATILE ACIDS

Method 8218**
Digital Titrator Method using Sodium Hydroxide†

* *Standard Methods for the Examination of Water and Wastewater,* 16th Ed., p. 505–506, 1985.
** Compliments of HACH Company.
† *Analytical Chemistry,* 38, 791(1966).

1. Distill the sample and collect 150 mL of distillate.
Note: Use the HACH Volatile Acids Procedure, Sample Distillation, accompanying the General Purpose Distillation Apparatus Set or the distillation procedure described in "*Standard Methods for the Examination of Water and Wastewater.*"

2. Attach a clean delivery tube to a 0.9274 N Sodium Hydroxide Titration Cartridge. Attach the cartridge to the titrator body.

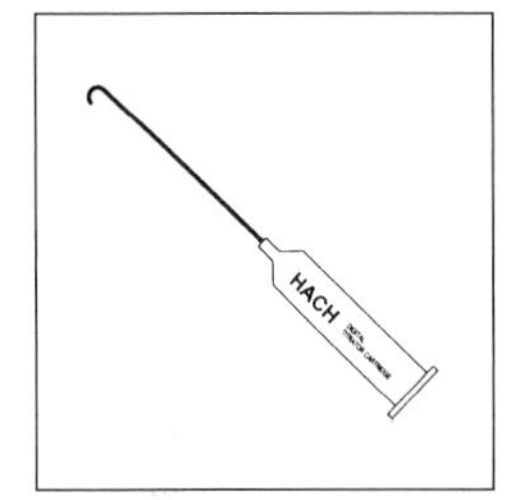

3. Flush the delivery tube by turning the delivery knob to eject a few drops of titrant. Reset the counter to zero and wipe the tip.
Note: For added convenience use the TitraStir® stirring apparatus.

4. Select the distillate volume corresponding to the expected volatile acids concentration as acetic acid from Table 2.41. Using a graduated cylinder, transfer the distillate volume into a clean 250-mL erlenmeyer flask and dilute to about the 150-mL mark with deionized water.

TABLE 2.41

Range (mg/L as CH_3COOH)	Volume (mL)	Titration cartirdge (N NaOH)	HACH Catalog number	Digit multiplier
100–400	150	0.9274	14842-01	1
200–800	75	0.9274	14842-01	2
600–2400	25	0.9274	14842-01	6

5. Add the contents of one Phenolphthalein Indicator Powder Pillow and swirl to mix.

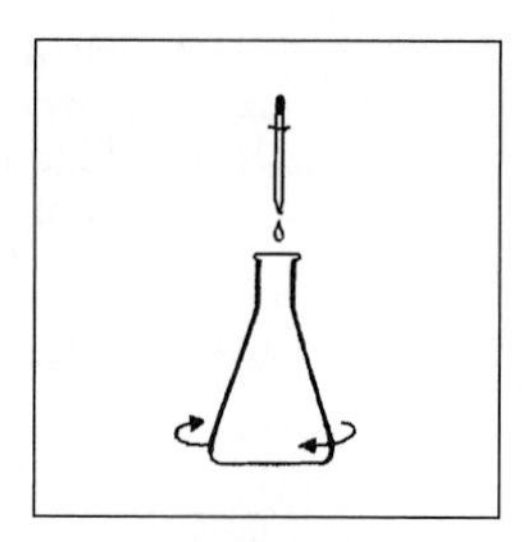

6. Place the delivery tube tip into the solution and swirl while titrating with sodium hydroxide until a light pink color appears. Record the number of digits required.

7. Calculate:

   Digits Required × Digit Multiplier = mg/L Volatile Acids (as acetic acid, CH_3COOH)

   Note: Approximately 70% of the volatile acids in the sample will be found in the distillate. This has been accounted for in the calculation.

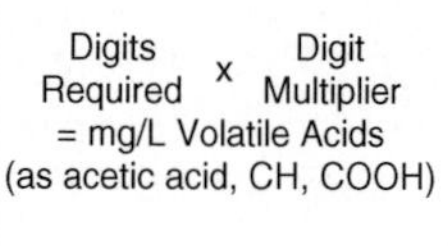

SAMPLING AND STORAGE

Collect samples in plastic or glass bottles. Analyze samples as soon as possible after collection. Samples can be stored up to 24 hours by cooling to 4°C (39°F) or below. Warm to room temperature before running the testing.

QUESTIONS AND ANSWERS

1. What factors affect recovery rate of a distillation?
   Recovery rates are affected by temperature fluctuations (heating rate) during distillation, length of time for distillation, operating conditions, the presence and the hydrolysis of sludge solids, final distillate volume, and reactions of materials in the mixture. During distillation, the same properly cleaned equipment should be used under the same conditions of operation during the experiment.
2. How accurate is this experiment?
   If the variables in the experiment are consistent during the determinations of volatile acids and a recovery factor, f, can be accurately determined, then the measurement and comparison of volatile acids concentrations will be consistent and meaningful.

3. What are some of the chemical interferences in the experiment?
Chemical interferences of H_2S *and carbon dioxide liberated during the experiment can provide a positive error during the experiment. Discarding the first 15 mL of distilled sample and accounting for the distillate in the recovery factor can eliminate the error.*

REFERENCES

1. *HACH Water Analysis Handbook,* 3rd Ed., (Loveland, CO.: HACH Company, 1997).
2. *Hawley's Condensed Chemical Dictionary,* 11th Ed., rev'd by N. Irving Sax and Richard J. Lewis (New York, N.Y.: Van Nostrand Reinhold, 1987).
3. *Standard Methods for the Examination of Water and Wastewater:* ORGANIC AND VOLATILE ACIDS, 16th Ed., Method 504B, A. E. Greenberg, R. R. Trussell, L. S. Clesceri and Mary Ann H. Franson (eds.), pp. 505–506, Washington, DC, American Public Health Association, 1985.

ZINC

Zinc is a beneficial element in human growth and is often added to foods such as cereals. The zinc concentration in drinking water averages about 1.3 mg/L. Concentrations at 5 mg/L display no harmful physiological effects on humans. Zinc enters the water supply by deterioration of brass and galvanized iron and through waste pollution.

Zinc metal and zinc compounds are used in a variety of applications including polymers, medicine, batteries, fungicides, animal repellant, insecticide and rodenticide, preservatives, food ingredient, cross-linking polymers, pharmaceuticals, bleaching of foods and textiles, surgical dressings, antiseptic and deodorant applications, dental cements, and ceramics. Additional uses include galvanized materials, automotive parts, roofing and gutters, organ pipes, batteries and fuses, soldering flux, rayon manufacturing, catalyst fire proofing, electroplating, pigments, semiconductors, piezoelectric devices, photoconductor, rubber compounding, curative, metal coating, phosphors, paint pigment, organic synthesis, resin curing, catalyst, wood preservative, insulating materials and many other applications. Zinc and compounds of zinc are used extensively in our society.

Zinc metal and zinc compounds are sometimes recycled and reused as collected. In other cases, zinc compounds are made into other complexes and materials. Some of the materials are placed in landfills and places that are inappropriate for long term storage and decomposition.

A number of interferences include aluminum, cadmium, copper, ferric iron, manganese, and nickel. Large amounts of organic material may interfere and require mild digestion. In the experiment, zinc and other metals in the sample form complexes with cyanide. Zinc is the only metal released by the addition of cyclohexanone. Then, zinc reacts with the "Zincon" indicator called 2-carboxy-2′-hydroxy-5′-sulfoforamazyl. The zinc concentration is proportional to the blue color developed in solution and is measured by a field spectrometer at 620 nm. Also, samples containing 0 to 3.00 mg/L Zn can be analyzed at a different wavelength by the Zincon Method using the DR/850 and the DR/890 Colorimeters. The Zincon Method is USEPA approved for wastewater analysis when the samples have been digested. Otherwise, sample digestion is not normally required. Both HACH and LaMotte Companies publish a similar "Zincon" procedure for determination of zinc. In both cases, the Zincon Method can be used to determine "free" zinc in drinking and surface waters and industrial water.

ZINC (0 to 2.00 mg/L Zn)

Method 8009 for water and wastewater*
Zincon Method **
USEPA approved for wastewater analysis†(digestion needed; see Section 1)

1. Enter the stored program number for Zinc.
   Press: **7 8 0 ENTER.** The display will show: **Dial nm to 620.**
   Note: The Pour-Thru Cell cannot be used.

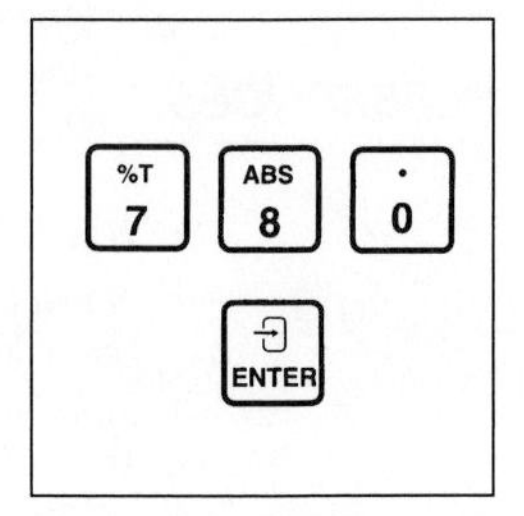

2. Rotate the wavelength dial until the small display shows: **620 nm.** When the correct wavelength is dialed in the display will quickly show: **Zero sample,** then: **mg/L Zn.**

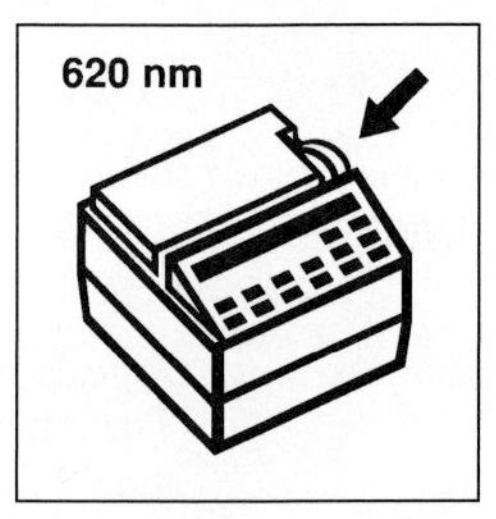

3. Insert the 10-mL Cell Riser into the cell compartment.

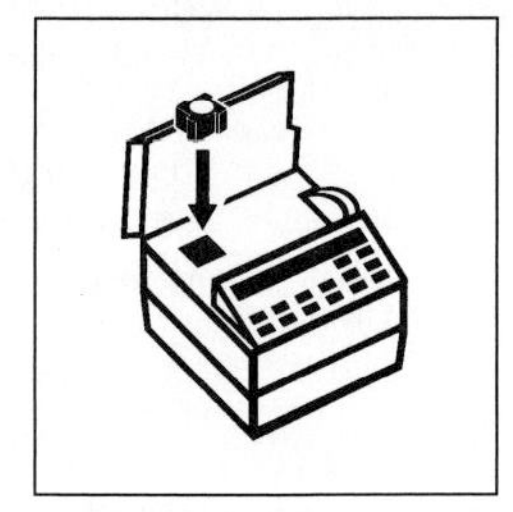

4. Fill a 25-mL graduated mixing cylinder with 20 mL of sample.
   Note: Use only glass stoppered cylinders. Rinse with 1:1 hydrochloric acid and deionized water before use.

* Compliments of HACH Company. Analysis may be performed by the DR/850 and DR/890 Colorimeters and the DR/2010 Field Spectrophotometer.

** Adapted from *Standard Methods for the Examination of Water and Wastewater, 328 D.*

† Federal Register, 45 (105) 36166 (May 29, 1980).

5. Add the contents of one ZincoVer 5 Reagent Powder Pillow. Stopper. Invert several times to completely dissolve the powder.
Note: Powder must be completely dissolved.
Note: The sample should be orange. If it is brown or blue, dilute the sample and repeat the test
***Caution:* ZincoVer 5 contains cyanide and is very poisonous if taken internally or inhaled. Do not add to an acidic sample. Store away from water and acids.**

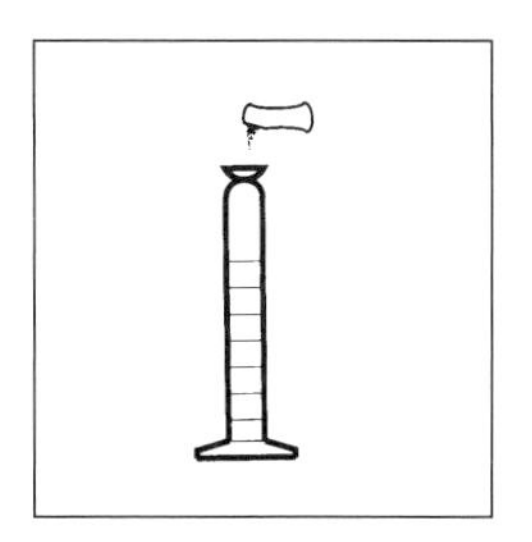

6. Measure 10 mL of the solution into a sample cell (the blank).

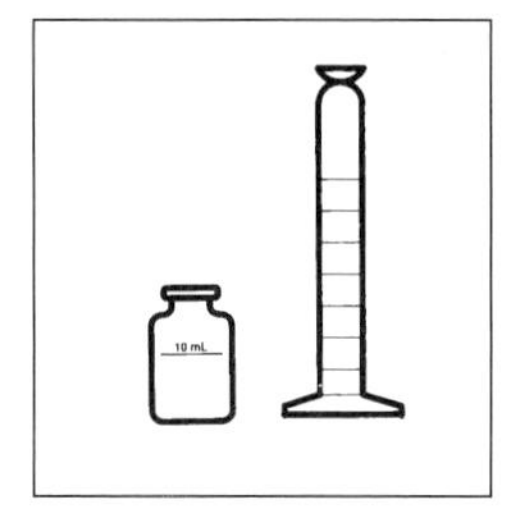

7. Add 0.5 mL of cyclohexanone to the remaining solution in the mixing cylinder.
Note: Use a plastic dropper as rubber bulbs may contaminate the cyclohexanone.

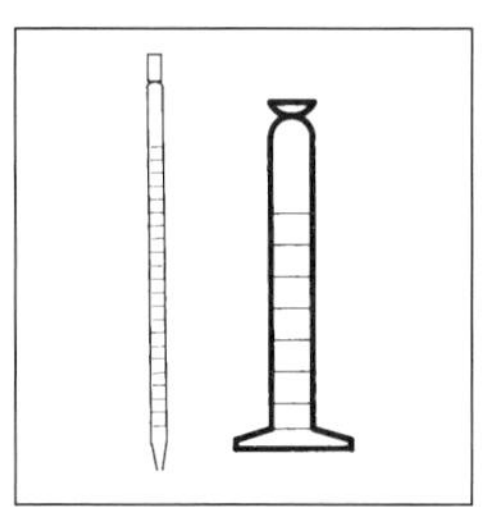

8. Stopper the cylinder. Shake vigorously for 30 seconds (the prepared sample).
Note: The sample will be red-orange, brown, or blue, depending on the zinc concentration.

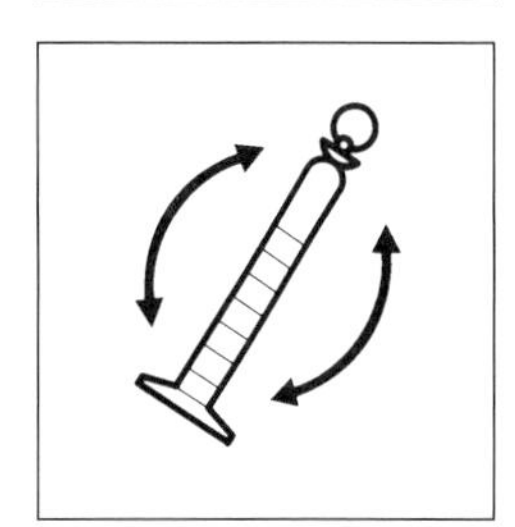

9. Press: **SHIFT TIMER.** A three-minute reaction period will begin.
Note: During this period, complete Step 10.

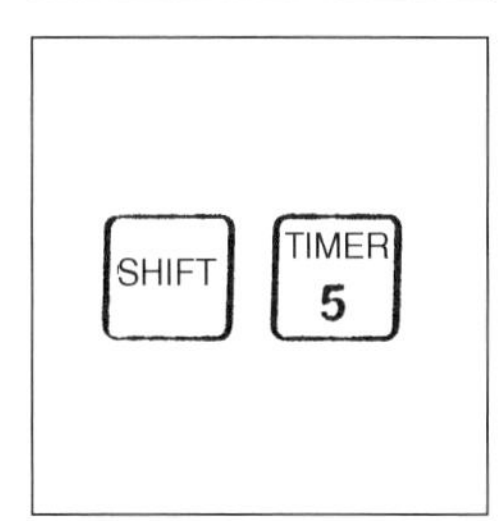

10. During the reaction period, pour the solution from the cylinder into a sample cell.

11. When the timer beeps, place the blank into the cell holder. Close the light shield.

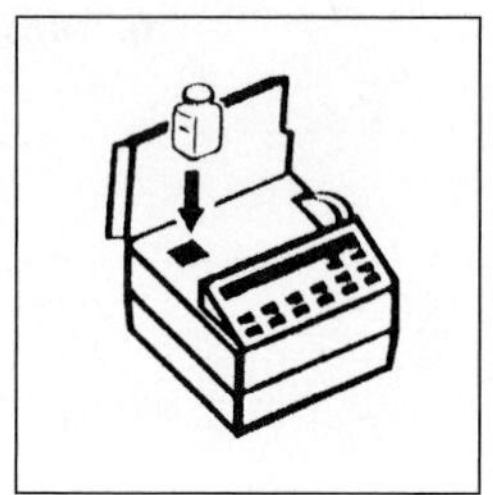

12. Press: **ZERO.** The display will show: **Zeroing ...** then: **0.00 mg/L Zn.**

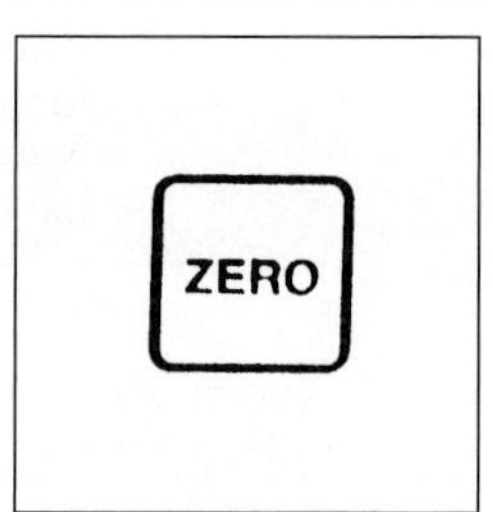

13. Place the prepared sample into the cell holder. Close the light shield.

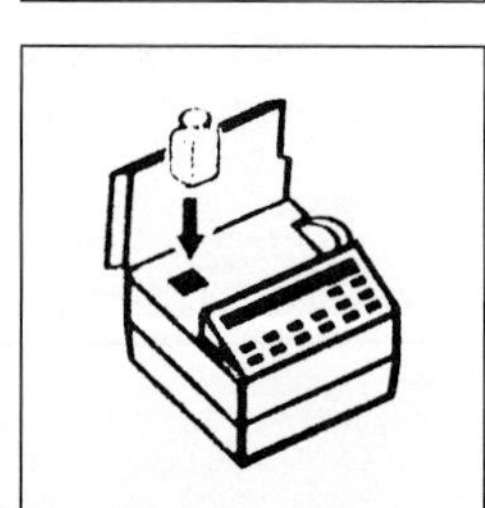

14. Press: **READ.** The display will show: **Reading ...** then the result in mg/L zinc will be displayed.

INTERFERENCES

Table 2.42 identifies the metals that may interfere when present in concentrations exceeding those listed below.

Large amounts of organic material may interfere. Perform the mild digestion (Section 1) to eliminate this interference.

Highly buffered samples or extreme sample pH may exceed the buffering capacity of the reagents and require sample pretreatment (see "pH Interference" in Section 1).

POLLUTION PREVENTION AND WASTE MANAGEMENT

ZincoVer 5 reagent contains potassium cyanide. Cyanide solutions are regulated as hazardous wastes by the Federal RCRA. Cyanide should be collected for disposal as reactive (D003) waste. Be sure that cyanide solutions are stored in a caustic solution with pH > 11 to prevent the release of hydrogen cyanide gas.

In the event of a spill or release, clean up the area by following these steps:

a) Use a fume hood or supplied-air or self-contained breathing apparatus.

b) While stirring, add the waste to a beaker containing a strong solution of sodium hydroxide and calcium hypochlorite or sodium hypochlorite (household bleach).

c) Maintain a strong excess of hydroxide and hypochlorite. Let the solution stand for 24 hours.

d) Neutralize and flush the solution down the drain with a large excess of water.

TABLE 2.42

Substance	Concentration
Aluminum	6 mg/L
Cadmium	0.5 mg/L
Copper	5 mg/L
Iron (ferric)	7 mg/L
Manganese	5 mg/L
Nickel	5 mg/L

QUESTIONS AND ANSWERS

1. In this experiment, what is the primary safety issue?
ZinconVer 5 contains cyanide which is very poisonous if taken internally or inhaled. Store the material away from acids and water. ***Do not add*** *the material to an acidic sample. When testing water outdoors, be sure that the wind does not blow in your face to avoid odors from the material.*

2. What are the interferences in this experiment?
The metals of aluminum, cadmium, copper, ferric iron, manganese and nickel can interfere in the analysis of zinc. Large amounts of organic material may interfere and require mild digestion of the sample.

3. What do you do with the waste or spills containing cyanide?
The ZincoVer 5 reagent contains potassium cyanide. Cyanide solutions have been regulated as hazardous wastes by the Federal RCRA and can be collected for disposal as reactive wastes (DOO3). The cyanides can be oxidized to cyanate by the addition of a strong solution of sodium hydroxide and sodium hypochlorite (household bleach).

REFERENCES

1. *HACH Water Analysis Handbook,* 3rd Ed. (Loveland, CO.: HACH Company, 1997).
2. *Hawley's Condensed Chemical Dictionary,* 11th Ed., rev'd by N. Irving Sax and Richard J. Lewis (New York, N.Y.: Van Nostrand Reinhold, 1987).
3. *Standard Methods for the Examination of Water and Wastewater:* ZINC, 16th Ed., A. E. Greenberg, R. R. Trussell, L. S. Clesceri and Mary Ann H. Franson (eds.), pp. 254, 259–260, Washington, DC, American Public Health Association, 1985.

CHAPTER 3
SOLIDS, GELS, OILS, SLUDGES AND NONAQUEOUS SUBSTANCES

SOLIDS TESTING

Environmental analyses of solids may require a variety of sampling techniques depending on the goals of the testing. A single test at specific times may be appropriate for quality control and process control purposes. Repetitive tests on a representative sample is appropriate for testing of many solids. The type of tests, the purpose of the tests, and the variability of the samples will determine the frequency and numbers of samples to be tested.

For example, soil sampling techniques vary depending on the test and purpose of the test. Soil fertility tests in a five acre field require a number of randomly selected composite soil samples which are mixed in a bucket and then analyzed. If the field has different soil characteristics or has variable elevation, the area may have to be divided into zones or areas of analyses and the individual sites may be tested as single composite samples or as representative samples. The variability of the field and the purpose of the test will determine the number of soil testing sites, the type of tests, depth of sampling, and the type of sampling. The mixing characteristics of the sample and the size of the sample are important features to consider when analyzing for a specific contaminant or element.

The determination of the pH of soil in order to add lime has different goals than determining whether a toxic substance was illegally dumped at a site. A crime scene investigation, such as a robbery and an assault of a person, may require soil sampling and testing at various depths and zones of investigation. Thus, the sampling techniques for solids depend on the goals of the project, the condition or state of the sample, and the type of test on the sample.

EXTRACTION OF MATERIALS FROM SOLIDS

Many cations, anions and neutral species are found in solids. Solids include soil, residues, ash, cinders, slag, tars, gels, sand, grease, waxes, sediments, metals, sawdust, mulch, shale, gravel and rock, dust, ores, plant tissue, organs of humans and animals, wastes, solid contaminants, sludge, supplies, chemical compounds, fertilizers, feed and forage, foods (cheese, cereal, etc.), ion exchange resins, and many additional manufactured products and byproducts of manufacturing. Solids are ionic and covalent compounds and polymeric materials found in suspended materials in solution and in crystalline materials and non-crystalline materials. For analysis purposes, oils can also be included as a solid even though a different sample size is required for analysis purposes.

Solids in waste materials or in water sediments, volatile solids, nonvolatile solids, and insoluble compounds require a number of different procedures for complete extraction. Samples containing a slurry, gel, sludge, oils or a semisolid are sometimes nearly impossible to analyze because of difficulty in extracting a tightly bound metal, ion (anion or cation), or organic compound from the matrix.

Extraction procedures on soil and other types of solids are appropriate for testing nutrients and bioavailable minerals in the ionic form. Extraction procedures are applied to materials where the species being analyzed is not confined to a lattice structure, organic matrices, or a unique structure.

Physical and chemical characteristics of soil or other solid materials can influence the pH of a sample. At a basic pH, some ions such as calcium precipitate as calcium carbonate and impact the pH of the soil or material. At a pH below 6.5, the low pH of the system may be modified by the dissolution of the calcium carbonate and shifts in the equilibrium of organic/inorganic complexes.

Digestion procedures are required for the analyses of materials that do not dissolve readily and freely in solution. Ions and organic molecules are often tied up or bound by a number of bonding types, crystal lattice structures, and organic matrices. During analysis for a metal, the digestion process is required to break apart the crystal lattice structure and the organic matrices to form the metal, carbon dioxide, water and additional free ions in an acidic solution. Then, analysis of the free ion or species can be performed on the sample.

Digestion procedures and test methods for evaluating solids, sludges and other materials are listed in USEPA SW-846, *Test Methods for Evaluating Solid Waste, Physical Chemical Methods,* 3rd Ed., Update III, December 1996.* Some of the materials are listed on the Internet under USEPA SW-846. Methods are identified for the extraction and determination of metals and specific organics in stack emissions, stationary sources, aqueous samples, oils, sediments, sludge, soils, and other materials.

Digestion methods for specific sample types are provided in the *DR/2010 Spectrophotometer Procedures Manual,* 1998, pp. 59–65 and in the *Wastewater and Biosolids Analysis* text published by HACH Company. In most cases, the Digesdahl Digestion System digests 0.25 to .50 grams of sample of material for 4 minutes containing 4 to 6 mL of sulfuric acid and 10 ml of 50% hydrogen peroxide. After a sample portion is diluted and pH adjusted, the sample can be analyzed for a specific ion or compound.

SAMPLE PREPARATION

Samples of soil, rocks and other inorganic materials do not usually require drying. Samples can be ground with a mortar and pestle or broken apart with a "stone crusher" or some other device. For soil and other ground samples, free minerals are extracted either with a 3% acetic acid solution or with a Melich Solution composed of a mixture of 0.5 N HCL and 0.025 N H_2SO_4. A particular solid may require additional research to identify the best method for dissolution of the "unknown."

The Melich I Extracting Solution or "Acid Extracting Solution" extracts free metals and additional anions and compounds from soils and other ground samples.** Extracted samples using the Melich I Extracting Solution (Acid Extracting Solution) include the following: 1) Nitrate Nitrogen; 2) Phosphorus (Phosphates); 3) Potassium; 4) Calcium; 5) Magnesium; 6) Ammonia Nitrogen; 7) Nitrite Nitrogen; 8) Manganese; 9) Copper; 10) Zinc; 11) Iron; and 12) Aluminum. Most of the extracted samples are neutralized with base before spectrophotometric analysis.

*Compliments of Mary Ann Detlefsen, Ph.D., Environmental Chemist, Latham, N.Y.

**Compliments of LaMotte Company.

For many solid samples, uniform grinding and proper sample mixing are critical in obtaining accurate analytical results. Some solid samples, such as plant tissue samples, can be dried at 80° C. These solid samples should be reduced to 0.5 to 1.0-mm particle size to ensure sample homogeneity and to facilitate organic matter destruction. If the sample aliquot being analyzed is less than 0.50 grams, the dried sample should be passed through a 40-mesh screen before analysis.* For samples larger than 0.50 grams, a 20-mesh screen should be used to prepare the dried sample for analysis. Most solid samples should be stored in a cool (4° C), dark, and dry environment.

There are three types of procedures used to prepare solid samples for analysis. The selection of the procedure will depend upon the method preference for determining the pollutant or sample constituent. The wet ashing procedure generally requires less time and less costly equipment.

Wet ashing procedures using a concentrated acid (sulfuric acid, or nitric acid or perchloric acid) or a combination of acids followed by hydrogen peroxide treatment of the sample completes the digestion process at high temperatures (440° C) on a hot plate or in a digestion system.** Digested samples containing aluminum, nickel and iron may require additional heating to completely dissolve the metal. After digestion is complete, the sample is cooled and is diluted with either water or acidified water (nitric acid or hydrochloric acid) to dissolve the digest residue and to bring the sample to final volume. After a pH adjustment on the sample, the final sample volume is used for the determination of the specific metal or neutral species.

Dry ashing of ground samples (plants, leaves, food, paper, organic products) is conducted in a muffle furnace at temperatures of 500° C to 550° C for 4 to 8 hours.* After ashing the dried sample to a clean white ash, the ash is cooled and the contents of the ash are dissolved in dilute nitric acid, or dilute hydrochloric acid, or a mixture of both (aqua regia). The final solution is diluted to meet the range requirements of the analysis procedure.

An accelerated wet digestion procedure can be used on plant materials and other organic matter. Organic matter destruction can occur in microwave ovens when the system is properly ventilated. This procedure will shorten the time for digestion of samples.

SOIL EXTRACTION PROCEDURE

EXTRACTION†

The following method of extraction is employed for obtaining the soil filtrate for the tests for nitrate nitrogen, phosphorous, potassium, calcium, magnesium, ammonia nitrogen, nitrite nitrogen, manganese, aluminum, copper, zinc, and iron. Separate individual extractions are required for the chloride and sulfate tests and are not included in this procedure. Consult the LaMotte Soil Handbook for additional information on "Sampling and Preparation of Sample for Testing."

---

* *Handbook of Reference Methods for Plant Analysis,* Soil and Plant Analysis Council, Yash P. Kalra (ed.) (Boca Raton, Fla.: CRC Press, 1998), pp. 40–47.

** *HACH Water Analysis Handbook,* 3rd Ed., HACH Company, 3rd Edition, 1997.

† Modified Text-Compliments of LaMotte Company.

PROCEDURE*

1. Use the 1 ml pipet to add 5 mL of the Acid Extracting Solution to the 100 mL graduated cylinder. Add deionized water to 75 mL graduation.

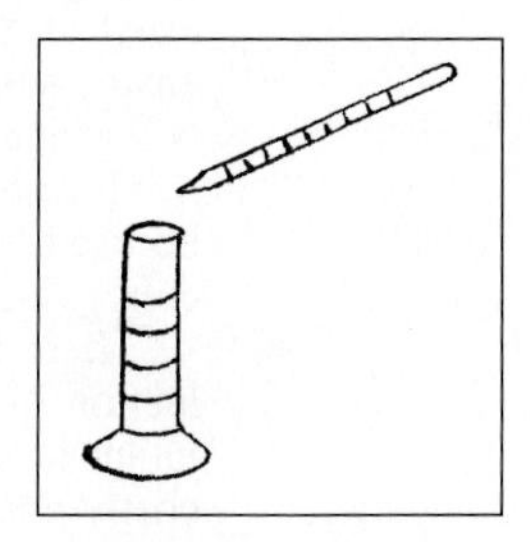

2. Pour this solution into the 100 mL bottle.

3. Use the Soil Measure to add 15 g (one level measure) of the soil sample to the bottle.

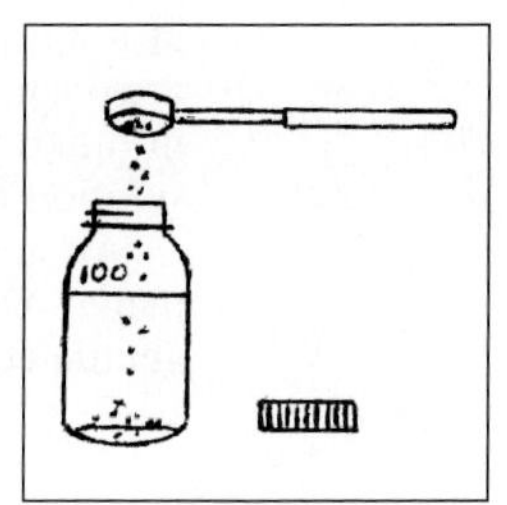

4. Cap the bottle and shake for a period of 5 minutes.

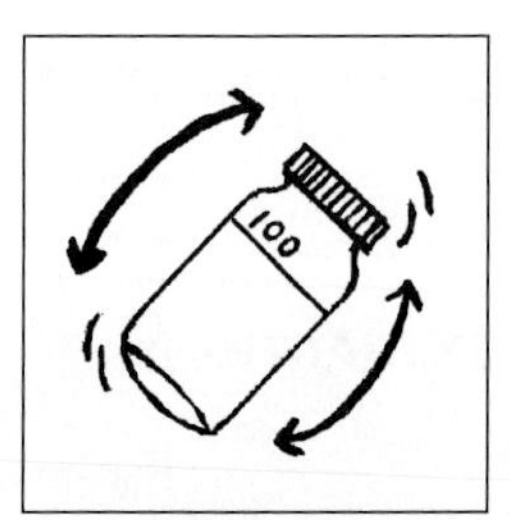

5. Use the funnel and filter paper to filter and collect all of the soil filtrate in a 100 mL bottle.

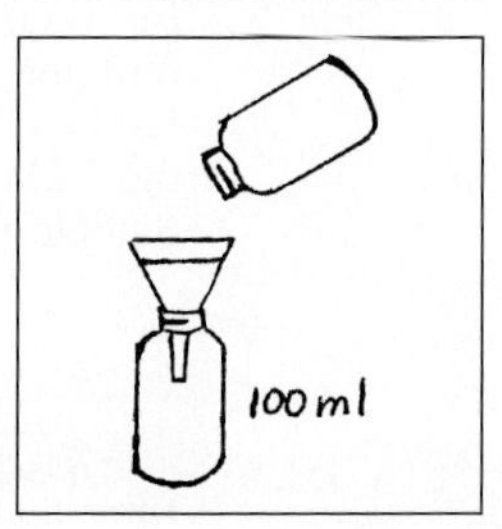

*Compliments of LaMotte Company.

6. The soil filtrate is used for all of the tests listed above, except chloride and sulfate.

*SINGLE TEST PROCEDURE**

1. Use the 1 mL pipet to add 1 mL of the Acid Extracting Solution to the graduated vial, then add deionized water to the graduation.

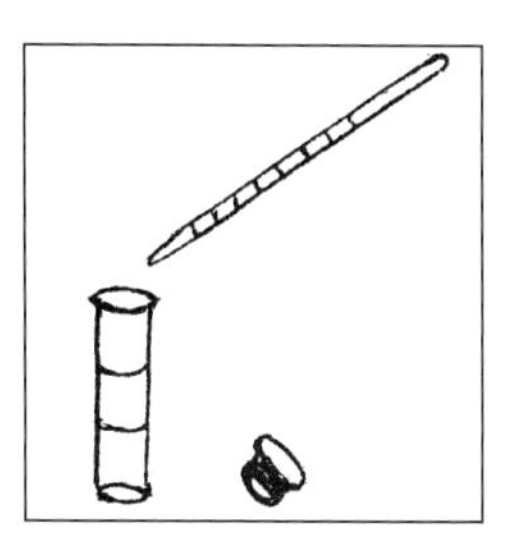

2. Using the 1.0 g spoon (0697), add 3 g of soil to the extracting solution in the vial.

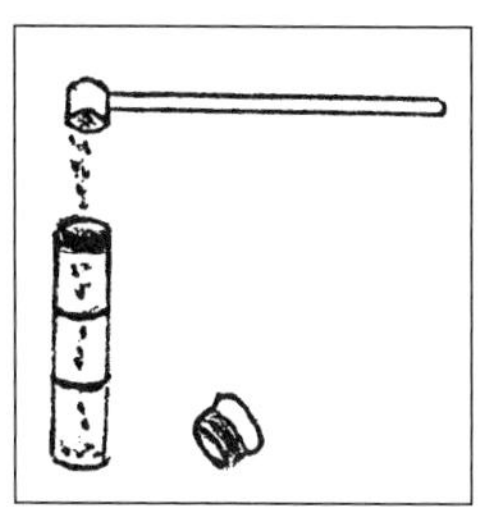

3. Cap the vial and shake for a period of 5 minutes.

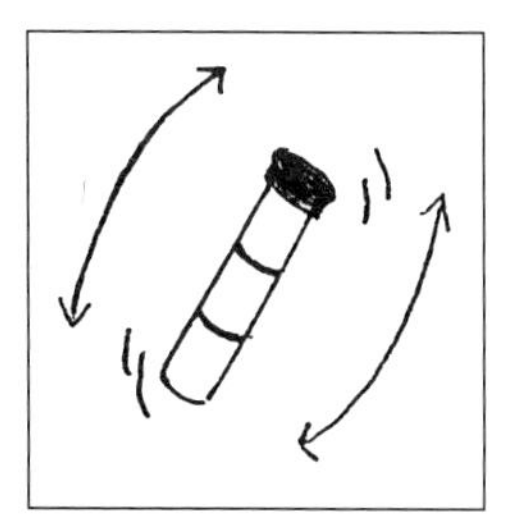

4. Filter, using the funnel and filter paper and collect all of the soil filtrate.

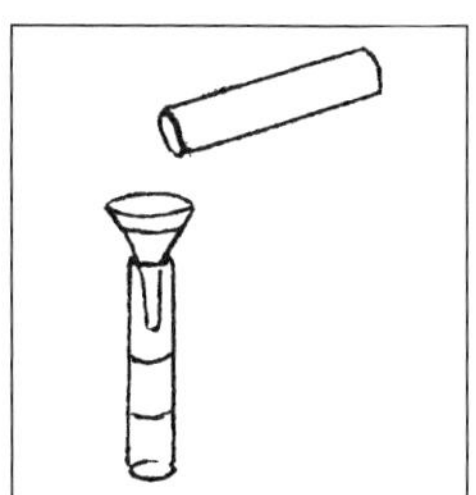

5. The soil filtrate can be used for all of the tests except chlorides and sulfates.

*Compliments of LaMotte Company.

DIGESTION PROCEDURE FOR SOLIDS AND OILS*

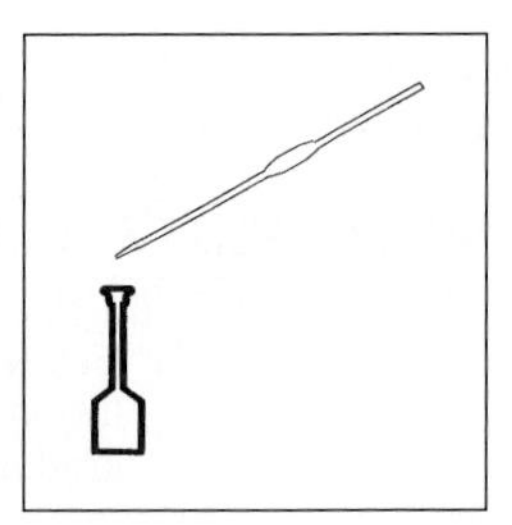

1. Transfer 0.50 g or less of solid sample into a 100-mL Digesdahl digestion flask; see Sample and Analysis Volume Tables for Solids following this procedure.
   Note: Be sure you have a homogenous sample. Solid sample should be finely ground or chopped and mixed well.
   Note: If a metal is being extracted from an oil, transfer 0.25 g of oil sample into a 100-mL Digesdahl digestion flask and proceed with the same extraction experiment beginning with Step 2 below. See Sample and Analysis Volume Tables for Oils following this procedure.

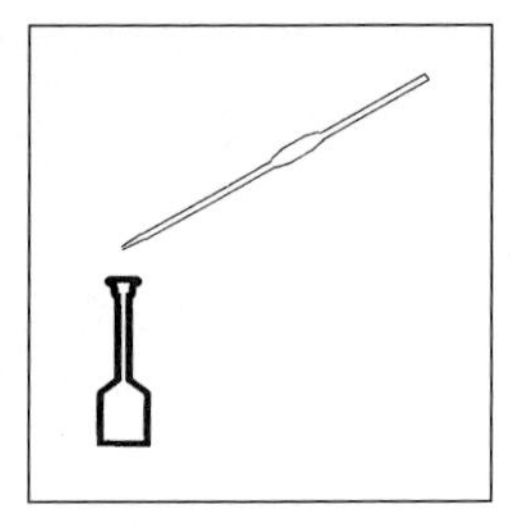

2. Add 4 mL concentrated sulfuric acid to the digestion flask.
   Note: Use only HACH Digesdahl digestion flasks. Volumetric flasks with concave bottom should not be used. Safety glasses and a safety shield placed between the operator and the Digesdahl are required.
   Note: Add 6 mL concentrated sulfuric acid to the flask for soil or fuel analysis and 15 ml of acid for ion exchange resins.

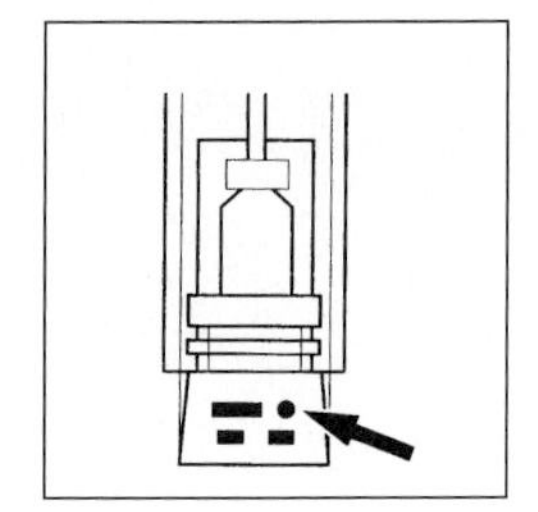

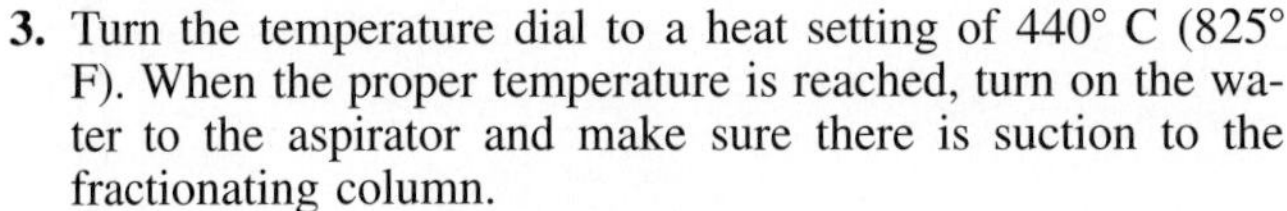

3. Turn the temperature dial to a heat setting of 440° C (825° F). When the proper temperature is reached, turn on the water to the aspirator and make sure there is suction to the fractionating column.
   Note: Wait for the proper temperature to be reached before sample is placed on the heater.

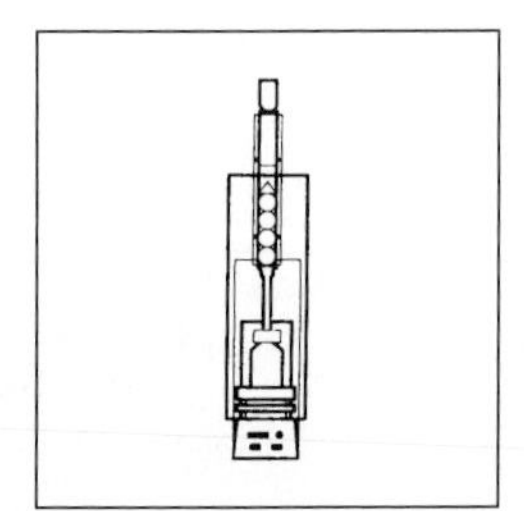

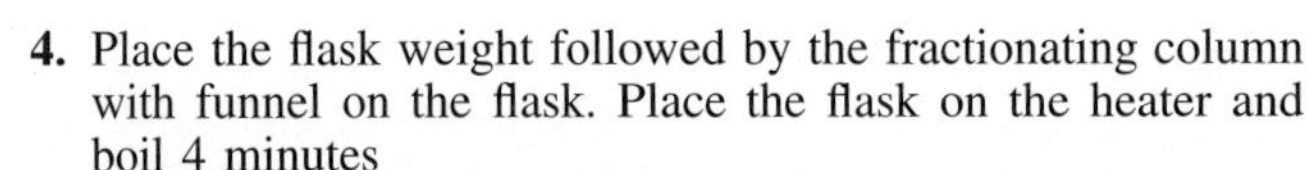

4. Place the flask weight followed by the fractionating column with funnel on the flask. Place the flask on the heater and boil 4 minutes
   Do not boil to dryness! If sulfuric acid is not present in the flask after the boiling period, do not proceed to Step 5! Discard the sample and use more sulfuric acid for the digestion procedure in Step 2. Or choose a smaller amount from the Sample and Analysis Volume Tables for Solids.
   Note: If sample foams up into the neck of the flask, lower temperature to 335° C (635° F). Continue heating at lower temperature until all water is evaporated. Then return to original digestion temperature. White acid vapors accompanied with a reflux line indicate that the sulfuric acid is boiling. Some organic samples may need more than 5 minutes for complete digestion.

---

*Compliments of HACH Company.

5. **Do not proceed if sulfuric acid is not visible in the flask.** Add 10 mL of 50% hydrogen peroxide to the charred sample via the funnel on the fractionating column.
   Note: Do not heat to dryness. Visually confirm the presence of sulfuric acid in the flask before adding hydrogen peroxide. If the digest does not turn colorless, add 5 mL increments of peroxide until the digest becomes clear or does not change color.
   Note: If sample foams during hydrogen peroxide addition, stop the peroxide flow and remove the digestion flask and fractionating column (use finger cots). Cool for 30 seconds and return apparatus to the heating block. Start peroxide addition with 2 mL, then follow with the remaining peroxide.

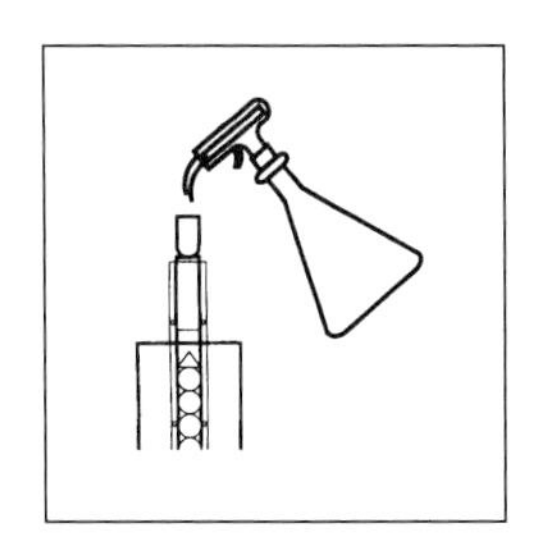

6. After addition of hydrogen peroxide is complete, boil off excess hydrogen peroxide by heating for one more minute. **Do not heat to dryness.**
   Note: If the sample goes to dryness, turn off the Digesdahl and air cool to room temperature. Add water to flask before handling. Repeat the digestion from the beginning using a new sample.

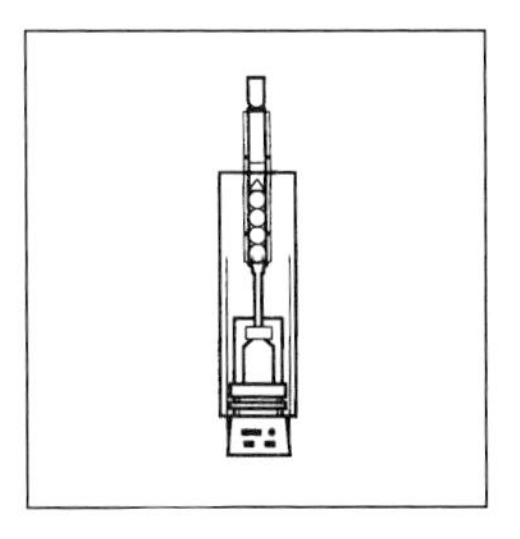

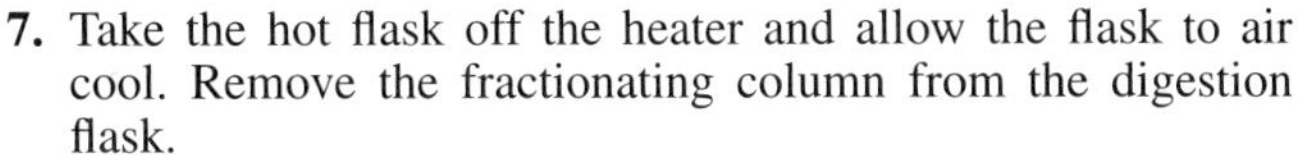

7. Take the hot flask off the heater and allow the flask to air cool. Remove the fractionating column from the digestion flask.
   Note: Use finger cots to remove the digestion flask. Place it on a cooling pad for at least one minute. Then remove the column. Do not add water to the flask until it has cooled.

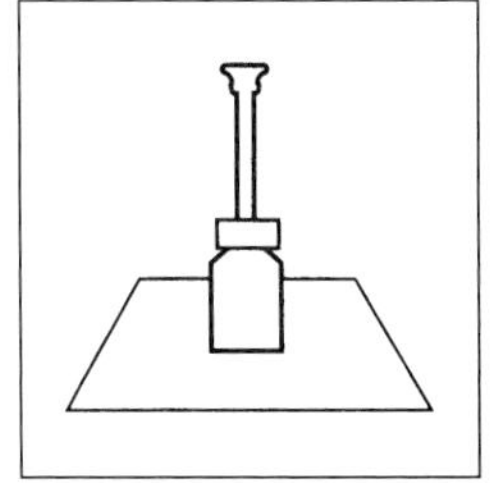

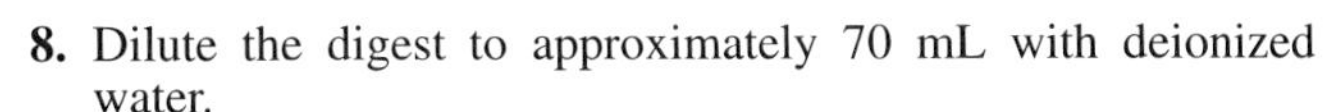

8. Dilute the digest to approximately 70 mL with deionized water.
   Note: Add deionized water slowly at first. Cool the flask if necessary for handling.

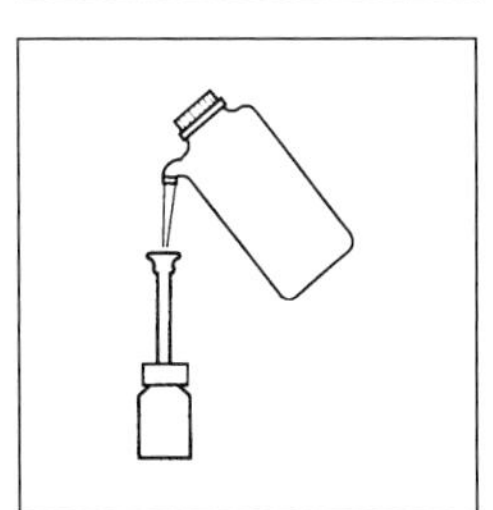

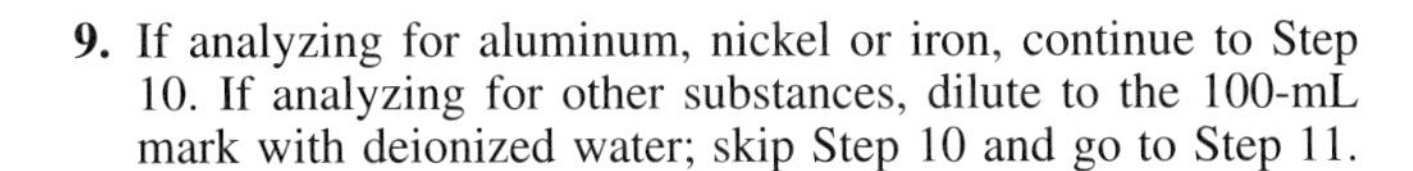

9. If analyzing for aluminum, nickel or iron, continue to Step 10. If analyzing for other substances, dilute to the 100-mL mark with deionized water; skip Step 10 and go to Step 11.

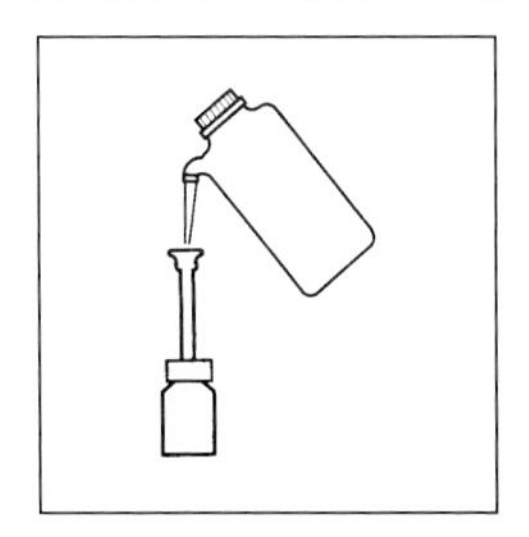

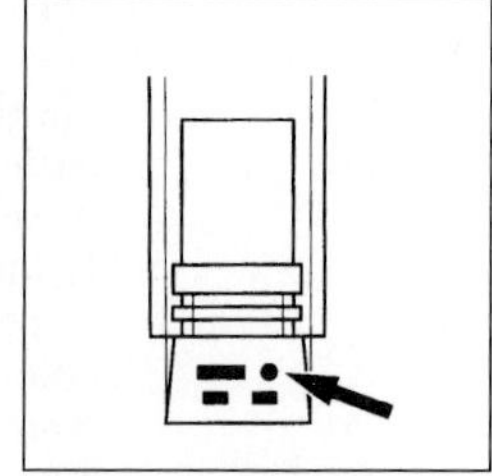

10. Turn the temperature dial to a heat setting of 204° C (400° F). Add 150 mL of water to a 400 mL beaker. Place the beaker on the heater. Place the flask in the beaker and boil for 15 minutes. Air cool to room temperature and dilute to the mark with deionized water. Invert several times to mix.
Note: When using a Digesdahl Digestion Apparatus System without temperature control dials, reset to a lower setting that gently boils the water.

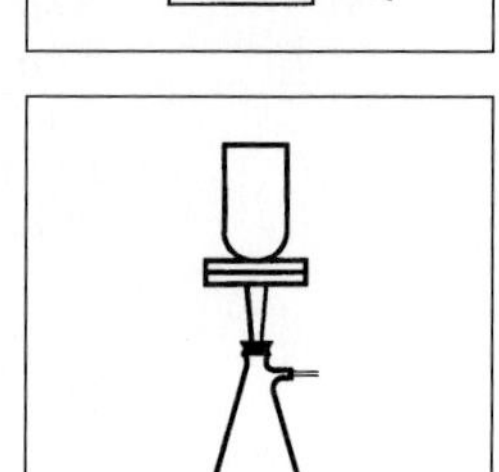

11. If the sample has visible turbidity, filter or wait until the turbidity settles, and the upper portion of the sample is clear.
Filter as follows:
a). *Place a filter paper into the filter holder with wrinkled surface upward.*
b). *Place the filter holder assembly in the filtering flask and wet the filter with deionized water to ensure adhesion to the holder.*
c). *While applying a vacuum to the filtering flask, transfer the sample to the filtering apparatus.*
d). *Slowly release the vacuum from the filtering flask and transfer to another container.*
Note: If small aliquots (≤ 1.0 mL) are analyzed, filtration is not needed.

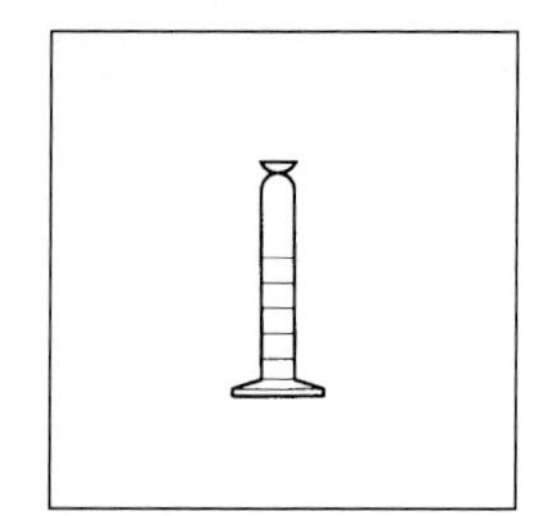

12. Continue with the analysis using the appropriate procedure following this procedure.

PH ADJUSTMENT FOR METALS METHOD (*p. 125,* HACH Water Analysis Handbook, *3rd Ed., 1997*)

Note: If analyzing aliquots smaller than 0.5 ml, pH adjustment is not necessary.

1. Pipet the appropriate analysis volume into the appropriate mixing graduate cylinder. See **Sample and Analysis Volume Tables for Solids** following the specific digestion to determine the analysis volume. For oils, see **Sample and Analysis Volume Tables for Oils** following the specific digestion to determine the analysis volume.
Note: Some methods require pipetting into a volumetric flask or a regular graduate cylinder.

2. Dilute to about 20 mL with deionized water.

3. Add 1 drop of 2,4-Dinitrophenol Indicator Solution.

4. Add 1 drop of 8 N Potassium Hydroxide (KOH) Standard Solution, swirling between each addition, until the first flash of yellow appears (pH 3). Do not use a pH meter if analyzing for silver.

5. Add 1 drop of 1 N KOH. Stopper the cylinder and invert several times to mix.
Note: Use pH paper to insure the pH is 3. If it is higher than 4, do not readjust with acid; start over with a fresh aliquot volume.
6. Continue to add 1 N KOH in this manner until the first permanent yellow color appears (pH 3.5 to 4.0).
Note: High iron content will cause precipitation (brown cloud) which will coprecipitate other metals. Repeat this procedure with a smaller aliquot volume.
7. Add deionized water to the volume indicated in the colorimetric procedure for the parameter you are analyzing. Fill a second graduated cylinder to the same volume with deionized water.
8. Continue with the colorimetric procedure for the parameter you are analyzing.

PH ADJUSTMENT FOR TOTAL KJELDAHL NITROGEN (TKN), COLORIMETRIC METHODS

To complete the TKN analysis, consult the spectrophotometer or colorimeter procedure. Consult the procedure on p. 115 for **oils** and p. 126 for **solids** in the HACH Water Analysis Handbook, 3rd Ed., 1997 if a guide is not available.

SAMPLE AND ANALYSIS VOLUME TABLES FOR SOLIDS*

Solids include carbohydrates (cellulose, gums, starch, sugars), inorganics, proteins, soil, sludges (slurries or dry), wood products (paper and pulp), stone, plastic, metals, and many additional substances. Digestion procedures are required to ensure all organic-metallic bonds are broken. The aliquot size of the digest to be used in the analysis is of primary importance. Tables are provided for determining the amount of initial sample to be digested. Equations have been developed in order to optimize the specific tests being performed.

ALUMINUM, Aluminon (Method 8012)

Expected AL conc. (mg/kg)	Sample amount (g)	Analysis volume (mL)	Dilute to
10–470	0.500	20.0	50.0 mL
25–1200	0.400	10.0	50.0 mL
65–3100	0.300	5.00	50.0 mL
480–23000	0.200	1.00	50.0 mL
1900–93000	0.100	0.50	50.0 mL

$$\text{Total mg/kg AL} = \frac{A \times 5000}{B \times C}$$

A = mg/L reading from the instrument

B = mL (g) sample amount from table

C = mL analysis volume from table

*Compliments of HACH Company.

CADMIUM, Dithizone (Method 8017)

Expected Cd conc. (mg/kg)	Sample amount (g)	Analysis volume (mL)	Dilute to
5–420	0.500	20.0	250.0 mL
12–1000	0.400	10.0	250.0 mL
30–2800	0.300	5.00	250.0 mL
240–21000	0.200	1.00	250.0 mL
960–85000	0.100	0.500	250.0 mL

$$\text{Total mg/kg Cd} = \frac{A \times 25}{B \times C}$$

A = μg/L reading from instrument

B = mL (g) sample amount from table

C = mL analysis volume from table

CHROMIUM, Total (Method 8024)

Expected Cr conc. (mg/kg)	Sample amount (g)	Analysis volume (mL)	Dilute to
4.0–165	0.500	20.0	25.0 mL
10–410	0.400	10.0	25.0 mL
25–1100	0.300	5.00	25.0 mL
190–8200	0.200	1.00	25.0 mL
750–33000	0.100	0.500	25.0 mL

$$\text{Total mg/kg Cr} = \frac{A \times 2500}{B \times C}$$

A = mg/L reading from instrument

B = mL (g) sample amount from table

C = mL analysis volume from table

COBALT (Method 8078)

Expected Co conc. (mg/kg)	Sample amount (g)	Analysis volume (mL)	Dilute to
10–450	0.500	20.0	25.0 mL
20–1200	0.400	10.0	25.0 mL
50–3000	0.300	5.00	25.0 mL
500–20000	0.200	1.00	25.0 mL
2000–100000	0.100	0.500	25.0 mL

$$\text{Total mg/kg Co} = \frac{A \times 2500}{B \times C}$$

A = mg/L reading from instrument

B = mL (g) sample amount from table

C = mL analysis volume from table

COPPER, Bicinchoninate (Method 8026)

Expected Cu conc. (mg/kg)	Sample amount (g)	Analysis volume (mL)	Dilute to
20–1200	0.500	20.0	25.0 mL
50–3000	0.400	10.0	25.0 mL
120–7500	0.300	5.00	25.0 mL
1000–60000	0.200	1.00	25.0 mL
4000–240000	0.100	0.500	25.0 mL

$$\text{Total mg/kg Cu} = \frac{\text{A} \times 2500}{\text{B} \times \text{C}}$$

A = mg/L reading from instrument

B = mL (g) sample amount from table

C = mL analysis volume from table

IRON, 1,10 Phenanthroline (Method 8008)*

Expected Fe conc. (mg/kg)	Sample amount (g)	Analysis volume (mL)	Dilute to
12–700	0.500	20.0	25.0 mL
35–1700	0.400	10.0	25.0 mL
85–4500	0.300	5.00	25.0 mL
650–35000	0.200	1.00	25.0 mL
2500–140000	0.100	0.500	25.0 mL

* Total Iron

$$\text{Total mg/kg Fe} = \frac{\text{A} \times 2500}{\text{B} \times \text{C}}$$

A = mg/L reading from instrument

B = mL (g) sample amount from table

C = mL analysis volume from table

LEAD, Dithizone (Method 8033)

Expected Pb conc. (mg/kg)	Sample amount (g)	Analysis volume (mL)	Dilute to
8–700	0.500	20.0	250.0 mL
20–1800	0.400	10.0	250.0 mL
50–4700	0.300	5.00	250.0 mL
400–35000	0.200	1.00	250.0 mL
1500–140000	0.100	0.500	250.0 mL

$$\text{Total mg/kg Pb} = \frac{\text{A} \times 25}{\text{B} \times \text{C}}$$

A = μg/L reading from instrument

B = mL (g) sample amount from table

C = mL analysis volume from table

MANGANESE, PAN (Method 8149)

Expected Mn conc. (mg/kg)	Sample amount (g)	Analysis volume (mL)	Dilute to
4–180	0.500	20.0	25.0 mL
10–450	0.400	10.0	25.0 mL
25–1200	0.300	5.00	25.0 mL
200–9300	0.200	1.00	25.0 mL
750–37000	0.100	0.500	25.0 mL

$$\text{Total mg/kg Mn} = \frac{A \times 2500}{B \times C}$$

A = mg/L reading from instrument

B = mL (g) sample amount from table

C = mL analysis volume from table

NICKEL, PAN (Method 8150)

Expected Ni conc. (mg/kg)	Sample amount (g)	Analysis volume (mL)	Dilute to
4–225	0.500	20.0	25.0 mL
10–585	0.400	10.0	25.0 mL
25–1500	0.300	5.00	25.0 mL
200–11500	0.200	1.00	25.0 mL
750–45000	0.100	0.500	25.0 mL

$$\text{Total mg/kg Ni} = \frac{A \times 2500}{B \times C}$$

A = mg/L reading from instrument

B = mL (g) sample amount from table

C = mL analysis volume from table

NITROGEN, Total Kjeldahl (Method 8075)

Expected nitrogen conc. (mg/kg)	Sample amount (g)	Analysis volume (mL)	Dilute to
42–1400	0.500*	20.0	25.0 mL*
106–3500	0.400*	10.0	25.0 mL*
350–9200	0.300*	5.00	25.0 mL*
1000–70000	0.200*	1.00	25.0 mL*
4200–275000	0.100*	0.500	25.0 mL*

* These are guidelines only. See the spectrophotometric procedure manual.

$$\text{Total mg/kg TKN} = \frac{A \times 75}{B \times C}$$

A = mg/L reading from instrument

B = mL (g) sample amount from table

C = mL analysis volume from table

PHOSPHORUS, Ascorbic Acid (Method 8048)

Expected PO_4 conc. (mg/kg)	Sample amount (g)	Analysis volume (mL)	Dilute to
10–450	0.500	20.0	25.0 mL
25–1150	0.400	10.0	25.0 mL
67–3100	0.300	5.00	25.0 mL
500–23500	0.200	1.00	25.0 mL
2000–93500	0.100	0.500	25.0 mL

$$\text{Total mg/kg phosphate} = \frac{A \times 2500}{B \times C}$$

A = mg/L reading from instrument

B = mL (g) sample amount from table

C = mL analysis volume from table

SILVER (Method 8120)

Expected Ag conc. (mg/kg)	Sample amount (g)	Analysis volume (mL)	Dilute to
6–325	0.500	20.0	50.0 mL
15–820	0.400	10.0	50.0 mL
40–2200	0.300	5.00	50.0 mL
300–16000	0.200	1.00	50.0 mL
1200–65000	0.100	0.500	50.0 mL

$$\text{Total mg/kg Ag} = \frac{A \times 5000}{B \times C}$$

A = mg/L reading from instrument

B = mL (g) sample amount from table

C = mL analysis volume from table

ZINC (Method 8009)

Expected Zn conc. (mg/kg)	Sample amount (g)	Analysis volume (mL)	Dilute to
16–1200	0.500	20.0	50.0 mL
40–2900	0.400	10.0	50.0 mL
100–7800	0.300	5.00	50.0 mL
800–58000	0.200	1.00	50.0 mL
3000–230000	0.100	0.500	50.0 mL

$$\text{Total mg/kg Zn} = \frac{A \times 5000}{B \times C}$$

A = mg/L reading from instrument

B = mL (g) sample amount from table

C = mL analysis volume from table

SAMPLE AND ANALYSIS VOLUME TABLES FOR OILS*

Oils include animal fats (lard, fish oil, tallow), edible oils (corn oil, peanut oil, olive oil, Cod liver oil, Halibut liver oil, food dressing oils), terpene-based oils, mineral oils (motor oils, transmission oils, machinery oils, grease, cutting and drilling oils, grinding oils, tool and dye- making oils, and crude oil), organic solvents, and vegetable oils (castor oil, coconut oil, cottonseed oil, linseed oil, palm and safflower oils, soybean oil, tung oil). Digestion procedures must be completed to ensure all organic-metallic bonds are broken. The aliquot size of the digest to be used in the analysis is of primary importance. Tables are provided for determining the amount of initial sample to be digested. Equations have been developed in order to optimize the specific tests being performed.

ALUMINUM, Aluminon (Method 8012)

Expected AL conc. (mg/kg)	Sample amount (g)	Analysis volume (mL)	Dilute to
15–950	0.250	20.0	50.0 mL
40–2500	0.200	10.0	50.0 mL
100–6200	0.150	5.00	50.0 mL
1000–45000	0.100	1.00	50.0 mL

$$\text{Total mg/kg AL} = \frac{A \times 5000}{B \times C}$$

A = mg/L reading from instrument

B = mL (g) sample amount from table

C = mL analysis volume from table

CADMIUM, Dithizone (Method 8017)

Expected Cd conc. (mg/kg)	Sample amount (g)	Analysis volume (mL)	Dilute to
10–850	0.250	20.0	250.0 mL
25–2000	0.200	10.0	250.0 mL
65–5500	0.150	5.00	250.0 mL
480–40000	0.100	1.00	250.0 mL

$$\text{Total mg/kg Cd} = \frac{A \times 25}{B \times C}$$

A = μg/L reading from instrument

B = mL (g) sample amount from table

C = mL analysis volume from table

*Compliments of HACH Company.

CHROMIUM, Total (Method 8024)

Expected Cr conc. (mg/kg)	Sample amount (g)	Analysis volume (mL)	Dilute to
8–330	0.25	20.0	25.0 mL
20–820	0.20	10.0	25.0 mL
50–2200	0.15	5.00	25.0 mL
350–16000	0.10	1.00	25.0 mL

$$\text{Total mg/kg Cr} = \frac{A \times 2500}{B \times C}$$

A = mg/L reading from instrument

B = mL (g) sample amount from table

C = mL analysis volume from table

COBALT (Method 8078)

Expected Co conc. (mg/kg)	Sample amount (g)	Analysis volume (mL)	Dilute to
15–950	0.250	20.0	25.0 mL
45–2400	0.200	10.0	25.0 mL
120–6200	0.150	5.00	25.0 mL
880–46000	0.100	1.00	25.0 mL

$$\text{Total mg/kg Co} = \frac{A \times 2500}{B \times C}$$

A = mg/L reading from instrument

B = mL (g) sample amount from table

C = mL analysis volume from table

COPPER, Bicinchoninate (Method 8026)

Expected Cu conc. (mg/kg)	Sample amount (g)	Analysis volume (mL)	Dilute to
40–2300	0.250	20.0	25.0 mL
100–6000	0.200	10.0	25.0 mL
320–15000	0.150	5.00	25.0 mL
2000–110000	0.100	1.00	25.0 mL

$$\text{Total mg/kg Cu} = \frac{A \times 2500}{B \times C}$$

A = mg/L reading from instrument

B = mL (g) sample amount from table

C = mL analysis volume from table

IRON, 1,10 Phenanthroline (Method 8008)*

Expected Fe conc. (mg/kg)	Sample amount (g)	Analysis volume (mL)	Dilute to
25–1400	0.250	20.0	25.0 mL
60–3500	0.200	10.0	25.0 mL
160–9300	0.150	5.00	25.0 mL
1200–70000	0.100	1.00	25.0 mL

* Total Iron

$$\text{Total mg/kg Fe} = \frac{A \times 2500}{B \times C}$$

A = mg/L reading from instrument

B = mL (g) sample amount from table

C = mL analysis volume from table

LEAD, Dithizone (Method 8033)

Expected Pb conc. (mg/kg)	Sample amount (g)	Analysis volume (mL)	Dilute to
15–1400	0.250	20.0	250.0 mL
38–3500	0.200	10.0	250.0 mL
100–9500	0.150	5.00	250.0 mL
750–70000	0.100	1.00	250.0 mL

$$\text{Total mg/kg Pb} = \frac{A \times 25}{B \times C}$$

A = μg/L reading from instrument

B = mL (g) sample amount from table

C = mL analysis volume from table

MANGANESE, PAN (Method 8149)

Expected Mn conc. (mg/kg)	Sample amount (g)	Analysis volume (mL)	Dilute to
7–370	0.250	20.0	25.0 mL
20–900	0.200	10.0	25.0 mL
50–2500	0.150	5.00	25.0 mL
400–19000	0.100	1.00	25.0 mL

$$\text{Total mg/kg Mn} = \frac{A \times 2500}{B \times C}$$

A = mg/L reading from instrument

B = mL (g) sample amount from table

C = mL analysis volume from table

NICKEL, PAN (Method 8150)

Expected Ni conc. (mg/kg)	Sample amount (g)	Analysis volume (mL)	Dilute to
7–470	0.250	20.0	25.0 mL
20–1200	0.200	10.0	25.0 mL
50–3100	0.150	5.00	25.0 mL
350–23000	0.100	1.00	25.0 mL

$$\text{Total mg/kg Ni} = \frac{A \times 2500}{B \times C}$$

A = mg/L reading from instrument

B = mL (g) sample amount from table

C = mL analysis volume from table

NITROGEN, Total Kjeldahl (Method 8075)

Expected Nitrogen conc. (mg/kg)	Sample amount (g)	Analysis volume (mL)	Dilute to
85–5600	0.25	20.0	25.0 mL*
210–14000	0.20	10.0	25.0 mL*
2100–140000	0.10	5.00	25.0 mL*

* These are guidelines only. See the spectrophotometric procedure manual.

$$\text{Total mg/kg TKN} = \frac{A \times 75}{B \times C}$$

A = mg/L reading from instrument

B = mL (g) sample amount from table

C = mL analysis volume from table

PHOSPHORUS, Ascorbic Acid (Method 8048)

Expected PO_4 conc. (mg/kg)	Sample amount (g)	Analysis volume (mL)	Dilute to
20–900	0.250	20.0	25.0 mL
50–2300	0.200	10.0	25.0 mL
130–6200	0.150	5.00	25.0 mL
1000–45000	0.100	1.00	25.0 mL

$$\text{Total mg/kg phosphate} = \frac{A \times 2500}{B \times C}$$

A = mg/L reading from instrument

B = mL (g) sample amount from table

C = mL analysis volume from table

SILVER (Method 8120)

Expected Ag conc. (mg/kg)	Sample amount (g)	Analysis volume (mL)	Dilute to
12–650	0.250	20.0	25.0 mL
30–1600	0.200	10.0	25.0 mL
80–4500	0.150	5.00	25.0 mL
620–32000	0.100	1.00	25.0 mL

$$\text{Total mg/kg Ag} = \frac{A \times 5000}{B \times C}$$

A = mg/L reading from instrument

B = mL (g) sample amount from table

C = mL analysis volume from table

ZINC (Method 8009)

Expected Zn conc. (mg/kg)	Sample amount (g)	Analysis volume (mL)	Dilute to
30–2300	0.250	20.0	25.0 mL
80–5800	0.200	10.0	25.0 mL
200–15000	0.150	5.00	25.0 mL
1600–115000	0.100	1.00	25.0 mL

$$\text{Total mg/kg Zn} = \frac{A \times 5000}{B \times C}$$

A = mg/L reading from instrument

B = mL (g) sample amount from table

C = mL analysis volume from table

DRY BASIS WEIGHT

The drying a solid can lead to weight loss attributed to volatilization of organic matter, volatilization of gases by heat-induced chemical decomposition, evaporation of water of crystallization, and loss of water enclosed within pores of solids (mechanically occluded water). Residues dried at 103 to 105° C may retain some water of crystallization and mechanically occluded water. Samples can also gain weight by oxidation processes during the drying process.

Drying a solid sample depends on the temperature and time of heating. Depending on the sample, attainment of a constant sample weight can be a very slow process. For consistency, a sample is heated in the oven for two hours at a temperature of 103 to 105° C. The procedure can provide the amount or the percentage of moisture lost during the drying procedure.

Some solids such as plants, leaves, sludge and other materials require a drying process before analysis. Some of these samples must be dried to a constant weight before analysis. The experiment will discuss the drying procedure before analysis.

In some cases, a sample may require additional drying, cooling and desiccating until a constant weight is obtained. This process should be repeated until the weight loss is less than 4% of the previous weight or 0.5 milligram difference, whichever is less.

DRY BASIS WEIGHT

1. Weigh an aluminum dish and record the weight. Record as "A."

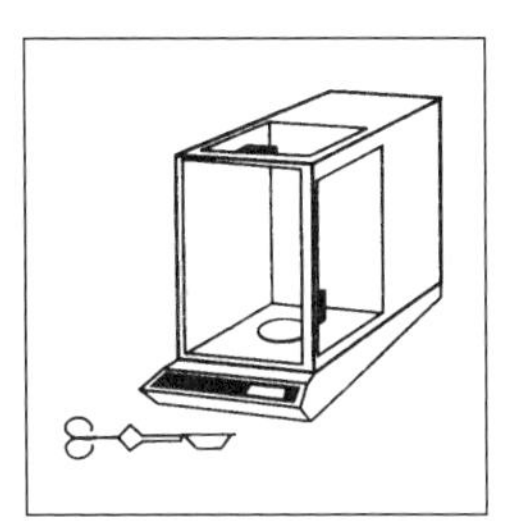

2. Weigh out approximately 2 g of solid sample onto the dish. Record the exact weight as "B."

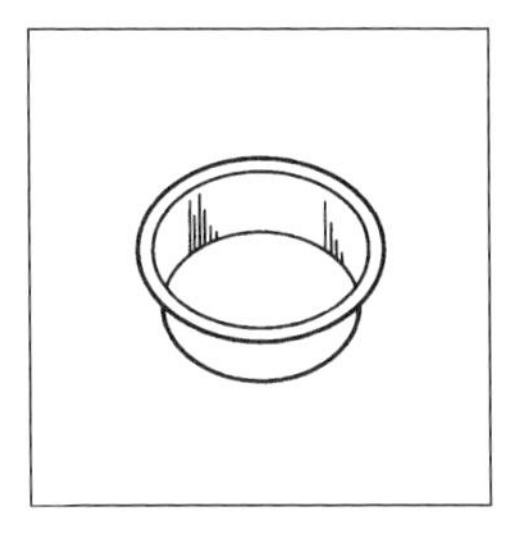

3. Place the dish in the oven (103 to 105° C) for two hours.

4. Cool to room temperature by placing in a desiccator.

5. Weigh the aluminum dish with oven dried sample. Record as "C."
   Note: The oven dried material generally is unsuitable for additional testing and should be discarded.

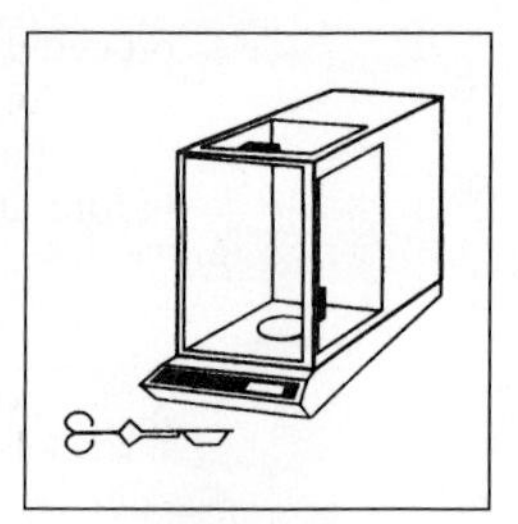

6. Use the following formula in calculating your sample on a "dry basis." Test result (dry basis) $= \dfrac{C - A}{B - A}$
   Note: Multiply the test result on an "as is" basis, by the factor above, to report as "dry basis."

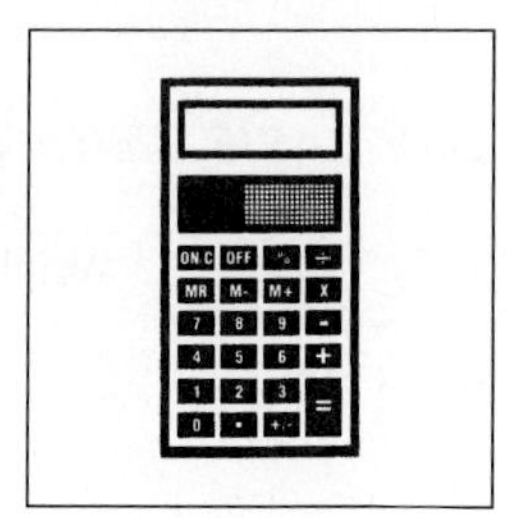

QUESTIONS AND ANSWERS

1. **Identify the problems with the process of drying materials?**
   Water of crystallization and mechanically occluded water can remain within the material particles. Loss of carbon dioxide through decomposition processes and loss of organic matter by volatilization can impact sample weights. In specific cases, a sample can gain weight due to the oxidation process.
2. **What types of samples require drying before analyses?**
   Wet soil-type samples, mulch, and other solids may require air drying before analyses. Samples of plants, leaves and other moisture-containing living materials will require drying before analyses. Wet sludge will require filtration and appropriate amount of drying to a constant weight before analyses. Generally, the procedure will provide sample preparation directions.
3. **When should an aluminum dish not be used for weighing materials?**
   Most solid samples can be dried and weighed in an aluminum dish. However, substances that react with aluminum by oxidation-reduction processes under heated conditions should be dried in a porcelain dish or on a glass evaporating dish.
4. **Why should a dried sample be allowed to cool in a desiccator before weighing?**
   Due to air buoyancy, a warm sample and container will weigh less than a thoroughly cooled sample and container.

REFERENCES

1. *HACH Water Analysis Handbook,* 3rd Ed. (Loveland, CO.: HACH Company, 1997).
2. *Standard Methods for the Examination of Water and Wastewater:* SOLIDS. Section 209, 16th Ed., A. E. Greenberg, R. R. Trussell, L. S. Clesceri and Mary Ann H. Franson (eds.), pp. 92–94, Washington, DC, American Public Health Association, 1985.

DISCUSSIONS AND PROCEDURES

ALUMINUM

Aluminum is an abundant element occurring in rocks, minerals and clay. Colloids and both insoluble and soluble aluminum compounds are found in soil, water, mud and other materials. A few aluminum compounds are decomposed by water. Aluminum metal generally increases solubility in the environment by changes in pH (above pH of 7) and excessive electrical charge. Changes in acidic and basic characteristics of the environment can increase or decrease the solubility of aluminum compounds.

Aluminum metal usually forms a hard exterior surface and has extensive uses in our society. Aluminum compounds have a number of practical uses in society such as the following: ceramics, catalyst, organic synthesis, defoamer, hydrophobic agent, drying agent, reducing agent, metallurgy, water repellant, lubricant, paint and ink thickener, organic reducing agent, ceramics, electroless coatings, dyeing, water purification, paints, stabilizer and many additional applications.

Some of the industries utilize closed loop systems or partial closed loop systems to recover aluminum wastes. Aluminum wastes are manifested for burial purposes and for recovery and reuse. Industries and metal recycling sites collect and recycle aluminum metal.

A known volume of acid is added to a solid or sludge sample weighing 0.100 to 0.500 grams in a digestion flask. The digestion procedure for oils requires 0.100 to 0.250 grams of material diluted to a specific volume. The sample is boiled for 4 to 5 minutes and the sample is allowed to cool. Then, 10 mL of 50% hydrogen peroxide is added to the charred sample and the sample is boiled for 1 minute. The digested sample is allowed to cool to room temperature and is diluted to 70 mL with deionized water. The sample is heated an additional 15 minutes in a water bath. The digested sample is cooled to room temperature, and the sample is diluted to 100 mL with deionized water. After selecting the appropriate sample size and diluting the solution to 50.0 mLs, the clear sample can be analyzed by the Aluminon (HACH Method 8012) procedure provided in Chapter 2. A pH adjustment for the sample will be required before analysis for aluminum.

The Aluminon method of determination has some interferences which can be eliminated by additional treatment of the solution. Also, fluoride interferes at all levels by complexing with aluminum. Aluminum can be determined accurately by using the Fluoride Interference Graph when the fluoride concentration is known in the sample.

The Aluminon method for determination of aluminum can be measured in the field by the DR/820, DR/850, and the DR/890 Colorimeters and the DR/2010 Spectrophotometer. Additional methods to determine aluminum by both colorimetric and atomic absorption methods are available in the literature.

QUESTIONS AND ANSWERS

1. After digestion, what is the primary interference with the aluminum test?
Fluoride interferes at all levels and accurate aluminum results depend on the interpretation of the Fluoride Interference Graph, which can be found in the HACH Water Analysis Handbook.

2. Are aluminum compounds dangerous to handle?
Many aluminum compounds are relatively safe to handle. However, some of the compounds in air evolve dangerous gases and burn easily, while other compounds are cor-

rosive, toxic, explosive, dangerous when wet, strong irritants, and are inhalation problems.

3. Does iron interfere with the aluminum test?
   Ascorbic acid is added to remove iron interference.
4. What will happen if AluVer 3 Aluminum Reagent powder does not totally dissolve during the test for aluminum?
   Inconsistent results will be obtained if any powder is undissolved.

REFERENCES

1. *Standard Methods for the Examination of Water and Wastewater:* ALUMINUM, 16th ed., A. E. Greenberg, R. R. Trussell, L. S. Clesceri and Mary Ann H. Franson (eds.), pp. 182 and 959–961, Washington, DC, American Public Health Association, 1985.
2. *HACH Water Analysis Handbook,* 3rd ed. (Loveland, CO.: HACH Company, 1997).
3. *Hawley's Condensed Chemical Dictionary,* 11th ed., rev'd by N. Irving Sax and Richard J. Lewis (New York, N.Y.: Van Nostrand Reinhold, 1987).

AMMONIA NITROGEN TEST IN SOILS

Nitrogen (N) deficiency in soil is evident by observing the characteristics of plants. Plant growth is slow and the plant will appear stunted and spindly in appearance.* The plant will mature early without the production of significant quantities of fruit or grain. The plant and leaves will appear light green in color and older leaves will develop a V-shaped yellowing. Eventually, the leaf will turn brown and die.

With excess Nitrogen (N), the plant will be dark bluish green and new growth will be evident. Plants can be stressed by disease, insect infestation, and drought. Blossom abortion and lack of fruit set may be observed.

Plants fertilized with too much ammonium-nitrogen may exhibit carbohydrate depletion and reduced plant growth. Cation imbalance and low potassium and magnesium are often the result of too much ammonium nitrogen. Lesions may be observed on plant stems and decay of the conductive tissue at the base of the stem may occur. Blossom-end rot of fruit will be evident. Consequently, grain and fruit yields will decrease.

A fertile soil may be expected to give a low ammonia nitrogen test reading, unless there has been a recent application of nitrogenous fertilizer in forms other than the nitrate.** After fertilizer application, ammonia rapidly disappears by being transformed by the nitrification process to the more available nitrate compounds. In some cases, volatilization of ammonia can occur near the surface of the soil. In forest soils, ammonia is the most abundant available form of nitrogen.** The humus layers of a forest soil can produce very high concentrations of ammonia nitrogen.

The Nesslerization Method tests samples for ammonia nitrogen in the range of 0 to 150 pounds per acre. Ammonia forms a colored complex with Nessler's Reagent in proportion to the amount of ammonia present in the sample. Rochelle salt is added to prevent precipitation of calcium or magnesium in undistilled samples. Sample interferences during the test

*Robert D. Munson, *Handbook of Reference Methods for Plant Analysis,* Yash P. Kalra (ed.) (Boca Raton, Fla.: CRC Press, 1998).

**Ammonia Nitrogen Test, *Instruction Manual,* (Chestertown, Md.: LaMotte Company, 1997), p. 11.

can be sample turbidity and color. Turbidity may be removed by filtration procedure. Color interferences may be eliminated by adjusting the instrument to 100%T with a sample blank.

AMMONIA NITROGEN TEST

Nesslerization Method*—Procedure

1. Use the 1 mL pipet to transfer 2 mL of soil filtrate into a clean colorimeter tube and dilute to the 10 mL mark with deionized water.

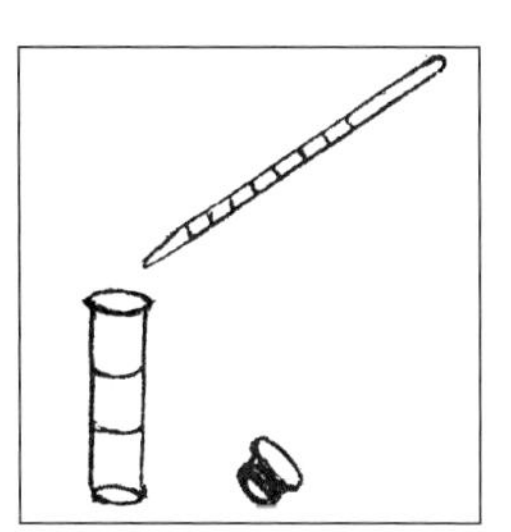

2. Mix and neutralize by adding the 15% solution Sodium Hydroxide to the soil filtrate, one drop at a time while stirring with the plastic rod. The stirring rod is touched to the Bromthymol Blue test paper after the addition of each drop of 15% Sodium Hydroxide until the color changes from yellow to green or blue.

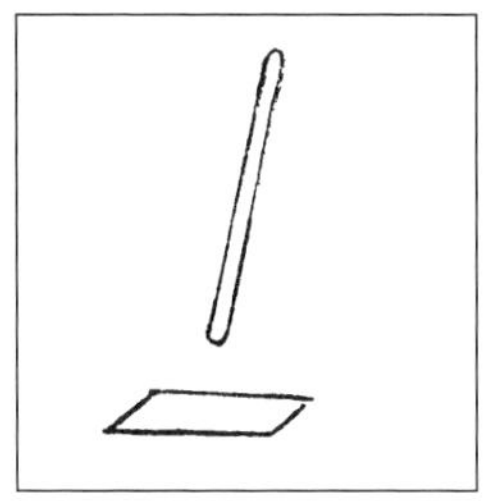

3. Select setting 1 on the "Select Wavelength" knob.

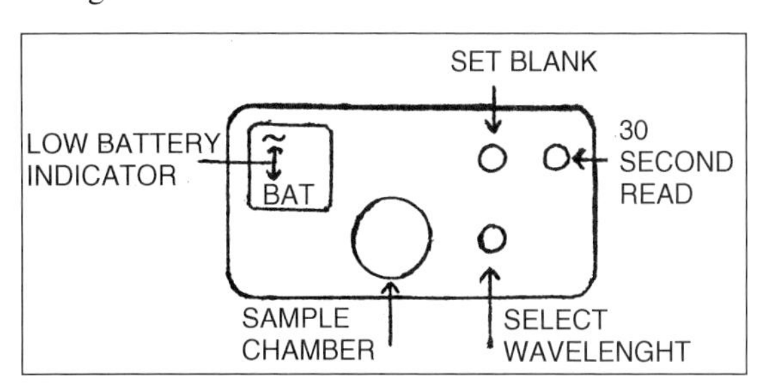

4. Insert sample into chamber and press the "30 Second Read" button. Adjust "100"%T with the "Set Blank" knob.

*Compliments of LaMotte Company.

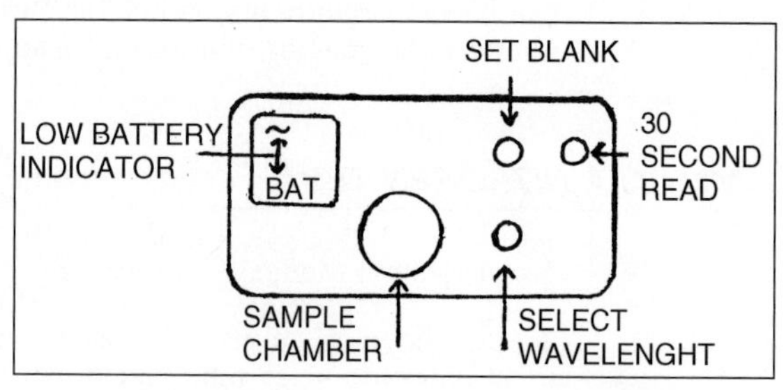

5. Remove sample tube and add 12 drops of Ammonia Nitrogen Reagent #1, cap and mix.

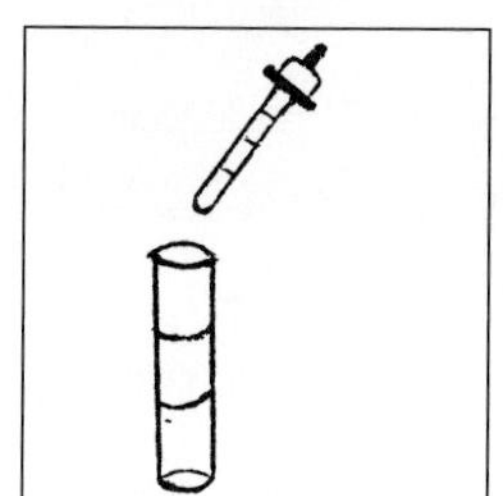

6. With the 1.0 mL pipet, add one measure of Ammonia Nitrogen Reagent #2 to the tube, cap and mix.

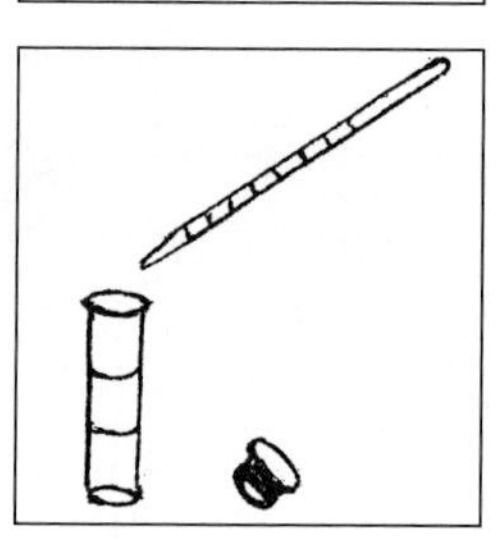

7. Allow 5 minutes for maximum color development.

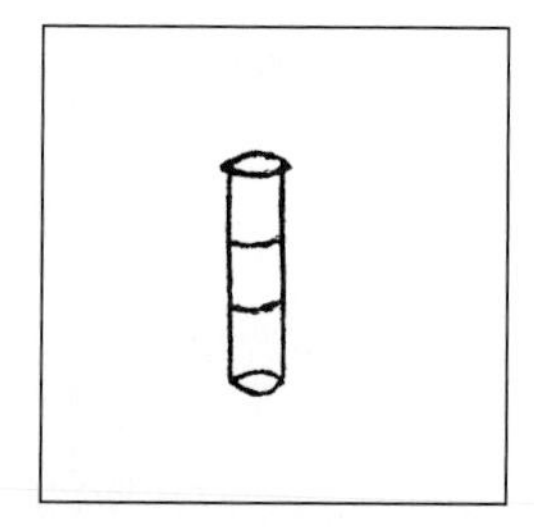

8. At the end of a 5 minute waiting period, insert sample into colorimeter, press "30 Second Read" button and measure %T as soon as reading stabilizes.

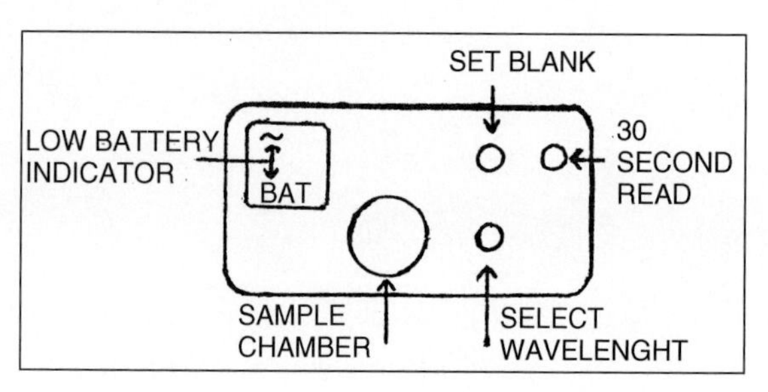

9. Consult the charts below to determine the concentration of Ammonia Nitrogen in pounds per acre.

Ammonia Nitrogen Calibration Chart (Lb/Acre)

%T	9	8	7	6	5	4	3	2	1	0
80			0.0	0.5	1.5	2.0	3.0	3.5		
70	4.5	5.5	6.0	7.0	8.0	8.5	9.5	10.5	11.5	12.0
60	13.0	14.0	15.0	16.0	17.0	18.0	19.0	20.0	21.0	22.0
50	23.0	24.0	25.5	26.5	27.5	28.5	30.0	31.0	32.5	33.5
40	35.0	36.0	37.5	39.0	40.5	42.0	43.0	44.5	46.5	48.0
30	49.5	51.0	53.0	54.5	56.5	58.0	60.0	62.0	64.0	66.0
20	68.5	70.5	73.0	75.5	78.0	80.5	83.5	86.0	89.0	92.0
10	95.5	99.0	102.5	106.5	110.5	115.0	120.0	125.0	130.5	137.0
0	143.5	151.0								

Ammonia Nitrogen Concentration Chart

%T	Range	Pounds per acre
58–100%	Low	0–24 lbs/acre
29–57%	Medium	25–68 lbs/acre
0–28%	High	Over 71 lbs/acre

HAZARDOUS SUBSTANCES

Read the MSDS before using the chemicals. The following substance is a hazardous substance and should be disposed of and handled properly. The hazardous substance is Ammonia Nitrogen Reagent #2.

RANGE OF MEASUREMENT

The range of measurement is 0 to 150 lbs/acre.

QUESTIONS AND ANSWERS

1. What forms of nitrogen are in soil?
Organic nitrogen from special fertilizers and decay products is present under a variety of conditions. Ammonia and ammonium ions are available as fertilizer additives. High concentrations of ammonia can be available in forest areas and in some forms of decaying matter. Ammonia is rapidly oxidized to nitrate by the nitrification process. Plants are generally exposed to varying proportions of both inorganic N-forms of ammonia and

nitrates. A number of nitrate compounds are converted to nitrites by a variety of reduction processes.

2. What is Rochelle salt?
Rochelle salt is a tartrate of sodium or potassium.
3. How does too much ammonium-nitrogen fertilizer affect plants?
High ammonium-nitrogen concentrations can deplete carbohydrates and reduce plant growth. Lesions on plant stems, decay of the conductive tissue at the base of the stem, and blossom-end rot of fruit will probably be evident. Grain and fruit yields will be low.

REFERENCES

1. *Handbook of Reference Methods for Plant Analysis,* Yash P. Kalra (ed.), (Boca Raton, Fla.: CRC Press, 1998).
2. *LaMotte Soil Handbook* (Chestertown, Md.: LaMotte Company, 1994).

CADMIUM

Cadmium dust is highly toxic and is a known carcinogen (OSHA). Cadmium can destroy the tissue lining the lungs and cause bone disease and kidney damage. Plating of food and beverage containers with cadmium has resulted in a number of cases of food poisoning. Some cadmium compounds are sparingly soluble in water. A number of cadmium compounds are increasingly soluble in water as the temperature of the water increases above 0°C.

Cadmium may enter water through the deterioration of galvanized pipes. Even though cadmium is fairly stable as a metal, cadmium metal exposed to acid, bases, and electrical current has a tendency to dissolve or deteriorate over a period of time. Cadmium metal has practical applications in metal coatings, alloys, batteries, transmission wire, enamels, pigments and glazes, fungicides, photoelectric cells, photography, and several electrical devices. As sparingly soluble or insoluble compounds, cadmium can be found in reagent chemicals, accelerator for butyl rubber, cadmium plating, cadmium salts, electrodes, electronic and optical applications, plating baths, glazes, phosphors, nematocides, pigments, ceramics, semiconductors, polymer stabilizers, fungicides, optical applications, lubricants, pyrotechnics, solar cells, and several additional applications.

Industries manufacturing cadmium and cadmium compounds must protect the water bodies by eliminating deposition of the metal and compounds in the environment. A number of industries have implemented closed loop systems to recover and reuse waste, while other facilities manifest waste for landfilling purposes. Some cadmium materials are sold to other companies to produce additional items. A number of materials containing cadmium are discarded in the garbage and at inappropriate sites.

A known volume of acid is added to a solid or sludge sample weighing 0.100 to 0.500 grams in a digestion flask. The digestion procedure for oils requires 0.100 to 0.250 grams of material diluted to a specific volume. The sample is boiled for 4 to 5 minutes and the sample is allowed to cool. Then, 10 mL of 50% hydrogen peroxide is added to the charred sample and the sample is boiled for 1 minute. The digested sample is allowed to cool to room temperature and is diluted to 100 mL with deionized water. After selecting the appropriate sample size and diluting the solution to 250.0 mL, the clear sample can be analyzed by the Dithizone (HACH Method 8017) procedure provided in Chapter 2. A pH adjustment for the sample will be required before analysis for cadmium.

Some of the chemicals used in the determination of cadmium should be stored away from light and heat. A treatment procedure can eliminate the interferences caused by the presence of copper, bismuth, mercury and silver in the experiment. During the analysis, cadmium ions react with dithizone in basic solution to form a pink to red complex that is extracted with chloroform. The cadmium-dithizonate complex in chloroform is measured photometrically at 515 nm.

QUESTIONS AND ANSWERS

1. **What are some features that affect cadmium compound solubility?**
   A number of cadmium compounds can become more soluble by changes in mixture pH, the presence of small electrical currents, the presence of bacterial action, and increases in water temperature above 0°C. Several cadmium compounds are solubilized by ammonia from decomposition processes. The deterioration of metal pipes and containers can add cadmium to water and contents of the metal container.
2. **What are some unique characteristics of cadmium metal?**
   Cadmium metal has several stable isotopes. The metal tarnishes in moist air and has poor corrosion resistance in industrial atmospheres. The metal becomes brittle at 80°C and is a high neutron absorber. The metal lowers the melting point of certain alloys when used in low percentage in the alloy. The metal is also combustible. In powder form, the metal is flammable and a known carcinogen.
3. **Why should the DithiVer powder pillows be stored in a cool and dark place in the laboratory?**
   DithiVer powder pillows tend to decompose slowly in light and heat. Even chloroform solutions of the powder should be stored in an amber glass bottle.
4. **What are some interferences in the experiment?**
   The interferences are copper, mercury and silver. At a higher level, bismuth can interfere.

REFERENCES

1. *Standard Methods for the Examination of Water and Wastewater:* CADMIUM 16th ed., A. E. Greenberg, R. R. Trussell, L. S. Clesceri and Mary Ann H. Franson (eds.), pp. 193–196, Washington, DC, American Public Health Association, 1985.
2. *HACH Water Analysis Handbook,* 3rd ed. (Loveland, CO.: HACH Company, 1997).
3. *Hawley's Condensed Chemical Dictionary,* 11th ed., rev'd by N. Irving Sax and Richard J. Lewis, (New York, N.Y.: Van Nostrand Reinhold, 1987).

CALCIUM AND MAGNESIUM

The Schwarzenbach EDTA titration method provides the parts per million or pounds per acre of both calcium and magnesium in the soil. Additional ground samples of a solid or semisolid can be tested by this method with final results recorded as parts per million. The laboratory process involves two different titrations. The first titration provides calcium and magnesium content and the second titration gives calcium results.

Calcium is a component of cell walls in plants and is known to stimulate root and leaf development as well as activate several enzyme reactions involved in plant metabolism.*

*Calcium and Magnesium Test, *Instruction Manual* (Chestertown, Md.: LaMotte Company, 1998), pp. 12–13.

Calcium influences crop yields by reducing soil acidity and by reducing the toxicity of several other minerals such as manganese, zinc, and aluminum. The amount of total calcium in soils may range from 0.1% to 25%.

During calcium deficiency, leaves on plants look ragged and the margins of emerging leaves tend to stick together.* The growing tips of roots and leaves tend to turn brown and die. The lower stem conductive tissue may display decay.

Older leaves will display yellowing between the veins and some reddish purpling during magnesium deficiency. Sections of the stripped area will die within the chlorotic stripe. Plant growth becomes slow. Disease infestation and blossom-end rot of fruit may become problems.

Calcium and magnesium deficiency can be reversed by adding lime (Dolomite-type) to the soil. Additional types of lime products (agricultural slag, carbon lime) seldom contain significant quantities of magnesium. The addition of lime raises the pH to the proper range for optimum plant growth.

Excess calcium and magnesium can lead to additional problems. Excess calcium can lead to plants exhibiting a magnesium deficiency or a potassium deficiency. Excess magnesium produces a cation imbalance with the plant showing signs of either a calcium and/or potassium deficiency.

A number of reagents in the test procedure are hazardous substances and should be treated accordingly. Sample color and turbidity may interfere with the endpoint. The experiment determines calcium in the range of 0 to 400 lbs/acre and magnesium in the range of 0 to 240 lbs/acre.

CALCIUM AND MAGNESIUM TEST

DILUTION OF SOIL EXTRACT**

Measure 10 mL of the soil extract using the 30 mL graduated cylinder and transfer it to a 50 mL beaker. Add 10 mL of deionized water, mix and neutralize by adding 15% Sodium Hydroxide Solution to the soil filtrate, one drop at a time while stirring with the plastic rod. The stirring rod is touched to the Bromthymol Blue test paper after the addition of each drop of 15% Sodium Hydroxide until the color changes from yellow to green or blue.

*Robert D. Munson, *Handbook of Reference Methods for Plant Analysis,* Yash P. Kalra (ed.) (Boca Raton, Fla.: CRC Press, 1998), pp. 7–8.

**Compliments of LaMotte Company.

*TITRATION A, CALCIUM AND MAGNESIUM**

1. Fill the test tube to the 5 mL line with the soil extract from above. Dilute to the 10 mL line with deionized water.

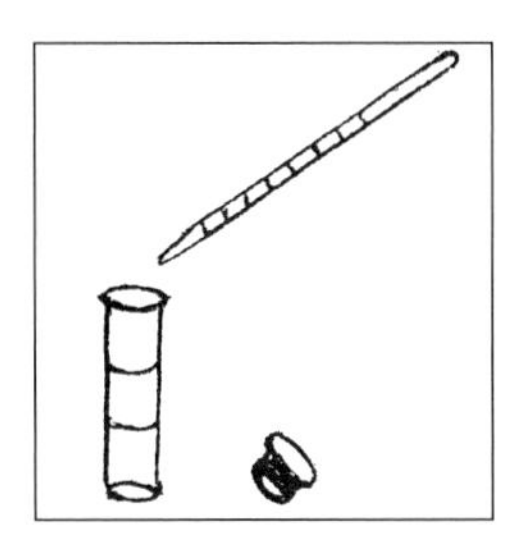

2. Add 5 drops of Calcium Magnesium Inhibitor Reagent.

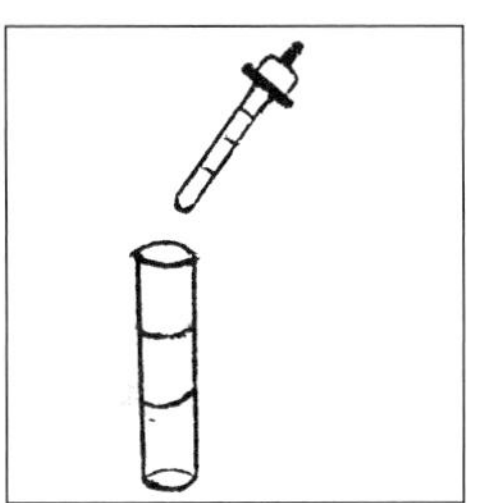

3. Wait 5 minutes.

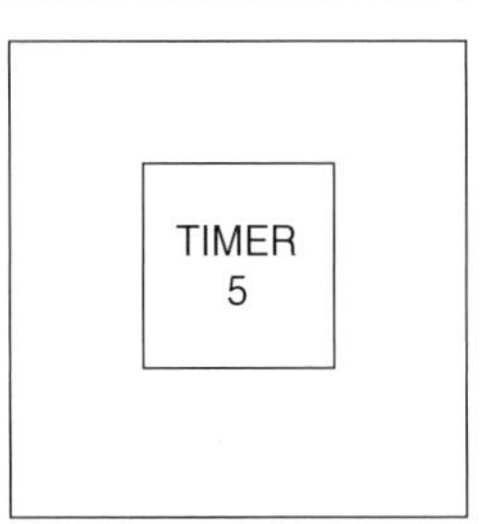

4. Use a transfer pipet to add 5 drops of Calcium and Magnesium Buffer.

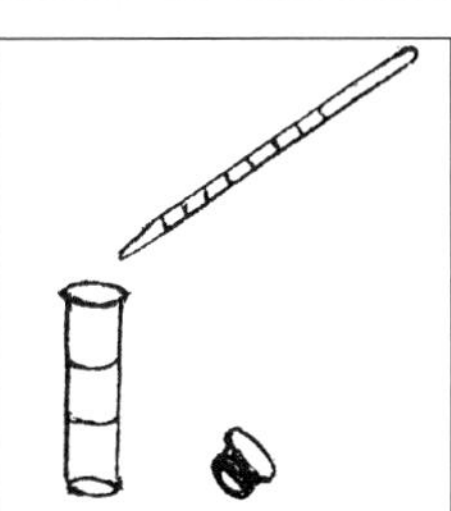

5. Add 10 drops of CM Indicator.

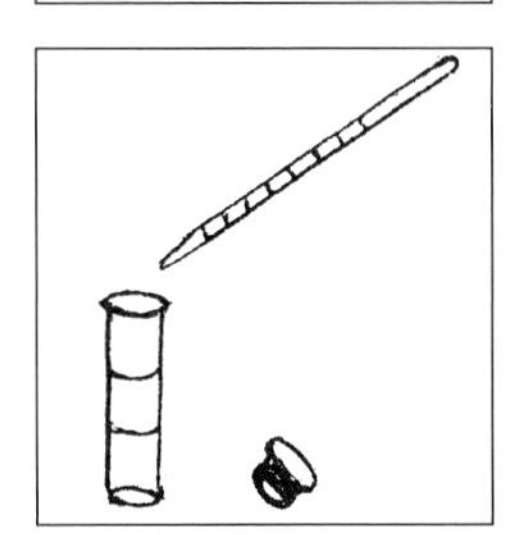

6. Fill the Direct Reading Titrator with the Standard EDTA Reagent. Insert the tip of the Titrator into the center hole of the test tube cap.

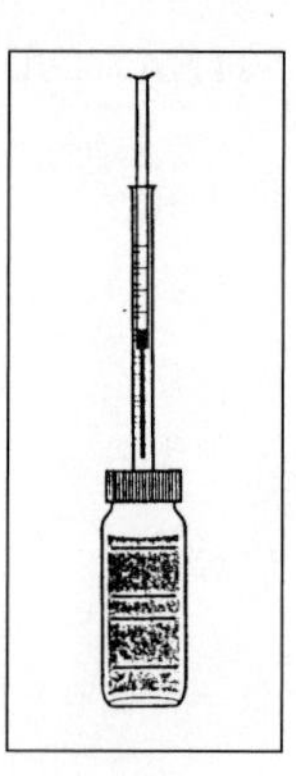

7. While gently swirling the tube, slowly press the plunger to titrate until the color changes from red to blue.

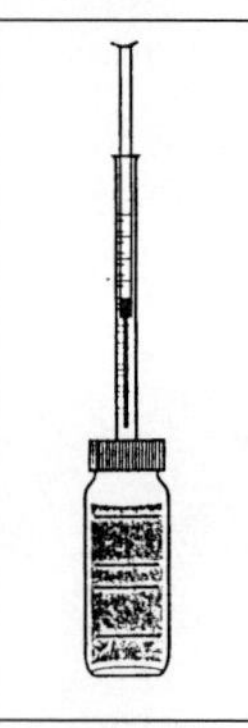

8. Read the Titrator scale at the tip of the plunger and multiply by 5.16. This is Titration Value A.

$$\text{Titration Value A} = \text{Titrator Reading} \times 5.16$$

*TITRATION B, CALCIUM**

1. Fill the test tube to the 5 mL line with the diluted soil extract. Dilute to 10 mL with deionized water.

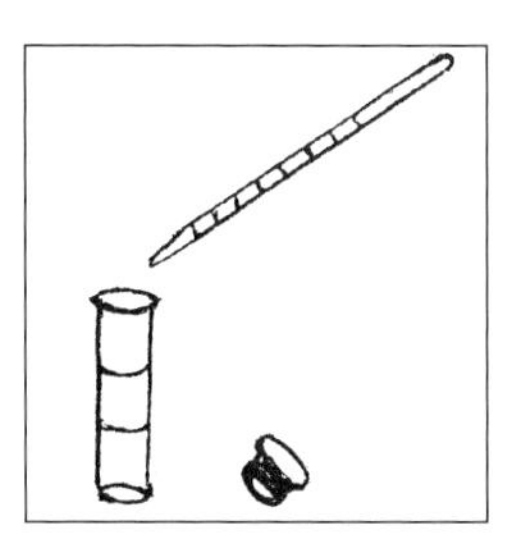

2. Add 2 drops of Inhibitor Solution.

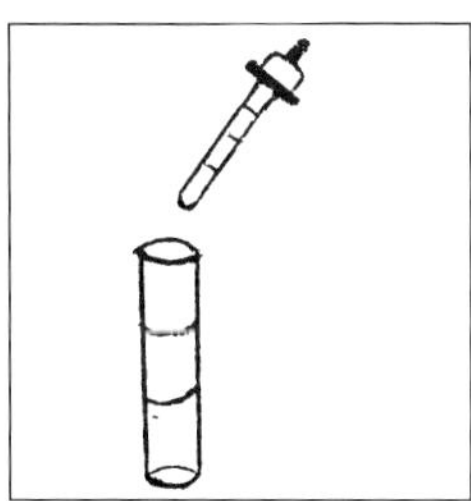

3. Add 2 drops of TEA Reagent.

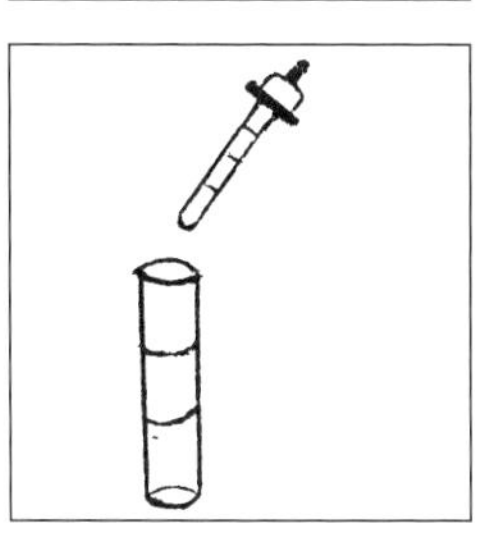

4. Add 8 drops of Sodium Hydroxide w/Metal Inhibitors.

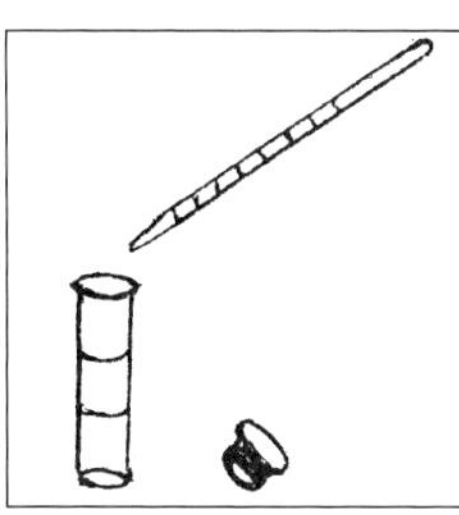

5. Add 1 Calcium-Hardness Indicator Tablet to the test sample. Cap and shake to dissolve the tablet. A red color will develop.

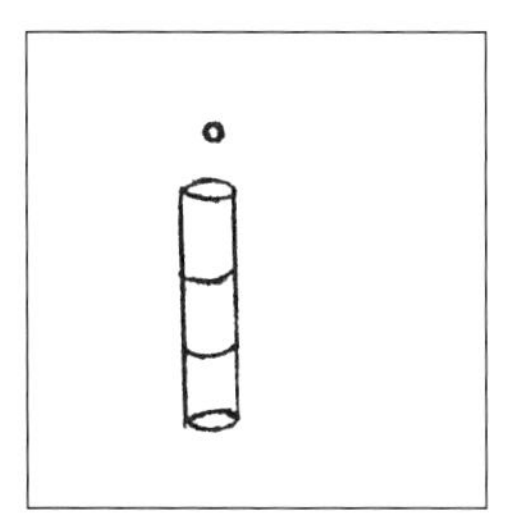

*Compliments of LaMotte Company.

Soluble chloride is classified as a plant micronutrient at levels below 100 mg CL/kg.* Some plants can accumulate a much higher concentration of chloride up to 20,000 mg CL /kg. Fertilizer applications may increase chloride levels.** High chloride levels in soil can poison plants and result in stunted growth. High chloride levels can poison a variety of living species.

A chloride deficiency in soil will produce plants with chlorotic leaves and wilting problems. Disease can infest grain-producing plants when chloride is deficient. With excess chloride, lower leaves tend to prematurely yellow and display burning of leaf margins and tips. Plants can easily wilt with excess chloride concentrations. Leaf abscission is evident in most woody plants.

Chloride concentration can be high in soils along salt water bodies and in some areas of "salt mines." This test provides valuable results on saline soils or when contamination from sea water or spray is evident. Normal soils of humid regions do not normally provide readable results unless the soil has been recently fertilized heavily.

The testing range for chlorides is 0 to 1000 lbs/acre. Potassium dichromate indicates the endpoint of the 0.141 N silver nitrate titration in a neutral or slightly alkaline solution. Interferences in the experiment are bromide, iodide, and cyanide which register as equivalent chloride concentrations.

CHLORIDE TEST

Direct Reading Titrator Method—Procedure†

1. Fill a clean extracting tube to the mark with deionized water.

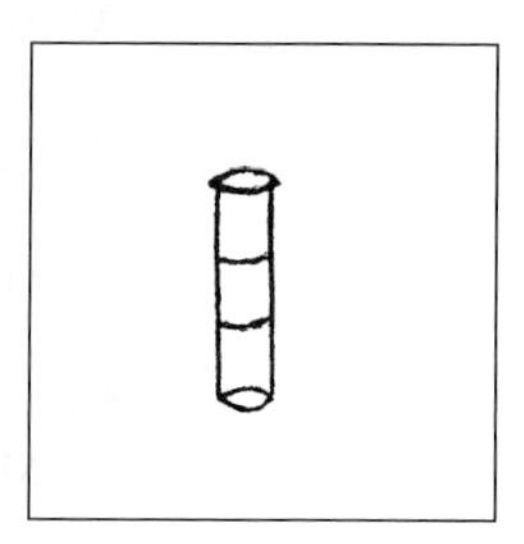

2. Add 3 measures of soil using the 1 g spoon. Cap tube and shake for 5 minutes.

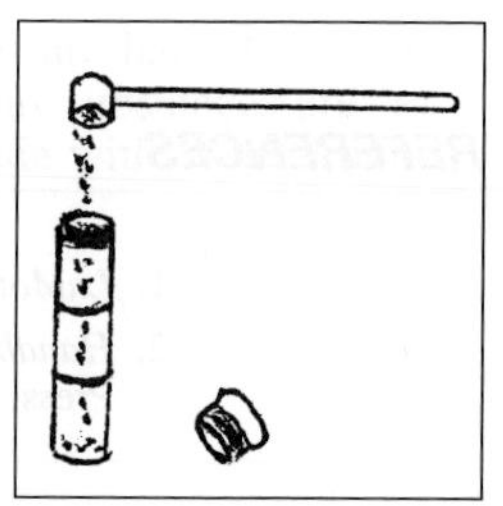

*Liangxue Liu, *Handbook of Reference Methods for Plant Analysis,* Yash P. Kalra (ed.) (Boca Raton, Fla.: CRC Press, 1998), p. 111.

**Chloride Test, *Instruction Manual,* (Chestertown, Md.: LaMotte Company, 1998), p. 14.

†Compliments of LaMotte Company.

3. Filter and collect all of the soil filtrate using the funnel and filter paper. The filtrate does not have to be clear since a slight turbidity does not interfere in the test.

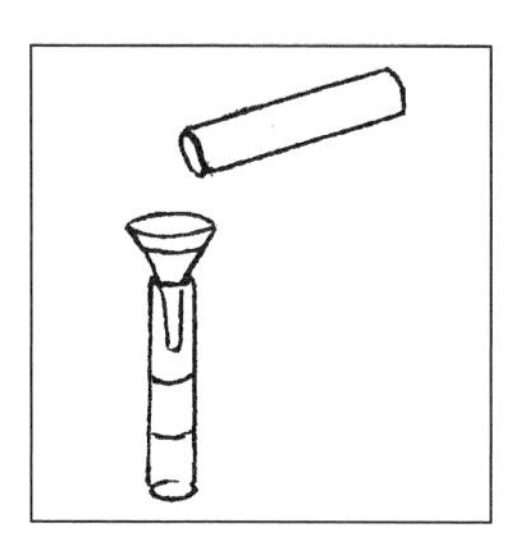

4. Fill the Titration Tube to the 10 mL line with the filtrate.

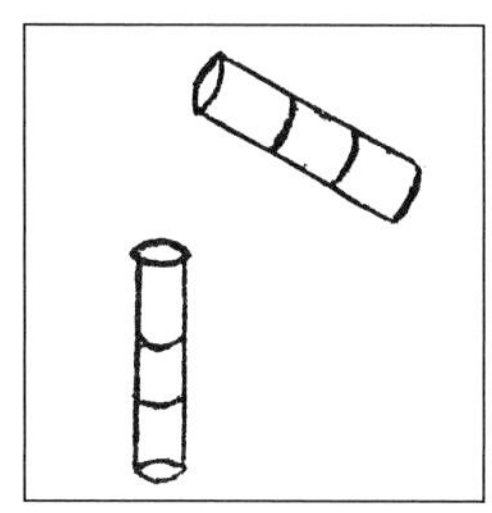

5. Add 3 drops of Chloride Reagent #1 to the sample. Cap and shake to mix. A yellow color will result.

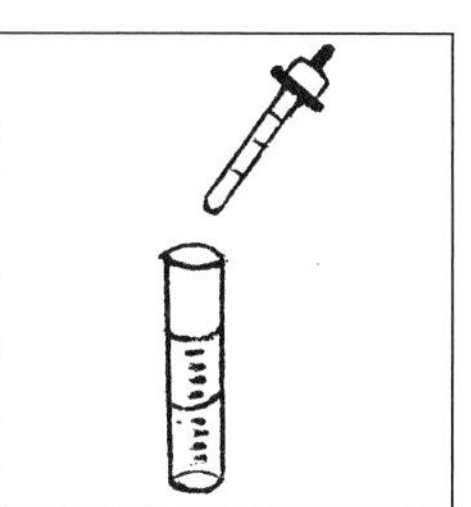

6. Fill the Direct Reading Titrator with 0.141 N Silver Nitrate Solution. Follow the procedure described in the instruction manual.

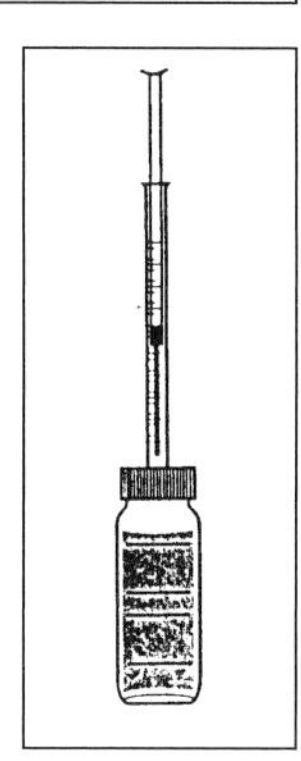

7. Titrate the test sample with 0.141 N Silver Nitrate until the yellow color changes permanently to pink. Record the Titrator reading.
   Note: If the plunger reaches the bottom mark (1000 ppm) on the Titrator scale before the endpoint color change occurs, refill the Titrator and continue the titration procedure. Be sure to include the value of the original amount added (1000 ppm) when recording the final result.
   Note: Each minor division on the Direct Reading Titrator Scale (0384) equals 20 ppm.

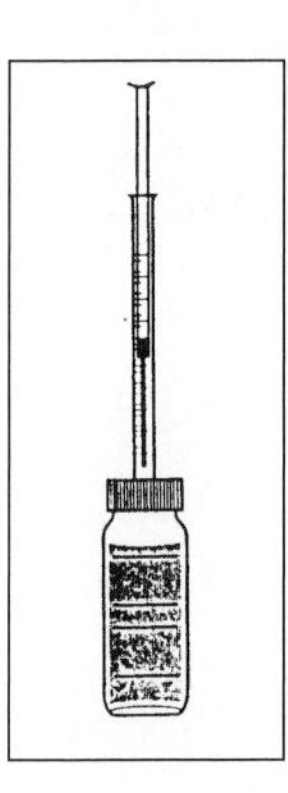

HAZARDOUS SUBSTANCES

Read the MSDS before using the chemicals. The following substances are hazardous substances and should be disposed of and handled properly. The hazardous substances are Chloride Reagent #1 and 0.141 N Silver Nitrate.

QUESTIONS AND ANSWERS

1. How important are chlorides to plants?
   Chlorides are classified as a plant micronutrient at levels below 100 mg CL/kg. Some plants can accumulate a much higher concentration of chloride in the range of 2000 to 20,000 mg CL/kg.
2. Where are high chloride levels found?
   High chloride levels are found in salt water, salt beds, heavily fertilized areas, industrial effluent, human waste effluent, and polluted streams passing through populated areas.
3. What are the major interferences in the experiment?
   Interferences are bromides, iodides and cyanides. All are soluble and react in the same way as chlorides, especially with silver nitrate.
4. If the Direct Reading Titrator uses 12.5 minor divisions on the Titrator scale to titrate a sample with silver nitrate, what is the concentration of chloride in the sample in parts per million (ppm)?
   Each minor division on the Direct Reading Titrator (0384) equals 20 ppm. Therefore, the chloride concentration in the sample is 250 ppm or (12.5×20 ppm).

REFERENCES

1. *LaMotte Soil Handbook* (Chestertown, Md.: LaMotte Company, 1994).
2. *Handbook of Reference Methods for Plant Analysis,* Yash P. Kalra (ed.), (Boca Raton, Fla.: CRC Press, 1998).

CHROMIUM

Chromium compounds have varying degrees of toxicity. A number of specific chromium compounds are strong oxidizing agents, dangerous chemicals, and toxic by ingestion and inhalation. Hexavalent chromium compounds are carcinogenic (OSHA) and corrosive on tissue. Chromium in lower oxidation states tend to be less of a problem to humans and animals.

Partially soluble and insoluble chromium compounds have several industrial and practical uses including the following: pigments, steel-making, plating, catalysts, waterproofing, textile mordant, mothproofing, printing, dyeing, asphalt roofing, abrasives, ceramics, electrical conductors, coatings, gasoline additive, and additional applications. A few of the compounds used in these applications are stable at high temperatures, but tend to dissolve under slightly acidic or basic conditions. Additional insoluble chromium compounds decompose in air and are sensitive to light. Increases in the water solubility of chromium compounds found in soils, gels, sludge, slag, and other solid materials will increase the chance of exposure of humans and animals to toxics.

Some industries have implemented loop systems to address the recovery and reuse issues of insoluble chromium compounds. Additional industries are manifesting the chromium waste stream for burial purposes. Slag and sludge from metal manufacturing processes often contain significant amounts of toxic forms of chromium. Slag and sludge are sometimes deposited near water resources such as springs and flowing streams. A significant amount of the chromium waste generated in small industries and in society has been poorly characterized and improperly treated.

A known volume of acid is added to a solid, gel or semisolid sample weighing 0.100 to 0.500 grams in a digestion flask. The digestion procedure for oils requires 0.10 to 0.25 grams of material diluted to a specific volume. The sample is boiled for 4 to 5 minutes and the sample is allowed to cool. Then, 10 mL of 50% hydrogen peroxide is added to the charred sample and the sample is boiled for 1 minute. The digested sample is allowed to cool to room temperature and is diluted to 100 mL with deionized water. After selecting the appropriate sample size and diluting the solution to 25 mL, the clear sample can be analyzed by the alkaline hypobromite oxidation (HACH Method 8024) procedure provided in Chapter 2. A pH adjustment for the sample will be required before analysis for chromium.

In order to determine total chromium in a sample, all chromium must be converted to the hexavalent state by treatment with an alkaline hypobromite oxidation method or by digestion of the sample with a sulfuric-nitric acid mixture followed by oxidation with potassium permanganate. The total chromium content can be determined colorimetrically by reaction of chromium with a 1,5-diphenylcarbohydrazide complex in an acidic buffer.

QUESTIONS AND ANSWERS

1. What form does chromium usually assume in solids, sludge and ash?
   Chromium (+2) has a tendency to oxidize to Cr (+3) in air and in solution. Cr (+6) has a tendency to be reduced to Cr (+3). However, chromium metal (0), chromium (+2, +3), and chromium (+6) are found in solids that have not been exposed to air and other materials. Sometimes the metal is bound in the structure of the material and not readily available for oxidation or reduction reactions.
2. What happens to a number of the chromous (+2 oxidation state) compounds in a bottle?
   Some chromous compounds polymerize on exposure to light, whereas other compounds decompose or oxidize in moist air. Some chromous compounds fume in air.

3. How dangerous are chromium compounds?
Chromium dust and fumes are known carcinogens. Chromium metal in stainless steels and plastics and as a protective coating is safe and has many applications. However, a number of chromium compounds are toxic by inhalation and ingestion and are irritants to skin and tissue. Additional compounds are radioactive and are potential explosives.
4. Why is oxidation required to determine total chromium?
Oxidation of chromium metal (Cr) and chromium ions (+2, +3) to Chromium (+6) and subsequent reaction with 1,5-diphenylcarbohydrazide in acidic solution allows the formation of the purple compound, which can be determined spectrophotometrically.

REFERENCES

1. *Standard Methods for the Examination of Water and Wastewater:* CHROMIUM. 16th ed., Method 312B, A. E. Greenberg, R. R. Trussell, L. S. Clesceri and Mary Ann H. Franson (eds.), pp. 201- 202, Washington, DC, American Public Health Association, 1985.
2. *HACH Water Analysis Handbook,* 3rd ed. (Loveland, CO.: HACH Company, 1997).
3. *Hawley's Condensed Chemical Dictionary,* 11th ed., rev'd by N. Irving Sax and Richard J. Lewis (New York, N.Y.: Van Nostrand Reinhold, 1987).

COBALT

Cobalt compounds and the metal have many practical uses. Cobalt metal corrodes readily in air. Cobalt is an important trace element in soil and is necessary for animal nutrition. Cobalt compounds have variable toxicity levels and a few compounds decompose or melt below hot water temperatures. Partially soluble and insoluble compounds at room temperatures have been used in pigments, cosmetics, whitener, paint and varnish drier, catalyst, ceramics, coloring enamels and glass, glazing pottery, oxygen stripping agent, temperature indicator, semiconductors, mineral supplement, foam stabilizer, antiknock agents, fertilizer additive, manufacture of vitamin B_{12}, storage batteries, hair dyes, hygrometers, nonionic surfactants (CTAS), bonding rubber to metals, electronic research, medicine and many additional applications.

Cobalt consists of three useful radioactive substances with half lives of 72 days to 5.3 years. Radioactive cobalt has applications in biological and medical research, radiation therapy for cancer, radiographic testing of welds and castings, gas discharge tubes, portable radiation units, gamma radiation for wheat and potatoes, quality of marketable products, wool dyeing, oil consumption in internal combustion engines, and for locating buried telephone and electrical conduits.

Cobalt easily forms coordination complexes with ammonia, water, and several other species. These colored compounds display almost none of the ordinary properties of cobalt. Also, cobalt forms a blue to green pigment named "cobalt blue" or "Thenard's blue." The cobalt blue pigment is a very durable blue pigment resistant to both chemicals and weathering. This pigment is used in oil and water and as a cosmetic for eye shadows and grease paints.

Some of the cobalt waste is recovered and recycled to a variety of sources for use in many applications. Some facilities manifest the waste for either recovery or burial purposes, while other wastes are poured down the drain. In most cases, cobalt wastes are not very dangerous to humans and animals.

A known volume of acid is added to a solid, gel or semisolid sample weighing 0.100 to 0.500 grams in a digestion flask. The digestion procedure for oils requires 0.100 to 0.250 grams of material diluted to a specific volume. The sample is boiled for 4 to 5 minutes and the sample is allowed to cool. Then, 10 mL of 50% hydrogen peroxide is added to the charred sample and the sample is boiled for 1 minute. The digested sample is allowed to cool to room temperature and is diluted to 100 mL with deionized water. After selecting the appropriate sample size and diluting the solution to 25 mL, the clear sample can be analyzed by the PAN (HACH Method 8078) procedure provided in Chapter 2. A pH adjustment for the sample may be required before analysis for cobalt by the PAN method.

A number of cations and anions interfere with the 1-(2-pyridylazo)-2-naphthol (PAN) test for cobalt. Cations of iron, copper, cadmium, chromium, manganese, lead, zinc, and aluminum, and the fluoride anion at concentrations below 40 mg/L (ppm) interfere with cobalt. After color development, EDTA is added to destroy all metal PAN complexes except cobalt and nickel. The PAN method is capable of detecting 0.1 mg/L of cobalt in the sample. Both nickel and cobalt can be determined on the same sample. The wavelength to determine the cobalt-PAN complex is 620 nm, whereas the wavelength to determine nickel in the sample is 560 nm. The analyses of both metals are noted in Method 8078 (Cobalt) and Method 8150 (Nickel) provided by the HACH Company.

QUESTIONS AND ANSWERS

1. What is "cobalt blue"?
Cobalt blue is a blue pigment which is extremely durable and non-toxic. The pigment resists weathering and chemicals and is used for eye shadowing and painting of the body for display in shows.

2. Discuss the solubility and stability of cobalt compounds.
A number of useful organo-compounds have limited solubility in water, but tend to dissolve in slightly acidic or basic conditions. Insoluble cobalt compounds tend to form soluble cobalt-ammine and hydrated ammine- complexes at waste decomposition sites and in water. In addition, a few insoluble cobalt complexes melt and/or decompose at relatively low temperatures. A number of cobalt compounds are stable at high temperatures.

3. Since cobalt compounds are important as vitamins and trace elements, how toxic are cobalt compounds?
Some cobalt compounds are toxic by ingestion while other substances are highly toxic. A few compounds are irritants while other substances are carcinogenic. A number of cobalt compounds in foods are relatively safe.

4. What is the purpose of pyrophosphate in the experiment?
Pyrophosphate masks Fe^{+3} in the experiment.

REFERENCES

1. *Standard Methods for the Examination of Water and Wastewater:* Nonionic Surfactants as CTAS, 16th ed., A. E. Greenberg, R. R. Trussel, L. S. Clesceri and Mary Ann H. Franson (eds.), pp. 585–586, Washington, DC, American Public Health Association, 1985.
2. *HACH Water Analysis Handbook,* 3rd ed. (Loveland, CO.: HACH Company, 1997).
3. *Hawley's Condensed Chemical Dictionary,* 11th ed., rev'd by N. Irving Sax and Richard J. Lewis (New York, N.Y.: Van Nostrand Reinhold, 1987).

COPPER

Copper compounds that are insoluble and partially soluble in water sometimes become more soluble with pH changes and during bacterial processes. Copper compounds usually increase solubility levels in the environment by forming colorful coordination complexes with ammonia, water, and a number of "environmental ligands" from products of decomposition and discards. In general, copper compounds lack the toxicity levels of other complexes of the same type.

Copper metal and copper compounds have numerous uses. Copper metal is more resistant to atmospheric corrosion than iron. The metal is essentially nontoxic in elemental form. The conversion of copper ores to compounds can present some environmental problems. Partially soluble and insoluble copper compounds are found in paint, catalysts, insecticides, antifouling paints, thermal composites, preservative, wood preservative, dental cement, mordant in dyeing, pigments, pyrotechnics, coloring brass black, feed additive, electroplating, seed treatment, osmotic membranes, ceramics, enamels, metallurgy flux, fluorinating agent, pesticides, staining paper, fire retardant, corrosion inhibitor, fish net preservative, emulsifying agent, photocells, phosphors, welding fluxes, brazing preparations, and fungus control. A number of the insoluble copper compounds do decompose or are altered to a more soluble form in the environment. Most organic-copper complexes are moderately toxic.

Copper is a micronutrient in soil and in many solid samples. Copper is added to soil at 0.2 to 25 pounds per acre (lbs/A) to correct soil deficiencies. Copper plays an important role during the formation of the chlorophyll molecule and like other metals, e.g. iron, manganese and zinc, acts as a catalyst.* The amount of available copper varies considerably with the type of soil. Well drained sandy soils are usually low in copper while heavily clay-type soils contain an abundant supply of copper. Copper may be unavailable in soils that have a high organic make-up because it readily forms insoluble complexes with organic compounds.

A known volume of acid is added to a solid, gel or semisolid sample weighing 0.100 to 0.500 grams in a digestion flask.** The digestion procedure for oils requires 0.100 to 0.250 grams of material diluted to a specific volume. The sample is boiled for 4 to 5 minutes and the sample is allowed to cool. Then, 10 mLs of 50% hydrogen peroxide is added to the charred sample and the sample is boiled for 1 minute. The digested sample is allowed to cool to room temperature and is diluted to 100 mL with deionized water. After selecting the appropriate sample size and diluting the solution to 25 mLs, the clear sample can be analyzed by the Bicinchoninate (HACH Method 8506) procedure provided in Chapter 2. A pH adjustment for the sample may be required before analysis.

In the experiment, a number of interferences can occur. A precipitate or sample turbidity may form if the solution is extremely acidic. Turbidity can be dissolved with the addition of a base such as KOH solution. Additional interferences including silver, cyanide, and high levels of iron, hardness, or aluminum can be found in solid or gel samples. Each of these interferences can be minimized or eliminated by the addition of specific chemicals.

Copper ion tends to adsorb on the surface of sample containers. Therefore, samples should be analyzed as soon as possible after collection. If storage of a water sample is necessary, addition of a dilute solution of HCL will prevent adsorption. Copper exists in solution as cuprous (Cu^{+}) ion and cupric (Cu^{+2}) ion. Cuprous ion (Cu^{+}) is sufficiently stable in solution to react with 1, 10-phenanthroline-type compounds to form colored complexes which can be analyzed spectrophotometrically. In this experiment, copper in the sample reacts with a salt of bicinchoninic acid to form a purple colored complex in proportion to the copper concentration. The copper concentration can be determined by a spectrophotometer at 560 nm.

---

*Compliments of LaMotte Company.

**Compliments of HACH Company.

The analysis for copper (0 to 5.00 ppm) can be applied to sludges and other solid materials in wastewater, seawater, and industrial wastes. Strip tests have been developed to obtain quick and reliable approximations of total copper ion in solution. The strip tests have a range of measurement of 0 to 3 ppm and can be applied to many areas of testing after the metal has been dissolved from solids or sludges.

The "Diethyldithiocarbamate Method" is a test developed by LaMotte for copper in soils and other solids. The test determines copper in the range of 0 to 25 ppm. Cupric ions form a yellow colored chelate with diethyldithiocarbamate around pH 9 to 10, in proportion to the concentration of copper in the sample.* Bismuth, cobalt, mercury, nickel and silver ions and chlorine (6 ppm or greater) are significant interferences in the experiment and must be absent.

COPPER TEST

Diethyldithiocarbamate Method—Procedure*

1. Fill a clean colorimeter tube to the 10 mL line with the soil filtrate.

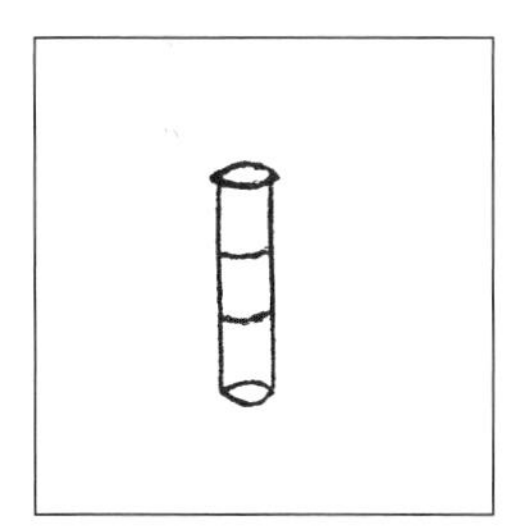

2. Neutralize by adding the 15% Sodium Hydroxide solution to the soil filtrate, one drop at a time while stirring the plastic rod. The stirring rod is touched to the Bromthymol Blue test paper after the addition of each drop of 15% Sodium Hydroxide until the color changes from yellow to green or blue.

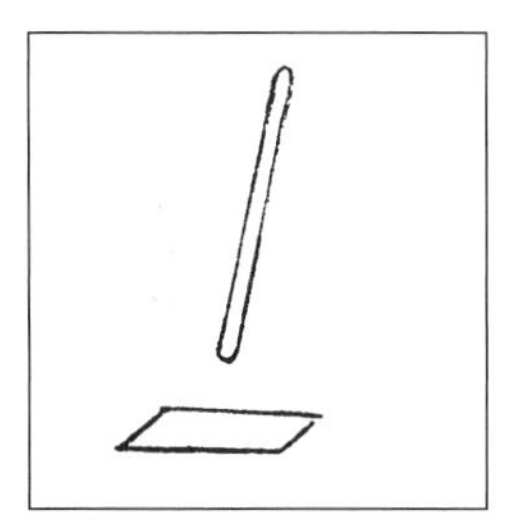

3. Select setting 2 on the "Select Wavelength" knob.

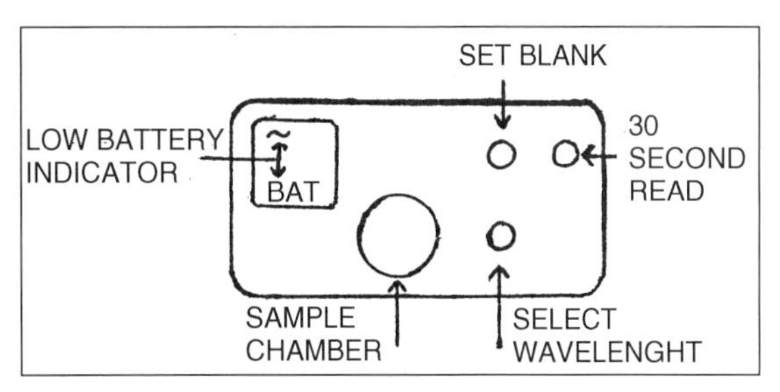

4. Insert tube containing sample into chamber and press the "30 Second Read" button. Adjust "100"%T with the "Set Blank" knob.

*Compliments of LaMotte Company.

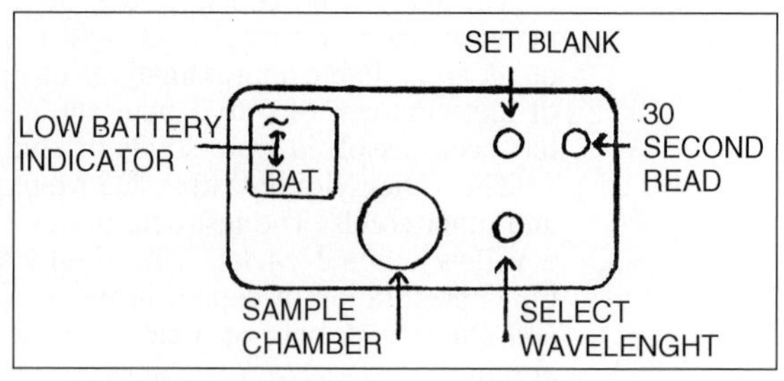

5. Remove the tube and add 5 drops of Copper Reagent. Cap and mix contents. A yellow color indicates the presence of copper.

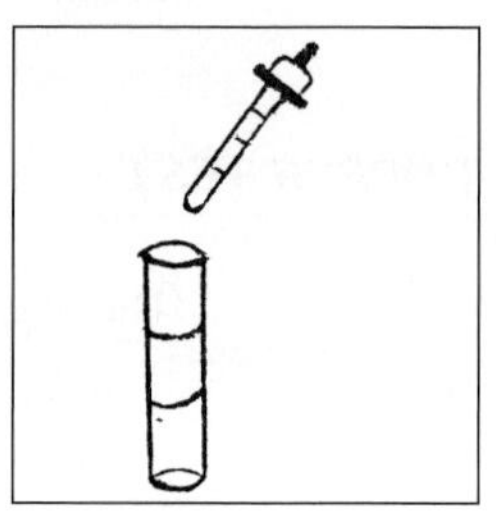

6. Insert sample into colorimeter chamber, press the "30 Second Read" button and measure %T as soon as reading stabilizes.

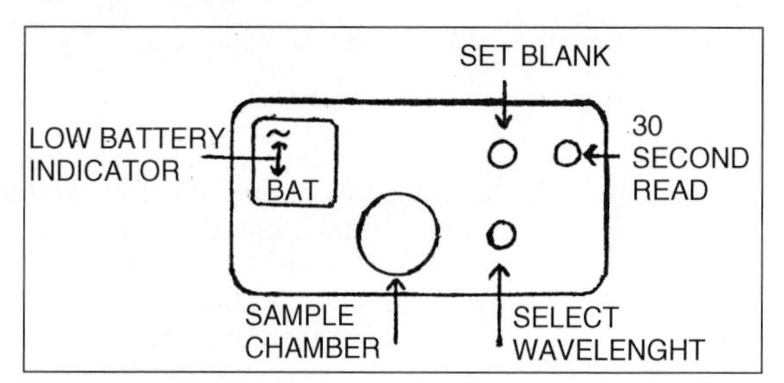

7. Consult the chart below to find the concentration of Copper in parts per million.
   Note: There is a tendency for the meter to drift with the use of the blue filter (415 nm) as a result of the photocell's response to blue light. For best results, after sample has been inserted into chamber and covered, allow approximately 10 seconds before taking the reading.

Copper Calibration Chart (ppm)

%T	9	8	7	6	5	4	3	2	1	0
90	0.0	0.1	0.2	0.3	0.4	0.5	0.6	0.7	0.8	0.9
80	1.0	1.1	1.2	1.3	1.4	1.5	1.6	1.7	1.8	1.9
70	2.0	2.2	2.3	2.4	2.5	2.6	2.8	2.9	3.0	3.1
60	3.3	3.4	3.5	3.7	3.8	3.9	4.1	4.2	4.4	4.5
50	4.7	4.8	5.0	5.2	5.3	5.5	5.6	5.8	6.0	6.2
40	6.4	6.5	6.7	6.9	7.1	7.3	7.5	7.7	8.0	8.2
30	8.4	8.6	8.9	9.1	9.4	9.6	9.9	10.2	10.5	10.8
20	11.1	11.4	11.7	12.1	12.4	12.8	13.2	13.6	14.0	14.4
10	14.9	15.4	15.9	16.4	17.0	17.6	18.3	19.0	19.8	20.7
0	21.6	22.7	23.9	25.3						

Copper Concentration Chart

%T	Range	Parts per million
89–100%	Low	0–1 ppm
71–88%	Marginal	1–3 ppm
63–70%	Adequate	3–4 ppm

HAZARDOUS SUBSTANCES

Read the MSDS before using the chemicals. The following substance is a hazardous substance and should be disposed of and handled properly. The hazardous substance is copper reagent.

RANGE OF MEASUREMENT

The range of measurement is 0 to 25 ppm.

INTERFERENCES

Bismuth, cobalt, mercury, nickel, and silver ions and chlorine (6 ppm or greater) interfere and must be absent during measurements.

COPPER (0 to 5.00 mg/L)*

Method 8506**

Bicinchoninate Method† using Powder Pillows

Digestion required for USEPA reporting of wastewater analysis‡

*Compliments of HACH Company. Analysis may be performed by DR/890 Colorimeter and DR/2010 Spectrophotometer.

**Pretreatment required; see Interferences (Using Powder Pillows) in Section 1.

†Adapted from S. Nakano, *Yakugaku Zasshi,* 82 486-491 (1962) [*Chemical Abstracts,* 58 3390e (1963)].

‡Powder Pillows only: *Federal Register,* 45 (105) 36166 (May 29, 1980)

1. Enter the stored program number for copper (Cu), bicinchoninate powder pillows. Press: **1 3 5 ENTER.** The display will show: **Dial nm to 560.**
Note: The Pour-Thru Cell can be used for 25-mL reagents only.

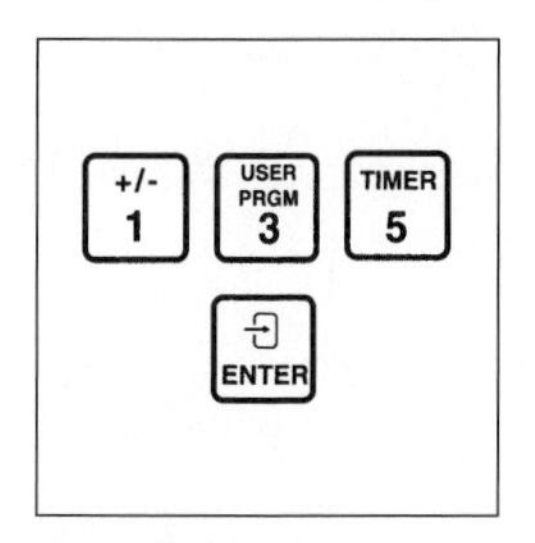

2. Rotate the wavelength dial until the small display shows: **560 nm.** When the correct wavelength is dialed in, the display will quickly show: **Zero sample,** then: **mg/L Cu Bicn.**
Note: Determination of total copper needs a prior digestion.

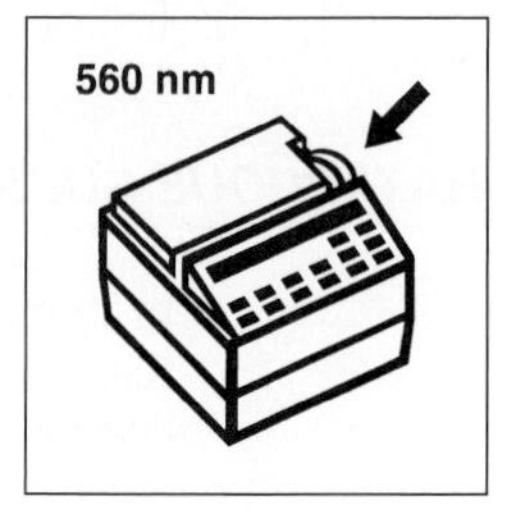

3. Insert the 10-mL Cell Riser into the cell compartment.

4. Fill a 10-mL sample cell with 10 mL of sample.
Note: Determine a reagent blank for each new lot of reagent. Use deionized water in place of the sample in the procedure. Subtract this value from each result obtained.
Note: Adjust pH of stored samples to between 4 and 6 before analysis.

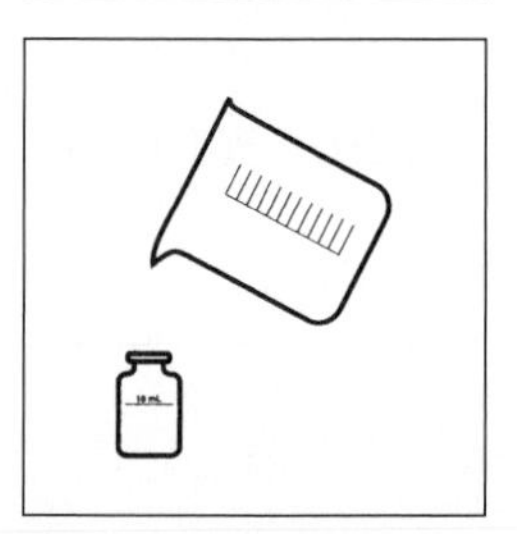

5. Add the contents of one CuVer 1 Copper Reagent Powder Pillow to thc sample cell (the prepared sample). Swirl to mix.
Note: A purple color will develop if copper is present.

6. Press: **SHIFT TIMER.** A two-minute reaction period will begin.
   Note: Accuracy is not affected by undissolved powder.

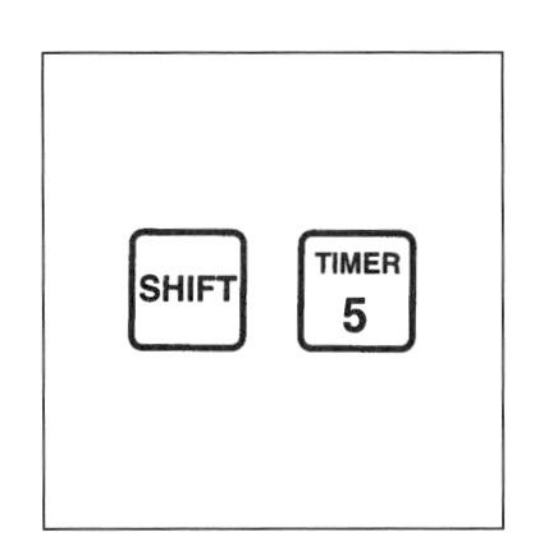

7. When the timer beeps, the display will show**: mg/L Cu Bicn.** Fill a second 10-mL sample cell (the blank) with 10 mL of sample.

8. Place the blank into the cell holder. Close the light shield.

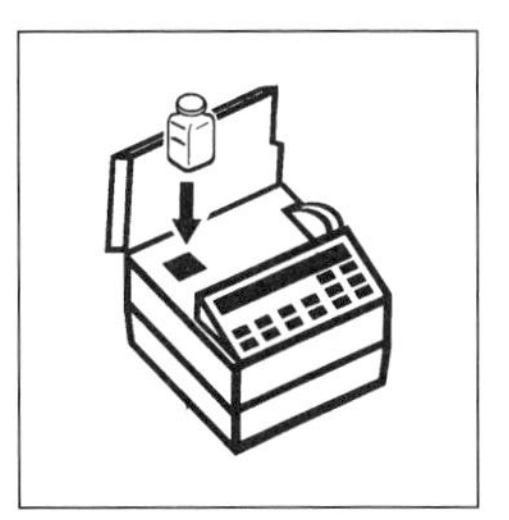

9. Press: **ZERO**. The display will show: **Zeroing . . .** then: **0.00 mg/L Cu Bicn.**

10. Within 30 minutes after the timer beeps, place the prepared sample into the cell holder. Close the light shield.

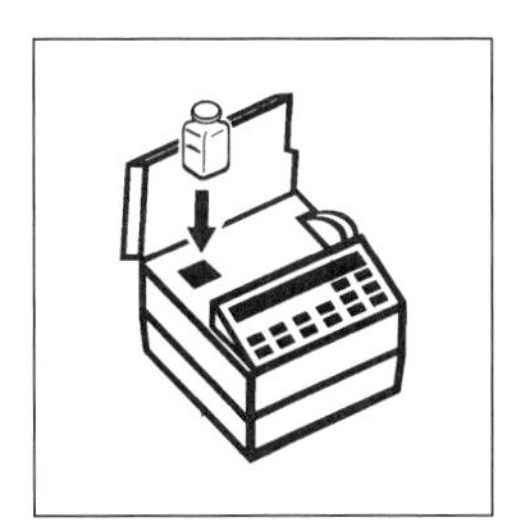

11. Press: **READ.** The display will show: **Reading . . .** then the results in mg/L copper will be displayed.

SAMPLING AND STORAGE

Samples that have been dissolved should be stored in acid-cleaned glass or plastic containers. Adjust the pH to 2 or less with nitric acid (about 2 mL per liter). Store preserved samples up to six months at room temperature. Before analysis, adjust the pH to 4 to 6 with 8 N potassium hydroxide. Do not exceed pH 6, as copper may precipitate. Correct the test result for volume additions; see "Correcting for Volume Additions" in Section 1 for more information. If only dissolved copper is to be determined, filter the sample before acid addition.

ACCURACY CHECK

See p. 533 of the HACH Water Analysis Handbook, 3rd Ed., 1997.

INTERFERENCES (USING POWDER PILLOWS)

If the sample is extremely acidic (pH 2 or less) a precipitate may form. Add 8 N Potassium Hydroxide Standard Solution drop-wise while swirling to dissolve the turbidity. Read the mg/L Cu. If the turbidity remains and turns black, silver interference is likely. Eliminate silver interference by adding 10 drops of saturated Potassium Chloride Solution to 75 mL of sample, followed by filtering through a fine or highly retentive filter. Use the filtered sample in the procedure. For additional interferences, see the copper analysis procedure in the water analysis section.

QUESTIONS AND ANSWERS

1. Why is digestion required?
Digestion is required for wastewater reporting to the USEPA. Digestion is required to break any bonds or forces that may hold substances together in the solid.
2. How is accuracy affected by undissolved powder?
In this experiment, accuracy is not affected by undissolved powder.
3. Why is 50% hydrogen peroxide added to the charred sample during the digestion process in the HACH Process?
Hydrogen peroxide is added to oxidize all copper forms to Cu^{+2} *and to free the copper ion from any other material in the matrix.*

4. What interferences can occur in the HACH experiment during analysis for copper?
If the sample is acidic (pH 2 or less), a precipitate or turbidity may form. Silver and cyanide interferences must be addressed. High levels of iron and aluminum in the sample can be a problem.

5. What interferences can occur in the LaMotte experiment during determination of copper?
Bismuth, cobalt, mercury, nickel and silver ions and chlorine (6 ppm or greater) are significant interferences and must be absent.

REFERENCES

1. *HACH Water Analysis Handbook,* 3rd Ed. (Loveland, CO.: HACH Company, 1997).
2. *LaMotte Soil Handbook* (Chestertown, Md.: LaMotte Company, 1994).
3. *Hawley's Condensed Chemical Dictionary,* 11th Ed., rev'd by N. Irving Sax and Richard J. Lewis (New York, N.Y.: Van Nostrand Reinhold, 1987).
4. *Standard Methods for the Examination of Water and Wastewater:* COPPER, 16th Ed., Method 313, A. E. Greenberg, R. R. Trussell, L. S. Clesceri and Mary Ann H. Franson (eds.), pp. 204–208, Washington, DC, American Public Health Association, 1985.

DIOXINS AND FURANS

Polychlorinated dibenzo-para-dioxins (PCDDs) and polychlorinated dibenzofurans (PCDFs) are composed of 210 individual congeners (isomers) of which 17 have a significant toxicity level. The most toxic form of dioxin, the 2,3,7,8-tetrachloro-dibenzo-p-dioxin (TCDD) congener, has been isolated in significant quantities from the organs and the blood of humans and animals. This toxic congener has adverse effects on the immune system of humans. TCDD and TCDF compounds bioaccumulated in tissue lipid cause a number of health problems at low levels of exposure over a long period of time. Breast feeding infants are at significant risk from these contaminates in human milk. Health affects from "agent orange spraying" will provide new information on human exposure limits to dioxins/furans. Toxic forms of dioxins have contaminated water, sediments and the organs of fish in some areas of Europe, Vietnam and other locations. Dioxins and furans are often measured in terms of the toxic equivalent (TEQ) of 2,3,7,8-TCDD (picograms/gram of material).

The normal exposure of humans to dioxins throughout the world is reflected in the blood levels in the range of 5 to 15 picograms of toxic-equivalent (TEQ) 2,3,7,8-TCDD per gram of blood.[1] Workers exposed to dioxins for 20 years were found to have blood levels of about 350 pg/g. They did not show indications of negative health effects, but those with blood levels of about 650 pg/g showed some indications of cancers above the normal rate.

The chlorinated furans are less toxic than the 2,3,7,8-TCDD (Dioxin) complex. Adverse reproductive and developmental effects in humans, wildlife, and laboratory animals have been observed after exposure to these agents. The furan complexes may interfere with endocrine system function and appear to mimic the action of hormones.

Dioxins and furans are often grouped together under the heading of "Dioxins." Dioxins have been studied extensively, whereas less attention has been granted to chlorinated furans. Chlorinated dioxins and furans can be easily formed from the combination of chlorinated phenols, chlorinated benzenes and polychlorobiphenyls (PCBs). The formation of PCDDs and PCDFs may easily occur during condensation or "cooling off" processes during high temperature burning of wastes or manufacturing of products. The formation rate of PCDDs and PCDFs generally increases as the level of chlorine (above 1%) in the fuel increases during combustion processes.[2]

Chlorinated dioxins and furans are common by-products of everyday processes such as the burning of coal, timber, coke, garbage, paper and charcoal, where trace amounts of chlorine are present during combustion. The open burning of household wastes (2 to 40 households) in barrels produces an average PCDD/PCDF emissions comparable to burning 200 tons of waste per day in a modern municipal waste combustor equipped with high efficiency flue gas cleaning technology.[3] Back-yard burning of general wastes that produce quantities of black smoke are natural dioxin and furan generators attributed to incomplete combustion processes. Significantly higher levels of PCDD/PCDF are emitted in the air during the open barrel burning of normal household wastes produced by the avid recycler as compared to the same burning process of wastes produced by the non-recycler.[4]

PCDDs and PCDFs enter the air from various types of engines, poorly operated and designed waste incinerators, wood-treating plants, steel plants, and metal production plants. Common burn processes with low combustion efficiencies produce soot, particulates, unburnt hydrocarbons, and toxic forms of dioxin and furan complexes.

Various materials have a small but significant toxic equivalent level of 2,3,7,8-TCDD. Some of these items are the following: plastic packaging (4.7 pg/g), clothes dryer lint (2.4 to 6.0 pg/g), vacuum cleaner dust (8.3 to 12 pg/g), room air filter (27 to 29 pg/g), car air filter (84 pg/g) and furnace air filter (170 pg/g).[5] Additional materials contaminated with chlorinated dioxins/furans include paper, soot, soils, sewage sludge, foods, settled dust, paper coffee filters, cigarette and cigar smoke, shopping bags, vehicle emissions, and human adipose tissue.[6] Dioxins have also been identified in waste oils, wood stains, plant sprays, remote lakes, fish, seals, herbicides, recycled paper and cardboard, composting processes, emissions from apartment furnaces and boilers, and during burning of paper and plastics in wood stoves, barbecue pits and fireplaces.

The toxicity level and amount of toxic dioxins formed during a fire will depend upon the amount of chlorine (chloride) available in the burning medium, the type of organic precursors, the availability of metal catalysts, and the temperature of the burn. For instance, the global abundance of PCDD and PCDF can be attributed to forest fires and brush fires. Fires at plastic-recycling centers and wood structures (buildings) can produce significant levels of toxic forms of chlorinated dioxins. Volcanic eruptions produce significant amounts of PCDD and PCDF. Trash burning in metal barrels and landfill fires are sources of large amounts of chlorinated dioxins and furans.

In general, the higher order hepta- and octa-chlorodioxins (PCDD) and furans (PCDF) compounds are less volatile and are predominately found in ash, soil, and on plants. The lower chlorinated forms are more soluble in water, are easily dispersed by wind, and altered by sunlight. Generally, the furan compounds are less soluble in water than the dioxin complexes.

Toxic forms of chlorinated dioxins and furans produced by municipal, hazardous and medical waste combustion can be effectively controlled by maintaining the combustion temperature in the range from 1700° F to 1800° F and mixing the combustion gases effectively.[1] Cooling of combustion products to below 350° F appears to reduce dioxin levels. Activated carbon injected prior to gas cleaning in a specific type of fabric filter (baghouse) has been found to be effective in adsorbing dioxins, hence is now generally used to reduce TEQ dioxins and furans to the European Union guideline of 0.3 nanograms per standard cubic meter. A recently developed alternative to carbon injection is the use of a catalytic layer on the fabric filters to cause the conversion of dioxins and furans to harmless forms, effectively destroying their toxic characteristics.

The formation of ferric chloride during incineration can promote the formation of dioxins during incineration processes.[7] During incineration, the formation of chlorinated dioxins and furans can be significantly inhibited by the addition of Diox-Blox Sorbent, EDTA, sodium sulfide or other agents.[7,8] Congener shifts of toxic substances to much less toxic materials can reduce many environmental problems. However, a number of older burning and heating operations, including incinerators and industrial units, must be modified or updated to operate more efficiently and to include specific emission control devices.

Levels of dioxin emissions from old models of gasoline-driven cars have been greatly reduced by the use of catalytic converters. Dioxins are still formed while the converter is being warmed up, and levels increase as the engine malfunctions. Dioxin emissions from diesel engines are still significant.

The important affect of dioxins on humans may be the impact on the Ah receptor, rather than the increase in cancer incidence. Immunoassays can directly indicate the affect of dioxins and dioxin-like substances on the Ah receptor.[10] The toxicity of the dioxin congeners can vary by one-thousand-fold and can be correlated to each congener's ability to transform the Ah-receptor. The Ah-IMMUNOASSAY Method is used for screening dioxin-like toxicity in tissues, ash, soils, smoke particles, and many types of environmental samples. The method provides total toxicity data of the dioxin mixture as 2,3,7,8-TCDD equivalents.

Samples to be tested by the Ah-IMMUNOASSAY Method are added to a mixture containing Ah-receptor and other components in a special ELISA plate and allowed to incubate near 30° C for two hours.[10] After incubation, any Ah-receptor transformed by dioxin-like compounds is bound to the plate media and excess material is washed away. Antibodies are added to the ELISA to detect the transformed, bound Ah-receptor. The changes in the Ah-receptor can be determined colorimetrically. Color development is proportional to the amount of transformed Ah-receptor. The detection limit is at 1pg (picogram) 2,3,7,8-TCDD equivalents.

Quantitative tests for chlorinated dioxin and furan compounds are usually difficult and expensive to perform. A number of PCDD/F congeners have been isolated by soxhlet extraction from tissue and fly ash samples and identified by a gas chromatograph (GC) with special columns and a mass spectrometer.[11] GC/MS analysis coupled with FT-IR spectroscopy is a viable method of determination.

Molecular imprinting of polymers creates recognition elements for the environmental toxin 2,3,7,8-tetrachlorodibenzodixin. The toxic dioxin molecule is identified after reacting with a primary amine complex to form a diurea template structure. In the future, this polymer imprinting technique of identification may simplify procedures for quantitative determination of toxic dioxins.[12]

DIOXIN EQUIVALENTS

AH-IMMUNOASSAY PROCEDURE*

1. Remove the kit from cold storage and warm to room temperature.
2. Plan the experiment and the placement of the samples on the plate. The detailed instructions provide a guideline for testing 12 items (including standard and solvent blank) in duplicate and at four dilutions, which will probably consume the entire kit. Steps 5 and 6 will change slightly with other experimental designs. Refer to "Representative Experimental Design" section in the kit for details in designing dilutions in Steps 5 and 6.

*GESS Environmental, 2575 Old Arcata Rd., Bayside, CA. 95524 (Phone: 707-822-4377) is the worldwide exclusive agent for the Paracelsian (Ah-IA) Kit.

3. Thaw the three vials of Sample Diluent in a beaker of tepid water. Immediately pool these three vials to create 30 mL of Sample Diluent and place the container on ice.
4. Remove the coated **ELISA Plates** from the plastic bag.
5. Add the samples, dissolved in DMSO, to aliquots of the Sample Diluent at a rate of 0.01 mL sampler per 1.0 mL Diluent. This constitutes the treated sample (1:100 dilution). Refer to "Representative Experimental Design" provided in the kit for details about designing dilutions in Steps 5 and 6.
6. Aliquot 0.4 mL of the Treated Samples into the **ELISA** plate wells representing the most concentrated sample. Serially dilute 0.2 mL into wells containing 0.2 mL untreated Sample Diluent using a multichannel analyzer. Refer to "Representative Experimental Design" section details about making serial dilutions in Steps 5 and 6.

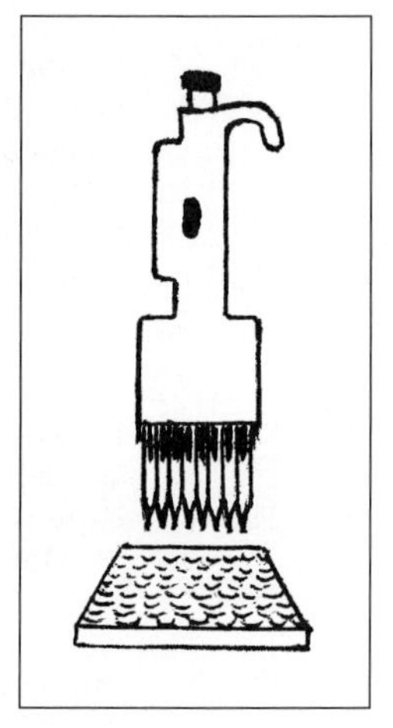

7. The plate is covered with the **ELISA Plate Cover** and incubated at 30° C for two hours.
   Note: Incubation at room temperature is satisfactory but will result in slightly lower responses in the assay. Incubation at higher temperature (i.e. 37° C) will also result in lower responses in the assay.

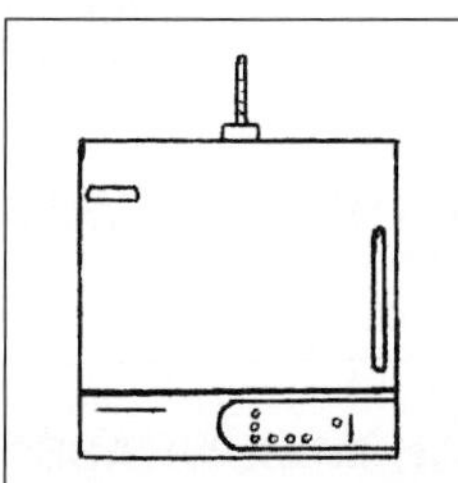

8. During the incubation of Step 7 (and prior to Step 11), prepare the wash buffer stock. To make this solution, dilute the contents of 20× Wash Buffer (25 mL) to 500 mL with reagent grade water. Label this wash buffer stock solution as Wash buffer.

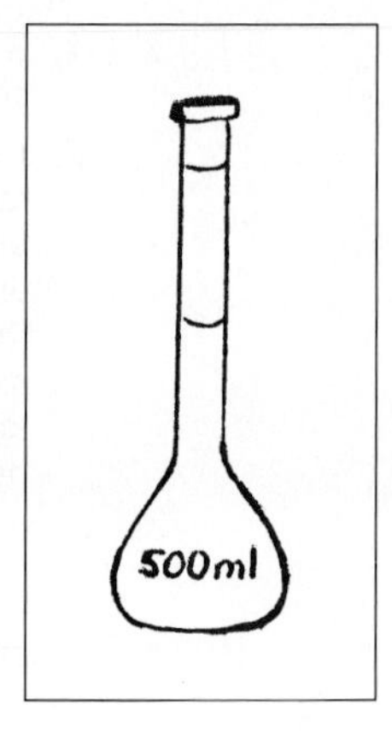

9. Prepare the primary antibody stock. Dissolve the contents of the AB 1 Vial in one 25 mL bottle of AB diluent using several rinses of the vial with the diluent. Mix gently. Set aside at room temperature.
   Note: This 25 mL solution is the Primary Antibody.

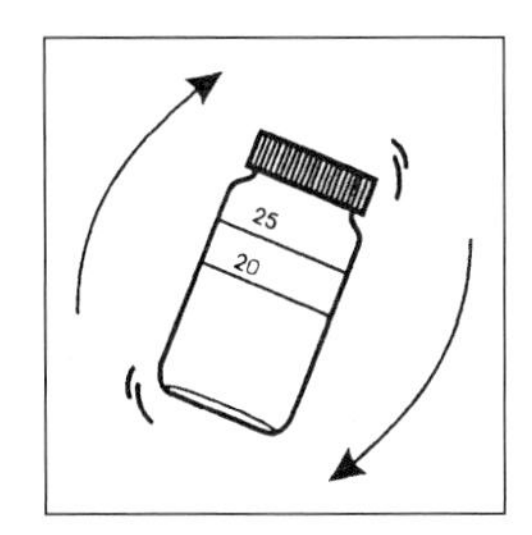

10. When the Incubation of Step 7 is complete, perform Steps 11 to 13.
11. Remove the treated samples from the **ELISA** plate wells.
    Note: Contents of the wells may need to be discarded as hazardous waste.
12. Add 0.4 mL of Wash Buffer to each well of the Reagent Reservoir using a multichannel pipettor. Wait two minutes, then remove the first wash. Repeat this wash twice form a total of three washes. Blot the inverted plate against paper toweling to remove residual wash buffer.

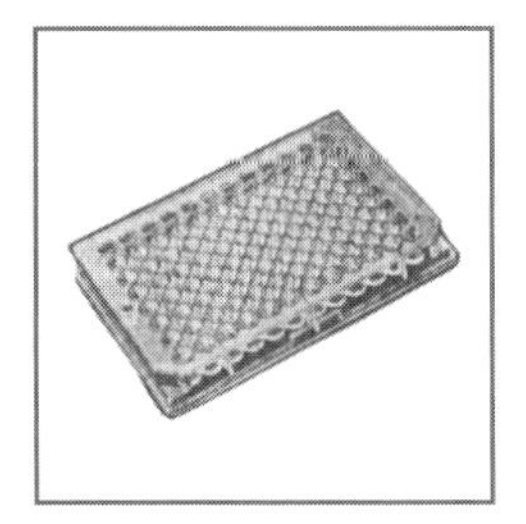

13. Pour the Primary Antibody (prepared in Step 9) into a clean reagent reservoir. Using a multichannel pipette, add 0.2 mL of Primary Antibody to each well.

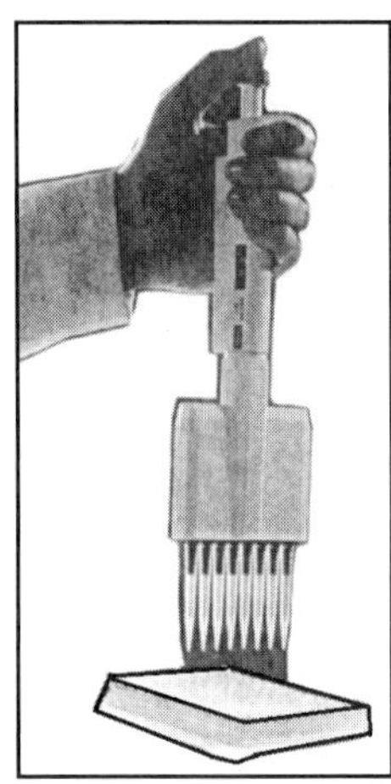

14. Incubate the samples for one hour at 30° C.

15. Before the end of incubation of Step 14 and prior to Step 17, prepare the secondary antibody stock. Dissolve the contents of the AB 2 vial in one 25 mL bottle of AB diluent using several rinses of the vial with the diluent. Mix gently. Set aside at room temperature.
Note: This 25 mL solution is the Secondary Antibody and should be properly labelled.

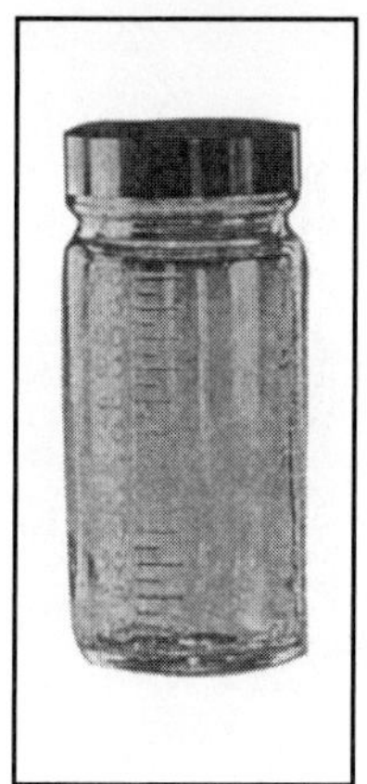

16. When incubation of Step 14 is complete, perform Steps 17 to 18.

17. Remove the contents of the **ELISA** plate by inverting over the sink. Blot the inverted plate against paper toweling to remove residual fluid. Add 0.4 mL wash buffer to all wells. Wait two minutes, then remove the first wash. Repeat this wash for a total of three washes. Blot or strike the inverted plate against paper toweling to remove residual wash buffer.

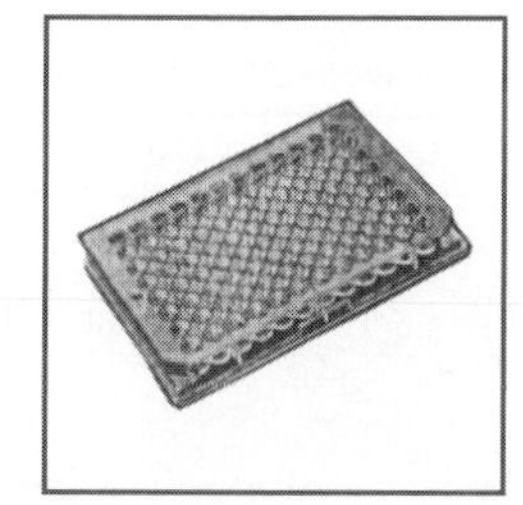

18. Pour the Secondary Antibody (prepared in Step 15) into a clean reagent reservoir. Using a multichannel pipette, add 0.2 mL of secondary antibody to each well.

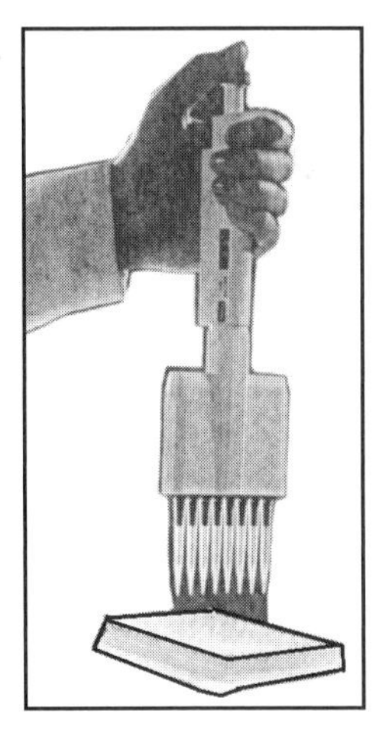

19. Incubate one hour at 30° C.

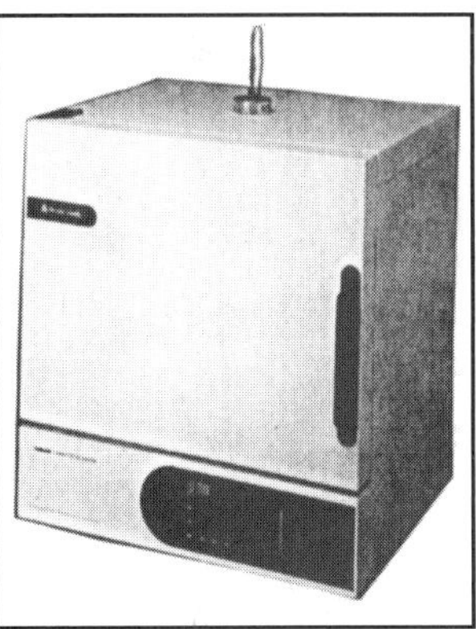

20. Before the end of incubation of Step 19 and prior to Step 22, prepare the Detection Reagent. Dissolve 3 Detection Tablets in the 30 mL bottle of Detection Buffer with gentle mixing. Label this mixture as Detection Reagent.
Note: Store this reagent away from light until use.

21. When incubation of Step 19 is complete, perform Steps 22 to 23.

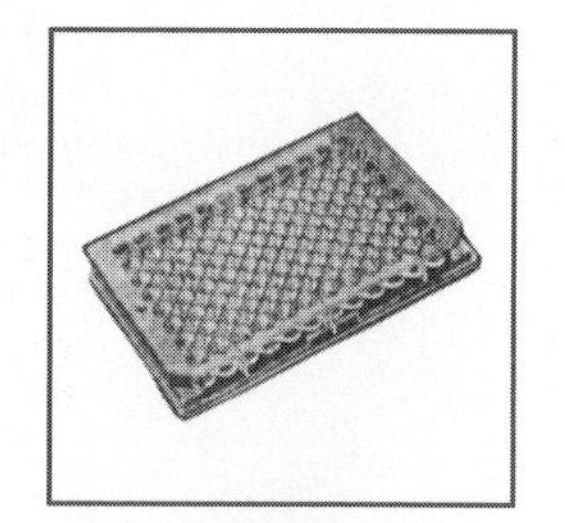

22. Remove the contents of the **ELISA** plate by inverting over the sink. Blot or strike the inverted plate against paper toweling to remove residual fluid. Add 0.4 mL Wash Buffer to all wells. Wait two minutes, then remove this first wash. Repeat this wash twice for a total of three washes. Strike inverted plate against paper toweling to remove residual wash.

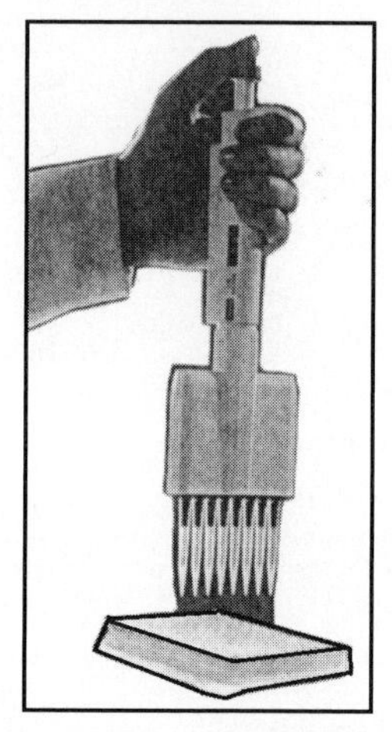

23. Pour the Detection Reagent, prepared in Step 20, into a clean reagent reservoir. Add 0.2 mL of the Detection Reagent to each well using a clean multi-channel pipettor.

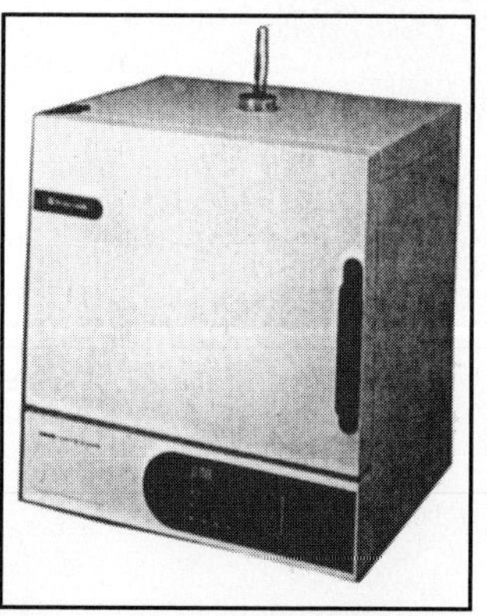

24. Incubate one hour at 30° C.

25. At the end of the one hour incubation, read the plate with an **ELISA Reader** at 405 nm.

26. Determine the amount of TCDD in the unknown sample by performing data analysis using TCDD as the Reference Standard. See the **Data Reduction Method** below.

REPRESENTATIVE EXPERIMENTAL DESIGN AND DATA REDUCTION METHOD

The experimental design including serial dilution information and data reduction method can be obtained through Andrew Jolin, GESS Environmental, 2575 Old Arcata Road, Bayside, California 95524 or through Dr. T. Colin Campbell, Cornell University, Nutritional Science Department, 305 Savage Hall, Ithaca, New York.

SAMPLING AND STORAGE

The components of the Ah-IMMUNOASSAY test kit are stable for at least three months. The "Sample Diluent" should be frozen in a freezer. The remainder of the kit can be stored in a refrigerator.

WARNINGS AND PRECAUTIONS

Dioxin contaminated materials should be handled by trained personnel. Observe all regulations for disposal.

MATERIALS REQUIRED (NOT PROVIDED)

1. Samples of dioxin in dimethyl sulfoxide (DMSO), preferably in the range of 0.5 pg/μL to 32 pg/μL TCDD equivalents. About 2 μL of sample is required per 200 μL ELISA plate well.
2. Adjustable multi-channel pipettor capable of accurately dispensing 200 μL volumes. Adjustable pippettors covering a range of 0.01 mL to 1.0 mL are also useful.
3. Disposable pipet tips for pipettor. Aerosol resistant tips are recommended to decrease the risk of cross contamination.
4. Incubator set at 30° C. (Recommended, but not required).
5. **ELISA Reader** capable of reading at 405 nm.

QUESTIONS AND ANSWERS

1. Are all PCDD and PCDF complexes significantly toxic?
   Approximately 17 complexes of 210 chlorinated-PCDD and PCDF compounds display significant levels of toxicity. The 2,3,7,8 -TCDD complex is a stable and most toxic congener. Other symmetric dioxin complexes are fairly stable, but none appear to be as toxic. Dioxin-like compounds have some degree of toxicity because of the ability of each congener to interact with a cellular protein called "Ah-receptor."[9]
2. Is the 2,3,7,8-TCDD complex very toxic?
   Some public reports of the toxicity level of 2,3,7,8- TCDD compound are exaggerated. A study of the Material Safety Data Sheets will reveal that many other common compounds are more dangerous (explosive, ignitable, etc.) and have a greater toxicity level. Chemical

gases such as HCN, H_2S and H_2Se appear to be much more life threatening to humans than the most toxic dioxin congener. The effects of some compounds on a person are immediate, whereas the most toxic form of TCDD requires a "small level" of exposure for an extended period of time to effectively harm an individual. People exposed to "high levels" of the 2,3,7,8- TCDD congener over a long period of time (20 to 30 years) have demonstrated health problems. At this time, we still do not know the effects of short term exposure (8 hours to 1 week) of the most toxic dioxin congener on a person's health. However, safety precautions must be implemented during collection and analysis of dioxin samples. All analysis of the substances containing dioxins should be performed in a laboratory hood that is properly working.

3. Where are chlorinated forms of dioxins and furans found in our society?
Chlorinated dioxins/furans are found in many places around the home and office. These compounds have been identified in remote regions on the surfaces of plants and on soil. Dioxins/furan complexes are found in cigarette and cigar smoke, wood burning stoves and fireplaces, vehicle exhausts, forest fires and brush burning, building burning, vacuum cleaning dust, various types of air filters, packaging, dryer lint, incinerator exhausts, industrial furnaces, plane exhausts, plastic recycling center fires, soot, fly ash, waste composting, landfills, chlorinated plastic producing industries, trash burning in barrels and landfills, etc. Many of the dioxin and furan emissions are not extremely toxic, whereas other sources (such as burn barrels, cigarette smoke, and burning landfills) generate more toxic forms of dioxins and/or furans than suspected.

REFERENCES

1. Floyd Hasselriis, P. E., Consulting Engineer, Forest Hills, New York.
2. C. Rappe, S. Marklund, et al., *Influence of Level and Form of Chlorine on the Formation of Chlorinated Dioxins, Dibenzofurans, and Benzenes During Combustion of an Artificial Fuel in a Laboratory Reactor,* Environmental Science & Technology, 30 (5): 1637–1638 (May 1996).
3. Paul M. Lemieux, Judith A. Abbott, Kenneth M. Aldous, et al., *Emissions of Polychlorinated Dibenzo-p- dioxins and Polychlorinated Dibenzofurans from the Open Burning of Household Waste in Barrels,* Environmental Science & Technology, 20 (February 1, 2000).
4. *Evaluation of Emissions From The Open Burning of Household Waste in Barrels,* Vol. 1, Technical Report, United States Environmental Protection Agency, EPA-600/R-97-134a, Control Technology Center, November 1997.
5. R. Berry, C. Luthe, and R. Voss, *Ubiquitous Nature of Dioxins: A Comparison of the Dioxin Content of Common Everyday Materials with That of Pulps and Papers,* Environmental Science & Technology, 27 (6): 1164–1168 (1993).
6. Donald A. Drum and Floyd Hasselriis, *What Do We Know About Dioxins and Furans in 1994?,* Paper 94-MP 17.05, Air and Waste Management Association, 87th Annual Meeting and Exposition, Cincinnati, OH, June 19–24, 1994.
7. B. J. Lerner, *Prevention of Dioxin Formation in Medical Waste Incineration,* Paper 97-P165-A30, Air and Waste Management Association, 90th Annual Meeting and Exposition, Toronto, Ontario, Canada, June 8–13, 1997.
8. K. Olie, et al., *Prevention of Polychlorinated Dibenzo-p-dioxins/dibenzofurans Formation on Municipal Waste Incinerator Fly Ash Using Nitrogen and Sulfur Compounds,* Environmental Science & Technology, 30 (7): 2350–2355 (July 1996).
9. Donald A. Drum and Jack D. Lauber, *Waste Management and Treatment or Disguised Disposal?* Air and Waste Manaement Association, 85th Annual Meeting and Exhibition, Kansas City, Mo., June 21–26, 1992.
10. Andrew Jolin, *Paracelsian Immunoassay* (*Ah-I A*), GESS Environmental, Bayside, CA., February 2000.

11. Ward Stone, Wildlife Pathologist, New York State Department of Environmental Conservation, Wildlife Research Center, Delmar, New York.

12. M. Lubke, M. J. Whitcombe, and E.N. Vulfson, *Molecular Imprinting of Polymers Creates Recognition Elements for the Environmental Toxin 2,3,7,8-tetrachlorodibenzodioxin (TCDD)*, J. Am. Chem. Soc., 120: 13342–13348 (1998).

13. Ben Pierson, P. E., Senior Sanitary Engineer, New York State Department of Health, Troy, New York.

IRON

Iron exists as a metal and as compounds of ferrous (+2) ion and ferric (+3) ion in organic and inorganic complexes. Iron as ferrous and ferric ions may exist in a colloidal state or form coarse suspended particles. Many ferrous and ferric compounds are soluble in water. Iron can be identified in many foods, plants, soils, silt, metals, stone and many additional items.

Silt and clay soils often contain significant amounts of iron. Iron has a tendency to form a colloidal suspension in solution. Adsorption of iron on various types of materials and deposition of iron on pipes walls are common occurrences. Bacterial growth, hydration processes, changes in pH, and oxidation-reduction processes will solubilize iron metal and insoluble iron compounds in the environment. These changes normally require a considerable amount of time.

In plants, iron is essential during the formation of chlorophyll. The deficiency of iron in plants causes chlorosis and slow growth. In excess, a bronzing of leaves and tiny brown spots become evident. A fraction of the abundant iron in soil is soluble and is available to the growing plant in neutral or alkaline soils. Acid soils usually contain significantly higher levels of available iron.

A known volume of acid is added to a solid, gel or semisolid sample weighing 0.100 to 0.500 grams in a digestion flask. The digestion procedure for oils requires 0.100 to 0.250 grams of material diluted to a specific volume. The sample is boiled for 4 to 5 minutes and the sample is allowed to cool. Then, 10 mL of 50% hydrogen peroxide is added to the charred sample and the sample is boiled for 1 minute. After cooling the mixture and adding 70 mL of deionized water to the digest, the sample is boiled in a hot water bath for 15 minutes. The digested sample is allowed to cool to room temperature and is diluted to 100 mL with deionized water. After selecting the appropriate sample size and diluting the solution to 25 mL, the clear sample can be analyzed by the procedure provided below. A pH adjustment on the sample will be required before analysis for iron.

The determination of the ferrous ion requires special precautions and immediate attention because ferrous iron is easily oxidized to the ferric form. In the phenanthroline test for ferrous iron (Fe^{+2}), three molecules of phenanthroline chelate a single ferrous ion to form an orange-red complex. The aqueous digested samples of the ferrous ion should be limited to light exposure. Ferric iron (Fe^{+3}) does not react with the phenanthroline.

Total iron includes both the soluble and insoluble forms of iron in the sample. In the HACH procedure, the FerroVer Iron Reagent reacts with these different forms of iron to produce soluble ferrous ion.* Then, the ferrous ion reacts with the 1,10-phenanthroline indicator to form an orange color solution in proportion to the "total" iron concentration, which can be easily measured by a Colorimeter or a Spectrophotometer. At times, a test strip can be used to measure total iron at low concentrations (0 to 3.00 mg/L). Digestion of the sample frees all of the iron for a total iron test by the FerroVer Method (Method 8008).

*Compliments of HACH Company.

Digestion of the sample before determination of total iron is required for reporting wastewater analysis to USEPA.

The Bipyridyl Method (LaMotte) can be used to determine soluble iron in the range of 0 to 25 ppm. In this experiment, ferric iron is reduced to ferrous iron and subsequently forms a colored complex with bipyridyl for a quantitative measure of total iron.* Interferences include copper and cobalt in excess of 5.0 mg/L. Strong oxidizing agents interfere with the test for ferrous ion.

IRON TEST

Bipyridyl Method*—Procedure

1. Fill a clean colorimeter tube to the 10 mL line with the soil filtrate.

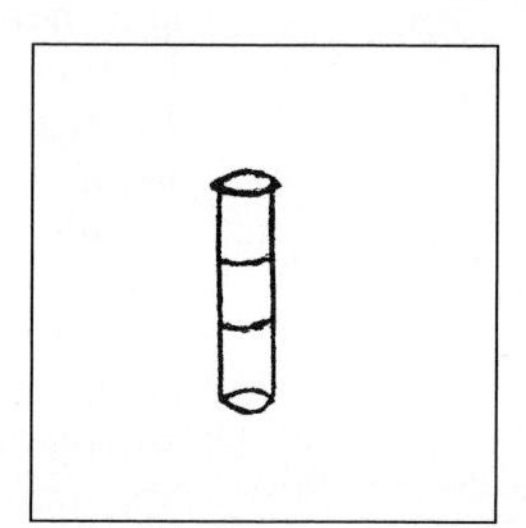

2. Neutralize the soil filtrate by adding 15% Sodium Hydroxide Solution to the soil filtrate, one drop at a time while stirring the plastic rod The stirring rod is touched to the Bromthymol Blue test paper after the addition of each drop of Sodium Hydroxide until the color changes from yellow to green or blue.

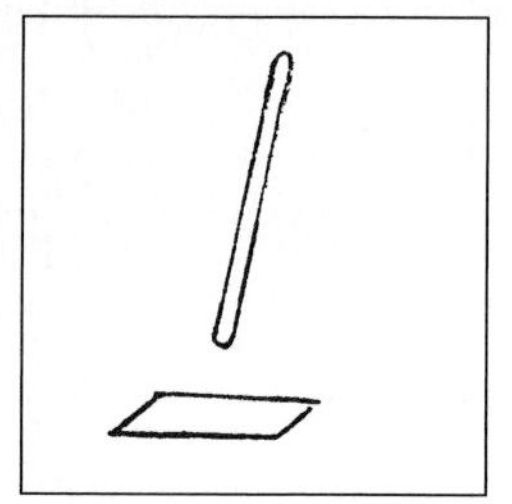

3. Select setting 3 on the "Select Wavelength" knob. Press the "30 Second Read" button.

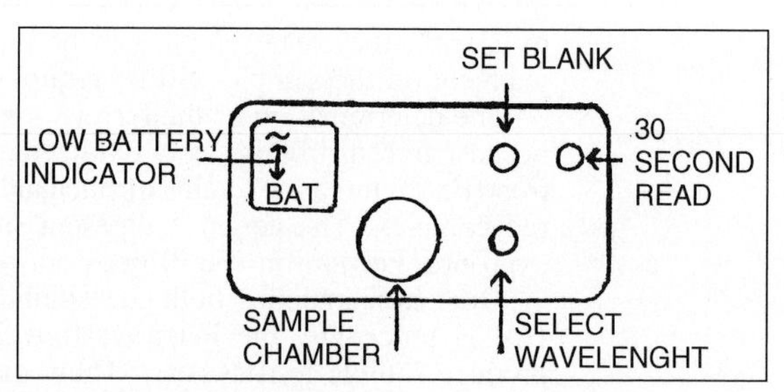

4. Insert tube containing sample into chamber and adjust to "100"%T with the "Set Blank" knob. This is the 100%T blank.

* Compliments of LaMotte Company.

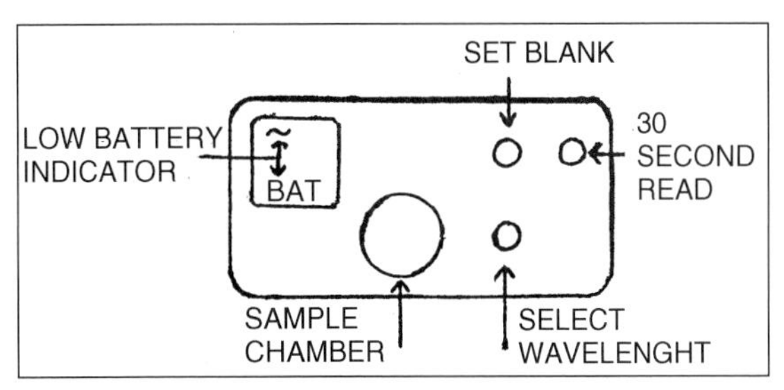

5. Remove the tube from the chamber. With 0.5 mL pipet, add one measure of Iron Reagent #1 to sample. Cap and mix.

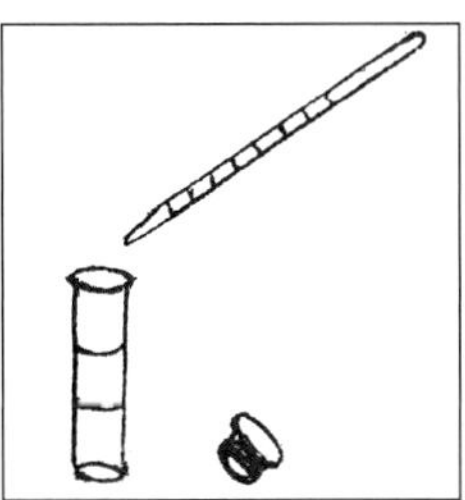

6. With the 0.1 g spoon, add 1 level measure of Iron Reagent #2 Powder to sample.

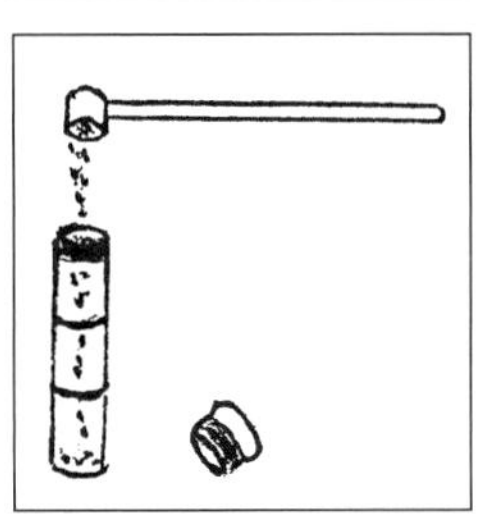

7. Cap and shake vigorously for 30 seconds.

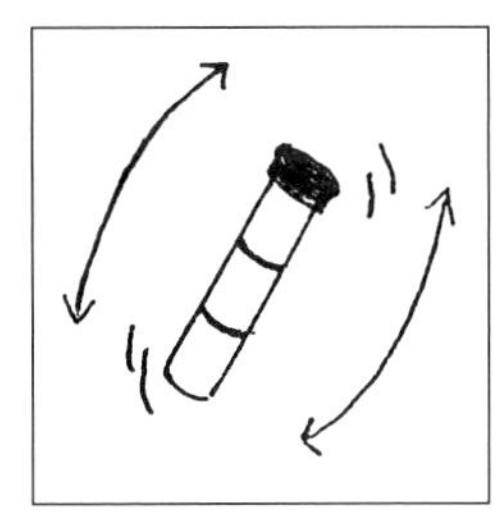

8. Allow 3 minutes for maximum color development.

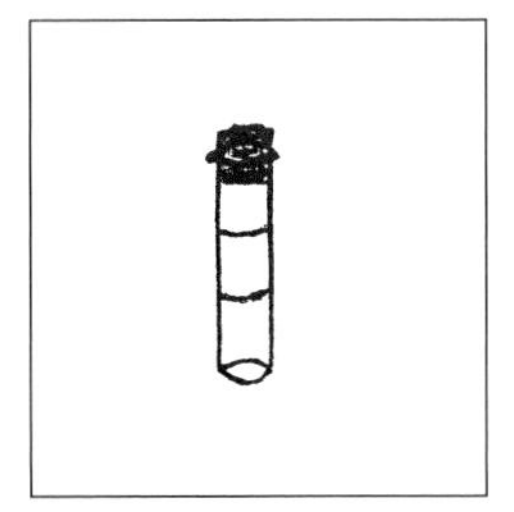

9. After 3 minutes, insert sample into colorimeter chamber, press the "30 Second Read" button and measure %T as soon as reading stabilizes.

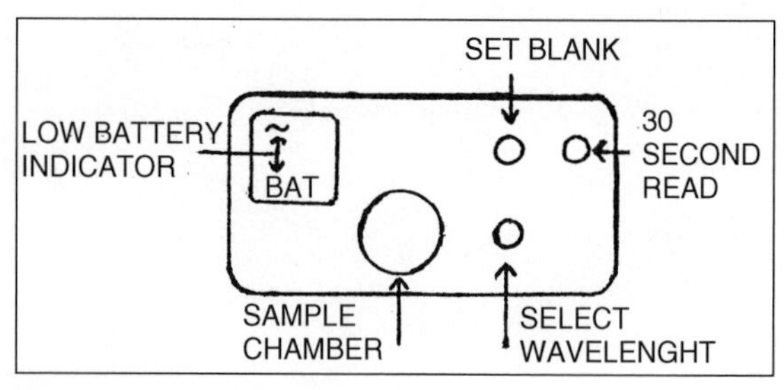

10. Consult the calibration chart to find the concentration of Iron in parts per million (ppm).

Iron Calibration Chart (PPM)

%T	9	8	7	6	5	4	3	2	1	0
90	0.1	0.2	0.3	0.4	0.4	0.5	0.6	0.7	0.8	0.9
80	1.0	1.1	1.2	1.3	1.4	1.5	1.6	1.7	1.8	2.0
70	2.1	2.2	2.3	2.4	2.5	2.6	2.8	2.9	3.0	3.1
60	3.3	3.4	3.5	3.7	3.8	3.9	4.1	4.2	4.4	4.5
50	4.6	4.8	5.0	5.1	5.3	5.4	5.6	5.8	6.0	6.1
40	6.3	6.5	6.7	6.9	7.1	7.3	7.5	7.7	7.9	8.1
30	8.4	8.6	8.9	9.1	9.4	9.6	9.9	10.2	10.5	10.8
20	11.1	11.4	11.8	12.1	12.5	12.9	13.3	13.7	14.1	14.6
10	15.0	15.6	16.1	16.7	17.3	17.9	18.7	19.4	20.3	21.2
0	22.2	23.4	24.7	26.2	28.0	30.3	33.2	37.4	44.7	

Iron Concentration Chart

% T	Range	Parts per Million
86–100%	Very Low	0–1.3 ppm
70–85%	Low	1.4–3 ppm
57–69%	Medium	3–5 ppm
32–56%	Medium High	5–10 ppm
0–31%	High	Above 10–25 ppm

HAZARDOUS SUBSTANCES

Hazardous substances are Iron Reagent #1 and Iron Reagent #2 Powder. Consult the material safety data sheet (MSDS) for proper handling and disposal procedures.

RANGE OF MEASUREMENT

The range of measurement is 0 to 25 ppm.

INTERFERENCES

Strong oxidizing agents interfere, as well as copper and cobalt in excess of 5.0 mg/L.

IRON, TOTAL (0 to 3.00 mg/L)

Method 8008*

FerroVer Method** using Powder Pillows

Digestion is USEPA accepted for reporting wastewater analysis†

1. Enter the stored program number for iron (Fe) FerroVer, powder pillows. Press: **2 6 5 ENTER.** The display will show: **Dial nm to 510.**
   Note: Adjust pH of stored samples before analysis.
   Note: The Pour-Thru Cell may be used with 25-mL reagents only.

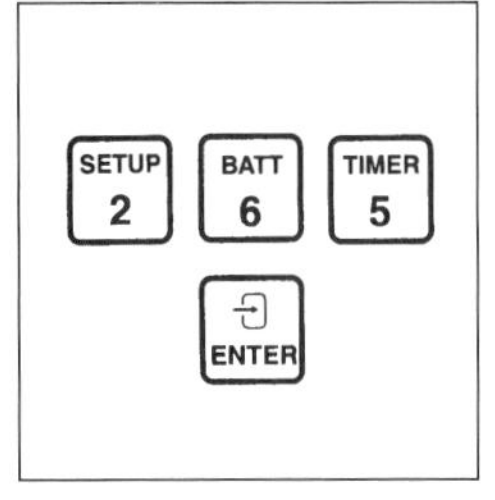

2. Rotate the wavelength dial until the small display shows: **510 nm.** When the correct wavelength is dialed in, the display will quickly show: **Zero sample,** then: **mg/L Fe FV.**

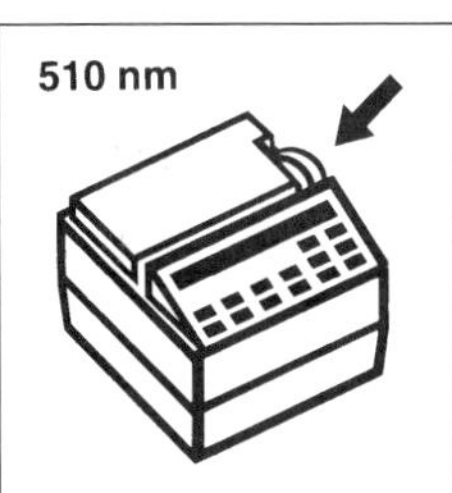

3. Insert the Cell Riser for 10-mL sample cells.

*Compliments of HACH Company. Analysis may be performed by DR/820, DR/850, and DR/890 Colorimeters and DR/2010 Spectrophotometer.

**Adapted from *Standard Methods for the Examination of Water and Wastewater.*

†Federal Register, 45 (126) 43459 (June 27, 1980).

4. Fill a clean sample cell with 10 mL of sample.
Note: Determination of total iron needs a prior digestion; use the mild, vigorous or Digesdahl disgestion in Section 1.

5. Add the contents of one FerroVer Iron Reagent Powder Pillow to the sample cell (the prepared sample). Swirl to mix.
Note: An orange color will form if iron is present.
Note: Accuracy is not affected by undissolved powder.

6. Press: **SHIFT TIMER.** A three-minute reaction period will begin.
Note: Samples containing visible rust should be allowed to react at least five minutes.

7. When the timer beeps, the display will show: **mg/L Fe FV.** Fill another sample cell with 10 mL of sample (the blank).

8. Place the blank into the cell holder. Close the light shield.
Note: For turbid samples, treat the blank with onc 0.1-g scoop of RoVer Rust Remover (use 0.2 g for 25-mL samples). Swirl to mix.

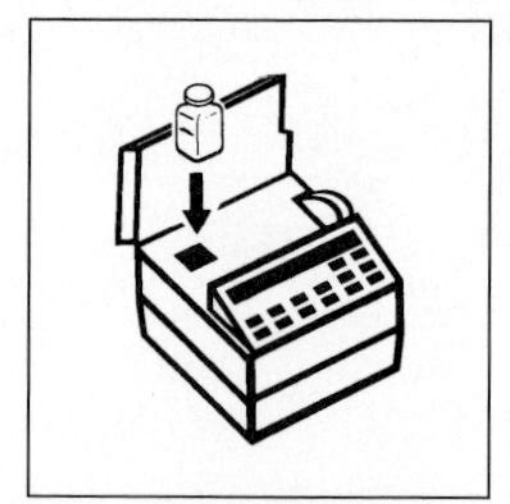

9. Press: **ZERO.** The display will show: **Zeroing . . .** then: **0.00 mg/L Fe FV.**

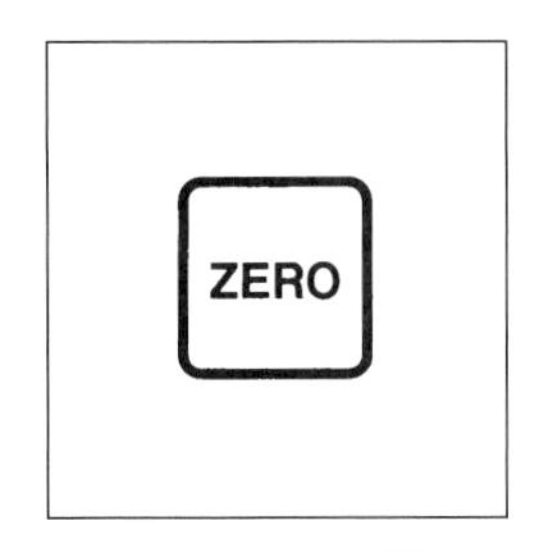

10. Within 30 minutes after the timer beeps, place the prepared sample into the cell holder. Close the light shield.

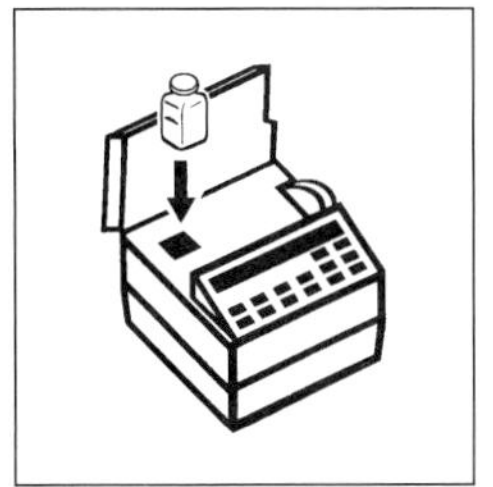

11. Press: **READ.** The display will show: **Reading . . .** then the result in mg/L iron will be displayed.

SAMPLING AND STORAGE

Collect samples in acid-cleaned glass or plastic containers. To preserve digested samples, adjust the pH to 2 or less with nitric acid (about 2 mL per liter). Preserved samples may be stored up to six months at room temperature. Adjust the pH to between 3 and 5 with 5.0 N Sodium Hydroxide Standard Solution before analysis. Correct the test result for volume additions; see "Correcting for Volume Additions" in Section 1 for more information.

If only dissolved iron is to be determined, filter the sample before acid addition.

ACCURACY CHECK

See p. 671 of the HACH Water Analysis Handbook, 3rd Ed., 1997.

INTERFERENCES

The following will not interfere below the levels shown:

Chloride	185,000 mg/L
Calcium	10,000 mg/L as $CaCO_3$
Magnesium	100,000 mg/L as $CaCO_3$
Molybdate Molybdenum	50 mg/L as Mo

A large excess of iron will inhibit color development. A diluted sample should be tested if there is any doubt about the validity of a result.

FerroVer Iron Reagent Powder Pillows contain a masking agent which eliminates potential interferences from copper.

Samples containing some forms of iron oxide require the digestion. After digestion adjust the pH to between 2.5 and 5 with ammonium hydroxide.

Samples containing large amounts of sulfide should be treated as follows in a fume hood or well ventilated area: Add 5 mL of hydrochloric acid to 100 mL of sample and boil for 20 minutes. Adjust the pH to between 2.5 and 5 with 5 N sodium hydroxide and readjust the volume to 100 mL with deionized water. Analyze as described above.

Highly buffered samples or extreme sample pH may exceed the buffering capacity of the reagents and require sample pretreatment; see "pH Interference" in Section 1.

QUESTIONS AND ANSWERS

1. What are the primary differences between ferrous ion and ferric ion in this experiment?
   Ferric compounds in solution have a more intense and bright color. Ferrous ions (Fe^{+2}) tend to oxidize to ferric (Fe^{+3}) ion in air or in the presence of oxidizing agents, whereas the ferric ion tends to be more stable. The ferrous cation complexes with 1,10-phenanthroline whereas ferric ion does not.
2. Why would a sample remain cloudy or have high turbidity after digestion?
   The sample may contain polymeric compounds or oxidized matter that will not completely dissolve in the solution.
3. Why is iron in solids, semisolids, gels, and oils sometimes difficult to detect?
   Iron exists as ferrous ion and ferric ion in colloidal states, as suspended particles, and as organic and inorganic complexes. Various types of iron compounds are found in silt, clay, ash and many other media. Some of the iron is free for analyses purposes, while additional iron compounds are bound and require digestion procedures.
4. The total iron concentration was determined to be 0.7 mg/L, while the ferrous iron concentration is 0.3 mg/L. What is the concentration of the ferric ion?
   Ferric ion concentration is 0.4 mg/L.

REFERENCES

1. *HACH Water Analysis Handbook,* 3rd Ed. (Loveland, CO.: HACH Company, 1997).
2. *LaMotte Soil Handbook* (Chestertown, Md.: LaMotte Company, 1994).
3. *Hawley's Condensed Chemical Dictionary,* 11th Ed., rev'd by N. Irving Sax and Richard J. Lewis (New York, N.Y.: Van Nostrand Reinhold, 1987).
4. *Standard Methods for the Examination of Water and Wastewater:* IRON, 16th Ed., A. E. Greenberg, R. R. Trussell, L. S. Clesceri and Mary Ann H. Franson (eds.), pp. 214–219, Washington, DC, American Public Health Association, 1985.

LIME

Lime has different concentrations of constituents based on the source. Lime or ground and pulverized limestone can contain calcium and magnesium with traces of iron, manganese, potassium, zinc, copper, and other ions. The concentration of the minor elements and magnesium in the limestone will depend on the site from which the limestone is obtained. Carbon lime generally lacks higher concentrations of magnesium and contains a limited number of the trace elements important for plant growth. In most cases, agricultural slag has a lower concentration of limestone and the minor elements in the mix than either carbon lime or lime.

Plants can exhibit a calcium deficiency. The growing tips of leaves and roots will turn brown and die.* The leaf edges tend to stick together and leaves may not fully emerge. Blossom-end rot on fruits will affect fruit quality. The decay of lower stem conductive tissue may produce plants that wilt easily.

Calcium excess can produce plants exhibiting a deficiency in magnesium and potassium.* A potassium deficiency is noted by burned edges of leaves, disease infestation, and stunted plants. Poor seed and fruit production leads to poor quality product. Magnesium deficiency is noted by interveinal chlorosis (yellowing between the veins). A beaded-streaking appearance can appear within the chlorotic stripe. The growth of plants will decrease and a number of plants will become diseased.

Iron deficiency can be observed by interveinal chlorosis on young leaves and bleaching of new growth.* The plant may turn light green in color and growth may be slowed. A manganese deficiency can display interveinal chlorosis of young leaves, but the leaves remain green in color. Plant growth will slow and white streaks will appear on the leaves. Manganese deficiency affects forage legumes such as alfalfa. Excess manganese can be identified by small black spots on the stems and fruit, commonly referred to as *measles.* Brown spots in a chlorosis zone are commonly observed.

Copper and zinc excesses may lead to iron deficiency in plants.* Plants will be stunted and die. Deficiencies in copper and zinc will lead to slow plant growth. With copper deficiencies, the plant will appear to be limp with distorted leaves. Zinc deficiencies will display interveinal chlorosis and whitening on the side of the leave mid-rib.

The outline of mineral deficiencies and excesses is applicable to many types of plants. Some plants require control of added nutrients. Additional information about mineral deficiencies and excesses in plants can be obtained by reference to the *Handbook of Reference Methods for Plant Analysis,* edited by Y. P. Kalra.

---

*Robert D. Munson, *Handbook of Reference Methods for Plant Analysis,* Yash P. Kalra (ed.) (Boca Raton, Fla.: CRC Press, 1998), pp. 7–10.

LIME REQUIREMENT—WOODRUFF METHOD

Procedure*

1. Use the 10g soil measure to add one level measure of the soil sample to a 50 mL beaker. Then, add 10 mL deionized water using a graduated cylinder to the beaker. Stir thoroughly.

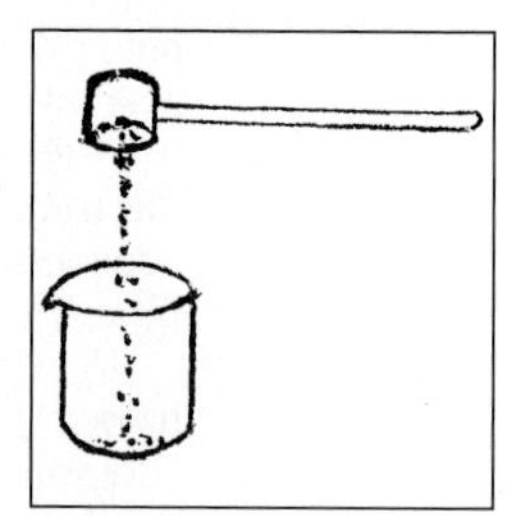

2. Let stand for at least 15 minutes.

3. Add 20 mL of Woodruff Buffer Solution. Mix well, and let stand for at least 20 minutes, stirring two or three times.

4. Read on pH meter. Stir mixture just prior to making reading. *Note:* Each 0.1 pH unit drop from pH 7.0 indicates a lime requirement equivalent to 1000 lbs calcium carbonate ($CaCO_3$).

QUESTIONS AND ANSWERS

1. What is lime?
Lime is generally calcium oxide and calcium hydroxide. Besides calcium, quality lime

---

* Modified Text-Compliments of LaMotte Company.

products tend to be a "grayish" color and can contain magnesium and traces of iron, manganese, potassium, molybdenum, zinc, copper, and other soluble ions. Several types of lime are almost pure white and do not contain a number of the elements. Agricultural slag has a granular structure and contains less calcium than lime. There are a variety of grades of lime dependent on the source and the manufacturing process.

2. Why is lime so important in soil?
Lime is primarily used to adjust the pH of soil. Many crops require a specific pH range which is satisfactory for quality plant growth and high crop yields. A calcium deficiency in soil results in poor plant growth, brown leaves and discolored root tips, stunted leaf growth, leaves that are ragged and stick together, decay of the lower stem conductive tissue resulting in wilting problems, blossom-end rot and poor fruit production. Excess calcium can exhibit magnesium and potassium deficiencies.

3. How much calcium is required to neutralize a soil with a pH of 6.8?
Each 0.1 pH unit drop below pH = 7.0 indicates a lime requirement equivalent to 1000 lbs of high quality lime. The pH (drop) = 0.2 and the lime requirement equivalent is 2000 lbs of high quality lime. About 1 to 2 tons of lime per acre would be recommended for many crops. Some plants may require lime containing a specific element for quality growth.

REFERENCES

1. *LaMotte Soil Handbook,* (Chestertown, Md.: LaMotte Company, 1994).
2. *Handbook of Reference Methods for Plant Analysis,* Yash P. Kalra (ed.), (Boca Raton, Fla.: CRC Press, 1998).

LEAD

Lead accumulates in the body and is a poison. It is absorbed through the GI tract and pulmonary system. Lead and some lead compounds tend to accumulate in the bone structure when ingested in excessive amounts. It affects the central and peripheral nervous systems and the kidneys. Lead is deposited in blood and inhibits the synthesis of hemoglobin. Accumulation of lead in the body may cause permanent brain damage, convulsions and death. Toxic effects of lead are attributed to enzyme inhibition.

As a result of the potential for lead poisoning, a national program to reduce the concentration of lead in consumer products has been implemented. Some lead compounds are toxic and have a tolerance (as Pb) of 0.15 mg per cubic meter of air. Calcium disodium EDTA salts complex lead in the body. The salts can be administered to a person who has consumed lead paint and any additional lead- contaminated substance.

Lead exists as a metal, inorganic compounds and as organometallic complexes. Sources of lead are industrial mines, industrial ash, smelter discharges, and old pipes used in plumbing. Lead compounds are used in varnish driers, waterproofing paints, rust inhibitors, paint primer component and pigments, stained glass, porcelain, insecticides, herbicides, rubber, explosives, radiation shielding, cable coverings, ammunition, piping and tank linings, reaction equipment, solders, alloys, bearings, ceramic glazes and coatings, battery components, electrodes, electronic and optical applications, infrared detector, bronzing, high pressure lubricants, semiconductor, radiation detectors, photoconductors, electronic and optical applications, lasers, matches, industrial wastes, and printing. Additional uses of lead include stabilizers, organic preparations, plastic, metallurgy, analytical reagents, medicine, and exterior paints. In addition, some types of paper and other materials adsorb lead compounds from aqueous solutions. Many lead compounds are not soluble in water or have very limited solubility in water. A number of lead compounds are slightly soluble under either acidic or basic conditions. A number of insoluble lead compounds can form more soluble compounds in the human body and under decaying environmental conditions.

Some industries manifest wastes containing lead for burial or reuse by other industries. A few industries utilize wastes containing small amounts of lead to produce other products. Organic wastes containing low levels of lead are sometimes burned in emission-controlled incinerators to recover energy. Wastes contaminated with lead are often disposed in landfills and by inappropriate procedures.

Lead from solid samples, gels, sludge and semisolid samples weighing 0.100 to 0.500 grams can be digested in a known volume of sulfuric acid. The digestion procedure for oils requires 0.100 to 0.250 grams of material diluted to a specific volume. After boiling for 4 to 5 minutes, 10 mL of 50% hydrogen peroxide is added to the charred sample and the sample is heated for 1 minute. The digested mixture is allowed to cool and is diluted to 100 mL with deionized water. After selecting the appropriate analysis volume and diluting the solution to 250.0 mL, the clear sample can be analyzed by the HACH procedure (Method 8033 or Method 10064) provided in Chapter 2. A pH adjustment on the sample will be required before analysis for lead.

Metals such as bismuth, copper, mercury, silver and tin can interfere with the analysis for lead. These interferences can be minimized by an extraction procedure with chloroform. In the experiment, chloroform (D002) and cyanide (D003) solutions are regulated as hazardous wastes by the Federal RCRA. These wastes should be collected for disposal with laboratory solvents.

Lead ions in basic solution react with dithizone to form a pink to red complex which can be extracted with chloroform. The pink lead-dithizone complex is somewhat light sensitive and should be kept out of direct sunlight. The spectrophotometric measurement for lead at 515 nm is read in micrograms per liter. Digestion of wastewater sludge samples is required for USEPA reporting.

The amount of lead in samples of sediments, sludges, soils, paint and other materials can also be dissolved by the Lead Extraction Method and various digestion procedures. After using the lead extractor, a HACH colorimetric analysis using the DR/2010 can be performed on the solution containing the lead. This procedure may be more appropriate for higher concentrations of tightly bound lead in mixtures.

QUESTIONS AND ANSWERS

1. What are some of the sources of lead and lead compounds in our environment?
The sources of lead and lead compounds are varied. The nonsoluble forms of lead compounds can spread in our environment by being exposed to slightly acidic or basic conditions and by solubilizing under inappropriate disposal conditions. Additional forms of lead are found after paint components chip or scale from items. Industrial slag containing significant quantities of lead has been used improperly. Improper disposal and the lack of recovery of lead contaminated sludges and other wastes contribute to lead contamination.

2. How do you avoid problems with cyanide in the experiment?
Before cyanide is added to the mixture, 5.0 N sodium hydroxide is added to the solution until the pH is a minimum of 11.0. Adding 5 extra drops of 5.0 N sodium hydroxide solution provides an extra sense of security.

3. Why should filter paper be avoided when filtering a cloudy or turbid sample?
Filter paper adsorbs lead. Instead, use a glass membrane filter or a sintered glass filter.

4. Why should this experiment be performed in a hood or with proper ventilation?
This experiment uses chloroform and potassium cyanide which require protection and open air. Both chemicals are regulated as hazardous wastes by the Federal RCRA. Do not pour these chemicals down the drain. The chemicals should be collected for disposal with laboratory solvent waste.

5. Using the Dithizone Method, what units are used to describe the reading from the DR/2010 Spectrophotometer?
The units of measurement from the DR/2010 Spectrophotometer are micrograms per liter.

REFERENCES

1. *HACH Water Analysis Handbook,* 3rd ed. (Loveland, CO.: HACH Company, 1997).
2. *Hawley's Condensed Chemical Dictionary,* 11th ed., rev'd by N. Irving Sax and Richard J. Lewis (New York, N.Y.: Van Nostrand Reinhold, 1987).
3. *Standard Methods for the Examination of Water and Wastewater:* LEAD, 16th ed., A. E. Greenberg, R. R. Trussell, L. S. Clesceri and Mary Ann H. Franson (eds.), pp. 221–223, Washington, DC, American Public Health Association, 1985.

MANGANESE

Manganese dust or powder is flammable. Quadrivalent manganese is usually in a suspension in water. Manganese has a tendency to be oxidized to a higher oxidation state and precipitates or becomes adsorbed to the collection container walls. The determination of total manganese usually requires a digestion.

Manganese compounds are generally not very toxic, and they can be easily modified to a less toxic form by oxidation-reduction processes. Some forms of manganese products are recovered and are recycled waste materials. Many of the compounds are not soluble in water or have limited solubility in water. Naturally occurring bacterial processes, pH changes, and oxidation-reduction processes tend to solubilize some manganese compounds in the environment.

Manganese appears to play an important role in some enzyme reactions and is required for the formation of chlorophyll in the plant. The amount of manganese available to the plant is dependant upon the soil pH, the quantity of organic matter present, and the degree of aeration.* Manganese deficiency is most likely to occur in neutral or alkaline soils because it is less soluble at elevated pH levels. Since manganese is more soluble in acid soils, increases in toxic levels of manganese may reduce crop yields to some degree. Also, manganese forms insoluble organic complexes in some soils that have high humus content. Tests can determine the level of soluble manganese in soil. After dry ashing techniques, the same tests can determine the amount of manganese in plant tissue and additional substances.

Manganese compounds are used in varnish and oil driers, deoxidizers, paints, ceramics, bleaching tallow, metal conversion coatings, dyeing processes, catalysts, pharmaceutical preparations, fertilizer, feed additives and dietary supplements, pyrotechnics, matches, oxidizing agent, decolorizers, scavenger, colored glass, bleaching tallow, coatings of metals, textile printing, medicines, fungicides, antiknock agents and oxidizing agents. Manganese metal is primarily used as a ferroalloy (steel manufacture) and a nonferrous alloy (improves

*Compliments of LaMotte Company.

hardness and corrosion resistance). The metal is a purifying and scavenging agent in metal production. A number of the processes used in preparing these products deposit more soluble manganese in the environment than the products themselves.

Manganese from a solid sample, gel or semisolid sample weighing 0.100 to 0.500 grams can be digested in a known volume of acid. The digestion procedure for oils requires 0.100 to 0.250 grams of material diluted to a specific volume. After boiling for 4 to 5 minutes, 10 mL of 50% hydrogen peroxide is added to the charred sample and the sample is heated for 1 minute. The digested mixture is allowed to cool and is diluted to 100 mL with deionized water. After selecting the appropriate volume and diluting the solution to 25.0 mL, the clear sample can be analyzed by either of the two procedures provided below. A pH adjustment for the sample will be required before metal analysis.

The PAN Method (HACH, Method 8149) is a sensitive, rapid procedure for determining low levels of manganese at 0 to 0.700 mg/L.* An ascorbic acid reagent reduces all oxidized forms of manganese to Mn^{+2}. An alkaline-cyanide reagent masks interferences and the PAN Indicator forms an orange-colored complex with Mn^{+2}. For levels higher than 0.700 mg/L manganese, the sample will have to be diluted appropriately or the periodate oxidation method will have to be implemented on the sample. The amount of manganese in the samples and the goals of the measurements will dictate which field instrument or method is most appropriate. For a number of measurements, the DR/2010 Spectrophotometer and the DR/890 Colorimeter will be required for the HACH procedure provided in Chapter 2.

In the LaMotte experiment, the Periodate Method for determining free manganese in solids, gels and other samples ranges from 0 to 100 ppm.** Interferences are reducing substances capable of reacting with periodate or permanganate. The interferences must be eliminated in the experiment. Small amounts of chlorine can be oxidized by periodate.

MANGANESE TEST

Periodate Method**—Procedure

1. Fill a clean colorimeter tube to the 10 mL line with the soil filtrate.

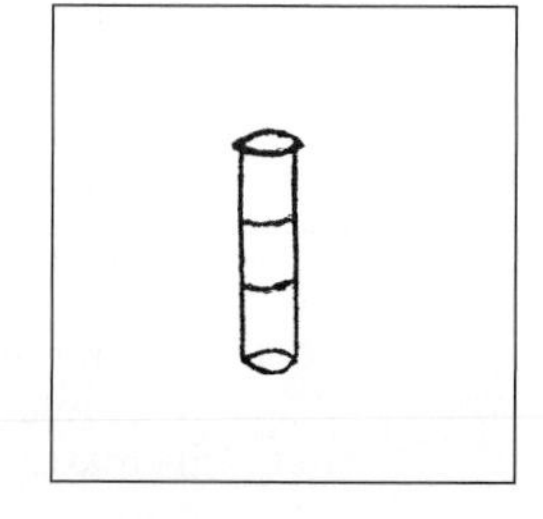

2. Neutralize the soil filtrate by adding 15% Sodium Hydroxide Solution to the soil filtrate, one drop at a time while stirring the plastic rod. The stirring rod is touched to the Bromthymol Blue test paper after the addition of each drop of sodium hydroxide until the color changes from yellow to green or blue.

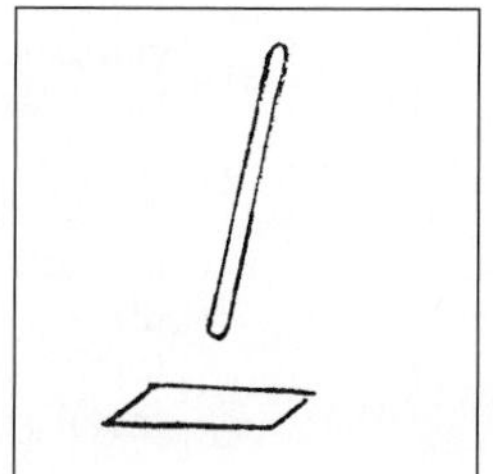

*Compliments of HACH Company.

**Compliments of LaMotte Company.

3. Select setting 4 on the "Select Wavelength" knob.

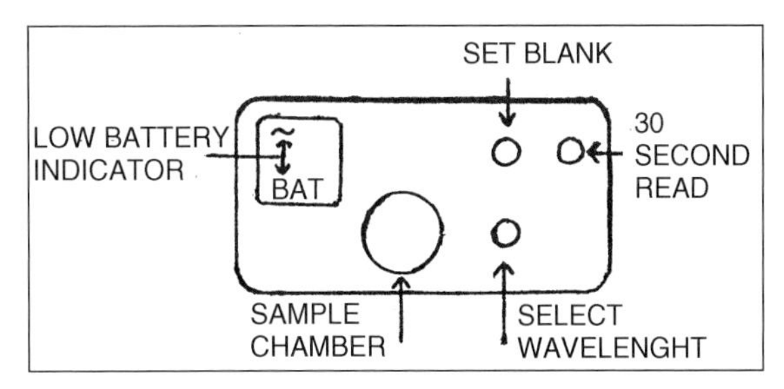

4. Insert the tube containing sample into chamber and press the "30 Second Read" button. Adjust to "100"%T with the "Set Blank" knob.

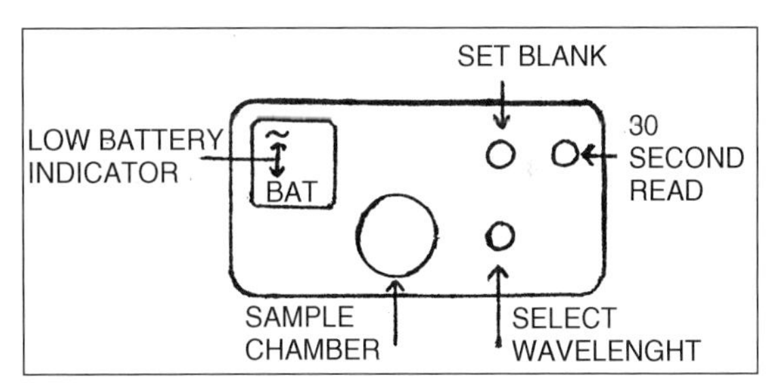

5. Remove the tube and add 2 level measures of Manganese Buffer Reagent with the 0.1 g spoon. Cap and mix to dissolve the powder.

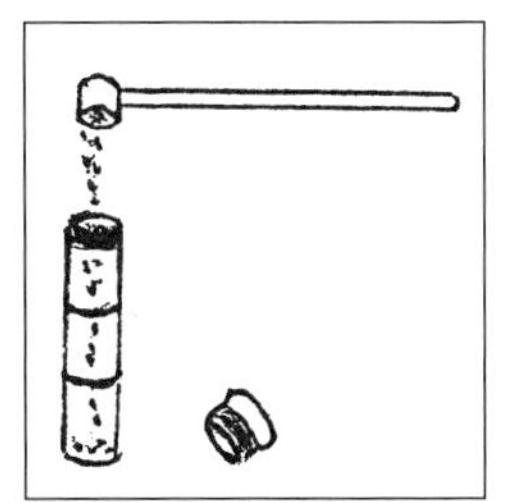

6. With the a clean 0.1 g spoon, add 1 heaping measure of Manganese Periodate Reagent to the contents of the tube, cap and mix. An undissolved portion of the reagents may remain in the bottom of the tube without adversely affecting the results.

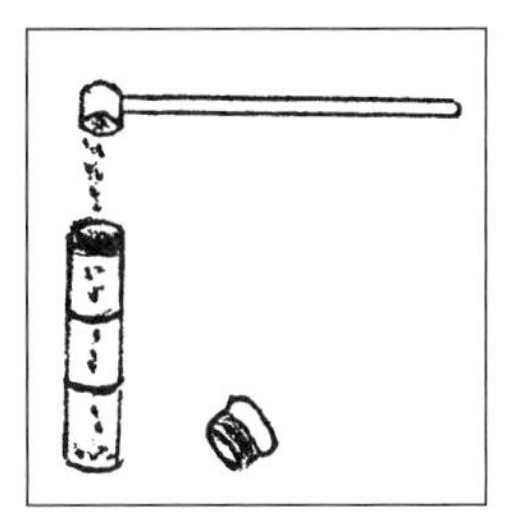

7. Allow approximately 2 minutes for the pink color to develop if manganese is present.

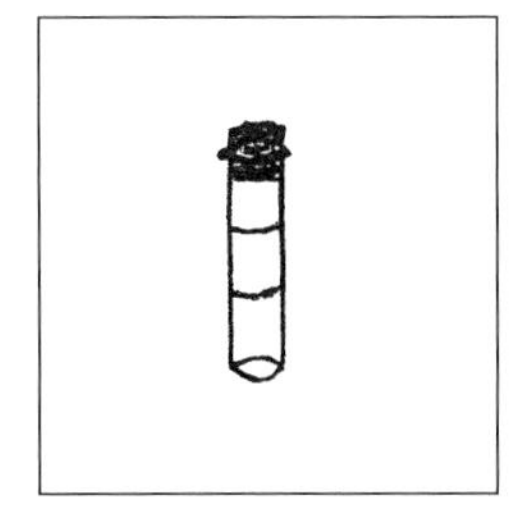

8. Insert test sample into colorimeter chamber, press the "30 Second Read" button, and measure %T as reading stabilizes.

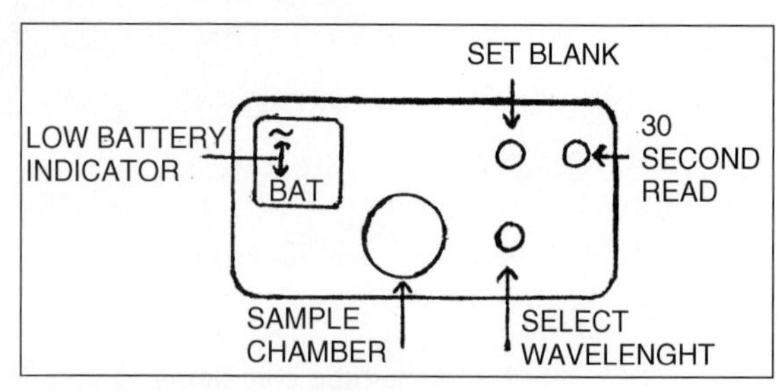

9. Consult the chart below to find the concentration of manganese in parts per million (ppm).

Manganese Calibration Chart (PPM)

%T	9	8	7	6	5	4	3	2	1	0
90		0.0	0.4	0.8	1.2	1.6	2.0	2.4	2.8	3.2
80	3.6	4.0	4.5	4.9	5.3	5.8	6.2	6.7	7.1	7.6
70	8.0	8.5	9.0	9.5	9.9	10.4	10.9	11.4	12.0	12.5
60	13.0	13.5	14.1	14.6	15.2	15.7	16.3	16.9	17.5	18.1
50	18.7	19.3	19.9	20.6	21.2	21.9	22.6	23.2	23.9	24.6
40	25.4	26.1	26.8	27.6	28.4	29.2	30.0	30.8	31.7	32.5
30	33.4	34.3	35.3	36.2	37.2	38.2	39.2	40.3	41.4	42.5
20	43.7	44.9	46.1	47.4	48.7	50.1	51.5	53.0	54.6	56.2
10	57.9	59.7	61.6	63.6	65.7	67.9	70.3	72.8	75.6	78.6
0	81.9	85.5	89.6	94.3	99.7					

Manganese Concentration Chart

%T	Range	Parts per million
86–100%	Low	0–5 ppm
71–85%	Medium	5–12 ppm
51–70%	Medium High	13–24 ppm
32–50%	High	25–40 ppm
0-31%	Very High	Over 40 ppm

HAZARDOUS SUBSTANCES

Read the MSDS before using the chemicals. The following substance is a hazardous substance and should be disposed of and handled properly. The hazardous substance is manganese periodate reagent.

RANGE OF MEASUREMENT

The range of measurement is 0 to 100 ppm.

INTERFERENCES

Reducing substances capable of reacting with periodate or permanganate must be eliminated. Chloride in small amounts can be oxidized by periodate.

MANGANESE, LR (0 to 0.700 mg/L)

Method 8149*

PAN Method**

1. Enter the stored program number for manganese (Mn). Press: **2 9 0 ENTER.** The display will show: **Dial nm to 560.** *Note:* The Pour-Thru Cell can be used with 25-mL reagents only.

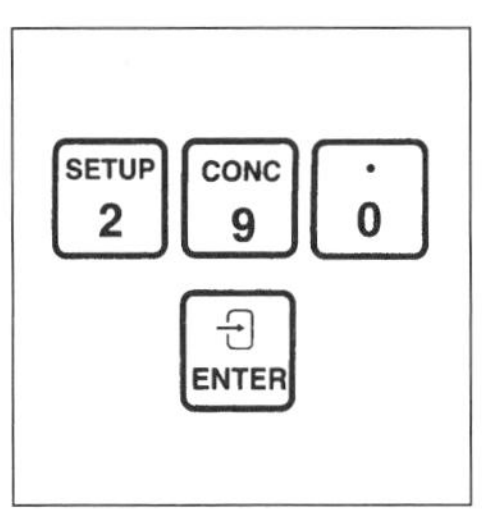

2. Rotate the wavelength dial until the small display shows: **560 nm.** When the correct wavelength is dialed in, the display will quickly show: **Zero sample,** then: **mg/L Mn LR (Low Range).**

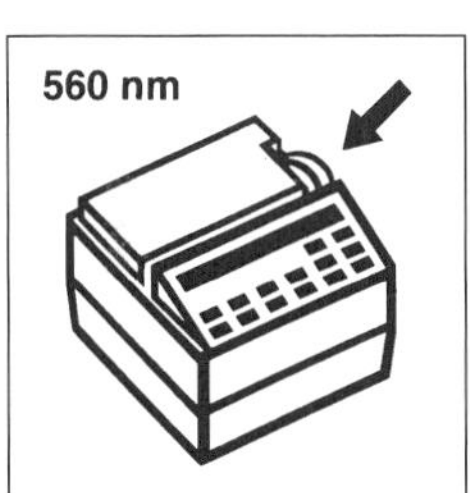

3. Insert the 10-mL Cell Riser into the cell compartment. *Note:* Total manganese determination requires a prior digestion.

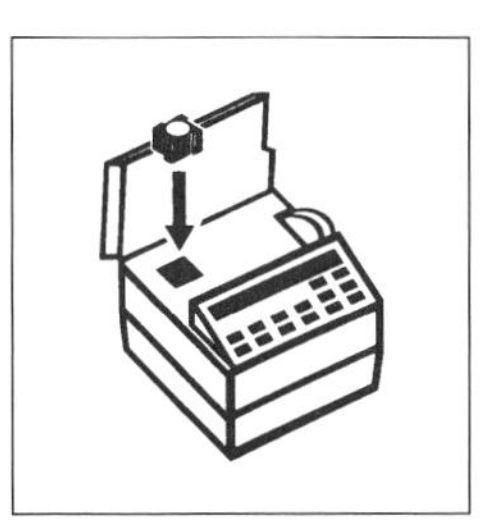

*Compliments of HACH Company. Analysis may be performed by DR/890 Colorimeter and DR/2010 Spectrophotometer.

**Adapted from K. Goto, et al., Talanta, 24, 752–3 (1977).

4. Fill a 10-mL sample cell with 10 mL of deionized water (this will be the blank).
Note: Rinse all glassware with 1:1 Nitric Acid Solution. Rinse again with deionized water.

5. Fill another 10-mL sample cell with 10 mL of sample (this will be the prepared sample).

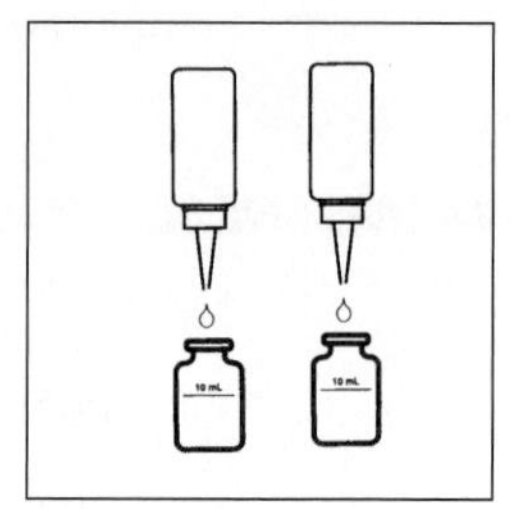

6. Add the contents of one Ascorbic Acid Powder Pillow to each cell. Swirl to mix.
Note: For samples containing hardness greater than 300 mg/L $CaCO_3$, add four drops of Rochelle Salt Solution to the sample after addition of the Ascorbic Acid Powder Pillow.

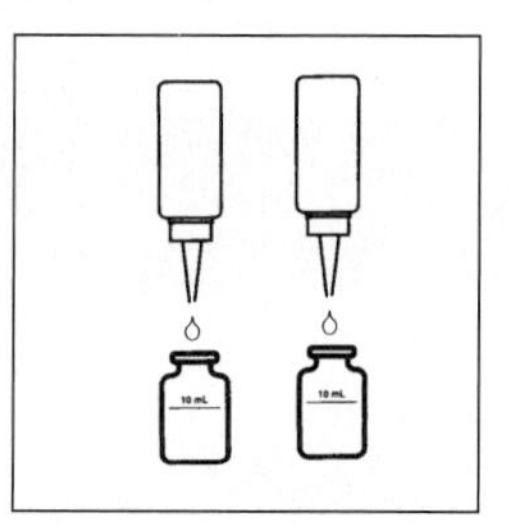

7. Add 15 drops of Alkaline-Cyanide Reagent Solution to each cell. Swirl to mix.
Note: A cloudy or turbid solution may form in some samples after addition of the Alkaline-Cyanide Reagent Solution. The turbidity should dissipate after Step 8.

8. Add 21 drops of PAN Indicator Solution, 0.1%, to each sample cell. Swirl to mix.
Note: An orange color will develop in the sample if manganese if present.
Note: For 25-mL reagents, use 1 mL of each liquid reagent in Steps 7 and 8.

9. Press: **SHIFT TIMER.** A two-minute reaction period will begin.
Note: If the sample contains more than 5 mg/L iron, allow ten minutes for complete color development. To set the timer for 10 minutes, press 1000 SHIFT TIMER.

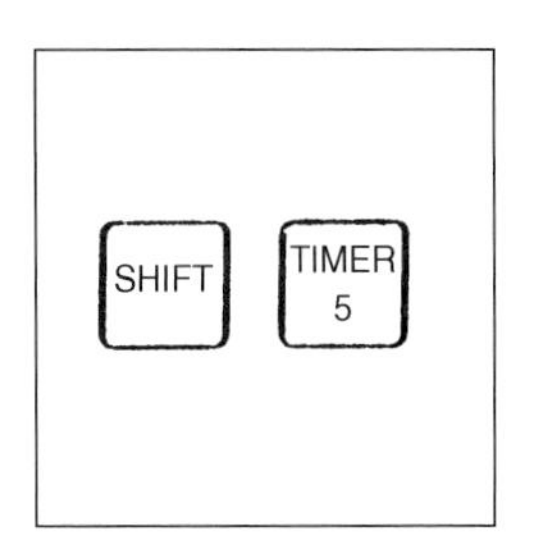

10. When the timer beeps, the display will show: **mg/L Mn LR.** Place the blank into the cell holder. Close the light shield.

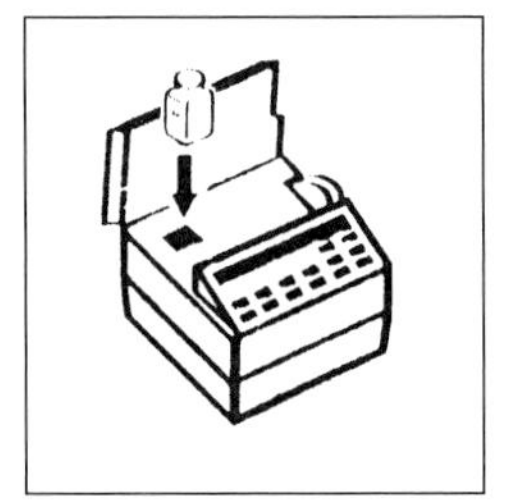

11. Press: **ZERO.** The display will show: **Zeroing . . .** then: **0.00 mg/L Mn LR.**

12. Place the prepared sample into the cell holder. Close the light shield. Press: **READ.** The display will show: **Reading . . .** then the result in mg/L manganese will be displayed.
Note: See Waste Disposal below for proper disposal of cyanide containing wastes.

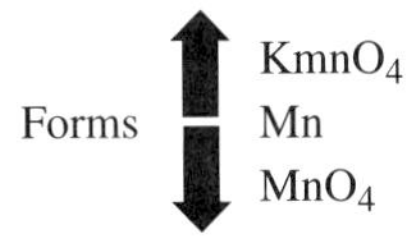

SAMPLING AND STORAGE

Collect samples in a clean glass or plastic container. A solution containing dissolved manganese should be adjusted to a pH of 2 or less with nitric acid (about 2 mL per liter) before storage. Preserved samples can be stored up to six months at room temperature. Adjust the pH to 4.0 to 5.0 with 5.0 N sodium hydroxide before analysis. Correct the test result for volume additions; see "Correction for Volume Additions" in Section 1.

ACCURACY CHECK

See p. 715 of the HACH Water Analysis Handbook, 3rd Ed., 1997.

INTERFERENCES

The following do not interfere up to the indicated concentrations:

Aluminum	20 mg/L
Cadmium	10 mg/L
Calcium	1000 mg/L as $CaCO_3$
Cobalt	20 mg/L
Copper	50 mg/L
Iron	25 mg/L
Lead	0.5 mg/L
Magnesium	300 mg/L as $CaCO_3$
Nickel	40 mg/L
Zinc	15 mg/L

WASTE MANAGEMENT

The alkaline cyanide solution contains cyanide. Cyanide solutions should be collected for disposal as reactive (D001) waste. Be sure cyanide solution are stored in a caustic solution with pH > 11 to prevent release of hydrogen cyanide gas.

QUESTIONS AND ANSWERS

1. If a dissolved sample contains 10 ppm manganese, can the analysis be performed by the PAN Method?
The sample will have to be diluted with distilled water to a level below 0.700 mg/L (1 ppm = 1mg/L). Alternatively, the Periodate Oxidation Method will have to be used for analysis.

2. Why is a digestion required for total manganese?
In solution, manganese can form a stable soluble complex, a partially soluble suspended material, and a precipitate. Also, some manganese compounds can adsorb to the walls of a container.

3. What are some of the uses of manganese?
Manganese has four allotropic forms, of which alpha is the most important. Manganese metals are used in thermocouples, electrical instrument parts, alloys, and metal coatings. Manganese can be used to form ferroalloys and nonferrous alloys. The metal improves

corrosion resistance and hardness of nonferrous alloys. Manganese is a purifying and scavenging agent in metal production. Manganese is used to manufacture aluminum by the Toth Process.

4. What is the purpose of the alkaline-cyanide reagent during the determination of manganese by the PAN Method?
The alkaline-cyanide reagent masks any potential interferences.

5. What is the purpose of the ascorbic acid reagent during the determination of manganese?
The ascorbic acid reagent reduces all forms of manganese to Mn^{+2}.

REFERENCES

1. HACH Water Analysis handbook, 3rd Ed. (Loveland, CO.: HACH Company, 1997).
2. *LaMotte Soil Handbook* (Chestertown, Md.: LaMotte Company, 1994).
3. *Hawley's Condensed Chemical Dictionary,* 11th Ed., rev'd by N. Irving Sax and Richard J. Lewis (New York, N.Y.: Van Nostrand Reinhold, 1987).
4. Standard Methods for the Examination of Water and Wastewater: MANGANESE, 16th Ed., A. E. Greenberg, R. R. Trussell, L. S. Clesceri and Mary Ann H. Franson (eds.), pp. 228–229, Washington, D.C., American Public Health Association, 1985.

MERCURY

Large quantities of mercury metal, inorganic mercury (I) and mercury (II) compounds are used annually. Mercury exists in a variety of chemical forms such as liquid mercury metal, simple inorganic and organic solids, and complexes of mercury. Mercury metal and many mercury compounds are highly toxic by ingestion, inhalation, and absorption. Mercury causes neurological damage, mutations and, at high levels, death of humans. Microscopic droplets of mercury can easily hide in crevices and cracks. Normal vibrations in the work environment will increase the vaporization and distribution of mercury metal.

Precious metals (i.e., gold, silver) have been extracted from ores by treatment with mercury. A number of metals (i.e., copper dust) form insoluble amalgams with mercury and are used to clean-up spilled mercury*. Amalgams, used for filling teeth, are quite stable and do not appear to degrade significantly. If the amalgam falls out of the tooth, the solid mercury containing the filling passes out of the body without any significant harm to the body. When a person ingests mercury, the doctor prescribes "Penicillamine" or "Polythiol Resins" which chelates mercury metal to form an insoluble mercury complex. The insoluble mercury complex will pass harmlessly through the body. In addition, EDTA has been previously used to complex mercury in the body to form a less toxic substance.

Most organic and inorganic mercury compounds are very toxic and should be monitored closely in the environment. Many of the mercury compounds remain as stable complexes in our environment and do not solubilize easily. The nonsoluble mercury compounds do not affect humans and animals as much as soluble mercury compounds. However, insoluble mercury compounds tend to solubilize under the appropriate conditions with time.

Mercury metal has been identified in the organs and tissues of fish, turtles and other sea life. Mercury has a tendency to accumulate in plant roots and may accumulate in other plant parts**. Mercury metal appears to become more soluble in water through decay processes, increases in temperature, pH changes, and various organic reaction mechanisms. Sewage

---

* Gershon Shugar, *Chem. Tech. Ready Reference Handook,* 4th ed., 1996.
** *Handbook of Reference Methods for Plant Analysis,* CRC Press, p. 187.

effluent sometimes contain up to ten times the level of mercury found in typical natural waters. Fossil fuel and lignite contain mercury at levels near 100 parts per billion, which creates some concern with increased use of these fuels as energy resources. Compounds of mercury, including mercury metal, have a tendency to cling to ash and particulates from emission processes during the burning of coal and other solid fuels. Mercury metal and a number of mercury compounds can be found in many types of waste sites and as throw-away items from industry, commercial business, schools, laboratories and homes. Products and processes containing mercury compounds and mercury metal are numerous. Based on recent studies, mercury levels continue to increase in our environment.

The solubility of mercury metal in water is extremely small, but increases by 15-fold from 0°C to 30°C. In addition, a number of mercury compounds that have limited solubility in water become somewhat more soluble as the temperature of the water increases above 0°C. Mercury metal becomes more solubilized in water through the formation of complexes from the reaction of products of decomposition, such as methane, to form methylated mercury compounds in the environment. The high concentration of mercury found in water and in fish tissue can be attributed to the formation of soluble monomethylmercury ion and volatile dimethylmercury. These compounds are produced by anaerobic bacteria in sediments. In addition, a number of insoluble and partially soluble mercury compounds can slowly form more soluble compounds by reaction with organic complexes and weak acidic solutions. Basic conditions will decompose some of the mercury compounds and allow a number of reactions and processes to occur in the environment.

Mercury compounds and liquid mercury have been used in the following areas: pharmaceutical products, batteries, thermoscopy, analytical reagents, polishing compound, cosmetics, perfumery, germicide, antiseptic, medicine, waterproof paints, pigments, sterilant, fungicide, instruments, catalyst, pyrotechnics, amalgam tooth fillings, photography, electrodes, antifouling paints, maggot control, silvering mirrors, slimicide, mold retardant, explosives, infrared detectors, vapor lamps, arc lamps, nuclear power plants, ultrasonic amplifiers, ceramics and semiconductors. Topical dressings often containing insoluble mercury compounds are used to treat wounds and to prevent infection. Equipment measuring pressure, temperature, flow rates and a variety of parameters primarily use liquid mercury. Items such as mercury-containing switches and lamps are used every day.

In industry, mercury metal and several toxic mercury compounds require a closed loop system for recovery and reuse. A number of partially soluble mercury wastes have been buried in the ground and recycled. Some wastes containing partially soluble and insoluble mercury compounds have been dumped in undesirable places in the past. Environmental monitoring must be extensive and necessary because many insoluble and slightly soluble mercury compounds are considerably toxic and do not change or decompose easily.

Inhalation of mercury vapor is hazardous to animals and people. Mercury metal should be used in areas of adequate ventilation. Exposed mercury metal in a container should be covered with a layer of water to inhibit mercury vaporization. If a mercury spill occurs, a mercury spill kit containing a powder, such as "Cinnasorb," can be mixed with water to form a paste, which forms an amalgam with mercury droplets. A powder called "Resisorb" can be sprinkled into cracks and crevices to minimize mercury vapors. In any case, mercury spills should be collected and cleaned up.

Mercury is difficult to analyze and analysis procedures are expensive compared to other metals. Due to the sensitivity of the procedure, dedicated digestion glassware and sample cells are suggested for the determination of mercury. Glassware must be cleansed thoroughly and rinsed with 1:1 HCL solution followed by several deionized water rinses. In this experiment, extra care must be taken to avoid chemical exposure. Proper protection of personnel and appropriate handling procedures must be implemented at all times. Digestion and extraction procedures must be performed in a properly vented laboratory hood. Also, waste disposal of the mercury waste can be a major problem. Any solution containing above 0.2 mg/L mercury must be destroyed as a hazardous waste as defined by RCRA.

Samples containing water insoluble mercury compounds require an extensive level of digestion before mercury analysis. Most routine methods for the determination of total mercury in biological samples involve a two-step procedure: 1) the conversion of all bound Hg in the sample to Hg (II) by wet oxidation, and (2) the reduction of Hg (II) to mercury vapor for analysis.[1] A successful digestion method for a variety of biological samples uses an HNO_3-H_2SO_4-V_2O_5 mixture to dissolve the metal. An oxidizing agent such as permanganate is added to minimize loss of mercury. To reduce losses of mercury, digestions should be done in closed vessels.

The amount of solid or sludge used for extraction purposes depends upon the estimated concentration of total mercury compounds in the sample. Generally, a dried 0.25 to 0.50 gram solid sample (plant tissue, soil, fish, peat, etc.) ground to a 0.5-mm particle size is added into a Folin-Wu digestion tube or a pre-weighed 100 mL digestion flask. Then, 50 ($\pm$5) mg V_2O_5 powder followed by 5 mL HNO_3 is carefully added into the digestion tube. After foaming has subsided, the closed tube is heated in a block digester (preheated to 160°C) for 5 minutes in a vented hood. The closed tube is removed from the block digester and allowed to cool. Then, 5 mL of concentrated sulfuric acid is added to the cooled digestion tube, and the closed tube is heated in the block digester for an additional 15 to 20 minutes. The tube is removed from the block digester and allowed to cool. The final digested material is made up to 50 mL total volume with deionized water.

With the hydride generation technique, a small amount of $NaBH_4$ in an acidic medium reduces the formation of mercury metal during the cold vapor separation technique. For additional sample digestion and analysis procedures about total mercury in solids, see the *Handbook of Reference Methods for Plant Analysis,* CRC Press, ISBN 1574441248. After cold vapor separation and preconcentration of mercury, the digested sample can be analyzed by colorimetric analysis. Alternatively, the sample can be determined by atomic absorption methods.

Solubilized forms of mercury in solids (soil, peat, fish, gels, plants, animal and human organs) can be analyzed by the same digestion procedures used for the liquid after the mercury is extracted from the solid or gel sample. In this case, a solid sample is ground or chopped to a 0.5-mm particle size on a surface that will not react with, absorb or adsorb free mercury. A mortar and pestle are recommended for reducing sample size. All sample preparation surfaces should be rinsed three times with 1.0 N nitric acid, and the rinse added to the glass stoppered flask before sample dilution to 1.0 liter. Then, 0.5 grams of a "ground" representative sample of solid material are carefully extracted in a 1.0 liter solution (including rinse waters) of 1.0 N nitric acid solution (55.8 mL conc. HNO_3 diluted to 1.0 liter final volume with distilled or deionized water) by stirring for a minimum of 12 hours in a glass stoppered flask at room temperature. After dissolution of the mercury, pour 450 (448) mL of the acidic solution into a separate flask and dilute the solution with deionized or distilled water to 1.0 liter. Carefully add 50 mL of concentrated sulfuric acid to 1.0 liter of the acidified solution and continue with step 4 of the sample digestion procedure (Phase 1) by adding 4.0 to 5.0 grams of potassium persulfate to the sample and stirring. Continue with the remaining steps of sample digestion, cold vapor separation and preconcentration of mercury, and colorimetric analysis at 412 nm.

This experiment determines the soluble forms of mercury in solids, semisolids, and sludges. Sample digestion converts various forms of soluble mercury compounds to mercuric (+2) ions. The mercuric ions in the digested sample are changed to mercury vapor, which is converted to mercuric chloride by a chemically active absorber column. After elution of the mercuric chloride from a specific column, a sensitive indicator is added to the solution to form a mercury indicator complex. Later in the experiment, this complex is broken apart to form the indicator and the metal. The calorimeter is zeroed using the absorbance peak (412 nm) of the unreacted indicator. The measurement of the solution where the indicator and the metal has been broken apart provides evidence of the presence of mercury. The increase in solution absorbance is directly attributed to the concentration of mercury in the solution.

After mercury is released from materials by the digestion procedures in a closed bubbler system, the solution can also be analyzed by a MAS-50D Mercury Analyzer.* The MAS-50D Analyzer utilizes the EPA-approved Hatch and Ott Cold Vapor method and features a sensitivity of 0.01 micrograms mercury. The Mercury Analyzer provides reproducible results at 253.7 nm.

MERCURY IN SOLIDS (0.1 to 2.5 μg/L)

Cold Vapor Mercury Concentration Method**
Adaptation of Method 10065†

Sample Preparation‡

Grind or crush the solid sample with a mortar and pestle to a 0.5 to 1.0 mm particle size. Some samples will require special devices to minimize particle size. If necessary, add the solid sample to a glass plate and dry the sample in a dessicator containing silica gel or an appropriate drying agent at room temperature. Weigh 0.500 grams of the dried, ground sample on a glass plate. Add the weighed solid material to 400 mL of 1.0 N HNO_3 acid solution (Solution A) in a glass stoppered flask. Carefully rinse the mortar, pestle, and glass plate a minimum of three times with 1.0 N nitric acid and add the rinses directly to Solution A. Add 1.0 N HNO_3 acid to bring the final volume of Solution A to 500 ml.

Phase 1 Sample Digestion: Must be done in a hood! Toxic gases may be produced

1. Dilute Solution A to 1.0 liter total volume with 1.0 N nitric acid solution. Place a glass stopper on the flask. Label the 1.0 liter solution as Solution B.
Note: This procedure must be done in a fume hood. Toxic gases may be produced.
Note: HACH recommends using dedicated digestion glassware and sample cells for this procedure.
2. Gently stir the solution (Solution B) in the *stoppered* flask with a magnetic stirring hot plate (stirring bar should be covered with a material that will not adsorb mercury) overnight to allow the mercury to completely dissolve at room temperature.
Note: The flask should be 1.0 liter in volume to minimize any possible volatilization of mercury.
3. After dissolution of the mercury, pour 552 mL of the acidic solution containing soluble mercury into a separate flask labelled "D." Dilute the remaining 448 mL solution with distilled or deionized water to 1.0 liter total volume and add the solution to a large erlenmeyer flask (Solution C).
Note: Stopper and save the remaining 552 mL of acidic solution (Solution D) of mercury for later analysis, if necessary.

---

*Compliments of Bacharach, Inc.

**Patents pending.

†Method 10065, Compliments of HACH Company.

‡Mercury metal forms amalgams and clings to surfaces of Fe, Cu, Ag, Ti and other metals. When grinding or crushing a sample, clean the metal tool surface properly or use another method to reduce particle size.

Note: If transferring the 448 mL solution to a larger flask labelled C, rinse the empty 1000-mL flask with small portions of deionized or distilled water three times and add the washings to flask C. Then dilute sample C to 1.0 liter total volume.

4. Carefully add 50 mL of concentrated sulfuric acid to 1.0 liter of the acidified solution (Solution C).
   Note: Determine a reagent blank for each new lot of reagent by running the entire procedure, including the digestion, using one liter of deionized water instead of sample. Add the same amount of potassium permanganate as required by the sample. Subtract the reagent blank from each test result.
5. Stopper the flask and stir the sample on the magnetic stirring hot plate for 10 minutes.
6. Add 5.0 grams of potassium persulfate to the sample. Stir until dissolved.
7. Add 7.5 g of potassium permanganate to the sample. Stir until dissolved.
   Note: Alternatively, add a 10-gram measuring scoop of potassium permanganate to the sample.
8. Cover the flask with a watch glass. Begin heating the sample to a temperature of 90°C *after* the reagents have dissolved. **AVOID BOILING.**
   Note: For a mercury standard or reagent blank in distilled water, the heat step is not necessary.
9. Continue to stir and heat the sample at 90°C for two hours.
   Note: A dark purple color must persist throughout the two-hour digestion. Some samples (high in organic matter or chloride concentration) require additional permanganate. It may be difficult to see a dark purple color if the sample contains a black/brown manganese dioxide precipitate. You may add more potassium permanganate if the solution is not dark purple.
10. Cool the digested sample to room temperature. A brown/black precipitate of manganese dioxide may settle during cooling. If the digested sample does not have a purple color, the digestion may be incomplete. Add more potassium permanganate. Return the sample to the magnetic stirring hot plate and continue digestion until a purple color persists.
11. Return the cool, digested sample to the cool, magnetic stirring hot plate. Turn the stirrer on.
12. Using a 0.5-gram measuring spoon, add 0.5 g additions of hydroxylamine-hydrochloride until the purple color disappears. Wait 30 seconds after each addition to see if the purple disappears. Add hydroxylamine-hydrochloride until all the manganese dioxide is dissolved.
13. Remove the stir bar.
14. The digested sample is now ready for processing by cold vapor separation and preconcentration.
    Proceed to Phase 2.

Phase 2 Cold Vapor Separation and Preconcentration of Mercury

Follow the mercury procedure provided in Chapter 2 or on pages 737 to 740 in the HACH Water Analysis Handbook, 3rd ed., 1997.

Phase 3 Colorimetric Analysis

Follow the procedure provided in Chapter 2 or on pages 740 to 741 in the HACH Water Analysis Handbook, 3rd ed., 1997. Step 10 in the procedure has been modified to account for dilution factors and is provided below.

10. Return the sample cell to the cell holder. Close the light shield. Results in $\mu g/L$ mercury will be displayed.
Note: Since the solid was digested and the solution was diluted, multiply the results in $\mu g/L$ mercury by 2.23 to obtain the concentration of mercury in the original sample.

$$\mu g/L \text{ mercury (actual concentration)} = 2.23 \times \mu g/L \text{ mercury}$$

Sample Handling and Analysis

Carefully follow all procedures for sampling, handling, storage, standard solution methods, start-up standard, storage and maintenance of the Cold Vapor Mercury Apparatus, dedicated glassware and apparatus storage, maintaining the system, safety, interferences, and waste disposal.

HACH recommends that the analyst perform a few analyses on mercury standards and blanks for system equilibration before beginning sample testing. This allows the system to stabilize before processing samples.

QUESTIONS AND ANSWERS

1. Where do we find mercury compounds?
Liquid mercury can be found in thermometers, manometers and several other devices to measure pressure, temperature, flow rates and environmental conditions. Medical facilities and different types of laboratories often use mercury containing devices to measure specific characteristics. Solid forms of mercury compounds have been used in topical dressings to treat wounds and to prevent infection. Partially soluble and insoluble compounds are in antifouling paints, explosives, electronic devices, infrared detectors, catalysts and intermediates in chemical manufacturing, maggot control, lamps, batteries, polishing compounds, silvering mirrors, fungicides, pigments, ceramics, cosmetics, perfumery, waterproof paints, and additional applications in the fields of chemistry, medicine, photography, explosives, and pyrotechnics. There are many applications using insoluble and partially soluble forms of mercury and mercury compounds.
2. How can insoluble forms of mercury get into the environment?
Many people have played with liquid mercury and do not really believe that liquid mercury metal is a problem. Spilled liquid mercury requires some special clean up techniques. Mercury metal and some mercury compounds combine with ammonia, methane, and other nitrogen-based organics in the environment to form soluble complexes. In some cases, a few mercury complexes can be decomposed by heat and form compounds which are more soluble in water. Some mercury compounds have a tendency to react or decompose in heat, light, and/or basic conditions to form more soluble compounds.
3. Why are mercury compounds and mercury so difficult to control in the environment?
Mercury and mercury compounds have so many practical uses and cannot be replaced easily. Many of the mercury compounds are insoluble or partially soluble in water and stable at high temperatures and pressure. Some mercury compounds are light sensitive and are soluble in lipids. A number of the mercury compounds and mercury metal do not react or decompose under normal environmental conditions and remain in matrix mixtures with other substances in the environment for extensive periods of time. Also, solubilized mercury tends to accumulate in the tissues and roots of some plants and in the organs and tissues of animals, sea life, and man.

REFERENCES

1. *Handbook of Reference Methods for Plant Analysis,* Yash P. Kalra (ed.) (Boca Raton, Florida: CRC Press, 1998).
2. *Standard Methods for the Examination of Water and Wastewater:* MERCURY 16th ed., A. E. Greenberg, R. R. Russell, L. S. Clesceri and Mary Ann H. Franson (eds.), p. 232, Washington, DC, American Public Health Association, 1985.
3. *HACH Water Analysis Handbook,* 3rd ed. (Loveland, CO.: HACH Company, 1997).
4. *Hawley's Condensed Chemical Dictionary,* 11th ed., rev'd by N. Irving Sax and Richard J. Lewis (New York, N.Y.: Van Nostrand Reinhold, 1987).

NICKEL

Nickel metal and a number of nickel compounds are flammable. Nickel metal is relatively nontoxic to humans and most animals. Nickel metal displays some level of carcinogenic characteristics as a dust in air. Some nickel compounds are toxic and known carcinogens (OSHA).

Nickel metal is primarily used in metal alloys, coatings, battery, fuel cells and catalytic processes. Nickel compounds are used as catalysts, ceramic colors and glazes, coatings on metals, electroplating, metallurgy, glazes, antioxidants, paints, cosmetics, fuel cell electrodes, capacitors, storage batteries, reflectors, electrodes, reagents, porcelain painting, and complexing agents in organic solvents. A number of processes preparing these products deposit more soluble nickel in the environment than the products themselves.

Hydration processes, pH changes, oxidation-reduction processes, and the formation of coordination compounds during decomposition processes will solubilize nickel metal and insoluble forms of the compound in the environment. These changes do not generally produce compounds of high toxicity and usually require a considerable amount of time during the decomposition or reaction process.

Some industries manifest the waste stream especially if the nickel waste is mixed with other types of wastes that can not be easily separated. Manifested wastes containing nickel compounds are often recovered for applications in other industries or buried in containers. Nickel wastes are often found as discards or in landfills.

A known volume of acid is added to a solid, gel or semisolid sample weighing 0.100 to 0.500 grams in a digestion flask. The digestion procedure for oils requires 0.100 to 0.250 grams of material diluted to a specific volume. After 4 mL of sulfuric acid is added to the digestion flask, the sample is boiled for 4 to 5 minutes and the sample is allowed to cool. Then, 10 mL of 50% hydrogen peroxide is added to the charred sample and the sample is boiled for 1 minute. After cooling the mixture and adding 70 mL of deionized water to the digest, the sample is boiled in a hot water bath for 15 minutes. The digested sample is allowed to cool to room temperature, and diluted to 100 mL with deionized water. After selecting the appropriate analysis volume and diluting the sample, the clear sample can be analyzed by the procedure provided below. A pH adjustment will be required before analysis by HACH Method 8150 in Chapter 2.

Nickel has a tendency to form complexes with chelating agents, such as EDTA. Vigorous digestion on the sample will eliminate these interferences. Common interferences in concentrations less than 50 mg/L (ppm) are ions of aluminum, cadmium, chromium, copper, fluoride, iron (+2, +3), manganese (+2), lead and zinc. Most of the metals react with 1-(2-pyridylazo-2-naphthol (PAN) to form colored complexes. Iron (+3) is masked by adding pyrophosphate. EDTA is added to destroy all metal-PAN complexes except those of nickel and cobalt.

This method can be used to determine nickel at relatively low concentration levels (0 to 1.000 mg/L) in samples from industrial operations and wastes. Solid samples or oil samples with high concentrations of nickel will have to be digested and diluted appropriately.

The wavelength to determine the nickel-PAN complex is 560 nm, whereas the wavelength to determine cobalt is 620 nm. A correction for cobalt interference can be applied to the spectrophotometer readings. The sensitive PAN procedure for detecting nickel and cobalt at concentrations near 1 mg/L can be made on the same sample by adjusting the wavelength and zeroing the instrument before both measurements.

QUESTIONS AND ANSWERS

1. If a sample of incinerator ash is believed to contain approximately 1.57 mg/L of nickel, what method should be used to determine nickel in the sample?
Samples containing high concentrations of nickel can be determined by diluting the digested sample and proceeding with the Pan Method. Also, higher concentrations of nickel in a sample can be determined by the Heptoxime procedure.

2. An analysis of a flyash sample gives 8.4 mg/L nickel at 560 nm. At 620 nm, after zeroing the instrument, the same sample reads 2.0 mg/L. What is the concentration of the nickel?
To correct for cobalt interference,

$$\text{mg/L Ni} = 8.4 - (0.89) \times 2.0 \quad \text{or essentially 6.6 mg/L of nickel}$$

3. What are some of the potential interferences during the nickel test?
Interferences below 60 ppm each are ions of aluminum, cadmium, chromium, copper, fluoride, zinc iron (ferous and ferric), manganese, lead, and molybdenum. Chloride and ions of calcium, magnesium, sodium and potassium can interfere at much higher levels.

4. How are interferences by metals avoided during the test for nickel?
Iron is masked by adding pyrophosphate. Metals, except nickel and cobalt, that form the metal-PAN complexes are destroyed by adding EDTA to the solution.

REFERENCES

1. *HACH Water Analysis Handbook,* 3rd Ed. (Loveland, CO.: HACH Company, 1997).
2. *Hawley's Condensed Chemical Dictionary,* 11th Ed. rev'd by N. Irving Sax and Richard J. Lewis (New York, N.Y.: Van Nostrand Reinhold, 1987).
3. *Standard Methods for the Examination of Water and Wastewater:* NICKEL. 16th Ed. A. E. Greenberg, R. R. Trussell, L. S. Clesceri and Mary Ann H. Franson (eds.), p. 234, Washington DC, American Public Health Association, 1985.

NITRATE

As solids, nitrates are found in many different foods and materials. Nitrates have many practical uses such as latex coagulant, mordant, fertilizer, glass manufacture, matches, black powders, pyrotechnics and rocket propellants, dynamite, pharmaceuticals, enamel, indelible inks, silvering mirrors, ceramics, insecticide, steel and metal processing, explosives, coatings, and many additional applications.

Plants absorb most of their nitrogen in the forms of ammonium ion and nitrate anion. Plant growth is often improved by nourishment of plants with both ions rather than either individually. Nitrate uptake by plants occurs by active absorption and is favored at low pH. When plant uptake levels of nitrate are high, organic synthesis within the plant increases and the level of inorganic cations (Ca, K, Mg) tends to increase.

Healthy plants contain nitrogen which is the basis for their rich green color. Nitrogen improves the quality of the leaf crops and stimulates the utilization of phosphorus, potassium and other essential nutrient elements. Nitrogen influences fruit sizing.

Extractable nitrate in plant tissue can be determined by the ion-selective electrode method.* Alternatively, extractable nitrate can be determined by spectrophotometric methods at 520 nm after extraction of the plant tissue with 2% acetic acid solution. Nitrate can be extracted from a number of solids with dilute acetic acid solution. In most cases, grinding of solid samples with a mortar and pestle or chopping of samples with a blender will allow dissolution of the nitrate by the dilute acid solution. Samples with high levels of nitrate may have to be diluted before analysis.

Both the LaMotte and HACH Methods perform nitrate analysis by the cadmium reduction method. The LaMotte test provides results in the range of 0 to 150 lbs/A. Powdered cadmium is used to reduce nitrate to nitrite. The nitrite and reduced nitrate forms are determined by diazotizing sulfanilide and coupling with N-(1-naphthyl)-ethylenediamine dihydrochloride to form a highly colored azo-dye which is measured colorimetrically. High concentrations of iron and copper may provide low results. Strong oxidizing and reducing substances interfere.

NITRATE NITROGEN TEST

Cadmium Reduction Method Procedure**

1. Select setting 4 on the "Select Wavelength" knob.

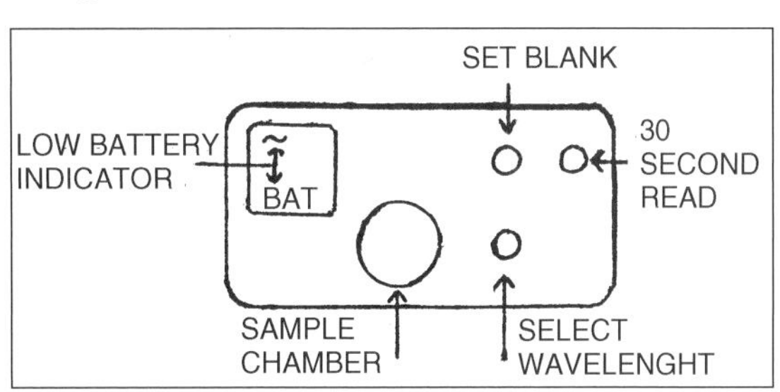

2. Use the 1 mL pipet to add 1 mL of soil filtrate to a clean colorimeter tube and dilute to the line with deionized water. Cap tube and mix.

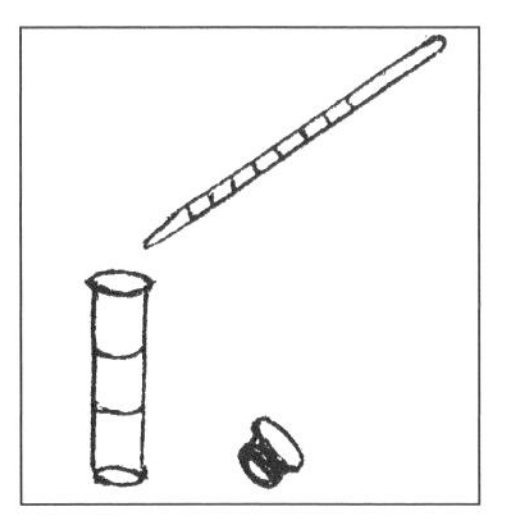

*Robert O. Miller, *Handbook of Reference Methods for Plant Analysis,* Yash P. Kalra (ed.) (Boca Raton, Fla.: CRC Press, (1998) at 115–117.

**Compliments of LaMotte Company.

3. Measure 5 mL of the diluted soil filtrate to another colorimeter tube, then add 5 mL of Mixed Acid Reagent. Cap the tube and mix.
   Note: Use the 10 mL graduated cylinder for these measurements.

4. Insert the tube into the colorimeter chamber and press the "30 Second Read" button. Adjust "100"%T with the "Set Blank" knob.

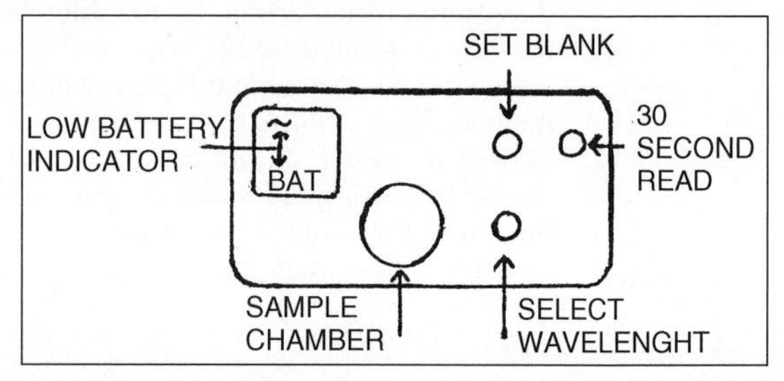

5. Remove the tube from the chamber. Use the 0.1 g spoon to add 2 level measures of Nitrate Reducing Reagent to the contents of tube. Cap the tube.

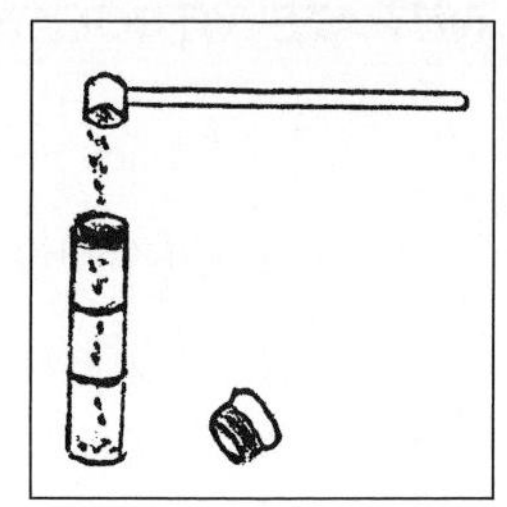

6. Hold tube by index finger and thumb and mix by inverting approximately 50 to 60 times in a minute, then let stand 10 minutes for maximum color development.
   Note: At the end of the waiting period an undissolved portion of the Nitrate Reducing Reagent may remain in the bottom of the tube without affecting the results.

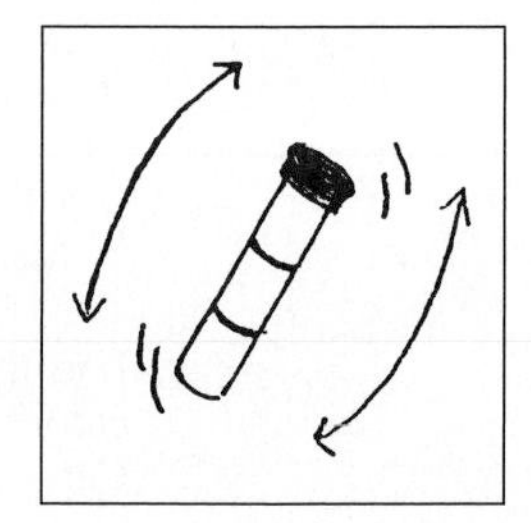

7. At the end of the 10 minute waiting period, insert tube into chamber of colorimeter, press the "30 Second Read" button and measure %T as soon as reading stabilizes.

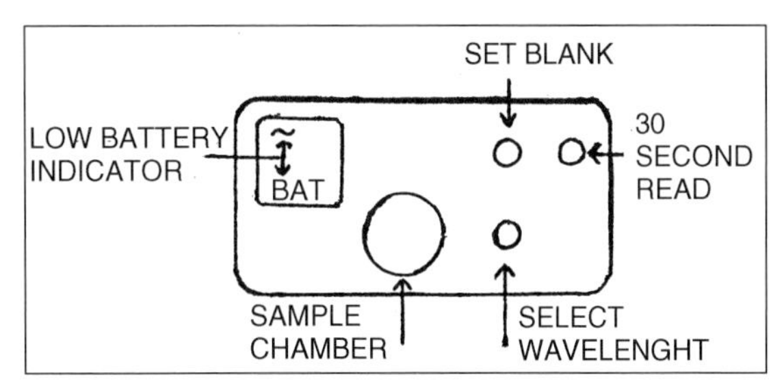

8. Consult the chart below to find the concentration, in pounds per acre, of nitrate nitrogen in the soil.

Nitrate Nitrogen Calibration Chart (lbs/acre)

%T	9	8	7	6	5	4	3	2	1	0
90	2	3	5	6	7	8	10	11	12	14
80	15	16	18	19	20	22	23	25	26	27
70	29	30	32	34	35	37	38	40	42	43
60	45	47	48	50	52	54	56	58	60	62
50	64	66	68	70	72	74	76	78	81	83
40	86	88	90	93	96	98	101	104	107	110
30	113	116	119	122	125	129	132	136	140	144
20	148	152								

Nitrate Nitrogen Concentration Chart

%T	Range	Pounds per acre
93–100%	Low	0–9.0 lbs/acre
79–92%	Medium	11–29 lbs/acre
65–78%	Medium High	33–51 lbs/acre
43–64%	High	53–100 lbs/acre
0–42%	Very High	Over 100 lbs/acre

HAZARDOUS SUBSTANCES

Hazardous substances are the mixed acid reagent and the nitrate reducing reagent. The substances must be handled appropriately. Consult the Material Safety Data Sheet (MSDS) for handling and disposal purposes.

RANGE OF MEASUREMENT

The range of measurement is 0 to 150 lbs/acre.

INTERFERENCES

Strong oxidizing and reducing substances interfere. Low results may be observed for samples that contain high concentrations of iron and copper.

QUESTIONS AND ANSWERS

1. What predictions can be made about nitrogen in soil?
Some types of bacteria in soil can tie up nitrogen and convert the nitrogen to other forms so that the nitrogen can no longer be used by plants. Oxidation and bacterial action can also create nitrates from other nitrites and organic-nitrogen compounds. The changes of nitrogen in soil are generally unpredictable.
2. Nitrogen exists in many forms in soil, plants and organic material. What forms does nitrogen take?
Nitrogen in soil, plants, and most organic material takes the preferred forms of ammonium ion and nitrate ion. However, nitrite and organic-nitrogen compounds are often available depending on pH, temperature, waste products, degradation processes, available oxygen, and bacteria.
3. What role does nitrogen play in plant development?
Nitrogen improves the quality of leaf crops and appears to stimulate the utilization of phosphorus, potassium and other essential elements. Nitrogen, applied correctly, enhances plant growth and is influential in fruit sizing.
4. How are ammonia, amines, amides, hydroxylamines oximes and other reduced nitrogen compounds converted to nitrate in soil and different media?
The biological oxidation of ammonia and nitrogen-organics to nitrate is a two-step process known as nitrification. At first, ammonium or ammonia is converted to nitrite ($N0_2^-$) primarily by heterotropic organisms and autotrophic bacteria known as Nitrosomonas. The conversion from nitrite to nitrate is accomplished by autotrophic bacteria, fungi, and a few other bacterial strains. In both steps, oxygen is required.

REFERENCES

1. *Handbook of Reference Methods for Plant Analysis,* Yash P. Kalra (ed.), (Boca Raton, Fla.: CRC Press, 1998).
2. *LaMotte Soil Handbook,* (Chestertown, Md.: LaMotte Company, 1994).
3. *HACH Water Analysis Handbook,* 3rd Ed. (Loveland, CO.: HACH Company, 1997).
4. *Handbook of Reference Methods for Plant Analysis,* Yash P. Kalra (ed.), (Boca Raton, Fla.: CRC Press, 1998).
5. Samuel Tisdale, Werner L. Nelson, and James D. Beaton. *Soil Fertility and Fertilizers,* (Macmillan Publishing Company, 1985) at 112–149.
6. *Hawley's Condensed Chemical Dictionary,* 11th Ed., rev'd by N. Irving Sax and Richard J. Lewis (New York, N.Y.: Van Nostrand Reinhold, 1987).
7. *Standard Methods for the Examination of Water and Wastewater:* NITROGEN (NITRATE), 16th Ed., A. E. Greenberg, R. R. Trussell, L. S. Clesceri and Mary Ann H. Franson (eds.), pp. 391, 394–396, Washington, DC, American Public Health Association.

NITRITE

Nitrite anion along with nitrate anion and ammonia is one of the primary nitrogen sources available to plants. Specific types of bacteria convert ammonia under aerobic conditions to nitrites. The conversion of ammonia and organic-nitrogen compounds to nitrite is normally accomplished by a group of autotropic bacteria known an *Nitrosomonas.* Additional organisms such as bacteria, actinomycetes, and fungi can convert reduced nitrogen compounds to nitrites.

A number of factors affect nitrite loss in soil and in substances. Nitrite concentrations can decrease by decomposition at pH values below 5.0, catalytically react with metals (Cu, Fe, Mn), and undergo fixation by soil organic matter. In the transformation of nitrite in soil, a portion of the nitrogen becomes organically bound and most of the remaining nitrogen is evolved as gaseous forms of nitrogen and oxides of nitrogen. Excessive nitrites are toxic to plants and remain stable in soils with inadequate aeration. High nitrites can be found in soils containing high levels of nitrates. Nitrate nitrogen can decompose to form nitrites in a reducing environment.

Nitrites are used as food curing agents, analysis reagents, rubber accelerators, color fixative, pharmaceuticals, dye manufacture, food additive, corrosion inhibitor, meat preservative, antidote for poisoning, and in many additional applications. Nitrites are easily oxidized in air and in water to nitrates. Nitrites have been used to prevent oxidation of other substances. Many nitrites are fairly strong oxidizing agents under the appropriate conditions. Nitrites are a fire and explosion risk when heated or shocked, or in contact with some types of organic materials.

The Nitrite Nitrogen Test can be performed by the diazotization method in the range of 0 to 53 lbs/acre.* The diazotization of sulfanilamide by nitrite in water under acid conditions produces a diazonium compound, which subsequently reacts with N-(1-naphthyl)-ethylenediamine to form a reddish-purple complex in solution. A colorimeter measures the intensity of the color. The presence of oxidants or reductants can affect the nitrite concentrations. High alkalinity near 600 mg/L can shift the pH of the solution and produce low nitrite results.

NITRITE NITROGEN TEST

Diazotization Method—Procedure*

1. Use the 1 mL pipet to add 2 mL of soil filtrate to a clean colorimeter tube and dilute to the line with deionized water. Cap tube and mix.

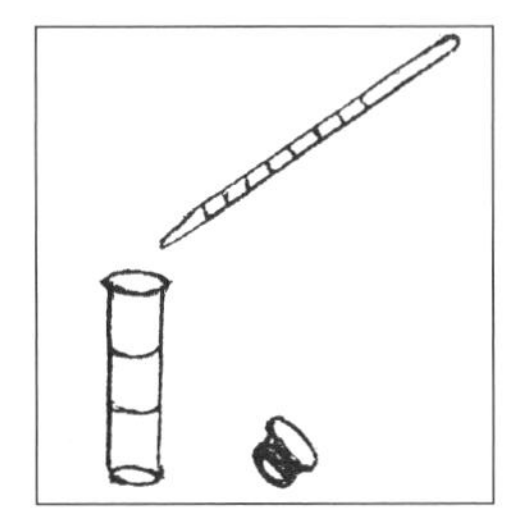

2. Select setting 4 on the "Select Wavelength" knob.

*Compliments of LaMotte Company.

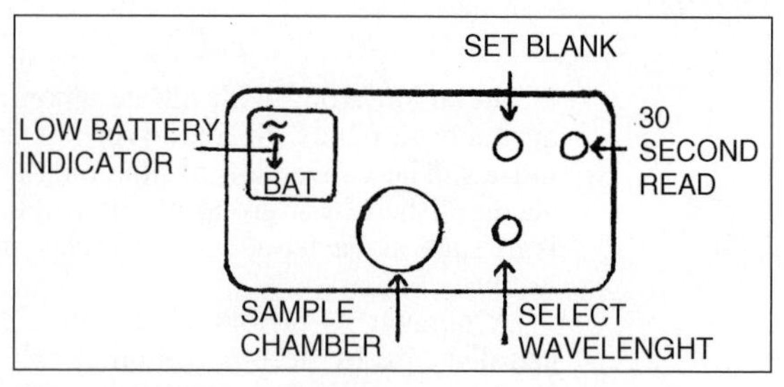

3. Measure 5mL of diluted soil filtrate into another colorimeter tube, then add 5 mL of "Mixed Acid Reagent. Use the small graduated cylinder for these measurements.

4. Cap the tube and mix.

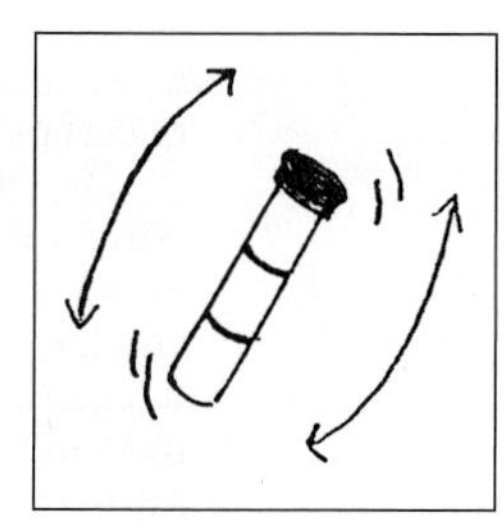

5. Insert the tube into colorimeter chamber and press the "30 Second Read" button. Adjust "100"%T with the "Set Blank" knob.

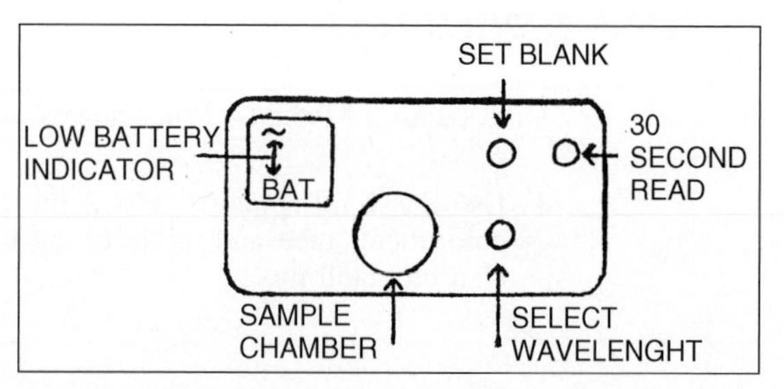

6. Remove the tube from the chamber. Using the 0.1 g spoon, add two level measures of "Color Developing Reagent" to the contents of tube and cap.

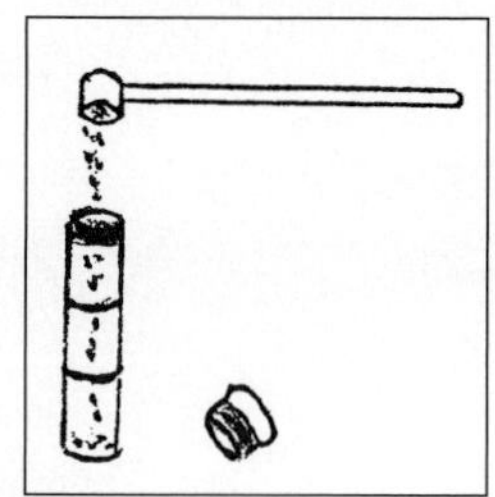

7. Shake tube for approximately one minute to dissolve the powder, then let stand for 5 minutes for maximum color development.

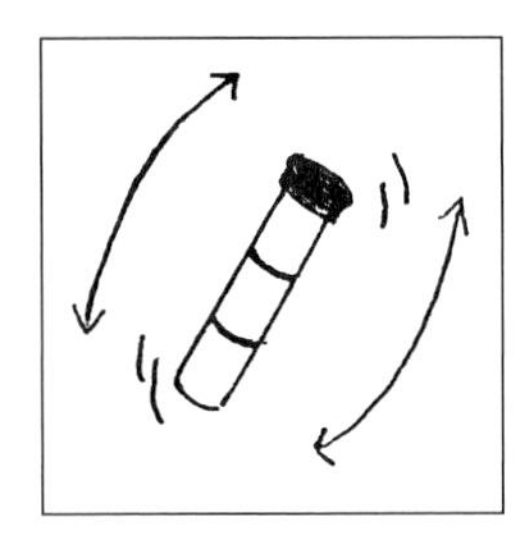

8. At the end of the 5 minute waiting period, insert tube into chamber, press the "30 Second Read" button and measure %T as soon as reading stabilizes.

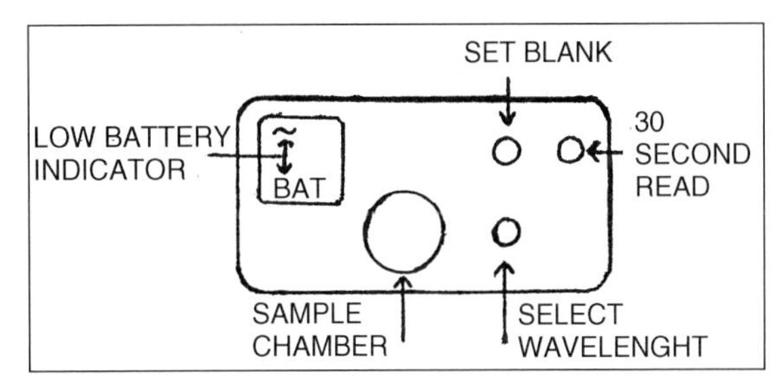

9. Consult chart to find the nitrite nitrogen concentration in pounds per acre.

Nitrate Nitrogen Calibration Chart (Lbs/A)

%T	9	8	7	6	5	4	3	2	1	0
100								0.0	0.1	0.2
90	0.3	0.5	0.6	0.7	0.8	1.0	1.1	1.2	1.3	1.5
80	1.6	1.7	1.9	2.0	2.1	2.3	2.4	2.6	2.7	2.8
70	3.0	3.1	3.3	3.4	3.6	3.7	3.9	4.1	4.2	4.4
60	4.5	4.7	4.9	5.0	5.2	5.4	5.6	5.8	5.9	6.1
50	6.3	6.5	6.7	6.9	7.1	7.3	7.5	7.7	7.9	8.2
40	8.4	8.6	8.8	9.1	9.3	9.6	9.8	10.1	10.3	10.6
30	10.9	11.2	11.4	11.7	12.0	12.4	12.7	13.0	13.4	13.7
20	14.1	14.4	14.8	15.2	15.6	16.1	16.5	17.0	17.5	18.0
10	18.5	19.1	19.7	20.4	21.0	21.8	22.6	23.4	24.3	25.3
0	26.5	27.7	29.2	30.9	32.9	35.5	38.8	43.8	53.0	

Nitrate Nitrogen Concentration Chart

%T	Range	Pounds per acre
86–100%	Low	0–2 lbs/acre
73–85%	Medium	2.5–4 lbs/acre
42–72%	High	4.5–10 lbs/acre
0–41%	Very High	Over 10 lbs/acre

HAZARDOUS SUBSTANCES

Hazardous substances are mixed acid reagent and color developing reagent. Follow the Material Safety Data Sheet (MSDS) for disposal and handling processes.

RANGE OF MEASUREMENT

The range of measurement is 0 to 53 lbs/acre.

INTERFERENCES

The presence of strong oxidants or reductants may readily affect the nitrite concentrations. There are few known interferences of substances at concentrations less than 1000 times that of nitrite. High alkalinity (above 600 mg/L) will give low results due to a shift in pH.

QUESTIONS AND ANSWERS

1. How stable is nitrite compared to nitrate?
   Nitrites tend to oxidize to nitrates in air and in surface waters. In closed bottles, nitrites can remain as fairly stable compounds.
2. Which compounds interfere during the test for nitrite?
   Oxidizing agents can oxidize nitrites to nitrates.
3. How does high alkalinity affect the nitrite determination?
   High alkalinity will shift the pH of the solution and a low nitrite value should be observed.
4. Are nitrites toxic and dangerous?
   Excessive nitrites can be toxic to plants. An excessive amount of nitrites can be somewhat toxic to humans. The nitrites are explosive when heated or shocked. Many nitrites are a danger when in contact with organic materials.

REFERENCES

1. *LaMotte Soil Handbook* (Chestertown, Md.: LaMotte Company, 1994).
2. Samuel Tisdale, Werner L. Nelson, and James D. Beaton. *Soil Fertility and Fertilizers* (Macmillan Publishing Company, 1875) at 112–149.
3. *Hawley's Condensed Chemical Dictionary,* 11th Ed., rev'd by N. Irving Sax and Richard J. Lewis (New York, N.Y.: Van Nostrand Reinhold, 1987).
4. *Standard Methods for the Examination of Water and Wastewater:* NITROGEN (NITRITE), 16th Ed., A. E. Greenberg, R. R. Trussell, L. S. Clesceri and Mary Ann H. Franson (eds.), p. 404, Washington, DC, American Public Health Association, 1985.

NITROGEN, TOTAL KJELDAHL

The Total Kjeldahl Method determines the amount of ammonia and organic nitrogen in a sample. The ammonia and organic nitrogen is attributed to decomposition of organic matter.

This method determines the amount of nitrogen in the trinegative state. The method fails to account for nitrogen in the form of nitrate, nitrite, nitrile, nitro, nitroso, azide, azine, azo, hydrazone, oxime, and semi-carbazone.

An analysis for ammonia and organic nitrogen (Total Kjeldahl Nitrogen) requires a digestion procedure. The digestion process oxidizes carbon compounds to carbon dioxide and converts organic forms of nitrogen (amino acids, proteins, peptides, etc.) to ammonia. The digestion procedure can be performed on sludge, oils, fats, and dry samples such as soil, plants, nuts and other products containing organic nitrogen. Samples of dried materials often require chopping or grinding into small pieces before the digestion process.

A known volume of acid is added to a specific weight (usually 0.100 to 0.500 grams) of a solid, gel, sludge, or semisolid sample in a digestion flask. For determination of organic nitrogen and ammonia in oils, 0.10 to 0.25 grams of material are usually required for dilution to a specific volume. After boiling for four minutes, 10 mL of 50% hydrogen peroxide is added to the charred sample and the sample is heated for one minute. The digested mixture is allowed to cool and is diluted to 100 mL with deionized water. After selecting the appropriate sample size and diluting the sample to 25.0 mL or a volume satisfactory for analysis, the clear sample can be analyzed for Total Kjeldahl Nitrogen (Nessler Method) by the procedure (HACH Method 8075) provided in Chapter 2. A pH adjustment on the sample will be required before analysis for total nitrogen.

Organically-bound nitrogen reacting with hydrogen peroxide and sulfuric acid is converted into ammonium salts. A Mineral Stabilizer is added to the sample to complex calcium and magnesium. Polyvinyl Alcohol Dispersing Agent helps to develop the color in the reaction of Nessler Reagent with ammonium ion. In the Modified Nessler Method Test, a yellow color proportional to the ammonia concentration in solution is measured at 460 nm by a spectrometer.

QUESTIONS AND ANSWERS

1. Why is a digestion process required in this experiment?
   Digestion of samples is required to determine the amount of bound nitrogen called organic nitrogen. The bound nitrogen would not be released for analysis if digestion of the sample was not implemented.
2. What types of samples require digestion before analysis?
   Aqueous samples, oils and fats, soil samples, hay and grass samples, specific types of fertilizers, and anything that contains organic nitrogen would require a digestion process.
3. What is the purpose of the dispersing agent?
   A dispersing agent is added to the solution to aid in the development of color for analysis purposes.
4. Will this experiment measure all of the nitrogen in the sample?
   This experiment does not determine the all of the nitrogen in the ionic form and the organic form. Nitrogen as nitrate, nitrite, nitrile, nitroso, nitro, azide, azine, azo, hydrazone, oxime, and semi-carbazone are not included in the determination.

REFERENCES

1. *HACH Water Analysis Handbook,* 3rd ed., Loveland, CO.: HACH Company.
2. *Standard Methods for the Examination of Water and Wastewater:* Macro-Kjeldahl Method. 16th Ed. A. E. Greenberg, R. R. Trussell, L. S. Clesceri and Mary Ann H. Franson (eds.), pp. 408–409, Washington, DC, American Public Health Association, 1985.

OIL AND GREASE

"Oil and grease" has been defined as any material recovered as a substance soluble in tichlorotrifluoroethane. Oils and greases are defined by the method used for their determination. Even though the definition includes other materials such as organic dyes, chlorophyll and some sulfur compounds, the limitation in definition omits these substances as oils and greases.

The term "HEM" is defined as "*n*-hexane extractable materials" which indicates that this method may be applied to materials other than oils and greases. The term "TPH" was traditionally used to characterize aliphatic hydrocarbon materials. An additional term "SGT-HEM" has been defined as "silica gel treated *n*-hexane extractable materials." This method may be applied to materials other than aliphatic petroleum hydrocarbons that are not adsorbed by silica gel.

Oil and Grease and Total Petroleum Hydrocarbons (TPH) include any material that is soluble in *n*-hexane extractant and can be recovered as a substance. Substances such as relatively non-volatile hydrocarbons, animal fats, waxes, greases, soaps, vegetable oils, plant extracts, and soluble additives are included in the definition. Special chemicals or additives are sometimes added to oil and grease compounds to improve wear characteristics of the product under stress.

To measure oil and grease (HEM) gravimetrically, the soluble materials are extracted from the sample with *n*-hexane. After evaporation of the *n*-hexane at 98°C, the residue is weighed to determine the concentration of oil and grease in mg/L.

To determine the Total Petroleum Hydrocarbons (SGT-HEM) gravimetrically, the soluble materials are extracted from the sample with *n*-hexane. The extracted material is mixed with silica gel to absorb non-TPH compounds. After evaporation of the *n*-hexane, the remaining residue is weighed to determine the concentration of total petroleum hydrocarbons.

OIL AND GREASE (15 to 3000 mg/L HEM and SGT-HEM)

Method 10056 for water and wastewater*

Hexane Extractable Gravimetric Method**

*Compliments of HACH Company.

**Equivalent to USEPA Method 1664.

1. Collect 350 mL of sample in a clean 500-mL separatory funnel.
   Note: Do not pre-rinse the collecting vessel with sample, or results may be increased.
   Note: If the sample is not collected in the separatory funnel, set the empty container and lid aside for use in step 4.
   Note: The sample must be at room temperature before analyzing.
   Note: Determine a blank value (350 mL of distilled or deionized water) with each new lot of reagents. If the blank result is greater than 5 mg, resolve the source of error or remove the interferences before performing this procedure.

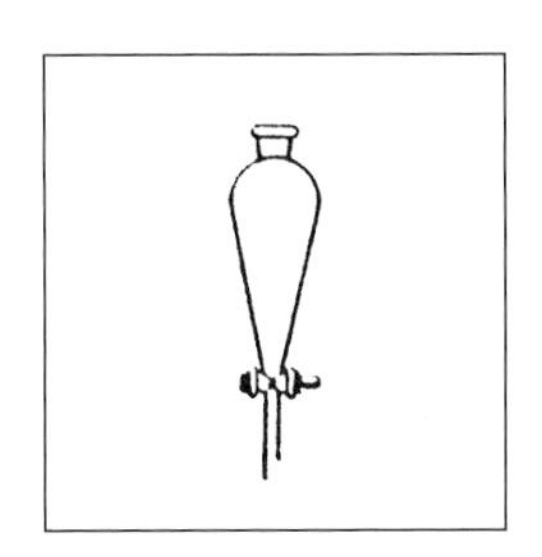

2. Using a pipette and pipette filler, add 4 mL of 1:1 Hydrochloric Acid solution to the separatory funnel. Mix well. The pH must be ≤2.
   Note: Check sample pH after acid addition by dipping a glass rod into the sample and allowing a few drops to touch the pH paper. Do not dip the pH paper into the sample. Rinse the glass rod with a small portion of hexane back into the separatory funnel to remove any grease/oil on the rod.
   Note: A pH of ≤2 is required to hydrolyze some oils and greases. A pH of greater than 2 dissolves the sodium sulfate used in step 8 and causes an interference.
   Note: Use the equivalent amount of acid to determine the blank and all samples from each sampling source.

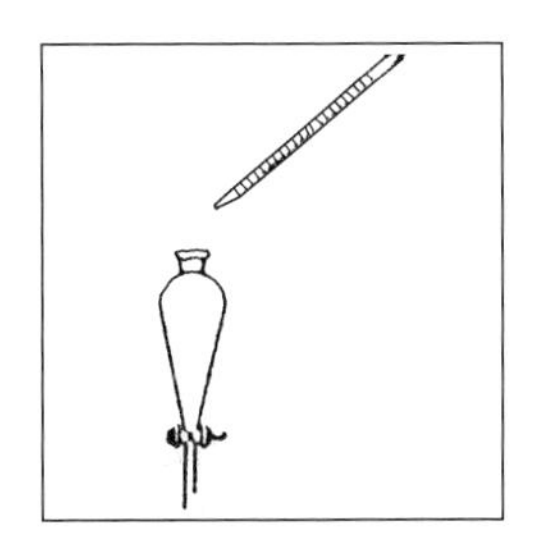

3. Using an analytical balance, weigh a previously dried and cleaned 125-mL distillation flask containing 3 to 5 boiling chips to the nearest 0.1 mg. Record the weight of the flask.
   Note: If determining both the HEM and the SGT-HEM, clean and dry two distillation flasks (one for each procedure) in advance.

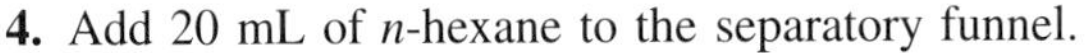

4. Add 20 mL of *n*-hexane to the separatory funnel.
   Note: Spilled reagent will affect test accuracy and is hazardous to skin and other materials.
   Note: If the sample was collected in a separate container or if repeating this step, rinse the collecting vessel/volumetric flask which contained the sample/water layer with the 20 mL of *n*-hexane, then add the 20 mL *n*-hexane rinse to the separatory funnel.

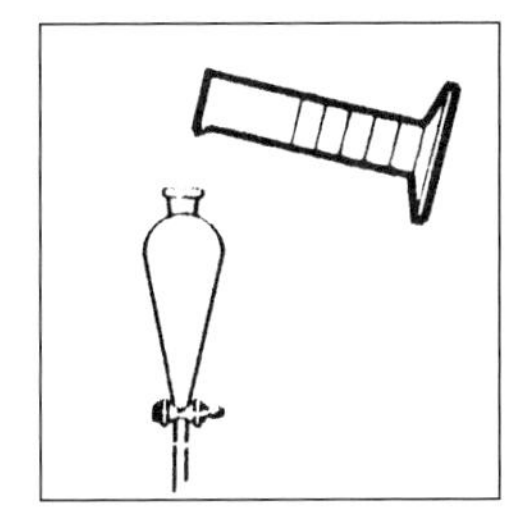

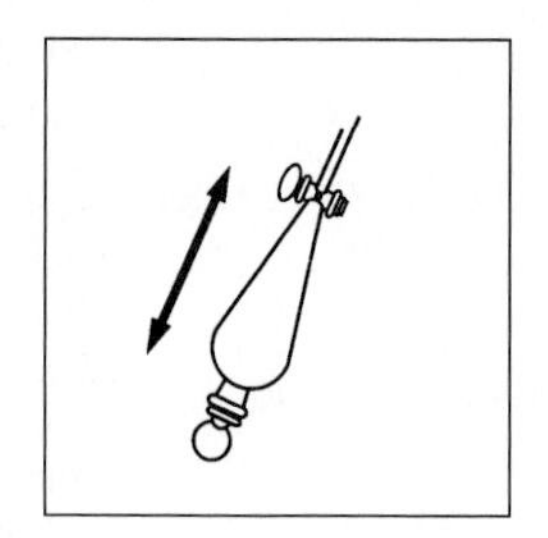

5. Stopper the funnel. Invert the funnel and release the gases through the stopcock. Then vigorously shake the funnel for two minutes.
Note: To release gases from the separatory funnel, invert it and shake it once very hard (support the stopper with your hand). Under a hood, point the delivery tube in a safe direction and SLOWLY open the stopcock to release any gas. Close the stopcock. Repeat the venting procedure until you no longer hear the release of gases.
Note: Do not count the venting time as part of the two-minute shaking time. Shaking for less than 2 minutes may decrease the results.

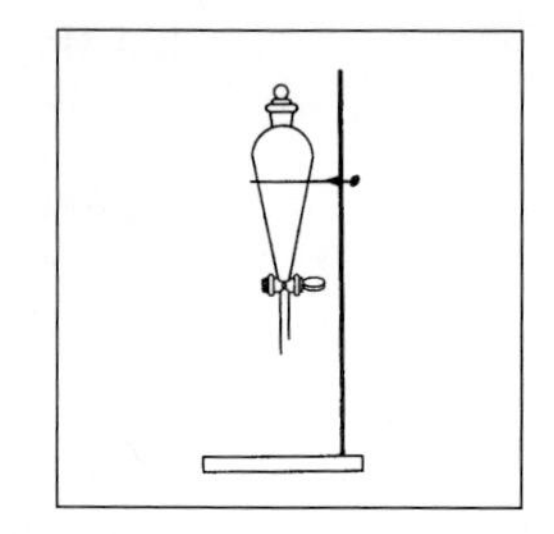

6. Let the funnel stand undisturbed for **at least** 10 minutes to ensure separation of the lower water layer and the upper solvent layer.
Note: The solvent layer may be brown if a colored oil is present.
Note: If an emulsion forms, see Interferences following this procedure. An emulsion is a bubbly layer between the aqueous and solvent layer.
Note: If you repeat this step the third time and the water layer is cloudy, allow the separatory funnel to stand undisturbed for 20 minutes for better separation of the water and solvent layers.

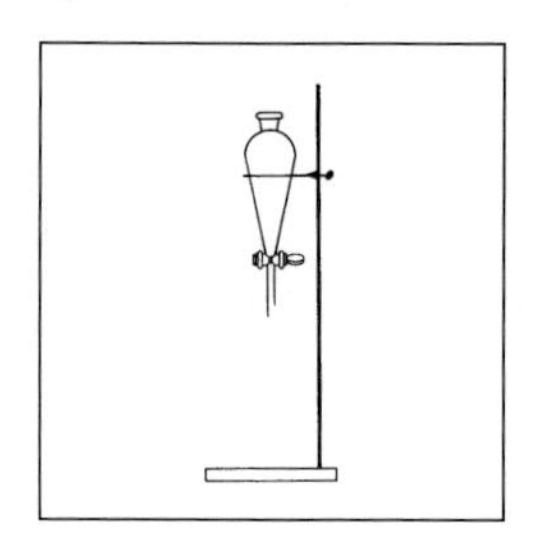

7. **SLOWLY** drain the lower water layer from the separatory flask into the original sample container or 500-mL volumetric flask. This should take about 3 to 4 minutes. Save the water layer for Step 9.
Note: To ensure water is not transferred in Step 8, allow several drops of solvent layer to drain into the water layer until the solvent layer is visible on top of the water.
Note: If the water layer drains too quickly, excess water will be present in the solvent layer. This causes sodium sulfate and water interference.

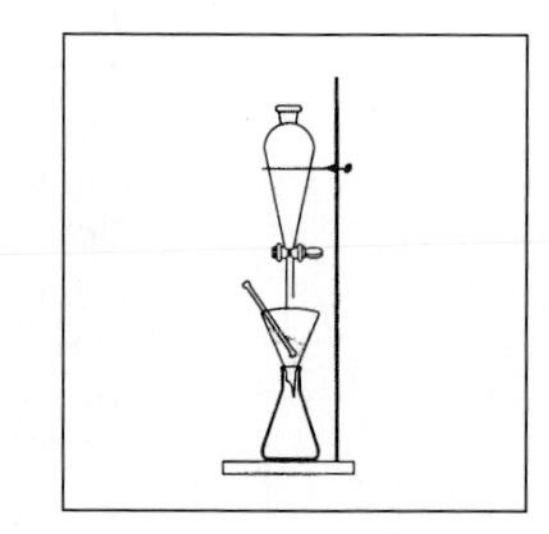

8. Drop-drain the solvent layer into the pre-weighed boiling flask through a funnel containing filter paper and 10 g anhydrous sodium sulfate. Gently stir the sodium sulfate with a glass stirring rod while the solvent layer is draining. Be careful not to rip the filter paper.
Note: Use the same filter, funnel, and sodium sulfate when repeating this step for the second and third extractions. Remove large, hard sodium sulfate chunks between extractions to reduce sodium sulfate contamination.
Note: To set up the filtering funnel, put the glass funnel in the neck of the distillation flask. Place a folded 12.5 cm filter paper in the funnel. Add 10 grams of anhydrous sodium sulfate to the filter paper. Rinse the sodium sulfate with a small amount of the hexane. Discard the hexane properly.
Note: Do not use any plastic tubing to transfer the solvent between containers.

9. Return the water layer to the separatory funnel.
Note: Use the same glass funnel for the second and third extraction (referred to in Step 10).
Note: A second funnel can be used to pour the water layer into the separatory funnel to reduce spillage.

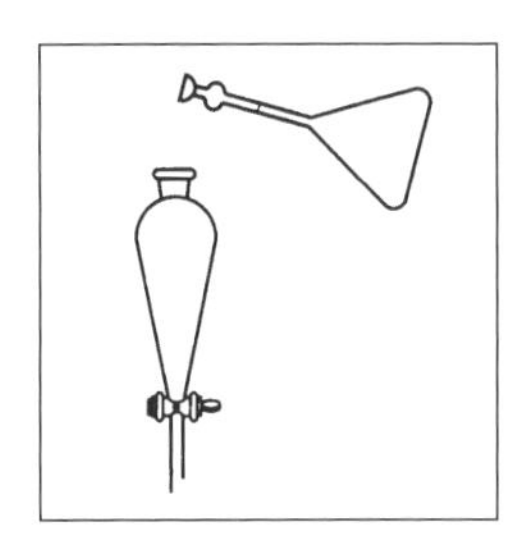

10. Repeat Steps 4 to 9 two more times. After the third extraction, discard the water layer.
Note: There may be small amounts of acetone and/or *n*-hexane in the water layer. Implement proper disposal methods.

11. Rinse the separatory funnel with three separate 5-mL aliquots of fresh *n*-hexane to remove any oil film left on the funnel walls. Drain each aliquot through the funnel containing the sodium sulfate into the distillation flask.

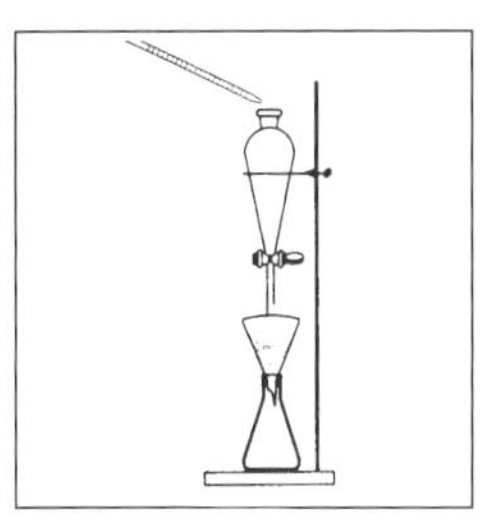

12. Rinse the tip of the glass funnel with 5 mL of *n*-hexane while removing it from the distillation flask. Check for sodium sulfate contamination.
Note: Sodium sulfate contamination will appear as cubic crystals at the bottom of the distillation flask. If present, refilter the solvent layer through filter paper without sodium sulfate. You must re-clean, dry and weigh the boiling flask and boiling chips or have an extra flask ready in case this is necessary.

13. If only the SGT-HEM is to be determined *and* the HEM is known, go to Step 21. If HEM is to be determined, go to Step 14.
Note: If only the SGT-HEM is to be analyzed, the HEM is needed to determine the amount of silica gel needed for the SGT-HEM. For each group of samples from a discharge, determine the HEM before the SGT-HEM.

For SGT-HEM
go to Step 21.
For HEM
go to Step 14.

14. Using a distillation apparatus, distill off the *n*-hexane. Distillation is complete when there are no boiling bubbles or the distillation flask appears dry.
Note: Use a steam bath or a hot plate to maintain a water bath at the proper temperature for the distillation. Do not place the flask directly on a hot plate. This will cause low results and is dangerous because *n*-hexane is volatile.
Note: Evaporation will be faster if the long vertical arm of the connector is wrapped with insulation (paper towel, cloth, or asbestos insulating tape). The distillation should take less than 30 minutes.

See Figure 1 for Distillation Assembly

15. Disconnect the condenser/connector portion of the distillation assembly at the pinch clamp and remove the distillation flask from the heat source with an anti-lint cloth or tongs.
Note: The distilled *n*-hexane may be re-used in future HEM extractions, but is not recommended for SGT-HEM due to the potential increased water content of the solvent.

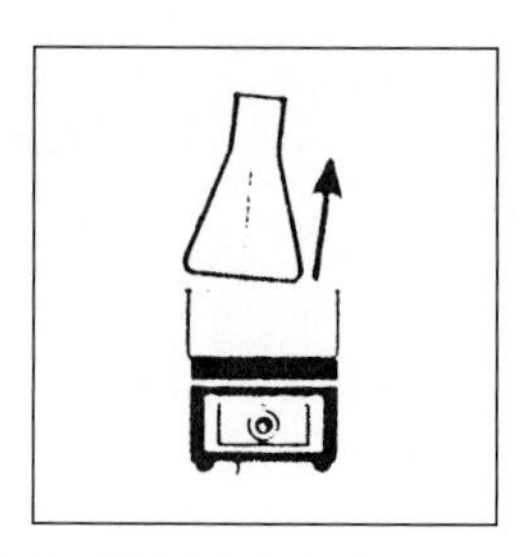

16. Remove the remaining solvent vapors from the distillation flask by attaching the vacuum connector/gas inlet adapter to the flask. Apply a vacuum for 1 to 2 minutes or until all *n*-hexane solvent vapors have been removed.
Note: Crystals on the bottom of the flask indicate that sodium sulfate may have been dissolved in the extraction steps. Re-dissolve the extract in *n*-hexane, filter into another pre-weighed flask and repeat Steps 13 to 15. This is not necessarily true for the "standard" extraction since stearic acid is crystalline below 69° C. If sodium sulfate is present in the standard, big cubical crystals (not the flattened stearic acid crystals) will be visible. Also, you will calculate an unusually high yield compared to the expected value.

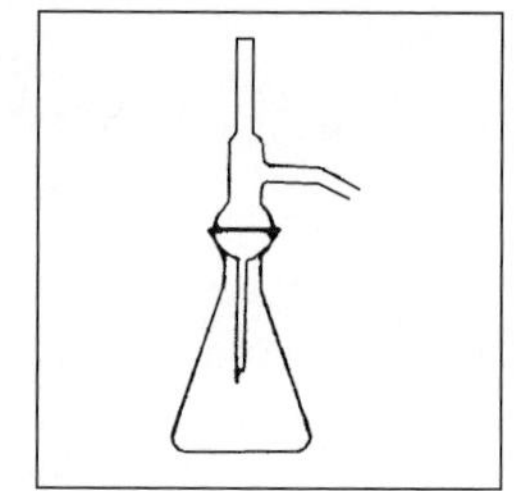

17. Place the flask in a desiccator for 30 minutes (or longer if necessary) until it cools to room temperature.
Note: If the silica gel indicator has turned red, replace the silica gel.

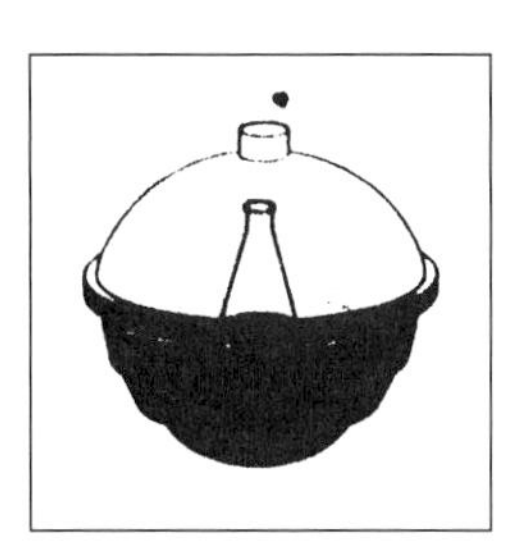

18. Using an analytical balance, weigh the flask to the nearest 0.1 mg. Record this weight. Do not touch the flask after weighing; fingerprints will add weight.
Note: Always use a tong or a lint-free wipe when handling the flask. If you touch the flask, clean the flask with line-free wipes.
Note: Precise weighing is necessary for accurate results; multiple weight measurements are recommended. Re-wipe the flask before each measurement to ensure all contaminants are removed. Record each weight; use the lowest repeatable value for calculations.

19. Calculate the test results:

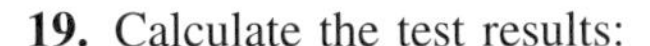

$$\frac{A - B}{\text{Sample Volume}} = \text{mg/L HEM}$$

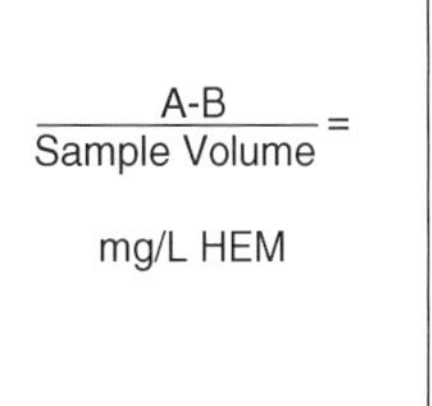

A = Weight (mg) of residue
B = Weight (mg) of flask with boiling chips.
Note: If yield is less than 15 mg/L and additional precision is needed, use a 1-liter sample.
Example: *A = 92.4659 mg*
B = 92.4206 mg
Sample volume = *0.350 L*

$$\frac{92.4659 - 92.4206}{.350\ L} = 129.4\ mg/L$$

20. If only calculating the HEM, stop here. If continuing with SGT-HEM at Step 21, re-dissolve residue with approximately 85 mL of **fresh** *n*-hexane. Heat slightly to ensure re-dissolving of all HEM materials.
Note: For a 350-mL water sample, dilution is necessary if the HEM is above 2850 mg/L (for a 1-liter sample, dilute if the HEM is greater than 1000 mg/L).
Note: To dilute to a 1000 mg/L sample, pour the re-dissolved HEM into 100 mL volumetric flask. Rinse the distillation flask 3 to 4 times with 2 to 3 mL of *n*-hexane. Fill the volumetric flask to volume with *n*-hexane. Mix well. Into a 100 mL beaker, volumetrically pipet the amount (V_a) determined by this equation:

$$V_a = \frac{100000}{W_h}$$

where V_a = Volume of aliquot to be withdrawn (mL) to get 1000 mg of HEM.
W_h = Weight of HEM (A-B) in Step 19 (mg). Dilute to about 100 mL with *n*-hexane.

21. Put a magnetic stir bar and the correct amount of silica gel based on the equation below into the flask with the solvent product from Step 20.

$$\frac{3 \times \text{mg of HEM}}{100} = \text{silica gel (g} \pm 0.3)$$

22. Stir a solution on a magnetic stirrer for a minimum of 5 minutes.

23. Pre-clean, dry and weigh a distillation flask with 3 to 5 boiling chips in it. Place a funnel on the distillation flask. Place a 12.5 cm filter paper in the funnel. Pre-moisten the filter paper with fresh *n*-hexane. Filter the solution through filter paper. Rinse the beaker containing remaining silica gel 3 times with 5-mL aliquots of fresh *n*-hexane and pour the aliquots into the distillation flask.

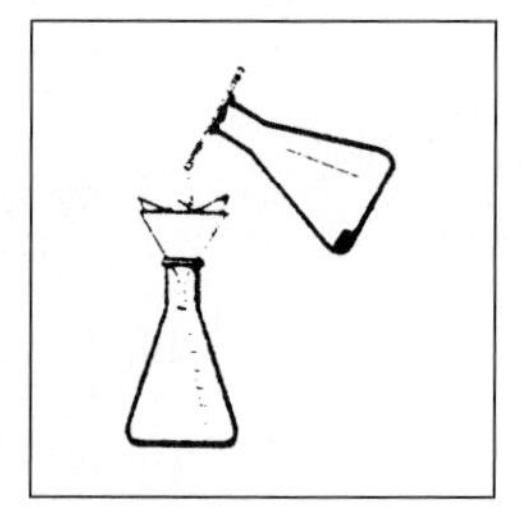

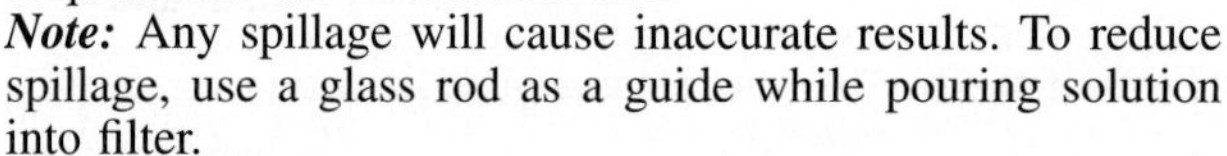

Note: Any spillage will cause inaccurate results. To reduce spillage, use a glass rod as a guide while pouring solution into filter.

24. Follow Steps 14 to 19. Weigh the product left in the bottom of the flask and calculate your results using the equation below:

$$\frac{A - B}{\text{Sample Volume}} = \text{mg/L SGT-HEM}$$

where: A = Weight (mg) of residue
B = Weight (mg) of flask with boiling chips

Perform
Steps 14-19

Sampling

Collect samples in wide-mouth glass bottles or directly in the separatory funnel for immediate analysis. Collection of sample may be done directly into the separatory funnel. Measure 350 mL of water with a graduated cylinder. Pour this into the separatory funnel. Use a laboratory pen to mark the 350-mL level. Fill with sample to this mark. Do not pre-rinse the bottle or separatory funnel with the sample.

Handling Glassware

Before analysis, careful cleaning and drying of the glassware and boiling chips is necessary. Clean the chips and distillation flask by washing with hot water and detergent, rinsing with distilled water, and then rinsing with acetone or *n*-hexane. Place the cleaned flask and boiling chips in a drying oven at 105 to 115° C for 2 hours. Cool to room temperature in a desiccator for at least 30 minutes. Store in the desiccator until needed.

To eliminate errors, always handle the flask with tongs or an anti-lint wipe. If the same flasks are used repeatedly, record their weights after drying in the oven without boiling chips. The drying step may be skipped if the flasks weigh the same after the acetone or *n*-hexane rinse as it does after drying. Boiling chips will vary in weight; their weight should be added to the flask weight.

Interferences

Substances extracted from samples will vary from source to source, depending upon the diversity of the site being sampled. Some samples may contain high amounts of detergents or particulates that can interfere with the extraction procedure. For these samples, it may be necessary to use a 350-mL sample size rather than 1 liter (which is optional). In this circumstance, the 350-mL sample size is EPA accepted for reporting. Wash all glassware in hot water with detergent, rinse with tap and distilled water and rinse with *n*-hexane or acetone.

If an emulsion forms between the two phases (at Step 6) and is greater than one-third the volume of the solvent layer, filter the emulsion and solvent layer through a funnel with glass wool in it. If an emulsion still exists, other possible solutions include: stirring the solvent and emulsion layer with a stir bar, using solvent phase separation paper, centrifugation, using an ultrasonic bath with ice, addition of NaCl, or other physical methods. (Solid phase or other extraction techniques would fall under performance based modifications).

A milky solvent/product layer in the distillation flask indicates water in the solvent layer. Let the flask stand one hour to allow the water to settle. Re-filter the solvent layer through sodium sulfate to remove remaining water.

Extremely low yields could mean a poor extraction (Steps 5 through 8) and a high yield could indicate a problem in the solvent drying (Step 8). Follow procedure Steps 5 to 8 very carefully and run your blank before you run samples in order to identify any possible interference due to these steps. If your blank indicates a yield above 1 mg per test, you should identify the source of contamination before continuing. Likely sources are sodium sulfate contamination and improperly rinsed glassware.

EPA Monitoring and Testing Procedures and Modifications

See p. 886 in the HACH Water Analysis Handbook, 3rd Edition, 1997.

Oil and Grease Reporting to EPA

See p. 887 in the HACH Water Analysis Handbook, 3rd Edition, 1997.

QUESTIONS AND ANSWERS

1. Define "HEM" and "SGT-HEM."
   HEM is defined as "n-hexane extractable materials." SGT-HEM is defined as "silica gel treated n-hexane extractable materials."
2. Define "oil and grease."
   Any material recovered as a substance soluble in trichlorotrifluoroethane. Additional substances are soluble in the organic solvent but are excluded from the definition. Oils and greases are defined by the method used for their determination.
3. Why are special chemicals added to oil?
   Chemicals and plant extracts are added to oil to improve wear and lifetime. Most of these oils are used in vehicles, train engines, and heavy equipment. The special oils are used on equipment such as drill presses, cutting, tool and dyes and other equipment which create high temperatures due to friction during operation.

REFERENCES

1. *HACH Water Analysis Handbook,* 3rd Ed. (Loveland, CO.: HACH Company, 1997).
2. *Standard Methods for the Examination of Water and Wastewater:* OIL and GREASE. 16th Ed., A. E. Greenberg, R. R. Trussell, L. S. Clesceri and Mary Ann H. Franson (eds.), pp. 496–497, Washington, DC, American Public Health Association, 1985.

PCBs

PCBs are highly toxic, colorless liquids with a specific gravity of 1.4 to 1.5. They have limited solubility in water and have very high chemical, thermal and biological stability. The highly chlorinated compounds have low vapor pressures and high dielectric constants.

Specific congeners of this compound tend to exhibit widespread environmental problems including toxicity, very slow biodegradation, and the ability to bioaccumulate within organisms. The decrease in water solubility of specific chlorinated PCBs tends to pose greater aquatic environmental problems. In water bodies, PCBs tend to deposit on the river bottom and concentrate in areas where the water flow is minimized. PCBs remain fairly stable in soil under normal conditions and do not decompose easily. However, reductive dechlorination of highly chlorinated PCBs can occur through microbial enrichment processes in the natural environment or in the laboratory. Anaerobic bacteria slowly convert toxic forms of PCBs to less toxic forms of dicloro-substituted and monochloro-substituted PCBs. The less toxic forms of PCBs can undergo degradation by aerobic bacteria to form non-toxic substances. In addition, PCBs can react under heat and oxygenated conditions to form polychlorinated furans (PCDFs).

In the past, PCBs have been used as dielectrics in transformers and capacitors. The compounds have also been used in plasticizers, hydraulic fluids, heat transfer fluids, lubricants and petroleum additives, impregnation of cotton and asbestos, epoxy paint additives, and in carbonless copy paper. Formerly, askarel, the PCB-containing dielectric fluid found in transformers, contained about 60% PCBs and approximately 40% trichlorobenzenes. As of 1994, most askarel- containing transformers have been replaced in the USA. The extensive use of PCB-type compounds has led to a widespread distribution of these substances in the environment. Because of their extreme persistence, bioaccumulation, and potential adverse health effects, these substances were no longer produced in the USA as of 1977. Significant amounts of PCBs have been disposed of in landfills and deposited in waterways and the oceans. Still, some amounts of the more persistent PCB compounds continue to cycle through the environment.

Polychlorinated biphenyls (PCBs) are less potent but toxicologically similar to TCDD (dioxins). PCBs tend to occupy the Ah-receptor. PCBs and dioxins can act similarly in certain mechanisms. The PCB compounds appear to mimic the actions of hormones. Therefore, adverse reproductive and developmental effects in humans, wildlife and laboratory animals have been displayed after exposure to these compounds.

An excessive amount of polychlorobiphenyl (PCB) has been identified in the organs and fatty tissues of ducks and turtles.[1] Specific types of PCBs tend to concentrate in fish (especially lake trout). The death of a red- tail hawk and a great-horned owl has been attributed to PCBs. Death of various species consuming PCBs appears to be a relatively slow process. During at least two different occasions, thousands of people ingested gram quantities of PCBs by consuming contaminated cooking oil. Ectodermal defects, developmental delay and mortality from non-malignant liver disease have increased among children born to poisoned mothers.[2,3] In addition, PCB concentrations in eggs and chicken meat have been reported up to 250 times the tolerance level for humans.[4] Some contamination of pigs was also reported. Animal feed was contaminated by the addition of fat containing PCB. According to the report, the isolated episode of food contamination will probably have little impact on the health of the general population due to the isolation of the problem and the removal of food, pet-food and feed ingredients from the market. A similar group of compounds of polybrominated biphenyls (PBBs) has been used as a flame retardant, which contaminated livestock feed in Michigan during 1973. Persons ingesting food contaminated with PBB displayed an increase in the incidence of rashes, liver ailments, and headaches.

PCBs can be effectively destroyed by incineration at relatively high temperatures. The "residence time" or "dwell time" for the chemical in contact with the flame becomes an important issue for complete combustion. For a short dwell time, higher consistent temperatures will be required for complete combustion of PCB. During combustion, different types of scrubbing materials such as lime, cement particles, and various other materials must be used to trap or tie up the chloride released during the incineration process. Appropriate scrubbing materials will essentially eliminate emissions of hydrochloric, sulfuric and nitrous/nitric acids during the combustion process.

PCB analysis requires soil extraction procedures, sample and standards preparation, immunoassay procedures, color development and measurement. Samples, standard and color-development reagents are added to test tubes coated with an antibody specific for PCB.[5] The PCB concentration in the sample is determined by comparing the developed color intensity to that of a PCB standard. The PCB concentration is inversely proportional to the color development where a lighter color indicates a higher PCB concentration. Different PCBs (Arochlor Compounds) require various levels of concentration for a positive test at the 1-ppm threshold level.

PCB IN SOIL (1 or 10 ppm Threshold)

Method 10053 for Soil*
Immunoassay Method

*Compliments of HACH Company.

1. Enter the stored program for absorbance. Press: **0 ENTER.**
   Note: The Pour-Thru Cell cannot be used.

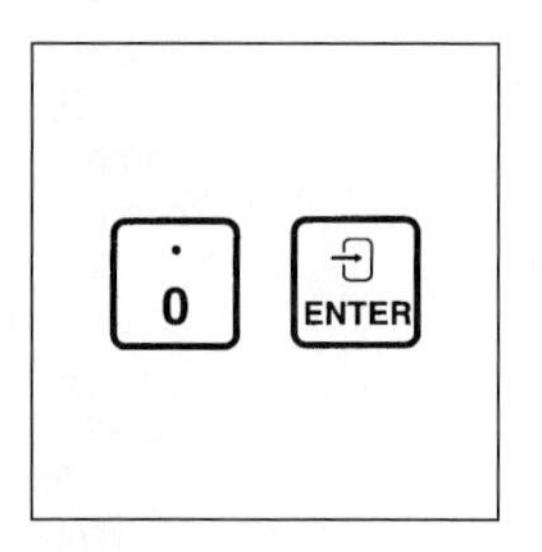

2. Rotate the wavelength dial until the small display shows: **450 nm.**

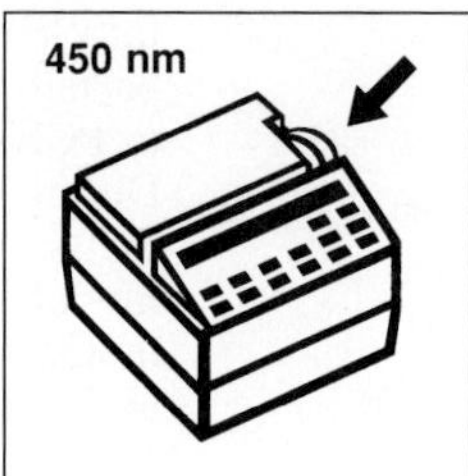

PCB PHASE 1: SOIL EXTRACTION

1. Fill the extraction vial to the 0.75-oz line with Soil Extractant Solution.
   Note: This is equivalent to adding 20 mL of Soil Extractant Solution.

2. Place a plastic weighing boat on an analytical balance. Tare the balance.
   Note: Use either the portable AccuLab Pocket Pro or a laboratory balance.

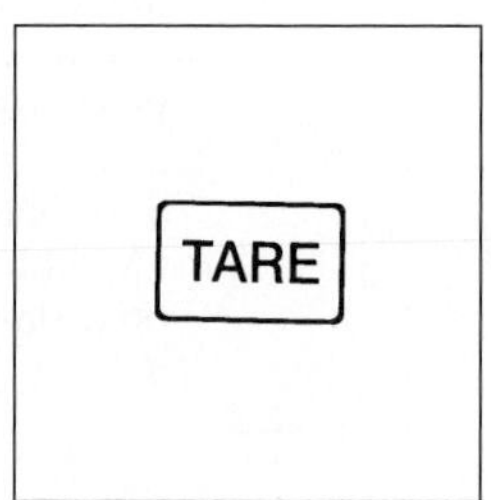

3. Weigh out 10 ± 0.1 g of soil in the plastic weighing boat. Carefully pour the soil from the weighing boat into the extraction vial.

4. Cap the extraction vial tightly and shake vigorously for 1 minute.

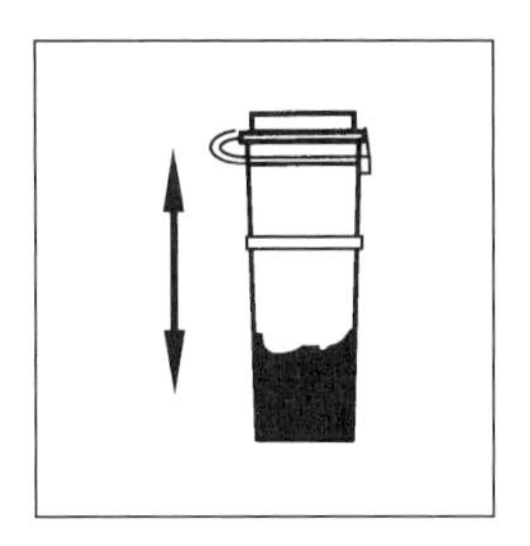

5. Allow to settle for 1 minute. Gently open the extraction vial.

6. Using the disposable bulb pipet, withdraw 1.0 to 1.5 mL from the liquid (top) layer in the extraction vial. Transfer this aliquot into the filtration barrel (the *bottom* part of the filtering assembly; the plunger is inserted into it).
Note: Do not transfer more than 1.5 mL into the barrel. The pipet is marked in 0.25-mL increments.

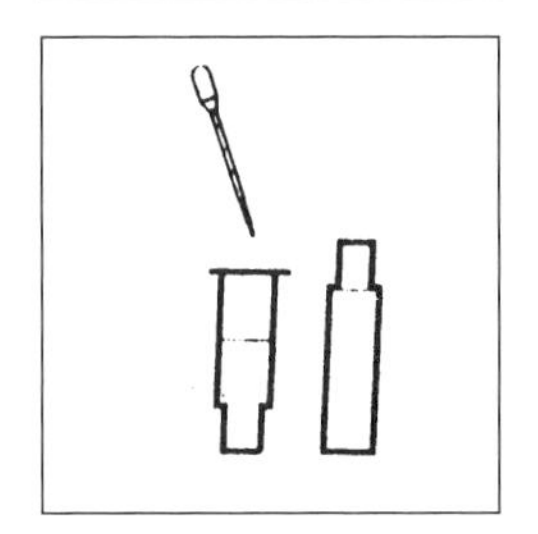

7. Insert the filtration plunger into the filtration barrel. Press firmly on the plunger until at least 0.5 mL of filtered sample is collected in the center of the plunger.
Note: The liquid will be forced up through the filter. The liquid in the plunger is the filtered extract.
Note: It may be necessary to place the filtration assembly on a table and press down on the plunger.

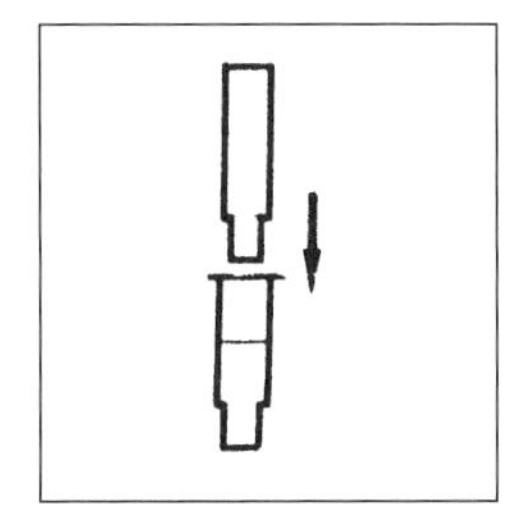

PCB PHASE 2: PREPARING SAMPLES AND STANDARDS

1. To prepare a 1-ppm threshold dilution, snap open a 1-ppm Dilution Ampule. Label the Dilution Ampule with appropriate information.

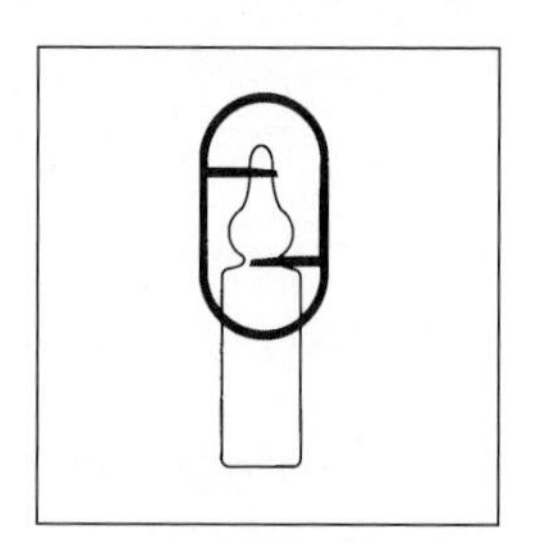

2. Using the WireTrol® pipet, withdraw 100 μL (0.1 mL) of sample extract from the filtration plunger and add it to the 1-ppm Dilution Ampule. Swirl to mix. Discard the capillary tube.

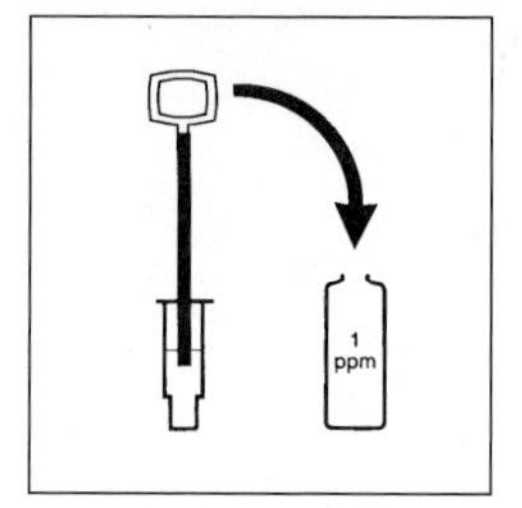

3. To prepare a 10-ppm threshold, snap open a 10-ppm Dilution Ampule. Label the Dilution Ampule. Using a TenSette Pipet, withdraw 1.0 mL from the 1-ppm Dilution Ampule (Step 2), and add it to the 10-ppm Dilution Ampule. Swirl to mix.

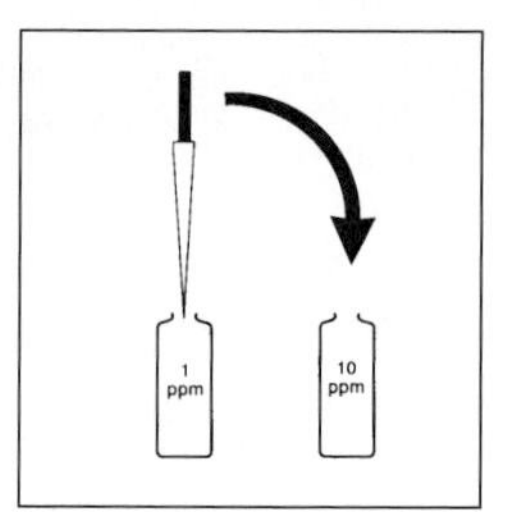

4. To prepare the standard, snap open a PCB Standard Ampule. Snap open a 1-ppm Dilution Ampule. Label the Dilution Ampule as "Standard."

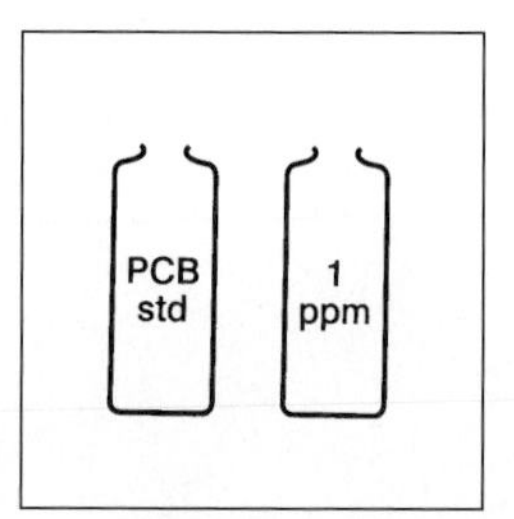

5. Using the WireTrol pipet withdraw 100 μL (0.1 mL) of the standard and add it to the 1-ppm Dilution Ampule. Swirl to mix thoroughly.
Note: Dispense standard and sample below the level of the solution in the Dilution Ampules.
Note: Use the standard dilution prepared above for both 1-ppm and 10-ppm thresholds. Do not further dilute the standard.

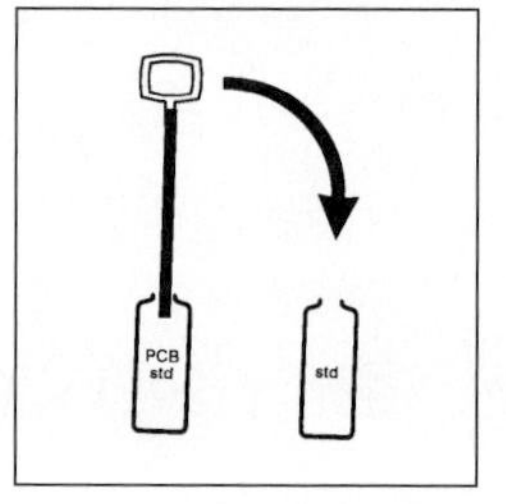

PCB PHASE 3: IMMUNOASSAY (STEPS IN THIS PHASE REQUIRE EXACT TIMING)

1. Label two PCB Antibody Tubes for each dilution ampule. Label two PCB Enzyme Conjugate Tubes for each dilution ampule.
Note: The PCB Conjugate and PCB Antibody Tubes are matched lots. Mixing with other lots will cause erroneous results.

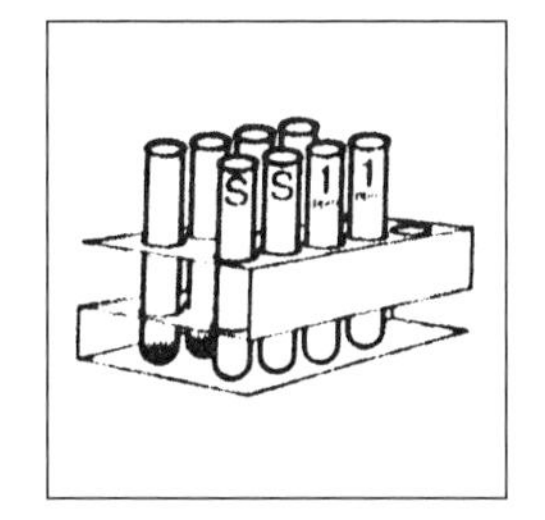

2. Use a TenSette Pipet to add a 1.0-mL aliquot from each dilution ampule prepared (1-ppm or 10-ppm) to the bottom of each appropriately labeled PCB Antibody Tube. Do this for each sample and standard. Use a new pipet tip for each solution.
Note: Do not touch the inside walls of the tubes.

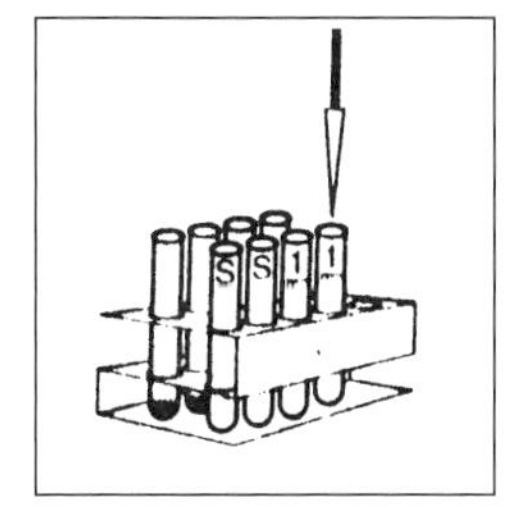

3. Begin a 10-minute reaction period.

4. At the end of the 10-minute reaction period, decant the solution from the Antibody Tubes into the respective Enzyme Conjugate Tubes.

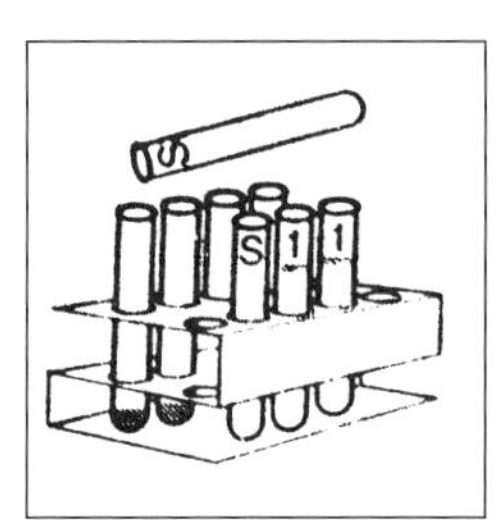

5. Invert and place the Antibody Tubes over the Enzyme Conjugate Tubes until they fit tightly onto the Enzyme Conjugate Tubes.

6. Begin a five-minute reaction period.
Note: Immediately proceed with the next step.

7. Immediately invert the solution repeatedly until the Antibody Tube has been filled four times and the Enzyme Conjugate has been dissolved. After the last inversion make sure that all of the solution is in the Antibody Tube and that it is upright.

8. Place the Antibody Tube in the rack and remove the Enzyme Conjugate Tube from the mouth of the Antibody Tube. Discard the used Enzyme Conjugate Tube.

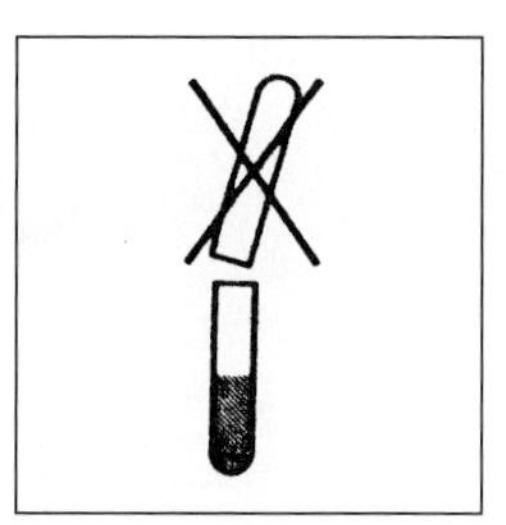

9. After the five-minute period, discard the contents of the PCB Antibody Tubes into an appropriate waste container.

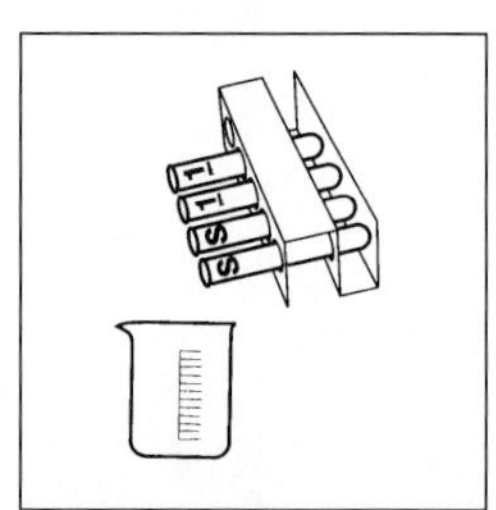

10. Wash each tube thoroughly and forcefully 4 times with Wash Solution. Empty the tubes into an appropriate waste container. Shake well to ensure most of the Wash Solution drains after each wash.
Note: Wash Solution is a harmless detergent.

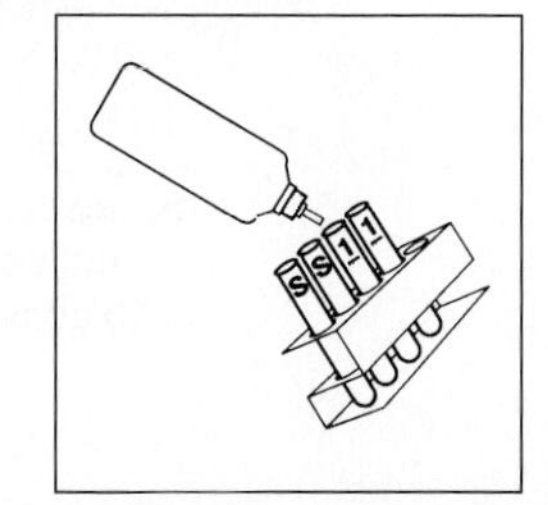

11. Continue to the next phase immediately.
Note: Ensure most of the Wash Solution is drained from the tubes by turning the tubes upside down and gently tapping them on a paper towel to drain. Some foam may be left from the Wash Solution; this will not affect results.

Continue to Phase 4

PCB PHASE 4: COLOR DEVELOPMENT (Check reagent labels carefully. Reagents must be added in proper order!)

1. Add 5 drops of **Solution A** to each tube. Replace the bottle cap.
Note: Hold all reagent bottles vertically for accurate delivery or erroneous results may occur.

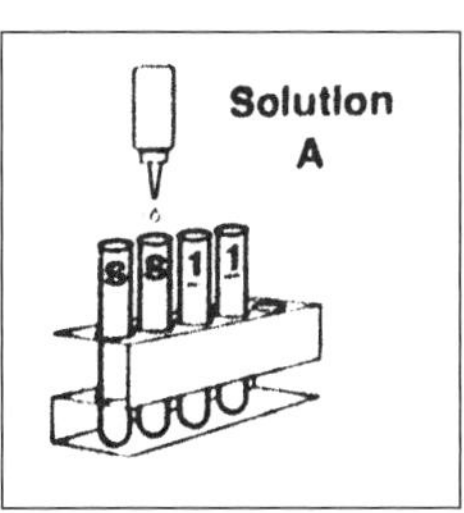

2. Begin a 2.5-minute period and immediately add 5 drops of **Solution B** to each tube. Swirl to mix. Replace the bottle cap.
Note: Add drops to the tubes in the same order to ensure proper timing (i.e., left to right). Solution will turn blue in some or all of the tubes.

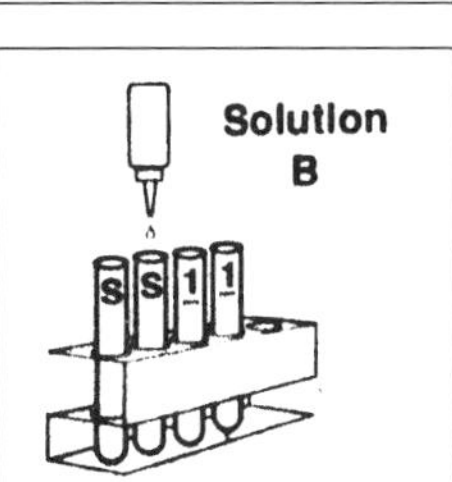

3. Let each tube react for exactly 2.5 minutes. Then add 5 drops of **Immunoassay Stop Solution** to each tube. Replace the bottle cap.
Note: Blue solutions will turn yellow when Stop Solution is added. PCB concentration is inversely proportional to color development; a light color indicates higher levels of PCB.

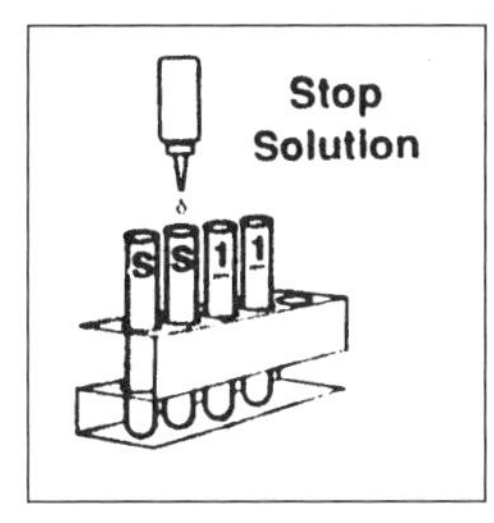

4. Using the TenSette Pipet and a new tip, add 0.5 mL of deionized water to each tube. Swirl to mix.

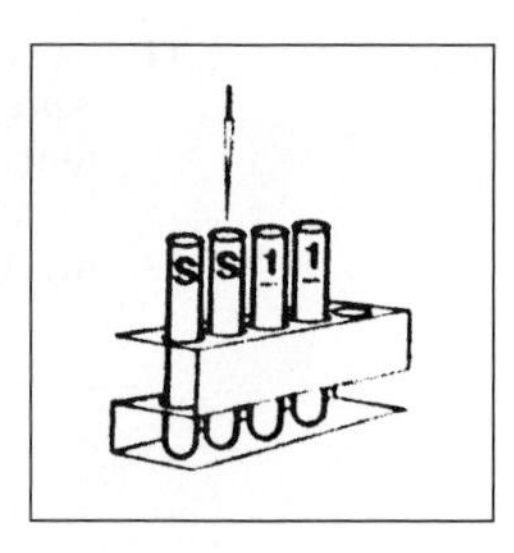

PCB PHASE 5: MEASURING THE COLOR

1. Fill a Zeroing Tube with deionized water (the blank). Wipe the outside of al the tubes with a tissue to remove smudges and fingerprints.

2. Insert the Immunoassay adapter into the sample cell compartment.
   Note: Align the adapter so the light beam openings face the sides of the DR/2010. Press firmly on the adapter to seat it.

3. Place the blank in the cell holder. Place the cover on the adapter.

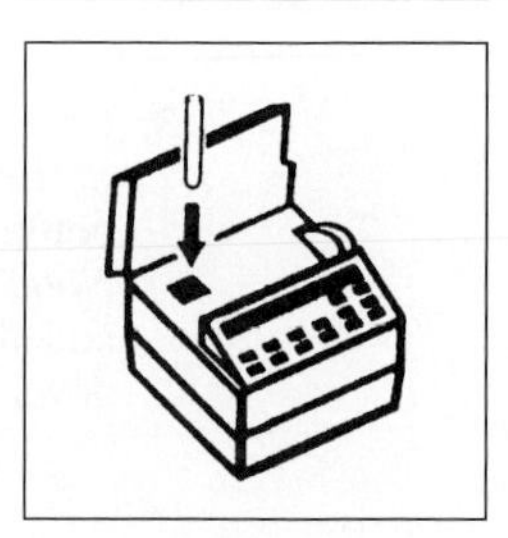

4. Press: **ZERO.** The display will show: **0.000 ABS.**

5. Insert Standard #1 Antibody Tube into the cell holder. Place the cover on the adapter.

6. Record the absorbance reading.

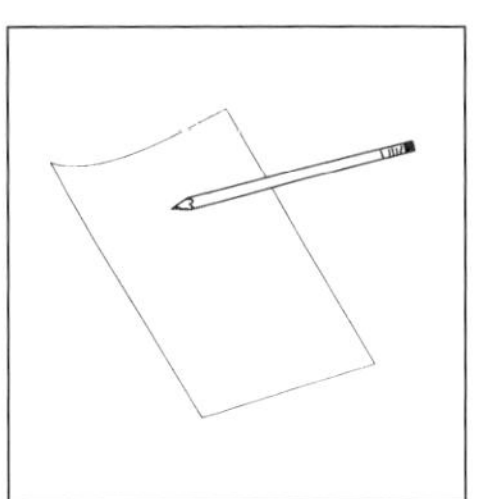

7. Repeat Steps 5 and 6 for the Standard #2 Antibody Tube. ***Note:*** If Standard 1 and Standard 2 are more than 0.350 absorbance units apart, repeat the test beginning at Phase 2, Standard Preparation.

8. Insert the Sample #1 Antibody Tube into the cell holder. Place the cover on the adapter.
Note: PCB concentration is inversely proportional to the color intensity (or absorbance value). More color means less PCB in the sample.

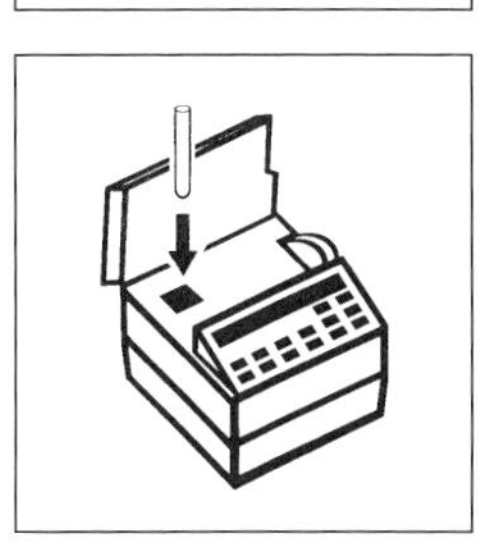

9. Record the absorbance reading.

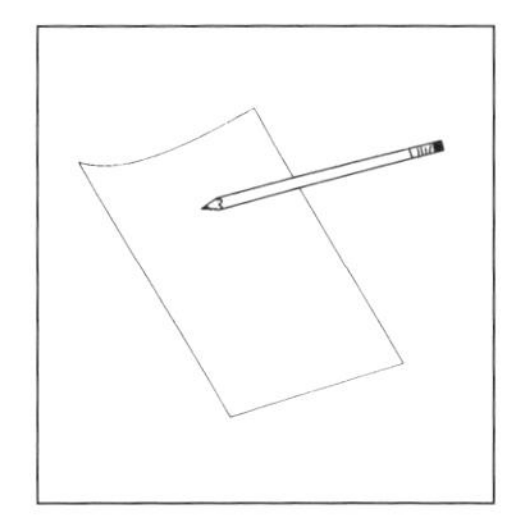

10. Repeat Steps 8 and 9 for the Sample #2 Antibody Tube. See Table 3.1 below to interpret results.

Repeat Steps 8 and 9 for Standard #2

INTERPRETING THE RESULTS

Use Table 3.1 to interpret the results.

TABLE 3.1

If the sample absorbance is . . .	1 ppm threshold	10 ppm threshold
. . . less than the highest standard absorbance	Sample PCB is greater than 1 ppm	Sample PCB is greater than 10 ppm
. . . greater than the highest standard absorbance	Sample PCB is less than 1 ppm	Sample PCB is less than 10 ppm

Sensitivity

Compound	Concentration required to result in positive test at 1-ppm threshold
PCBs	
Arochlor 1260	0.4
Arochlor 1254	0.4
Arochlor 1248	1
Arochlor 1242	2
Arochlor 1016	4
Arochlor 1232	4
Other Halogenated Compounds	
2,4,6-trichloro-p-terphenyl	>10,000
Halowax 1013	10,000
Halowax 1051	1,000
o,p-DDT	>10,000
2,4-D	10,000
Silvex	1,000
Bifenox	1,000
Tetradifon	100
Dicofop, methyl	1,000
Dichlorofenthion	10,000
Trichloroethylene	>10,000
1,2,4-trichlorobenzene	10,000
2,4-dichloro-1-naphthol	50
2,4-dichlorophenyl benzene sulfonate	1,000
1-chloronaphthalene	>10,000
Pentachlorobenzene	>10,000
Hexachlorobenzene	>10,000
2,5-dichloroaniline	>10,000
Miscellaneous Compounds	
Toluene	>10,000
Naphthalene	>10,000
DIALA(R) Oil AX	>10,000
R-Temp fluid	>10,000
Envirotemp 200 fluid	>10,000
Diesel fuel	>10,000
Gasoline	>10,000

STORING AND HANDLING REAGENTS

- Wear protective gloves and eye wear.
- Store reagents at room temperature and out of direct sunlight (less than 80° F or 27° C).
- Keep aluminized pouch that contains antibody-coated tubes sealed when not in use.
- If Stop Solution or liquid from the extraction jar comes in contact with eyes, wash thoroughly with cold water and seek immediate medical help.
- Operational temperature of the reagents is 40 to 90° F (5 to 32° C).

POLLUTION PREVENTION AND WASTE MANAGEMENT

The soil extractant (methanol) is an ignitable (D001) waste regulated by the Federal RCRA. Collect this material with laboratory solvents for disposal. If the soil samples being analyzed are contaminated with hazardous waste, the samples and resulting test waste may also need to be disposed of in accordance with RCRA.

QUESTIONS AND ANSWERS

1. Do PCBs decompose in water bodies and in soil?
PCBs are stable in water bodies and most soils and remain for years without major decomposition. PCBs can dechlorinate through microbial action in the laboratory and in the field if sufficient population of the microbes are present and the conditions for microbe survival are appropriate. Based on additional research, the microbes exist in highly polluted water and soil. PCBs appear to react under heat to form polychlorinated furans.

2. How does the toxicity of PCBs compare to chlorinated-dioxins?
PCBs are less soluble in water and less polar solvents than chlorinated dioxins. The most toxic form of dioxin (2,3,7,8-TCDD) is believed to be more toxic than any of the PCBs compounds. Many of the 210 dioxin/furan compounds are not toxic to any significant degree. However, the PCBs are more stable and exist in the environment for longer periods of time and tend to concentrate in fatty tissues and organs.

3. Are PCBs one of the most toxic substances known?
PCBs are relatively toxic, but certainly not one of the most toxic substances known. A careful review of Material Safety Data Sheets will indicate that many other substances are more toxic than PCBs and have an immediate impact on humans and animals. Also, as the chlorines change position on the PCB ring (i.e., ortho, meta and para positions) and the numbers of chlorines on the biphenyl rings vary, the toxicity level and solubility in water are both modified. The lower chlorinated PCB compounds (1 to 2 chlorines per structure) are not very toxic and decompose more readily under aerobic conditions.

4. How can PCBs be effectively destroyed?
PCBs can be effectively destroyed by incineration at high temperatures in a system with a high residence time (dwell time) and effective scrubbing. An effective means of destroying solvents containing PCBs is a rotary kiln producing cement. The temperature required to produce cement is sufficient to destroy PCBs and the residence time is more than adequate. Cement dust will neutralize any chlorinated acids formed during the process.

REFERENCES

1. Communications with Ward Stone, Wildlife Pathologist, New York State Department of Environmental Conservation, Wildlife Research Center, Delmar, New York.
2. W. J. Rogan, et al., *Increased Mortality from Chronic Liver Disease and Cirrhosis 13 Years after the Taiwan Yucheng ("Oil Disease") Incident* Am. J. Ind. Med., 31:172–175 (1997).
3. W. J. Rogan, et al., *Congenital Poisoning by Polychlorinated Biphenyls and their Contaminates in Taiwan.* Science, 241:334–336 (1998).
4. A. Bernard, et al., *Food Contamination by PCBs and Dioxins,* Nature, 401:231 (Sept. 16, 1999).
5. *DR/2010 Spectrophotometer Procedures Manual, PCB in Soil,* 3rd Ed. (Loveland, CO.: HACH Company, 1998).

6. Roger N. Reeve, *Environmental Analysis,* (ACOL), John D. Barnes (ed.), (New York, N.Y.: John Wiley & Sons, Inc., 1994).
7. Greg G. Wilber, J. L. Malanchuk, L. Smith, M. Swanson, R.A. Hites, et al., *Fate of Pesticides and Chemicals in the Environment,* Jerald L. Schnoor (ed.), Chapters 2 and 3 (New York, N.Y.: John Wiley & Sons, Inc., Wiley-Interscience Publication, 1992) at pp. 46–47 and 58–74.
8. Stanley E. Manahan, *Environmental Chemistry,* 6th Ed. (Boca Raton, Fla.: Lewis Publishers, 1994).
9. Communications with Jon Powell, Columbia-Greene Community College, Hudson, New York.

pH VALUE

A solid or gel can be slightly basic in water due to the presence of soluble forms of carbonate and bicarbonate and other anions. Additional ammonium and specific organic-complexes can demonstrate basic characteristics. Acidic characteristics are observed with solids containing specific protonated ions and complexes dissolved in water. Some solids composed of metals, such as aluminum, are hydrated with water and can easily produce a slightly acidic solution upon dissolution.

A definition for pH is $pH = -\log [H^+]$. The term pH or hydrogen ion activity represents the intensity of the acidic or basic character of a solution. At pH = 7.0, the activities of hydrogen ion and hydroxyl ion are equal, and as a result a neutral system exists at 25° C. A sample with a pH above 7.0 is considered basic, whereas a pH below 7.0 is acidic.

The pH of a sample may be measured by a pH strip (paper), pH indicator, or a pH meter. The pH meter can be a single or double electrode model. For routine analysis, a pH meter accurate and reproducible to 0.1 pH unit with a 0 to 14 pH range and a temperature adjustment system is appropriate for measurements of pH on soluble compounds.

Buffer solutions of pH 4, 7, and 10 are used to calibrate the meter before measurements. For reliable results, it is important to rinse the electrode with distilled water or a portion of the sample and gently blot the electrode dry with a kimwipe or another type of "soft tissue." For most samples in water, the calibration of the meter against the pH 7 buffer solution and rinsing and gently drying of the electrode(s) before measuring the sample pH is standard practice. When several samples are measured in a few hours, calibration of the pH meter before measuring each sample is not necessary. Over a one-hour period, the calibration of the pH meter should be rechecked with buffer solution after the measurement of 10 samples of water. However, it is necessary to properly rinse the electrode(s) after and before each pH measurement.

A pH electrode should be stored in a buffer solution (pH = 7) when not in use. The glass sensing membrane needs proper care to function properly. Proper storage instructions are provided in the manual provided with the pH meter or the electrode.

A colorimetric pH determination using phenol red indicator is practical for field studies. This method uses the DR/820, DR/850 or DR/890 Colorimeters to determine the pH of solids dissolved in water samples in the pH range of 6.5 to 8.5. This method is appropriate for a number of environmental applications, but has limited applicability to solids from industrial sites and hazardous waste sites.

pH

Procedure* **

1. Use the 10g Soil Measure to add one level measure of the soil sample to a 50 mL beaker†.

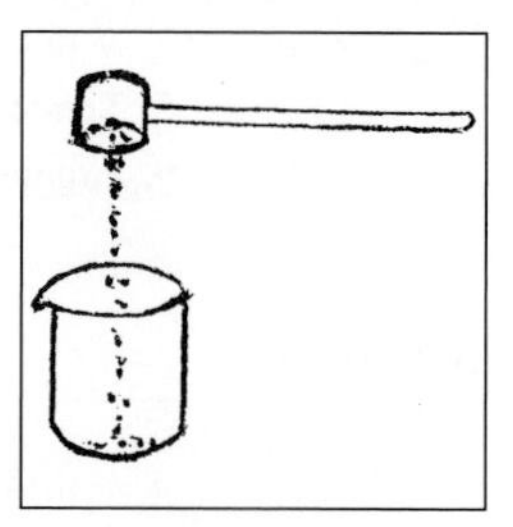

2. Add 10 mL of deionized water in a graduated cylinder to the soil sample in the beaker. Stir thoroughly.

3. Let stand for at least 30 minutes, stirring two or three times.

4. Properly rinse off the pH meter electrode(s) with distilled or deionized water. Blot the electrodes dry with kimwipes or an appropriate tissue.

---

* Modified Text, Compliments of LaMotte Company.

** Additional solid samples may be tested if ground properly and mixed with deionized or distilled water.

† Preferably use polyethylene or Teflon beakers or equivalent.

5. Read the pH of the solution on a pH meter. Stir the mixture just prior to making the pH reading.

QUESTIONS AND ANSWERS

1. What procedures should be standard practice when measuring the pH of several solutions?
The electrode of the pH meter must be rinsed with distilled water or sample water and gently blotted with a kimwipe or another soft tissue before each measurement. After each measurement of pH, the same routine should be implemented. Calibration of the pH meter should be checked after every 10 samples measured over a one-hour period. If the pH meter is used intermittently, check the calibration of the pH meter for every sample.

2. How is the pH of a solid or gel measured?
The pH of the soluble substances in solid materials can be measured by adding a small sample of solid to water, stirring the sample every few minutes (5 to 10 minutes), and again stirring the sample before reading the pH on a properly calibrated pH meter. At times, the solid material will have to be broken apart or crushed in order to obtain an accurate pH of the material. Insoluble materials will not change the pH unless bacterial action, heating, decomposition or some other process occurs.

3. What are some precautions when using a pH meter?

a) *Properly rinse the electrode(s) before and after use with distilled water. Blot the wet electrode with a kimwipe or appropriate material.*

b) *Calibrate the pH meter at appropriate times with the pH = 7 standard solution. Adjust the temperature setting if one is available.*

c) *Do not damage the electrode(s) by careless activity.*

d) *Store the electrode tip in distilled water in a beaker or cover the electrode tip with the rubber guard if not in use.*

e) *Examine the internal and external parts of the electrode occasionally to note any significant changes, such as cracked glass, salt solution needs to be upgraded, crusty or dirty surface needs to be cleaned, or wires disconnected. Check for broken wire or improper plug.*

REFERENCES

1. *LaMotte Soil Handbook* (Chestertown, Md.: LaMotte Company, 1994).

2. *Standard Methods for the Examination of Water and Wastewater:* pH VALUE, 16th Ed., A. E. Greenberg, R. R.Trussell, L. S. Clesceri and Mary Ann H. Franson (eds.), pp. 429–432, Washington, DC, American Public Health Association, 1985.

PHOSPHATE

Phosphates occur in many different forms in our environment and exhibit various degrees of solubility and physical properties. They are classified as orthophosphates, condensed phosphates (pyrophosphates, metaphosphates, and polyphosphates), and organically bound phosphates.

Many transition metal phosphates, alkaline earth phosphates and organic phosphates are insoluble or partially soluble in water. A number of the phosphate chemicals are soluble in weak solutions of acids, ammonia, and other bases. These compounds can be found as polymers in water and wastewater treatment facilities. Partially soluble and insoluble phosphate compounds are found in fertilizers, feed and food additives, textile processing, metal conversion coatings, pharmaceuticals, fireproofing materials, medicine, research, emulsifing agents, toothpaste additive, dental cements, phosphors, photographic developers, gasoline purification processes, organic catalysts, curing agents, rust remover, galvanoplastics, ceramic glazes, resins, coatings, and solder flux. Most human and animal wastes contain both soluble and nonsoluble forms of phosphate material.

A few industries treat phosphates as a waste which can be recovered and recycled as an usuable material. Some phosphate wastes are manifested for reuse or burial. Some phosphate materials enter our water sources through the use of fertilizer, sewage dumping, and garbage disposal in landfills. The organic and inorganic phosphates can be solubilized by bacterial action, pH changes, and any process that dissolves or alters various forms of phosphates in soil.

Phosphates exist under slightly varying conditions of pH, organic materials, and industrial wastes. Reactive phosphorus, acid-hydrolyzable phosphorus and organic phosphorus can occur in the suspended fractions in a sample, resulting in difficult analyses and possibly erroneous results. Therefore, phosphate must be in the appropriate form for accurate results during analysis. In addition, commercial detergents containing phosphate must be avoided when cleaning glassware used for phosphate analysis.

Phosphorus promotes plant growth, rapid cell growth, and increases in crop yield.* Phosphorus increases the ratio of grain to straw and stimulates the formation of fats, convertible starches and healthy seed. Phosphorus increases the resistance to disease in plants. Analyses of plant samples and similar samples require heating the sample in a muffle furnace to form an ash, dissolution of the ash contents, diluting the sample to volume and analyzing the sample.

Phosphate from a solid sample, gel or semisolid sample weighing 0.100 to 0.500 grams can be digested in a known volume of acid. The digestion procedure for oils requires 0.100 to 0.250 grams of material diluted to a specific volume. After boiling for 4 to 5 minutes, 10 mL of 50% hydrogen peroxide is added to the charred sample and the sample is heated for 1 minute. The digested mixture is allowed to cool and is diluted to 100 mL with deionized water. After selecting the appropriate volume and diluting the solution to 25.0 mL, the clear sample can be analyzed by the Ascorbic Acid (Method 8048) procedure. A pH adjustment on the sample will be required before analysis for phosphate.

With the HACH procedures, phosphates can be determined by the PhosVer 3 (Ascorbic Acid) Method 8048 using the DR/2010 Spectrophotometer, the DR/850 or the DR/890 Colorimeters in the field.** The PhosVER 3 experiment (HACH method 8048) for testing orthophosphates is presented in Chapter 2. A number of interferences are noted in this procedure. In addition, test strips can provide fast and "approximate" results in the range of 0 to 50 ppm phosphate.

---

* Robert O. Miller, *Handbook of Reference Methods for Plant Analysis,* Yash P. Kalra (ed.), (Boca Raton, Fla.: CRC Press, 1998).

** Compliments of HACH Company.

In a few cases, the total phosphorus content of a sample will be necessary. Total Phosphorus (organic and acid hydrolyzable) can be determined by similar field methods after persulfate digestion of the bound phosphate to orthophosphate by HACH Method 8190. Pretreatment of the sample with acid and heat provides the conditions for hydrolysis of the condensed (meta-, pyro- and polyphosphates) inorganic forms. Organic phosphates are converted to orthophosphates by heating with acid and persulfate. Organically bound phosphates can be determined indirectly by subtracting the result of an acid hydrolyzable phosphorus test from the total phosphorus result.

With the LaMotte procedure, the phosphate reacts with a solution of ammonium molybdate and antimony potassium tartrate in a filtered acid medium to form an antimony-phosphomolybdate complex. Ascorbic acid reduces the complex to an intense color, which is proportionate to the amount of orthophosphate present. Polyphosphates and some organic phosphorus compounds can be converted to the orthophosphate form by sulfuric acid digestion. Organic phosphorus compounds can be converted to orthophosphate by persulfate digestion. High iron content in the solid can cause precipitation of phosphates and loss of phosphates in the experiment.

PHOSPHORUS TEST

Ascorbic Acid Reduction Method*—Procedure**

1. Use the 1 mL pipet to add 1 mL of the soil filtrate to a clean colorimeter tube and dilute to the 10 mL line with deionized water.

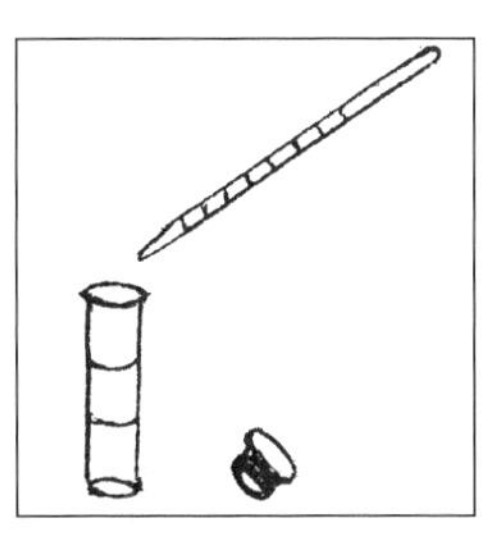

2. Select setting 4 on the "Select Wavelength" knob.

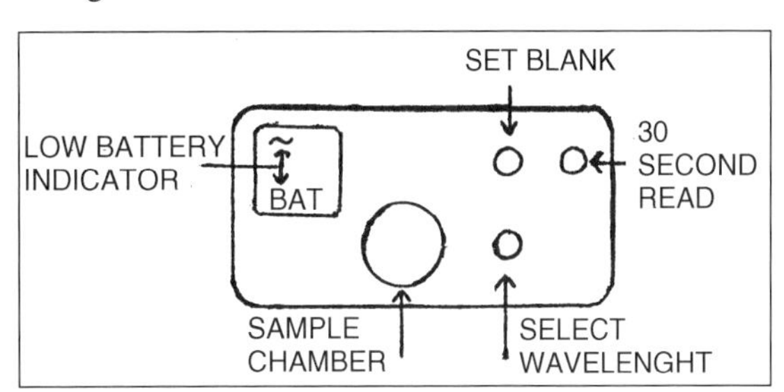

3. Insert the tube containing the diluted soil filtrate into chamber and press the "30 Second Read" button. Adjust "100"%T with the "Set Blank" knob.

---

*Compliments of LaMotte Company.

**A special extraction procedure is provided below for soil of pH above 7.0.

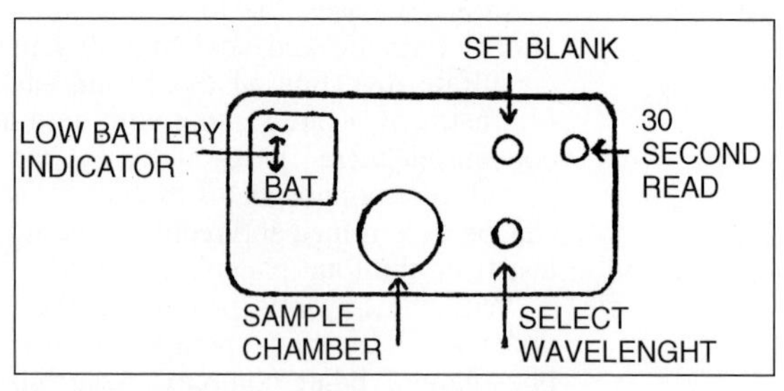

4. Remove the tube containing the diluted soil filtrate and add one measure Phosphate Acid Reagent with the 1 mL pipet, cap the tube and mix.

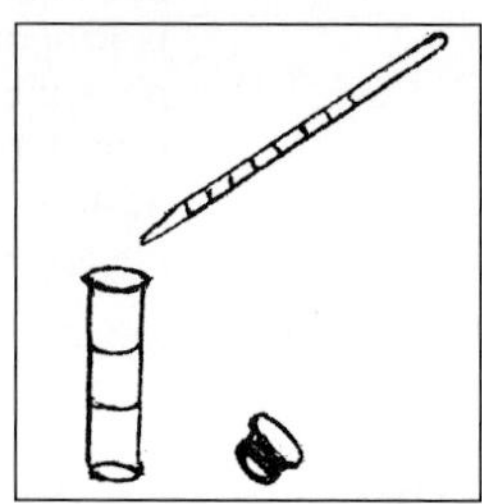

5. With the 0.1 g spoon, add 1 level measure of Phosphate Reducing Reagent and shake until dissolved. Allow 5 minutes for full color development.
   Note: Phosphates exhibit a clear blue color.

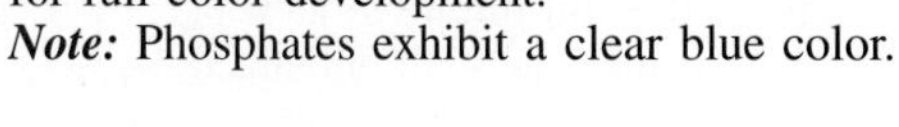

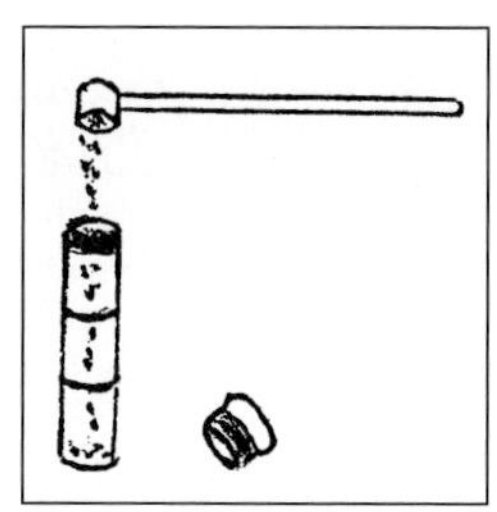

6. At the end of the waiting period, insert test sample into the colorimeter chamber, press the "30 Second Read" button and measure %T as soon as reading stabilizes.

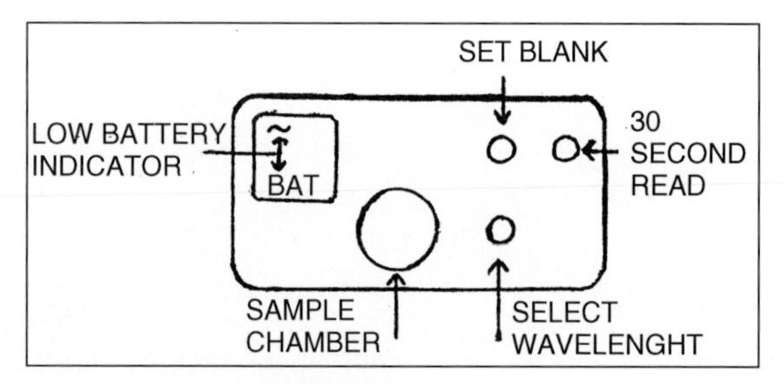

7. Consult the chart below to find the concentration, in pounds per acre, of Phosphorus (P) in the soil.

PHOSPHORUS IN ALKALINE SOILS

A special extraction procedure is used for determining the available phosphorus content of Western U. S. alkaline soils where the pH value is above 7.0.

EXTRACTION PROCEDURE

1. Use the 1 mL pipet to add 1 mL of the Special NF Extracting Solution to the graduated vial, then add deionized water to the graduation.

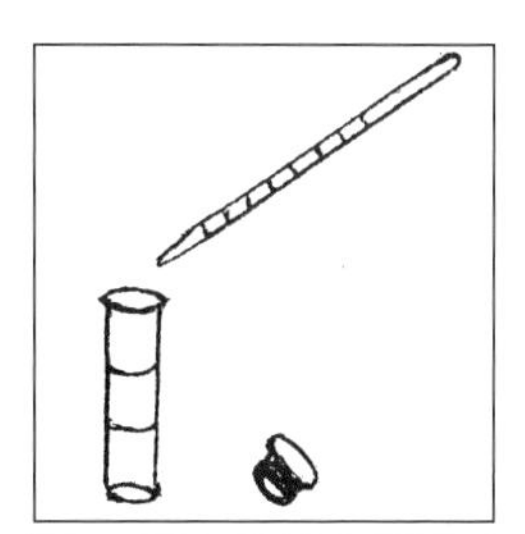

2. Add 3 of the 1 g measures of soil using the 1 g spoon to the extracting solution in the vial.

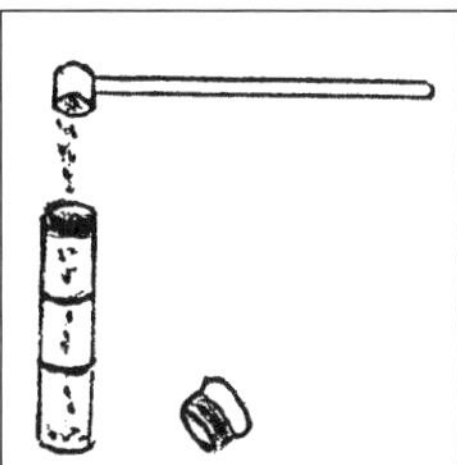

3. Cap the vial and shake for a period of 5 minutes.

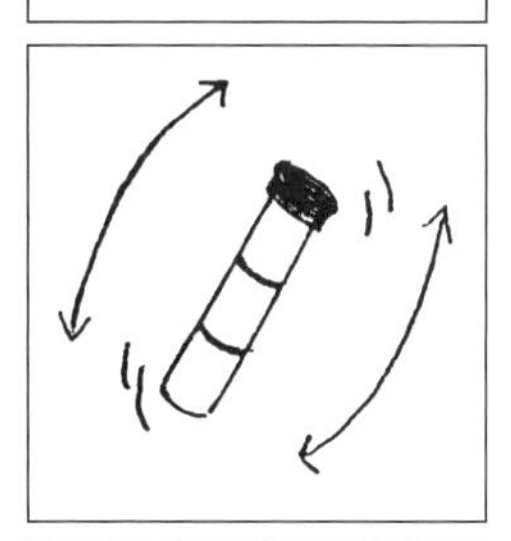

4. Filter using the funnel and filter paper. Collect all of the filtrate.

5. Perform the Phosphorus test according to the Phosphorus procedure provided above.

Phosphorus Calibration Chart (Lbs/A)

%T	9	8	7	6	5	4	3	2	1	0
90	0.2	1.0	1.7	2.4	3.1	3.9	4.6	5.4	6.2	6.9
80	7.7	8.5	9.3	10.1	10.9	11.8	12.6	13.5	14.3	15.2
70	16.1	17.0	17.9	18.8	19.7	20.7	21.6	22.6	23.6	24.6
60	25.6	26.7	27.7	28.8	29.9	31.0	32.1	33.2	34.4	35.6
50	36.8	38.0	39.3	40.5	41.8	43.2	44.5	45.9	47.3	48.8
40	50.2	51.8	53.3	54.9	56.5	58.2	59.9	61.7	63.5	65.4
30	67.3	69.3	71.3	73.4	75.6	77.9	80.2	82.6	85.1	87.8
20	90.5	93.3	96.3	99.4						

Phosphorus Concentration Chart

%T	Range	Pounds per acre
81–100%	Very Low	0–14 lbs/acre
61–80%	Low	16–34 lbs/acre
39–60%	Medium	35–67 lbs/acre
0–38%	High	Over 70 lbs/acre

HAZARDOUS SUBSTANCES

Hazardous substances are the phosphate acid reagent and the phosphate reducing agent. Solutions should be handled appropriately and disposed of in accordance with the Material Safety Data Sheets (MSDS).

RANGE OF MEASUREMENTS

The range of measurements is 0 to 99 lbs/acre.

INTERFERENCES

High iron concentrations can cause precipitation of phosphate and subsequent loss of phosphorus.

QUESTIONS AND ANSWERS

1. Which forms of phosphate require digestion or pretreatment?
   All forms, except the orthophosphate form, require digestion and pretreatment before analysis.
2. What forms of phosphate are used in industrial processes and water and wastewater treatment processes to protect the inside surfaces of pipes and equipment?
   Polymeric phosphate compounds have often been used to line pipes for protection purposes. In sewage and water plants, the polyphosphates also aid in the precipitation process.
3. During collection and analysis of samples, why must samples be collected in glass bottles?
   Phosphates tend to adsorb onto the surfaces of plastic containers. Before use, glass containers should be rinsed with dilute HCL followed by several distilled water rinses.
4. What type of detergent should be used to rinse equipment?
   Commercial detergents containing phosphate must be avoided when cleaning glassware used for phosphate analysis.

REFERENCES

1. *HACH Water Analysis Handbook,* 3rd Ed. (Loveland, CO.: HACH Company, 1997).
2. *LaMotte Soil Handbook* (Chestertown, Md.: LaMotte Company, 1994).
3. *Handbook of Reference Methods for Plant Analysis,* Yash P. Kalra (ed.), (Boca Raton, Fla.: CRC Press, 1998).
4. *Standard Methods for the Examination of Water and Wastewater:* PHOSPHORUS, 16th Ed., Method 424, A. E. Greenberg, R. R. Trussell, L. S. Clesceri and Mary Ann H. Franson (eds.), pp. 437–438, Washington, DC, American Public Health Association, 1985.
5. *Hawley's Condensed Chemical Dictionary,* 11th Ed., rev'd by N. Irving Sax and Richard J. Lewis (New York, N.Y.: Van Nostrand Reinhold, 1987).

POTASSIUM

Potassium forms a variety of compounds which have practical applications in decorative coloring of metals, medicine, antibiotics, dehydrating agent, textile conditioning and printing, crystal glass, synthetic flavoring, rubber formulations, soaps and detergents, driers, insecticides, silvering mirrors, inks, baking powders, alkylation agent, etching glass, silver solder flux, brazing, laboratory agent, pigments, soaps, stabilizer, purification of various compounds, food preservative, food additive, foaming agent for plastics, and special spectroscopy applications. Additional applications for potassium-type compounds include enamels, disinfectant, explosives, matches, fertilizer, special plant nutrients, herbicide, fungicide, manufacture of organic chemicals and drugs, treatment for sickle cell anemia, painting porcelain and glass, fumigant, batteries, synthetic mica, bactericides, emulsifying agents, purification of gasoline, determination of nitrogenous matter in water, oxidizer in solid rocket propellants, pyrotechnics, peptizing agent, flares, sequestrant, buffer, fat emulsifier, flame retardant, photography, and many additional applications.

Potassium plays a vital role in the physiological and biochemical functions of plants, animals, and other types of species. Potassium chloride is a common form of fertilizer, called muriate of potash. In plants, potassium enhances disease resistance by strengthening stalks and stems. Also, it activates specific enzyme systems, and contributes to a thicker cuticle which guards against disease and water loss.* Potassium controls the turgor pressure within plants to prevent wilting, and enhances fruit size, texture, flavor, and development. Potassium is involved in the production of amino acids, chlorophyll formation, starch formation, and sugar transport from leaves to roots. The plants' ability to take up potassium appears to provide a resistance to stresses of disease, temperature, and moisture.

Plants with a potassium deficiency will display older leaves with burned edges.** The plants are sensitive to disease infestation and will appear stunted. Fruit and seed production will be of poor quality. With potassium deficiency, the seasonal duration of leaf photosynthesis is limited, transport of sugars and nutrients can be severely hampered, starch formation is hindered, and nitrogen plays little role in the development of the plant. With an excess of potassium, plants will exhibit a cation imbalance with magnesium, and possibly calcium deficiency symptoms.

Potassium is present in most solids and gels as a soluble ionic compound. Potassium can be trapped in the crevices and cavities of a rock or solid material. A substance such as

---

*Potassium Test, *Instruction Manual* (Chestertown, Md.: LaMotte Company, 1998), pp. 25–26.

**Robert D. Munson, *Handbook of Reference Methods for Plant Analysis,* Yash P. Kalra (ed.), (Boca Raton, Fla.: CRC Press, 1998), pp. 7–8.

potassium chloroplatinate used in photography is slightly soluble in water. Complexes like potassium penicillin G, commonly used as an antibiotic, are extremely soluble in various types of solutions. Because of the solubility of most potassium compounds, potassium ion can be easily determined in solids by dissolution in a 2% weak acid solution.

During the soil test, a high ammonia nitrogen test result can lead to a false high reading in the potassium test. If significant quantities of ammonia-type fertilizer have been recently applied or if the soil pH is below 5.0, the ammonia test should be performed before the potassium test. In the test, potassium reacts with sodium tetraphenylboron to form a colloidal white precipitate in quantities proportional to the potassium concentration measured as turbidity.* Very high concentrations of calcium and magnesium can interfere during the test.

A tissue testing kit can estimate the amount of potassium in plant tissue.* The semi-quantitative test is inexpensive and easy to run. Usually potassium is determined by the more expensive methods of atomic absorption and flame photometry.

POTASSIUM TEST

Tetraphenylboron Method—Procedure**

1. Use the 1 mL pipet to add 2 ml of the soil filtrate to a clean colorimeter tube and dilute to the 10 mL line with deionized water.

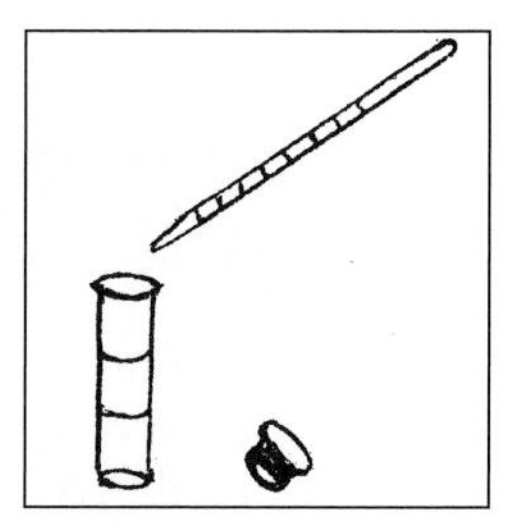

2. Select setting 1 on the "Select Wavelength" knob.

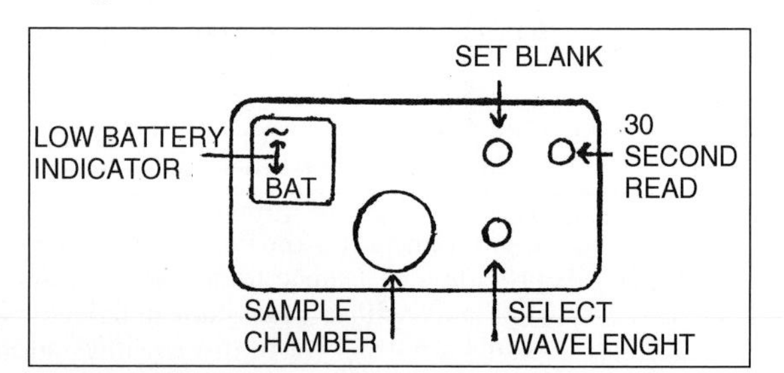

3. Insert the tube containing the diluted soil filtrate into chamber and press the "30 Second Read" button. Adjust "100"%T with the "Set Blank" knob.

*J. Benton Jones, Jr. and Denton Slovacek, *Handbook of Reference Methods for Plant Analysis,* Yash P. Kalra (ed.), (Boca Raton, Fla.: CRC Press, 1998), pp. 122–124.

**Compliments of LaMotte Company.

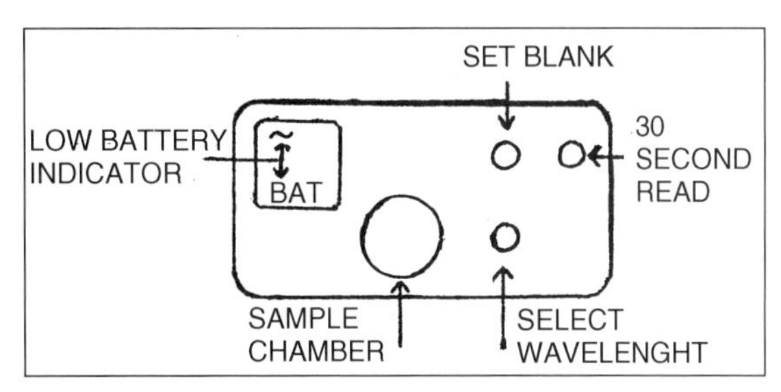

4. Remove the tube containing the diluted soil filtrate and add 4 drops of 1.0 N sodium hydroxide and mix.

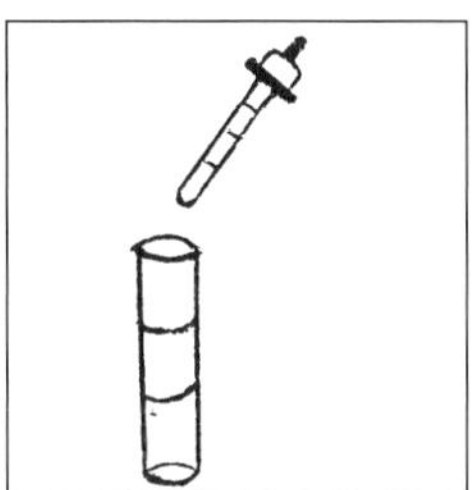

5. With the 0.05 g spoon, add 1 level measure of tetraphenylboron powder. Cap the tube and shake vigorously until all of the powder has dissolved.

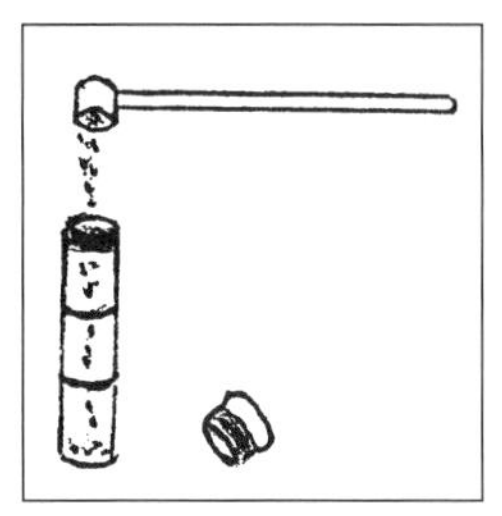

6. After standing 5 minutes, shake the tube to suspend any settled precipitate and immediately place it in the colorimeter chamber, press the "30 Second Read" button and measure the %T as soon as the reading stabilizes.

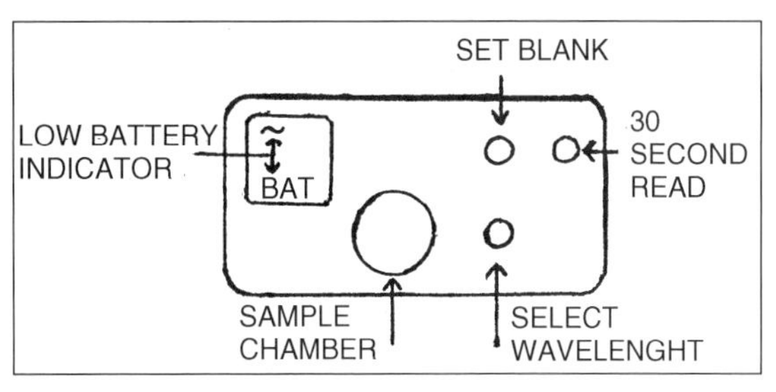

7. Consult the chart to find the concentration, in pounds per acre, of potassium in the soil.

Potassium Calibration Chart (Lbs/A)

%T	9	8	7	6	5	4	3	2	1	0
90	22.9	26.5	30.1	33.6	37.1	40.5	43.9	47.3	50.5	53.8
80	57.0	60.1	63.2	66.2	69.3	72.1	75.0	77.8	80.6	83.3
70	86.0	88.6	91.2	93.7	96.2	98.6	100.9	103.2	105.5	107.7
60	109.8	111.9	114.0	116.0	118.0	119.9	121.7	123.6	125.4	127.1
50	128.8	130.5	132.1	133.7	135.3	136.8	138.4	139.9	141.4	142.9
40	144.3	145.8	147.3	148.9	150.4	152.0	153.7	155.3	157.1	159.0
30	160.9	163.0	165.2	167.5	170.1	172.8	175.8	179.1	182.7	186.6
20	191.0	195.8	201.1	207.0	213.5	220.9	229.1	238.2	248.6	260.2
10	273.2	288.0	304.7	323.7	345.4	370.1	398.3	430.8	468.3	511.6

Potassium Concentration Chart

%T	Range	Pounds per acre
93–100%	Very Low	0.44 lbs/acre
82–91%	Low	50–76 lbs/acre
50–81%	Medium	82–143 lbs/acre
18–80%	High	144–281 lbs/acre
0–17%	Very High	Over 294 lbs/acre

HAZARDOUS SUBSTANCES

Hazardous substances are 0.1 N Sodium Hydroxide Solution and tetraphenylboron powder. The substances are considered hazardous substances and must be handled appropriately. Consult the material data sheet for additional information.

RANGE OF MEASUREMENT

The range of measurement is 0 to 500 lbs/acre.

INTERFERENCES

Calcium and magnesium at very high concentrations interfere with the measurement.

QUESTIONS AND ANSWERS

1. What chemicals are considered to be hazardous in the experiment?
   Sodium hydroxide and tetraphenylboron powder are hazardous substances. The sodium hydroxide can be effectively neutralized by the addition of an equal volume of 0.1 N HCL.

2. What is the range of determination for potassium in the experiment?
For the experiment, the range of determination is 0 to 500 lbs/acre.

3. Where do we find potassium?
Potassium, like sodium, is found in many common items from medicine to pigments and explosives. Potassium is necessary for quality living in both plants and humans. Potassium enhances disease resistance functions.

4. Why should the ammonia (ammonium) test be performed before the potassium test?
The ammonium cation acts very much like the potassium ion. When present in large amounts, the ammonium cation will produce a precipitate similar to that produced by potassium. If fertilizer containing high levels of ammonia (or ammonium compounds) has been recently applied to soil, the ammonia test should be performed before the potassium test. A high ammonia nitrogen test result will probably result in a high reading for potassium. The actual results for potassium should be somewhat lower.

REFERENCES

1. *LaMotte Soil Handbook* (Chestertown, Md.: LaMotte Company, 1994).
2. *Handbook of Reference Methods for Plant Analysis,* Yash P. Kalra (ed.), (Boca Raton, Fla.: CRC Press, 1998).

SILVER

Silver metal tends to tarnish in air. Compounds, such as silver chloride, are light sensitive and become dark on exposure to light. A number of silver compounds are soluble in organic solvents and are decomposed by acids and other solvents. Many insoluble silver compounds will dissolve in the presence of normal decomposition products such as ammonia. Some silver compounds are shock irritants to skin and tissue, while others are an explosion risk.

Silver compounds have been used for detonators, laboratory reagents, photographic film and plates, x-ray film, plating processes, photochromic glass, light bulb coating, jewelry, cutlery, electrical contacts, bearing metal, magnet windings, dental amalgams, organic synthesis, batteries, photometry, optics, catalysts, silver plating, antiseptics, absorption cells, cloud seeding, silvering mirrors, germicide, alloys, medicine, ceramics, pharmaceuticals, reaction vessels, and polishing glass. The metal is used as a liner for chemical vats and equipment and as a coating on numerous items used every day. Colloidal silver is a nucleating agent in photography and during its combination with protein in medicine.

Since silver has significant value, many industries recover silver for investment purposes and for industrial applications. The amount of silver buried in hazardous waste sites is usually quite small. Small quantities of silver in complex mixtures are normally recovered by various separation procedures.

Silver from a solid sample or semisolid sample weighing 0.100 to 0.500 grams can be digested in a known volume of acid. The digestion procedure for oils requires 0.100 to 0.250 grams of material diluted to a specific volume. After boiling for 4 to 5 minutes, 10 mL of 50% hydrogen peroxide is added to the charred sample and the sample is heated for 1 minute. The digested mixture is allowed to cool and is diluted to 100 mL with deionized water. After selecting the appropriate volume and diluting the solution to 50.0 mL (25 mL for oils), the clear sample can be analyzed by a colorimetric (HACH Method 8120) procedure after the pH of the solution has been adjusted. Method 8120 for silver analysis is provided in Chapter 2.

When determining silver, there is a significant number of negative interferences by metals, chloride and ammonia at relatively high concentrations. A few metals, especially mercury at

2 mg/l, can provide a positive interference. During the digestion procedure, most interferences will probably be eliminated. If the interferences are not eliminated, then proceed with the alternative digestion procedure listed at the end of the experiment (HACH Method 8120).

Silver ions in basic solution react with an additive to form a colored complex. Sodium thiosulfate performs as a decolorizing agent in the blank while other additives act as buffer, indicator and masking agents. Organic extractions are avoided in this procedure. The HACH colorimetric procedure measures the concentration, in mg/L, of silver at 560 nm.

QUESTIONS AND ANSWERS

1. Why should the pH meter not be used for samples containing silver?
   The pH electrodes containing silver could contaminate the sample.
2. Why is sodium thiosulfate added to the blank?
   Sodium thiosulfate acts as a decolorizing agent.
3. What are colloidal silver and argyrol?
   Colloidal silver is a nucleating agent used primarily in photography and in medicine. Argyrol, a colloidal silver complex, is a silver and protein organic compound used in medicine for its specific antiseptic and bacteriostatic action. The compound has low toxicity.
4. What happens to many of the silver effluents from industrial processes?
   A considerable amount of silver is recovered from industrial effluents and made into bars. Silver bars have considerable value.

REFERENCES

1. *HACH Water Analysis Handbook,* 3rd Ed. (Loveland, CO.: HACH Company, 1997).
2. *Standard Methods for the Examination of Water and Wastewater:* SILVER, 16th Ed., A. E. Greenberg, R. R.Trussell, L. S. Clesceri and Mary Ann H. Franson (eds.), p. 242, Washington, DC, American Public Health Association, 1985.
3. *Hawley's Condensed Chemical Dictionary,* 11th Ed., rev'd by N. Irving Sax and Richard J. Lewis (New York, N.Y.: Van Nostrand Reinhold, 1987).

SULFATE

Metal sulfates of alkali metals and alkaline earth metals have many practical uses in our society and exhibit very little toxicity. The sulfates of mercury, copper, lead, nickel, and other metals have a much higher toxicity level and are somewhat less soluble in water. Insoluble metal sulfate cakes form in oil fields when two or more types of water are mixed. Some applications of sulfate compounds are in dietary supplement, preservative, pharmaceuticals, cosmetic lotions, flour enrichment, and medicine. Additional applications of some of the more insoluble forms of sulfate include analytical reagents, cement, alum manufacture, pigments, glass manufacture, chemical manufacturing, coatings, ceramics, paperboard manufacture, fibers, fireproofing, catalyst, fertilizers, laboratory reagent, plating, metal priming paints, rust inhibitor in paints, lubricants, plastic and rubber products, rayon manufacturing, herbicide, metallurgy, drying industrial gases, desiccant, wallboard, dyes, surgical casts, and many additional applications.

Sulfur, in sulfate, affects plant metabolism and is essential for the formation of protein.* Sulfur-deficient plants are pale green in color. Plants grow slowly, delay in maturity, and have low yields. Excess sulfur may develop a premature aging of the leaves.

Negatively charged sulfate ions from soluble sulfates are easily leached in soil. The major sources of soluble soil sulfates are fertilizer, wastewater, and atmospheric sulfur dioxide carried into the soil by precipitation. Some of the more insoluble forms of sulfate tend to dissolve more readily in heated conditions and with changes in pH and bacterial action.

Insoluble sulfates can be determined by grinding solid samples to a constant particle size and then extracting the sulfate with hot acid. Plants and additional organic materials will require drying before analysis. Soluble sulfates can be extracted from most solids by a "Sulfate Extracting Solution" provided by LaMotte Company.** Sulfate ion can be precipitated in an acid medium with barium chloride to form barium sulfate in proportion to the amount of sulfate present. The Barium Chloride Method for determining sulfate dissolved in solution has a range of 0 to 172 ppm. Suspended matter may be removed by a filtration step. Silica present above 500 mg/L will interfere. Color interference can be a problem.

SULFUR (SULFATE)

Barium Chloride Method—Procedure†

1. Use the 1 mL pipet to add 1 mL of Sulfate Extracting Solution to the graduated vial.

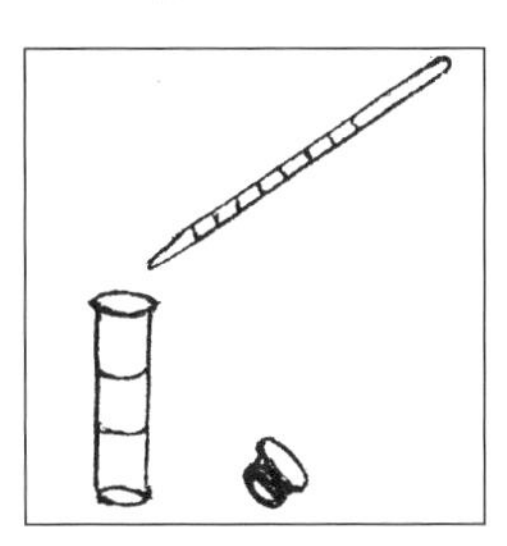

2. Add deionized water to the graduation.

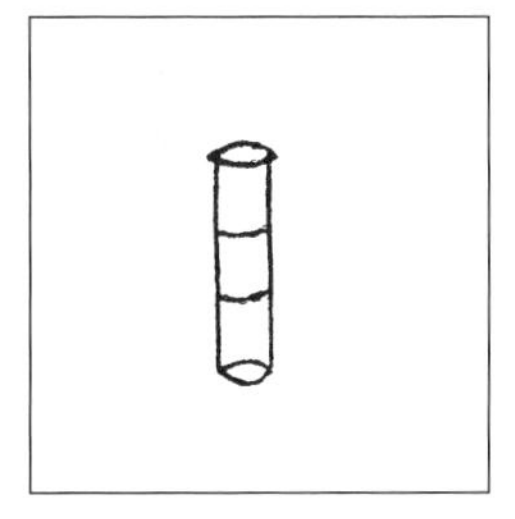

3. Add 3 of the 1 gram measures of soil using the 1 g spoon. Cap vial and shake for 5 minutes.

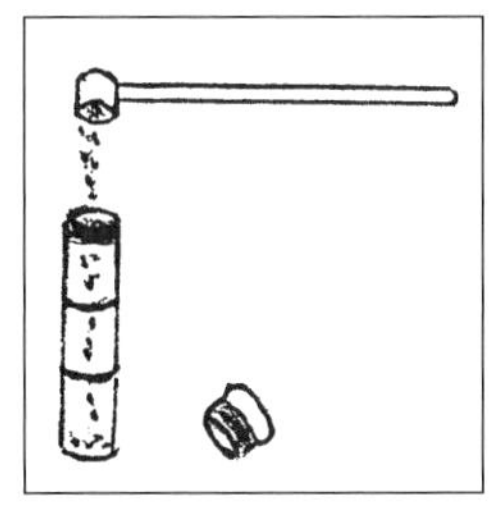

* Sulfur Test, *Instruction Manual,* (Chestertown, Md.: LaMotte Company, 1998), pp. 27–28.

** Robert D. Munson, *Handbook of Reference Methods for Plant Analysis,* Yash P. Kalra (ed.), (Boca Raton, Fla.: CRC Press, 1998), p. 8.

† Compliments of LaMotte Company.

4. Filter and collect all of the soil filtrate using the funnel and filter paper. If the filtrate is not clear, filter a second time.

5. Fill a clean colorimeter tube to the 10 mL line with the soil filtrate.

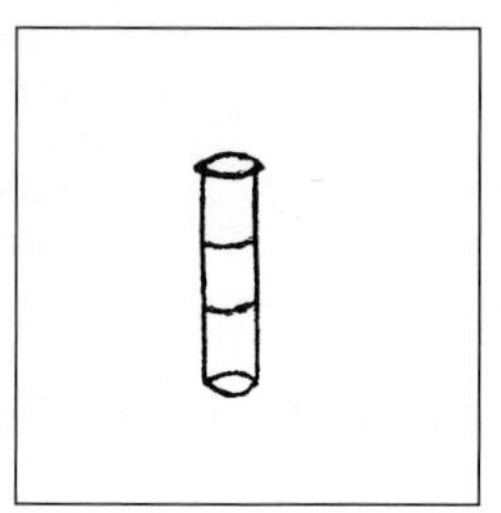

6. Select setting 1 on the "Select Wavelength" knob.

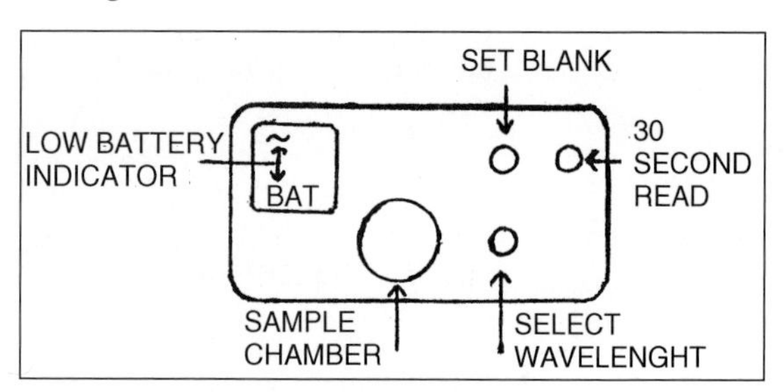

7. Insert tube containing sample into chamber and press the "30 Second Read" button. Adjust "100% T with the "Set Blank" knob.

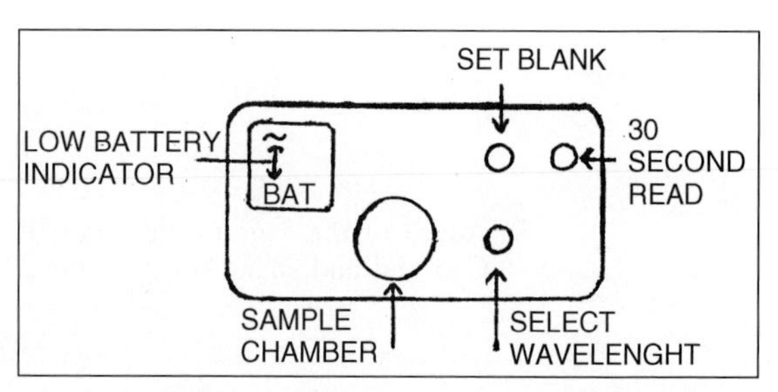

8. Remove the tube from the chamber. Use the 0.1 g spoon to add one level measure of Sulfate Reagent to the sample tube.

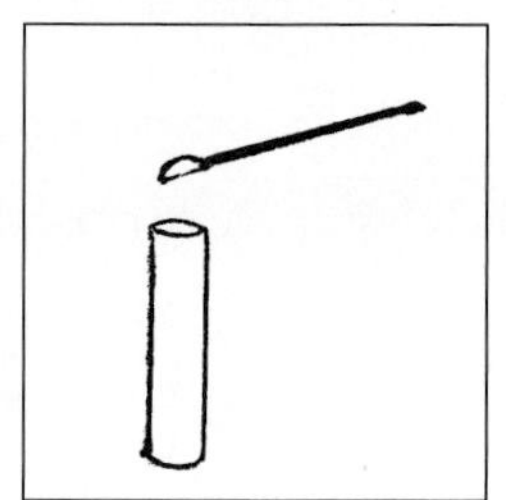

9. Cap the tube and shake until powder has dissolved.
Note: A white precipitate will develop if sulfates are present.

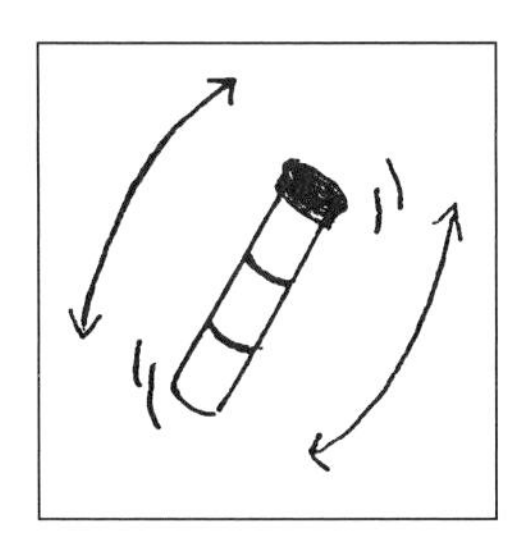

10. Allow the reaction to proceed for 5 minutes, then mix again before inserting tube into chamber of colorimeter. Cover and measure %T as soon as reading stabilizes.

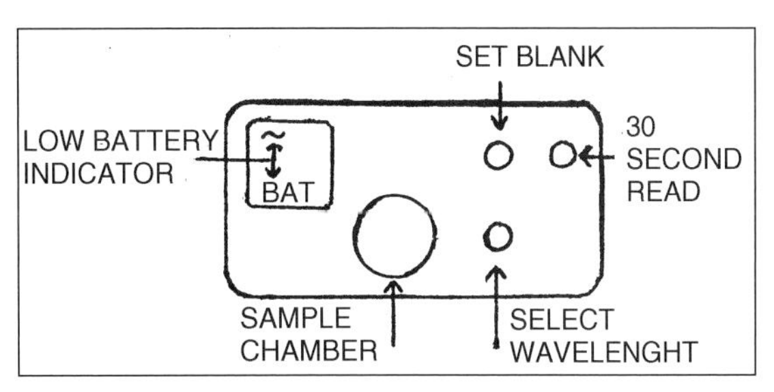

11. Consult chart to find the concentration in parts per million (ppm) of sulfur.
Note: A white film is deposited on the inside of test tubes as a result of the sulfate test. Thoroughly clean and rinse test tubes after each test.

Sulfur Calibration Chart (ppm)

%T	9	8	7	6	5	4	3	2	1	0
90	6.7	7.2	7.6	8.1	8.5	9.0	9.4	9.9	10.3	10.8
80	11.2	11.7	12.2	12.6	13.1	13.6	14.0	14.5	15.0	15.5
70	16.0	16.5	17.0	17.5	18.0	18.5	19.1	19.6	20.1	20.7
60	21.2	21.8	22.3	22.9	23.5	24.1	24.6	25.3	25.9	26.5
50	27.1	27.8	28.5	29.1	29.8	30.5	31.3	32.0	32.8	33.5
40	34.4	35.2	36.0	36.9	37.8	38.8	39.7	40.7	41.8	42.9
30	44.0	45.2	46.4	47.7	49.1	50.5	52.0	53.6	55.2	57.0
20	58.9	60.9	63.0	65.3	67.8	70.4	73.3	76.4	79.8	83.5
10	87.6	92.0	97.0	102.6	108.8	115.9	124.0	133.3	144.1	156.9
0	172.1									

Sulfur Concentration Chart

%T	Range	Parts per million
79–100%	Low	0–16 ppm
55–78%	Medium Low	17–30 ppm
35–54%	Medium	31–50 ppm
0–33%	High	52–75 ppm

HAZARDOUS SUBSTANCES

Hazardous substances are the Sulfate Extracting Solution and the Sulfate Reagent. Material Safety Data sheets must be available. Solutions should be handled safely and disposed of properly.

RANGE OF MEASUREMENT

The range of measurement is 0 to 172 ppm.

INTERFERENCES

Suspended matter and color interference may be removed by a filtration step. Silica in excess of 500 mg/L will interfere.

QUESTIONS AND ANSWERS

1. Which sulfates are more toxic and more insoluble?
The sulfates that are more toxic are mercury and lead. The sulfates that have some toxicity are copper and nickel. The least toxic sulfates are iron, and the alkaline-earth and alkali sulfates. Mercuric sulfate tends to decompose in water with time and mercurous sulfate has limited solubility in water. Various hydrated lead sulfate compounds are somewhat soluble in hot water, but are nearly insoluble in water at room temperature. Barium sulfate forms an insoluble suspension in solution.
2. What are the interferences during sulfate measurements?
The primary interferences are highly turbid samples and colored samples. Silica can interfere at levels above 500 mg/L.
3. How can insoluble sulfates in natural forms be dissolved for determination purposes?
Insoluble sulfates in concrete, glass and other materials can be dissolved by hot acid after the solid sample has been crushed or reduced to particles with size of 0.5 to 1.0 mm.
4. Why is sulfur (sulfate) important to plants?
Sulfur affects plant metabolism and is essential during the formation of protein. Plants deficient in sulfur develop a pale green color with thin stems, grow slowly, and do not mature properly.

REFERENCES

1. *LaMotte Soil Handbook* (Chestertown, Md.: LaMotte Company, 1994).
2. *Handbook of Reference Methods for Plant Analysis,* Yash P. Kalra (ed.), (Boca Raton, Fla.: CRC Press, 1998).
3. *Hawley's Condensed Chemical Dictionary,* 11th Ed., rev'd by N. Irving Sax and Richard J. Lewis (New York, N.Y.: Van Nostrand Reinhold, 1987).

4. *Standard Methods for the Examination of Water and Wastewater:* SULFATE, 16th Ed., A. E. Greenberg, R. R. Trussell, L. S. Clesceri and Mary Ann H. Franson (eds.), pp. 464–468, Washington, DC, American Public Health Association, 1985.

SUSPENDED SOLIDS

Suspended solids, sometimes called nonfilterable residue, are normally determined by filtering a well mixed sample through a weighed standard glass fiber filter and the residue retained on the filter is dried to a constant weight at 103–105° C. The increase in the weight of the filter represents the total suspended solids.

This photometric method is easy to perform and is often used for checking in-plant processes. This method of determining suspended solids does not require filtration, ignition of the sample, and sample weighing used in gravimetric analyses. The USEPA specifies the gravimetric method for solids determinations.

This method can be used to study metal distribution coefficient models which predict metal sorption to suspended solids in natural waters. This technique may find application in recovery of viruses from suspended solids in water and wastewater.

This photometric method can be calibrated against the gravimetric technique on sewage samples from a municipal sewage plant.* When higher accuracy is required, run parallel photometric and gravimetric determinations with portions of the same sample that are appropriately stirred and mixed.

The photometric method is limited to a concentration of 750 mg/L for determining suspended solids. The estimated detection limit using the photometric method is about 22 mg/L.

SUSPENDED SOLIDS (0 to 750 mg/L)

Method 8006** for Water and Wastewater
Photometric Method†

*DR/2010 Spectrophotometer Procedures Manual, 1998, pp. 352–353.

**Compliments of HACH Company.

†Adapted from *Sewage and Industrial Wastes*, 31, 1159 (1959).

1. Enter the stored program number for suspended solids. Press: **PRGM** The display will show: **PRGM?**

2. Press: **94 ENTER** The display will show **mg/L, SuSld** and the **ZERO** icon.

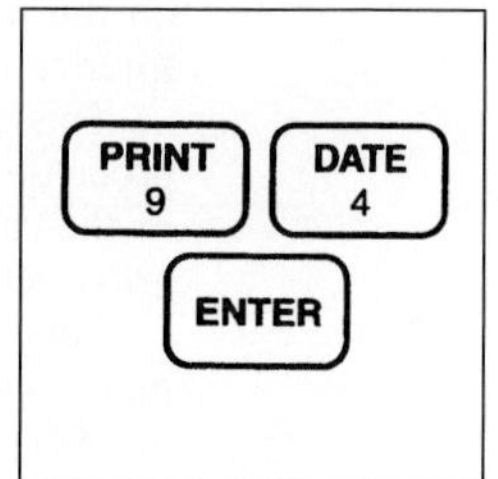

3. Blend 500 mL of sample in a blender at high speed for exactly 2 minutes.

4. Pour the blended sample into a 600-mL beaker.

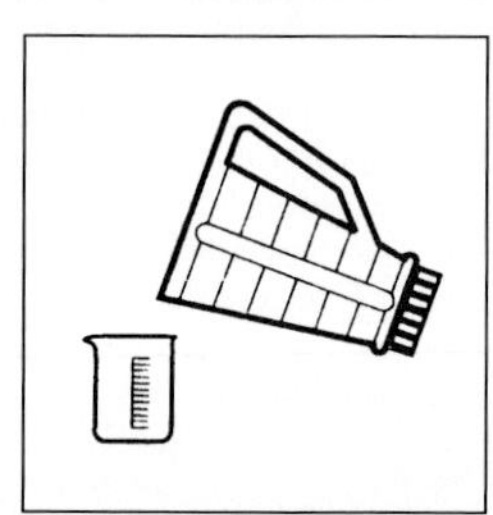

5. Fill a sample cell with 25 mL of tap water or deionized water (the blank).
Note: Remove gas bubbles in the water by swirling or tapping the bottom of the cell on the table.

6. Place the blank in the cell holder. Tightly cover the sample cell with the instrument cap.

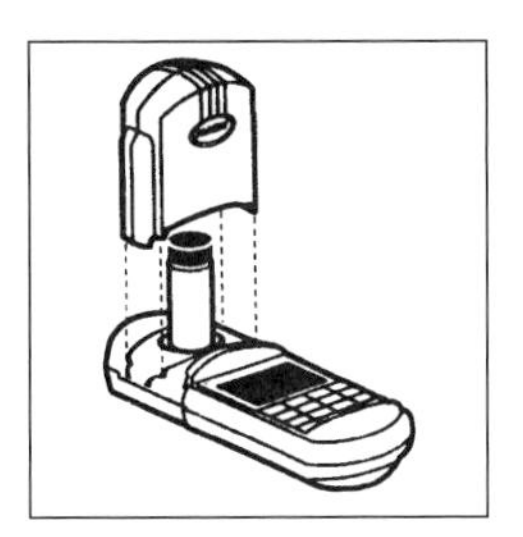

7. Press: **Zero** The cursor will move to the right, then the display will show: **0 mg/L SuSld**.

8. Stir the sample thoroughly and immediately pour 25 mL of the blended sample into a sample cell (the prepared sample).

9. Swirl the prepared cell to remove any gas bubbles and uniformly suspend any residue.

10. Place the prepared sample into the cell holder. Tightly cover the sample cell with the instrument cap.

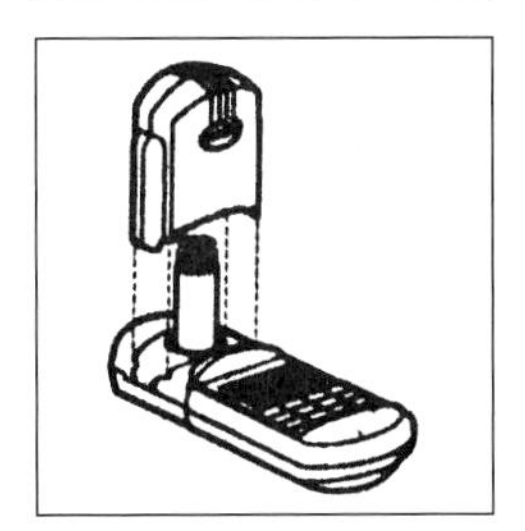

11. Press: **READ** The cursor will move to the right, then the result in mg/L suspended solids will be displayed.

SAMPLING AND STORAGE

Collect samples in clean plastic or glass bottles. Analyze samples as soon as possible after collection. The sample may be stored seven days by cooling to 4° C (39° F).

INTERFERENCES

Calibration for this test is based on parallel samples using gravimetric technique on sewage samples from a municipal sewage plant. For most samples, this calibration will provide satisfactory results. When higher accuracy is required, run parallel photometric and gravimetric determinations with portions of the same sample. The new calibration should be made on your particular sample using a gravimetric technique as a basis.

QUESTIONS AND ANSWERS

1. What are the advantages of this photometric test?
This test can be run in the field. The test is easier to run than the gravimetric method and results are obtained quickly.

2. What are the disdvantages of this photometric test?
The USEPA does not accept this method of suspended solid determination. The photometric test is not as accurate as the gravimetric method. However, technique and material drying errors are more likely in the gravimetric procedure.

3. How can high accuracy be obtained by the photometric method?
The photometric method can be run against the gravimetric technique using portions of the same sewage sample from a wastewater treatment facility. The sample should be stirred and mixed appropriately before dividing the sample into aliquots.

4. What are some new applications of this method?
The photometric method could be used to study metal concentrations and metal distribution coefficient models in suspended solids. This equipment may also be used to recover specific types of viruses from suspended solids in water and wastewater.

REFERENCES

1. DR/2010 Spectrophotometer Procedures Manual, 3rd Ed. (Loveland, CO.: HACH Company, 1998).
2. *Standard Methods for the Examination of Water and Wastewater:* TOTAL SUSPENDED SOLIDS DRIED AT 103–105° C. 16th Ed., Method 209C, A. E. Greenberg, R. R. Trussell, L. S. Clesceri and Mary Ann H. Franson, (eds.) pp. 96–97, Washington, DC, American Public Health Association, 1985.

TOTAL PETROLEUM HYDROCARBONS (TPH) IN SOILS

Total Petroleum Hydrocarbons (TPH) such as gasoline, kerosene, diesel fuel, #2 fuel oil, Jet Fuel A, Jet Fuel JP-4, #6 fuel oil, mineral spirits (a grade of naphtha), toluene, styrene, hexanes, heptanes, iso-octane, 2-methylpentane, benzene, ethylbenzene, xylenes (o, m, p-forms), dichlorobenzene, hexachlorobenzene, naphthalene, acenaphthene, biphenyl, creosote, used motor oil, and various sources of crude oil will provide a positive test at a variety of concentration levels in soils. Additional formulated petroleum products such as grease, mineral oil, unused motor oil, brake fluid and machine oil at 1000 ppm each will respond to the Total Petroleum Hydrocarbon Test in soil. Aliphatic compounds including undecane, trichloroethylene, and MTBE at 1000 ppm in soil each will also respond to the TPH Test.

The TPH components are primarily substituted aromatic organic compounds and alkane-type compounds consisting of straight chain paraffins, branched chain paraffins and cycloparaffins. Styrene is an phenyl alkene-type compound whereas acenaphthene is a tricyclic naphthalene compound used as an insecticide and fungicide. A number of compounds such as creosote, biphenyl, hexachlorobenzene, dichlorobenzene, naphthalene and xylenes (ortho, meta, para) are used as insecticides, fungicides, disinfectants, wood preservatives, repellants, fumigants, moth repellant, and for control of plant diseases.

Specific petroleum products are used as fuels, petroleum chemicals, paint and varnish thinners, dry-cleaning fluid, asphalt and road tar solvents, rubber cement solvents, and for domestic heating. A number of the alkanes, benzene, xylene and other solutions are used as cleaning solvents, reaction media and in various types of sprays. Ethylbenzene is used as a solvent and as an intermediate in the production of styrene.

Most of the materials are flammable and a dangerous fire risk. Many of the substances are moderately toxic by ingestion, inhalation and skin absorption. Human tolerance levels are limited and somewhat variable depending on the type of hydrocarbon or chlorinated hydrocarbon.

An analytical technique using a TPH-analyzer can quantify Total Petroleum Hydrocarbons at a specific wavelength as specified in EPA Method 418.1. The analyzer is a wavelength-specific, portable infrared instrument. After extracting the solid sample, the concentration of the materials can be measured by determining the intensity of the characteristic infrared absorption bands.

This semi-quantitative test for TPH is an immunoassay procedure.* The TPH Stabilizing Agent stabilizes the substituted aromatic organic compounds in the samples. Sample, standard and color development reagents are added to test tubes that are coated with an antibody specific for petroleum fuels. The TPH concentration in a sample is determined by comparing the developed color intensity to the color of the TPH standard. The final TPH concentration is inversely proportional to the intensity of color development. Lighter color development indicates higher levels of TPH in the sample.

TOTAL PETROLEUM HYDROCARBON (TPH) in SOIL (10 and/or 100 ppm TPH thresholds)**

Method 10050† for Soil
Immunoassay Method

---

*DR/2010 Spectrophotometer Procedures Manual, 1998, pp. 787–795.

**Test is semi-quantitative. Results are expressed as greater or less than the threshold value used.

†Compliments of HACH Company.

1. Enter the stored program for absorbance. Press: **0 ENTER.**
   Note: The Pour-thru Cell cannot be used.

2. Rotate the wavelength dial until the small display shows: **450 nm**

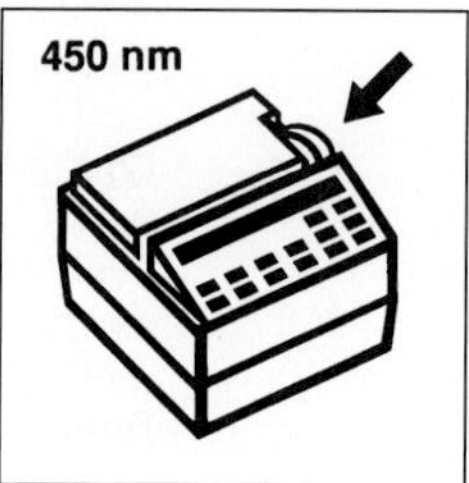

TPH PHASE 1: SOIL EXTRACTION

1. Fill the extraction vial to the 0.75 oz line with Soil Extractant Solution.
   Note: This is equivalent to adding 20 mL of Soil Extractant Solution.

2. Place a plastic weighing boat on an analytical balance. Tare the balance.
   Note: Use either a laboratory balance or the portable AccuLab Pocket Pro.

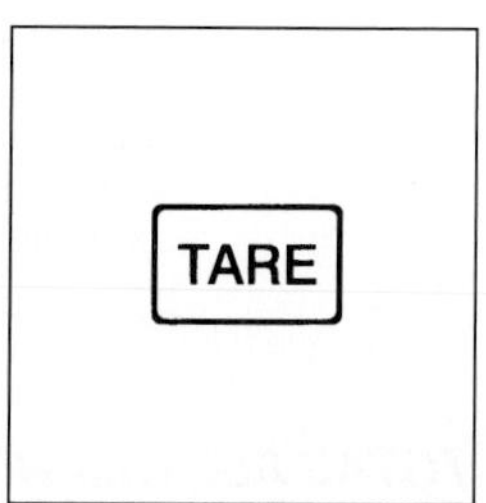

3. Weigh out 10 ± 0.1 g of soil in a plastic weighing boat. Carefully pour the soil into the extraction vial.

4. Cap the extraction vial tightly and shake vigorously for 1 minute.

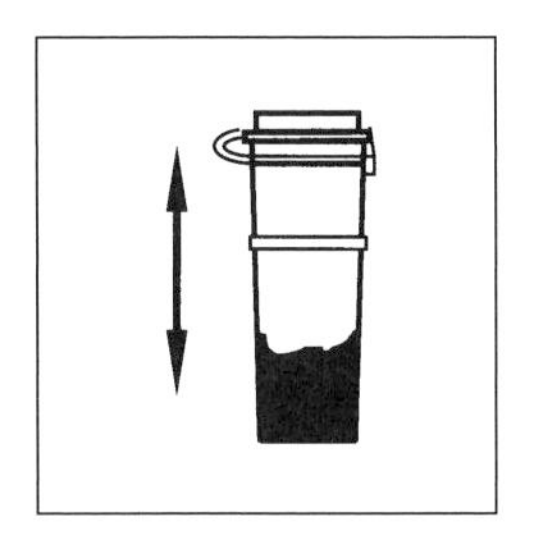

5. Allow to settle for 1 minute. Gently open the extraction vial.

6. Using the disposable bulb pipet, withdraw 1.0 to 1.5 mL from the liquid (top) layer in the extraction vial. Transfer this aliquot into the filtration barrel (the *bottom* part of the filtering assembly: the plunger is inserted into it).
   Note: Do not use more than 1.5 mL. The pipet has 0.25-mL increments on it.

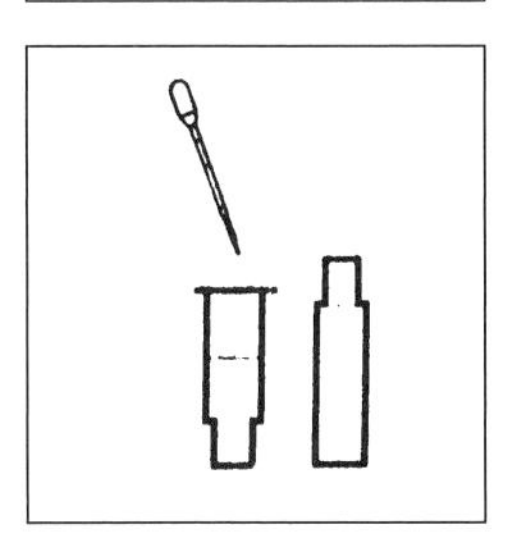

7. Insert the filtration plunger into the filtration barrel. Press firmly on the plunger until at least **0.5 mL** of filtered sample is collected in center of the plunger.
   Note: The liquid will be forced up through the filter. The liquid in the plunger is the filtered extract.
   Note: It may be necessary to place the filtration assembly on a table and press down on the plunger.

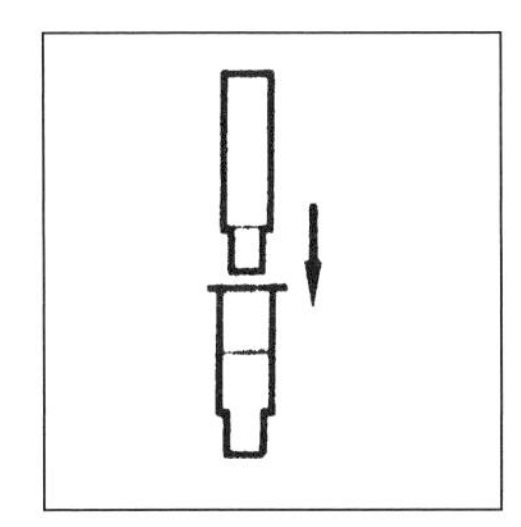

TPH PHASE 2: PREPARING SAMPLES AND STANDARDS

1. Label four TPH Enzyme Conjugate Tubes Standard 1, Standard 2, Sample 1 and Sample 2.
Note: In Phase 2, choose the desired threshold level and perform Step 6 or Steps 7 to 9, but not both. Additional tubes are required to perform analyses at both threshold levels.
Note: The TPH Conjugate and Antibody Tubes are matched lots. Mixing with other lots will cause erroneous results.

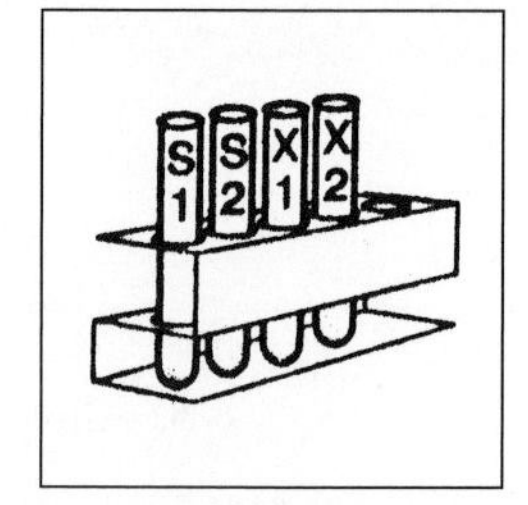

2. Label one TPH Antibody Tube for each sample and standard. Be sure to label the 10 and 100 ppm samples appropriately.

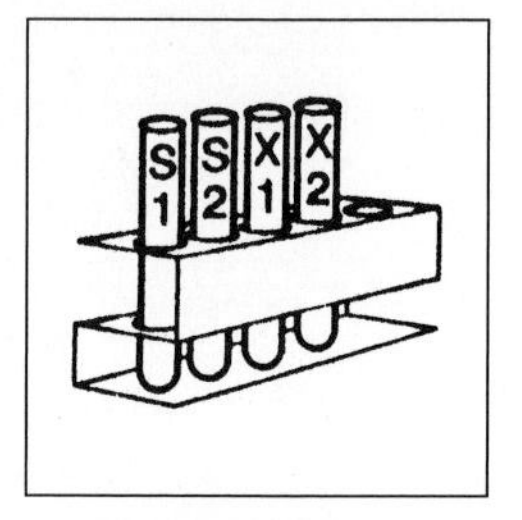

3. Pour the contents of one TPH Buffer Pillow into each Enzyme Conjugate Tube.
Note: Squeezing the top portion of the pillow will help empty the pillow contents. If necessary, clip open the top corner of the pillow to release the liquid.

Preparing the TPH Standard

4. Snap open a TPH Standard Ampule.

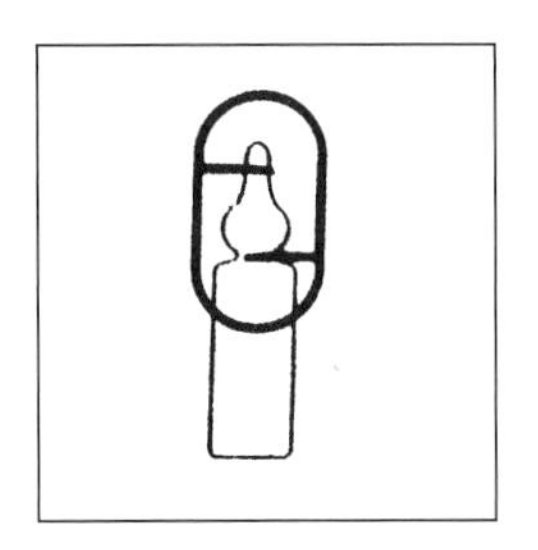

5. Using the WireTrol pipet, add 50 μL of TPH Standard to the Standard 1 and Standard 2 TPH Enzyme Conjugate Tubes. Swirl to mix thoroughly.
Note: Dispense standard below the level of the solution in the Enzyme Conjugate Tubes.

Preparing Samples for Measuring at 10-ppm Threshold

6. Using the WireTrol pipet, add 50 μL of the sample extract from the filter unit to the tubes labeled Sample 1 and Sample 2. Swirl each to mix.

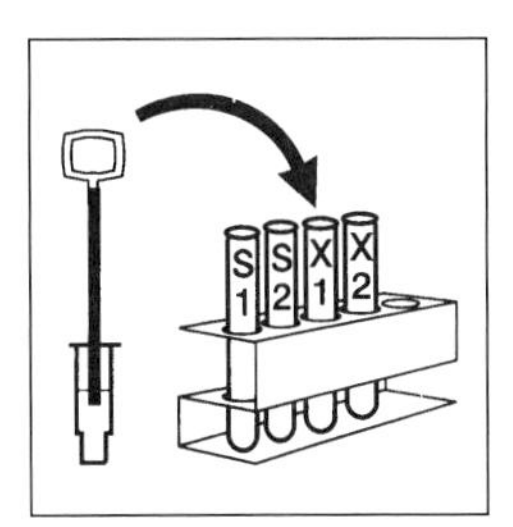

Preparing Samples for Measuring at the 100-ppm Threshold

7. Label a 100-ppm TPH Dilution Ampule for each sample. Snap open each Dilution Ampule.

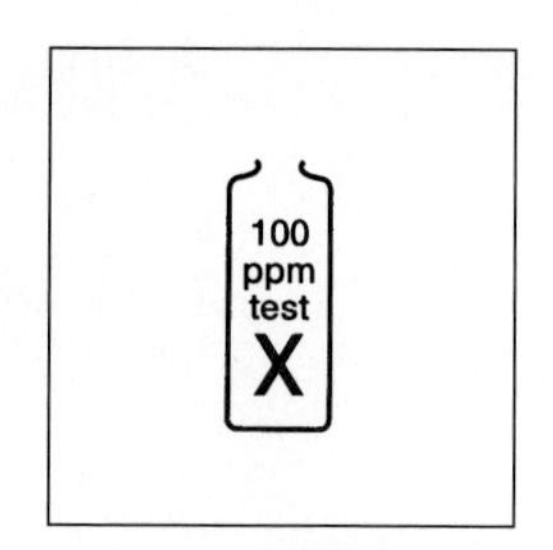

8. Using the WireTrol pipet, add 100 μL of the sample extract from the filtration plunger to the labeled 100 ppm Dilution Ampule. Mix by repeatedly drawing 100 μL into the pipet and dispensing back into the ampule.
Note: The upper line on the WireTrol capillary tubes is equal to 100 μL.

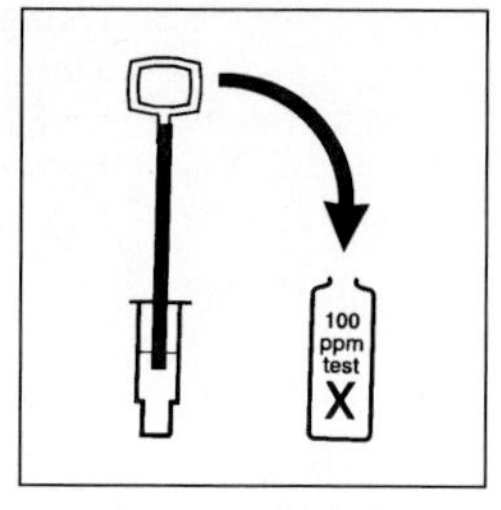

9. Using the WireTrol with a new pipet tip, add 50 μL of the diluted sample extract from the 100-ppm Dilution Ampule to the TPH Enzyme Conjugate tubes labeled Sample 1 and Sample 2. Swirl to mix.

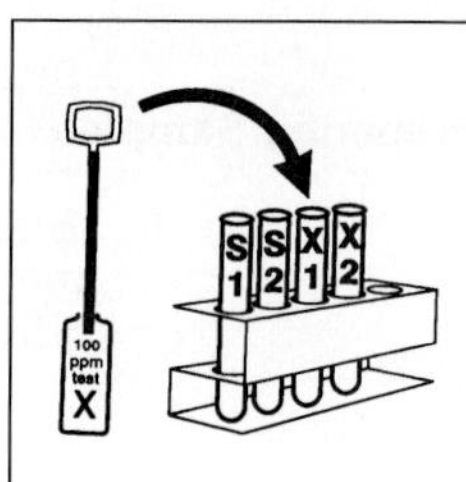

TPH PHASE 3: IMMUNOASSAY-STEPS IN THIS PHASE REQUIRE EXACT TIMING

1. Pour the contents of the Enzyme Conjugate Tubes into the correct TPH Antibody Tubes. Swirl to mix.
Note: The TPH Conjugate and Antibody Tubes are matched lots. Mixing with other lots will cause erroneous results.

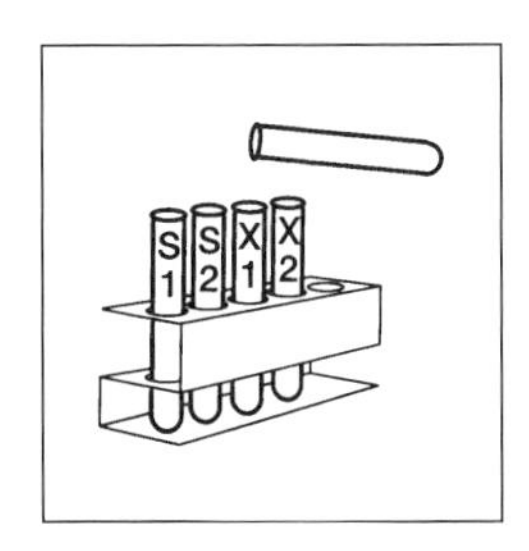

2. Begin a 10-minute reaction period.
Note: During this phase, analytes in the sample compete with the enzyme conjugate for a limited number of binding sites on the inside of the antibody tubes.

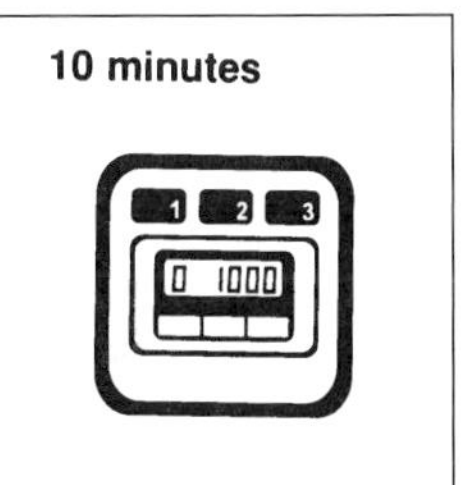

3. After the 10-minute period, discard the contents of the TPH Antibody Tubes into an appropriate waste container.

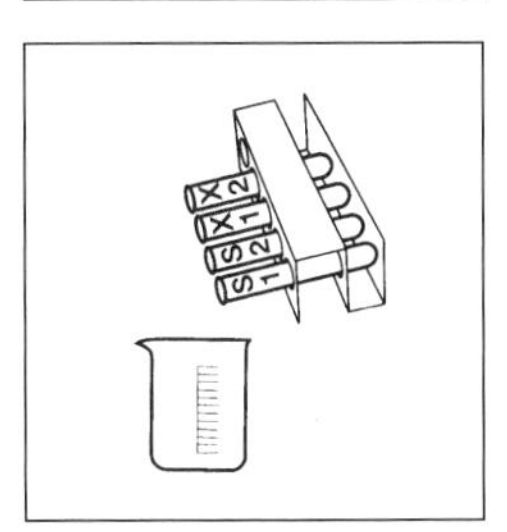

4. Wash each tube thoroughly and forcefully 4 times with Wash Solution. Empty the tubes into an appropriate waste container. Shake well to ensure most of the Wash Solution drains after each wash.
Note: Wash Solution is a harmless detergent.

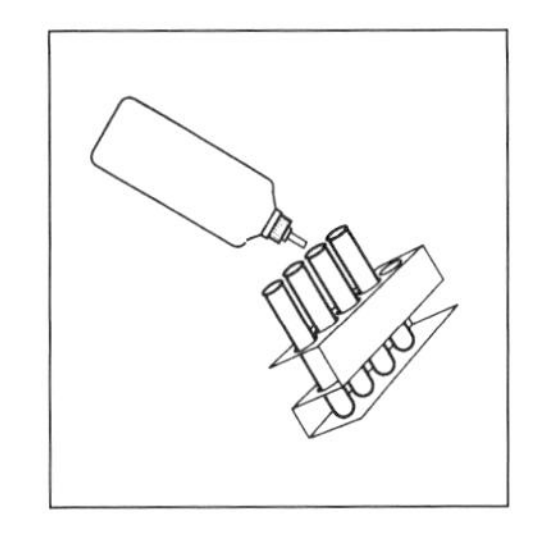

5. Continue to the next phase immediately.
Note: Ensure most of the Wash Solution is drained from the tubes. Turn the tubes upside down and gently tap on a paper towel to drain. Some foam may be left from the Wash Solution; this will not effect results.

Go To next Phase immediately!

TPH PHASE 4: COLOR DEVELOPMENT

Check reagent labels carefully! Reagents must be added in proper order for valid test results.

1. Add 5 drops of Solution A to each tube. Replace the bottle cap.
   Note: Hold reagent bottles vertically for accurate delivery or erroneous results may occur.

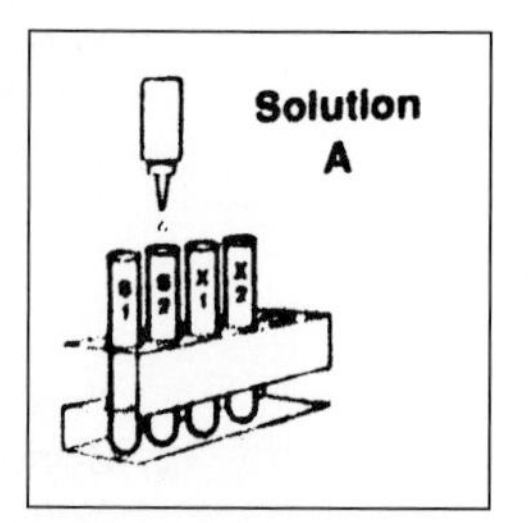

2. Begin a 2.5-minute period and immediately add 5 drops of Solution B to each tube. Swirl to mix. Replace the bottle cap.
   Note: Add drops to the tubes in the same order to ensure proper timing (i.e. left to right). Solution will turn blue in some or all of the tubes.

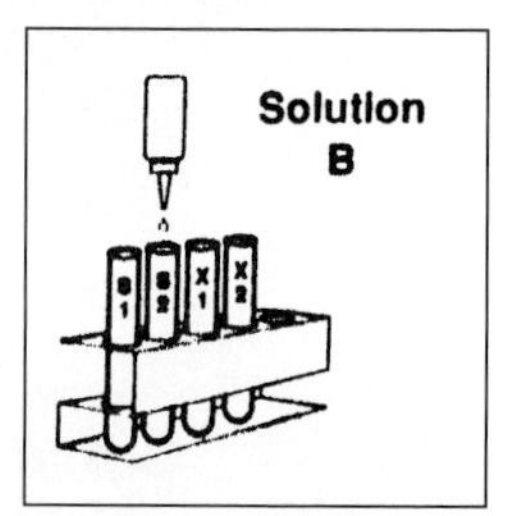

3. Let each tube react for exactly 2.5 minutes. Then add 5 drops of Immunoassay Stop Solution to each tube. Replace the bottle cap.
   Note: Blue solutions will turn yellow when Stop Solution is added. TPH concentration is inversely proportional to color development; a lighter color indicates higher levels of TPH.

4. Using the TenSette Pipet with a new tip, add 0.5 mL of deionized water to each tube. Swirl to mix.

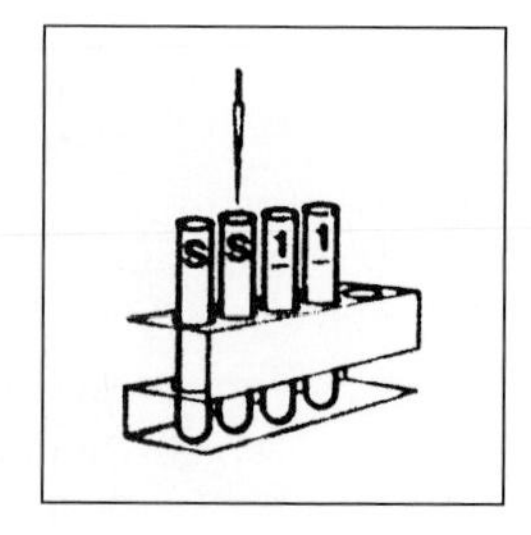

TPH PHASE 5: MEASURING THE COLOR

1. Fill a Zeroing Tube with deionized water (the blank).

2. Insert the Immunoassay adapter into the sample cell compartment.
Note: Align the adapter so the light beam openings face the sides of the DR/2010. Press firmly on the adapter to seat it.

3. Place the blank in the cell holder. Place the cover on the adapter.

4. Press: **ZERO.** The display will show: **0.000 Abs.**

5. Insert Standard 1 sample cell into the cell holder. Close the light shield.

6. Record the absorbance reading.

7. Repeat Steps 5 and 6 for the Standard 2 tube.
Note: If Standard 1 and Standard 2 are more than 0.350 absorbance units apart, repeat the test beginning at Phase 2, Standard Preparation.

8. Insert the Sample 1 tube into the cell holder. Place the cover on the adapter.
Note: TPH concentration is inversely proportional to the color intensity (or absorbance value). More color means less TPH in the sample.

9. Record the absorbance reading.

10. Repeat Steps 9 and 10 for the Sample 2 tube. See Table 3.2 below to interpret results.

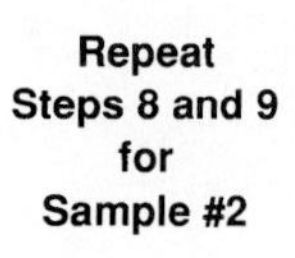

INTERPRETING THE RESULTS

Use Table 3.2 to interpret the results.

TABLE 3.2

If sample absorbance is . . .	10 ppm threshold	100 ppm threshold
. . . less than the highest standard absorbance	Sample TPH is greater than 10 ppm	Sample TPH is greater than 100 ppm
. . . greater than the highest standard absorbance	Sample TPH is less than 10 ppm	Sample TPH is less than 100 ppm

SENSITIVITY

The TPH in the Soil Test has sensitivities to the following chemicals at the stated levels.

Compound	Concentration required to give a Positive Result (ppm)*	Compound	Concentration required to give a Positive Result (ppm)*
Petroleum Fuels		**Aromatic Compounds**	
Gasoline	10	Acenaphthene & Naphthalene	0.5 & 0.8
		Creosote	1.5
Diesel fuel, # 2 Fuel Oil, Jet Fuels A & JP-4, Kerosene	15	1,2-Dichlorobenzene	2.5
		Styrene & Ethylbenzene	7
		m-Xylene & o-Xylene	8 & 8.5
#6 Fuel Oil	25	Hexachlorobenzene & Biphenyl	10
		Toluene	40
Formulated Petroleum Products		p-Xylene	45
Mineral Spirits	40	Benzene	400
Unused Motor Oil	50		
Machine Oil, Brake Fluid, Grease, Unused Motor Oil, & Mineral Oil	>1000	**Aliphatic Compounds**	
		iso-Octane	8.5
		2-Methylpentane	35
Crude Oil**		Hexanes, mixed	65
Ingram	60	Heptane	130
Walker, Louisiana, & Main Pass	100	Undecane, Trichloroethylene & MTBE	>1000
Vermilion	130		

* Samples with stated concentration will give positive result greater than 95% of the time when tested at stated concentration level.
** Variable, depending on source.

SAFETY AND STORING AND HANDLING OF REAGENTS

- Wear protective gloves and eye wear.
- Store reagents at room temperature and out of direct sunlight (less than 80° F or 27° C).

- Keep aluminized pouch that contains antibody-coated tubes sealed when not in use.
- If Stop Solution or liquid from the extraction jar comes in contact with eyes, wash thoroughly with cold water. Seek immediate medical help.
- Operational temperature of the reagents is 40 to 90° F (5 to 32° C).
- The soil extractant (methanol) is an ignitable (D001) waste regulated by Federal RCRA.
- Consult the material Data Safety Sheets for information specific to the reagents used.
- Implement good safety habits and laboratory techniques throughout the experiment.

MEASURING HINTS

- Timing is crucial; follow instructions carefully.
- Handle the antibody tubes carefully. Scratching the inside or outside may cause erroneous results. Wipe tubes with a paper towel to remove smudges or fingerprints before measuring.
- Hold all dropper bottles vertically and direct drops to the bottom of the tube.
- Antibody Tubes and Enzyme Conjugate Tubes are made in matched lots. Do not mix with other reagent lots.

QUESTIONS AND ANSWERS

1. What is the soil extractant?
   The soil extractant is methanol, which is an ignitable (D001) waste regulated by the Federal RCRA.
2. Why does the analyst use a WireTrol pipet in the experiment?
   A WireTrol capillary tube (pipet) is used to transfer microliters of sample.
3. During color development, why is it necessary to hold the reagent delivery bottle vertically over each solution tested?
   Hold the reagent bottle vertically to minimize inaccuracies in drop size and to reduce erroneous results.
4. During color development, why are 5 drops of Solution B added to the test tubes in the same order (left to right) as the previous step (A)?
   Reagents must be added in proper order for valid test results. The drops must be added in the same order as Step 1 to ensure proper timing and proper color development.

REFERENCES

1. DR/2010 Spectrophotometer Procedures Manual, 3rd Ed. (Loveland, CO.: HACH Company, 1998).
2. *Hawley's Condensed Chemical Dictionary,* 11th Ed., rev'd by N. Irving Sax and Richard J. Lewis (New York, N.Y.: Van Nostrand Reinhold, 1987).

ZINC

Zinc forms many inorganic and organometallic compounds which are used extensively in our society. It is an essential element, which appears to be toxic to plants at higher concentrations. Zinc enters the water supply by deterioration of brass and galvanized iron and through industrial waste pollution. Some zinc organometallic compounds, such as diethyl zinc, decompose under heated conditions to form a toxic substance. A number of organozinc compounds have a variety of solubilities in water or decompose in water.

Zinc metal and zinc compounds are used in a variety of applications, including polymers, medicine, batteries, fungicides, insecticide, metal plating, fuses, rodenticide, catalyst, preservatives, food ingredient, cross-linking polymers, pharmaceuticals, surgical dressings, antiseptic, suppositories, mildew inhibitor, deodorant applications, dental cements, welding, soldering, complex pigments, and ceramics. Additional uses include fire proofing textiles, electroplating, pigments, electrical devices, rubber compounding, curative, metal coating, phosphors, animal repellant, organic synthesis, resin curing, wood preservative, insulating materials, accelerator in rubber cements, and many other applications. Zinc metal and zinc compounds are sometimes recycled and reused. Some of the materials are placed in landfills and places that are inappropriate.

Zinc is essential in promoting specific enzymes in the soil and is required for the production of chlorophyll and the formation of carbohydrates in plants.* As the soil pH increases, the availability of zinc in soil tends to decrease. Soils with high levels of both phosphate and zinc tend to uptake phosphorus instead of zinc. In addition, zinc and phosphate ions can also react to form insoluble zinc phosphate. Excess zinc appears to create an iron deficiency. After applications of zinc to soil, zinc is readily absorbed by organic matter and is relatively immobile in the soil. Zinc deficiency in plants will show interveinal chlorosis with an eventual whitening on either side of the mid-rib in the upper leaves.** Leaves may be small and distorted. Short internodes and stunted plants demonstrate zinc deficiency.

A solid sample, gel or semisolid sample weighing 0.100 to 0.500 grams can be digested in a known volume of acid. The digestion procedure for oils requires 0.100 to 0.250 grams of material added to a specific volume of acid. After boiling for 4 to 5 minutes, 10 mL of 50% hydrogen peroxide is added to the charred sample and the sample is heated to boiling for 1 minute. The digested mixture is allowed to cool and is diluted to 100 mL with deionized water. After selecting the analysis volume and diluting the solution to the appropriate volume (25.0 mL or 50.0 mL), the clear sample can be analyzed by either of the Zincon procedures provided by HACH or LaMotte. A pH adjustment on the sample will be required before metal analysis.

A number of interferences include aluminum, cadmium, copper, ferric iron, manganese, and nickel. Large amounts of organic material may interfere in the test provided by HACH. In the experiment, zinc and other metals in the sample form complexes with cyanide. Zinc is the only metal released by the addition of cyclohexanone. Then, zinc reacts with the "Zincon" indicator called 2-carboxy-2′-hydroxy-5′-sulfoforamazyl. The zinc concentration is proportional to the blue color developed in solution and is measured by a field spectrometer at 620 nm†. Also, samples containing 0 to 3.00 mg/L Zn can be analyzed at a different wavelength by the HACH Zincon Method using the DR/850 and the DR/890 Colorimeters.

The LaMotte Method for the determination of zinc is slightly different than the Zincon Method provided by HACH. Zinc forms a blue colored complex with Zincon in a solution buffered at pH = 9.0.* Most interfering metals are complexed by cyanide and the zinc cyanide complex is released by the addition of formaldehyde. Manganese interference is reduced by the addition of sodium ascorbate. The sample should be analyzed soon after collection. A number of cations and a few anions can interfere at relatively low concentrations.

---

*Compliments of LaMotte Company.

**Robert D. Munson, *Handbook of Reference Methods for Plant Analysis*, Yash P. Kalra (ed.), (Boca Raton, Fla.: CRC Press, 1998).

†Compliments of HACH Company.

ZINC

Zincon Method—Procedure*

1. Fill a clean colorimeter tube to the 10 mL line with the soil filtrate.

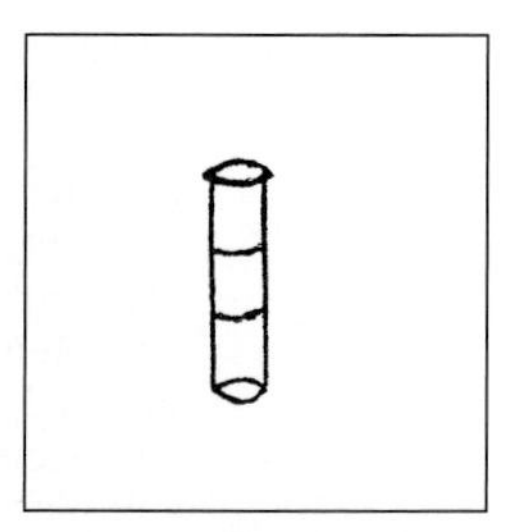

2. Neutralize the soil filtrate by adding 15% Sodium Hydroxide solution to the soil filtrate, one drop at a time while stirring the plastic rod. The stirring rod is touched to the Bromthymol Blue test paper after the addition of each drop of Sodium Hydroxide until the color changes from yellow to green or blue.

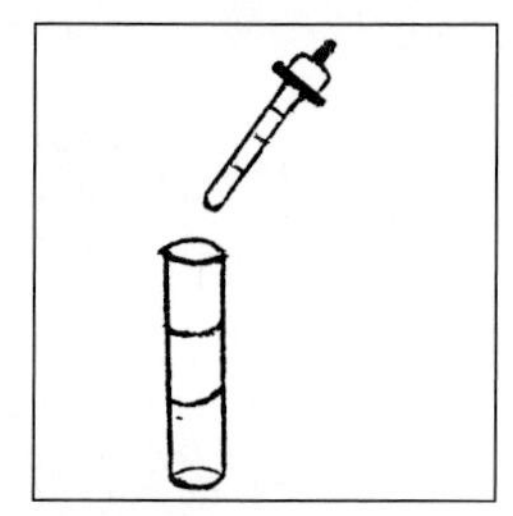

3. Select setting 6 on the "Select Wavelength" knob. Press the "30 Second Read" button.

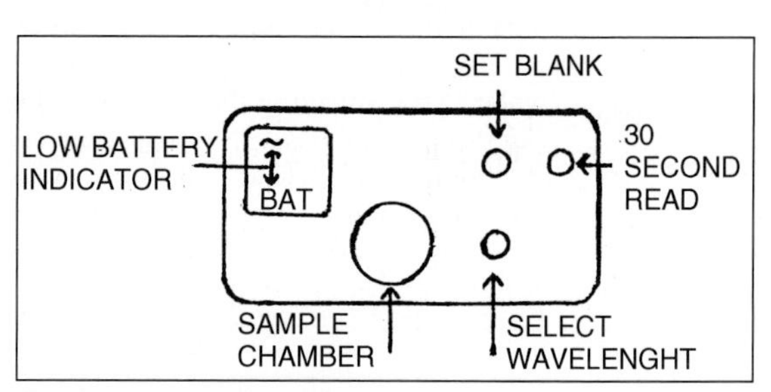

4. Insert the tube into the colorimeter chamber and adjust to "100"%T with the "Set Blank" knob. This is the 100%T blank.

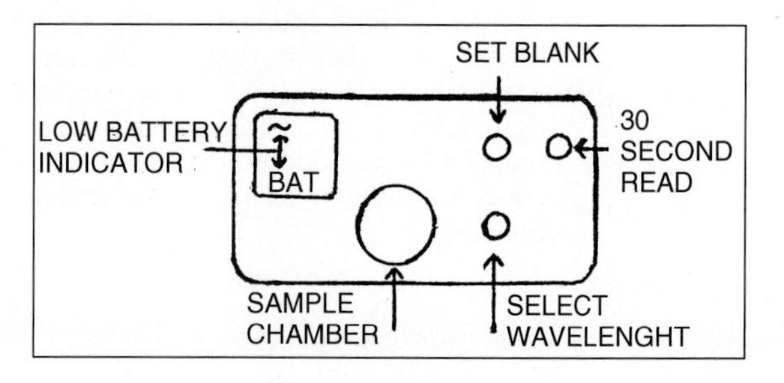

*Compliments of LaMotte Company.

5. Remove the tube, add 0.1 g of Sodium Ascorbate with the 0.1 g spoon.

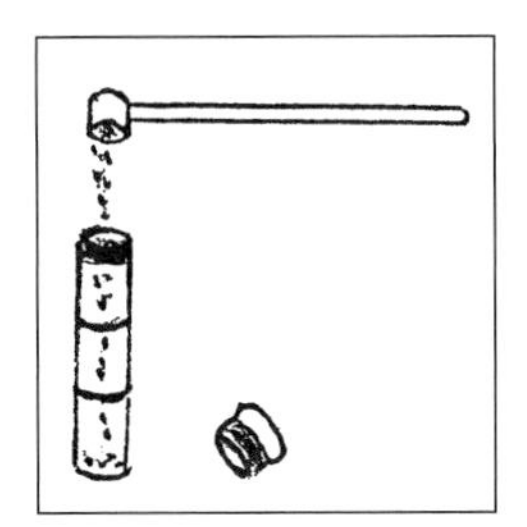

6. Add 0.5 g of Zinc Buffer Reagent with 0.5 g spoon, cap and shake vigorously for 1 minute.

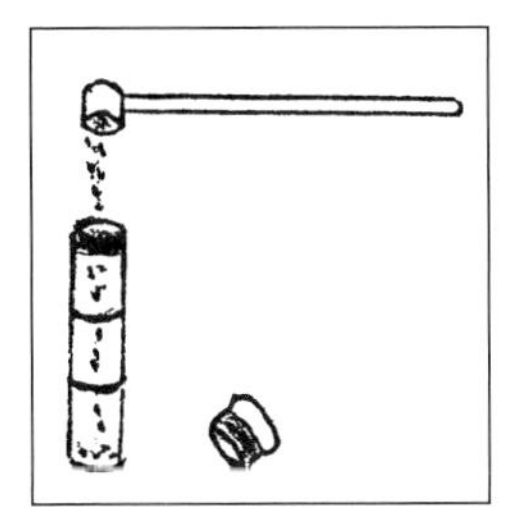

7. Add 3 drops of sodium cyanide solution with a clean, dry eye dropper. Cap and mix contents of the tube.

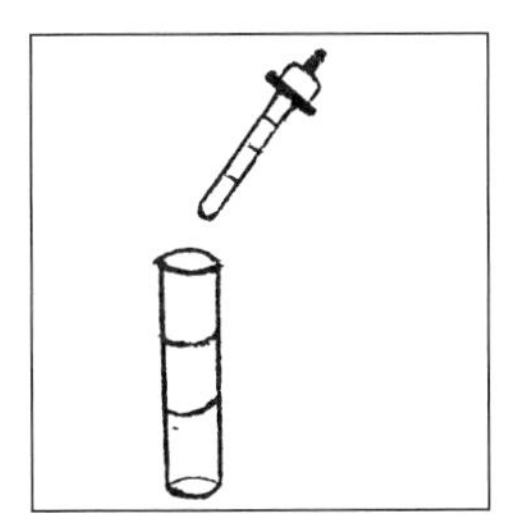

8. Using the 1 mL pipet assembly, add 1 mL of "Diluted Zinc Indicator Solution." Cap and mix contents.
Note: The "Diluted Zinc Indicator Solution" is prepared before the experiment begins and is stored in a capped bottle. Note the preparation procedure below.

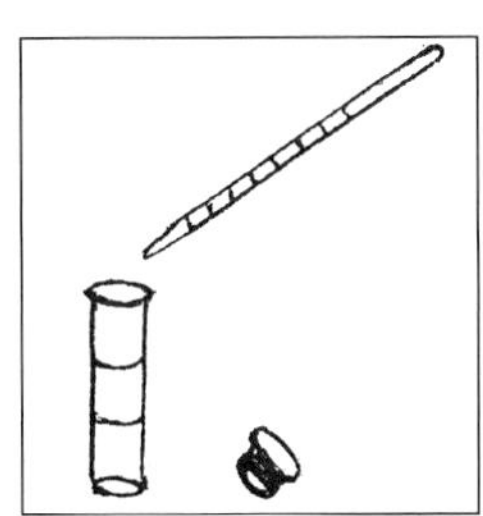

9. Using another clean pipet, add 4 drops of 37% Formaldehyde Solution. Cap and mix by inverting 15 times.

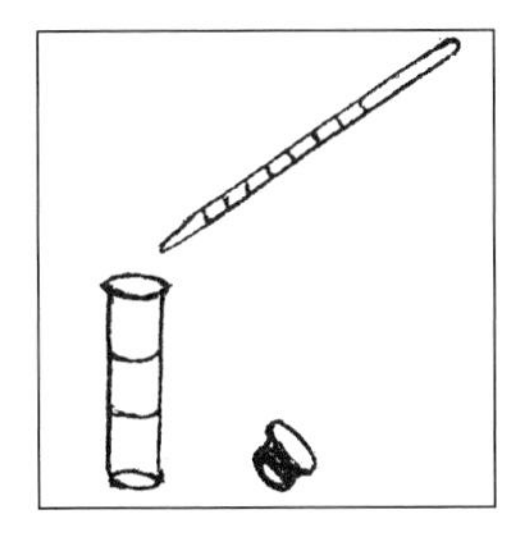

10. Press the "30 Second Read" button and insert the tube into the colorimeter chamber. Measure the test result as soon as the reading stabilizes.

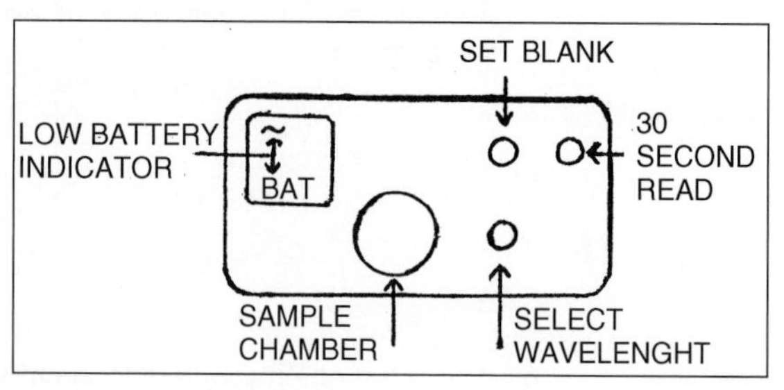

11. Consult the calibration chart to find the concentration of zinc in parts per million.

PREPARATION OF DILUTED ZINC INDICATOR SOLUTION

1. Using a clean pipet, measure exactly 5.0 mL of Zinc Indicator Solution into a 10 mL graduate cylinder. Pour the solution into the bottle labeled "Diluted Zinc Indicator Solution." ***Note:*** The bottom of the curved surface (meniscus) of the liquid should be at the top portion of the 5.0 mL line on the graduate cylinder.

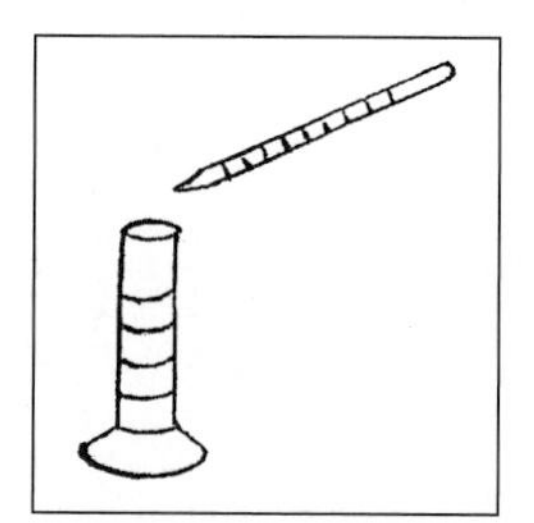

2. Using the unrinsed graduated cylinder, add 10.0 mL of methyl alcohol and then 7.8 mL of methyl alcohol (total of 17.8 mL) to the bottle labeled "Diluted Zinc Indicator Solution." Cap and mix the ingredients in this bottle. Do not leave the bottle uncapped.

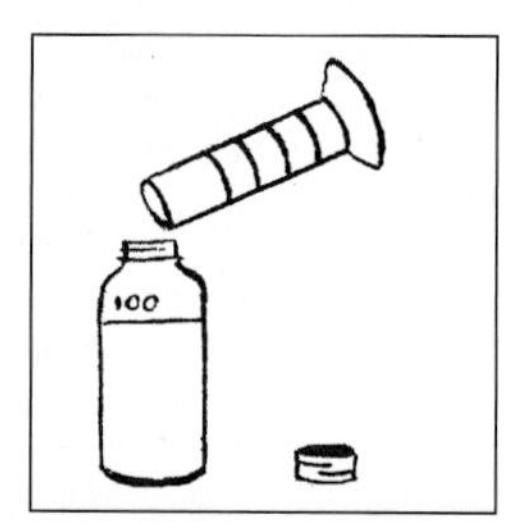

Zinc Calibration Chart (ppm)

%T	9	8	7	6	5	4	3	2	1	0
90										
80				0.0	0.1	0.1	0.2	0.2	0.3	0.3
70	0.4	0.4	0.5	0.5	0.6	0.6	0.7	0.7	0.8	0.9
60	0.9	1.0	1.0	1.1	1.2	1.2	1.3	1.4	1.4	1.5
50	1.6	1.7	1.7	1.8	1.9	2.0	2.0	2.1	2.2	2.3
40	2.4	2.5	2.6	2.7	2.8	2.9	3.0	3.1	3.2	3.3
30	3.4	3.5	3.6	3.8	3.9	4.0	4.2	4.3	4.4	4.6
20	4.8	4.9	5.1	5.3	5.5	5.6	5.9	6.1	6.3	6.5
10	6.8	7.0	7.3	7.6	8.0	8.3	8.7	9.1	9.5	10.0
0	10.6	11.2	11.9	12.8	13.8	15.1	16.8	19.2	23.6	

Zinc Concentration Chart

%T	Range	Parts per million
82–100%	Low	0–0.5 ppm
77–81%	Marginal	0.6–1.0 ppm
69–76%	Adequate	1.1- 2 ppm

HAZARDOUS SUBSTANCES

Read the MSDS before using the chemicals. The following substances are hazardous substances and should be disposed of and handled properly. Hazardous substances are: a) zinc indicator solution; b) methyl alcohol; c) zinc buffer powder; d) 10% sodium cyanide; and e) 37% formaldehyde solution.

RANGE OF MEASUREMENT

The range of measurement is 0.01 to 3.0 ppm.

INTERFERENCES

The following ions interfere in concentrations greater than those listed.*

Ion	mg/L (ppm)	Ion	mg/L (ppm)
Cd (II)	1	Cr (III)	10
AL (III)	5	Ni (II)	20
Mn (II)	5	Co (II)	30
Fe (III)	7	CrO_4^{2-}	50
Fe (II)	9		

ZINC (0 to 2.00 mg/L Zn)

Method 8009**
Zincon Method†

*Compliments of LaMotte Company.

**Compliments of HACH Company. Analysis may be performed by the DR/850 and DR/890 Colorimeters and the DR/2010 Field Spectrophotometer.

†Adapted from *Standard Methods for the Examination of Water and Wastewater, 328 D.*

Digestion Required for Solids, Oils and Wastewater*

1. Enter the stored program number for Zinc.
Press: **7 8 0 ENTER.** The display will show: **Dial nm to 620.**
Note: The Pour-Thru Cell cannot be used.

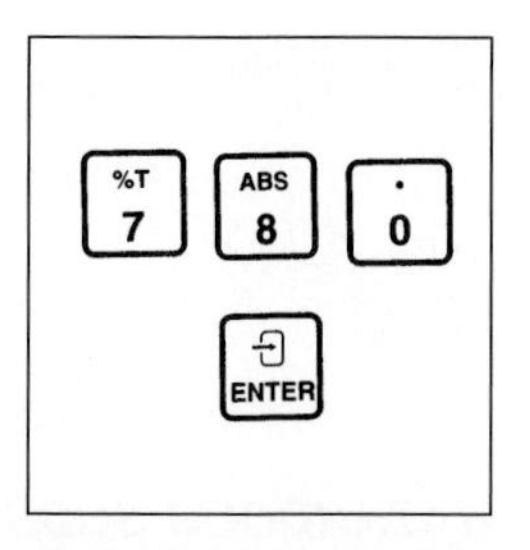

2. Rotate the wavelength dial until the small display shows: **620 nm.** When the correct wavelength is dialed in the display will quickly show: **Zero sample,** then: **mg/L Zn.**

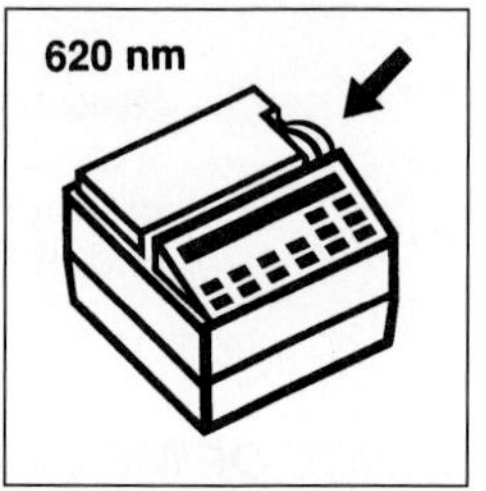

3. Insert the 10-mL Cell Riser into the cell compartment.

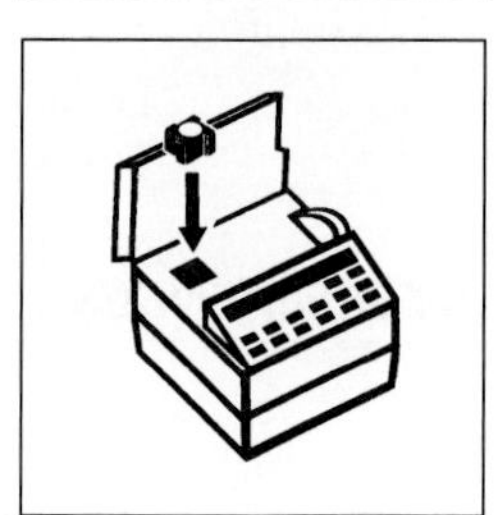

4. Fill a 25-mL graduated mixing cylinder with 20 mL of sample.
Note: Use only glass stoppered cylinders. Rinse with 1:1 hydrochloric acid and deionized water before use.

5. Add the contents of one ZincoVer 5 Reagent Powder Pillow. Stopper. Invert several times to completely dissolve the powder.
Note: Powder must be completely dissolved.
Note: The sample should be orange. If it is brown or blue, dilute the sample and repeat the test.
Caution: *ZincoVer 5 contains cyanide and is very poisonous if taken internally or inhaled. Do not add to an acidic sample. Store away from water and acids.*

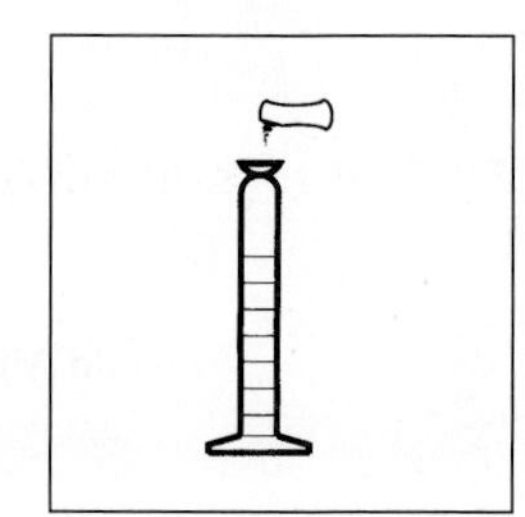

* Federal Register, 45 (105) 36166 (May 29, 1980).

6. Measure 10 mL of the solution into a sample cell (the blank).

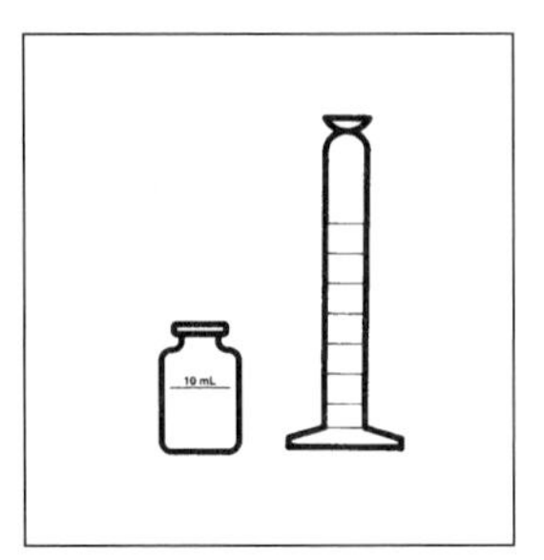

7. Add 0.5 mL of cyclohexanone to the remaining solution in the mixing cylinder.
Note: Use a plastic dropper as rubber bulbs may contaminate the cyclohexanone.

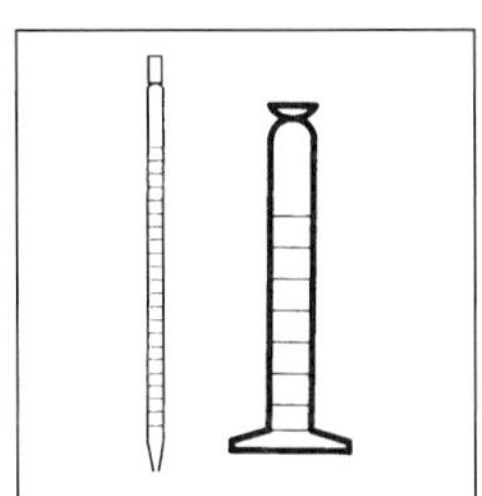

8. Stopper the cylinder. Shake vigorously for 30 seconds (the prepared sample).
Note: The sample will be red-orange, brown, or blue, depending on the zinc concentration.

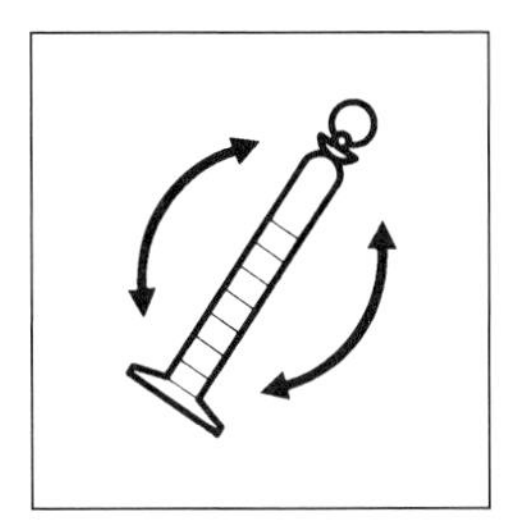

9. Press: **SHIFT TIMER.** A 3-minute reaction period will begin.
Note: During this period, complete Step 10.

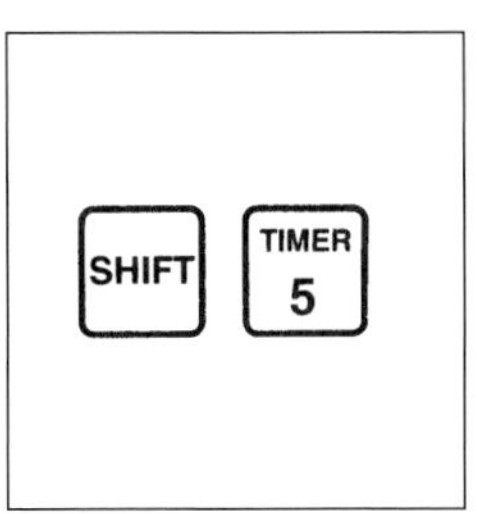

10. During the reaction period, pour the solution from the cylinder into a sample cell.

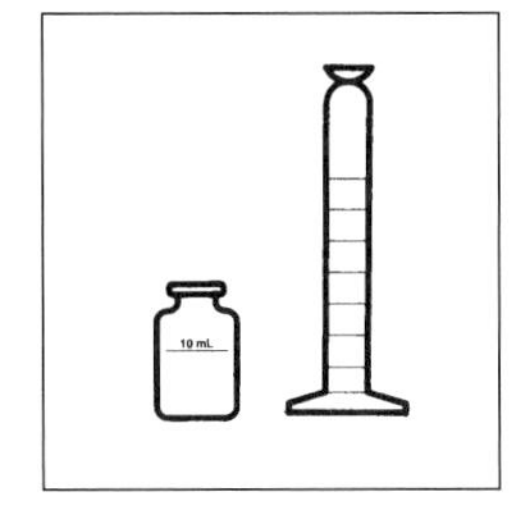

11. When the timer beeps, place the blank into the cell holder. Close the light shield.

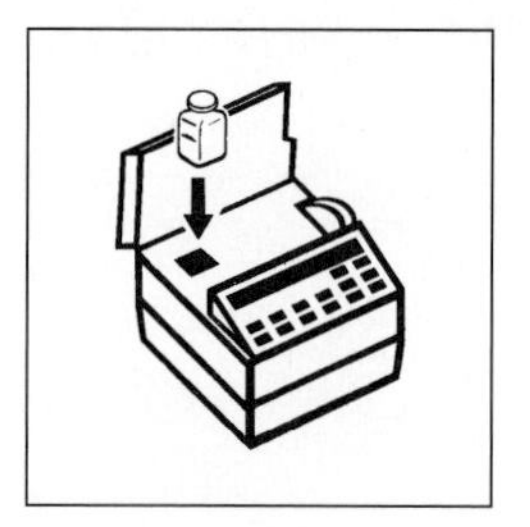

12. Press: **ZERO.** The display will show: **Zeroing . . .** then: **0.00 mg/L Zn.**

13. Place the prepared sample into the cell holder. Close the light shield.

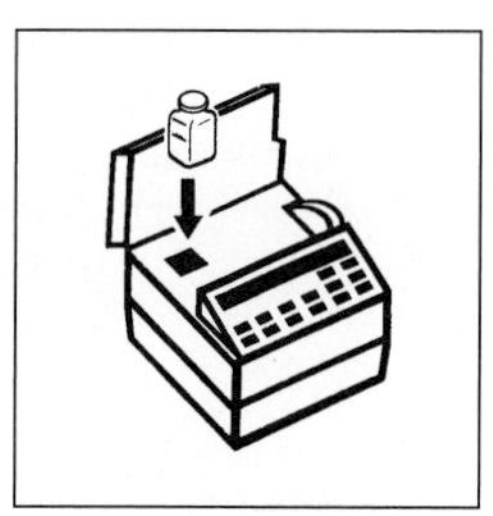

14. Press: **READ.** The display will show: **Reading . . .** then the result in mg/L zinc will be displayed.

SAMPLING AND STORAGE

Collect samples in acid-washed plastic bottles. For storage of zinc in solution, adjust the pH to 2 or less with nitric acid (about 2 mL per liter). The preserved samples can be stored up to six months at room temperature.

Adjust the pH to 4 to 5 with 5.0 N sodium hydroxide before analysis. Do not exceed pH 5, as zinc may be lost as a precipitate. Correct the test result for volume additions; see "Sample Collection, Preservation and Storage, and Correcting for Volume Additions," in Section 1 for more information.

If only dissolved zinc is to be determined, filter the sample before the acid addition.

ACCURACY CHECK

See p. 1186 of the HACH Water Analysis Handbook, 3rd Ed., 1997.

INTERFERENCES

The following may interfere when present in concentrations exceeding those listed below.

Substance	Concentration
Aluminum	6 mg/L
Cadmium	0.5 mg/L
Copper	5 mg/L
Iron (ferric)	7 mg/L
Manganese	5 mg/L
Nickel	5 mg/L

Large amounts of organic material may interfere. Perform the mild digestion (Section 1) to eliminate this interference.

Highly buffered samples or extreme sample pH may exceed the buffering capacity of the reagents and require sample pretreatment (see "pH Interference" in Section 1).

POLLUTION PREVENTION AND WASTE MANAGEMENT

ZincoVer 5 Reagent contains potassium cyanide. Cyanide solutions are regulated as hazardous wastes by the Federal RCRA. Cyanide should be collected for disposal as reactive (D003) waste. Be sure that cyanide solutions are stored in a caustic solution with pH > 11 to prevent the release of hydrogen cyanide gas.

In the event of a spill or release, clean up the area by following these steps:

a) Use a fume hood or supplied-air or self-contained breathing apparatus.

b) While stirring, add the waste to a beaker containing a strong solution of sodium hydroxide and calcium hypochlorite or sodium hypochlorite (household bleach).

c) Maintain a strong excess of hydroxide and hypochlorite. Let the solution stand for 24 hours.

d) Neutralize and flush the solution down the drain with a large excess of water.

QUESTIONS AND ANSWERS

1. In this experiment, what is the primary safety issue?
ZinconVer 5 contains cyanide which is very poisonous if taken internally or inhaled. Store

the material away from acids and water. ***Do not add*** *the material to an acidic sample. When testing water outdoors, be sure that the wind does not blow in your face to avoid odors from the material.*

2. What are the interferences in the HACH experiment and in the LaMotte experiment?
   In the HACH experiment, the metals of aluminum, cadmium, copper, ferric iron, manganese and nickel can interfere in the analysis of zinc. In the LaMotte experiment, interferences are cadmium, aluminum, manganese, iron (III & II), and chromium at less than 10 ppm and nickel, cobalt, chromate, and zinc at 20 to 50 ppm.
3. What do you do with the waste or spills containing cyanide?
   The ZincoVer 5 Reagent contains potassium cyanide. Cyanide solutions have been regulated as hazardous wastes by the Federal RCRA and can be collected for disposal as reactive wastes (DOO3). The cyanides can be oxidized to cyanate by the addition of a strong solution of sodium hydroxide and sodium hypochlorite (household bleach).
4. What types of zinc compounds are known?
   Zinc forms many inorganic and organometallic compounds. Some of the compounds are antibacterial agents, tracer metals, metallic complexes, multi-metal complexes, simple and complex inorganic compounds, resinates, and many organo-zinc complexes.

REFERENCES

1. *HACH Water Analysis Handbook,* 3rd Ed. (Loveland, CO.: HACH Company, 1997).
2. *LaMotte Soil Handbook* (Chestertown, Md.: LaMotte Company, 1994).
3. *Handbook of Reference Methods for Plant Analysis,* Yash P. Kalra (ed.), (Boca Raton, Fla.: CRC Press, 1998).
4. *Hawley's Condensed Chemical Dictionary,* 11th Ed., rev'd by N. Irving Sax and Richard J. Lewis (New York, N.Y.: Van Nostrand Reinhold, 1987).
5. *Standard Methods for the Examination of Water and Wastewater:* ZINC, 16th Ed., A. E. Greenberg, R. R. Trussell, L. S. Clesceri and Mary Ann H. Franson (eds.), pp. 254, 259–260, Washington, DC, American Public Health Association, 1985.

CHAPTER 4
AIR AND GAS POLLUTANTS

INTRODUCTION

Safety is an important issue when addressing air pollution and various types of spills. A number of gases and vapors from spilled liquids are extremely toxic to humans. Many gases are explosive and fire hazards in the appropriate environment. During spills of liquids and releases of dangerous gases, people must wear the appropriate safety equipment and respond to the incident in a safe and professional manner.

In a building, incident respondents must be aware of the flowing air currents and drafts created by fans and open doors. When outdoors, the direction of the wind will play an important role during the emergency response to a spill or release of dangerous vapors in open spaces or outside of a building. In each case, the wind or air currents should be *behind* (blowing from behind to the front) the person addressing the escaping gases or volatile vapors.

Air contaminated with specific types of gases from releases will require the responder to wear appropriate breathing apparatus, safety clothing, and equipment. A number of the gases are hazardous substances, explosive, poisons, reactive, and extremely toxic to humans. Responders to releases must be aware of the area, tight spaces, traffic patterns, air currents, air flow direction, and potential problems with the released chemical. If possible, know what chemicals are present during the release or spill and respond appropriately. Read the Material Safety Data Sheets, the labels and shipping documents, the Emergency Response Guide, and additional appropriate materials. Other important information is available through the local HAZMAT Team or through a number of Federal and State agencies, fire departments, and emergency coordinators.

AIR AND GAS CONTAMINATES

Air pollutants are emitted from mobile sources (vehicles, lawn mowers), stationary sources (industrial and power plants) , and natural activities (volcanoes). Some primary air pollutants can react with one another or with another chemical in the air to form secondary air pollutants. In general, primary air pollutants are carbon monoxide (CO), carbon dioxide (CO_2), sulfur dioxide (SO_2), nitric oxide (NO), nitrogen dioxide (NO_2), hydrocarbons, and suspended particles. Pollutants from volcanoes normally contribute to the list of primary pollutants. Secondary air pollutants include sulfur trioxide (SO_3), nitric acid (HNO_3), sulfuric acid (H_2SO_4), hydrogen peroxide (H_2O_2), ozone (O_3), peroxyacyl nitrates (PANs), and most nitrate and sulfate salts.

There is a degree of uncertainty regarding the health effects of air pollution. For example, the synergistic effect of the more than 800 single substances in cigarette smoke and the physiological condition of the smoker are all factors in determining the toxicological influence of cigarette smoke on the individual*.

The monitor or measurement method used to determine air pollution quantities depends upon the objective of the measurement. There is no universal monitor for all possible scenarios. Direct reading colorimetric indicators have many practical applications and are relatively easy to perform. The companies of Mine Safety Appliances, Dräger, and Sensidyne produce air monitoring equipment available in the form of tubes filled with direct reading colorimetric indicators. Long-term and short-term measurements using different types of pumps and procedures produce reliable data.

A vial contains a number of layers comprised of different reagent systems, which react with a gaseous substance by changing color. The reactive layers in a tube can react with selective chemicals (such as CO_2) or selective substance groups (chlorinated hydrocarbons) and class selective groups such as easily oxidizable substances. The length-of-stain discoloration is an indicator of the concentration of the measured substance. A printed scale on the vial allows a direct reading of the contaminant concentration. The calibration scale is prepared in the engineering units of ppm or volume percent. At times, humidity can cause erroneous measurements.

Gas measuring devices are used to perform measurements in the range of the occupational exposure limits during investigations in the work place. The devices are also used for measurement of emission concentrations from stacks and other sources during technical analysis. In addition, gas measuring devices determine contaminants, such as CO, CO_2, water, and oil.

Short-term tubes are used to evaluate the contaminant concentration variations in the work place and measure the contaminants in the worker's breathing zone. Short-term tubes have necessary applications during investigation of confined spaces prior to entry and are used to check process pipe lines for gas leaks.

Long-term tubes provide integrated measurements that represent the average concentration during the sampling period (1 to 8 hours). They can be used as personal monitors or area monitors to determine the time waited average concentration. Direct reading diffusion tubes and badges can also be used for long-term measurement.

Chemical components in a mixture can be chemically similar (methanol, ethanol, and propanol) and have the same sensitivity to a specific tube. A colorimetric reaction system based on a chromate indicator cannot distinguish between the three alcohol types and indicates a sum of the concentration.

Discoloration of colorimetric chemical sensors in a tube is proportional to the mass of the reacting gas. Generally, the length of the stain indicates the concentration or mass of the reacting gas in a tube. In some cases, the length of a stain in a tube must be interpreted in terms of color and intensity according to a given reference standard or a set of standards.

AIR CONTAMINATE MEASUREMENTS

A number of different tubes containing a variety of chemicals are used to detect the presence of many types of air pollutants. Short-term measurements of air contaminated with a gaseous substance can be performed with a Dräger tube and similar devices in confined spaces (grain silos, chemical tanks, sewers,) and near leaking pipelines*. Short-term tubes are designed for immediate measurements at a particular location over a relatively short period of time (10 seconds to 15 minutes). The air pollutant changes the color of the tube media to indicate the level of contaminate in the air. The various tubes are limited by a standard range of

* Compliments of Dräger.

measurement, specific interferences (chemical cross sensitivity), and ambient operating conditions, i.e., absolute humidity and temperature range.

Long-term measurements of air are required for TLVs (Threshold Limit Values) and represent the average concentration of the contaminate during the sampling period. Long-term tubes can be used as personal monitors or area monitors to determine the time weighted average concentration of a pollutant. A low flow pump that can be calibrated in the 10 to 20 cc/minute range is required for detection of air pollutants by specific types of Dräger tubes. Long-term measurements over a 2 to 10 hour period of time can be recorded for a number of toxic and dangerous chemicals such as ammonia, benzene, carbon monoxide, carbon dioxide, carbon disulfide, chlorine, hydrogen sulfide, hydrogen cyanide, hydrochloric acid, hydrogen fluoride, nitrogen dioxide, perchloroethylene, sulfur dioxide, trichloroethylene, vinyl chloride, toluene and other volatile substances. Again, each tube contains a reaction medium that changes color to indicate the concentration of a specific air contaminate.

"Diffusion Tubes with Direct Indications" consist of sampling tubes which are exposed to the air contaminate. After breaking off one end of the diffusion tube, air is allowed to diffuse through the tube. The same type of diffusion principles apply to "Direct indications badges." In both cases, the color of the reaction medium in the tube or on the badge will change if a specific level of contaminate is present in the atmosphere. Direct Indication Diffusion Tubes are used to detect the presence of relatively toxic substances in the air for a period of 1 hour or 8 hours at 20° C and 1.013 hPa. Direct indication badges are primarily used to determine the presence of an air contaminate for a minimum or maximum measurement time.

Dräger tubes have been developed to determine the components of compressed air at a constant flow. When the pressure in a line is reduced to about 150 psi or less, the tube is compatible to the flow index. Also, additional tubes and procedures have been developed to address special needs of hazardous material responders in the company and in many communities.

Portable soil vapor monitors can detect both methane and non-methane hydrocarbon vapors at very low part per million (ppm) vapor levels. Some monitors can distinguish between methane or natural gas and other volatile hydrocarbon vapors. The modern combustible sensor is not affected by changes in relative humidity (RH). Monitors can measure a variety of gaseous components emitting from soil and solid matrices.

TESTING INFORMATION

Long-term tubes and short-term tubes require both ends of the Dräger tube to be broken off during the sampling phase. However diffusion tubes only require one end open during testing of air contaminates and do not require a pump. During long-term testing procedures, a low flow pump calibrated at 10 to 20 cc/minute will be required during sample analysis. For additional information, read the insert in the tube case or the instrument operations manual.

MEASUREMENT OF AIR CONTAMINATES

Procedure*

1. Break off both tips of the tube in the tube opener.

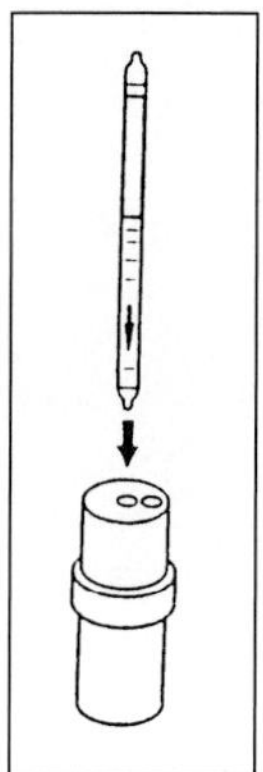

2. Insert the tube tightly in the pump. Arrow on the tube point towards the pump.
Note: Refer to the Operating Instruction Sheet in the case of tubes or the Dräger Handbook to select the Number of Strokes (n) for each tube.

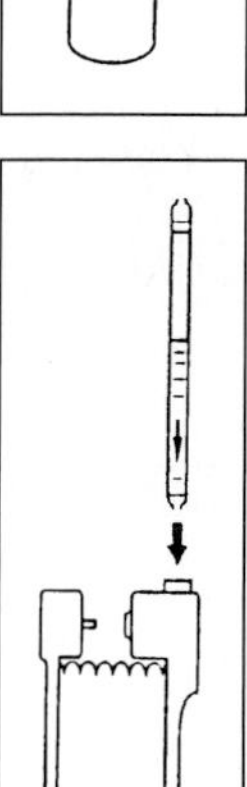

3. Suck air or gas sample through the tube.
Note: Measuring period is approximately 3 minutes for 10 strokes. The sampling time and the number of strokes will usually vary for each type of chemical tested. Refer to the Operating Instruction Sheet or Dräger Handbook.

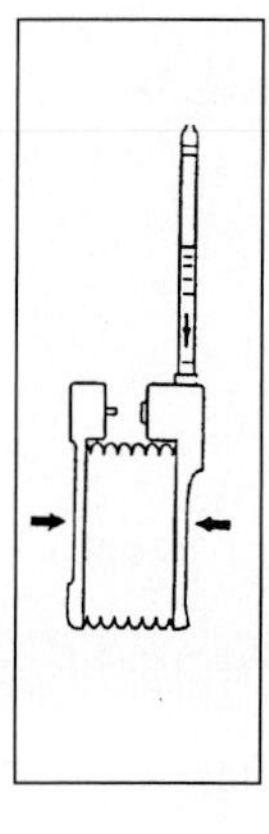

*Compliments of Dräger.

4. Read the entire length of the tube medium discoloration.
   Note: The discoloration will depend upon the tube medium and the chemical being tested.

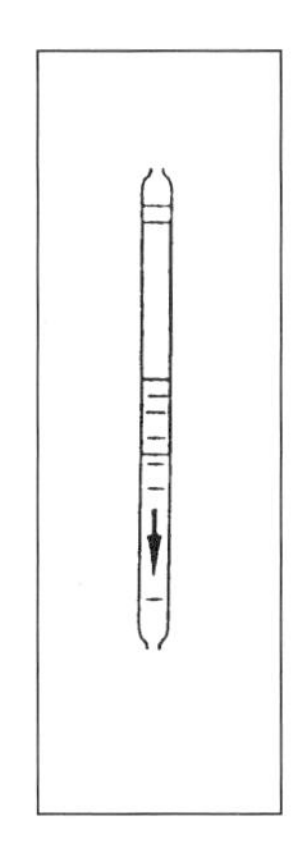

5. Record the temperature in °C and the humidity level. Compare the recorded values with the required experimental ranges. Record the atmospheric pressure and correct the reading by using one of the following formulas:

$$F = \frac{1013 \text{ millibars}}{\text{actual atmospheric pressure (hPa)}} \quad \text{or}$$

$$F = \frac{760 \text{ mm Hg}}{\text{actual atmospheric pressure (mm Hg)}}$$

6. Multiply the value by "Factor F" for correction of the atmospheric pressure. Enter the result in the measurement record. Record the standard deviation of the measurement. Observe or note the possible cross sensitivities.

$$\text{Reading} \times \text{Factor } F = \text{Concentration (in units of ppm, mg/l, mg/m}^3\text{, or Vol.\%)}$$

7. Flush the pump with air after operation.

DRÄGER TUBES REFERENCE TABLE

Dräger tubes and other types of tubes perform a number of measurements of gases and vapors under a variety of conditions. The table below identifies the name of the compound, a basic outline of the standard range of measurement, and the number of tubes for each type of measurement. Types of measurement included in the table are the following: a) short-term measurements, b) long-term measurements, c) diffusion tubes with direct indications, and d) measurements of compressed air. Badges with direct indications are available to detect the presence of specific gases. For additional details about operation and measurement techniques, calculations, measuring range extension and other details, contact the Dräger-Tube Handbook, 11th Edition, 1998 or read the information sheets and manuals in the tube kits and equipment cases.

Dräger tube*	No. of tubes, short term	Standard range of measurement (short term)	No. of tubes, long-term	No. of tubes, diffusion	No. of tubes, compressed air
Acetaldehyde	1	100 to 1,000 ppm	—	—	—
Acetic Acid	1	5 to 80 ppm	1	1	—
Acetone	1	100 to 12,000 ppm	1	—	—
Acid Test	1	qualitative	—	—	—
Acrylonitrile**	2	0.5 to 10 ppm/5 to 30 ppm	—	—	—
Alcohol**	2	25 to 2,000 ppm/50 to 4,000 ppm	—	—	
Amine Test	1	qualitative	—	—	—
Ammonia**	5	0.25 to 3 ppm/5 to 100 ppm and 0.5 to Vol.-%	1	1	—
Aniline	2	0.5 to 10 ppm/1 to 20 ppm	—	—	—
Arsenic Trioxide	1	0.2 mg/m^3 as As	—	—	—
Arsine	1	0.05 to 3 ppm	—	—	—
Benzene**	6	0.5 to 10 ppm/15 to 420 ppm	1	—	—
Butadiene	—	—	—	1	—
Carbon Dioxide**	6	100 to 3,000 ppm and 0.5 to 10 Vol.-%/5 to 60 Vol.-%	1	2	1
Carbon Disulphide**	3	3 to 95 ppm and 0.1 to 10 mg/L	1	—	—
Carbon Monoxide**	7	2 to 60 ppm/100 to 3,000 ppm and 0.001 to 0.03 Vol.-%/0.3 to 7 Vol.-%	2	1	1
Simultaneous Tube: (Carbon Monoxide 200/a+ Carbon Dioxide 2%/a)	1	200 to 2,500 ppm CO and 2 to 12 Vol.-% CO_2	—	—	—
Carbon Tetrachloride**	3	0.2 to 10 ppm/10 to 70 ppm	—	—	—
Chlorine**	3	0.2 to 3 ppm/50 to 500 ppm	1	—	—
Chlorobenzene	1	5 to 200 ppm	—	—	—
Chloroform	1	2 to 10 ppm	—	—	—
Chloroformates	1	0.2 to 10 ppm	—	—	—
Chloroprene	1	5 to 60 ppm	—	—	—
Chromic Acid	1	0.1 to 0.5 mg/m^3	—	—	—
Cyanide	1	2 to 15 mg/m^3	—	—	—
Cyanogen Chloride	1	0.25 to 5 ppm	—	—	—
Cyclohexane	1	100 to 1,500 ppm	—	—	—
Cyclohexylamine	1	2 to 30 ppm	—	—	—
Diethyl Ether	1	100 to 4,000 ppm	—	—	—
Dimethyl Formamide	1	10 to 40 ppm	—	—	—
Dimethyl Sulfate	1	0.005 to 0.05 ppm	—	—	—
Dimethyl Sulphide	1	1 to 15 ppm	—	—	—
Epichlorohydrin	1	5 to 50 ppm	—	—	—
Ethanol	—	500 to 8,000 ppm/62.5 to 1,000 ppm	1	1	—
Ethyl Acetate	1	200 to 3,000 ppm	—	1	—
Ethyl Benzene	1	30 to 400 ppm	—	—	—
Ethyl Glycol Acetate	1	50 to 700 ppm	—	—	—
Ethylene**	2	0.2 to 5 ppm/50 to 2,500 ppm	—	—	—
Ethylene Glycol	1	10 to 180 mg/m^3	—	—	—
Ethylene Oxide	2	1 to 15 ppm/25 to 500 ppm	—	—	—
Fluorine	1	0.1 to 2 ppm	—	—	—
Formaldehyde**	2	0.2 to 2.5 ppm/2 to 40 ppm	—	—	—
Formic Acid	1	1 to 15 ppm	—	—	—
Halogenated Hydrocarbons**	1	200 to 2,600 ppm	—	—	—
Hexane	1	100 to 3,000 ppm	—	—	—

Dräger tube*	No. of tubes, short term	Standard range of measurement (short term)	No. of tubes, long-term	No. of tubes, diffusion	No. of tubes, compressed air
Hydrazine	2	0.25 to 3 ppm/0.5 to 10 ppm	—	—	—
Hydrocarbon**	2	3 to 23 mg/L and 0.5 to 1.3 Vol.-%	1	—	—
Hydrochloric Acid	2	1 to 10 ppm/500 to 5,000 ppm	1	1	—
Hydrochloride Acid/Nitric Acid	1	H.A.-1 to 10 ppm N.A.-1 to 15 ppm	—	—	—
Hydrocyanic Acid	1	2 to 30 ppm	1	1	—
Hydrogen	2	0.2 to 2.0 Vol.-%/0.5 to 3.0 Vol.-%	—	—	—
Hydrogen Fluoride	1	1.5 to 15 ppm	1	—	—
Hydrogen Peroxide	1	0.1 to 3 ppm	—	—	—
Hydrogen Sulphide**	11	0.2 to 5 ppm/100 to 2,000 ppm and 0.2 to 7 Vol.-%/2 to 40 Vol.-%	1	1	—
Hydrogen Sulphide + Sulfur Dioxide	1	0.2 to 7 Vol.-%	—	—	—
Mercaptan	2	0.5 to 5 ppm/20 to 100 ppm	—	—	—
Mercury Vapor	1	0.05 to 2 mg/m^3	—	—	—
Methanol	1	50 to 3,000 ppm	—	—	—
Methyl Acrylate	1	5 to 200 ppm	—	—	—
Methyl Bromide**	3	0.5 to 5 ppm/10 to 100 ppm	—	—	—
Methylene Chloride	1	100 to 2,000 ppm	1	—	—
Natural Gas Test	1	qualitative	—	—	—
Nickel	1	0.25 to 1 mg/m^3	—	—	—
Nickel Tetracarbonyl	1	0.1 to 1 ppm	—	—	—
Nitric Acid	1	1 to 15 ppm/5 to 50 ppm	—	—	—
Nitrogen Dioxide**	2	0.5 to 10 ppm/5 to 100 ppm	1	1	—
Nitroglycol	1	0.25 ppm	—	—	—
Nitrous Fumes**	4	0.5 to 10 ppm/500 to 5,000 ppm	2	—	1
Oil	—	0.1 to 10 mg/m^3	—	—	1
Oil Mist	1	1 to 10 mg/m^3	—	—	—
Olefine	1	Propylene, 0.06 to 3.2 Vol.-% Butylene, 0.04 to 2.4 Vol.-%	—	1	—
Organic Arsenic Compounds and Arsine	1	0.1 ppm as Arsine	—	—	—
Organic Basic Nitrogen Compounds	1	1 mg/m^3 threshold value	—	—	—
Oxygen	1	5 to 23 Vol.-%	—	—	—
Ozone	2	0.05 to 0.7 ppm/20 to 300 ppm	—	—	—
Pentane	1	100 to 1,500 ppm	—	—	—
Perchloroethylene**	4	0.1 to 1 ppm/500 to 10,000 ppm	1	1	—
Petroleum Hydrocarbons	2	10 to 300 ppm/100 to 2,500 ppm	—	—	—
Phenol	1	1 to 20 ppm	—	—	—
Phosgene**	3	0.02 to 0.06 ppm/0.25 to 15 ppm	—	—	—
Phosphine**	5	0.1 to 4 ppm/200 to 10,000 ppm	—	—	—
Phosphoric Acid Ethers	1	0.05 ppm Dichlorvos	—	—	—
Polytest	1	qualitative	—	—	—
Pyridine	1	5 ppm	—	—	—
Simultaneous Test-Set I	1	—	—	—	—
Simultaneous Test-Set II	1	—	—	—	—
Simultaneous Test-Set III	1	—	—	—	—
Styrene**	3	10 to 200 ppm/50 to 400 ppm	—	—	—
Sulphur Dioxide**	5	0.5 to 5 ppm/400 to 8,000 ppm	2	1	—

Dräger tube*	No. of tubes, short term	Standard range of measurement (short term)	No. of tubes, long-term	No. of tubes, diffusion	No. of tubes, compressed air
Sulphuric Acid	1	1 to 5 mg/m^3	—	—	—
Tetrahydrothiophene	1	1 to 10 ppm	—	—	—
Thioether	1	1 mg/m^3 threshold value	—	—	—
Toluene**	3	5 to 80 ppm/50 to 400 ppm	1	1	—
Toluene Diisocyanate	1	0.02 to 0.2 ppm	—	—	—
O-Toluidine	1	1 to 30 ppm	—	—	—
Trichloroethane	1	50 to 600 ppm	—	—	—
Trichloroethylene	2	2 to 50 ppm/50 to 500 ppm	1	1	—
Triethylamine	1	5 to 60 ppm	—	—	—
Vinyl Chloride**	4	0.5 to 3 ppm/100 to 3,000 ppm	1	—	—
Water Vapor**	4	0.1 to 1 mg/L/1 to 40 mg/L	—	1	1
Xylene	1	10 to 400 ppm	—	—	—

*Compliments of Dräger.
**Not all ranges indicated.

LAMOTTE EQUIPMENT*

Air studies can be performed by using LaMotte equipment and chemicals. During air pollution studies, some type of vacuum equipment is normally used to perform the experiment. The applied vacuum draws an air sample through a chamber containing a special absorbing solution. An absorbing solution in a special glass bubbling tube called an impinger is used to trap or "bind" the gas for analysis purposes. A variety of absorbing solutions with different chemical properties are used to trap the gaseous pollutants.

For quantitative testing of air samples, the amount of air pulled through the absorbing solution is determined when an adjustable flowmeter is attached to the vacuum portion of the air sampling train. A flowmeter with an adjustable flow device (like a needle valve) can be preset to control the sampling rate of the vacuum pump. For most air pollution studies, the flowmeter is calibrated to measure the rate of air flow in liters per minute (Lpm). If air is drawn through the absorbing solution at 1.5 Lpm for 10 minutes, then 15 liters of air have been sampled.

Air samples collected from outdoors require battery operated and completely portable equipment. The pump must maintain a particular flow rate for a pre-determined time to regulate the amount of air sampled. Proper sampling techniques require that the air sample be pulled through the impinger prior to contact of the air with the pump*. Air sampling equipment must be completely portable, possess a regulating device for sampling at different rates, and be able to maintain a particular sampling rate for a known amount of time.

Some precision impingers consist of a fitted glass end which is immersed into the absorbing solution to disperse many minute bubbles. The smaller bubbles provide greater surface contact between the gas and absorbing solution, which results in a higher efficiency of gas absorption. A fitted glass bubbler producing small bubbles of gas is used for nitrogen

*Compliments of LaMotte Company, Chestertown, Md.

dioxide determinations*. For other gas sampling, a glass bubbler assembly with a 1-mm opening at the end disperses the air sample into the absorbing solution.

During some of the air tests, an indicator is added directly to the absorbing solution to cause a color reaction. In other systems, the indicator is added to a pre-treated absorbing solution The reaction color changes are measured by a visual comparator or an electronic colorimeter.

Visual comparators are used to match the color of the test sample to color standards of known values. The appropriate index number is recorded after the color of the solution is matched as closely as possible with the eight permanent color standards of the comparator. After referring to the calibration chart, the index number will provide the concentration of the air pollutant in units of milligrams per cubic meter (mg/M^3) or an appropriate unit.

The pollution test equipment can be used for testing air pollution levels in confined areas or for monitoring atmospheric pollutants. The equipment can also be used to detect gas leaking or escaping from large tanks, bottled gas containers, and various types of vehicles. In these cases, prompt and cautious actions are required to implement appropriate responses and correct analytical techniques. Safety procedures must be fully implemented.

ESTABLISHING SAMPLING SITES*

Air pollution monitoring programs require a complete survey of the area and preparation of a sampling map to select appropriate sampling sites. Heavy industrial areas, high traffic patterns, hidden pollution sites, and relatively large population centers contribute significantly to the total air pollution in an area. Areas with excessive numbers of high pollution sources such as lawn mowers, boilers, fireplaces, barbecue pits, tire fires, outdoor burning, forest fires, and high vehicle traffic patterns should be tested extensively.

A number of tests can accurately monitor emissions from these sites. When sampling at various sites and recording data, the dispersion of air pollutants will be impacted by meteorological data such as the prevailing winds, temperature inversions, rainfall and other factors. For control purposes, a remote sheltered location away from the normal air pollution sources is necessary to obtain normal background data for comparison purposes.

Any observations which might adversely affect the result should be recorded during data collection. Air pollution investigations should include proper sample labeling, sampling site location, sampling and test procedures, and determining the concentration of the pollutant. Meteorological data (prevailing winds, temperature, atmospheric pressure, relative humidity; duration, date, and time of sampling) should be included in the report.

AIR SAMPLING APPARATUS*

The Model BD Air Sampling Pump is designed especially for use with the LaMotte Air Pollution Test Equipment. The air sampling pump consists of an adjustable flowmeter, intake and exhaust connections, impinger holder, on-off switch, four "C" cell batteries and AC jack. For proper maintenance and to guarantee the life and usefulness of the air sampling pump, thc following operating procedures are recommended.

*Compliments of LaMotte Company, Chestertown, Md.

INSTRUCTIONS*

1. Check battery capacity by turning switch to the "ON" position. Adjust flowrate to sample at 2.0 Lpm. If indicator float of flowmeter vibrates up and down with a deflection of more than 0.2 Lpm, or if air sampling pump cannot be adjusted to sample at 2.0 Lpm, replace batteries. (See procedure for replacement of batteries in the LaMotte manual titled "Air Sampling & Measurement".) Alternatively the pump may be operated with an AC adapter.
2. Fill the impinging tube to the designated line with the absorbing solution. Insert stopper assembly, thereby immersing the long tube into the absorbing solution. Attach one end of the flexible tubing to the pump intake fitting. The other end of the tubing is attached to the outlet connection of the impinger.
   Note: Several air sampling kits use the Midget Impinger or the Gas Bubbler Impinger instead of the General Purpose Impinger. In Step 2, the flexible tubing from the pump is attached to the glass side tube of the impinger by placing the end of the tubing onto the glass tube. This is a tight connection which requires some care and patience to make. The outlet connection draws air from above the liquid in the impinger chamber. **DO NOT** connect to the inlet tube of the impinger which is below the surface of the liquid.
   Note: Check to insure all connections are tight and that the tubing is connected to correct connection of the impinging tube.

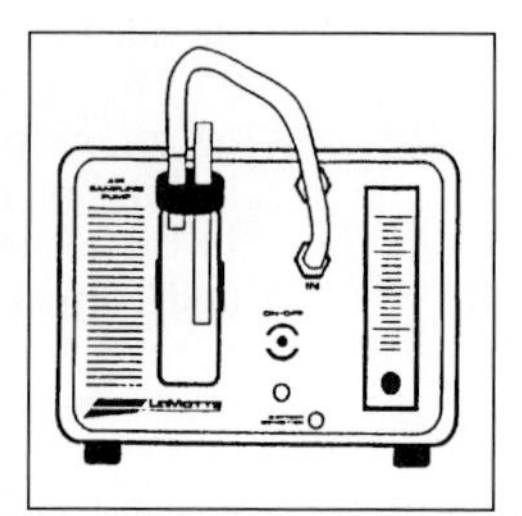

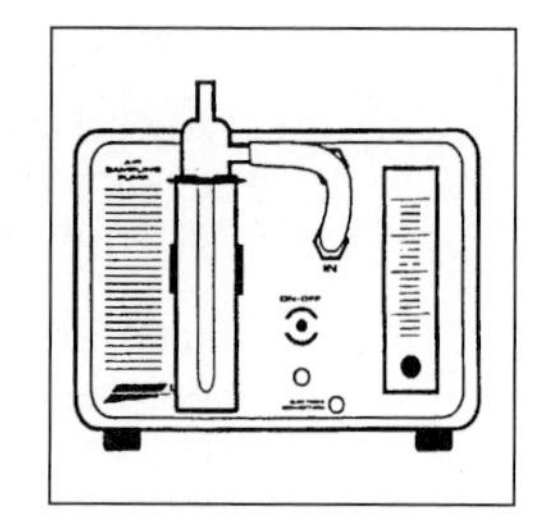

3. Turn switch to "ON" position.
4. Adjust flowmeter to sample at the designated rate (according to instructions of testing unit). Turn knob clockwise to reduce flow, counterclockwise to increase flow. Do not unscrew or withdraw valve stem beyond threaded section except for maintenance. The flowmeter is read by aligning the reader's eye with the center of the black indicator float and the scale. Graduations on the scale are in 0.1 Lpm increments.
   Note: To avoid introducing foam or residue into the pump, a separate trap or empty impinger may be inserted in series or down-stream and next to the pump.
5. At the end of the sampling period, turn switch to the "OFF" position. Disconnect impinging apparatus and remove from holder.
6. The absorbing solution is then subjected to the testing procedure as outlined in the individual air pollution test instructions.

*Compliments of LaMotte Company, Chestertown, Md.

Maintenance and Storage of Flowmeter*

Required instrument maintenance includes an occasional cleaning to ensure reliable operation and good float visibility. For disassembly, cleaning, and reassembly of the flowmeter, refer to the Instruction Manual titled "Air Sampling & Measurement."

To reduce the effect of possible variation introduced by the interval of collection and analysis, the time period should be kept as short as possible, and preferably of the same duration. The sample should be kept from exposure to heat and light. If the analysis cannot be made immediately after the sample collection, the sample may be stored at a low temperature.

Instructions for the Model LG Hand Operated Syringe Pump*

Introduction This manually operated syringe pump is capable of drawing 50 cc of the test atmosphere through the impinging apparatus by one complete stroke of the syringe. The pumping mechanism relies upon the use of a special check valve which ensures that a positive vacuum is formed on the upstroke of the syringe. The check valve also allows the expulsion of the air sample as the plunger is depressed. Therefore, this pump allows a continuous sampling of the test atmosphere as the plunger of the syringe is manually pulled out and depressed.

A general purpose impinging apparatus is held in place by a special holder on the barrel of the syringe. The short plastic tube of the impinging apparatus is connected by flexible tubing to the intake connection of the syringe. As the syringe plunger is pulled out, air is drawn through the longer tube and bubbled into the absorbing solution. The vacuum formed on the upstroke of the plunger draws air from above the absorbing solution into the syringe chamber. When the syringe plunger is depressed, the air is forced through the outlet or exhaust of the check valve. After the syringe is in the downstroke position it is ready for another 50 cc air sample.

INSTRUCTIONS

1. Depress plunger to "0" position.

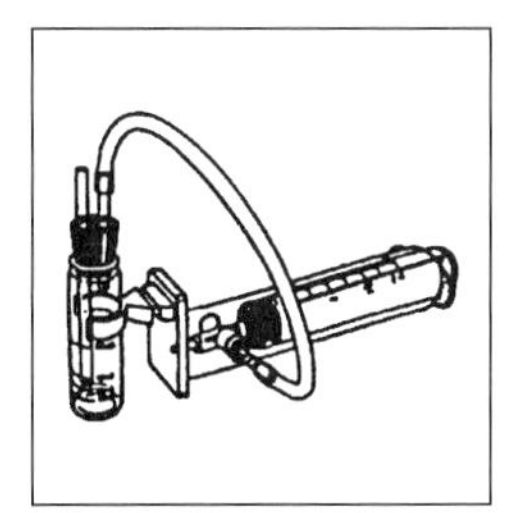

2. Remove stopper assembly from impinging tube and pour the designated amount of absorbing solution into the glass tube. ***Note:*** See instructions of individual test kits to determine what amount of absorbing solution is added to impinging apparatus.
3. Replace stopper assembly. Connect flexible tubing to short tube (outlet) of impinger assembly and to intake connections of syringe.

*Compliments of LaMotte Company, Chestertown, Md.

4. Begin to sample the test atmosphere by completing one complete cycle (pulling out-depressing plunger). Twenty complete strokes on the syringe per minute is equivalent to sampling the air at one liter per minute (Lpm). Continue sampling until the appropriate volume of air has been sampled (See instructions of individual test procedures).
5. At the end of the sampling period, disconnect impinging apparatus and remove from holder. At this point, the absorbing solution can be tested by the procedure of the individual air pollution test instructions.

Air Pollution Test Equipment*

The testing methods used in each LaMotte experiment follow the accepted procedures of the Environmental Protection Agency, which recommends collection of the gas in a special absorbing solution and the subsequent chemical analysis. Each analytical method has been appropriately modified for testing in the field or in confined areas. Both precision impinging units and general purpose impinging units are readily available for testing samples outdoors or in confined spaces.

CONVERSIONS*

The individual air pollution test sets are calibrated with read-outs or results in parts per million (ppm). To convert concentrations of gases and vapors from part per million (ppm) by volume to milligram per cubic meter (mg/m^3) and vice versa at any given temperature and pressure, the following expression is useful:

$$\text{C2 (ppm)} = \frac{\text{C1} \times 24.450 \times \text{T} \times 760}{\text{mol. wt.} \times 298 \times \text{P}}$$

$$\text{C1 (mg/M}^3\text{)} = \frac{\text{C2} \times \text{mol. wt.} \times 298 \times \text{P}}{24.450 \times \text{T} \times 760}$$

C1 = Concentration of gas or vapor in mg/M^3
C2 = Concentration of gas or vapor in ppm
T = Absolute temperature in °Kelvin (°C + 273)
P = Absolute pressure in Torr of sample air stream (mm of mercury)

Note: It is usually necessary to convert the measure of air flow to the temperature and pressure of air stream.

FLOW EQUIVALENT CONVERSION TABLE*

1 Cu Ft/Hr	**1 Lpm**
0.016 Cu Ft/Min	**60 Lph**
0.471 Lpm	**0.035 Cu Ft/Min**
28.317 Lph	**2.118 Cu Ft/Hr**

*Compliments of LaMotte Company, Chestertown, Md.

CONCENTRATION SPECIFICATIONS AND CONVERSIONS*

The measurement of contaminants in air is expressed in terms of concentration or amount of substance per volume of air. A specific engineering unit (gram per liter, milligram per cubic meter, etc.) indicates a concentration level of contaminate.

High concentrations are generally given in volume percent (Vol.-%), i.e., 1 part of a substance in 100 parts of air. If air consists of 10 Vol.-% of a contaminate, then 100 parts of air contain 10 parts of the contaminate. At lower pollution levels, the engineering units of ppm (parts per million or mL/m^3) and ppb (parts per billion) have many applications. The concentration ppm is defined as 1 part of a substance in 1 million parts of air, whereas ppb refers to 1 part of a substance in 1 billion parts of air.

The conversion of very small concentration units to Vol.-% is as follows:

$$1 \text{ Vol.-\%} = 10{,}000 \text{ ppm} = 10{,}000{,}000 \text{ ppb}$$

In addition to gaseous components, the air also contains solid particulates or liquid droplets called aerosols. Since an indication in volume percent is not very useful due to the small size of the droplets or particles, the concentration of the aerosols is given in mg/m^3. Tables of conversion units or calibration factors are provided below.

Calibration Factors*

	Vol.-%	ppm	ppb
Vol.-% = $10\ L/m^3$, 1 cL/L	1	10^4	10^7
ppm = mL/m^3, $\mu L/L$	10^{-4}	1	10^3
ppb = $\mu L/m^3$, nL/L	10^{-7}	10^{-3}	1

	g/L	mg/L	mg/m^3
g/L = kg/m^3, mg/mL	1	10^3	10^6
mg/L = g/m^3, $\mu g/mL$	10^{-3}	1	10^3
mg/m^3 = ng/mL	10^{-6}	10^{-3}	1

Since each volume is related to a corresponding mass, the volume concentrations of gaseous substances can be converted into mass per unit volumes and vice versa. These conversions must be done at a specified temperature and pressure since the gas density is a function of temperature and pressure. For standard measurements at work places, the reference parameters are 20° C and 1013 hPa*. Examples of calculations are provided below.

a) Conversion from mg/m^3 to ppm

$$c[\text{ppm}] = \frac{\text{mole volume [L]}}{\text{molar mass [g]}} \times c[\text{mg/m}^3]$$

The mole volume of any gas is 24.1 L at 20° C and 1013 hPA, the molar mass (molecular weight) is gas specific.

Example for acetone:

*Compliments of LaMotte Company, Chestertown, Md.

Mole volume	24.1 L/mole
Molar mass	58 g/mole
Assumed concentration	876 mg/m³

$$c[\text{ppm}] = \frac{24.1\ [\text{L}]}{58[\text{g}]} \times 876\ [\text{mg/m}^3]$$

Concentration in ppm: c = 364 ppm or 364 mL/m³.

b) Conversion from ppm to mg/m³

$$c[\text{mg/m}^3] = \frac{\text{molar mass [g]}}{\text{mole volume [L]}} \times c[\text{ppm}]$$

with the assumed concentration of 364 ppm it is:

$$c[\text{mg/m}^3] = \frac{58\ [\text{g}]}{24.1\ [\text{L}]} \times 364\ [\text{ppm}]$$

Concentration in mg/m³: c = 876 mg/m³.

*GENERAL PRECAUTIONS**

A. Store chemicals in a cool, dry place to prolong reagent shelf-life.

B. Read all instructions to familiarize yourself with the test procedure before you begin. Note any precautions in the instructions.

C. Read the label on each LaMotte reagent container prior to use. Some containers include precautionary notices and first aid information.

D. Keep all equipment and reagent chemicals out of the reach of young children.

E. In the event of an accident or suspected poisoning, immediately call the Poison Center phone number in the front of your local telephone directory or call your physician. Be prepared to give the name of the reagent in question and its LaMotte code number. LaMotte reagents are registered with **POISINDEX,** a computerized poison control information system available to all local poison control centers.

F. Avoid contact between reagent chemicals and skin, eyes, nose, and mouth.

G. Wear safety goggles or glasses when handling reagent chemicals.

H. Use the test tube caps or stoppers, not your fingers, to cover test tubes during shaking or mixing.

I. When dispensing a reagent from a plastic squeeze bottle, hold the bottle vertically upside-down (not at an angle) and gently squeeze it (if a gentle squeeze does not suffice, the dispensing cap or plug may be clogged).

J. Wipe up any reagent chemical spills, liquid or powder, as soon as they occur. Rinse area with wet sponge, then dry.

K. Thoroughly rinse test tubes before and after each test. Dry your hands and the outside of the tube.

L. Tightly close all reagent containers immediately after use. Do not interchange caps from different containers.

---

*Compliments of LaMotte Company, Chestertown, Md.

M. Avoid prolonged exposure of equipment and reagents to direct sunlight. Protect them from extremely high temperatures, and protect them from freezing.

DISCUSSIONS AND PROCEDURES

ACID RAIN AND ACID VAPORS

A significant number of sulfur and nitrogen oxides entering the atmosphere are converted to strong and weak acids on water droplets. The formation of these acids in the atmosphere and the emission of hydrochloric acid and other acidic compounds can cause major environmental problems in some areas. Changes in the acidity of the atmosphere can create a number of problems identified below. Excessive acid vapors and/or acid concentrations[1] can cause

- Severe health problems and death at high exposure levels
- Phytotoxicity to plants and other living species
- Liberation of metals from soil and rock creating an increase in metal toxicity
- Diminished productivity of forests and crops by leaching of soil nutrients
- Destruction of sensitive forests and sensitive vegetation
- Decreases in alkaline earth metal ions in leaves
- Respiratory effects on humans and animals
- Acidification of waters with toxic effects to flora, fauna, and fish
- Dissolution of limestone building materials, stone and decorations
- Reduction of visibility by formation of aerosols and acidic fog
- Corrosion to exposed structures such as metals in piping, equipment and electrical devices

Some compounds accidentally released to the atmosphere can hydrolyze and cause acid vapors. Silicon tetrachloride can react with atmospheric water to form a fog of hydrochloric acid droplets. Aluminum chloride will react with rain or high levels of moisture to form HCL gas. Leaking or accidentally punctured tanks containing these materials can produce significant volumes of acid vapors. Acid vapors of sulfur dioxide can be produced by metal sulfide smelting processes and the burning of coal. Emission of NO_x from a variety of sources creates acid vapors. Also, specific types of pine trees create an acid environment. In many cases, these changes have localized impacts. Acidic mist particles in the Los Angeles area during 1982 have been measured at a pH = 1.7[1]. Acid precipitation values can vary widely but are normally in the pH range of 4.0 and above.

Acid conditions of collected rain water can be measured by a pH meter or estimated by pH indicators or litmus paper. A Dräger test using an "acid test" sampling tube can qualitatively identify an acidic air environment at a pH near 5.0. The test procedure is provided in the experiment titled "Measurement of Air Contaminates." The tube contents react to some acids and form a pink to yellow reaction product depending on the concentration and type of acid in the air. The indicator tube is primarily used for indoor air, vehicle accidents, accidental spills, and industrial leaks from tanks or other equipment. The recorded pH ranges for some common substances are provided in the following diagram.

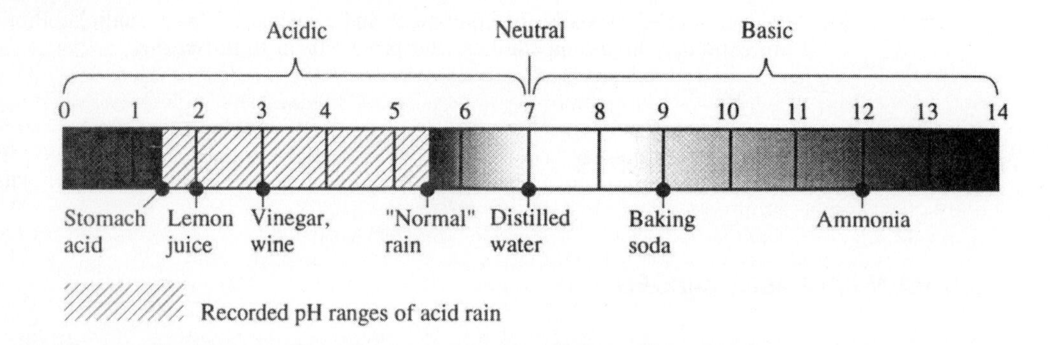

QUESTIONS AND ANSWERS

1. What are the major acids that cause environmental damage?
   The acids that cause the most environmental damage are sulfuric acid, nitric acid and hydrochloric acid.
2. How would you detect a leak of an acid?
   An air monitoring system with the appropriate tube would qualitatively indicate the presence of acid gases. A dramatic release of acid vapors would require individuals to use breathing apparatus and wear special gear or "Hazmat" suits.
3. If an acid spill occurs, what should a person do without appropriate equipment?
   If the spill is minor, the acid can be neutralized with baking soda or sodium carbonate in water. For major spills or accidents, call the Hazmat Response Team and stay away from the area. Read the Emergency Response Guidebook for further instructions. Isolate the hazard area and remain upwind from the spill or gas. Stay out of low areas.
4. What is the lowest pH of an acidic vapor?
   The pH of an acidic moisture in fog has been previously measured at 1.7 However, the acid vapor in a relatively confined space can reach levels well below a pH = 1.

REFERENCES

1. Stanley E. Manahan, *Environmental Chemistry,* 6th Ed. (Boca Raton, Fla.: Lewis Publishers, (1994).
2. *Dräger Tube Handbook,* 11th Ed. (Lübeck: Dräger Sicherheitstechnik GmbH., 1998).
3. Roger N. Reeve, *Environmental Analysis,* (ACOL), John D. Barnes (ed.), (New York, NY: John Wiley & Sons, 1994).

AMMONIA

Sources of gaseous ammonia are animal waste decay, sewage treatment processes, microorganism activities, and other biochemical processes. Additional sources include coke and ammonia manufacturing, refrigeration systems, chemical manufacturing, laboratories, and from soil fertilization processes with anhydrous ammonia machines. Fairly concentrated ammonia solvents are used as cleansing agents in many homes and industry. Ammonia-based reaction solvents can create significant air pollution problems in industry. Numerous releases of ammonia occur on a daily basis.

Ammonia gas is poisonous to most life. When ammonia combines with a proton in acidic water to form the ammonium ion, the toxicity of ammonia in water decreases significantly. Ammonia has an affinity for water and acts as a base in water. Ammonium salts are fairly corrosive substances in atmospheric aerosols. Ammonia or ammonium salts can be found in cleaning solutions, solid and liquid fertilizers, anhydrous ammonia fertilizer, refrigerants, surface stripping solutions, and as products of decomposition and sewage. In polluted environments, ammonia neutralizes and reacts with nitrate and acidic sulfate aerosols to form non-harmful products and useful compounds.

Ammonia can be determined spectrophotometrically by bubbling the gas into a solution of alkaline mercury (II) iodide to produce a colloidal orange brown compound of $NH_2Hg_2I_3$. The compound absorbs light between 400 to 500 nm. This procedure requires relatively expensive equipment.

Alternatively, the compound can be determined in the gaseous state by Dräger tubes. A portion of the pH indicator media, bromophenol blue, reacts with ammonia and changes color from yellow or yellow-orange to either a blue or violet reaction product. The "short-term" tube standard measurement ranges are 0.25 to 3 ppm, 2 to 30 ppm, 5 to 70 ppm, 5 to 100 ppm, and 0.5 to 10 Volume %. The Dräger test procedure for ammonia is provided in the experiment titled "Measurement of Air Contaminates." Both basic and acidic gases can be measured for ammonia concentration by the tube method. Basic organic amines will interfere with the reaction of ammonia in the tube media.

Ammonia can be reacted with bromophenol blue to form a blue reaction product in a Dräger tube during a long-term measurements of 1 hour, 2 hours, or 4 hours. A low flow pump that can be calibrated in the 10 to 20 cc/minute range pumps the contaminated air into the tube. Additional basic gases respond to the tube reactions at different sensitivities. Isopropylamine and triethylamine are indicated with approximately the same sensitivity. In addition, ammonia can be measured for 1 to 8 hours by a "Diffusion Tube with Direct Indications," where the gas molecules automatically move into the tube without a pump.

Ammonia can be quantitatively determined by the LaMotte Method. Air containing ammonia is pumped into an Ammonia Nitrogen Absorbing Solution at 1.0 Lpm for a specific time period. After two different reagents are added to the solution, the mixture in a test tube is shaken until a yellow to brown color develops indicating the presence of ammonia. The sample color is matched to the eight permanent color standards and an index number is recorded. From an Ammonia in Air Calibration Chart, the index number can be matched with the time of sampling (probably 10 minutes) to provide the concentration of ammonia in ambient air in parts per million (ppm).

AMMONIA

Procedure*

1. Pour 10 mL of Ammonia Nitrogen Absorbing Solution into impinging apparatus. Connect impinging apparatus to intake of air sampling pump.
   Note: Make sure the long tube is immersed in the absorbing solution.

2. Set flowmeter to sample at a rate of 1.0 Lpm for 10 minutes or until a measurable amount of ammonia is absorbed.

*Compliments of LaMotte Company, Chestertown, Md.

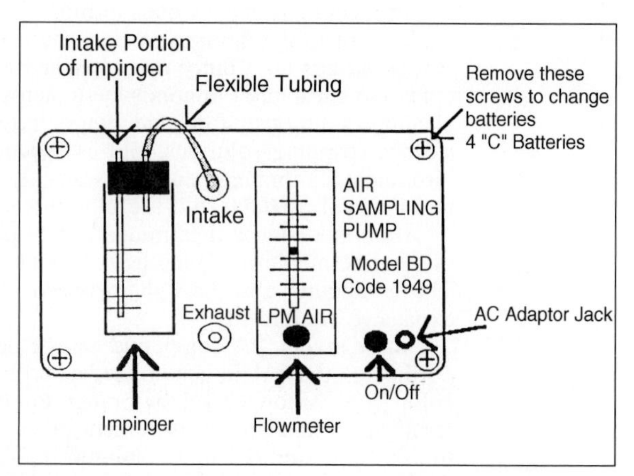

3. At the end of the sampling period, add contents of impinger to the 5-mL line of the test tube.

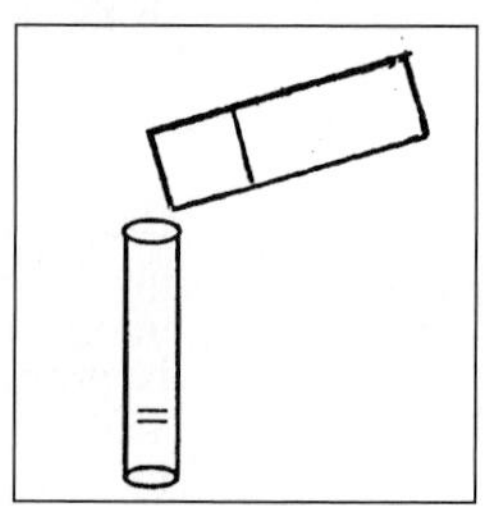

4. Add 2 drops of Ammonia Nitrogen Reagent #1. Cap and mix.

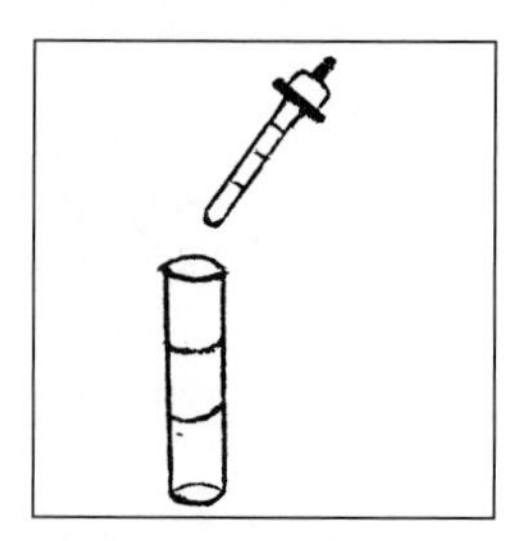

5. Add 8 drops of Ammonia Nitrogen Reagent #2. Cap and mix. Solution will turn yellow to brown color if ammonia is present.

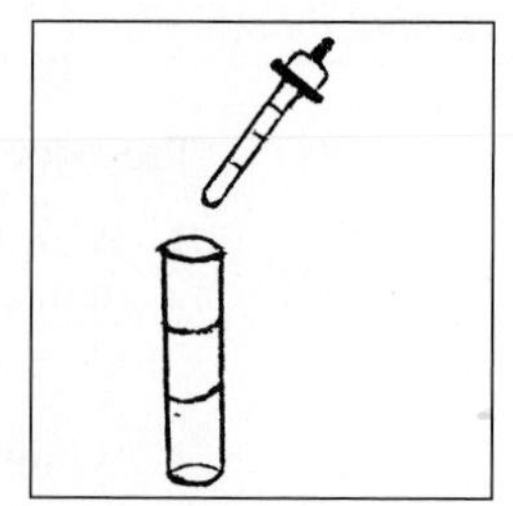

6. Place test tube into Ammonia in Air Comparator. Match sample color to index of color standards. Note the index number which gives the proper color match.

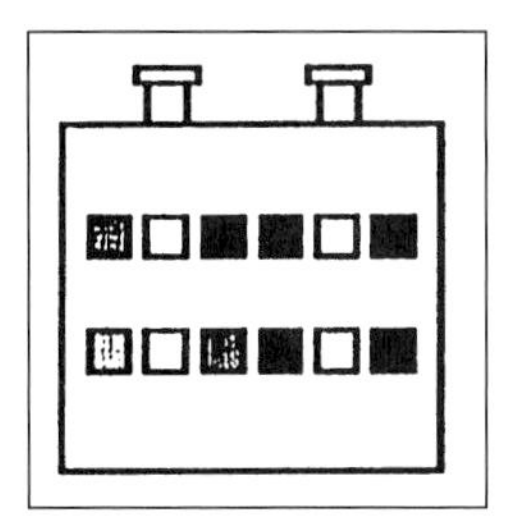

Ammonia in Air Calibration Chart*

Time (min)	Comparator index number 1	2	3	4	5	6	7	8
5	3.43	6.86	10.30	13.73	17.16	20.59	24.02	27.46
10	1.72	3.43	5.15	6.86	8.58	10.30	12.01	13.73
15	1.15	2.28	3.43	4.58	5.71	6.86	8.02	9.14
20	0.86	1.72	2.58	3.43	4.30	5.15	6.01	6.86

* values in ppm

Note: The sampling period specified in the directions corresponds to concentrations which may be encountered in ambient air conditions. High concentrations of pollutants in the atmosphere will require shorter sampling periods, while low concentrations will require longer sampling periods.

USE OF THE OCTET COMPARATOR

The Octet Comparator contains eight permanent color standards. To perform a test, the sample is inserted into one of the slots in the top of the comparator, where it can be compared to four color standards simultaneously. For optimum color comparison, the comparator should be positioned between the operator and light source, so the light enters through the special diffusing screen on the back of the comparator. Avoid viewing the comparator against direct sunlight or an irregularly lighted background.

SAFETY INFORMATION

The Ammonia Nitrogen Absorbing Soution and the Ammonia Nitrogen Reagent #2 are considered hazardous substances. Material Safety Data Sheets (MSDS) are available for these reagents. For your safety, read the labels and accompanying MSDS before using.

QUESTIONS AND ANSWERS

1. How can industry control ammonia emissions?
Ammonia gas has a high volatility rate. The scrubber technology and equipment to trap the gas and control emissions are fairly expensive to install. Bubbling ammonia into cooled water at a slow rate and reacting the ammonia in solution with other chemicals is probably the most effective method of controlling emission costs. Alternatively, reacting gaseous ammonia in a closed environment can be an effective means of controlling emissions.
2. What are some general characteristics of ammonia in the atmosphere?
Ammonia is a base and tends to neutralize acidic compounds in the atmosphere and in the soil. Ammonia has an affinity for water and easily reacts with and clings to water droplets.
3. What are the sources of ammonia?
Sources of ammonia are accidental releases from many different types of anhydrous ammonia equipment and vehicle accidents, evaporating ammonium hydroxide and ammonia solutions, coke and ammonia manufacturing facilities, refrigeration systems, chemical manufacturing and laboratory waste, animal waste decay, sewage treatment processes, and biochemical processes. Ammonia-type solvents are often used as cleaning solvents and reaction solvents. Some volatilization of ammonia from the ground occurs after injection of anhydrous ammonia between corn rows on farms.
4. What pH indicator is often used to determine ammonia concentrations in air?
Bromophenol blue reacts with ammonia vapors to form a blue reaction product in the Dräger tube or similar types of tubes.

REFERENCES

1. Stanley E. Manahan, *Environmental Chemistry,* 6th Ed. (Boca Raton, Fla.: Lewis Publishers, 1994).
2. Roger N. Reeve, *Environmental Analysis.* (ACOL), John D. Barnes (ed.) (New York, N.Y.: John Wiley & Sons, 1994).
3. *Dräger Tube Handbook.* 11th Ed. (Lübeck: Dräger Sicherheitstechnik GmbH., 1998).
4. *Air Quality Sampling & Measurement Equipment* (Chestertown, Md.: LaMotte Company, 1997).
5. *Ammonia in Air Test Kit* (Chestertown, Md.: LaMotte Company, 1996).

CARBON MONOXIDE

Carbon monoxide, CO, is an odorless, colorless gas and a poisonous inhalant. It is a toxicant which binds in the place of oxygen in the hemoglobin structure and prevents hemoglobin from carrying oxygen to the body tissues. In industry, the upper limit of exposure levels during an 8-hour period for healthy persons within certain age ranges is about 50 parts per million (ppm). Carbon monoxide exposure at 100 parts per million (ppm) causes headache, dizziness, drowsiness, irregular heartbeat, and weariness. Loss of consciousness occurs at 250 ppm and inhalation at 1000 ppm results in rapid death. Chronic long-term exposure to low levels of CO causes disorders of the heart and the respiratory system.

The concentration of CO in the earth's atmosphere is about 0.1 ppm. Much of the CO is formed as an intermediate during the oxidation of methane[1]. Most of the carbon monoxide pollution is the result of incomplete combustion of carbonaceous materials (including fossil fuels) from transportation and stationary sources. Concentrations of carbon monoxide may reach hazardous levels in poorly vented garages, tunnels, and in closed buildings. Inefficiently operating furnaces and vehicle exhausts can be a significant problem in poorly ventilated areas. Degradation of chlorophyll during the autumn months releases a considerable amount of CO. Additional sources of carbon monoxide in the atmosphere can be attributed to decay of plant materials, specific living plant organisms, forest and structure fires, burning of rubber tires and other materials, chemical reactions and processes, and unknown sources.

Carbon monoxide emissions in a home or any building are attributed to faulty furnaces, kerosene heaters, woodstoves, fireplaces, unvented gas stoves, and vehicle emissions[2]. The health risks from exposure to this chemical are magnified because people spend at least 70% or more of their time indoors. In addition, carbon monoxide levels inside of cars in traffic-clogged urban areas are many times higher than those in rural areas. Carbon monoxide levels from vehicles operating inside garages directly under a house can be a major problem for the residents.

Carbon monoxide emissions from internal combustion engines in congested urban areas may become as high as 50 to 100 ppm. Carbon monoxide emissions of vehicles can be significantly lowered by employing a lean fuel mixture approaching air-fuel (weight:weight) ratios of 16:1. Modern automobiles use a catalytic exhaust system to convert emissions from carbon monoxide to carbon dioxide.

The lifetime of carbon monoxide in the atmosphere is about 4 months. Carbon monoxide is removed from the atmosphere by reaction with hydroxyl radical, HO·, to produce carbon dioxide and a sequence of reactions to yield radical products[1]. In addition, specific types of fungi, soil microorganisms and other species tend to remove CO from the atmosphere.

Analysis for carbon monoxide often includes non-dispersive infrared spectrometry and flame ionization gas chromatography. However, the equipment in these methods are expensive and the procedures are often replaced with other experiments. The test procedures to detect the various levels of carbon monoxide in air are provided in an experiment titled "Measurement of Air Contaminates." During the experiments, each tube measures contaminates at different concentrations and with specific criteria.

The amount of carbon monoxide in a number of environments can be determined by Dräger gas detector tubes. The tube reaction converts iodine pentoxide to iodine by reaction with carbon monoxide. The contents of the tubes provide a color change of the media from white to blue-green or brown-green at a specific concentration of contaminate. Carbon monoxide can be measured in various short-term tubes at 2 to 60 ppm, 100 to 700 ppm/5 to 150 ppm, 8 to 150 ppm, 100 to 3000 ppm/10 to 300 ppm, 10 to 250 ppm, 0.01 to 0.3 volume % and 0.001 to 0.03 volume %, 0.3 to 7 volume %, and 200 to 2500 ppm carbon monoxide and 2 to 12 vol% CO_2.

Olefins can interfere with the Dräger tube reactions. Many of the gases and vapors that may cause interferences can be adsorbed by activated carbon in a pretube. Some compounds such as acetylene react with the tube media with a different sensitivity.

Carbon monoxide can also be determined by reaction with I_2O_5 and $SeO_2/H_2S_2O_7$ to form a brown stain during "long-term measurements" of 1 hour, 2 hours, 4 hours, or 8 hours. Acetylene interferes with the measurement. A constant low flow pump can be used for long-term measurements to determine the time weighted average concentration of the air pollutant. In addition, "Diffusion Tubes with Direct Indications" perform long-term CO measurements by allowing air or vapor to diffuse through the tube and react with a palladium salt to form a gray black palladium indicator mark on the tube. A Dräger tube can also measure the CO content in compressed air from pressurized containers if the pressure of the gas is somewhat reduced to an appropriate level.

Also, carbon monoxide can be determined by the LaMotte Method. This method bubbles air containing CO into an absorbing solution at a rate of 1.0 Lpm for an appropriate time. The LaMotte Method requires the use of an axial reader for the measurement of faint color reactions. The color of the unknown sample is matched to a set of color standards. The color match provides an index number which can be used to determine the concentration of carbon monoxide in parts per million (ppm).

In the LaMotte experiment, a blank determination is subtracted from the test measurement. For example, after sampling the atmosphere for 30 minutes and chemically developing a yellow solution, an index reading of 6.0 is observed on the comparator. From the calibration chart, the carbon monoxide concentration would be 120 ppm for the sample. If the 30 minute blank determination resulted in an index of 1.0, then the concentration of the blank is 20 ppm CO. Therefore, the final concentration of CO would be 100 ppm.

CARBON MONOXIDE

Procedure*

1. Pour 10 mL of Carbon Monoxide Absorbing Solution into impinging tube.

2. Connect impinging apparatus to intake of air sampling pump. *Note:* Make sure the long tube is immersed in the absorbing solution.

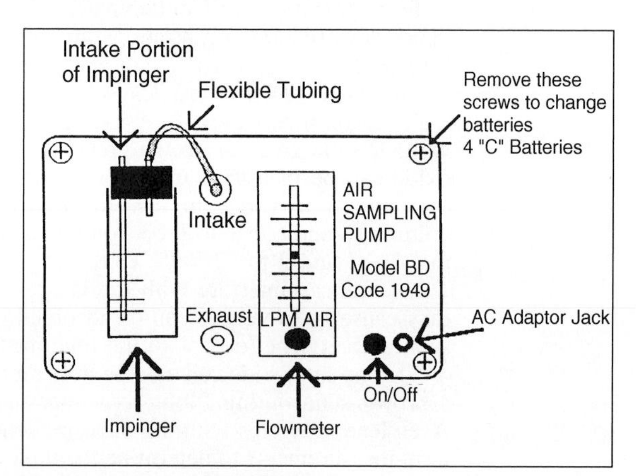

3. Adjust flowmeter to collect air at a rate of 1.0 Lpm for 30 minutes or until a measurable amount of carbon monoxide is absorbed.
Note: A measurable amount of carbon monoxide is indicated by the development of a yellow color in the absorbing solution.

---

*Compliments of LaMotte Company.

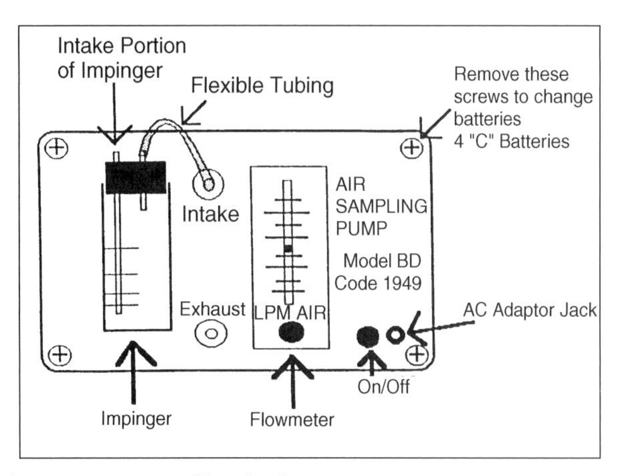

4. At the end of the sampling period, pour contents of impinging apparatus into a clean test tube.

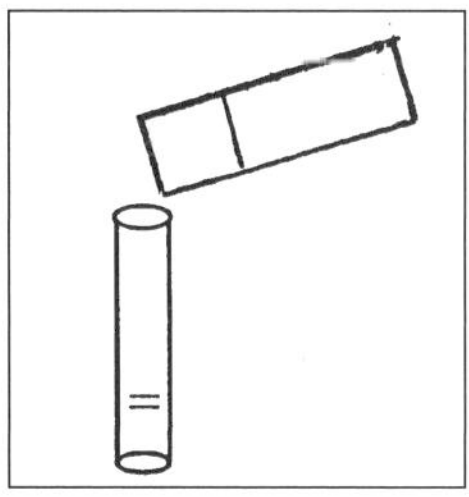

5. Insert test tube into Carbon Monoxide Comparator with the Axial Reader. Match sample color to color standards. Record the index number which gives the proper color match.

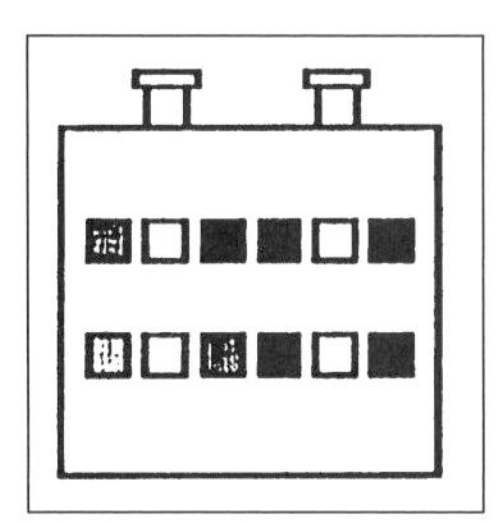

6. Use chart to convert index reading to a concentration. Record as ppm carbon monoxide.

Carbon Monoxide in Air Calibration Chart*

Time (min.)	Comparator index number 1	2	3	4	5	6	7	8
10	33	67	100	133	166	200	233	267
20	25	50	75	100	125	150	175	200
30	20	40	60	80	100	120	140	160
40	17	34	51	68	83	100	117	134
60	12.5	25	37.5	60	62.5	75	87	100

* values in ppm

Note: The sampling period specified in the directions corresponds to concentrations which may be encountered in ambient air conditions. High concentrations of pollutants in the atmosphere will require shorter sampling periods, while low concentrations will require longer sampling periods.

SAFETY INFORMATION

Carbon Monoxide Absorbing Solution is a hazardous substance. A Material Safety Data Sheet (MSDS) is supplied for this reagent. For your safety, read the label and the accompanying MSDS before using the materials. **Read the Axial Reader Instruction Manual before proceeding.**

QUESTIONS AND ANSWERS

1. What levels of CO cause problems to humans?
   Carbon monoxide can cause memory problems and visual perception problems at 10 ppm. Exposure at 100 ppm causes headache, dizziness, drowsiness, irregular heartbeat, and weariness. Loss of consciousness occurs near 250 ppm and inhalation at 1000 ppm causes death.
2. What are the sources of carbon monoxide?
   Carbon monoxide is generated by incomplete combustion of fossil fuels and other carbonaceous materials. Forest fires, material decay, chlorophyll degradation, specific organisms, vehicle engines, burning of buildings and materials, and many other activities are sources of carbon monoxide.
3. How is carbon monoxide depleted or consumed on earth?
   Carbon monoxide can have a varied life which can extend to several months. Carbon monoxide reacts with oxygen to form carbon dioxide. Also, CO reacts with the hydroxyl radical to produce carbon dioxide in the atmosphere. The byproducts of the reaction can react further to produce additional radical products in the atmosphere. Additional reactions with chemicals in the environment are very limited. Specific types of fungi, soil microorganisms and other species consume carbon monoxide from the atmosphere.
4. How does carbon monoxide harm a person?
   Carbon monoxide bonds to hemoglobin in the place of oxygen and prevents hemoglobin from carrying oxygen to the body tissues. Thus, the body starves for oxygen and death can result.

REFERENCES

1. Stanley E. Manahan, *Environmental Chemistry,* 6TH Ed. (Boca Raton, Fla.: Lewis Publishers, 1994).
2. G. Tyler Miller, *Living in the Environment,* 9th Ed. (Florence, Ky.: Wadsworth Publishing Co., 1999).
3. Roger N. Reeve, *Environmental Analysis,* (ACOL), John D. Barnes (ed.), (New York, N.Y.: John Wiley & Sons, 1994).
4. *Dräger Tube Handbook,* 11th Ed. (Lübeck: Dräger Sicherheitstechnik GmbH., 1998).
5. *Air Quality Sampling & Measurement Equipment* (Chestertown, Md.: LaMotte Company, 1997).
6. *Carbon Monoxide in Air Test Kit* (Chestertown, Md.: LaMotte Company, 1998).

CHLORINATED HYDROCARBONS

Halogenated hydrocarbons of trichloroethane, trichloroethylene, perchloroethylene, and other chlorinated organic compounds have extensive applications as cleaning solvents and as degreasing solvents used in industry. The solvents are sometimes used as cleaning agents in the home, laundry facilities, garages, and offices.

Perchloroethylene, also known as tetrachloroethylene, is a colorless liquid with an ether-like odor. The nonflammable solvent is extremely stable and insoluble in water. The compound is moderately toxic and irritates the eyes and skin. Human tolerance level is 100 ppm in air. Tetrachloroethylene has applications as a dry-cleaning solvent, vapor-degreasing solvent, drying agent for solids and metals, veterinary medicine to eliminate intestinal worms, and a heat transfer medium. The solvent is used extensively in the dry-cleaning industry. After clothes are dry cleaned, the fumes from the dry-cleaning solvent remains on the clothes for an extended period of time. The chlorinated organic compound can cause nerve disorders and damage to the liver and kidneys, and is possibly carcinogenic. Tetrachloroethylene is one of the most important indoor air pollutants in the home.

The solvent, 1,1,1-trichloroethane, is insoluble in water and nonflammable. The compound is irritating to the eyes and tissue. Humans have a tolerance level of 350 ppm in air. The solvent is used for cleaning precision instruments, metal degreasing, pesticide manufacturing, textile processing, and tool and die applications. This chemical is used in industry and around the home as a common aerosol spray. The chlorinated compound can cause dizziness and irregular breathing when inhaled. This chemical is an important indoor air pollutant in the home.

The compound, 1,1,2-trichloroethane, is a nonflammable solvent and is insoluble in water. The toxic compound is an irritant that is easily absorbed by the skin. The human tolerance level is 10 ppm in air. The solvent is used for isolation of fats, oils, waxes, resins, and other products and in organic synthesis. Generally, this chemical is seldom used in the home.

Trichloroethylene is nonflammable and slightly soluble in water. The stable liquid has a chloroform-like odor and is a photoreactive substance that is toxic by inhalation. Human tolerance level is 50 ppm in air. Specific applications of trichloroethylene are not permitted in some states. The Food and Drug Administration (FDA) has prohibited its use in foods, drugs and cosmetics. However, trichloroethylene has applications during metal degreasing, extraction solvent of materials, solvent dyeing, heat exchange liquid and refrigerant, diluent in paints and adhesives, textile processing, aerospace operations, and as a chemical intermediate. At present, trichloroethylene cannot be used for decaffeination of coffee in the USA. However, the solvent is still used for coffee decaffeination in several foreign countries and the coffee is often shipped to many different countries.

Landfill gases measured from samples at more than 40 landfills contain averages of methylene chloride at 21,200 parts per billion (ppb), tetrachloroethylene (perchloroethylene) at 7810 ppb, 1,1,1-trichloroethane at 410 ppb, and trichloroethylene at 3650 ppb[1]. Dry-cleaning operations are excellent sources of tetrachloroethylene at 5000 to 16,500 parts per billion (ppb) by volume. Vapors from chlorinated solvents used by laundries to remove spots on clothes can volatilize into the air and impact apartments, food markets, and restaurants located nearby[2]. A number of tests for vapors from these chlorinated solvents can result in improved control of fugitive emissions from chlorinated solvents.

By comparison, a number of hazardous waste combustors have been tested for stack emissions of trichloroethane, trichloroethylene, perchloroethylene, and other chlorinated organic compounds[1]. Hazardous waste incinerators produce 2.8 ppb (parts per billion) of perchloroethylene by volume measured in the stack before dispersion. In MACT hazardous waste combustors, several additional compounds have been measured in the stacks before dispersion including 1,1,1-trichloroethane at 15.9 ppb, trichloroethylene at 4.1 ppb and other compounds which tested below 5.11 ppb. A study of concentrations of many chlorinated and nonchlorinated organic compounds indicates that the emissions in the stack of hazardous waste incinerators are significantly less toxic than the emissions of backyard burning, tire fires, wood burning, print shop, beauty salon, auto repair shops, paint mixing facilities, furniture refinishing, bathroom remodeling, dry cleaners, firearm cleaning, and wood stoves.

Perchloroethylene vapors within the standard measuring range of 0.5 to 4 ppm/0.1 to 1ppm and 20 to 300 ppm/2 to 40 ppm can be determined by observation of tube color change from yellowish white to gray-blue in a Dräger "short-term" tube. Permanganate ion oxidizes the chlorinated organic molecule to form chlorine, which reacts with diphenylbenzidine to produce a gray-blue reaction product. Interferences are halogenated hydrocarbons and free halogens during this measurement. Also, petroleum ethers vapors can reduce the perchloroethylene indications at higher concentrations.

Perchloroethylene at a concentration of 10 to 500 ppm in air can be determined by observing the development of an orange reaction product within the temperature range of 15 to 40° C in a short-term tube. After oxidation of the chlorinated hydrocarbon by permanganate ion, immediate reaction of o-tolidine with liberated chlorine produces the indicator product in the tube. This test also indicates halogenated hydrocarbons and free halogens. Also, petroleum vapors can reduce the perchloroethylene indication.

In addition, reactions of MnO_2 with perchloroethylene produces chlorine, which reacts with diphenylbenzidine to form a brown-blue reaction product. This brown-blue product identifies the concentration of perchloroethylene in the ranges of 500 to 10,000 ppm and 50 to 600 ppm during "short-term" tests. This measurement, which determines the concentration of perchloroethylene in the air sample, is temperature-dependent as noted on p. 194, Drager-Tube Handbook, 11th Ed., 1998. In this case, other chlorinated hydrocarbons, free halogens and mineral acids can also interfere with the tests.

During "long-term" measurements, the reaction of perchloroethylene with chromium (VI) to form HCL, followed by a reaction of the acid with bromophenol produces a yellowish white reaction product in a Dräger tube. This reaction sequence will determine the amount of perchloroethylene present in air by using prepared tubes during long-term measurements of 1 hour, 2 hours, or 4 hours. During the measurement, a constant low flow pump will be used to determine the time weighted average concentration of the air pollutant. Other chlorinated hydrocarbons such as trichloroethylene and 1,1-dichloroethylene will interfere with the determination of perchloroethylene.

Direct Indicating Diffusion Tubes can be used to perform long-term measurements over an 8-hour period. Perchloroethylene vapors react with chromium (VI) to form chlorine gas, which reacts with o-tolidine to produce a yellow-orange product on the tube medium. Various ranges of product concentrations are indicated on the tube during 1 hour, 2 hours, 4 hours, and 8 hours of tube reaction. Additional chlorinated hydrocarbons are indicated at different sensitivities. Trichloroethylene and 1,1,1-trichloroethane are indicated with nearly the same sensitivity.

Vapors of 1,1,1-trichloroethane can be detected at 50 to 600 ppm after tube oxidation with an acidified oxidizing agent followed by reaction with o-tolidine in a "short-term" Dräger tube. The brown-red reaction product will indicate the concentration of thc 1,1,1-trichloroethane in air. Additional chlorinated hydrocarbons are indicated at different sensitivities.

Trichloroethylene present in air at 20 to 250 ppm/2 to 50 ppm and 50 to 500 ppm combines with chromium (VI) ion on media in "short-term" Dräger tubes to form chlorine, which reacts directly with o-tolidine to form an orange reaction product. Again, other halogenated hydrocarbons are indicated at different sensitivities. Petroleum hydrocarbons can cause low readings. Free halogens and hydrogen halides interfere in these measurements. The amount of trichloroethylene in the air sample can be read directly from the tube.

"Long-term" measurements of trichloroethylene involve the same reaction principles as "short-term" measurements. The standard range of measurement varies with time of 1 hour, 2 hours, and 4 hours. The presence of other chlorinated hydrocarbons interferes with the determination of trichloroethylene. In addition, "Direct Indicating Diffusion Tubes" can be used to determine various concentrations of trichloroethylene levels at 1 hour, 2 hours, 4 hours, and 8 hours. The indicating layer in the tube changes from a white to a yellow-orange color during the gas reaction. Interferences by other chlorinated hydrocarbons can be indicated at different sensitivities.

All procedures for the "short-term" test are provided in an experiment titled "Measurement of Air Contaminates." The procedures for "long-term" tubes are an adaptation of the "Measurement of Air Contaminates" experiment. The tubes are open at both ends during both types of measurements. The detailed procedures for each experiment are enclosed with the equipment and the specific tubes in the cases. Also, "long-term" tubes require a constant low flow pump calibrated at 10 to 20 cc/minute to perform measurements of 1 hour or more.

Direct Reading Diffusion Tubes can be used for long-term measurements. However, the diffusion tubes are open on one end during measurements and do not require a pump or additional equipment. The detailed directions are include within the tube case.

QUESTIONS AND ANSWERS

1. Which one of the highly chlorinated solvents has been forbidden for use in foods, drugs and cosmetics by the FDA?
The Food and Drug Administration has forbidden the use of trichloroethylene in food, drugs and cosmetics.

2. What are the human tolerance levels for the major chlorinated hydrocarbons?
Tolerance levels for humans are: 350 ppm in air for 1,1,1-trichloroethane; 100 ppm in air for perchloroethylene; 50 ppm in air for trichloroethylene; and 10 ppm for 1,1,2-trichloroethane.

3. During the tube determination of a specific chlorinated hydrocarbons, what are the major interferences?
Similar compounds contain the same reactive groups even though the molecular weights vary. Chlorinated hydrocarbons are interferences and are sometimes indicated at nearly the same sensitivities. The interferences for each air pollutant is provided in the Cross Sensitivity Section of the Dräger-Tube Handbook.

4. Which of the chlorinated hydrocarbons has the greatest rate of usage?
Perchloroethylene has significant uses in dry cleaning, vapor-degreasing, drying agent for solids and metals, heat transfer medium, and veterinary medicine. Trichloroethylene has uses in metal-degreasing, extraction solvent, solvent dyeing, dry cleaning, heat exchange liquid and refrigerant, paint and adhesive diluent, textile processing, aerospace operations, chemical intermediate, and general solvent. Trichloroethylene probably has the greatest amount of application in industry, however perchloroethylene has the greatest usage rate in laundromats.

REFERENCES

1. James Cudahy, Chris McBride, Floyd Hasselriis, et al., *Comparison of Hazardous Waste Incinerator Trace Organic Emissions With Emissions From Other Common Sources,* Incineration and Thermal Treatment Technologies Conference, Portland, OR., May 10, 2000.
2. Jack Lauber, P.E., Consultant, Latham, New York.
3. Stanley E. Manahan, *Environmental Chemistry,* 6th Ed. (Boca Raton, Fla.: Lewis Publishers, 1994).
4. *Fate of Pesticides and Chemicals in the Environment,* J. L. Schnoor (ed.). (New York, N.Y.: John Wiley & Sons, Inc., 1992).
5. *Hawley's Condensed Chemical Dictionary,* 11th Ed., rev'd N. Irving Sax and Richard J. Lewis (New York, N.Y.: Van Nostrand Reinhold, 1987).
6. *Dräger Tube Handbook,* 11th Ed., (Lübeck: Dräger Sicherheitstechnik GmbH., 1998).

CHLORINE

Chlorine is a poisonous and toxic gas which irritates mucous membranes. The human tolerance level of chlorine has been established at 1 ppm in air. The substance is a strong oxidizing agent that is slightly soluble in cold water. Human exposure to 10 to 20 ppm of chlorine gas in air irritates the respiratory tract of humans. Human exposure to 1000 ppm of chlorine can be fatal. Chlorine can form a number of dangerous materials when in contact with turpentine, ether, common hydrocarbons, ammonia, hydrogen, powdered metals and reducing materials.

Chlorine gas has been used to manufacture carbon tetrachloride, trichloroethylene, chlorinated hydrocarbons, neoprene, polyvinyl chloride, ethylene dichloride, hydrogen chloride, hypochlorous acid, bleach solutions, metallic chlorides, chloroacetic acid, and chlorobenzene. Chlorine gas has been used in chlorinated lime, water and wastewater treatment, shrinkproofing wool, flame-retardant compounds, and special batteries. It has also been used to prepare chlorinated organics such as PCB, chlorinated phenols, dioxins, and many other compounds. Chlorine has applications in processing meat, fish, vegetables and fruit.

Accidents of vehicles carrying chlorine gas have resulted in a number of deaths attributed to inhaling of the gas. Confinement in enclosed spaces and in improperly ventilated areas with a leaking tank or leaking supply lines can be a major life threatening problem. Chlorine leaks at industrial sites, water and wastewater treatment facilities, or the local swimming pool are potential problems Chlorine dissolves in atmospheric water droplets to form hypochlorous acid and hydrochloric acid.

Chlorine gas can be measured by various types of electrochemical sensing personal monitors. The analyte gas reacts at an electrode to produce a current proportional to its gas-phase concentration. The monitors consist of a concentration display and an audible alarm to alert individuals to a specific dangerous exposure level.

A Dräger experiment to detect levels of chlorine gas in air by indicator tubes is provided under the title of "Measurement of Air Contaminates." A specific reactant in the tubes indicates the amount of chlorine gas in the air. Each tube has a specific standard deviation level for the measurement.

Gaseous chlorine can be determined by Dräger tubes at concentrations in the ranges of 0.2 to 3 ppm, 0.3 to 5 ppm, and 50 to 500 ppm. During short-term measurements, the o-tolidine in the tube reacts with chlorine vapors to form a yellow-orange reaction product at low concentration levels of chlorine gas or a dark brown product at high concentrations of chlorine. Interferences can be bromine, chlorine dioxide and nitrogen dioxide which react to provide a pale discoloration of the tube medium. In most cases, the tube sensitivity will vary due to the interferences.

In another type of Dräger tube, gaseous chlorine can also react with o-tolidine to form a yellow-orange reaction product during a long-term measurement. A constant low flow pump will be used in the experiment to determine the time weighted average concentration of thc air pollutant. As the time of gas measurement varies between 1 hour and 8 hours, the standard range of measurement (ppm) changes significantly. Bromine and nitrogen dioxide interfere with the measurement of chlorine.

Chlorine in air can also be determined by the LaMotte Method. Air containing chlorine gas is bubbled into a water solution containing a small amount of sodium hydroxide at 1.0 Lpm for a specific period of time. After diluting the sample with distilled water to exactly 10 mLs and adding a Chlorine DPD tablet, the sample is shaken until the tablet dissolves. If chlorine is available, the solution will turn pink at low chlorine concentrations and red at higher levels of chlorine. The quantity of unknown in the test tube can be determined by matching the sample color with an index of color standards in a Chlorine in Air Comparator. After recording the index number from the Comparator, the concentration of ppm chlorine can be obtained from the Chlorine in Air Calibration Chart by reading the number at the intersection of the Comparator Index number and the specific time required to collect the air sample at 1.0 Lpm.

CHLORINE

Procedure*

1. Fill a clean impinging tube to 10-mL line with deionized water.

2. Add 2 drops of 6% Sodium Hydroxide to the contents of the impinging tube and mix. Connect impinging apparatus to intake of air sampling pump. Make sure the long tube is immersed in the absorbing solution.

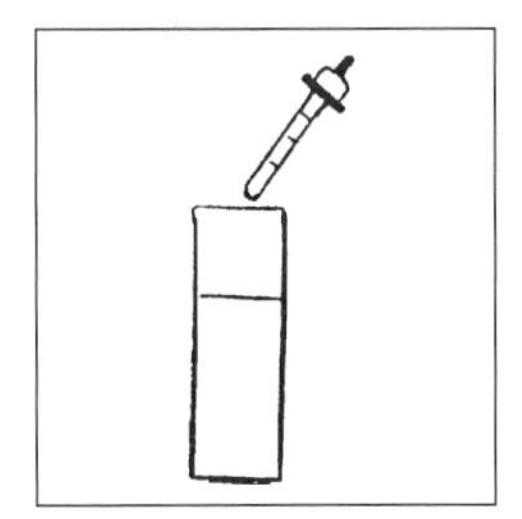

3. Adjust flowmeter of sampling pump to collect air at a rate of 1.0 Lpm. Sample for 30 minutes or until a measurable amount of chlorine is absorbed.

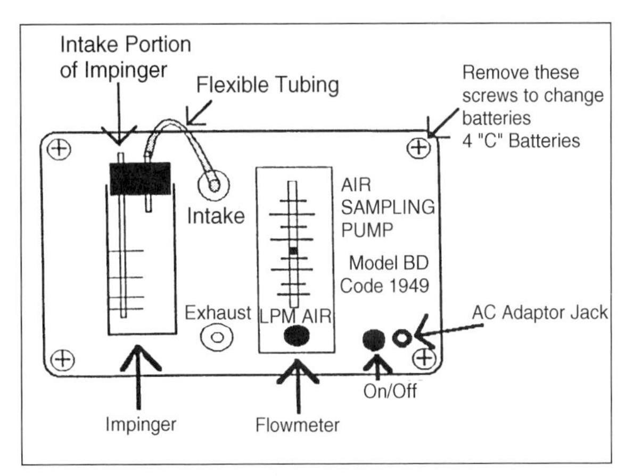

*Compliments of LaMotte Company, Chestertown, Md.

4. At the end of the sampling period, disconnect impinging tube and pour contents into a clean test tube. Dilute to 10-mL line with deionized water, if necessary, to replace any liquid lost through evaporation.

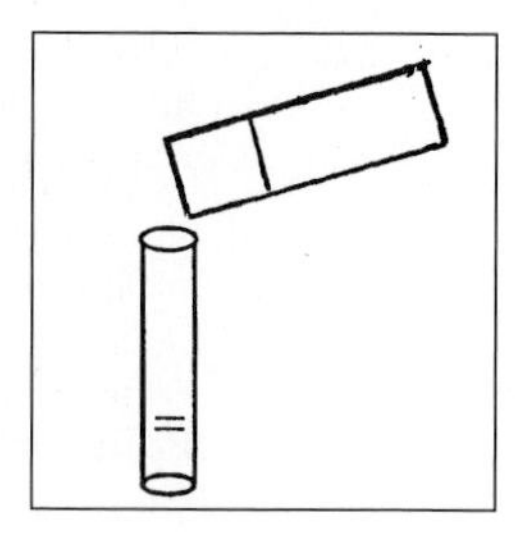

5. Add one Chlorine DPD Tablet #4 to the test tube. Cap and shake until tablet is dissolved. Solution will turn pink or red if chlorine is present.

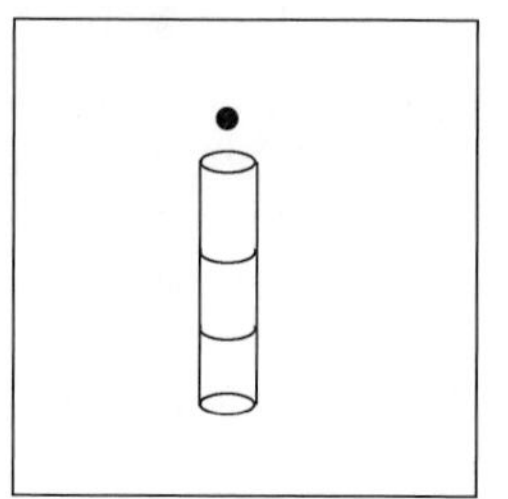

6. Insert test tube into Chlorine in Air Comparator. Match sample color to an index color standard. Record index number.

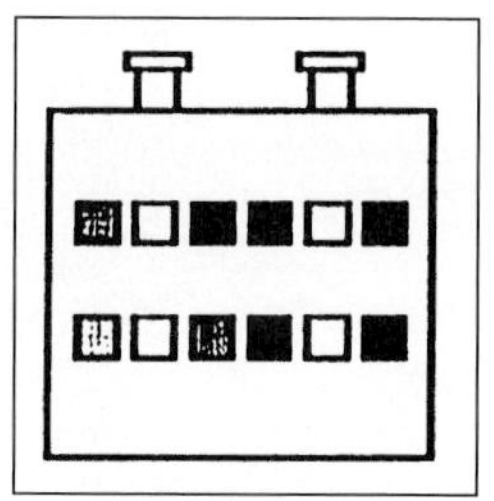

7. Use chart below to convert index reading to a concentration. Record as ppm chlorine.

Chlorine in Air Calibration Chart*

Time (min)	Comparator index number							
	1	2	3	4	5	6	7	8
15	0.04	0.09	0.13	0.18	0.22	0.33	0.44	0.67
30	0.02	0.04	0.07	0.09	0.11	0.17	0.22	0.33
60	0.01	0.02	0.03	0.04	0.06	0.08	0.11	0.17
120	0.005	0.01	0.02	0.02	0.03	0.04	0.06	0.08

* values in ppm

Note: The sampling period specified in the directions corresponds to concentrations which may be encountered in ambient air conditions. High concentrations of pollutants in the atmosphere will require shorter sampling periods, while low concentrations will require longer sampling periods.

USE OF THE OCTET COMPARATOR

The Octet Comparator contains 8 permanent color standards. To perform a test, the sample is inserted into one of the slots in the top of the comparator, where it can be compared to 4 color standards simultaneously. For optimum color comparison, the comparator should be positioned between the operator and light source, so the light enters through the special diffusing screen on the back of the comparator. Avoid viewing the comparator against direct sunlight or an irregularly lighted background.

SAFETY INFORMATION

The 6% Sodium Hydroxide Solution and Chlorine DPD #4 Tablets are hazardous substances. Material Safety Data Sheets (MSDS) are supplied for these reagents. For your safety, read the labels and accompanying MSDS before using the chemicals.

QUESTIONS AND ANSWERS

1. Why is chlorine gas so dangerous?
   Chlorine gas is a greenish yellow gas with a pungent irritating odor, and is both toxic and an irritant. The gas reacts vigorously with turpentine, ether, ammonia, hydrocarbons, hydrogen, powdered metals and many reducing materials. Chlorine is an extremely strong oxidizing agent. The chemical can cause severe breathing problems and extensive property damage.
2. Why is chlorine used in swimming pools and in drinking water?
   A small amount of the strong oxidizing agent will kill harmful bacteria and oxidize materials that may cause disease or other problems.
3. What happens if chlorine vapors are accidentally mixed with ammonia gases?
   Chlorine mixed with ammonia can easily produce chloroamines. Chloroamines can easily form in the vapor state as well as in the liquid state. The chloroamine gases can be deadly to an individual without proper ventilation.
4. When chlorine escapes into the atmosphere, what are the products formed?
   Chlorine combines with atmospheric water droplets to form hypochlorous acid and hydrochloric acid.

REFERENCES

1. Stanley E. Manahan, *Environmental Chemistry,* 6th Ed. (Boca Raton, Fla.: Lewis Publishers, 1994).
2. Roger N. Reeve, *Environmental Analysis,* (ACOL), John D. Barnes (ed.) (New York, N.Y.: John Wiley & Sons, 1994).
3. *Hawley's Condensed Chemical Dictionary,* 11th Ed., rev'd N. Irving Sax and Richard J. Lewis (New York, N.Y.: Van Nostrand Reinhold, 1987).
4. *Dräger Tube Handbook,* 11th Ed. (Lübeck: Dräger Sicherheitstechnik GmbH., 1998).
5. *Air Quality Sampling & Measurement Equipment* (Chestertown, Md.: LaMotte Company, 1997).
6. *Chlorine in Air Test Kit* (Chestertown, Md.: LaMotte Company, 1996).

ALDEHYDES (FORMALDEHYDE, ACETALDEHYDE, ACROLEIN)

Formaldehyde is a colorless gas with a pungent suffocating odor. It is commonly called formalin, a 37 to 50% aqueous solution of formaldehyde containing some methanol. The compound is also commercially available in solutions containing n-butanol or urea. Inhaled formaldehyde vapors can cause hypersensitivity and irritate mucous membrane linings of both the respiratory and alimentary tracts. The toxicity of formaldehyde is often attributed to its oxidation product, formic acid. At high concentrations, formaldehyde appears to be a lung carcinogen in some types of animals.

Formaldehyde has many practical applications in everyday life. Often, solutions containing formaldehyde are dumped down the drain in research centers, laboratories, industry, ore recovery processes, laundries, and mortuaries. The volatile chemical is used to prepare various types of resins, plastics, lacquers, insulation, rubber accelerators, dyes, explosives, adhesives, chemical intermediates, ethylene glycol, disinfectants, germicides, preservatives, reducing agent for recovery of precious metals, and industrial sterilants. Formaldehyde is also used in pentaerythritol solutions, fertilizer, embalming fluids, mobile home construction, hardening agent, laboratory and industrial solvent, corrosion inhibitor, durable-press treatment of textile fabrics, and during the treatment of grain smut.

Emissions and exhaust streams containing formaldehyde can vary significantly over a short period of time from a variety of sources. Some forms of wall insulation have produced unexpected sources of formaldehyde gases. Formaldehyde spills from pipes can be easily trapped within a clay-soil layer. Formaldehyde can leach out from the soil and evaporate into the air because of its high volatility rate.

Formaldehyde in water can be easily oxidized with hydrogen peroxide to form a relatively nontoxic substance. Normally formaldehyde concentrations in water and soil should be below 100 ppm to minimize toxic effects. At this point, evaporation of formaldehyde will be minimized.

The lower molecular weight aldehydes, such as acetaldehyde and acrolein, are relatively water soluble. These compounds attack the eyes and moist mucous membrane of the upper respiratory tract. The presence of these low molecular weight aldehydes in photochemical smog produce irritating reactions in humans and animals.

Acetaldehyde is sometimes emitted by specific types of vegetation. Formaldehyde and acetaldehyde are readily produced by a variety of microorganisms. Since the carbonyl group is a **chromophore** (readily absorbs light in the near-ultraviolet region), aldehydes are significant sources of free radicals in the atmosphere. The photodissociation process is usually a two-step process that produces HCO radical, carbon monoxide and other species. Acrolein contains a double bond and a carbonyl group and is especially reactive in the atmosphere to form a variety of species during the photochemical dissociation process.

As the molecular weight of the aldehyde increases, aldehydes become less soluble in water and penetrate further into the respiratory tract and affect the lungs. Colorless liquid acetaldehyde is an irritant and acts as a narcotic to the nervous system. Acetaldehyde is highly flammable and has a fruity odor. Explosive limits in air are 4 to 57%, and the human tolerance level is 100 ppm in air.

Acetaldehyde is used primarily to manufacture acetic acid, acetic anhydride, glycols, plastics, and several additional chemicals in industry. The simple aldehyde compound is also used during the manufacture of various types of synthetic flavors. The compound volatilizes easily because of its low molecular weight.

Acrolein is an extremely irritating vapor which can cause severe damage to respiratory tract membranes. Acrolein vapor is a lachrymator which can be very hazardous to skin tissue and eyes. The compound is a dangerous fire risk and has explosive limits in air of 2.8 to 31%.

Acrolein liquid is very reactive and polymerizes readily unless inhibited. The compound is used to prepare synthetic glycerol, polyurethane, polyester resins, methionine, pharmaceuticals, and herbicides. Small amounts of acrolein are also used as a warning agent in various types of gases.

Formaldehyde in air can be determined by the LaMotte Method. The air sample is collected by a BD pump connected to a Midget Impinger. The flowmeter is adjusted to 1.0 liter per minute and the air is sampled for 30 minutes or until a measurable amount of formaldehyde is absorbed. After reacting formaldehyde with a series of specific reagents, the sample color is matched to an index of color standards in an Octet Comparator. Using a table, the index reading from the Octet Comparator is converted to a concentration value of ppm formaldehyde.

Both samples of formaldehyde and acetaldehyde in air can be determined by procedures listed in an experiment titled "Measurements of Air Contaminates." In the experiment, each tube changes medium color to indicate the concentration of the chemical component in the air sample after a set sampling time. Each measurement lists a number of interferences to influence color changes of the medium, tube response time, and ambient operating conditions.

Formaldehyde can be measured by a Dräger tube within the standard measuring ranges of 2 to 40 ppm and 0.5 to 5 ppm/0.2 to 2.5 ppm. The short-term measurements in the tubes produce a white to pink color change during the formation of the quinoid reaction products. During one of the measurements, the measuring range can be extended with a specific tube. Styrene, vinyl acetate, acetaldehyde, acrolein, diesel fuel, and furfuryl alcohol interfere to form a yellowish brown discoloration of the tube media during formaldehyde measurements.

Acetaldehyde and other short-chain aldehydes can be determined in the range of 100 to 1000 ppm in air using a medium containing Chromium (VI) in a Dräger tube applied to short-term measurements. Acrolein tends to polymerize when exposed to air and becomes non-detectable by a Dräger tube upon polymerization. The orange indicating layer changes to a brownish green color upon exposure to the different short-chain aldehydes. Esters, ethers, ketones, aromatics, and petroleum hydrocarbons are indicated at different sensitivities with the Dräger tube.

FORMALDEHYDE

Procedure*

1. Fill a clean impinging tube to 10-mL line with distilled water.

2. Connect BD Pump to Midget Impinger.
   Note: Make sure the long tube is immersed in the absorbing solution.

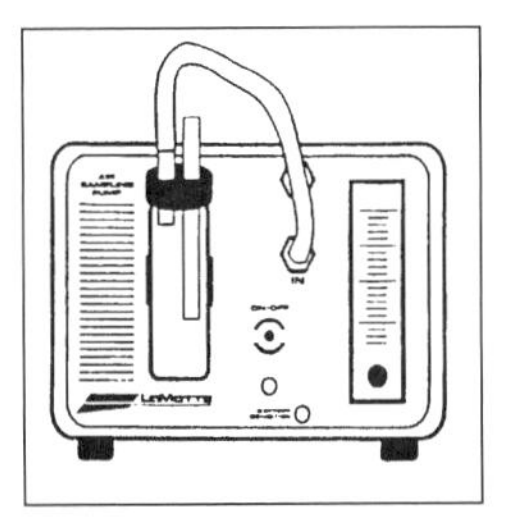

*Compliments of LaMotte Company, Chestertown, Md.

3. Adjust flowmeter to collect air at a rate of 1.0 Lpm. Sample for 30 minutes or until a measurable amount of formaldehyde is absorbed.

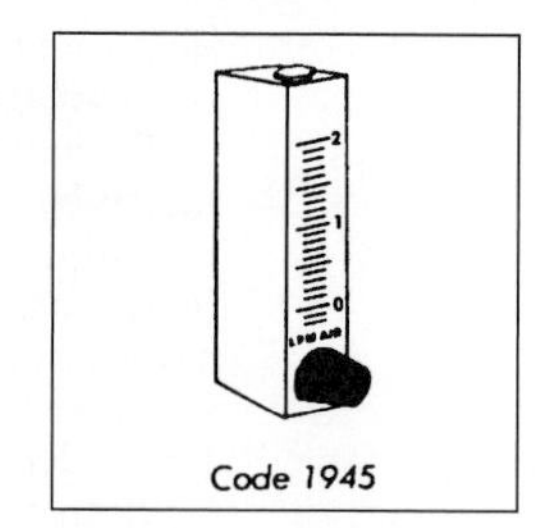

4. At end of sampling period, disconnect impinger and pour contents into clean 10-mL test tube. Dilute to mark if necessary with distilled water to replace any liquid lost through evaporation.

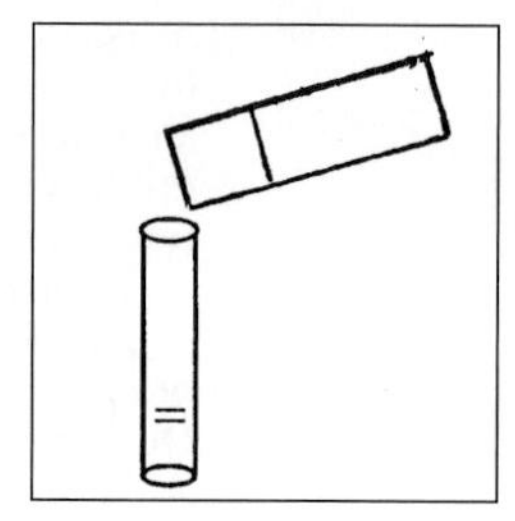

5. Carefully pour 12.5 mL of Formaldehyde Reagent #1 into the larger test tube.
   Caution: Formaldehyde Reagent #1 is highly poisonous. Read MSDS before use.

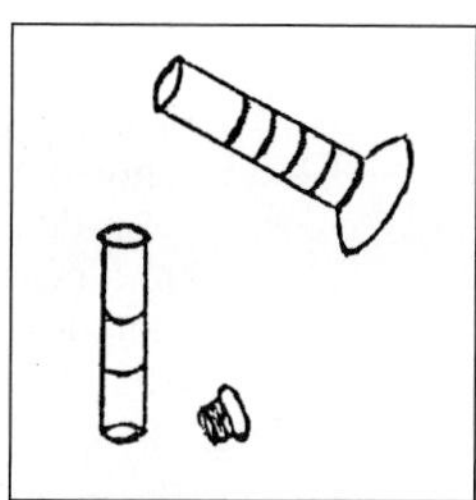

6. Use the 0.05-g spoon to add 0.05 g of Formaldehyde Reagent #2. Cap and mix until powder dissolves. This mixed reagent is stable for one day only.

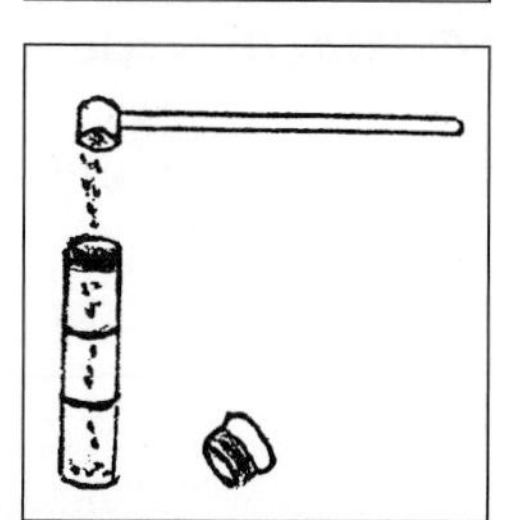

7. Use a 1-mL pipet to add 1.0 mL of reagent from Step 5 to the contents of the 10-mL test tube. Cap and mix.

8. Use a second pipet to add 1.0 mL of Formaldehyde Reagent #3. Cap and mix. Wait 20 minutes.

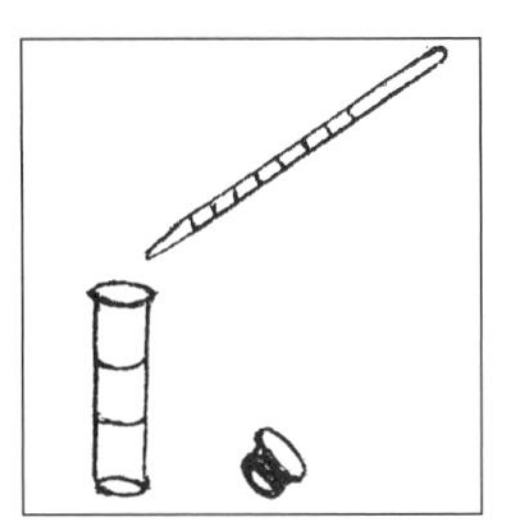

9. Insert test tube into Octet Comparator. Match sample color to index of color standard. Record the index number.

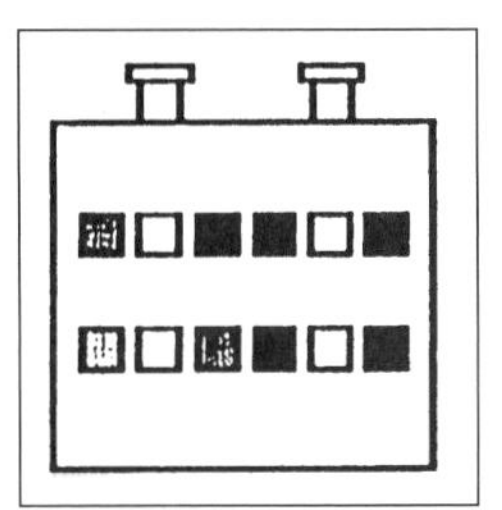

10. Use the chart below to convert index reading to a concentration. Record as ppm formaldehyde.

Formaldehyde in Air Calibration Chart*

Time (min)	Comparator reading							
	1	2	3	4	5	6	7	8
15	0	0.26	0.54	1.07	2.14	3.20	4.26	5.34
30	0	0.13	0.27	0.54	1.07	1.60	2.13	2.67
60	0	0.07	0.14	0.27	0.54	0.80	1.07	1.34
120	0	0.03	0.07	0.13	0.27	0.40	0.54	0.67

* values in ppm

Note:

A. The sampling period specified in the directions corresponds to concentration which may be encountered in ambient air conditions. High concentrations of pollutants in the atmosphere will require shorter sampling periods, while low concentrations will require longer sampling periods.

B. For applications requiring a high degree of accuracy, an external flowmeter should be used to verify the flow under actual working conditions. LaMotte Company can provide certification of the BD Pump's flow rate.

SAFETY AND WARNINGS

Formaldehyde Reagent #1 is considered to be a hazardous substance. Material Safety Data Sheets (MSDS) are supplied for all of these reagents. For your safety, read the label and accompanying MSDS before using the materials.

INTERFERENCES

To determine if there are any interferences present in the reagents, follow the procedure above on a reagent blank while eliminating Steps 2 and 3. If the amount of color which develops in the reagent blank is greater than zero, subtract this value from the test results for the actual concentration of formaldehyde.

QUESTIONS AND ANSWERS

1. How toxic are the three aldehyde-based substances discussed above?
The human tolerance level for acrolein is 0.1 ppm in air. Acrolein has a boiling point 52.7° C. Acetaldehyde has a boiling point of 20.2° C. The human tolerance of acetaldehyde is 100 ppm in air. The boiling point of an aqueous 37% solution of formaldehyde with 15% methanol is approximately 96° C and the human tolerance level is reported at 2 ppm in air. Pure formaldehyde has the smallest molecular weight and is the most volatile of the three liquids. However, most solutions containing formaldehyde in water are considerably less volatile and less toxic than the other two chemicals.
2. Which of the three liquids is the most dangerous?
Acrolein is the most reactive of the three compounds and easily polymerizes unless inhibited by hydroquinone or other chemical additives. The compound is a dangerous fire risk and has explosive limits in air of 2.8 to 31%. The compound in air can easily cause a disagreeable choking of humans and animals. Acrolein is the least soluble in water when compared to formaldehyde and acetaldehyde.
3. Which compound has been used to indicate the presence of odorless gases?
Small amounts of acrolein has been mixed with odorless gases to provide an odor for detection purposes.
4. Briefly explain the photodissociation process of acetaldehyde, formaldehyde and acrolein in sunlight?
Formaldehyde readily absorbs near-ultraviolet light and dissociates by a two-step process. The first step produces formyl radical (HCO) and an hydrogen atom (H), whereas the second step forms stable hydrogen gas and carbon monoxide. Carbon monoxide can undergo additional photochemical reactions. Acetaldehyde can participate in a similar two-step photodissociation process to form the methyl radical and formyl radical (HCO), which continue to react with other atmospheric species to form methane, carbon monoxide and several additional substances. In addition, acrolein forms a number of photochemically active substances in the atmosphere. The compound is especially photochemically sensitive due to the carbon-carbon double bond and the carbonyl group. However, acrolein has a higher boiling point than the other compounds and tends to polymerize in air. Unlike formaldehyde or acetaldehyde, acrolein will not be as available in the upper atmosphere because of its higher molecular weight and tendency to polymerization.

REFERENCES

1. Stanley E. Manahan, *Environmental Chemistry,* 6th Ed. (Boca Raton, Fla.: Lewis Publishers, 1994).
2. Roger N. Reeve, *Environmental Analysis,* (ACOL), John D. Barnes (ed.) (New York, N.Y.: John Wiley & Sons, 1994).
3. *Fate of Pesticides and Chemicals in the Environment,* J. L. Schnoor (ed.) (New York, N.Y.: John Wiley & Sons, Inc., 1992).
4. *Hawley's Condensed Chemical Dictionary,* 11th Ed., rev'd by N. Irving Sax and Richard J. Lewis (New York, N.Y.: Van Nostrand Reinhold, 1987).
5. *Dräger Tube Handbook,* 11th Ed., (Lübeck: Dräger Sicherheitstechnik GmbH., 1998).
6. *Air Quality Sampling & Measurement Equipment* (Chestertown, Md.: LaMotte Company, 1997).
7. *Formaldehyde in Air* (Chestertown, Md.: LaMotte Company, 1993).

GASEOUS FLUORINE COMPOUNDS

Gaseous fluorine compounds such as fluorine and hydrogen fluoride are dangerous substances that are generally corrosive. Of the fluorine-based gases, fluorine and hydrogen fluoride appear to be the most toxic. Brief exposure to any of the gaseous fluorine compounds may be fatal.

Additional gaseous compounds of silicon tetrafluoride, sulfur hexafluoride, sulfur tetrafluoride, sulfur monofluoride, and sulfuryl fluoride tend to decompose in moisturized air with time to form hydrogen fluoride or fluorine vapors. These fluorine-based gases are toxic to humans at higher levels. Almost all of the compounds are toxic by inhalation and are very strong irritants to body tissue. The least dangerous fluorine compound listed above is sulfur hexafluoride at 1000 ppm in air.

Aluminum metal, phosphoric acid, superphosphate fertilizer, steel and other metals are often manufactured by processes which produce hydrogen fluoride gas and silicon tetrafluoride. Coal-fired plants also have been known to produce small quantities of hydrogen fluoride near 20 ppm. Most of the fluorine by-products require expensive procedures to avoid severe pollution problems.

Sulfuryl fluoride is primarily used as an insecticide and fumigant, whereas sulfur tetrafluoride has applications as a lubricity agent and as a fluorinating agent for preparing water and oil repellants. Sulfur hexafluoride is a gaseous insulator for electrical equipment and radar wave guides. The compound is used as an atmospheric tracer. Each of these gaseous fluorine-based chemicals require special procedures and proper control to avoid unnecessary releases.

Fluoride poisoning in plants causes chlorosis, edge burn, and tip burn. Fluoride pollutants from aluminum plants in Norway have damaged areas of conifer forests. Plants can easily display the effects of gaseous fluorides at low exposure levels for a prolonged period of time.

Fluorine vapors can be detected at 0.1 to 2 ppm levels by a white to yellow change in medium color in a Dräger tube during short-term measurements. Nitrogen dioxide, chlorine and chlorine dioxide are indicated with different sensitivities.

During a short-term measurement, hydrogen fluoride can be detected at 1.5 to 15 ppm with a tube medium color change of pale blue to pink. In the presence of high humidity, some difficulty may be encountered in obtaining the reading. Other halogenated hydrocarbons do not interfere.

During a long-term measurement by a Dräger tube, hydrogen fluoride gas can react with bromophenol blue to form a yellowish reaction product. Other specific gases (HCL, CL_2, and NO_2) can interfere with the measurement, however sulfur dioxide at 5 ppm and hydrogen sulphide at 100 ppm do not interfere.

Sometimes, significant levels of the fluorine-based gases, i.e., sulfur hexafluoride, sulfuryl fluoride, etc., can decompose in moist air and be detected as hydrogen fluoride and fluorine. Some of these compounds are measured directly during off gas measurements by special equipment.

The test procedures to detect either hydrogen fluoride or fluorine in air are provide in an experiment titled "Measurement of Air Contaminates." Each tube requires a specific time for measurement of the contaminate. In a few cases, humidity levels can impact the level of detection.

QUESTIONS AND ANSWERS

1. What is the toxicity level of the more important gaseous fluorine compounds?
Sulfur hexafluoride, SF_6, has low toxicity compared to the other compounds. The human tolerance level is 1000 ppm in air. Sulfur tetrafluoride, SF_4, is highly toxic by inhalation and is a strong irritant to eyes and mucous membranes. Human tolerance level for sulfur tetrafluoride is 0.1 ppm in air. Sulfuryl fluoride, SO_2F_2, is toxic by inhalation and has a human tolerance level of 5 ppm in air. Silicon tetrafluoride is toxic by inhalation and is a strong irritant to mucas membranes. Humans have a tolerance (as F) of 2.5 mg per cubic meter of air. Sulfur monofluoride, S_2F_2, is a colorless gas with toxicity level similar to HF. However, fluorine is highly toxic by inhalation and is an extremely strong irritant to tissue. Fluorine is a strong oxidizing agent which reacts violently with inorganic and organic materials. Human tolerance level to fluorine is 1 ppm in air. Hydrogen fluoride is highly toxic by ingestion and inhalation. The compound is a strong irritant to eyes, skin and mucous membranes. Human tolerance level is 3 ppm in air. Many of the sulfur-fluorine compounds decompose in moist air to form HF and F_2. Heat tends to increase the decomposition rate and volatility rate of the compounds.

2. Which of the "fluoride" gases are the most dangerous?
Hydrogen fluoride has a light molecular weight and is a very volatile reactive acid. The human tolerance for hydrogen fluoride is 3 ppm compared to 1 ppm for fluorine. Fluorine is a powerful oxidizing agent and a dangerous fire risk in the presence inorganic and organic materials. Sulfur tetrafluoride is less volatile and humans have a lower tolerance level for the higher molecular weight compound. Sulfur monofluoride has a toxicity level similar to HF. All four gases are toxic by inhalation and are strong irritants.

3. Why are many of the fluoride gases so difficult to measure?
The fluoride gases have significantly different decomposition characteristics and are toxic at low concentrations. A few of the fluoride compounds in water vapor decompose to hydrogen fluoride and fluorine. The products of decomposition are generally varied.

REFERENCES

1. Stanley E. Manahan, *Environmental Chemistry,* 6th Ed. (Boca Raton, Fla.: Lewis Publishers, 1994).
2. *Dräger Tube Handbook,* 11th Ed., (Lübeck: Dräger Sicherheitstechnik GmbH., 1998).

3. *Hawley's Condensed Chemical Dictionary,* 11th Ed., rev'd by N. Irving Sax and Richard J. Lewis (New York, N.Y.: Van Nostrand Reinhold, 1987).
4. Roger N. Reeve, *Environmental Analysis,* (ACOL), John D. Barnes (ed.) (New York, N.Y.: John Wiley & Sons, 1994).

HYDROGEN PEROXIDE

Concentrated hydrogen peroxide solutions are highly toxic and strong irritants. The solutions are dangerous fire and explosion risks. Humans have a tolerance of 1 ppm in air to the strong oxidizing agent. Concentrated hydrogen peroxide solutions are often contaminated with impurities of iron, copper and heavy metals. An inhibitor of either acetanilide or sodium stannate is normally added to the concentrated hydrogen peroxide solution to inhibit the catalytic effects of the traces of metal impurities.

Fairly concentrated solutions of hydrogen peroxide (27.5 to 70%) are often shipped in large containers over the highways and rails. Hydrogen peroxide solutions are used to bleach and deodorize wood pulp, textiles, hair, fur and many additional items. The peroxide solutions are used for epoxidation, hydroxylation, viscosity control for starch and cellulose derivatives, cleaning and refining metals, bleaching and oxidizing foods, neutralizing agent in wine distillation, and in numerous oxidation processes. The compound is used in plasticizers, rocket fuel, foam rubber, glycerol, antichlor, dyeing, antiseptics, laboratory reagent, seed disinfectant and electroplating. Hydrogen peroxide is a substitute for chlorine in water and sewage treatment.

Hydrogen peroxide in the liquid or vapor forms has the ability to oxidize many types of cations, anions, organic species, and decaying matter. It can chemically oxidize and degrade many compounds, such as cyanides and sulfides, that are volatile and extremely toxic in wastewater treatment plants and at industrial waste sites. Hydrogen peroxide oxidizes free cyanide to cyanate, and then the cyanate slowly hydrolyses to ammonia and bicarbonate in alkaline solutions. The reaction of hydrogen peroxide with volatile sulfides forms sulfate, sulfites, and water. Hydrogen peroxide reacts with pungent forms of mercaptans and disulfides to generally form less noxious oils that are usually water insoluble. Peroxide is used to treat and destroy many toxic compounds in solutions containing phenol, creosol, benzene, aniline and several types of chlorinated phenols. For difficult-to-treat organics, hydrogen peroxide catalyzed with ferrous ion or ultraviolet light produces significant amounts of the hydroxyl radical.

The more concentrated forms of hydrogen peroxide in water can display significant volatility problems. Hydrogen peroxide is a natural source of free radicals (such as ·OH) and complexed radicals in air and in water. The radicals can effectively react with many different species including difficult-to-treat contaminants to produce less toxic substances.

Hydrogen peroxide in the air can be determined quantitatively by the LaMotte Method. After properly connecting equipment, the air is sampled at 1.0 liters per minute (Lpm) for 10 minutes or until a measurable amount of hydrogen peroxide is absorbed. Then, after disconnecting equipment and chemicals are added to the collected sample, a red solution develops with the shaking of the sample and allowing the sample to stand for two minutes. The color of the sample is matched with an index of color standards to obtain an index reading for the sample. From the calibration chart, the index value and the sampling time provide the concentration of hydrogen peroxide in air in parts per million (ppm).

HYDROGEN PEROXIDE

Procedure*

1. Fill a clean impinging tube to 10-mL line with deionized water. Connect impinging apparatus to intake of the air sampling pump. Make sure the long tube is immersed in the absorbing solution.

2. Collect air sample at a rate of 1.0 Lpm for 10 minutes or until a measurable amount of hydrogen peroxide is absorbed as is indicated by a light yellow color in the absorbing solution.

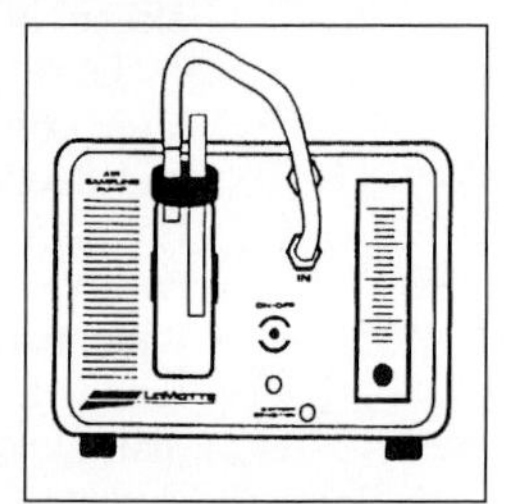

3. At the end of the sampling period, disconnect impinger and pour contents into a test tube. Dilute to 10-mL line, if necessary, with deionized water to replace any liquid lost through evaporation.

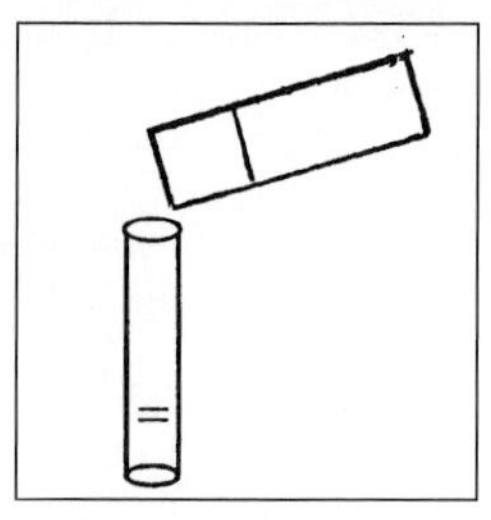

4. Add 4 drops of Hydrogen Peroxide Reagent.

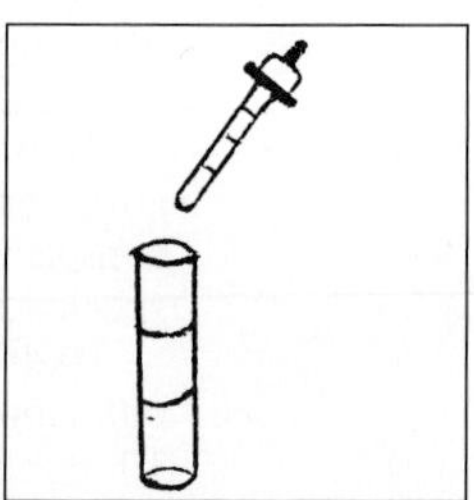

*Compliments of LaMotte Company, Chestertown, Md.

5. Add 1 hydrogen peroxide tablet to contents of tube, cap and shake to disintegrate tablet. Allow test sample to react for two minutes for full color development. A red color indicates the presence of hydrogen peroxide.

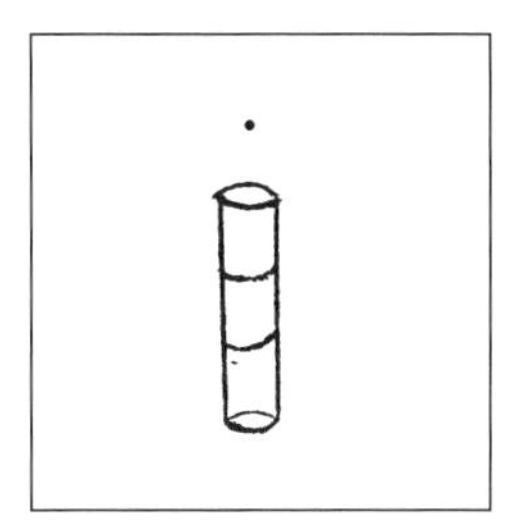

6. Place tube into comparator and compare color to index of color standards. Record the index which gives the proper color match.

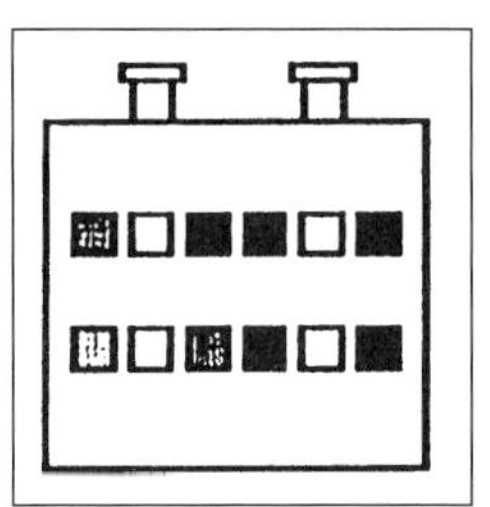

Hydrogen Peroxide in Air Calibration Chart*

Time (min)	Comparator index number 1	2	3	4	5	6	7	8
5	0.15	0.31	0.46	0.62	0.77	1.16	1.54	2.31
10	0.08	0.15	0.23	0.31	0.39	0.58	0.77	1.16
20	0.04	0.08	0.12	0.15	0.19	0.29	0.39	0.58
40	0.02	0.04	0.06	0.08	0.10	0.14	0.19	0.29

* values in ppm

Note: The sampling period specified in the directions corresponds to concentrations which may be encountered in ambient air conditions. High concentrations of pollutants in the atmosphere will require shorter sampling periods, while low concentrations will require longer sampling periods.

SAFETY INFORMATION

Hydrogen Peroxide Reagent #1 and hydrogen peroxide tablets are considered hazardous substances. Material Safety Data Sheets (MSDS) are available for these reagents. For your safety, read the labels and the accompanying MSDS before using the chemicals.

QUESTIONS AND ANSWERS

1. Why is hydrogen peroxide a potential problem?
Dilute solutions of 3% hydrogen peroxide are not a real problem unless taken internally.

However, the more concentrated solutions present a potential hazard attributed to releases of vapors and liquids.

2. What potential problems does hydrogen peroxide present?
Concentrated hydrogen peroxide solutions are highly toxic and strong irritants. The solutions are dangerous fire and explosion risks. Humans have a tolerance of 1 ppm in air to the strong oxidizing agent. Heated industrial processes using hydrogen peroxide and heated tanks containing the substance have the potential to volatilize liquid forms. In addition, the rate of oxidation reactions using hydrogen peroxide tends to increase with temperature and can present some interesting problems.

3. What are some basic applications for hydrogen peroxide?
Dilute solutions of hydrogen peroxide can kill many forms of bacteria on the body and in water and wastewater effluent. The compound has many home and industrial applications during bleaching and oxidation processes, viscosity control, cleaning, disinfectant processes, and compound preparation.

4. How can hydrogen peroxide be used to solve major problems?
Hydrogen peroxide will oxidize many toxic chemicals to form less toxic or non-toxic substances. Many poisonous or toxic substances such as cyanides, sulfides, mercaptans, disulfides, and many organic chemicals can be reacted with hydrogen peroxide to form essentially non-toxic or less toxic substances. Hydrogen peroxide can be used as a cleaner to attack specific type of molds, mildew, and rust problems.

REFERENCES

1. *Fate of Pesticides and Chemicals in the Environment,* J. L. Schnoor (ed.) (New York, N.Y.: John Wiley & Sons, Inc., 1992).
2. *Hawley's Condensed Chemical Dictionary,* 11th Ed., rev'd by N. Irving Sax and Richard J. Lewis (New York, N.Y.: Van Nostrand Reinhold, 1987).
3. *Air Quality Sampling & Measurement Equipment* (Chestertown, Md.: LaMotte Company, 1997).
4. *Hydrogen Peroxide in Air* (Chestertown, Md.: LaMotte Company, 1991).

HYDROGEN SULFIDE

Hydrogen sulfide poisoning irritates the respiratory tract and damages the central nervous system. The colorless gas is toxic by inhalation and is a strong irritant to the eyes and mucous membranes. The compound exhibits a "rotten egg" smell and is highly flammable. In some cases, inhalation of hydrogen sulfide kills faster than hydrogen cyanide. The tolerance level for humans is 10 ppm in air. Low doses of hydrogen sulfide can cause headache, dizziness, and damage to the central nervous system. Higher doses near 1000 ppm in air can cause rapid death due to asphyxiation from respiratory paralysis. Hydrogen sulfide has explosive limits in air of 4.3 to 46%.

Hydrogen sulfide and the mercaptan compounds can be produced by decaying matter. Microbiological processes reduce organic materials by the process of anaerobic decay to produce hydrogen sulfide, ammonia, and methane. In addition, hydrogen sulfide can be produced by microbial reduction of sulfate and from wood pulping operations and geothermal steam sources. Hydrogen sulfide is a by-product of petroleum refining. The recovery of sulfur from natural gas can produce significant levels of toxic hydrogen sulfide vapors. Deep well drilling for oil and natural gas can significantly increase the levels of hydrogen sulfide released in the local area. Hydrogen sulfide emissions can be converted to less toxic sulfur dioxide and sulfates in the atmosphere.

Hydrogen sulfide can be produced by rotting of certain types of vegetation without the presence of oxygen. Commonly called sewer gas, it is produced during the decomposition of sewage during oxygen deficient conditions. Hydrogen sulfide or sewer gas can back up into the home through bathtub drains, basement drains, and open water drains. A number of people have died from exposure to hydrogen sulfide as they entered large sewer lines through the manhole.

Hydrogen sulfide can destroy plant tissue, reduce plant growth, and create leaf lesions and defoliation. Paints containing lead pigments darken in its presence. Hydrogen sulfide reacts with exposed copper metal and silver metal to form black coatings of metal sulfide. With time, the metal sulfide can be oxidized to an inert layer of sulfate in air.

The test procedures to detect hydrogen sulfide in air are provided in an experiment titled "Measurement of Air Contaminates." A variety of reactants in the tubes are used to indicate the amount of hydrogen sulfide gas present at the test site. Each tube can sample at a specific temperature range and has a specific standard deviation level for the measurement. In a few cases, the reading has to be multiplied by a factor to correct for temperature.

Hydrogen sulfide can be determined by a number of Dräger tube chemistries and at a variety of concentrations. Some hydrogen sulfide detection tubes contain a colorless lead salt adsorbed on an inert medium such as silica gel. A tube of a lead compound develops a brown lead sulfide color at 0.2 to 5 ppm, 10 to 200 ppm and 1 to 20 ppm, 2 to 60 ppm, 5 to 60 ppm, and 100 to 2000 ppm. In another type of tube reaction, yellow or white mercuric compounds react with hydrogen sulfide to form a darkened colored mercuric sulfide and an acid in the standard measuring range of 0.2 to 6 ppm, 0.5 to 15 ppm, and 20 to 200 ppm/ 2 to 20 ppm. Additional short-term tubes react hydrogen sulfide with copper ion to form a brown to black color indication in the tube. Short-term tubes measuring Vol.% of hydrogen sulfide in air are also available. In each case, the printed scale on the tube allows a direct reading of the concentration of the contaminate in air. Few, if any, interferences are noted for these reactions.

During long-term measurements (1 hour to 8 hours) by a Dräger tube, hydrogen sulfide reacts with lead ion to form brown colored lead sulfide at 0 to 40° C. A constant flow pump will be used during long-term measurements to determine the time weighted average concentration of the air pollutant. In addition, "Diffusion Tubes with Direct Indications" perform long-term hydrogen sulfide measurements by allowing air containing hydrogen sulfide to diffuse through an open end of the tube. A brown discoloration of the tube medium indicates the concentration of hydrogen sulfide in the air. Interferences by other compounds are extremely limited.

Hydrogen sulfide in air can also be determined by the LaMotte Method. After the addition of Sulfide Reagent #1 to the impinging tube and preparing the equipment for analysis, the flowmeter is adjusted to collect air at a rate of 2.0 Lpm for 30 minutes, or until a measurable amount of hydrogen sulfide is absorbed. Then, various Sulfite Reagent Solutions (#2, #3 and #4) are appropriately added to the sulfide solution. After mixing, the solution develops a blue color indicating the presence of hydrogen sulfide. Sulfide Reagent #5 is added to the solution and then the amount of hydrogen sulfide in the solution is determined by a Hydrogen Sulfide in Air Comparator. After the sample color is matched to an index of color standards, the index is recorded and converted to the concentration of hydrogen sulfide in parts per million (ppm).

HYDROGEN SULFIDE

Procedure*

1. Add 7 mL of Sulfide Reagent #1 to the impinging tube.

2. Connect impinging apparatus to intake of air sampling pump. *Note:* Make sure the long tube is immersed in the absorbing solution.

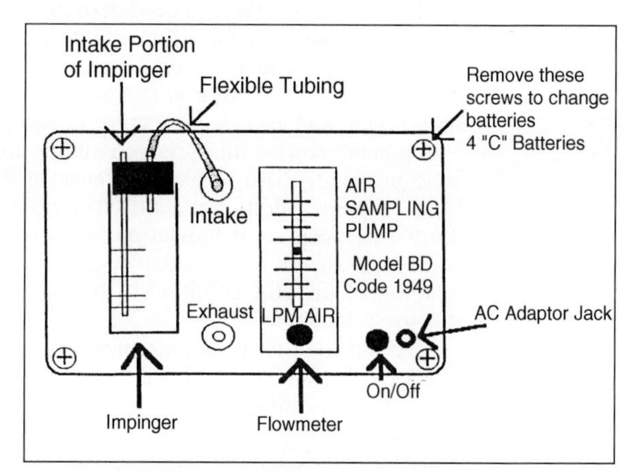

3. Set flowmeter to collect air at a rate of 2.0 Lpm for 30 minutes, or until a measurable amount of hydrogen sulfide is absorbed.

*Compliments of LaMotte Company, Chestertown, Md.

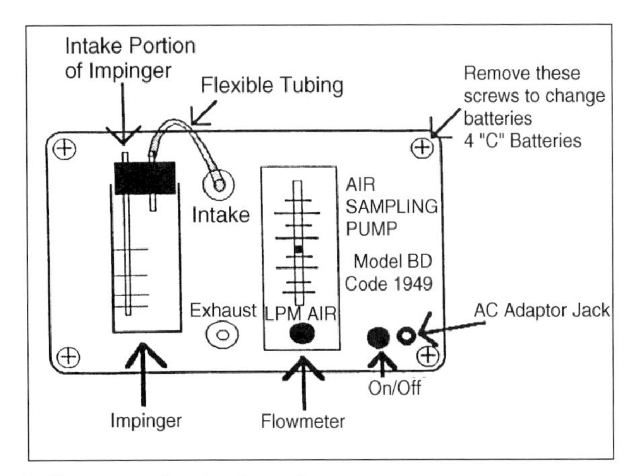

4. At the end of the sampling period, disconnect impinging tube from sampling pump. Use a 0.5-mL pipet to add 0.5 mL of Sulfide Reagent #2.

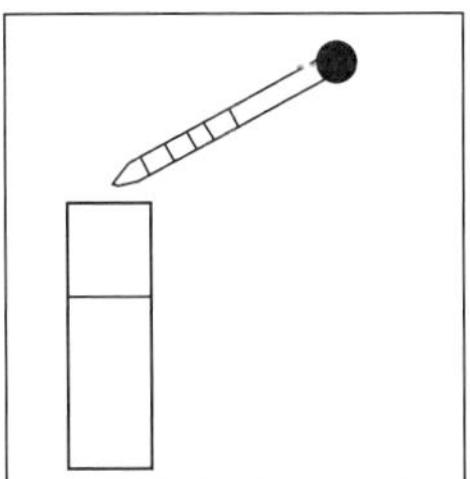

5. Use the second 0.5-mL pipet to add 0.5 mL of Sulfide Reagent #3. Mix.

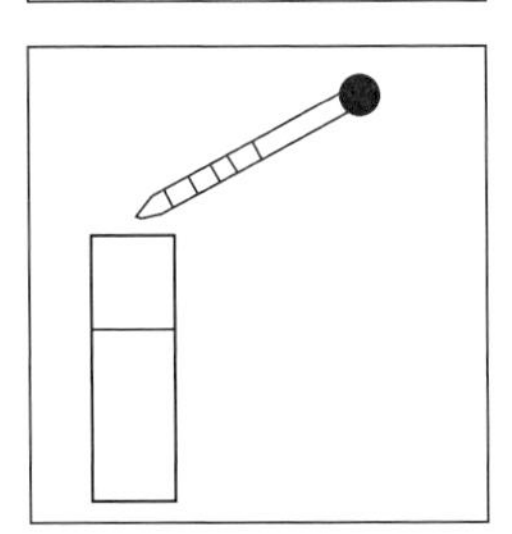

6. Add 4 drops of Sulfide Reagent #4. Mix. Solution will turn blue. Wait one minute.

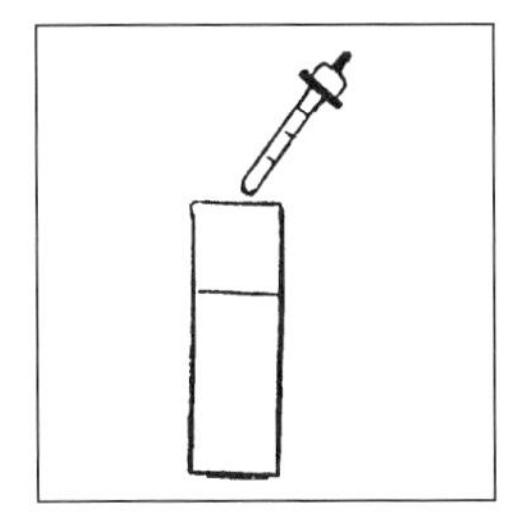

7. Use the 1.6-mL pipet to add 1.6 mL of Sulfide Reagent #5. Mix.

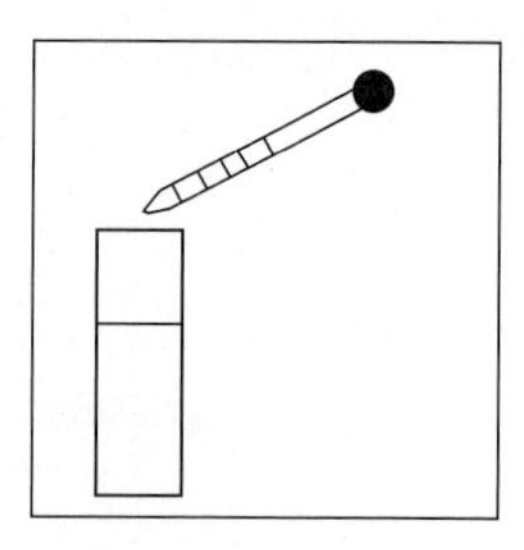

8. Transfer solution to a test tube. Insert test tube into the Hydrogen Sulfide in Air Comparator. Match sample color to the index of color standards. Record the index which gives the proper color match.

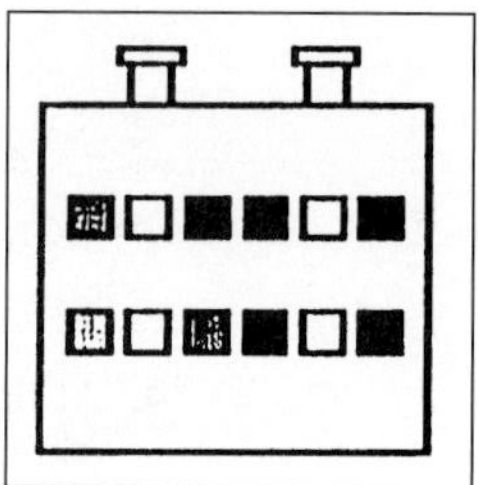

9. Use the chart below to convert index reading to a concentration. Record as ppm hydrogen sulfide.

Hydrogen Sulfide in Air Calibration Chart*

Time (min)	Comparator index number							
	1	2	3	4	5	6	7	8
10	0.06	0.14	0.28	0.55	1.11	1.66	2.22	2.77
20	0.03	0.07	0.14	0.28	0.55	0.83	1.11	1.39
30	0.02	0.05	0.09	0.19	0.37	0.55	0.74	0.92
60	0.01	0.02	0.05	0.095	0.19	0.28	0.37	0.46
90	0.01	0.02	0.03	0.06	0.12	0.19	0.25	0.30

*values in ppm

Note: The sampling period specified in the directions corresponds to concentrations which may be encountered in ambient air conditions. High concentrations of pollutants in the atmosphere will require shorter sampling periods, while low concentrations will require longer sampling periods.

SAFETY INFORMATION

Sulfide Reagents #1, #2, and #4 are considered to be hazardous substances. Material Safety Data Sheets (MSDS) are available for these reagents. For your safety, read the labels and accompanying MSDS before using.

QUESTIONS AND ANSWERS

1. What are some of the properties of hydrogen sulfide?
Hydrogen sulfide smells like "rotten eggs" and is a strong irritant to the eyes and mucous membranes. The compound is flammable and has explosive limits in air of 4.3 to 46%. The compound is toxic by inhalation.

2. What happens to hydrogen sulfide emissions in the atmosphere?
Eventually, hydrogen sulfide gas is converted to less toxic sulfur dioxide and sulfates in the atmosphere.

3. What damage can hydrogen sulfide create?
Hydrogen sulfide can damage the central nervous system and irritate the respiratory tract of individuals. It has killed a number of people exposed to sewer gas, natural gas explosions, and oil well drilling. Hydrogen sulfide can destroy plant tissue, reduce plant growth, and create leaf lesions and defoliation. It can discolor paints containing lead and react with copper and silver to form black coatings of metal sulfide.

4. How are mercaptans different from hydrogen sulfide?
Hydrogen sulfide has the formula of H_2S, whereas mercaptans are organic sulfides as in ethanethiol such as C_2H_5SH. Mercaptans are a group of organic compounds resembling alcohols, but having the oxygen of the hydroxyl group replaced by a sulfur. Mercaptans have a greater molecular weight and are less volatile than hydrogen sulfide.

REFERENCES

1. Stanley E. Manahan, *Environmental Chemistry,* 6th Ed. (Boca Raton, Fla.: Lewis Publishers, 1994).
2. *Hawley's Condensed Chemical Dictionary,* 11th Ed., rev'd by N. Irving Sax and Richard J. Lewis (New York, N.Y.: Van Nostrand Reinhold, 1987).
3. Roger N. Reeve, *Environmental Analysis,* (ACOL), John D. Barnes (ed.) (New York, N.Y.: John Wiley & Sons, 1994).
4. *Dräger Tube Handbook,* 11th Ed. (Lübeck: Dräger Sicherheitstechnik GmbH., 1998).
5. *Air Quality Sampling & Measurement Equipment* (Chestertown, Md.: LaMotte Company, 1997).
6. *Hydrogen Sulfide in Air Test Kit* (Chestertown, Md.: LaMotte Company, 1997).

LEAD

Lead and lead compounds have a number of toxic effects and inhibits the synthesis of hemoglobin in the body. Also, lead affects the central and peripheral nervous systems and the kidneys. The retardation of intellectual development in people has been related to lead pollution. Small dust particles containing lead are especially dangerous to children and pregnant women.

Lead bioconcentrates in marine organisms and has no known natural biological function. As a heavy metal, lead is significantly toxic. Lead compounds and lead metal tend to dissolve readily in acidic water and can be transported relatively easily after dissolution.

Solid particulate matter containing lead contaminates has been dispersed long distances in the atmosphere. Significant quantities of lead on atmospheric particulates have been identified near the North Sea off the coast of Britain. Lead has been identified in volcano ash, dust, lake waters, bogs, and other surface materials. It is generally distributed in the environment as metallic lead, inorganic compounds, and organometallic compounds. With the reduction of leaded fuels, atmospheric lead has decreased significantly in the environment.

Sources of lead include coal and oil combustion, lead smelting, improper solid waste and sewage sludge incineration, chemical manufacturing, battery manufacturing, lead product processing, and other industrial processes. Other sources of lead include ammunition, wear of equipment containing lead parts, solder and alloys, dyes, pigments, insecticides, paints, hair dyes, staining glass, varnish and paint dryers, waterproof paints, ceramic glazes, lead glass, curing agents, textile mordant, explosives, safety matches, lubricants, bronzing, printing, cloud seeding, medicine, wood preservative, catalysts, pipe joint packing materials, pyrotechnics, and stabilizers. Lighted candlewicks with lead metal cores can release particles containing lead into the air. Many of these items produce lead contaminated dust particles, volatilized lead, or water soluble lead during production and usage processes.

Air containing unknown lead quantities is pumped through a lead absorbing solution at 1.0 liter per minute for a specific recorded time (preferably 10 minutes) in the LaMotte Method. After adjusting the solution pH between 9 and 11, a sample of lead dithizone and a cyanide solution are appropriately added to the unknown lead solution. After the appropriate sample preparation procedure, a comparator index number is recorded by matching the sample color with an index of color standards. The lead concentration value (mg/M^3) in air can be obtained from a "Lead in Air Calibration Chart" by locating the comparator index number at the specific recorded time.

LEAD

Procedure*

1. Fill a clean impinging tube to the 10-mL line with Lead in Air Absorbing Solution. Connect impinging apparatus to intake of air sampling pump.
   Note: Make sure the long tube is immersed in the absorbing solution.

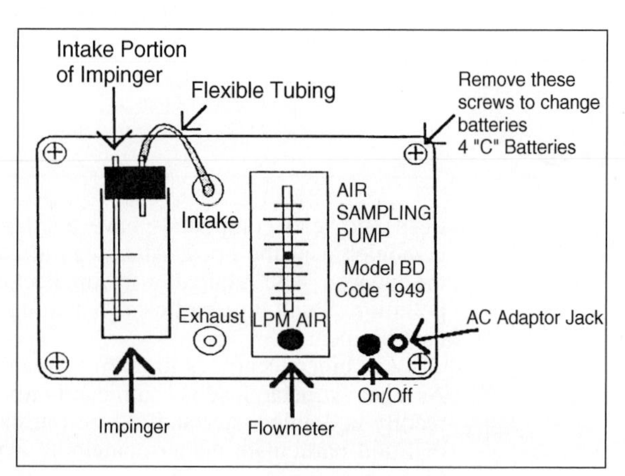

*Compliments of LaMotte Company, Chestertown, Md.

2. Adjust flowmeter to collect air sample at 1.0 Lpm for 10 minutes or until a measurable amount of lead is absorbed.

3. Using a glass stirring rod, test the pH of the solution. Alternatively, immerse the end of the pH strip into the solution. Compare the color of the strip to the color chart. If pH is not between 9 and 11, adjust by adding drops of 5N Sodium Hydroxide to raise the pH or drops of 1N Hydrochloric Acid to lower pH. **Make sure the pH is adjusted to between 9 and 11.**

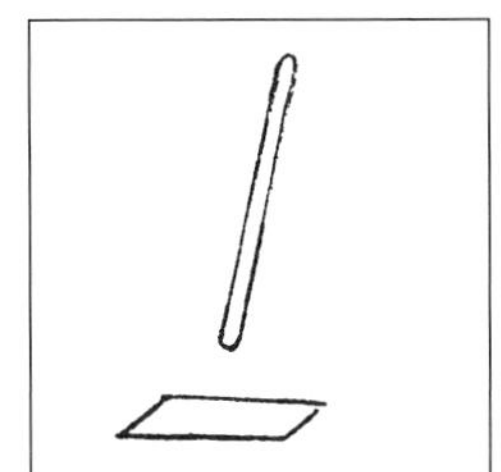

4. Using a glass pipet with a squeeze rubber bulb, insert pipet below clear preservative layer in Lead Dithizone Reagent. Release bulb and draw green solution into pipet until the solution measures 1.6 mL. Transfer the 1.6 mL of Dithizone Reagent to a clean test tube.
   Caution: Lead Dithizone Reagent contains chloroform as a solvent. Avoid spills and breathing vapor. Use with adequate ventilation.

5. Carefully add 5 drops of Lead Reagent #2.
   Caution: Lead Reagent #2 is an alkaline solution of cyanide and should be handled with extreme care. Do not dispose of cyanide solutions in presence of acids because the highly toxic HCN gas is evolved.

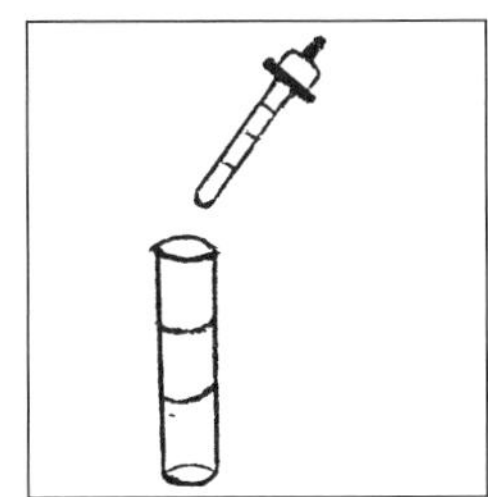

6. Cap and shake vigorously for 10 to 15 seconds. Remove cap and allow test tube to stand undisturbed for 1 minute. After 1 minute the layers will separate, producing a yellow to orange layer on top. No pink or red should be present in the bottom layer. If a pink or red color is present, rewash tube with dilute acid, rinse with distilled water and start the procedure over.

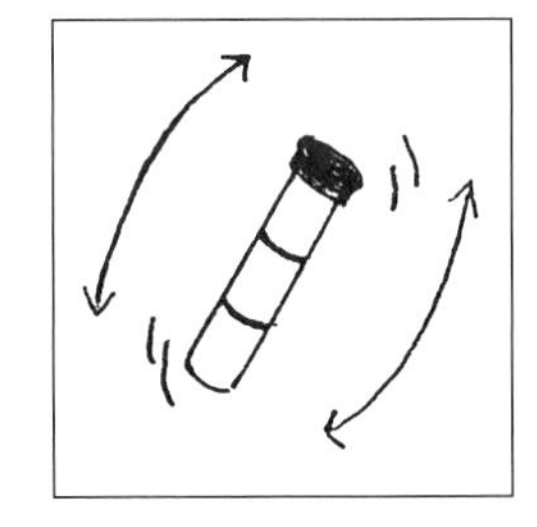

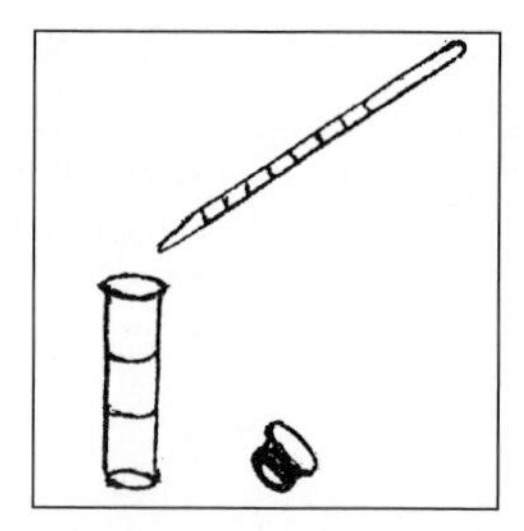

7. Dilute to 10-mL line with pH adjusted Absorbing Solution from Step 3. Cap and shake vigorously for 30 seconds.

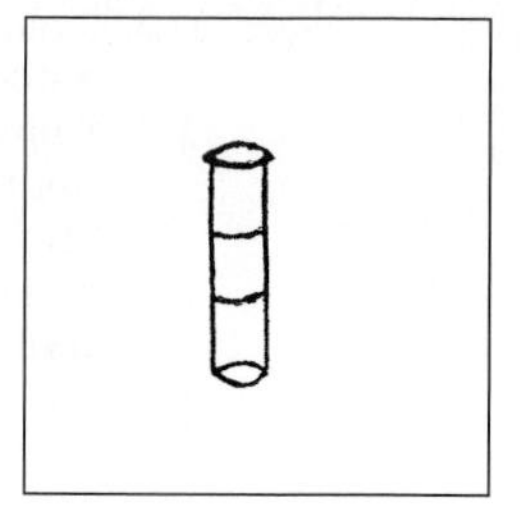

8. Allow test tube to stand undisturbed for 1 minute. If bottom layer remains cloudy, shake another 10 to 15 seconds. Check the pH to ensure pH is between 9 and 11.

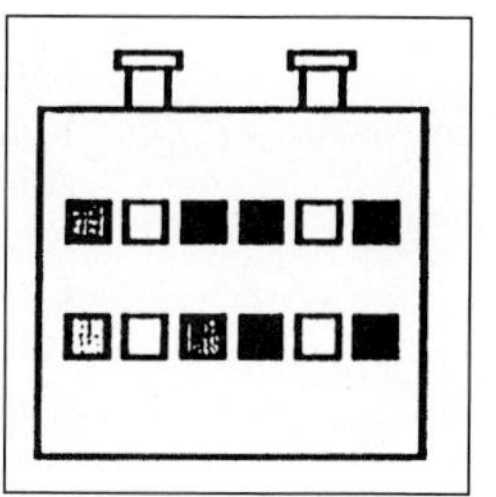

9. Insert test tube into the Lead in Air Comparator. Match sample color to index of color standards. Record the index number which gives the proper color match.

10. Repeat Steps 3 to 9 on a blank of deionized water and subtract result obtained from test result.
11. Use chart to convert index reading to a concentration. Record as mg/M^3 lead.

Lead in Air Calibration Chart*

Time (min)	Comparator index chart							
	1	2	3	4	5	6	7	8
5	0.0	0.2	0.4	0.8	1.2	1.6	2.0	3.0
10	0.0	0.1	0.2	0.4	0.6	0.8	1.0	1.5
15	0.0	0.07	0.13	0.27	0.4	0.53	0.6	1.0
20	0.0	0.05	0.1	0.2	0.3	0.4	0.5	0.75

* values in mg/M^3

Note: The sampling period specified in the directions corresponds to concentrations which may be encountered in ambient air conditions. High concentrations of pollutants in the atmosphere will require shorter sampling periods, while low concentrations will require longer sampling periods.

SAFETY INFORMATION

Hazardous substances are Lead in Air Absorbing Solution, Lead Dithizone Reagent, Lead Reagent #2, 5N Sodium Hydroxide, and 1N Hydrochloric Acid. The Material Safety Data Sheets (MSDS) are available for these reagents. For your safety, read the label and accompanying MSDS before using the materials.

QUESTIONS AND ANSWERS

1. What form does lead take in air?
   Lead can be found in air as metal and/or compounds attached to various sizes and types of particulate matter. Lead from lead strips in candles and other materials can be volatilized to a small extent in air by heating with a simple flame, i.e., candle flame, even though lead has a melting point near 327° C.
2. What happens to lead dissolved in acidic water?
   The lead can be consumed by humans and sea life and concentrate in specific organs. Lead dissolved in acidic water becomes more mobile and accessible to living species. At times, the soluble lead in slightly acidic water is deposited as an insoluble compound as the water becomes neutral or basic.
3. How does lead affect the human body?
   Lead and lead compounds inhibit the synthesis of hemoglobin in the body. Lead affects the central and peripheral nervous systems. Lead also affects the kidneys and slows intellectual development of people.
4. If the lead pollutant in air exists at significant concentrations, how could lead be detected in the air sample?
   Using this procedure, the relatively high lead concentration in air can be determined by collecting the air sample for 5 minutes at 1.0 liter per minute.

REFERENCES

1. Stanley E. Manahan, *Environmental Chemistry,* 6th Ed. (Boca Raton, Fla.: Lewis Publishers, 1994).
2. Roger N. Reeve, *Environmental Analysis,* (ACOL), John D. Barnes (ed.) (New York, N.Y.: John Wiley & Sons, 1994).
3. *Hawley's Condensed Chemical Dictionary,* 11th Ed., rev'd by N. Irving Sax and Richard J. Lewis (New York, N.Y.: Van Nostrand Reinhold, 1987).
4. *Fate of Pesticides and Chemicals in the Environment,* J. L. Schnoor (ed.) (New York, N.Y.: John Wiley & Sons, Inc., 1992).
5. *Air Quality Sampling & Measurement Equipment* (Chestertown, Md.: LaMotte Company, 1997).
6. *Lead in Air* (Chestertown, Md.: LaMotte Company, 1995).

MERCURY VAPORS

Mercury vapors from spilled liquid mercury are very widespread. Mercury can be found in switches, vapor lamps, fluorescent tubes, thermometers, barometers and thermostats found around the home and office. Even today, small amounts of mercury often remain deposited after a thermometer is broken or after an accidental spill from broken equipment. Containers of used mercury are still stored on the shelf at many homes and businesses. The clean up process for spilled liquid mercury is not taken very seriously by a number of people in the society.

Mercury droplets can be trapped in crevices in floor and table surfaces. Mercury vapor concentrations may exceed the recommended limit. Vibrations in the work environment can significantly increase the vaporization of mercury. Droplets of mercury can be picked up by a suction made from a filter flask, rubber stopper, and several pieces of rubber and glass tubing. Spilled mercury can immediately be amalgamated with a mixture of copper and iron dust powder, which can be quickly picked up by a magnet for disposal in a suitable container.

Mercury has a significant toxicity level and has some volatility and mobility. Mercury vapor and fumes are highly toxic by skin absorption and inhalation. It is absorbed by the respiratory system and intestinal tract. Mercury appears to bioconcentrate in humans, animals and in marine organisms and has no known natural biological function. Mercury can be easily converted into covalent organometallic compounds which will preferentially accumulate in fatty tissue. The tolerance level (as Hg) is 0.05 mg per cubic meter of air.

Volatile elemental mercury is primarily associated with particulate matter from coal combustion, volcanoes, and improper clean up of spills. In addition, mercury compounds and liquid mercury have been deposited in water bodies or in wet areas due to accidents or as discarded materials. Mercury forms water soluble organomercury compounds such as dimethyl mercury, $(CH_3)_2Hg$, and monomethylmercury salts, CH_3HgBr during normal decomposition processes in the presence of bacteria. These compounds are significantly volatile and are encountered in the atmosphere.

Mercury is emitted to the atmosphere from a variety of manufacturing processes that prepare paint additives, batteries, dental preparations, light fixtures, electrical devices, and laboratory products. The largest emission sources of mercury vapor appear to be those that use liquid mercury and the natural sources of mercury. Mercury vapor from vapor lamps, manufacturing processes and from spilled mercury can be detected in the air by a Dräger tube at 0.05 to 2 mg/m^3. The short-term tube medium in the tube changes color from a pale yellow gray to a pale orange. The color change is attributed to the reaction of mercury vapor with CuI to form a copper-mercury complex. The test procedures to detect mercury vapors in air are provided in an experiment titled "Measurement of Air Contaminates."

During the tube reaction, free halogens cause significant negative errors and can invalidate the mercury measurements. Substances such as arsine, phosphine, hydrogen sulfide, ammonia, nitrogen dioxide, sulfur dioxide, and hydrazine do not interfere in the TLV range.

QUESTIONS AND ANSWERS

1. What are the sources of mercury vapors?
   Spilled or uncontrolled mercury produces volatilization of mercury vapors. Improper clean-up of mercury metal droplets, particulate matter containing mercury, and volcanoes are primary sources of mercury. Bacterial decomposition of wastes generating methane forms volatile dimethyl mercury and monomethylmercury salts These compounds tend to solubilize mercury and make mercury much more mobile.
2. How do excess halogens affect the measurement for mercury vapor?
   Free halogens create significant minus errors. The measurement of mercury vapor in the presence of halogens is nearly impossible.
3. How does humidity affect the measurement of gaseous mercury?
   Absolute humidity must be below 20 mg water/ L for meaningful readings.
4. Are mercury vapors dangerous?
   Mercury vapors are highly toxic by skin absorption and inhalation. The human tolerance

level (as Hg) is 0.05 mg per cubic meter of air. Mercury is absorbed by the respiratory system and intestinal tract.

REFERENCES

1. Stanley E. Manahan, *Environmental Chemistry,* 6th Ed. (Boca Raton, Fla.: Lewis Publishers, 1994).
2. *Dräger Tube Handbook,* 11th Ed. (Lübeck: Dräger Sicherheitstechnik GmbH., 1998).
3. *Hawley's Condensed Chemical Dictionary,* 11th Ed., rev'd by N. Irving Sax and Richard J. Lewis (New York, N.Y.: Van Nostrand Reinhold, 1987).
4. *Fate of Pesticides and Chemicals in the Environment,* J. L. Schnoor (ed.) (New York, N.Y.: John Wiley & Sons, Inc., 1992).
5. Roger N. Reeve, *Environmental Analysis,* (ACOL), John D. Barnes (ed.) (New York, N.Y.: John Wiley & Sons, 1994).

NITROGEN OXIDES

Nitrogen oxides in the atmosphere are nitrous oxide (N_2O), nitric oxide (NO), and nitrogen dioxide (NO_2). Nitrous oxide, (N_2O), is an anesthetic known as "laughing gas" produced by microbiological processes. The gas does not appear to impact important reactant species in the lower atmosphere. The compound decomposes relatively easily in photochemical reactions to nitrogen and oxygen. Nitrous oxide, (N_2O), also reacts with singlet atomic oxygen to form nitrogen, oxygen and nitric oxide (NO), thus contributing to ozone layer depletion and to the greenhouse effect.[1]

NO_x gases include the pungent red-brown nitrogen dioxide (NO_2) and the colorless, odorless nitric oxide (NO). Additional gases of N_2O_5 and N_2O_3, and N_2O_4 are generally considered to be included as dimers and combinations of the basic NO_x gases. The NO_x gases enter the atmosphere through biological processes, lightning, vehicle emissions, supersonic transport planes, highly charged electrostatic precipitators, high voltage lines, and during fossil fuel combustion in engines and in high temperature furnaces. Large diesel engines are sometimes good sources of these compounds. Both NO and NO_2 have contributed to the acid rain problem.

In general, NO is a product of the combustion process of any nitrogen-containing fuel such as coal, leaves, and petroleum fuels. NO is the primary product of NO_x emissions. NO is relatively rapidly converted to nitrogen dioxide in the troposphere. Nitric oxide (NO) is less stable, less toxic and biochemically less active than nitrogen dioxide (NO_2). NO can attach to hemoglobin and reduce oxygen transport efficiency.[1]

Nitrogen dioxide increases a person's susceptibility to pneumonia and lung cancer. The NO_2 levels near 100 ppm causes inflammation of lung tissue. Exposure of humans to about 180 ppm of NO_2 can cause death of humans within weeks. Death is normally observed within days of human exposure to 500 ppm of NO_2.

Nitrogen dioxide poisoning can be generated by the fermentation of ensilage containing nitrate. Inhalation of NO_2-containing gases from burning celluloid and nitrocellulose film and from spillage of NO_2 oxidant used with liquid hydrazine fuel in rocket systems have resulted in a number of deaths.[1]

The NO_2 compound absorbs both ultraviolet and visible light and plays an important role in photochemical smog.[1] The compound undergoes photodissociation to form NO and excited molecules and reacts with other substances to produce a number of inorganic and organic species. The NO_2 gaseous compound causes paints and dyes to fade and combines directly with water molecules to form nitric acid or acid rain. Then, the nitric acid rain can be an environmental problem or it can react with bases (ammonia, particulate lime) to form particulate nitrates.[1]

The amount of nitrogen dioxide in air can be determined by diffusion tube measurements. The gas is absorbed by triethanolamine liquid and can be analyzed by field spectrometry methods. Remote sensing applications using ultraviolet absorptions have been used to identify the concentrations of NO_2 in air.

Both nitrogen dioxide and nitrous fumes (NO) in air can be determined by an experiment titled "Measurements of Air Contaminates." During the measurement, separate tubes are required for a specific concentration range of the pollutant in air. Each tube changes medium color to indicate the presence of the contaminate and requires a specific time for measurement under ambient operating conditions. Each tube application is generally discussed below.

Nitrogen dioxide reacts with diphenylbenzidine to form a blue-gray reaction product in short-term Dräger tubes. The tubes can easily measure nitrogen dioxide within the standard measuring ranges of 5 to 25 ppm/0.5 to 10 ppm and 5 to 100 ppm/2 to 50 ppm. Nitrogen dioxide cannot be measured when ozone and/or chlorine are present in levels above their threshold limit values (TLVs).

Nitrogen dioxide (NO_2) can be determined by long-term measurements (1 hour to 8 hours) using similar tube chemistry over the range of 1.25 ppm to 100 ppm. A constant flow pump will be used during long-term measurements to determine the time weighted average concentration of the air pollutant. Also, interferences by 10 ppm ammonia and 1 ppm ozone are not a problem. Nitrogen monoxide (NO) is not indicated during these measurements.

"Direct Indicating Diffusion Tubes" containing o-tolidine will provide a yellow-orange reaction product when nitrogen dioxide diffuses through an open end of a tube. The tube standard range of measurement for nitrogen dioxide gas is 1.3 to 200 ppm during a 1 hour to 8 hour time period. Chlorine and ozone are also indicated with approximately half the indicating sensitivity. Sulfur dioxide at 5 ppm and ammonia at 100 ppm display no influence on the tubes.

Nitrous fumes (NO) can be measured by sequential tube reactions with chromium (VI) and diphenylbenzidine to form blue-gray reaction products on the Dräger tube column during short-term measurements. The ranges of measurements are 0.5 to 10 ppm and 5 to 100 ppm/ 2 to 50 ppm. Ozone and/or chlorine in excess of their TLV's interfere during the measurements. Also, short-term tube oxidation of NO with chromium (VI) to form NO_2 and subsequent reaction with o-dianisidine produces a red-brown reaction stain on the column within the standard measuring ranges of 20 to 500 ppm and 100 to 1000 ppm/500 to 5000 ppm. Again, ozone and/or chlorine will interfere with the measurements in the same concentration range.

Nitrous fumes can also be determined by long-term measurements in a Dräger tube. Different reaction chemistries provide either a brown reaction stain or a blue-gray reaction stain on the tubes. The standard range of measurement is 1.25 to 50 ppm over a 1 to 4-hour period of testing as the brown indicator stain forms during the o-dianisidine test. The standard range of detection is 12.5 to 350 ppm over a 1 to 4-hour period for the blue-gray stain formed during the diphenylbenzidine test. Again, the interferences depend upon which method is used for the air test.

Nitrogen dioxide in air can also be determined quantitatively by the LaMotte Method. After adding Nitrogen Dioxide #1 Absorbing Solution into the impinging tube and connecting equipment, the air is sampled at 0.2 liters per minute (Lpm) for 10 to 20 minutes by a Model BD Sampling Pump equipped with an adapter for restricting the flow of air through a specific type of impinging apparatus. Then, a variety of chemicals are added to the collected sample to develop a colored solution. After waiting 10 minutes to develop a full color, the sample concentration of nitrogen dioxide can be determined by matching the sample color with permanent color standards in a comparator. The index reading from the comparator is converted to the concentration of the nitrogen dioxide in the atmosphere in parts per million (ppm) by finding the corresponding index value in the calibration chart and reading down the column until the sampling time is located.

NITROGEN DIOXIDE

AIR SAMPLING PROCEDURE*

The Model BD Air Sampling Pump is equipped with an adapter for restricting the flow of air through the impinging apparatus. This adapter consists of a small 27-gauge hypodermic needle which is fitted into a small plastic holder. A piece of plastic tubing is attached to the plastic holder. By joining one end of the tubing to the intake portion of the impinging apparatus, the flow of air is restricted to 0.2 Lpm. In order to sample the test atmosphere at this rate for nitrogen dioxide, the following procedure is outlined:

1. Attach adapter to intake of impinging apparatus.
2. Unscrew knob of flowmeter (counterclockwise—6 complete turns from closed position of flowmeter).
3. Turn switch to "ON" position. Follow recommended operating procedure for nitrogen dioxide test.
   Note: The sampling period should not exceed 20 minutes for a single determination of nitrogen dioxide when the adapter is in place as damage might result to the pumping mechanism.

Procedure*

1. Pour 10 mL of Nitrogen Dioxide #1 Absorbing Solution into the impinging tube. Connect impinging apparatus to intake of air sampling pump. Make sure the long tube is immersed in the absorbing solution.
   Note: For reference analysis of nitrogen dioxide, a gas bubbler impinger is recommended. For general applications and demonstration, a general purpose impinger or standard midget impinger may be used.

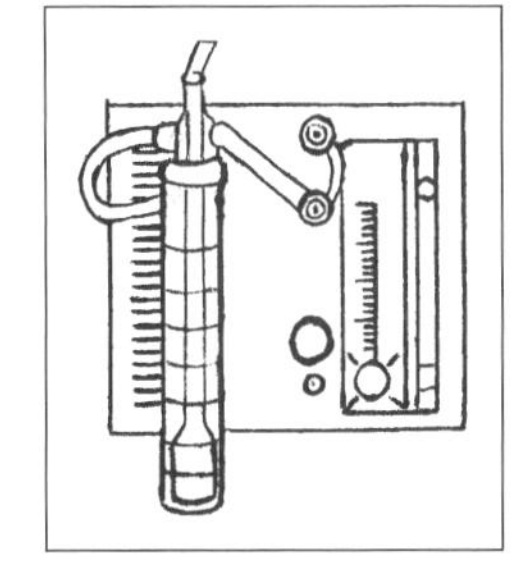

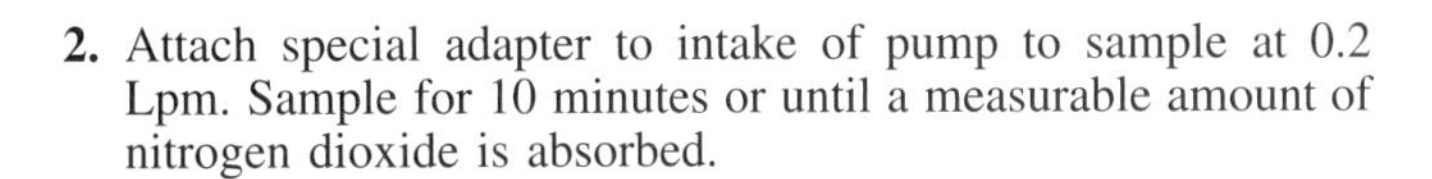

2. Attach special adapter to intake of pump to sample at 0.2 Lpm. Sample for 10 minutes or until a measurable amount of nitrogen dioxide is absorbed.

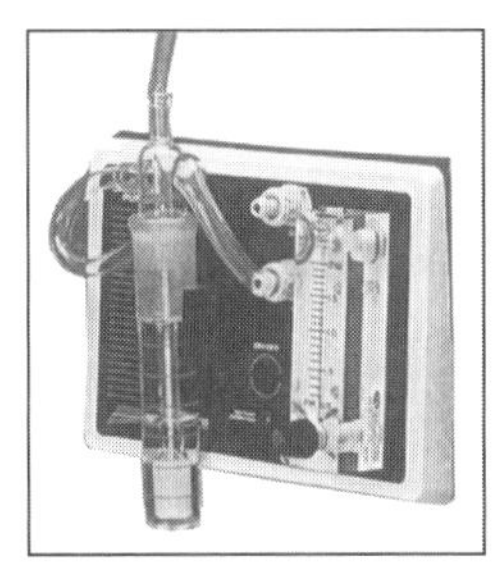

*Compliments of LaMotte Company, Chestertown, Md.

3. At the end of the sampling time, pour contents of impinging tube into test tube. Dilute to 10-mL volume with absorbing solution if any absorbing solution evaporated during sampling procedure.

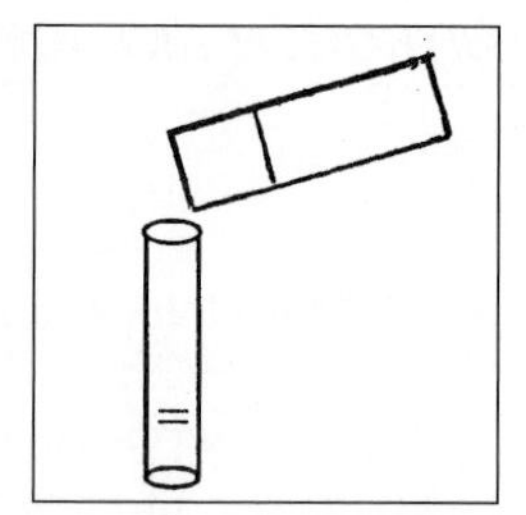

4. Use the pipet to add 1 drop of Nitrogen Dioxide Reagent #2. Cap and mix.

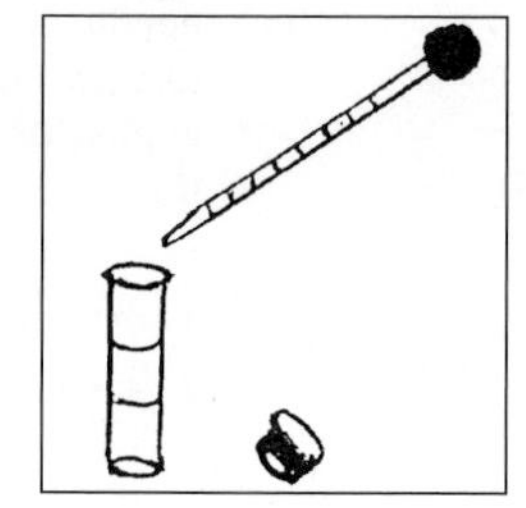

5. Use the 0.05-g spoon to add one level dipper of Nitrogen Dioxide #3. Cap and mix. Wait 10 minutes for full color development.

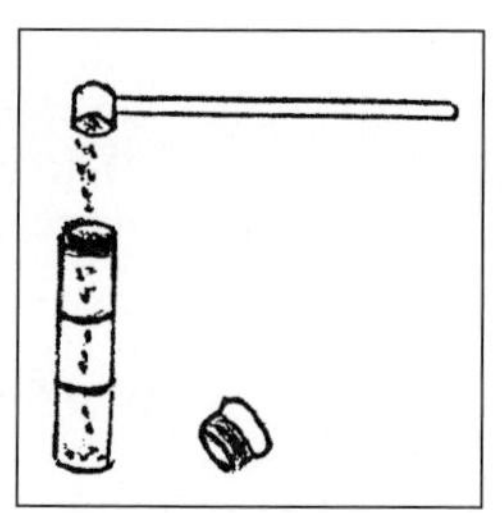

6. Place test tube into the comparator. Match test sample to index of color standards. Record the index number which gives the proper color match.

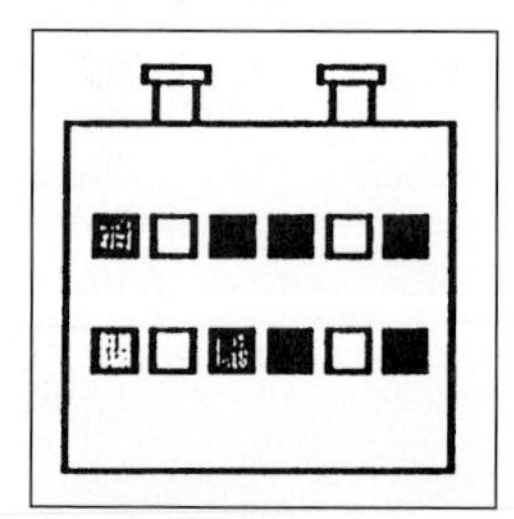

The following calibration chart is provided to convert the comparator index reading into the concentration of the contaminant in the atmosphere to ppm nitrogen dioxide. The chart is based upon the prescribed sampling period for individual tests. After the color match is made, find the corresponding index value on the calibration chart, then read down the line until the sampling time is found.

Nitrogen Dioxide in Air Calibration Chart*

Time (min)	Comparator index number							
	1	2	3	4	5	6	7	8
1	0.00	2.8	7.0	14.0	21.0	28.0	42.0	56.0
5	0.00	0.56	1.40	2.80	4.20	5.60	8.40	11.20
10	0.00	0.28	0.70	1.40	2.10	2.80	4.20	5.60
15	0.00	0.19	0.47	0.93	1.40	1.87	2.80	3.74
20	0.00	0.14	0.35	0.70	1.05	1.40	2.10	2.80

* values in ppm

Note: The sampling period specified in the directions corresponds to concentrations which may be encountered in ambient air conditions. High concentrations of pollutants in the atmosphere will require shorter sampling periods, while low concentrations will require longer sampling periods.

SAFETY INFORMATION

Nitrogen Dioxide #1 Absorbing Solution is considered to be a hazardous substance. Material Safety Data Sheets (MSDS) are supplied for any of the reagents. For your safety, read the labels and the accompanying MSDS before using the chemicals.

USE OF THE OCTET COMPARATOR

The Octet Comparator contains 8 permanent color standards. To perform a test, the sample is inserted into one of the slots in the top of the comparator, where it can be compared to 4 color standards simultaneously. For optimum color comparison, the comparator should be positioned between the operator and light source, so the light enters through the special diffusing screen on the back of the comparator. Avoid viewing the comparator against direct sunlight or an irregularly lighted background.

QUESTIONS AND ANSWERS

1. How toxic are N_2O, NO, and NO_2 to humans?
Nitrous oxide, N_2O, is known as laughing gas. The compound decomposes relatively easily to form oxygen and nitrogen in photochemical reactions. The gas, NO, can become attached to hemoglobin and reduce oxygen transport efficiency. The NO compound is biochemically less active and less toxic than NO_2. Human exposure to levels of nitric oxide near 180 ppm can cause death within weeks. However, NO is easily oxidized in air to NO_2. Nitrogen dioxide, NO_2, causes pneumonia and lung cancer at relatively low levels. Exposure of a human to 500 ppm NO_2 usually causes death within a few days.

2. How is nitrogen dioxide generated?
Nitrogen dioxide, NO_2, is generated by air oxidation of NO. Both components can be

generated by biological processes, lightening, vehicle and plane emissions, electrostatic precipitators, high voltage lines, fossil fuel combustion, and high temperature furnaces. Nitrogen dioxide can also be generated during fermentation of ensilage and during burning of celluloid fibers and nitrocellulose film. The compound is used as an oxidant in the fuel systems of rockets.

3. What problems does nitrogen dioxide create in the environment?
Nitrogen dioxide forms photochemical smog. The compound undergoes photodissociation to form NO and additional excited molecules, which react with other substances to produce a number of inorganic and organic molecules. The gaseous substance attacks paints and dyes and reacts directly with water droplets to form nitric acid rain. Nitric acid rain can be an environmental problem or it can be neutralized with basic chemicals to form nitrates.

4. How can the different nitrogen oxide fumes be distinguished from each other?
Nitrous oxide is "laughing gas" and is an anesthetic. Nitric oxide, NO, is colorless and odorless. NO is easily oxidized in air to nitrogen dioxide. Nitrogen dioxide, NO_2, is a pungent, red-brown gas which can cause severe problems.

REFERENCES

1. Stanley E. Manahan, *Environmental Chemistry,* 6th Ed. (Boca Raton, Fla.: Lewis Publishers, 1994).
2. Roger N. Reeve, *Environmental Analysis,* (ACOL), John D. Barnes (ed.) (New York, N.Y.: John Wiley & Sons, 1994).
3. *Dräger Tube Handbook*, 11th Ed. (Lübeck: Dräger Sicherheitstechnik GmbH., 1998).
4. *Air Quality Sampling & Measurement Equipment* (Chestertown, Md.: LaMotte Company, 1997).
5. *Nitrogen Dioxide in Air* (Chestertown, Md.: LaMotte Company, 1991).

OZONE

Ozone is a fire and explosion risk in contact with organic materials. The substance is toxic by inhalation and is listed as a strong irritant. The EPA standard for ozone in ambient air is 0.12 ppm.

Ozone at a concentration of 1 ppm by volume in air has a distinct odor. At this level it causes severe irritation and headaches during inhalation. Ozone irritates the eyes, upper respiratory system and lungs. Inhalation of too much ozone can cause fatal pulmonary edema, where there is an abnormal accumulation of fluid in the lung tissue.

Ozone can create chromosomal damage and generate free radicals in tissue of humans. The chemical can oxidize sulfhydryl (-SH) groups and other groups in living organisms. Also, ozone can cause lipid peroxidation in the body. Compounds that protect the body chemistry from the effects of ozone are antioxidants, radical scavengers, and other compounds.

Ozone is a photochemical oxidant that is an important component of urban smog. Ozone pollution can lead to deterioration of rubber, textiles, paints and other materials which can be easily oxidized. Ozone can impact vegetation by increasing leaf drop and producing premature fruit.

Ozone, O_3, is a strong oxidizing agent that is often used to deodorize air and sewage gases and to treat swimming pool waters, drinking water, and industrial process water. Ozone is also used to remove other gases from solvents and oxidize materials in several chemical processes. It also provides microbial sterilization and acts as a disinfectant.

The photolysis of ozone in sunlight at wavelengths below 308 nm produces oxygen in an excited state (O*) and oxygen in the ground state (O). Excited atomic oxygen emits visible light at wavelengths in the 558 to 636 nm region of the visible spectrum and intense radiation in the infrared region of the spectrum. This phenomena of emitted light in the visible and infrared regions of the spectrum is called **"airglow."**

Ozone absorbs harmful ultraviolet radiation in the stratosphere and provides a radiation shield for humans by protecting us on the Earth from the effects of excessive amounts of the radiation. Ozone is formed in the stratosphere by a two-step process reaction involving oxygen, light at a wavelength near 242 nm, and ground state oxygen (O). It is also destroyed by photodissociation and a series of reactions of ground state oxygen (O) with ozone to produce oxygen. A considerable amount of ozone is formed and destroyed daily.

Ozone absorbs or filters out dangerous forms of ultraviolet radiation. Ozone absorption of electromagnetic radiation converts the radiation to heat, which creates a temperature maximum at the boundary between the stratosphere and the mesosphere. A diminished shield of ozone will produce higher temperatures at the earth surface and create a number of problems with food production, skin cancer, cataracts, and cell growth. In addition, ozone depletion by freon chemicals has been addressed to a large extent by removing most of the freon from the marketplace. However, the reactions of chlorine and NO_x gases with ozone at cold temperature in the Arctic and Antarctic atmospheres could be a significant problem with time.

The test procedures to detect ozone in air are provided in an experiment titled "Measurement of Air Contaminates." In the experiment, ozone reacts with a pale blue or greenish blue indigo dye to form either a white or yellow stain on the column of a short-term tube. The standard measuring range of the Dräger tubes are 0.05 to 0.7 ppm or 20 to 300 ppm within the temperature range of 0 to 40° C with absolute humidity of 2 to 30 mg H_2O/L. The measuring range for ozone can be extended for lower concentrations of ozone. Interferences by low concentrations of sulfur dioxide, chlorine, and nitrogen dioxide are limited.

QUESTIONS AND ANSWERS

1. Why should the indigo reagent be stored in the dark?
   The indigo reagent is light sensitive and should be stored in the dark at all times.
2. How does ozone affect people?
   Ozone at 1 ppm by volume in air causes eye irritation and headaches during inhalation. Prolonged exposure to ozone can result in cardiovascular and respiratory illnesses. Ozone can create chromosomal damage and generate free radicals in human tissue. The chemical can oxidize a number of body chemicals and cause lipid peroxidation in the body under the right conditions. Pulmonary edema, an abnormal accumulation of fluid in the lung tissue, is an illness which has been attributed to high ozone levels.
3. Why is ozone important in the atmosphere?
   Ozone absorbs or filters out dangerous forms of ultraviolet radiation by providing a shield that protects humans from higher temperatures and problems with food production, cataracts, diseases, and plant growth. Most of this filtering process takes place in the ozone layer between the stratosphere and the mesosphere.
4. What are some compounds that will provide protection for the body from ozone?
   To some degree, the body can be protected from the effects of ozone by antioxidants, radical scavengers, and other compounds that act as reducing agents.

REFERENCES

1. Stanley E. Manahan, *Environmental Chemistry,* 6th. Ed. (Boca Raton, Fla.: Lewis Publishers, 1994).
2. Roger N. Reeve, *Environmental Analysis,* (ACOL), John D. Barnes (ed.) (New York, N.Y.: John Wiley & Sons, 1994).
3. *Hawley's Condensed Chemical Dictionary*, 11th Ed., rev'd by N. Irving Sax and Richard J. Lewis (New York, N.Y.: Van Nostrand Reinhold, 1987).
4. *Dräger Tube Handbook,* 11th Ed. (Lübeck: Dräger Sicherheitstechnik GmbH., 1998).

PHENOL

Phenol is toxic by ingestion, inhalation, and skin absorption. It can be absorbed through the skin and can lead to fatal overdose. Phenol poisoning can cause gastrointestinal disturbances, circulatory system failure, convolutions, kidney malfunction, and lung edema. Chronic exposure can damage the spleen, pancreas, and kidneys. Substituted phenols (such as chlorophenol) can, in some cases, have toxic effects similar to those of phenol.

The human tolerance level to phenol is approximately 5 ppm in air. The maximum admissible concentration (MAC) allowable for phenols is 0.5 micrograms per liter. Phenol is a nonpersistent compound which degrades in approximately 70 hours when exposed to different types of bacteria in water, soil, and particulates.

A number of phenol-based resins, such as bakelite, epoxy resins, nylon-6, and phenol-furfural resin, are fairly stable complexes that are used for many mechanical devices and items in our society. Phenol is used in 2,4-D, germicidal paints, pharmaceuticals, laboratory reagent, dyes, indicators, slimicide, and general disinfectant. Traditionally, traces of phenol have been used as disinfectants and antiseptic agents in salves, mouth washes, and cleaning solvents. Phenol is used as a selective solvent for refining lubricating oils and is often used in the preparation of a variety of chemicals such as adipic acid, salicylic acid, phenolphthalein, pentachlorophenol, acetophenetidine, and picric acid.

Phenol has also been used to determine the effectiveness of a disinfectant. The compound provides a standard of comparison to develop the phenol coefficient. In addition, phenol volatility in solution can be lowered by the addition of sodium hydroxide to the solution. Phenol and phenolic type compounds (such as cresol and nitrophenol) have significantly different properties than those of the aliphatic and olefinic alcohols. Nitro groups and other atoms attached to the aromatic phenolic rings can significantly affect the toxicological behavior of phenolic compounds.

Visible spectrophotometry is often used to detect the presence of phenols. Phenol can be measured by tube reactions of $Ce(SO_4)$ in the presence of sulfuric acid to form brown-gray reaction products in the Dräger tube during short-term measurements provided in an experiment titled "Measurement of Air Contaminates." The standard range of measurement is 1 to 20 ppm in the temperature range of 10 to 30° C and at an absolute humidity of 1 to 18 mg H_2O/L. At a temperature of 0° C, the tube reading must be multiplied by 1.3, whereas the final reading must be mutiplied by 0.8 at a temperature of 40° C.

Creosols are also indicated at different sensitivities during tube analysis. To determine m-cresol, multiply the indication by 0.8. Benzene, toluene, and other aromatic compounds without the hydroxyl group are not indicated in the tube. In addition, aliphatic hydrocarbons and alcohols do not interfere with the tube chemistry.

Phenols in air can also be determined quantitatively by the LaMotte Method. After adding 0.25 N ammonium hydroxide into the impinging tube and connecting equipment, the air is sampled at 1.0 liters per minute (Lpm) for 10 minutes or until a measurable amount of phenol is absorbed. Then, a variety of chemicals are added to the collected sample, and a colored solution develops after adding potassium ferricyanide solution. After inserting the test tubes of the colored sample and two blanks in an axial reader, the color of the sample is matched with an index of color standards to obtain an index reading for the sample. The index value and the sampling time provide the concentration of phenol in air in parts per million (ppm) from the calibration chart.

PHENOLS

Procedure*

1. Fill a clean impinging tube to the 10-mL line with 0.25 N Ammonium Hydroxide.
   Note: Read the Axial Reader Instruction manual before proceeding.

2. Connect impinging apparatus to intake of air sampling pump.
   Note: Make sure the long tube is immersed in the absorbing solution.

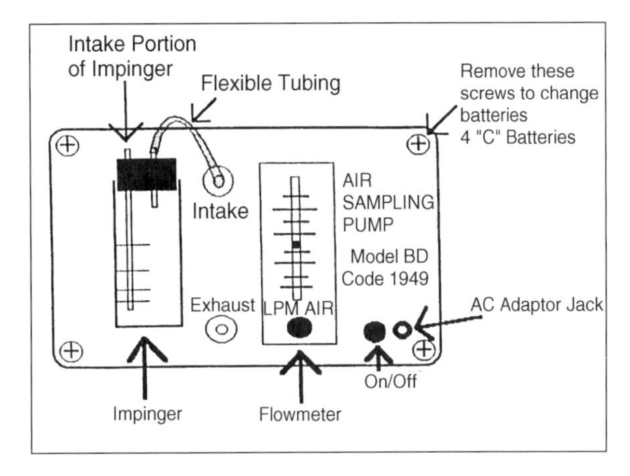

*Compliments of LaMotte Company, Chestertown, Md.

3. Adjust flowmeter of sampling pump to collect air at a rate of 1.0 Lpm for 10 minutes or until a measurable amount of phenol is absorbed.

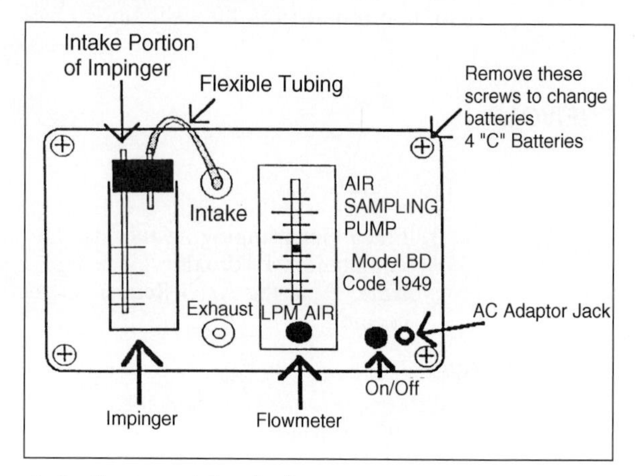

4. At the end of the sampling period, disconnect impinging tube. Dilute to 10-mL line with 0.25 N Ammonium Hydroxide, if necessary.

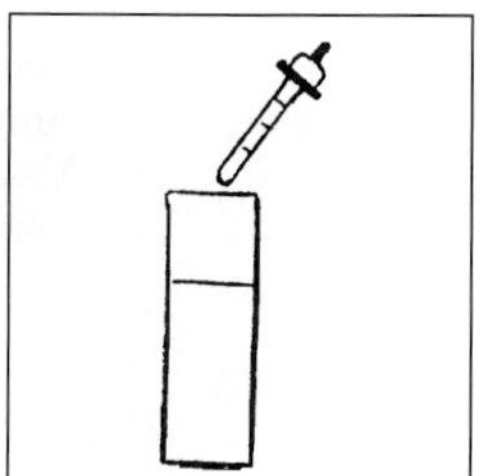

5. Use the 0.1-g spoon to add 1 level measure of Aminoantipyrine Reagent. Mix until powder dissolves.

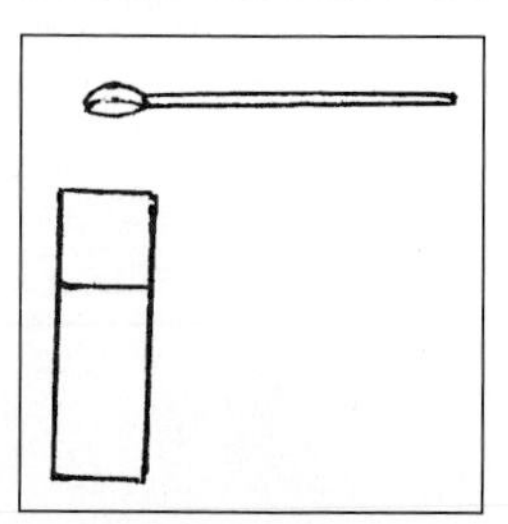

6. Use the plain pipet to add 4 drops of Ammonium Hydroxide Solution (Solution 7826). Mix.

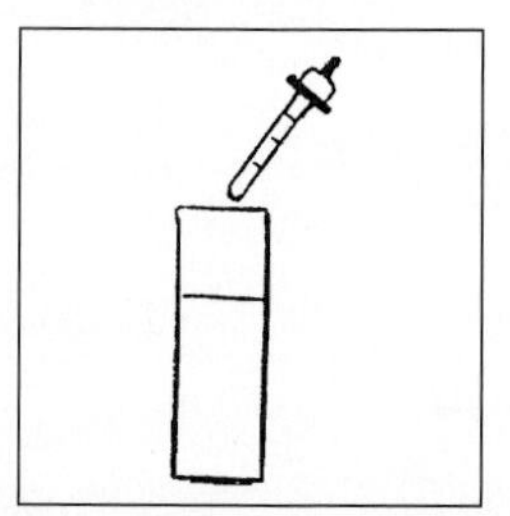

7. Use the 1-mL pipet to add 2.0 mL (2 measures) of Potassium Ferricyanide Solution. Mix.

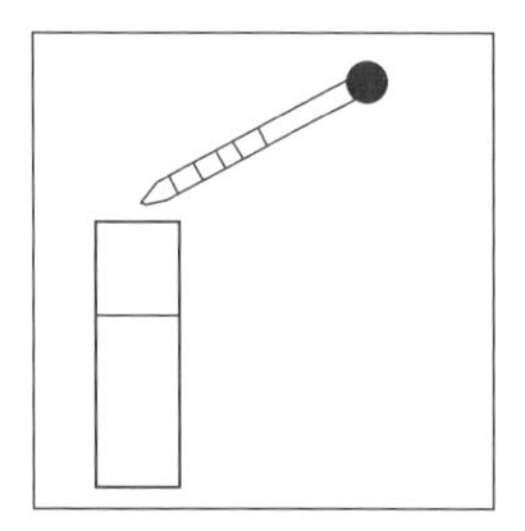

8. If phenols are present, the color develops almost immediately. Pour entire test sample into a test tube and insert test tube into Axial Reader. Place two untreated samples on either side of sample as blanks.

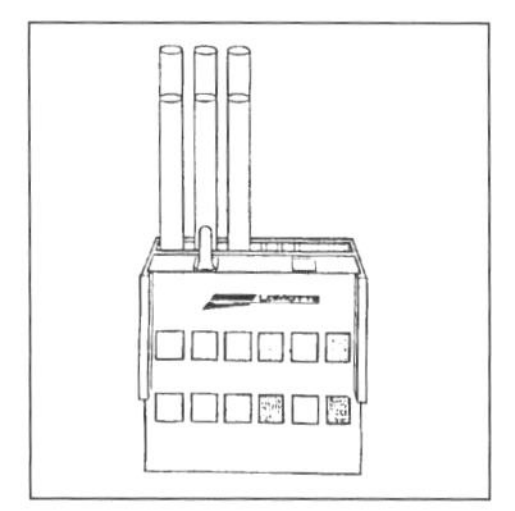

9. Match sample color to the index of color standards. Record the index number which gives the proper color match. ***Note:*** The reading must be taken within 2 to 5 minutes after the addition of the potassium ferricyanide.

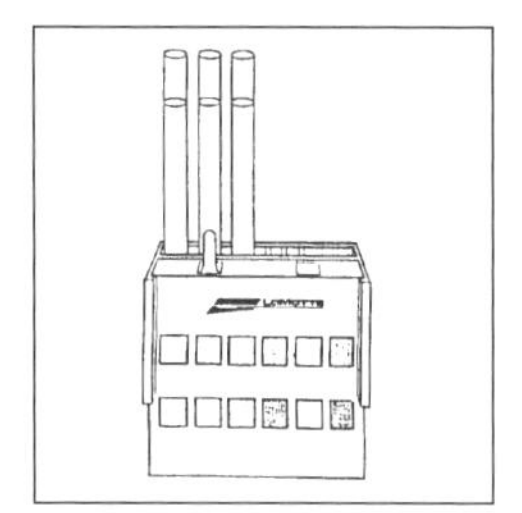

10. Use chart to convert index reading to a concentration. Record as ppm phenol.

Phenols in Air Calibration Chart*

Time (min)	Comparator reading							
	1	2	3	4	5	6	7	8
5	0.05	0.11	0.16	0.21	0.26	0.32	0.42	0.53
10	0.03	0.05	0.08	0.11	0.13	0.16	0.21	0.26
20	0.01	0.03	0.04	0.05	0.07	0.08	0.11	0.13
40	0.01	0.01	0.02	0.03	0.03	0.04	0.05	0.07

* values in ppm

Note: The sampling period specified in the directions corresponds to concentrations which may be encountered in ambient air conditions. High concentrations of pollutants in the atmosphere will require shorter sampling periods, while low concentrations will require longer sampling periods.

SAFETY INFORMATION

The Ammonium Hydroxide Solutions are considered to be hazardous materials. A Material Safety Data Sheet (MSDS) is available for these reagents. For your safety, read the label and the MSDS before using the chemicals.

QUESTIONS AND ANSWERS

1. How can the phenol volatility level be lowered in solution?
Phenol volatility in solution can be lowered by adding either sodium hydroxide or potassium hydroxide to the solution.
2. Are all phenol compounds toxic?
Resins that are comprised of phenolic structures are not toxic. Items made of phenoxy bakelite, epoxy resins, nylon-6, phenol-furfural resin, phenol-formaldehyde resin and several other types of polymeric phenolic-type structures are stable complexes used as machine parts, machine and instrument housings, electrical devices, brake linings, coatings, insulation, moldings and many additional applications. Some of the substituted phenol compounds such as ortho-nitrophenol or para-nitrophenol display a significanly different toxicological behavior than phenol. A number of substituted phenols are significantly less toxic than phenol.
3. Is phenol stable?
Some resins and substituted phenols are quite stable. Phenol in a bottle or container is stable for an extensive time. However, phenol tends to degrade in approximately 70 hours when exposed to different types of bacteria in water, soil and particulates.
4. How toxic is phenol?
Phenol can be absorbed through the skin and can lead to a fatal overdose. Phenol poisoning can cause circulatory system failure, gastrointestinal disturbances, convolutions, kidney malfunction, and lung edema. Chronic exposure to phenol can damage the spleen, pancreas, and kidneys. The human tolerance level to phenol is approximately 5 ppm in air. The maximum admissible concentration (MAC) allowable for phenols is 0.5 micrograms per liter.

REFERENCES

1. Stanley E. Manahan, *Environmental Chemistry,* 6th Ed. (Boca Raton, Fla.: Lewis Publishers, 1994).
2. Roger N. Reeve, *Environmental Analysis,* (ACOL), John D. Barnes (ed.) (New York, N.Y.: John Wiley & Sons, 1994).
3. *Fate of Pesticides and Chemicals in the Environment,* J. L. Schnoor (ed.) (New York, N.Y.: John Wiley & Sons, Inc., 1992).
4. *Hawley's Condensed Chemical Dictionary,* 11th Ed., rev'd by N. Irving Sax and Richard J. Lewis (New York, N.Y.: Van Nostrand Reinhold, 1987).
5. *Dräger Tube Handbook,* 11th Ed. (Lübeck: Dräger Sicherheitstechnik GmbH., 1998).
6. *Air Quality Sampling & Measurement Equipment* (Chestertown, Md.: LaMotte Company, 1997).
7. *Phenols in Air* (Chestertown, Md.: LaMotte Company, 1994).

POLYCYCLIC AROMATIC HYDROCARBONS (PAH)

Polycyclic aromatic hydrocarbons (PAH) are found in atmospheric particles and include compounds such as benzo(a)pyrene, benzo(a)anthracene, benzo(e)pyrene, benzo-(j)fluroanthene, indenol, chrysene, and benz(e)acephenanthrylene. The compounds are usually grouped together and referred to as an index of PAHs called "BaP." The compounds display various levels of carcinogenic effects. These compounds have been identified in black smoke and on soot particles produced during incomplete combustion processes. Inadequate burning promotes the formation of these compounds, whereas high combustion efficiency incineration processes destroy PAH compounds.

The PAH compounds are found in polluted urban environments and are products of natural fires, such as forest and prairie fires. Coal tars and asphalt-type compounds contain high levels of PAHs. PAHs are formed in gasoline and diesel engine exhausts, wood stove smoke, cigarette and cigar smoke, building fires, vehicle fires, coal and wood burning, oil burning, and burn barrels. Small amounts of PAHs are also formed during barbecues and food grilling. Larger amounts of PAHs are formed during poor waste incineration processes and during most incomplete combustion processes that produce gray or black smoke. Cigarette smoke contains almost 100 $\mu g/m^3$ of PAH compounds, whereas stack gas from a coal furnace may contain over 1000 $\mu g/m^3$ of PAH. Even though a number of the PAH products of combustion are fairly stable, a few PAH compounds adsorbed on soot particles can easily oxidize to form other organic products which are normally less toxic.

Benzo(a)pyrene and similar compounds can be metabolized by the body to form carcinogenic compounds. Some metabolites of PAH compounds, such as the 7,8-diol-9,10-epoxide of benzo(a)pyrene, are known to be potent mutagens and to cause cancer[1]. PAHs can be synthesized from low molecular weight hydrocarbons under oxygen-deficient conditions. PAHs can be formed relatively easily from cyclic compounds, unsaturated organic compounds, and higher alkanes present in fuels and plant materials.

Since the formation of soot particles and unburnt hydrocarbons is an indicator of PAH compounds, tests for PAH-type compounds can be performed with a PAH meter or specific types of hydrocarbon meters. Some common smoke detectors will provide evidence of smoke and soot particles formed during a fire. Each of these devices will detect the formation of smoke and, therefore, the potential formation of PAH-type compounds. These devices will *not* identify a specific PAH compound or the toxicity level of the compound. During the process of PAH detection, it is wise to remember that not all soot and black smoke contain significant levels of toxic forms of PAHs.

A number of PAH compounds and carcinogenic PAH compounds can be determined by a series of kit methods sold by Strategic Diagnostics Inc.[7] The U.S. EPA SW-846 approved methods can be used to detect the PAH toxics in soot, soil and other matrices. Most of the test kits are immunoassay methods for testing PAH substances at the parts per billion (ppb) level. The PAH compounds are detected as a group of compounds, called antigens. The antigen and an enzyme compete to combine with the antibody. Then, a color is developed based on the amount of enzyme remaining in the solution. The resulting color can be detected by a spectrophotometer. During the test, the darker the color, the less analyte in the tube. A tube or microliter well with a color lighter than the negative control is presumed positive. To quantitatively determine the concentration of the unknown, the analyst can run unknown samples alongside standards containing known concentrations of the antigen. Each compound within the group of PAH compounds has a different sensitivity level. Sample dilution techniques before analysis can extend the range of detection.

Analysis for specific PAH compounds include more expensive and elaborate methods of detection such as gas chromatography and mass spectrometry. Soot and other solids containing the PAH compounds must be heated to provide positive test results for these toxic forms.

QUESTIONS AND ANSWERS

1. **Where are PAHs found?**
   Polycyclic aromatic hydrocarbons, PAHs, are found on soot from natural fires during the burning of wood, coal, wood, and paper. PAHs are present in engine smoke, cigarette and cigar smoke, building fires, and charbroiled food. Black smoke exhausts from many types of fires and combustion processes contain various quantities of the PAH compounds. Coal tars and asphalt normally contain these compounds. PAHs are often found in polluted urban environments.
2. **Why are some PAH compounds considered dangerous?**
   These compounds can be metabolized by the body to form carcinogenic compounds. Some of the metabolites are potent mutagens and are known to cause cancer.
3. **Without a test for the specific chemicals, how would we know if the harmful substances are being formed?**
   Some of the PAH compounds are generally present during the formation of soot, smoke from smoking and fires, unburnt hydrocarbons, and improperly burned fuels. Paper, wood, plants, fuels, and other items which have been charred or partially burned are indicators of the presence of PAH compounds. The toxicity levels of the PAH compounds formed during incomplete combustion processes will remain unknown until more expensive and extensive tests can identify the specific compounds.
4. **What compounds are called PAHs?**
   The PAH-type compounds are benzo(a)pyrene, benzo(a)anthracene, benzo(e)pyrene, benzo(j)fluroanthene, indenol, chrysene, and benz(e)acephenanthrylene.

REFERENCES

1. Stanley E. Manahan, *Environmental Chemistry,* 6th Ed. (Boca Raton, Fla.: CRC Press, Inc., 1994).
2. *Fate of Pesticides and Chemicals in the Environment,* J. L. Schnoor (ed.) (New York, N.Y.: John Wiley & Sons, Inc., 1992).
3. James Cudahy, Chris McBride, Floyd Hasselriis, et al., *Comparison of Hazardous Waste Incinerator Trace Organic Emissions With Emissions From Other Common Sources,* Incineration and Thermal Treatment Technologies Conference, Portland, OR., May 10, 2000.
4. Ben Pierson, P. E., New York State Department of Health, Albany, New York.
5. Jack Lauber, P.E., Latham, New York.
6. Cathy McDonald, Mine Safety Appliances Company, Safety Products Division, Cranberry Township, PA.
7. James Eberts, Strategic Diagnostics Inc., 111 Pencader Drive, Newark, Delaware, 19702.

SULFUR DIOXIDE

Sulfur dioxide, SO_2, dissolves in water to form sulfurous acid, hydrogen sulfite ion, and sulfite ion. Because of its solubility in water, inhaled sulfur dioxide combines with the moisture in the upper respiratory system, irritating mucous membranes and the respiratory tract as well as eyes and skin. Exposure to the sulfur dioxide gas stimulates mucus secretion in humans and causes death near 500 ppm. Incidents of sulfur dioxide entrapment in a valley area and during temperature inversions have been the cause of the death of people at much lower concentrations.

Exposure of plants to high levels of the gas can kill leaf tissue (leaf necrosis). Chronic exposure to sulfur dioxide causes a bleaching or yellowing of the green portion of the leaf. Plant injury due to sulfur dioxide increases with the rise in relative humidity. Continual exposure to sulfur dioxide will reduce yields of grain crops. Sulfur dioxide can form sulfuric acid aerosols in the atmosphere and damage leaves of plants.

The primary sources of sulfur dioxide are the burning of coal and residual oil. Sulfur dioxide emissions from large coal-fired power stations without scrubbers can easily reach the 500 ppm level. During the burning of fuels, the primary sulfur product leaving the stack is sulfur dioxide with approximately 1 to 2% sulfur trioxide in the waste gases. Factors which influence the atmospheric chemical reactions of sulfur dioxide include humidity, temperature, light intensity, atmospheric transport, particulate matter surface characteristics and the presence of additional reactants. Sulfur dioxide has a tendency to react with particulate matter which settles from the atmosphere during rainfall or other processes. Much of the sulfur dioxide is oxidized and reacted with water to form sulfuric acid and sulfate salts of ammonium. In addition, soot particles produced during incomplete combustion of carbon-type fuels tend to catalyze the oxidation of the sulfur compound to sulfate.

Increased levels of sulfur dioxide can occur from chemical reaction processes and from fuel burning in unventilated areas. Additional sources of sulfur dioxide and sulfur-containing emissions include smelting of sulfur-bearing metal ores and petroleum refining. Significant sources of sulfur dioxide are tire burning and air pollution from vehicle exhausts. Diesel engines can be significant producers of sulfur oxides. Minor sources of sulfur dioxide are from the oxidation of carbonyl sulfide, COS, and carbon disulfide, CS_2, by the hydroxyl radical (HO·) in the atmosphere.

Sulfur dioxide pollution causes deterioration of building materials by attacking limestone, marble and dolomite materials, which are primarily calcium and magnesium carbonates. The reactions of these materials with SO_2 produce either water soluble compounds or solid crusts of hydrated forms of calcium and/or magnesium sulfate on walls of structures.

Some of the lakes in the USA appear to be acidified by pines of the area and sulfur dioxide emissions from metal sulfide roasting or smelting industry and from coal fired utilities. The operation of cement kilns and lime producing kilns emitting fine lime and cement particles made from limestone can partially neutralize the sulfur dioxide emissions of coal-fired plants and metal smelting processes. Sulfur dioxide and sulfur trioxide spills and accidents can be addressed by the appropriate application of lime-neutralizing materials.

Sulfur dioxide can react by a number of mechanisms in the atmosphere including: a) chemical processes in water vapor or droplets; b) photochemical reactions; c) photochemical and chemical reactions in the presence of alkenes, nitrogen oxides and other substances. A very important chemical reaction includes the reaction of sulfur dioxide with oxygen to form sulfur trioxide and subsequent reaction with water droplets or vapor to form sulfuric acid contaminated water droplets in the atmosphere.

In the atmosphere, hydrogen sulfide is rapidly converted to sulfur dioxide[1]. The hydroxyl radical, HO·, oxidizes SO_2 in the gas phase to form a reactive free radical, $HOSO_2$·, which can be easily converted to a sulfate. In water droplets, hydrogen peroxide reacts with SO_2 to form sulfuric acid. Also, the oxidation of SO_2 by ozone to form sulfate in photochemical smog appears to be significant. The presence of ammonia in water droplets favors the formation of bisulfite and sulfite ions during oxidation of sulfur dioxide. The metals, iron (III), and manganese (II) catalyze the oxidation of SO_2 in water. Dissolved nitrogen species, NO_2 and HNO_2, also tend to oxidize sulfur dioxide.

Sulfur dioxide removal processes in industry have become important to control emission levels. A number of methods are appropriate for sulfur removal from coal. Physical separation techniques and chemical methods can be used to remove sulfur from coal. Burning coal with finely ground limestone under a variety of conditions can effectively minimize or neutralize sulfur dioxide emissions. Major gas scrubbing operations using a lime slurry, calcium carbonate, magnesium hydroxide slurry, sodium sulfite scrubbing, or a combination of sodium hydroxide and lime can effectively significantly reduce sulfur dioxide emissions. Additional methods of sulfur dioxide conversion to sulfur can be implemented.

Sulfur dioxide can be monitored by an absorption train in an external atmosphere. An aqueous solution of hydrogen peroxide can be used as an absorbent during the procedure. In addition, electrochemical sensors can detect the sulfur dioxide gas at low levels by producing a current proportional to its gas-phase concentration at a gas sensitive electrode. Sulfur dioxide can also be analyzed by a portable spectrometer which measures gases in the ultraviolet range.

For quantitative analysis, sulfur dioxide gas or an air sample containing the gas is bubbled into a tetrachloromercurate solution. The compound is reacted with pararosaniline in formaldehyde to form a red-violet dye which can be measured spectrophotometrically at 548 nm.

Alternatively, a number of Dräger-tubes can be used to determine the concentration of sulfur dioxide in air. The tubes for short-term measurements have detection ranges of 0.1 to 3 ppm, 1 to 25 ppm/0.5 to 5 ppm, 1 to 25 ppm, 20 to 200 ppm, and 400 to 800 ppm/50 to 500 ppm. A variety of chemical reactions occur in the different tubes to provide colored products to indicate the level of sulfur dioxide sampled in the air. In one case, the measuring range of the tube can be extended. The interferences can be acidic gases, hydrogen sulfide, nitrogen dioxide, or hydrochloric acid depending on the tube components and the application.

During long-term measurements, sulfur dioxide will react with pH indicator to form a yellow product indicating the concentration of the air contaminate. A constant low flow pump will be used during long-term measurements to determine the time weighted average concentration of the air pollutant. A number of acid gases will respond to this test with different discolorations of the tube medium and different levels of sensitivities. Therefore, a sulfur dioxide measurement will probably not be possible in the presence of other acid gases.

A long-term measurement of sulfur dioxide using a mercury complex and methyl red complex on a tube medium forms a red indication in the tube. Again, a constant low flow pump will be used over a period of 1 to 4 hours to determine the concentration of the contaminate in air within the range of 1.25 to 50 ppm. The interferences are not a real problem, however sulfur dioxide in the presence of nitrogen dioxide is impossible to measure.

The test procedure to detect sulfur dioxide in an air sample is provided in an experiment titled "Measurement of Air Contaminates." Each tube of reactants forms a colored medium indicating the presence of a specific level of sulfur dioxide in the unknown sample. The reactions in the tubes occur within a specific range of temperature and absolute humidity. Each tube reaction is provided with a standard deviation for the measurement.

"Diffusion Tubes with Direct Indications" perform long-term measurements by allowing contaminated air to flow through an open end of a sampling tube. The blue-violet medium of the tube will turn light yellow to indicate the concentration (ppm) of contaminate in the sample. The standard range of measurement is 0.7 to 150 ppm of sulfur dioxide determined during a 1 to 8-hour time frame. Generally, it is nearly impossible to measure sulfur dioxide in the presence of other acidic substances. However, a sample containing 10 ppm HCL produces a pink color on the medium in 6 hours, and a sample containing 20 ppm acetic acid produces a yellow color on the tube medium. Nitrogen dioxide and chlorine also influence the development of tube medium color.

Sulfur dioxide in air can also be determined quantitatively by the LaMotte Method. After sampling the air containing SO_2 gas for a specific time at 1.0 liter per minute and developing a colored solution, the sample concentration of sulfur dioxide can be determined by matching the sample color with 8 permanent color standards in a comparator. Then, the comparator index reading is converted to the concentration of the contaminant in the atmosphere in parts per million (ppm) by finding the corresponding index value number in the calibration chart and reading down the column until the sampling time is located.

SULFUR DIOXIDE

Procedure*

1. Add 10 mL of Buffered Absorbing Solution to the impinging tube. Connect impinging apparatus to intake of air sampling pump. Make sure the long tube is immersed in the absorbing solution. Sample at 1.0 Lpm for 30 minutes or until a measurable amount of sulfur dioxide is absorbed. Cover impinging apparatus with aluminum foil to protect from light.
Note: Before collecting sample, read Air Pollution Sampling and Test Equipment Instruction Manual included with BD Pump.

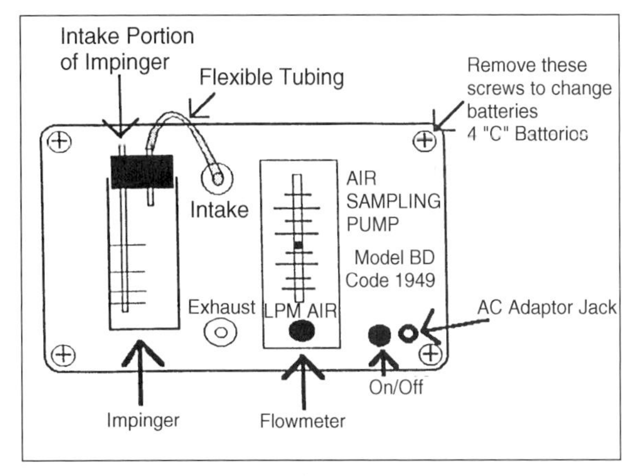

2. At the end of the sampling time, fill the small test tube to the line with the absorbing solution from the impinging tube. Add one level measure of Sulfur Dioxide Reagent 1 with the 0.25-g spoon. Cap test tube and shake vigorously to dissolve the powder.

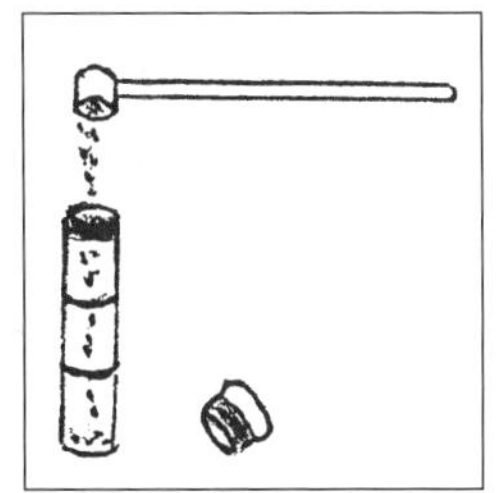

3. Use a 1-mL pipet to add 1 mL of 1N Sodium Hydroxide to the small test tube. Cap and invert several times to mix.

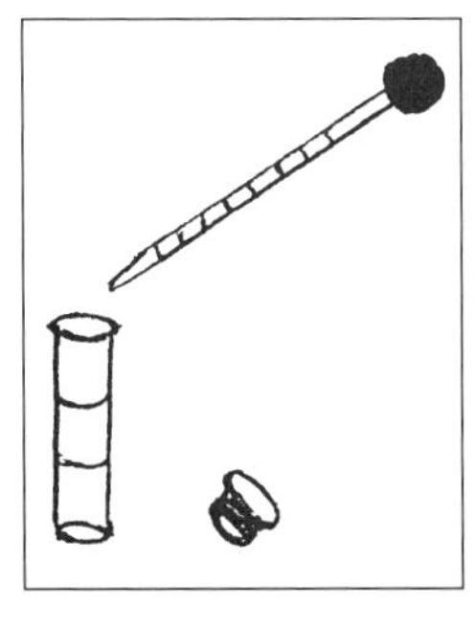

*Compliments of LaMotte Company, Chestertown, Md.

4. Use the other 1-mL pipet to add 2 mL (2 measures) of Sulfur Dioxide Passive Bubbler Indicator to a large test tube.

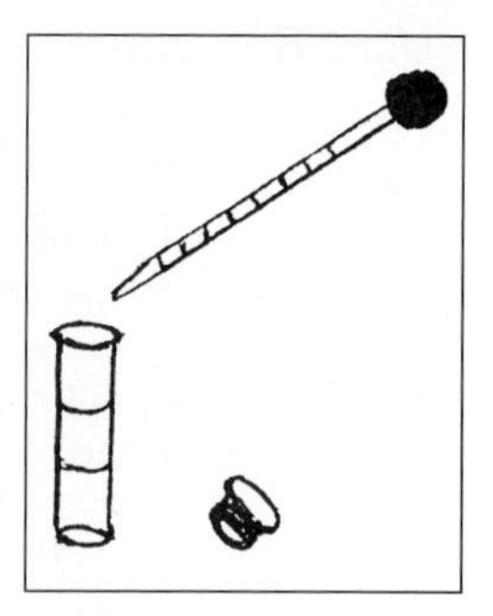

5. Pour the contents of the small test tube into the large test tube containing the indicator. Immediately cap tube and invert 6 times, holding cap firmly in place with index finger.

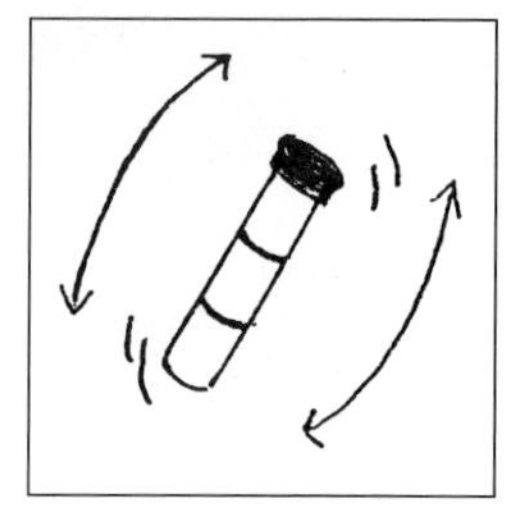

6. Wait 15 minutes. Place test tube into the Sulfur Dioxide Passive Bubbler Comparator. Match sample color to a color standard. Record index number from comparator.

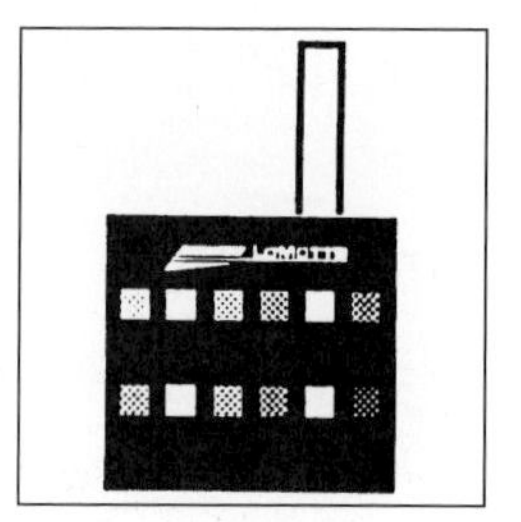

The following calibration chart is provided to convert the comparator index reading into the concentration of the contaminant in the atmosphere in parts per million (ppm). The chart is based upon the prescribed sampling period for individual tests. After the color match is made, find the corresponding index value on the calibration chart, then read down the line until the sampling time is found.

Sulfur Dioxide in Air Calibration Chart*

Time (min)	Comparator reading							
	1	2	3	4	5	6	7	8
10	0.00	0.19	0.29	0.38	0.48	0.57	0.67	0.76
30	0.00	0.06	0.10	0.13	0.16	0.19	0.22	0.25
60	0.00	0.03	0.05	0.06	0.08	0.10	0.11	0.13
90	0.00	0.02	0.03	0.04	0.05	0.06	0.07	0.08

* values in ppm

Note: The sampling period specified in the directions corresponds to concentrations which may be encountered in ambient air conditions. High concentrations of pollutants in the atmosphere will require shorter sampling periods, while low concentrations will require longer sampling periods.

SAFETY INFORMATION

Sulfur Dioxide Reagent 1, Sodium Hydroxide, 1N, and Sulfur Dioxide Passive Bubbler Indicator are considered hazardous substances. Material Safety Data Sheets (MSDS) are available for these reagents. For your safety, read the labels and accompanying MSDS before using.

USE OF THE OCTET COMPARATOR

The Octet Comparator contains 8 permanent color standards. A test sample is inserted into the openings in the top of the comparator. The sample can then be compared to 4 color standards at once, and the value read off the comparator. For optimum color comparison, the comparator should be positioned between the operator and a light source, so that the light enters through the special light-diffusing screen in the back of the comparator. Avoid viewing the comparator against direct sunlight or an irregularly lighted background.

QUESTIONS AND ANSWERS

1. What are some of the problems with sulfur dioxide?
   Sulfur dioxide vapors can cause breathing problems at relatively low concentrations. Death of individuals have been noted in areas measured at levels below 10 ppm of sulfur dioxide. The limiting level of human exposure is near 500 ppm. Sulfur dioxide is an irritant to the eyes, skin, mucous membranes, and respiratory tract. Exposure of plants and stone buildings to sulfur dioxide pollution can cause permanent damage. Sulfur dioxide which is oxidized and adheres to water vapor forms an acid which can dramatically affect the pH of lakes and water supplies.
2. How can the problem of sulfur dioxide emissions be addressed?
   Most industries have added emission control devices to regulate sulfur dioxide, sulfur trioxide, and sulfuric acid vapor emissions. The burning of coal with limestone-type materials will minimize sulfur dioxide emissions. A variety of alkaline-type or limestone-based scrubbers can be installed on coal burning operations and residual oil burning units. The emissions from cement and lime-producing processes can neutralize sulfur dioxide air emissions.
3. How is sulfur dioxide formed?
   Sulfur dioxide is usually formed by the burning and oxidation of sulfur-containing fuels such as coal and residual fuel oils. Volcanoes, biological decay of organic matter, and reduction of sulfates are significant sources of hydrogen sulfide and/or sulfur dioxide. Hydrogen sulfide is rapidly converted to sulfur dioxide in the atmosphere.
4. How much of the sulfur is converted to sulfur dioxide during the burning of coal and residual fuel oil?
   During the oxidation process of burning coal and residual fuel oil, most of the sulfur is converted to sulfur dioxide. Only 1 to 2% of the sulfur leaves the stack as SO_3*. In cement and lime kilns, metals (calcium, iron, sodium) can tie up sulfur (in the form of* SO_2 *&* SO_3*) as inert sulfates during incineration processes.*

REFERENCES

1. Stanley E. Manahan, *Environmental Chemistry,* 6th Ed. (Boca Raton, Fla.: Lewis Publishers, 1994).
2. Roger N. Reeve, *Environmental Analysis,* (ACOL), John D. Barnes (ed.) (New York, N.Y.: John Wiley & Sons, 1994).
3. *Dräger Tube Handbook,* 11th Ed. (Lübeck: Dräger Sicherheitstechnik GmbH., 1998).
4. *Air Quality Sampling & Measurement Equipment* (Chestertown, Md.: LaMotte Company, 1997).
5. *Sulfur Dioxide in Air Test Kit* (Chestertown, Md.: LaMotte Company, 1999).

TOTAL OXIDANTS

A compound that spontaneously evolves oxygen at room temperature or under slightly heated conditions is an oxidizing material. Such chemicals react vigorously at ambient temperatures when stored near or in contact with chemical reducing agents. Specific organic chemicals such as cellulose-based materials can also react vigorously with oxidants. At times, the chemical reaction will produce fire or an explosion.

Chemicals described as oxidants can exist as solids, liquids, and gases. Generally, oxidizing compounds dissolve readily in water and rapidly oxidize many materials containing metal ions, anions, organic chemicals, and bacteria. Oxidants will dissolve or destroy paper, clothing, skin, protein, fibers, and many other items. Some oxidizers mixed with specific fuels burn explosively, while other oxidants decompose relatively easily in air.

Wrecks of commercial transport vehicles loaded with oxidizing agents in drums, tanks, and packages can present a challenge to the Hazmat team. When tanks containing toxic gaseous forms of oxidizing agents are ruptured, they can cause severe problems. Oxidants can present fire problems, water runoff challenges, and potential human health hazards.

Some types of oxidants are regarded as cancer-causing chemicals when inhaled by humans or animals. Many cancer-causing oxidants, such as chromates, are not easily inhaled. Excess amounts of ozone can cause headaches, irritations, and damage to pulmonary functions, whereas nitric acid causes irritation, corrosion, and pulmonary edema in humans.

Various chemicals such as peroxides, chlorates, perchlorates, nitrates, chromates, dichromate, ozone, peroxyacetyl nitrate (PAN), chlorine, bromine, fluorine, nitric acid, and permanganates are categorized as oxidants. Oxidizing substances can exist as inorganic compounds, acids, organic complexes, salt-acid compounds, and organometallic complexes. A few hazardous waste mixtures are categorized as oxidizers and have the ability to evolve oxygen.

Substances such as oxygen and epoxides are generally considered to be weak oxidants at normal temperatures. However, if these substances are stored under pressure or at low temperatures, these types of compounds can easily react as an oxidant when warming to room temperature occurs. Generally, increases in temperature will increase the rate of the oxidative process.

A number of oxygen-rich compounds are not normally included as oxidizing agents or total oxidants. However, these substances can be oxidizing agents under the appropriate conditions, and can oxidize solution components or materials in particulates. The oxidizing substances can also attach to oxidized materials on particulates and be transported in air.

The test for total oxidants in air can be used in industrial applications and various environmental problems. The test can also be applied to fires, spills, air studies, and storage facilities containing oxidants. The test includes the addition of three different reagents to the impinging tube, adjusting the flowmeter rate to collect air at a rate of 1.0 liter per minute, and sampling the air for 10 minutes or more until a pink sample color develops. The pink color is matched to a color of the "Total Oxidants in Air Comparator" and an index value is obtained for the unknown. By using a calibration chart, the recorded index value is converted to concentration of the total oxidant contaminates in the atmosphere in parts per million.

TOTAL OXIDANTS

Procedure*

1. Pour 10 mL of Total Oxidents Reagent 1 into impinging tube.

2. Add 2 drops of Total Oxidents Reagent 2. Swirl to mix.

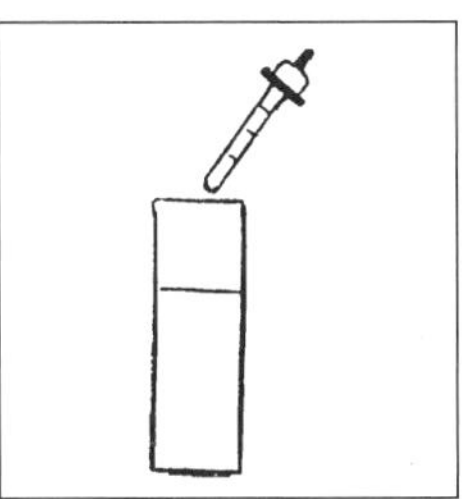

3. Use the pipet to add 2 drops of Total Oxidants Reagent 3. Swirl to mix.

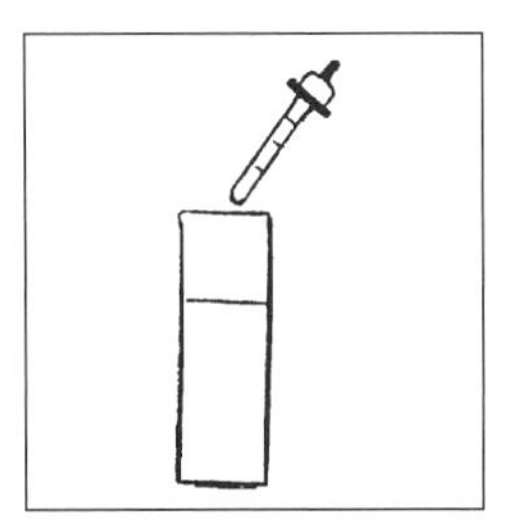

*Compliments of LaMotte Company, Chestertown, Md.

4. Connect impinging apparatus to intake of air sampling pump. Make sure the long tube is immersed in the absorbing solution. Cover impinging tube with foil or other opaque material to protect from light while sampling.
Note: An impinging tube and air sampling pump is also required for this experiment.

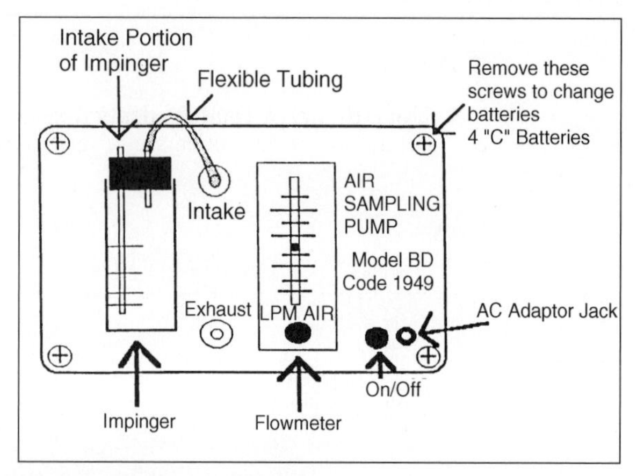

5. Adjust flowmeter of sampling apparatus to collect air at a rate of 1.0 Lpm. Sample for 10 minutes or until a measurable pink color develops. Note precise time of sampling.

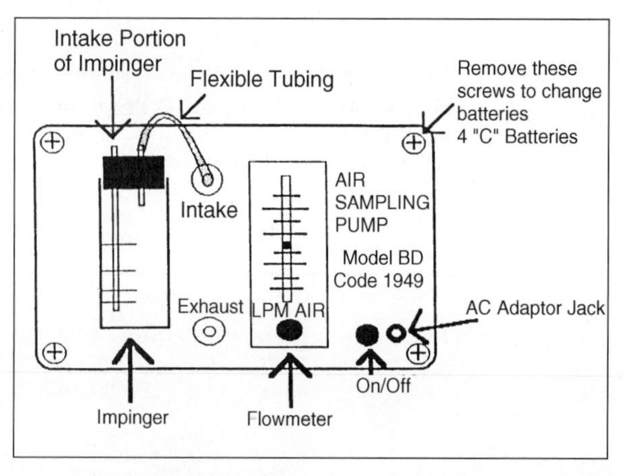

6. Disconnect impinging tube from pumping apparatus. Pour contents into a clean test tube.

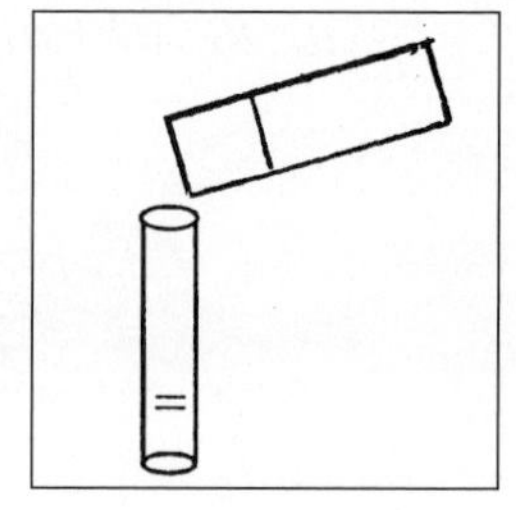

7. Insert test tube into the Total Oxidents in Air Comparator. Match sample color with an index value. Record index value.

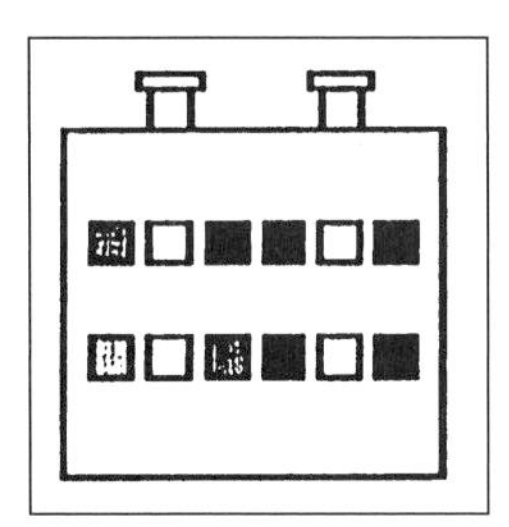

8. Use calibration chart to convert index reading into the concentration of the contaminant in the atmosphere in parts per million.

Total Oxidants in Air Calibration Chart*

Time (min)	Comparator index number							
	1	2	3	4	5	6	7	8
5	0.14	0.36	0.72	1.08	1.44	2.88	4.32	5.76
10	0.07	0.18	0.36	0.54	0.72	1.44	2.16	2.88
15	0.05	0.12	0.24	0.36	0.48	0.96	1.44	1.72

* values in ppm

Note: The sampling period specified in the directions corresponds to concentrations which may be encountered in ambient air conditions. High concentrations of pollutants in the atmosphere will require shorter sampling periods, while low concentrations will require longer sampling periods.

SAFETY INFORMATION

Total Oxidants Reagents 2 and Total Oxidants Reagents 3 are considered to be hazardous substances. Material Safety Data Sheets (MSDS) are available for these reagents. For your safety, read the label and accompanying MSDS before using the chemicals.

QUESTIONS AND ANSWERS

1. What is an oxidant?
   An oxidant can be defined as a compound that spontaneously evolves oxygen at room temperatures or under slightly heated conditions. Some chemicals will react vigorously at ambient temperatures when stored near or in contact with reducing agents.
2. Why should most oxidants be stored in closed containers?
   Some oxidants will decompose in air, while other active compounds will catch fire when exposed to inappropriate materials. Some oxidants are dangerous gases with the potential of causing death in humans and animals. Ozone can cause headaches and damage pul-

monary functions, whereas nitric acid fumes can cause pulmonary edema and blistered soft tissue in breathing passages. Most chromate compounds are solids and dissolve readily in water. The chromate compounds are cancer-causing to humans when taken internally.

3. Give a procedure to identify oxidized materials in water.
   Many metals in the oxidized state develop a fairly intense color in water. For instance, a solution of iron (II) in water has little color, whereas iron (III) in water will be a deeper yellow color. The oxidized metal will also react with specific chemicals. The reduced form of the metal will not generally react with the same chemical, but will react with other chemicals in solution.
4. An index value for total oxidants from the analysis of an air sample in a closed storage facility is 6.0 after the air was sampled for 10.0 minutes at 1.0 liter per minute. What is the concentration of the of the total oxidants in parts per million?
   The concentration of total oxidants in air is 1.44 ppm.

REFERENCES

1. Stanley E. Manahan, *Environmental Chemistry,* 6th Ed. (Boca Raton, Fla.: Lewis Publishers, 1994).
2. Roger N. Reeve, *Environmental Analysis,* (ACOL), John D. Barnes (ed.) (New York, N.Y.: John Wiley & Sons, 1994).
3. *Hawley's Condensed Chemical Dictionary,* 11th Ed., rev'd by N. Irving Sax and Richard J. Lewis (New York, N.Y.: Van Nostrand Reinhold, 1987).
4. *Air Quality Sampling & Measurement Equipment* (Chestertown, Md.: LaMotte Company, 1997).
5. *Total Oxidants in Air Test Kit* (Chestertown, Md.: LaMotte Company, 1999).

CHAPTER 5
RESIDENTIAL POLLUTION

TOXIC MATERIALS IN THE HOME ENVIRONMENT

A number of chemicals in products used around the home are present in small quantities. Some of these chemicals are significantly toxic, flammable, corrosive, poisonous, and dangerous. Chemicals isolated from other substances are sometimes fairly safe; however, mixing of a chemical with another material (chemical, cloth, etc.) can create gases, fires, heat, explosions, and extremely harmful situations. Always read the caution labels on containers of materials.

People need to read the labels, follow the warnings, and store the dangerous materials away from children. Some people are extremely sensitive to specific chemicals (e.g., acetates in clothing materials) used around the home and property. Individuals handling chemicals should avoid breathing of fumes and odors. Generally, each person should avoid or minimize contact with toxic chemicals by wearing protective clothing.

The primary chemicals and materials of significant problems found around the home include lead, asbestos, cleaning powders and solvents, molds, pollen, methane, chemical fumes from septic system, vehicle exhausts, and radon. Important substances which can lead to significant problems include motor oils, gasoline, kerosene, weed killer, lawn sprays, shrubbery and plant sprays, insecticides, pesticides, paints and stains, paint thinners, brush and roller cleaners, chemical shampoo used on rugs and materials, wall cleaners, floor cleaners, furniture cleaners, protective coatings, and chemicals used to open drains. Broken thermometers, barometers, light fixtures and other devices can lead to a mercury spill. Chemicals in solutions and pastes used on the human body can be a significant problem.

At times, noxious gases back up the basement floor drain or bathtub drain from the septic system. Exhausts from cars and other vehicles can contaminate the living quarters. Outdoor fires, leaf burning, and burning in barrels cause a significant amount of chemical pollution in the air. Commercial laundry facilities located next door can provide your home with noxious fumes from the use of chlorinated solvents. A number of cleaning solvents used to clean car parts and a variety of items can be fairly toxic to humans.

In addition, old transformers and other electrical devices hanging on electric poles or stored in the garage may still contain significant levels of PCBs. If a spill ever occurred, traces of PCB may still be located in areas where the transformer was once used. Oils from vehicles and a variety of sources such as electrical devices have been used to minimize dust on dirt roads. For years, a number of the used oils contained PCBs and other toxic materials. Toxic substances leaking from trucks driving along the road or trains passing an area can dramatically increase the level of toxicity along the roadside or railroad track. In fact, areas of the roadside in front of your property may be a very highly polluted site with toxic chemicals.

Some compounds used as insecticides, pesticides, weed killers, poisons, and compounds applied outdoors are usually not included in this list. Sprays applied to yards, gardens, and farm crops are generally not discussed. Compounds such as Diazinon, Carbonyl Insecticide, Methoxychlor, and Isopropylamine Salts usually require sophisticated methods of analysis procedures.

The Environmental Protection Agency (EPA) has linked pollutants found in buildings to dizziness, headaches, coughing, sneezing, nausea, burning eyes, chronic fatigue, and flu-like symptoms.[2] These symptoms are known as the "*sick building syndrome*." A building is considered "*sick*" when at least 20% of the occupants suffer persistent symptoms that disappear when they leave the building.[2] In most cases, new buildings are more commonly "*sick*" than old facilities because of new furniture, new paneling, and new carpeting. Also, buildings with reduced air exchange can cause severe problems for individuals who are more susceptible to the odors.

According to EPA studies, levels of pollution are generally higher inside homes and commercial buildings than outdoors.[2] Also, pollution inside cars on traffic-clogged U.S. urban highways can be considerably higher than the pollution outside cars. The health risks from exposure to toxic chemicals are significantly magnified because people spend most of their daily living indoors. Indoor air pollutants are a high-risk health problem for humans, especially in the area of cancer.[2] At greatest risk are infants and young children, smokers, the elderly, the sick, pregnant women, people with respiratory problems or heart problems, and factory workers.

The chart (Table A) below is a partial list of potentially dangerous materials commonly found around the home. Chemicals discussed in various chapters of the text are referenced in the last column. Some information about the chemical tolerance level in air and the source of the pollutant is also provided in the table.

REFERENCES

1. *Hawley's Condensed Chemical Dictionary,* 11th Ed., rev'd by N. Irving Sax and Richard J. Lewis (New York, N.Y.: Van Nostrand Reinhold, 1987).
2. G. Tyler Miller, *Living in the Environment,* 9th Ed. (Florence, Ky.: Wadsworth Publishing Co., 1999).

ASBESTOS

Asbestos is a group of impure magnesium silicate crystals which occur in fibrous forms. The multi-color fibers are noncombustible. The three types of asbestos of different structural compositions and components are called serpentine asbestos, amphibole asbestos, and Amosite. Some fibers are somewhat brittle and more resistant to heat and chemicals, while other forms of asbestos are composed of flexible and long fibers.

Asbestos is toxic by inhalation of dust particles, and is classified as an active carcinogen. The primary illnesses of humans exposed to asbestos fibers over a period of time are lung disease and lung cancer. Human tolerance levels to all forms of asbestos is 5 fibers per cubic centimeter with length greater than 5 μm.

Asbestos is primarily used in the home as pipe insulation and fire protection sheets. At times, the roof materials will also contain some asbestos, especially on older roofs. Vinyl ceiling and floor tiles containing asbestos are common in many homes and offices.

Asbestos is used in fireproof fabrics, vehicle brake linings, gaskets, roofing compositions, electrical insulation, insulation around pipes, fire protection barriers, paint filler, chemical filters, diaphragm cells, component of paper dryer felts, and as a reinforcing agent in rubber and plastics. Often, these materials do not appear to produce dust or particles that are a problem. However, individuals who work with these types of asbestos-rich items can come

TABLE A Residential Hazards

Chemical	Source/application	Tolerance level in air	Chapter
Acetaldehyde	Synthetic flavors	100 ppm	4
Acetates	Clothing, fibers	—	—
Acetic Acid	Drain cleaners, cleaning agent, acidifier, solvent	10 ppm	(4)
Acetone	Paints, spray paint, varnish & lacquer solvent, stain blocker, cleaning solvent, paint stripper, brush cleaner, wood cleaner	750 ppm	(4)
Acetylene	Welding, brazing and cutting metals	Asphyxiant	—
Acids, strong (HCL, H_2SO_4, HNO_3, etc.)	Cleaning metals, etchant, dissolving substances	2–5 ppm & variable	2 & (4)*
Alcohol(Methanol, Ethanol & Isopropanol)	Liquid fuel, cleaning solvents, wood cleaner, paint stripper, all purpose remover, windshield washer solution, home heating oil extender, rubbing alcohol	200 ppm	(4)
Ammonia or Ammonium Hydroxide	Cleaning solutions, fertilizers, refrigerant in limited cases, stripper, decomposition products & sewage	25 ppm	2 & 4
Ammonium Nitrate	Fertilizer, pyrotechnics, freezing mixture ingredient, rocket propellant, herbicides & insecticides	—	2 & 3 (Nitrate)
Arsenic Trioxide	Insecticides, weed killer, rodenticide, herbicide, wood preservative	0.05 mg /m^3	2
Asbestos	Pipe insulation, vinyl ceiling and floor tiles, fire wall insulation	5 fibers per cc > 5μm in length	5
Benzo(*a*) pyrene or Polycyclic Aromatic Hydrocarbon (PAH)	Cigarette and cigar smoke, wood stoves, engine exhausts, coal burning, burn barrels, vehicle and building fires, wood fires, agriculture burning, wildfires	Unknown	4
Carbon Monoxide**	Vehicle exhausts, furnace exhausts, fireplace fumes, woodstove exhausts, unvented gas stoves, kerosene heaters	35 ppm	4 & 5
Chlorinated Hydrocarbons of PERC, TCA, & TCE	Cleaning solvents, degreasers, heat transfer medium, veterinary medicine, dry-cleaning solvent	10–300 ppm in air	4
Chlorine***	Swimming pool additive, water purification, cleaning solutions	1 ppm	2 & 4
Chlorine Dioxide	Cleaners, oxidizers, swimming pool, odor control, biocide	0.1 ppm	—
Chloroform	Contaminate in chlorinated water, insecticide, fumigant	10 ppm	5
Chlorotoluene	All purpose remover (cleaner)	50 ppm	—
Gasoline, Kerosene	Fuels for various motors, degreasing & cleaning parts	—	(4)
Ethylene Glycol	Paint thinner, antifreeze, coolant	50 ppm	(4)
Formaldehyde	Urea-formaldehyde foam insulation, disinfectant, germicide, durable press treatment of fabrics, paneling, particle board, furniture stuffing , synthetic cushion materials	2 ppm	4 & 5
Hydrocarbons or Petroleum Distillates	Fuels, wood finish, charcoal lighter, cleaning solvent	—	(4)
Hydrogen Sulfide	Sewer gas, rotting vegetation in wet areas, bathtub drains, open water drains, basement drains	10 ppm	4
Lead & Lead Compounds	Hair dyes, solders, ammunition, paints, safety matches, wood preservative, pipe joint packing materials, candle wicks with metal cores, insecticides, stained glass	1.5 mg/m^3 (Variable)	2, 3, 4, & 5
Mercury	Thermometers, barometers, bottles of mercury in storage, mirror coating, mercury vapor lamps, electrical apparatus	0.05 mg/m^3	2, 3 & 4

TABLE A Residential Hazards (*Continued*)

Chemical	Source/application	Tolerance level in air	Chapter
Methane & Ethane	Exhausts from improper combustion of fuels, natural gas leaks	—	(4)
Methylene Chloride	Wood cleaner, paint strippers, degreasing, paint thinners	100 ppm	5
Methyl Ethyl Ketone	Spray paint, brush cleaner, paint remover, cleaning fluids, adhesive solvent	200 ppm	—
Monoethanolamine	Stripper, dry-cleaning, paints, polishes	3 ppm	—
Naptha (Petroleum Ether)	Paints, all purpose remover, liquid nails, paint and varnish thinner, brush cleaner, rubber cement solvent	—	—
Nitrogen Oxides - NO_x as NO & NO_2	Lightning, vehicle exhausts, burning leaves, burning grass clippings, unvented gas stoves, woodstoves, kerosene heaters, high voltage devices	25 ppm - NO 3 ppm - NO_2	4 & 5
Perchloroethylene or Tetrachloroethylene	Tar and grease removal, dry cleaning solvent	100 ppm	5
Phenol	Roof tars, patching materials, blacktop, tube tars, salves and skin products, mouthwash, disinfectant	5 ppm	2 & 4
Phosphate (Trisodium, Dibasic & Monobasic)	Surface cleaners on deck, siding & walls, water softening, detergent, food & feed additive, fertilizer, plant nutrients, boiler water treatment, baking powders, fireproofing materials, buffer, cheese	—	2 & 3
Propane & Butane	Cigarette lighter, fuel for stoves & furnaces, food grill	800 ppm Asphyxiant	—
Radon-222	Radioactive soil and soil surrounding foundation, water supply	5 picocuries per liter in water	5
Sodium Hydroxide	Deck wash, oven cleaner, drain opener, detergents	2 mg/m^3	2 (pH)
Sodium Hypochlorite	Deck wash, mildew stain remover, drain opener, bleach, swimming pool disinfectant, water purification, laundering	—	—
Stoddard Solvent (Petroleum Distillate)	Dry cleaning, spot & stain removal, paint thinner, paint stripper, brush cleaner	100 ppm	(4)
Styrene	Coatings, adhesives, radiator stop leak, plastics and resins, footwear, protective coatings, polyesters, toys, and furniture	50 ppm	5
Sulfur Dioxide & Sulfur Trioxide	Coal burning, tire burning, air pollution from vehicle exhausts, food additive to control bacterial growth	2 ppm	4
2,3,7,8-TCDD (Toxic Forms of Chlorinated Dioxins & Furans)	Garbage & waste burning in barrels and in the open, house & building fires, furnace air filter, clothes dryer lint, cigarette & cigar smoke, wood stove, fireplace, vehicle emissions, soot, brush fires, black smoke, incomplete combustion	—	3
Toluene	Wood cleaner, rust preventive, lacquer, adhesive solvent in plastic toys	100 ppm	5
1,1,1-Trichloroethane	Metal degreasing, pesticide, water and stain protection, aerosol spray	350 ppm	5
1,1,1-Trichloroethylene	Degreasing, paint & adhesive diluent, fumigant	50 ppm	4
Xylene	All purpose remover, paint thinner, brush cleaner, rust preventive, paint stripper, insecticides	100 ppm	5

*The numbers in parentheses provide the location of the chemical listed in a table within a chapter.

**Carbon monoxide tolerance level is 50 ppm in industrial workrooms.

***Chlorine in contact with ammonia forms poisonous chloroamines. Chlorine is dangerous when in contact with turpentine, ether, hydrocarbons, hydrogen, powdered metals and other reducing materials.

home to a family with asbestos fibers on their clothing. Limited exposure to asbestos and proper cleaning efforts by people exposed to these substances must be implemented to avoid health problems later in life.

Safety clothing and special breathing apparatus are required for asbestos removal from homes, schools and other buildings. Areas containing asbestos insulation must be isolated before asbestos removal can occur. In many areas, only licensed people can remove asbestos from a site.

Asbestos particles are tracked and collected by air filtration for a specific length of time. The fibers are analyzed by counting the number of fibers collected on the filter. The length of the fibers is noted as an important piece of information.

QUESTIONS AND ANSWERS

1. Are asbestos fibers dangerous?
   According to the EPA and public health officials, asbestos fibers are one of the four most dangerous indoor air pollutants.
2. How dangerous are asbestos fibers?
   Asbestos is toxic by inhalation of dust particles. Asbestos is classified as an active carcinogen. Lung disease and lung cancer are the primary diseases caused by breathing in too many asbestos fibers.
3. What is the best procedure for removing asbestos?
   A state licensed person should be hired to remove asbestos. Breathing apparatus and protective clothing should be required for removal of asbestos.
4. Where do you find asbestos around the home?
   Asbestos is primarily found in pipe insulation, vinyl ceiling tiles, floor tiles, older roofing materials, and fire protection panels.

REFERENCES

1. G. Tyler Miller, *Living in the Environment,* 9th Ed. (Florence, Ky.: Wadsworth Publishing Co., 1999).
2. *Hawley's Condensed Chemical Dictionary,* 11th Ed., rev'd by N. Irving Sax and Richard J. Lewis (New York, N.Y.: Van Nostrand Reinhold, 1987).

CARBON MONOXIDE

Carbon monoxide, CO, is an odorless, colorless gas and a poisonous inhalant. It is a toxicant which binds in the place of oxygen in the hemoglobin structure and prevents hemoglobin from carrying oxygen to the body tissues. Carbon monoxide exposure at 100 parts per million (ppm) causes headache, dizziness, drowsiness, irregular heartbeat, and weariness. Loss of consciousness occurs at 250 ppm and inhalation at 1000 ppm results in rapid death. Chronic long-term exposure to low levels of CO causes disorders of the heart and the respiratory system

Around the home, the carbon monoxide pollution is the result of incomplete combustion of fuels in furnaces and vehicles. Concentrations of carbon monoxide may reach hazardous levels in poorly vented garages, and in closed buildings. Additional sources of carbon monoxide pollution include forest and structure fires, and burning of rubber tires and other materials.

Carbon monoxide emissions in a home or any building are attributed to faulty furnaces, kerosene heaters, woodstoves, fireplaces, unvented gas stoves, and vehicle emissions.[2] The health risks from exposure to this chemical are magnified because people spend at least 70% or more of their time indoors. In addition, carbon monoxide levels inside cars in traffic-clogged urban areas are many times higher than those in rural areas. Carbon monoxide concentrations in congested urban areas with high traffic patterns may become as high as 100 ppm. Carbon monoxide levels from vehicles operating inside garages directly under a house can be a major problem for the residents.

The test procedures to detect the various levels of carbon monoxide in air are provided in an experiment titled "Measurement of Air Contaminates" provided in Chapter 4. Carbon monoxide can be measured in the "short term" tube within the standard measuring ranges of 100 to 700 ppm and 5 to 150 ppm within a few minutes of time. The white tube medium containing iodine pentoxide reacts with carbon monoxide to form a brown-green band on the tube constituents to indicate a specific concentration of contaminate. Additional tubes are available for CO measurements at various concentrations.

In addition, "Diffusion Tubes with Direct Indications" perform long-term CO measurements by allowing air or vapor to diffuse through an open end of the tube. Carbon monoxide reacts with a palladium salt to form a gray-black palladium indicator mark on the tube. This method can detect carbon monoxide in the range of 6 to 600 ppm in an exposed tube for an extended period of time of 1 to 8 hours. Interferences from other gases are very limited by this method.

Carbon monoxide can also be determined by the LaMotte Method identified in Chapter 4. During the experiment, air containing carbon monoxide must be bubbled through an absorbing solution at a rate of 1.0 Lpm for 30 minutes, or until a measurable amount of CO is absorbed. This method requires the use of an axial reader for the measurement of faint color reactions. The color of the unknown sample is matched to a set of color standards. The color match provides an index number which can be use to determine the concentration of carbon monoxide in parts per million (ppm). The complete procedure in the CO experiment is sufficient for analysis for carbon monoxide.

QUESTIONS AND ANSWERS

1. What levels of CO cause problems to humans?
Carbon monoxide can cause memory problems and visual perception problems at 10 ppm. Exposure at 100 ppm causes headache, dizziness, drowsiness, irregular heartbeat, and weariness. Loss of consciousness occurs near 250 ppm and inhalation at 1000 ppm causes death.

2. What are the sources of carbon monoxide?
Carbon monoxide is generated by incomplete combustion of fossil fuels and other carbonaceous materials. Forest fires, vehicle engines, improperly operating furnaces, inadequately vented heat sources, burning of buildings and materials, and congested traffic provide the greatest sources of carbon monoxide poisoning.

3. Of the analysis methods presented, which is the least expensive method for CO analysis?
The least expensive method for CO analysis is the "Diffusion Tube with Direct Indications." Unfortunately, the time required for a reading will take 1 to 8 hours. Either of the other methods would be more appropriate for quick readings. For safe response, call one of the following: 1) the furnace cleaning and repair company, 2) the local fire department, 3) a hazmat team member, or 4) a local company safety officer or hazmat responder.

4. How does carbon monoxide harm a person?
Carbon monoxide bonds to hemoglobin in the place of oxygen and prevents hemoglobin

from carrying oxygen to the body tissues. Thus, the body starves for oxygen and death can result.

REFERENCES

1. Stanley E. Manahan, *Environmental Chemistry,* 6th Ed. (Boca Raton, Fla.: Lewis Publishers, 1994).
2. G. Tyler Miller, *Living in the Environment,* 9th Ed. (Florence, Ky.: Wadsworth Publishing Co., 1999).
3. Roger N. Reeve, *Environmental Analysis,* (ACOL), John D. Barnes (ed.), (New York, N.Y.: John Wiley & Sons, 1994).
4. *Dräger Tube Handbook,* 11th Ed. (Lübeck: Dräger Sicherheitstechnik GmbH., 1998).
5. *Air Quality Sampling & Measurement Equipment* (Chestertown, Md.: LaMotte Company, 1997).
6. *Carbon Monoxide in Air Test Kit* (Chestertown, Md.: LaMotte Company, 1998).

CHLOROFORM

Chloroform, sometimes referred to as trichloromethane, is a colorless volatile liquid with a sweet taste and a characteristic odor. The compound is slightly soluble in water and is considered to be nonflammable. However, the compound will burn on prolonged exposure to flame or high temperature.

Chloroform is a narcotic and toxic by inhalation. Prolonged inhalation or ingestion of chloroform may be fatal. The human tolerance level is 10 ppm in air or 50 mg per cubic meter of air. Chloroform is a known carcinogen (OSHA) and has been prohibited by the FDA from use in drugs (cough medicine), cosmetics, food packaging, toothpastes and other substances.

Chloroform is used as a fumigant and insecticide, and is a popular solvent used as a wash and rinse solvent to clean organics and water-insoluble substances from surfaces. It has been identified in chlorine treated water in hot showers at microgram per liter concentrations. Chloroform can be produced in trace quantities during the disinfection of water by chlorination. In the past, chloroform has been identified in samples of drinking water from the water tap, shower water, swimming pool water, and bottled drinking water.

Chloroform in air can be determined by an experiment with the title of "Measurements of Air Contaminates" outlined in Chapter 4. In the experiment, chloroform reacts with chromium (VI) to form chlorine which continues to react with o-tolidine to form a yellow reaction product on the Dräger tube. The test uses a combination of two short term tubes containing a white oxidation layer, a brown oxidation layer and a white indicating layer to display the reaction concentration. The standard measuring range is 2 to 10 ppm within the normal temperature range. Other chlorinated hydrocarbons are indicated with different sensitivities.

QUESTIONS AND ANSWERS

1. What is the primary source of chloroform around the home?
Chloroform can be produced in trace quantities during chlorination of water.

2. Why is chloroform considered to be dangerous?
Chloroform is a known carcinogen (OSHA). The FDA has prohibited its use in drugs

(cough medicine, etc.), cosmetics, food and food packaging, toothpastes and other items used by humans. Chloroform is toxic by inhalation and is classified as a narcotic.

3. In the past, what types of water have contained traces of chloroform?
Chlorinated water used for showers, drinking, and swimming have contained traces of chloroform. In a couple of documented cases, bottled water has also contained low levels of chloroform.
4. What substances can interfere with the tube determination of chloroform?
Chlorinated hydrocarbons such as dichloromethane and chlorinated ethanes can interfere with the determination of chloroform. Usually, these chlorinated hydrocarbons are indicated with a different sensitivity.

REFERENCES

1. Roger N. Reeve, *Environmental Analysis,* (ACOL) John D. Barnes, (ed.) (New York, N.Y.: John Wiley & Sons, 1994).
2. G. Tyler Miller, *Living in the Environment,* 9th Ed. (Florence, Ky.: Wadsworth Publishing Co., 1999).
3. *Hawley's Condensed Chemical Dictionary,* 11th Ed., rev'd by N. Irving Sax and Richard J. Lewis (New York, N.Y.: Van Nostrand Reinhold, 1987).
4. *Dräger Tube Handbook,* 11th Ed. (Lübeck: Dräger Sicherheitstechnik GmbH., 1998).

FORMALDEHYDE

Formaldehyde is a colorless gas with a pungent suffocating odor. It is commonly called formalin, a 37 to 50% aqueous solution of formaldehyde containing some methanol. Formaldehyde has a high volatility rate. Its vapors can irritate the eyes, throat, skin, and lungs, and with low-level exposure over an extended period of time, the compound causes nausea, dizziness, sinus and eye irritation, sore throat, headaches, rash, and chronic breathing problems.

The toxicity of formaldehyde is often attributed to its oxidation product, formic acid. At high concentrations, formaldehyde appears to be a lung carcinogen in some types of animals. The EPA has estimated that approximately 2 out of every 10,000 people who live in a manufactured home for more than 10 years will develop cancer from formaldehyde exposure.[3] Formaldehyde is considered to be one of the four most dangerous indoor pollutants.

Formaldehyde is found in furniture stuffing, paneling, particleboard, carpet and wallpaper adhesives, and foam insulation. Small amounts of the substance have been identified in laminated wood such as plywood. Formaldehyde is used during durable-press treatment of textile fabrics and as a disinfectant and germicide. In water, it can be easily oxidized with hydrogen peroxide to form a relatively nontoxic substance. Normally formaldehyde concentrations in water and soil should be below 100 ppm to minimize toxic effects.

Formaldehyde is readily produced by a variety of microorganisms. Since the carbonyl group is a **chromophore** (readily absorbs light in the near-ultraviolet region), aldehydes are significant sources of free radicals in the atmosphere. The photodissociation process is usually a two-step process that produces HCO radical, carbon monoxide and other species.

Formaldehyde in air can be determined by the LaMotte Method presented in Chapter 4. The air sample is collected by a BD pump connected to a Midget Impinger. The flowmeter is adjusted to 1.0 liter per minute and the air is sampled for 30 minutes or until a measurable amount of formaldehyde is absorbed. After reacting formaldehyde with a series of specific reagents, the final sample color is matched to an index of color standards in an Octet Comparator. Using a table, the index reading from the Octet Comparator is converted to a concentration value of ppm formaldehyde.

Alternatively, various concentrations of formaldehyde vapors can be measured by a Dräger tube as noted in Chapter 4. The short-term measurements in the tubes produce a white to pink color change during the formation of quinoid reaction products. During one of the measurements, the measuring range can be extended with a specific tube. Interferences such as styrene, vinyl acetate, acetaldehyde, acrolein, diesel fuel, and furfuryl alcohol are indicated with a yellowish brown discoloration of the tube media.

QUESTIONS AND ANSWERS

1. How toxic is formaldehyde?
The boiling point of an aqueous 37% solution of formaldehyde with 15% methanol is approximately 96°C and the human tolerance level is reported at 2 ppm in air. Pure formaldehyde is very volatile. However, most solutions containing formaldehyde in water are considerably less volatile that pure formaldehyde.

2. How flammable is formaldehyde?
An aqueous solution of 37% formaldehyde and 15% methanol has a flash point of 122°C. Methanol-free formaldehyde appears to have a flash point of 185°C. Compared to ethers and methanol, formaldehyde does not burn as well.

3. Where do we find formaldehyde around the home?
Formaldehyde is found in furniture stuffing, foam cushions, paneling, particle board, and foam insulation. Formaldehyde is used as a germicide and disinfectant. Also, formaldehyde is used during durable-press treatment of textile fabrics and clothes.

4. Briefly explain the photodissociation process of formaldehyde in sunlight?
Formaldehyde readily absorbs near-ultraviolet light and dissociates by a two-step process. The first step produces formyl radical (HCO) and a hydrogen atom (H), whereas the second step forms stable hydrogen gas and carbon monoxide. Carbon monoxide can undergo additional photochemical reactions.

REFERENCES

1. Stanley E. Manahan, *Environmental Chemistry,* 6th Ed. (Boca Raton, Fla.: Lewis Publishers, 1994).
2. Roger N. Reeve, *Environmental Analysis,* (ACOL), John D. Barnes (ed.) (New York, N.Y.: John Wiley & Sons, 1994).
3. G. Tyler Miller, *Living in the Environment,* 9th Ed. (Florence, Ky.: Wadsworth Publishing Co., 1999).
4. *Hawley's Condensed Chemical Dictionary,* 11th Ed., rev'd by N. Irving Sax and Richard J. Lewis (New York, N.Y.: Van Nostrand Reinhold, 1987).
5. *Dräger Tube Handbook,* 11th Ed. (Lübeck: Dräger Sicherheitstechnik GmbH., 1998).
6. *Air Quality Sampling & Measurement Equipment* (Chestertown, Md.: LaMotte Company, 1997).
7. *Formaldehyde in Air* (Chestertown, Md.: LaMotte Company, 1993).

LEAD

Lead and lead compounds have a number of toxic effects and inhibit the synthesis of hemoglobin in the body. High lead levels can cause a variety of health problems including headaches, digestive problems, high blood pressure, kidney damage, mood changes, sleep disturbances, and muscle and joint pain. Lead can lead to serious problems with the central and peripheral nervous systems in children. The retardation of intellectual development in children and young adults has been related to lead pollution. In addition, small dust particles containing lead are especially dangerous to small children, pregnant women and workers.

Sources of lead and lead compounds include coal and oil combustion, ammunition, solder and alloys, dyes, pigments, insecticides, paints, hair dyes, staining glass, varnish and paint dryers, waterproof paints, ceramic glazes, lead glass, curing agents, textile mordant, explosives, safety matches, lubricants, printing, medicine, wood preservative, and pipe joint packing materials. Lighted candlewicks with lead metal cores can release particles containing lead into the air.

Many common items contain lead contaminated chips and dust particles, volatilized lead, or water soluble lead. A home built before 1978 is likely to have surfaces painted with lead-based paint.[1] Scraping, brushing, dry-sanding, or blasting lead-based paint can produce poisonous chips or dust. Burning lead-based paint with an open flame torch to loosen the paint can produce fumes containing lead and volatile chemicals that are poisonous when inhaled. When working with lead-based paints, the worker should wear protective clothing and a respirator, and should shield individuals from exposure to lead-based paints.

The U.S. Environmental Protection Agency (EPA) established an action level for lead in drinking water at 15 $\mu g/L$ (micrograms per liter) or 15 ppb (parts per billion) during 1991. Brass and bronze metal parts found in submersible pumps can potentially leach a small amount of lead into waters with a pH below 6.8. Soft, corrosive, or acidic water is more likely to cause leaching of lead from lead pipes or soldered joints. Water left standing in the pipes increases the possibility of lead leaching. Stray electrical currents from improperly grounded electrical outlets or equipment can cause an increase in the lead levels in drinking water. Lead pipes and pipes repaired with lead-containing materials carrying water from the source to the home can contribute lead to the water supply. Generally, older pumps and pipes with oxidized metals and carbonate coatings reduce the possibility of lead leaching.

Lead has been identified in many types of ash and clinker discarded from a number of industrial furnaces and used in driveways and for construction materials. In many cases, these lead compounds identified in the solids are bound and not accessible to children. Additional sources of lead include old toys and furniture, lead crystal, lead-glazed pottery or porcelain, lead smelters, and hobby projects (pottery, stained glass, furniture). Folk remedies such as "greta" and "azarcon" use materials that contain lead to treat an upset stomach.

Air containing unknown lead quantities is pumped through a lead absorbing solution at 1.0 liter per minute for a specific recorded time (preferably 10 minutes) in the LaMotte Method described in Chapter 4. After adjusting the solution pH between 9 and 11, a sample of lead dithizone and a cyanide solution are appropriately added to the unknown lead solution. After appropriate sample preparation procedures, a comparator index number is recorded by matching the sample color with an index of color standards. The lead concentration value (mg/M^3) in air can be obtained from a "Lead in Air Calibration Chart" by locating the comparator index number at the specific recorded time.

Solids containing lead can be analyzed by the Dithizone Method (HACH Method 8033) presented in Chapter 3 of this textbook. Oils containing lead can be analyzed by the same method. Free or available lead can be determined by extracting the lead with slightly acidified water (pH = 1.0 with nitric acid) and testing the extracted sample. Lead that has been physically and chemically bound by materials can be determined by the Dithizone Method (HACH Method 8033) after digestion of the sample. Lead in water can be determined by the same analysis procedure located in Chapter 2.

QUESTIONS AND ANSWERS

1. What form does lead take in air?
Lead can be found in air as metal and/or compounds attached to various sizes and types

of particulate matter. Lead from lead strips in candles and other materials can be volatilized to a small extent in air by heating with a simple flame, i.e., candle flame, even though lead has a melting point near 327°C. Burning lead-based paint with an open flame torch to loosen paint from surfaces can produce toxic fumes containing lead.

2. What happens to lead dissolved in acidic water?
The lead can be consumed by humans and sea life and concentrate in specific organs. Lead dissolved in acidic water becomes more mobile and accessible to living species. At times, the soluble lead in water is deposited as an insoluble compound under neutral or basic conditions.

3. How does lead affect the human body?
Lead and lead compounds inhibits the synthesis of hemoglobin in the body. Lead affects the central and peripheral nervous systems. Headaches, digestive problems, high blood pressure, mood changes, sleep disturbances, and muscle and joint pain are symptoms of lead exposure. Lead also affects the kidneys and slows intellectual development of children and young people.

4. If the lead pollutant in air exists at significant concentrations, how could lead be detected in the air sample?
Using this procedure, the relatively high lead concentration in air can be determined by collecting the air sample for 5 minutes at 1.0 liter per minute.

REFERENCES

1. *Reducing Lead Hazards When Remodeling Your Home.* U.S. EPA Reprint, Office of Pollution Prevention and Toxics, EPA 747-R-94-002, April 1994.
2. Stanley E. Manahan, *Environmental Chemistry,* 6th Ed. (Boca Raton, Fla.: Lewis Publishers, 1994).
3. Roger N. Reeve, *Environmental Analysis,* (ACOL), John D. Barnes (ed.), (New York, N.Y.: John Wiley & Sons, 1994).
4. *Hawley's Condensed Chemical Dictionary,* 11th Ed., rev'd by N. Irving Sax and Richard J. Lewis (New York, N.Y.: Van Nostrand Reinhold, 1987).
5. *Fate of Pesticides and Chemicals in the Environment,* J. L. Schnoor (ed.). (New York, N.Y.: John Wiley & Sons, Inc., 1992).
6. *Protect Your Family From Lead in Your Home.* U.S. EPA, U.S. CPSC, U.S. HUD, EPA 747-K-94-001, Washington, D.C, May 1995.
7. *Air Quality Sampling & Measurement Equipment* (Chestertown, Md.: LaMotte Company, 1997).
8. *Lead in Air* (Chestertown, Md.: LaMotte Company, 1995).

METHYLENE CHLORIDE

Methylene chloride is a colorless, volatile liquid with a penetrating ether-like odor. The organic compound is nonflammable and nonexplosive in air. The human tolerance level is 100 ppm in air. The compound is a narcotic at high concentrations, and it appears to cause nerve disorders and diabetes in humans.

Methylene chloride is used as a paint stripper, paint thinner, and wood cleaner. Methylene chloride is soluble in organic solvents and also slightly soluble in water. Because of these characteristics, methylene chloride is often used to clean surfaces contaminated with both organic materials and inorganic compounds. Grease, salts, dyes, and other substances can usually be wiped away from surfaces with methylene chloride. The organic compound is found in mixtures with other organic solvents.

The test procedures to detect various levels of methylene chloride in air are provided in an experiment titled "Measurement of Air Contaminates" in Chapter 4. Methylene chloride can be determined by a "short-term" measurement by a Dräger tube within the standard measuring range of 100 to 2,000 ppm. In the tube, methylene chloride reacts with Cr (VI) to form a gaseous cleavage product, which continues to react with I_2O_5 to form brown-green iodine on the column. The listed ambient operating conditions for the air tests are 10 to 30°C with the absolute humidity in the range of 3 to 15 mg H_2O/L. Additional halogenated hydrocarbons are indicated at different sensitivities. In the presence of petroleum hydrocarbons and carbon monoxide, methylene chloride is impossible to measure due to the same sensitivity levels.

Methylene chloride has similar tube chemistry in acid conditions to form a brownish green product of iodine in a long-term Dräger tube. The standard range of measurement varies between 12.5 to 800 ppm over a 1 to 4 hour time period. During the test, a constant low flow pump will be used to determine the time weighted average concentration of the air pollutant. In addition, perchloroethylene, vinyl chloride, 1,1-dichloroethylene and other chlorinated hydrocarbons interfere with the measurement of methylene chloride. Carbon monoxide displays a higher sensitivity on the tube, whereas butane responds with a lower sensitivity on the tube.

QUESTIONS AND ANSWERS

1. Where do you find methylene chloride around the home?
Methylene chloride is used as a paint remover and wood cleaner. The compound cleans organic materials and inorganic compounds from surfaces.
2. Why is this compound more of an environmental problem than other organic chemicals such as toluene and xylene?
Methylene chloride is slightly soluble in water, whereas the other compounds are less soluble in water. The compound can contaminate water resources easily.
3. What are some interferences during the determination of methylene chloride by the tube method?
Halogenated hydrocarbons are indicated at slightly different sensitivities. Petroleum hydrocarbons and carbon monoxide interfere directly with the measurement of methylene chloride during short-term measurements. During long-term measurements, perchloroethylene, vinyl chloride, 1,1-dichloroethylene and other chlorinated hydrocarbons interfere with the measurement of methylene chloride.
4. What is the primary difference between a short-term measurement and a long-term measurement?
A constant low flow pump is used to determine the time weighted average concentration of an air pollutant during long-term measurements.

REFERENCES

1. *Hawley's Condensed Chemical Dictionary,* 11th Ed., rev'd by N. Irving Sax and Richard J. Lewis (New York, N.Y.: Van Nostrand Reinhold, 1987).
2. *Dräger Tube Handbook,* 11th Ed. (Lübeck: Dräger Sicherheitstechnik GmbH., 1998).

NITROGEN OXIDES

The important nitrogen oxides around the home are nitric oxide (NO) and nitrogen dioxide (NO_2), which are commonly referred to as NO_x gases. Nitrogen dioxide is a pungent red-brown gas, while nitric oxide (NO) is colorless and odorless. Nitric oxide (NO) is less stable, less toxic and biochemically less active than nitrogen dioxide (NO_2).

Nitric oxide (NO) can attach to hemoglobin and reduce oxygen transport efficiency.[1] In most cases, NO is easily oxidized to nitrogen dioxide in an oxidizing atmosphere. Nitrogen dioxide will increase the probability of colds, headaches, and irritated lungs and damaged breathing passages in children and susceptible adults. Nitrogen dioxide increases a person's susceptibility to pneumonia and lung cancer. The NO_2 levels near 100 ppm cause inflammation of lung tissue. Exposure of humans to about 180 ppm of NO_2 can cause death within weeks. Death is normally observed within days of human exposure to 500 ppm of NO_2.

In the home, the largest sources of nitrogen oxides are unvented gas stoves, wood stoves, and kerosene heaters. Burning homes are good sources of the nitrogen oxide compounds. Vehicles emissions of NO_x gases can be significant in poorly vented garages under a house or in areas of high traffic patterns. If the home is near a major airport, exhausts from supersonic aircraft will contribute some level of NO and NO_2 gases to the atmosphere around the home. The burning of leaves and grass clippings generate significant amounts of NOx gases. Burning of specific types of nitrogen-containing fibers and film can also produce significant amounts of NO_2-containing gases in woodstove or fireplace emissions. In general, nitrogen oxides are products of the combustion process of any nitrogen-containing fuel such as coal, leaves, wood, and petroleum fuels. Also, nitrogen dioxide poisoning can be generated by the fermentation of ensilage containing nitrate on farms.

The NO_2 gaseous compound causes paints and dyes to fade and combines directly with water molecules to form nitric acid or acid rain. Then, the nitric acid rain can be an environmental problem or it can react with bases (ammonia, particulate lime) to form particulate nitrates.[1]

Both nitrogen dioxide and nitrous fumes (NO) in air can be determined by an experiment titled "Measurements of Air Contaminates," located in Chapter 4. During indoor measurements, a short-term Dräger tube can easily measure nitrous fumes within the standard measuring range of 5 to 100 ppm/2 to 50 ppm. During the measurement, air containing NO and NO_2 is pulled through a tube containing chromium (VI) and diphenylbenzidine on an inert medium. The tube medium changes from a yellow to a blue-gray color to indicate the presence of the contaminate within a short-time period. The concentration of the air contaminate can be read directly from the tube. The presence of ozone and/or chlorine in excess of their threshold limit values (TLVs) will interfere with the measurement.

Tubes and equipment to determine both NO and NO_2 are available through Dräger and other companies. For indoor air pollution problems, a specific tube may be more appropriate for a determination of the NO and NO_2 gases. Additional methods are available for the determination of the amounts of gases in air.

Nitrogen dioxide in air can also be determined quantitatively by the LaMotte Method provided in Chapter 4. After adding Nitrogen Dioxide #1 Absorbing Solution into the impinging tube and connecting equipment, the air is sampled at 0.2 liters per minute (Lpm) for 10 to 20 minutes by a Model BD Sampling Pump equipped with an adapter for restricting the flow of air through a specific type of impinging apparatus. Then, a variety of chemicals are added to the collected sample to develop a colored solution. After waiting 10 minutes to develop a full color, the sample concentration of nitrogen dioxide can be determined by matching the sample color with permanent color standards in a comparator. The index reading from the comparator is converted to the concentration of the nitrogen dioxide in the atmosphere in parts per million (ppm) by finding the corresponding index value in the calibration chart and reading down the column until the sampling time is located.

QUESTIONS AND ANSWERS

1. What levels of NO and NO_2 cause problems to humans?
The gas, NO, can become attached to hemoglobin and reduce oxygen transport efficiency. The NO compound is biochemically less active and less toxic than NO_2. However, NO is easily oxidized in air to NO_2. Nitrogen dioxide, NO_2, causes pneumonia and lung cancer at relatively low levels. Human exposure to levels of nitrogen dioxide (NO_2) near 180 ppm can cause death within weeks. Exposure of a human to 500 ppm NO_2 usually causes death within a few days.
2. How are nitrogen dioxide and nitric oxide generated?
Nitrogen dioxide, NO_2, is generated by air oxidation of NO. In the home, both compounds can be generated by the burning of wastes, coal, wood, coal, kerosene, leaves, grass and plants, building materials, and buildings. Vehicle exhausts are also a significant source of these pollutants. NO_2 can also be generated during fermentation of ensilage in farm silos and during the burning of celluloid fibers and nitrocellulose film.
3. Which is the most stable in air: NO_2 or NO?
NO is easily oxidized to NO_2 in air containing oxygen. Both compounds combine with water to form acids. NO_2 causes the greastest problems.
4. What problems does nitrogen dioxide create in the environment?
The gaseous substance attacks paints and dyes and reacts directly with water droplets to form nitric acid rain. Nitric acid rain can be an environmental problem or it can be neutralized with basic chemicals to form nitrates.
5. How can the different nitrogen oxide fumes be distinguished from each other?
Nitric oxide, NO, is colorless and odorless. NO is easily oxidized in air to nitrogen dioxide. Nitrogen dioxide, NO_2, is a pungent, red-brown gas which can cause severe problems.
6. How can the homeowner solve the problem of the NO_x gases in the home?
The venting system can be modified so that the fumes are vented outside the house. Fans blowing air out of the house can be installed. Garage doors, windows and normal doors can be left open for a period of time. The chimney and smoke exhaust systems should be open and properly adjusted.

REFERENCES

1. Stanley E. Manahan, *Environmental Chemistry,* 6th Ed. (Boca Raton, Fla.: Lewis Publishers, 1994).
2. Roger N. Reeve, *Environmental Analysis,* (ACOL), John D. Barnes (ed.), (New York, N.Y.: John Wiley & Sons, 1994).
3. G. Tyler Miller, *Living in the Environment,* 9th Ed. (Florence, Ky.: Wadsworth Publishing Co., 1999).
4. *Dräger Tube Handbook,* 11th Ed. (Lübeck: Dräger Sicherheitstechnik GmbH., 1998).
5. *Air Quality Sampling & Measurement Equipment* (Chestertown, Md.: LaMotte Company, 1997).
6. *Nitrogen Dioxide in Air* (Chestertown, Md.: LaMotte Company, 1991).

PERCHLOROETHYLENE (TETRACHLOROETHYLENE)

Perchloroethylene, also known as tetrachloroethylene, is a colorless liquid with an ether-like odor. The nonflammable solvent is extremely stable and insoluble in water. The compound is moderately toxic and irritates the eyes and skin. Human tolerance level is 100 ppm in air.

The chlorinated organic compound can cause nerve disorders and damage to the liver and kidneys. The compound is possibly carcinogenic.

Around the home, tetrachloroethylene has applications as a dry cleaning solvent, degreasing solvent, and a veterinary medicine to eliminate intestinal worms. The solvent is used extensively in the dry-cleaning industry. After dry cleaning clothing, the fumes from the dry cleaning solvent remains on the clothes for an extended period of time. Tetrachloroethylene is one of the most important indoor air pollutants in the home.

Landfill gases measured from samples at more than 40 landfills contain averages of tetrachloroethylene (perchloroethylene) at 7810 ppb.[1] Dry cleaner operations are excellent sources of tetrachloroethylene at 5000 to 16,500 parts per billion (ppb) by volume. These emission levels for perchloroethylene from these sources are 1800 to 5700 times higher than emission levels from operating hazardous waste incinerators. In addition, vapors from chlorinated solvents used by laundries to remove spots on clothes can volatilize into the air and impact apartments, food markets, and restaurants located nearby.[2] A number of tests for vapors from these chlorinated solvents can result in improved control of fugitive emissions from chlorinated solvents.

All procedures for the "short-term" test are provided in an experiment titled "Measurement of Air Contaminates" in Chapter 4. The tubes are open at both ends during short-term measurements. The detailed procedures for each experiment are enclosed with the equipment and the specific tubes in the cases.

Perchloroethylene vapors within the standard measuring range of 0.5 to 4 ppm/0.1 to 1 ppm and 20 to 300 ppm/2 to 40 ppm can be determined by observation of tube color change from yellowish white to gray-blue in a Dräger "short-term" tube. Permanganate ion oxidizes the chlorinated organic molecule to form chlorine gas, which immediately reacts with diphenylbenzidine to produce a gray-blue reaction product. Interferences are halogenated hydrocarbons and free halogens during this measurement. Also, petroleum ether vapors can reduce the perchloroethylene indications at higher concentrations.

Perchloroethylene at a concentration of 10 to 500 ppm in air can be determined by observing the development of an orange reaction product within the temperature range of 15 to 40°C in a short-term tube. After oxidation of the chlorinated hydrocarbon by permanganate ion, immediate reaction of o-tolidine with liberated chlorine produces the indicator product in the tube. Halogenated hydrocarbons and free halogens are interferences during this test. Also, petroleum vapors can reduce the perchloroethylene indication.

In addition, reactions of MnO_2 with perchloroethylene produce chlorine, which reacts with diphenylbenzidine to form a brown-blue reaction product. This brown-blue product identifies the concentration of perchloroethylene in the ranges of 500 to 10,000 ppm and 50 to 600 ppm during "short-term" tests at 20°C. This measurement is temperature dependent and requires interpretation of a table of values for measurements at higher temperatures. In this case, other chlorinated hydrocarbons, free halogens and mineral acids can also interfere with the tests.

During "long-term" measurements, the reaction of perchloroethylene with chromium (VI) to form HCL followed by a reaction of the acid with bromophenol produces a yellowish white reaction product in a Dräger tube. This reaction sequence will determine the amount of perchloroethylene present in air by using prepared tubes during long-term measurements of 1 hour, 2 hours, or 4 hours. During the measurement, a constant low flow pump will be used to determine the time weighted average concentration of the air pollutant. Other chlorinated hydrocarbons such as trichloroethylene and 1,1-dichloroethylene will interfere with the determination of perchloroethylene.

Direct Indicating Diffusion Tubes can be used to perform long-term measurements over an 8-hour period. The diffusion tubes are open on one end during measurements and do not require a pump or additional equipment. During the measurement, perchloroethylene vapors react with chromium (VI) to form chlorine gas, which reacts with o-tolidine to produce a yellow-orange product on the tube medium. Various ranges of product concentrations are indicated on the tube during 1 hour, 2 hours, 4 hours, and 8 hours of tube reaction. Additional chlorinated hydrocarbons are indicated at different sensitivities. Trichloroethylene and 1,1,1-trichloroethane are indicated with nearly the same sensitivity.

QUESTIONS AND ANSWERS

1. What is another name for perchloroethylene?
   Perchloroethylene is commonly called tetrachloroethylene or "perc."
2. What is the human tolerance levels for perchloroethylene in air?
   The tolerance level for perchloroethylene is 100 ppm in air.
3. What are the levels of air emissions of perchloroethylene at dry-cleaning facilities?
   Dry cleaner operations are excellent sources of tetrachloroethylene (perchloroethylene) air emissions at 5000 to 16,500 parts per billion (ppb) by volume.
4. What are the levels of air emissions of perchloroethylene at Hazardous Waste Incinerators?
   Hazardous Waste Incinerators emit 2.8 parts per billion (ppb) of perchloroethylene by volume measured in the stack before dispersion.
5. What are some uses for perchloroethylene or "perc"?
   Perchloroethylene or "perc" has significant uses in dry cleaning, vapor-degreasing, drying agent for solids and metals, heat transfer medium, and veterinary medicine.

REFERENCES

1. James Cudahy, Chris McBride, Floyd Hasselriis, et al., *Comparison of Hazardous Waste Incinerator Trace Organic Emissions With Emissions From Other Common Sources,* Incineration and Thermal Treatment Technologies Conference, Portland, Or., May 10, 2000.
2. Jack Lauber, P.E., Consultant, Latham, New York.
3. Stanley E. Manahan, *Environmental Chemistry,* 6th Ed. (Boca Raton, Fla.: Lewis Publishers, 1994).
4. *Hawley's Condensed Chemical Dictionary,* 11th Ed., rev'd by N. Irving Sax and Richard J. Lewis (New York, N.Y.: Van Nostrand Reinhold, 1987).
5. *Dräger Tube Handbook,* 11th Ed. (Lübeck: Dräger Sicherheitstechnik GmbH., 1998).

PESTICIDES

Pesticide categories are sometimes defined as insecticides, herbicides, rodenticides, fungicides, molluscicides, and nematodicides. In other cases, scientists have identified categories of pesticides by a less inclusive definition. Some of the more toxic pesticides have been identified in dead birds, fish and other wildlife. Some environmental contaminates interfere with endocrine system functions.

Specific pesticides and insecticides, such as DDT, have been banned in the United States since 1973. Since DDT is not truly biodegradable and is ecologically damaging, it can be used for a few specialized purposes under strict handling and application conditions. The application of DDT can be used to control the tussock moth. DDT is still manufactured for export purposes for a number of other countries.

Evolving climates and farming practices may hinder the efforts of scientists to predict the toxicity levels and persistence of some long-lived pollutants.[1] About one-quarter of all commercial pesticides are chiral or exist as enantiomers. Thus, the molecules are present as mirror-image twins and can significantly impact the life of a variety of species. A molecule of a pesticide, such as the levorotatory form, may kill pests and leave humans and other nontargeted species unharmed, whereas the chiral sibling known as the dextrorotatory form does the opposite. In addition, microbes may convert a single-twin of a herbicide or pesticide into its missing enantiomer. At this point, not much is understood about the conversion of one pesticide or herbicide enantiomer to the other form. Additional studies of the impacts of dextrorotatory (d) and levorotatory (l) forms on different species and the toxicity level of each enantiomer will be required.

Planar chromatography analyzes pesticides in drinking water. Depending on the type of chemical structure, some pesticides are more soluble in water than others. The system can screen for pesticides simultaneously with determination limits of 50 ng/L. In most cases, pesticides can be isolated from the environment by a solid phase carbon trap and the resulting fraction determined by gas chromatography/mass spectrophotometry.

A number of pesticides can be determined by a series of kit methods.[2] Some of the methods of detection are U.S. EPA SW-846 approved test procedures, depending on the application and whether the State and federal agencies will accept the test result. The tests can be used to detect specific pesticides and additional toxics in groundwater, soil and other matrices. Most of the test kits are immunoassay methods or extraction procedures for substances including DDT, Toxaphene, Chlordane, Triazine, Lindane 2,4-D, and other toxic substances. Several test methods must be used to test for pesticides present in soil, water or other media at various contaminate concentrations. The upper range of detection can be variable and extensive due to dilution techniques before analysis.

The pesticide compounds are detected as a group of compounds, called antigens. The antigen and an enzyme compete to combine with the antibody. Then, a color is developed based on the amount of enzyme conjugate reacting with the antibody. The color can be determined by reading the absorbance of a photometer. During the test, the darker the color, the less analyte in the tube. A tube or microliter well with a color lighter than the negative control is presumed positive. To quantitatively determine the concentration of the unknown, the analyst can run unknown samples alongside standards containing known concentrations of the antigen. Each pesticide has a different sensitivity level. Sample dilution techniques before analysis can extend the range of detection.

Experiments to determine Chlordane, Lindane, and Toxaphene in soil require a coated tube immunoassay procedure and a photometer to obtain final semi-quantitative results. Soil samples require prior extraction of the analyte using the SDI Extraction Kit. The minimum detection level (MDL) and measurement range for each analyte are the following: 20 ppb (parts per billion) at 20 to 600 ppb Chlordane, 0.4 ppm (parts per million) at 0.4 to 40.0 ppm Lindane, and 0.5 ppm at 0.5 to 10 ppm Toxaphene. A number of interferences including Endrin, Endosulfan I, Endosulfan II, Dieldrin, Heptachlor, Aldrin, and others interfere with each contaminate. Each experiment procedure can be obtained from Strategic Diagnostics Inc.[2]

A test for the triazine family of herbicides including atrazine provides semi-quantitative results during a coated tube immunoassay. The assay range of testing is 0.1 to 1.0 ppb and the lower limits of detection (LLD) is 0.053 ppb. A number of pesticides and substances including Ametryn, Propazine, Prometryn, Prometon, Terbutylazine, and substituted Atrazine compounds have lower limits of detection near the same level as the triazine family (atrazine). During the test procedure, the colored solution should not be exposed to direct sunlight. The test procedure is provided below.

TRIAZINE (ATRAZINE)

Procedure*

1. Add 160 μL of reference sample to a tube labeled 'R.' Add 160 μL of negative control to a tube labeled 'N.' Then add 160 μL of sample to a tube labeled 'S.'
   Note: Multiple samples (6) can be run if labeled properly.

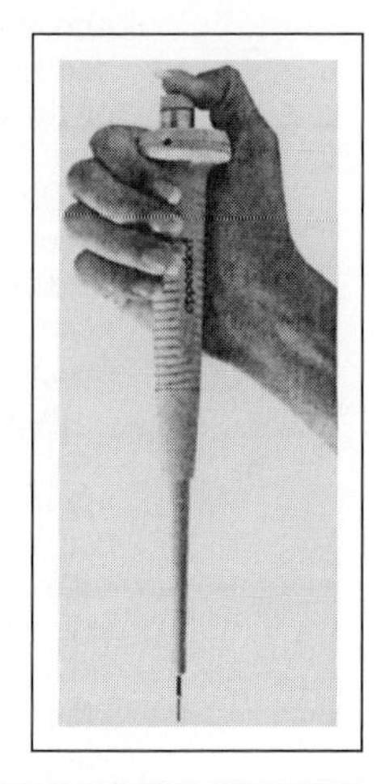

2. Add 4 drops of atrazine-enzyme conjugate to each of the tubes.
   Note: Gently swirl the test tubes to mix for 2 to 3 seconds.

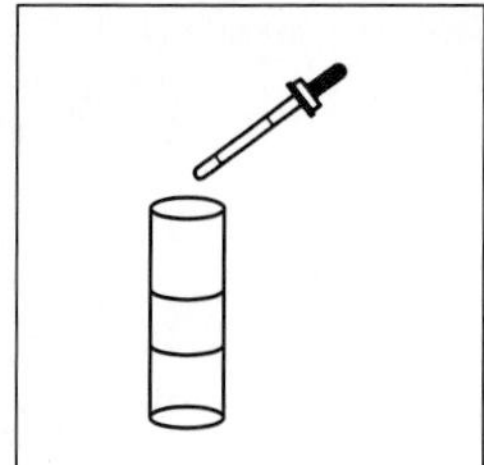

3. Allow the samples to stand undisturbed for 5 minutes.
4. Shake out the contents of each of the test tubes. Fill each test tube to overflowing with distilled water (or tap water), then decant and vigorously shake out the remaining water. Shake out as much water as possible during each wash. Repeat this wash step 3 more times for each test tube.

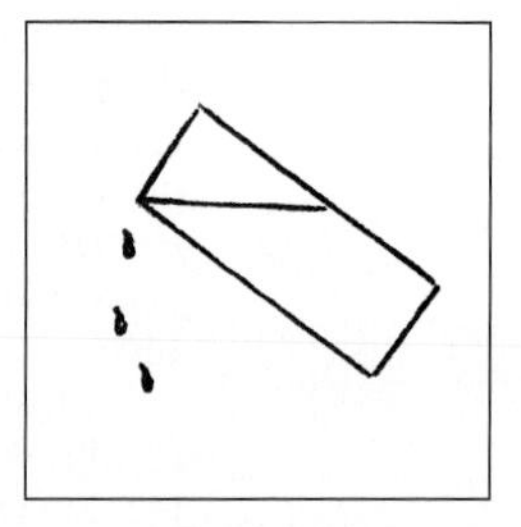

5. Add four drops of *substrate* to the each of the test tubes, and using a different clean eye dropper immediately add 4 drops of *chromogen* to each of the test tubes. Gently shake each of the tubes for a few seconds.
   Note: Do not reverse this order sequence. Add the substrate before the chromogen.

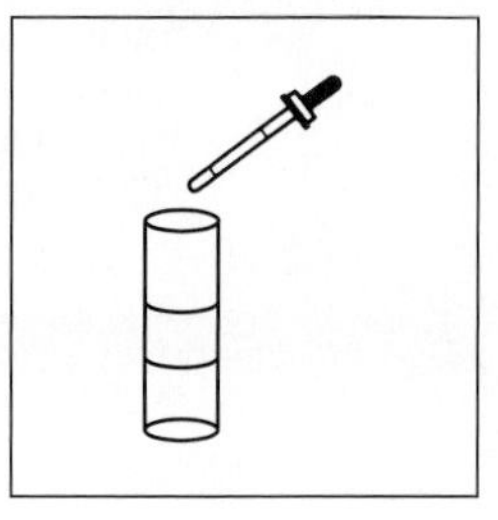

*Compliments of Strategic Diagnostics Inc.

6. Allow the sample and reference samples to stand for 2 minutes at room temperature and interpret the results.**
   Note: If a blue color does not develop in the negative control test tube within 2 minutes after adding the substrate and chromogen, the test is invalid and you must repeat it.
   Note: You can interpret the results visually within 2 minutes after adding the substrate and the chromogen to each test tube or you can perform a more precise analysis with a photometer. The photometer may have to be calibrated to measure the peak maximum for the blue colored sample (absorbance near 600 nm.).

QUESTIONS AND ANSWERS

1. Are all pesticides and insecticides biodegradable?
   Substances like DDT are not truly biodegradable. However, most pesticides and insecticides are considered to be biodegradable. The substances like DDT require strict handling and application regulations in some countries, whereas the substance can be freely applied in other countries.
2. What is an enantiomer?
   An enantiomer is a chemical substance of the same formulation that has two mirror-image twins. A twin rotates the plane of polarized light to the (left levorotatory, denoted as −), while the other twin rotates the plane of polarized light to the right (dextrorotatory, denoted as +).
3. Why are enantiomers important?
   Even though the enantiomers have identical properties, except "the way" that they rotate the plane of polarized light, one form can kill pests while the other form is non-toxic. The levorotatory form may kill pests and not have any impact on animals or humans, whereas the dextrorotatory form may have an impact on humans and animals but not kill pests. The preparation methods for the different forms of the same compound may have to be altered to avoid this problem or the final mixture of pesticide will have to be analyzed to be certain that the correct enantiomer is available in the product. Much more study has to be undertaken in this area.
4. What seems to be the problem with pesticides?
   Some pesticides seem to hang around for a considerable length of time, and appear to concentrate in the various organs of birds, fish and other wildlife. A number of pesticides appear to interfere with the endocrine system functions.

REFERENCES

1. J. Raloff, *Pesticides Change "Hands" and Risks,* Science News, 156(18), p. 276, Oct. 30, 1999.
2. Joe Dautlick, James Eberts and Kia Wyatt, Strategic Diagnostics Inc., 111 Pencader Drive, Newark, DE.
3. Ward Stone, Del Mar Wildlife Pathology Laboratories, Del Mar, N.Y.

**Compare the sample test to the negative control test tube against a white background. If the sample test tube contains less color than the negative control test tube, then the original sample contains triazine or members of the triazine family or a cross-reactant. For more information, refer to the specific directions supplied by SDI on the internet at www.SDIX.com or call 1-800-544-8881.

RADIOACTIVE PARTICLES

Atoms containing the same number of protons, but different numbers of neutrons in their nuclei are called *isotopes*. Isotopes have different masses and usually differ in their nuclear properties, but are chemically identical atoms of the same element. Some isotopes are *radionuclides* or *radioactive isotopes*. Radionuclides are unstable nuclei which give off charged particles and gamma rays in the form of *radioactivity*.[1]

Radionuclides are produced as fission products of heavy nuclei such as uranium and plutonium. They are also produced by the reaction of neutrons with relatively stable nuclei during nuclear power generation. Specific types of radionuclides are widely used as "*tracers*" in industry and in medical applications. Large coal plants lacking ash-control equipment can introduce significant quantities of radionuclides into the air. In addition, natural sources of radionuclides are abundant.

A number of hazardous substances are radioactive and can cause major environmental problems. An explosion and fire at a power reactor in Chernobyl during 1986 resulted in massive contamination of an area. The mass production of radioactive isotopes during the use and production of weapons and nuclear reactors has increased concern about the exposure levels in the environment. Excessive amounts of the wrong type of radioactivity can cause detrimental, or even fatal, health effects.

The ecological and health effects of radionuclides are dependent on a number of factors including the type and energy of radiation emitter and the half-life of the source. In addition, the degree to which a particular element is transported within the ecosystem and the ability of living species to absorb the radioactive element are important factors to consider.

Radionuclides emit ionizing radiation called *alpha particles, beta particles,* and *gamma rays*. An alpha particle is a helium nucleus consisting of two neutrons and two protons with atomic mass 4. Alpha production can occur by radioactive decay of substances such as uranium-238. Alpha particles can be stopped by cardboard, paper or skin. However, they are very dangerous when ingested. Alpha particles have a velocity of one-tenth the speed of light and have a high ionizing power.

Highly energetic, negative electrons or positive electrons, named positrons, are called *beta radiation.* Radioactive chlorine-38 is a typical beta emitter. The radioactive chlorine-38 loses a negative beta particle and becomes an argon-38 nucleus. Beta particles have a velocity approximately nine times the speed of alpha particles and are about 100 times more penetrating than an alpha particle. They produce much less ionization per unit path length than alpha particles. Beta particles can penetrate several millimeters into human tissue and can be stopped by a sheet of aluminum.

A chlorine-37 nucleus can absorb a neutron to produce chlorine-38 and gamma radiation. Gamma rays are electromagnetic radiation with shorter wavelengths than x-rays. Gamma rays are more energetic than x-rays and have a velocity equal to the speed of light. They can easily penetrate the human body, and in fact, they can penetrate several feet of concrete. The degree of penetration for a gamma ray is proportional to its energy. They are used for the qualitative and quantitative analyses of radionuclides.

Gamma rays are produced by nuclear energy transitions. For example, when cobalt-60 decays by beta emission, nickel-60 is produced. The process involves the emission of a beta particle and two gamma quanta. The decay process for radioactive elements is unique for each type of element.

X-rays are ionizing radiations that can penetrate approximately 15 cm into tissue in diagnostic tests. The depth of penetration is regulated by the voltage. X-rays pass more easily through tissue than through bones and teeth.

The intensity of radiation and the distance from a radiation source obeys the inverse square law. The exposure to radiation caused by changing the distance is inversely proportional to the square of the distance. For example, a person 5 cm from a radiation source will be exposed to a radiation of relative intensity 1/25 of the original amount. A person located

10 cm from the radiation source will be exposed to a radiation of relative intensity 1/100 of the original amount.

The decay of a radioactive species is represented by its half-life, represented as $t_{1/2}$. Half-life can be defined as the period of time during which half of a given number of atoms of a specific radionuclide decays. In general, 10 half-lives are required for the loss of 99.9% of a radionuclide.[1] The rate of decay (disintegration) of radioactive substances is constant and cannot be changed by any treatment method to which a substance may be subjected.

The term, *LD-50/30*, is the *30-day medium lethal dose equivalent,* which is a measure of radiation toxicity.[2] This term provides the dose (in *rems* or *millirems*) that will kill 50% of the exposed individuals within 30 days. Different species (animals, bacteria, viruses) have variable 30-day lethal dose equivalents, LD_{50} (roentgens).

Most people are exposed to higher levels of natural radiation than radiation from artificial sources. Radiation initiates harmful chemical reactions by breaking bonds in the tissue macromolecules of living organisms. In severe cases of radiation poisoning, the number of red blood cells are greatly diminished as the bone marrow is destroyed. Radiation can cause considerable amount of genetic damage which is normally evident years after exposure.

The above-ground detonation of nuclear devices can add large amounts of a variety of radioisotopes to particulate matter in the atmosphere. The processing of spent reactor fuels and the operation of nuclear power plants generate the radioactive noble gas, ^{85}Kr (half-life of 10.3 years). In general, additional radionuclides produced by reactor operation are usually removed from the reactor effluent by chemical reaction or have significantly short half-lives and decay before emission.

Strontium-90, a common waste product of nuclear testing, can be interchanged with calcium in bone. The radioactive element can fall onto pasture and crop land and be ingested by cattle. Eventually, strontium-90 can enter the bodies of people who drink milk. Strontium-90 will remain in a human body throughout the person's life, because the radioactive substance has a half-life of about 50 years. In addition, cesium-137 can replace quantities of sodium in the body and cause considerable damage even though its half-life is quite short at 140 days. Additional radionuclides, radium and potassium-40, from natural sources can easily leach into water. Many types of radioactive species are products and wastes that we are exposed to every day. Emissions from industry, medical centers, research facilities, nuclear power plants, and nuclear weapons sites are readily available for human exposure.

Radioactive decay can occur by a combination of several types of processes and stepwise procedures. Radon is a noble gas product of radium decay. It can enter the atmosphere in the form of two isotopes, ^{222}Rn (half-life of 3.8 days) and ^{220}Rn (half-life of 54.5 seconds). Both isotopes of radon emit alpha particles in decay chains that terminate with stable lead isotopes, which can adhere to atmospheric particulates. Also, the atmospheric particulate matter may already contain some radioactivity from natural origins. In addition, cosmic rays can act on atmospheric nuclei to form a number of additional radionuclides.

Radium leaches from minerals and is sometimes found in drinking water. Significant radium concentrations can be found in the areas of uranium-producing regions. The USEPA has established standards for the maximum contaminant level (MCL) for total radium (^{226}Ra and ^{228}Ra) in drinking water at 5 picocuries per liter. Many municipal water supplies exceed this standard and require water softening treatment processes to remove radium. A number of natural and artificial radionuclides have been identified in water. Many of these radionuclides are long-lived and highly toxic.

Radon has been identified in uranium mine tailings that have been used as soil conditioner, backfill, and building foundations. Radon enters buildings through cracks or openings in the foundation or basements, and has been identified in a number of basements of homes. In addition, radium containing significant quantities of radon has been used to make watches and other jewelry.

The *Curie* (Ci) was originally defined in terms of the disintegration rate associated with one gram of radium. The Curie is a measure of the activity and not the quantity of material.

Additional units of disintegration rate for a substance include megacuries, millicuries, and microcuries.

The *Roentgen* is a unit that measures the intensity level of x-ray or gamma radiation. This unit is limited to the effect of x-rays or gamma rays in air.

The *RAD* (Radiation Absorbed Dose) is the unit used to measure the absorbed dose. Because of its ionizing effect on body tissue, a total body dose of about 600 rads of gamma radiation is considered to be lethal to most people.

The *REM* (Roentgen Equivalent for Man) of any given radiation is that quantity that causes, when absorbed by man, an effect equivalent to the absorption of one roentgen.

The *RBE* (Relative Biological Effectiveness) of a radiation is defined as the ratio of the absorbed dose delivered by the gamma rays of Co-60 to the absorbed dose delivered by a particular radiation in question when both are compared in producing the same biological effect.[2] The RBE of radiation signifies its ability to produce a specific effect in a particular tissue.

The National Council on Radiation Protection has established the *Maximum Permissible Dose* (MPD) for the general population and for personnel engaged in radiology.[2] The permissible level for total body radiation for individuals subject to long term radiations is provided by the following formula:

$$\text{MPD} = 5(\text{N} - 18) \text{ rems}$$

where

MPD = maximum permissible dose
5 = 5 rems
N = person's age in years

The total-body radiation must be of sufficient strength and penetrating power to significantly affect certain important parts of the body, such as the genitals, head, blood forming organs, and upper and lower portions of the trunk to exceed the MPD. However, the MPD for the general population is reduced to one-tenth of that set for those actively working with radiations.

Today, many types of devices can detect various types of radiation for a number of purposes. Most of the monitors or badges can detect a type of radiation within a specific measuring range. Other types of equipment detect levels of radiation at a preset level of danger to humans. A number of devices that can determine the quantities of various types of radiation are identified below.

Radioactive contamination of water is normally detected by measurements of gross beta and gross alpha activity.[1] After evaporating water to obtain a very thin layer, the measurement of beta and alpha activity is performed with an *internal proportional counter*. For gamma detection, solid state detectors are usually employed to resolve closely spaced peaks in the sample spectra. A multichannel spectrometric data analysis can normally determine a number of radionuclides in a sample without chemical separation.

The best known device for the detection of radiation is the *Geiger counter* or *Geiger-Müller counter*. Radiation ionizes the gas in a tube, causing the flow of electric current that is amplified and measured as pulses by a sensing element. The recorder, called a scaler, acts as an adding machine by keeping a record of the total number of pulses emitted by the Geiger tube or another detector. The number of pulses is the measure of the radiation intensity. When using a Geiger counter, the background effect must be subtracted from the experimental result. A Geiger-Müller tube can be linked with a PC-microcomputer to collect and analyze data from field experiments and research. The Geiger-Müller counter is presented in Fig. R.1.

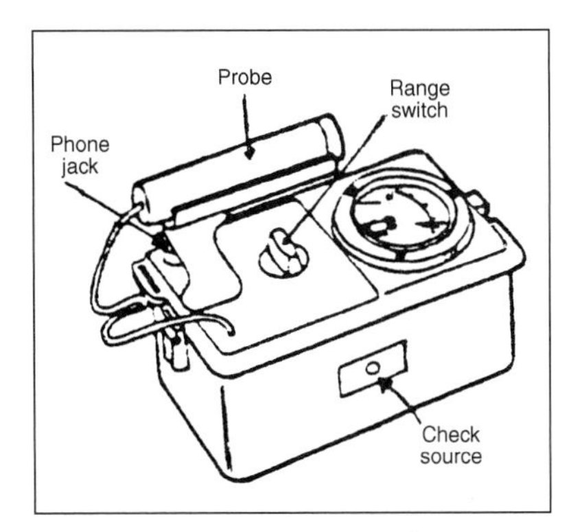

FIGURE R.1 Geiger Müller Counter.

The Geiger-Müller counter measures alpha, beta and gamma particles. Geiger tubes are not equally sensitive to alpha, beta and gamma radiations. Efficiencies of measuring each of these particles can be improved by changing the type of Geiger tube used for the measurement. A Geiger-Müller counter detects a small amount of background radiation. Background radiation can occur from natural sources or from artificially produced radioactive sources. When performing measurements, remove watches with radioactive dials and other radioactive materials that may influence the count. The background should be checked every hour if performing a series of measurements.

A *radon indoor air monitor* can provide air monitoring for radon over a variety of time periods. A variety of monitors purchased locally can offer 3-day, 7-day, 2 to 4 weeks or longer periods of monitoring. In each case, directions with the monitor provide ample guidance for sampling the air for radon. The monitors are normally charcoal canisters or alpha track detectors. Handling procedures, sample location, and amount of time required to collect the sample are presented in the directions. A picture of one type of radon indoor air monitor is provide below as R.2.

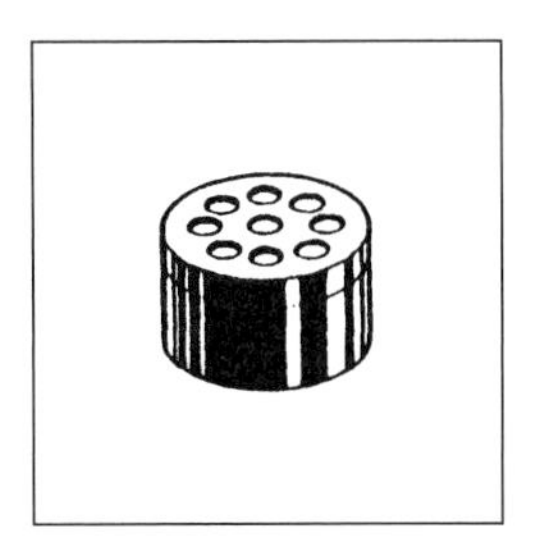

FIGURE R.2 Radon Indoor Air Monitor.

A *portable radiation meter* can detect minute amounts of gamma, beta or x-rays. The meter registers 0.1–10 mR/hour (milli-Roentgen) on a logarithmic scale. The instrument begins to beep at 20 mR/hour and increases frequency in proportion to the radiation level. The portable radiation meter is identified below as R.3.

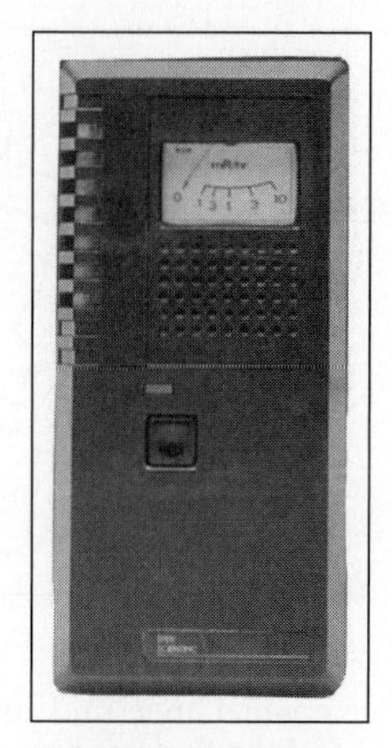

FIGURE R.3 Radiation Meter.

Monitoring dosimeters, or pin badges, are normally mounted on a pocket of a shirt or laboratory coat to measure "millirems." The diagram labeled R.4 displays the dosimeter.

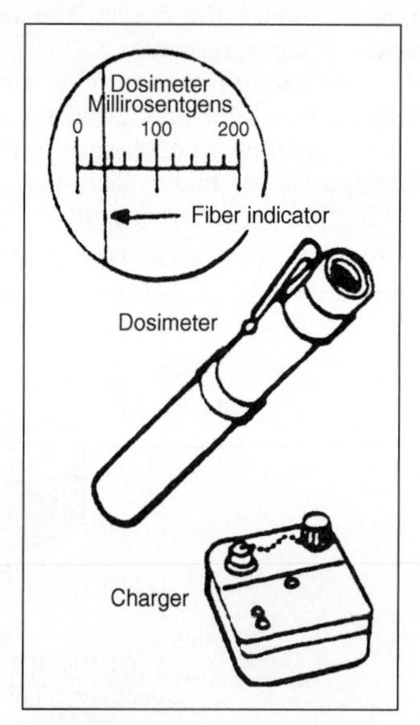

FIGURE R.4 Dosimeter and Charger.

Film badges measure absorbed radiation in "Rads." Film badges are normally mounted on the wrist or on the index finger. Film badges determine the small amount of energy which can have an ionizing effect on body tissue. The diagram labeled R.5 is a film badge mounted on the wrist for radiation monitoring.

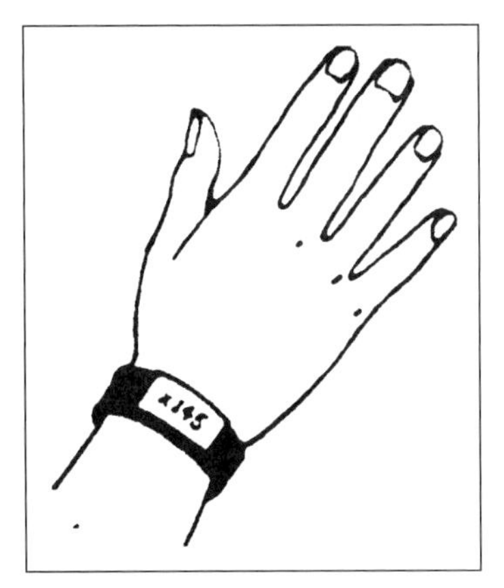

FIGURE R.5 Radiation Detection Bracelet.

A number of types of radiation can penetrate living tissue to a variety of depths. For example, microwave energy can leak through the door seals of microwave ovens after continued use and penetrate living tissue to a depth of about 1.18 inches. A *microwave detector,* identified as R.6, can detect the radiation leakage from an oven.

FIGURE R.6 Radiation Meter.

Certain substances (called phosphors or fluors) will emit light when subjected to radiation. For example, sodium iodide can be activated by gamma rays, whereas anthracene is activated by beta radiation. *Scintillation counters* can be used to amplify light emission. The output signal will provide a visual record of intensity on readout instruments. In this experiment, the background effect must be subtracted from the experimental result. The scintillation counter is illustrated in Fig. R.7.

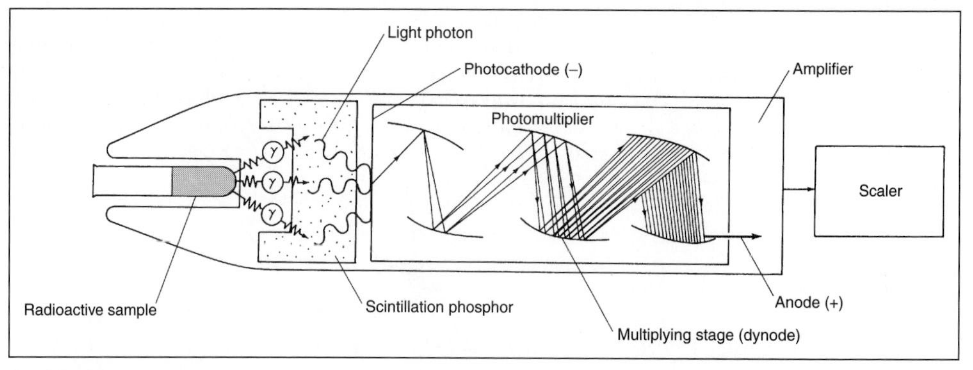

FIGURE R.7 Radiation Photomultiplier.

A radioactive standard set consisting of low levels of radionuclides can be used to check the operation of a Geiger tube, scintillation probe, and other equipment. Most standards consist of 1 microcurie (μCi) or an exempt quantity of a radioactive metal, such as Co-60 with half-life of 5.3 years, sealed in a small cavity in a 25 mm plastic disk.

Radioactive substances can become embedded in porous surfaces and make decontamination of equipment and areas more difficult. All working surfaces and equipment should be nonporous and resistant to chemicals. Stainless steel containers and acid resistant surfaces are easily decontaminated.

QUESTIONS AND ANSWERS

1. What are alpha, beta and gamma radionuclides?
An alpha particle is a helium consisting of two neutrons and two protons with atomic mass 4. Alpha particles have a velocity of one-tenth the speed of light and have a high ionizing power. In contrast, beta particles are highly energetic, negative electrons or positive electrons called positrons. Beta particles have a velocity approximately nine times the speed of an alpha particle and are 100 times more penetrating than the alpha particle. Beta particles have a lower ionizing power than alpha particles. However, gamma rays are electromagnetic radiation with shorter wavelengths than x-rays. Gamma rays are more energetic than x-rays and have a velocity equal to the speed of light. Gamma rays can penetrate the human body and several feet of concrete.

2. What is the relationship of the intensity of radiation on a body versus the distance of the body from the source?
The exposure to radiation caused by changing distance is inversely proportional to the square of the distance. If a person located 1 centimeter from a radiation source would have a relative intensity of 1 of radiation on the body, then a person located 5 centimeters from the radiation source would have a relative intensity of 1/25 of radiation on the body. Also, a person located 20 centimeters from a radiation source would have a relative intensity of 1/400 of radiation on the body.

3. What are the primary sources of radioactive particles?
The largest source of radionuclides is from naturally occurring reactions in the atmosphere. Radioactive particles are produced as fission products of heavy metal nuclei and during the reaction of neutrons with relatively stable nuclei. Some types of radionuclides are used extensively as tracers in industry, medicine and research. Large coal plants without proper ash-control equipment can introduce significant quantities of radionuclides

into the air while burning coal. Nuclear power plants, nuclear reactors, and the production of nuclear-based weapons can produce significant quantities of radioactive particles. Nuclear wastes also are a significant source of radioactive substances. Radioactive contamination of soil and water from uranium mine tailings and from uranium-producing areas has significantly increased the exposure of humans to both radium and radon. Jewelry, watches and other devices designed for wearing apparel sometimes contain small amounts of radioactive nuclides.

REFERENCES

1. Stanley E. Manahan, *Environmental Chemistry,* 6th Ed. (Boca Raton, Fla.: Lewis Publishers, 1994).
2. Dr. Gershon Shugar, *Health Sciences Chemistry,* 2nd Ed. (Dubuque, Iowa: Kendall Hunt Co., 1996).
3. *A Citizen's Guide to Radon,* OPA-86-004, EPA United States Environmental Protection Agency, U.S. Department of Health and Human Services, August 1986.
4. *Radon Reduction Methods,* OPA-87-010, United States Environmental Protection Agency, September 1987.

STYRENE

Styrene monomer is insoluble in water and polymerizes easily. The styrene monomer is toxic by ingestion and inhalation. The simple styrene compound can produce liver and kidney damage in humans. The styrene monomer has a human tolerance limit of 50 ppm in air. The substance is flammable and has explosive limits in air of 1.1 to 6.1 %. The styrene monomer must be inhibited during storage.

The styrene monomer is a primary constituent in many compounds and resins called polystyrene (polymerized styrene), SBR (styrene-butadiene rubber), ABS (acrylonitrile-butadiene-styrene), SAN (styrene-acrylonitrile polymer), latex and alkyds protective coatings, polyesters, copolymers, and intermediates. The SBR polymer is a strongly basic anion-exchange resin used for water treatment, while other polymers (ABS) have applications as moldings, packaging, and engineered plastics. Some of the compounds tend to discolor or slowly break down in outdoor light, while others are very stable in light. A number of the compounds dissolve and break down in the presence of common hydrocarbon solvents such as gasoline, kerosene, and alkanes. As the stable polymers tend to dissolve, decompose or break apart, the smaller units of dimer or monomer become much more toxic and considerably more volatile.

The compounds have practical uses including carpet backing, footwear, coatings, adhesives, sealants, engineered plastics (bottles, luggage, etc.), household wares, toys, furniture, radiator leak stop, and many additional household applications. Most of the styrene sources are carpets and plastic products. Generally, the compounds are safe in the polymerized stage, but become a greater problem as decomposition occurs. Burning these products in a wood stove, fireplace, outdoor barrel, open fire, or during a house fire can produce some major air pollution problems and dangerous air contaminates.

Styrene in air can be determined by an experiment titled "Measurement of Air Contaminates" in Chapter 4. Styrene monomer reacts with sulfuric acid to form a pale yellow reaction product in short-term tubes with standard measuring ranges of 10 to 200 ppm and 50 to 400 ppm. Butadiene-type compounds which tend to polymerize, interfere with the determination of monostyrene.

Another "short-term" Dräger tube can also be used to determine styrene in the standard measuring range of 10 to 250 ppm. A formaldehyde/sulfuric acid mixture on the tube medium reacts with styrene to form a red-brown reaction product characteristic of the amount of styrene in the air. Xylene, toluene, butadiene, and ethyl benzene can react with the formaldehyde/sulfuric acid indicating system and interfere with the results.

QUESTIONS AND ANSWERS

1. How toxic is the styrene monomer?
Styrene monomer is toxic by inhalation and ingestion. The compound can cause liver damage and kidney damage in humans. The monomer has a human tolerance limit of 50 ppm in air.

2. How stable are the resins made from the monomer?
Some of the polymeric resins are very stable to light, heat, and other conditions. The polystyrene materials are soluble in hydrocarbon solvents, while others types of polymers are more soluble in alcohol. Some polystyrene polymers tend to discolor in light. Many of these types of plastics break down and dissolve in hydrocarbons (gasoline, kerosene, etc.) commonly found around the home.

3. What are some of the sources of styrene pollution in the home?
The gaseous vapors from carpet backing, coatings, adhesives, sealants, and radiator stop leak are sources of styrene monomer pollution. Decomposition of these products and certain types of plastics can also produce sources of styrene pollution. In addition, burning of the styrene-based polymer in a wood stove, fireplace, outdoor barrel, open fire or during a house fire can produce some major air contaminates.

4. What compounds tend to interfere with the analysis of styrene?
Butadiene-type compounds tend to interfere with styrene analysis procedures. Also, xylene, toluene, and ethyl benzene can interfere with the analysis when using a formaldehyde/sulfuric acid tube for the determination of styrene.

REFERENCES

1. *Hawley's Condensed Chemical Dictionary,* 11th Ed., rev'd by N. Irving Sax and Richard J. Lewis (New York, N.Y.: Van Nostrand Reinhold, 1987).
2. *Dräger Tube Handbook,* 11th Ed. (Lübeck: Dräger Sicherheitstechnik GmbH., 1998).

TOLUENE

Toluene is a dangerous fire risk and has explosive limits in air of 1.27 to 7%. Toluene is toxic by ingestion, inhalation, and skin absorption. The human tolerance level is 100 ppm in air, however toluene can be tolerated in ambient air up to 200 ppm. Exposure at 500 ppm may cause headache, nausea, impaired coordination, and body weariness. Even though toluene is less toxic than benzene, high levels of human exposure to toluene can lead to a coma. The colorless liquid has a benzene-like odor and is insoluble in water.

Around the home, toluene is a thinner in lacquers and a solvent for various types of paints and coatings. The compound is an adhesive solvent in plastic toys and model airplanes. Toluene can be found in wood cleaners, rust preventive materials, and a variety of adhesives.

The test procedures to detect various levels of toluene in air are provided in an experiment titled "Measurement of Air Contaminates" in Chapter 4. A variety of reactants in the tubes indicate the amount of toluene in the air. Each tube can be used to sample within a specific temperature range and absolute humidity levels. A standard deviation is provided for each test.

Toluene can be determined by a Dräger tube in the standard measuring ranges of 50 to 400 ppm, and 50 to 300 ppm/5 to 80 ppm. During short-term measurements, toluene reacts with I_2O_5 in the presence of sulfuric acid on a tube medium to form brown-colored iodine. Xylene and benzene are indicated with the same sensitivity in the lower level of the multi-range tube, however different colors may develop with p-xylene and benzene. Phenol, acetone, ethanol, and octane can interfere with low level determinations of toluene. In the range of 50–400 ppm toluene, xylenes are indicated with a lower sensitivity, while benzene and petroleum hydrocarbons discolor the indicating layer in the tube. Methanol, ethanol, acetone, and ethyl acetate do not interfere at concentrations in the TLV range. The temperature ranges and the absolute humidity levels are not identical for each determination.

Toluene can be detected by reaction with SeO_2 in the presence of sulfuric acid on the tube medium to form a brown-violet reaction product within the standard measuring range of 100 to 1800 ppm during short-term measurements. Xylenes forms a bluish-violet color on the medium at approximately the same sensitivity, whereas benzene and petroleum hydrocarbons tend to discolor the entire indicating layer. Alcohols, acetone, and ethyl acetate do not interfere with the determination of toluene.

During long-term measurements of 1 to 8 hours, toluene reacts by a similar I_2O_5 reaction in sulfuric acid on a tube medium to form brown-colored iodine. A constant low flow pump will be used during long-term measurements to determine the time weighted average concentration of the pollutant in air. A number of aromatic hydrocarbons and petroleum hydrocarbons are indicated with different sensitivities. In addition, toluene reacts with a yellow tube medium to form a brown reaction product at 10 to 40°C in a "Direct Indicating Diffusion Tube." Additional aromatic hydrocarbons such as ethyl benzene and xylene can be determined after 6 hours of measurement. However, benzene does not interfere within the TLV range and aliphatic hydrocarbons are not indicated.

QUESTIONS AND ANSWERS

1. Where do you find toluene around the home?
   Toluene can be found in wood cleaners, rust preventive materials, and adhesives. It is often used as a thinner in lacquers, coatings and specific types of paints. Toluene has applications as a cleaning solvent for grease and other non-water soluble agents, and is also used as an adhesive solvent for plastic toys and model airplanes.
2. How toxic is toluene?
   Toluene is toxic by ingestion, inhalation, and skin absorption. The human tolerance level is 100 ppm in air.
3. What are some interferences during the determination of toluene by the tube method?
   During the determination of toluene at low concentrations in air, benzene and specific xylene compounds are indicated with the same sensitivity as toluene. Compounds such as phenol, acetone, ethanol, and octane at fairly high concentrations can interfere with the toluene determination by tube. In the range of 50 to 400 ppm toluene, xylenes can interfere at a lower sensitivity. Benzene and petroleum hydrocarbons tend to discolor the tube indicating layers of both the 50 to 400 ppm and 100 to 1800 ppm toluene sample ranges. Xylenes form a bluish layer on the medium at approximately the same sensitivity as high levels of toluene.
4. What primary requirements are necessary during long-term measurements that are not required during short-term measurements?
   Long-term measurements require a constant low flow pump to determine the time weighted

average concentration of the pollutant in the air over a period of time, usually 1 to 8 hours.

REFERENCES

1. Stanley E. Manahan, *Environmental Chemistry*, 6th Ed. (Boca Raton, Fla.: Lewis Publishers, 1994).
2. *Hawley's Condensed Chemical Dictionary*, 11th Ed., rev'd by N. Irving Sax and Richard J. Lewis (New York, N.Y.: Van Nostrand Reinhold, 1987).
3. *Dräger Tube Handbook*, 11th Ed. (Lübeck: Dräger Sicherheitstechnik GmbH., 1998).

1,1,1-TRICHLOROETHANE

The solvent 1,1,1-trichloroethane, called methyl chloroform or TCA, is insoluble in water and nonflammable. The compound is irritating to the eyes and tissue. Humans have a tolerance level of 350 ppm in air. The chlorinated compound can cause dizziness and irregular breathing when inhaled. This chemical is an important indoor air pollutant in the home and is highly resistant to biodegradation in the environment.

The solvent is used for cleaning precision instruments, metal degreasing, pesticide manufacturing, textile processing, and tool and die applications. This chemical is used in industry and around the home as a common aerosol spray. The compound is a good cleaning agent and provides a protective shield on specific materials. As a spray, the solvent provides water and stain protection of fabrics, leathers, and suedes. A warning on the can suggests spraying the 1,1,1-trichloroethane solvent in a well ventilated area.

Landfill gases measured from samples at more than 40 landfills (municipal and city) contain averages of 1,1,1-trichloroethane at 410 ppb.[1] By comparison, a number of hazardous waste combustors have been tested for stack emissions of trichloroethane and other chlorinated organic compounds. In MACT hazardous waste combustors, 1,1,1-trichloroethane has been measured in the stacks before dispersion at 15.9 ppb (parts per billion). Higher concentrations for this substance can be expected in homes and industry.

Vapors of 1,1,1-trichloroethane can be determined by an experiment titled "Measurements of Air Contaminates" in Chapter 4. A variety of reactants in the tube is used to indicate the presence of the nonflammable organic compound. The measurement time is approximately 2 minutes.

Vapors of 1,1,1-trichloroethane can be detected at 50 to 600 ppm after tube oxidation with an acidified oxidizing agent followed by reaction with o-tolidine in a "short-term" Dräger tube. The brown-red reaction product will indicate the concentration of the 1,1,1-trichloroethane in air. The range of detection is generally limited to temperatures of 15 to 40°C and an absolute humidity range of 5 to 15 mg H_2O/L. Additional chlorinated hydrocarbons are indicated at different sensitivities. Aromatic hydrocarbons will significantly lower the indication values.

QUESTIONS AND ANSWERS

1. What are some uses for 1,1,1-trichloroethane?
This chemical is used around the home as a common aerosol spray. The compound is a good cleaning agent and protects surfaces from water, stains, scratches, and outside attack by chemicals. The aerosol spray provides water and stain protection for fabrics, leathers

and suedes used for furniture, car seats, upholstery, jackets, vests, shoes, and other clothing and items found around the home.

2. What is the human tolerance level for 1,1,1-trichloroethane?
The tolerance level for humans is 350 ppm in air for 1,1,1-trichloroethane.

3. How does this chemical affect most people?
The chlorinated compound can cause dizziness and irregular breathing when inhaled.

4. What interferences create measurement problems during detection of 1,1,1-trichloroethane levels?
Other chlorinated hydrocarbons are indicated on the tube at different sensitivities. Aromatic hydrocarbons interfere with the test method by lowering the actual test results.

REFERENCES

1. James Cudahy, Chris McBride, Floyd Hasselriis, et al., *Comparison of Hazardous Waste Incinerator Trace Organic Emissions With Emissions From Other Common Sources,* Incineration and Thermal Treatment Technologies Conference, Portland, Or., May 10, 2000.
2. Jack Lauber, P.E., Consultant, Latham, New York.
3. *Fate of Pesticides and Chemicals in the Environment,* J. L. Schnoor (ed.) (New York, N.Y.: John Wiley & Sons, Inc., 1992).
4. *Hawley's Condensed Chemical Dictionary,* 11th Ed., rev'd by N. Irving Sax and Richard J. Lewis (New York, N.Y.: Van Nostrand Reinhold, 1987).
5. *Dräger Tube Handbook,* 11th Ed. (Lübeck: Dräger Sicherheitstechnik GmbH., 1998).

XYLENE

Dimethyl benzene is commonly called ortho-, meta- or para-xylene. Xylene solutions often contain different forms of xylene in various percentages, even though the meta-para mixtures are generally more common. Most xylene compounds are a moderate fire risk, however para-xylene is a dangerous fire risk. The human tolerance levels are 100 ppm in air for all three compounds. The compounds are toxic by ingestion and inhalation. The three compounds exist as a colorless liquid, which is insoluble in water.

Xylene is used as a paint thinner, brush cleaner, and paint stripper around the home. Xylene is classified as an "all purpose remover." Xylene can be used to clean organic materials from surfaces and is found in rust preventive materials. A variety of common insecticide mixtures contain xylene. Also, motor fuels contain some levels of ortho-xylene.

The test procedures to detect various levels of xylene in air are provided in an experiment titled "Measurement of Air Contaminates" in Chapter 4. Xylene can be determined by a Dräger tube in the standard measuring range 10 to 400 ppm. During a short-term measurement, xylene reacts with formaldehyde in the presence of sulfuric acid on a tube medium to form a red brown quinoid reaction product within the temperature range of 0 to 40°C and an absolute humidity range of 3 to 15 mg H_2O/L. Styrene, vinyl acetate, toluene, ethylbenzene, and acetaldehyde are indicated at different sensitivities. Octane, methanol, and ethyl acetate do not interfere at relatively significant concentrations.

QUESTIONS AND ANSWERS

1. Where do you find xylene around the home?
Xylene can be found in rust preventive materials, organic cleaning solvents, brush clean-

ing solvents, paint strippers, and insecticides. Xylene is used as a paint thinner and as an "all purpose remover." Ortho-xylene is also a contaminate in motor fuels.

2. How variable are the flash points of the various forms of xylene?
   The flash point of m-xylene is 85°F, while the flashpoint of p-xylene is 81°F. The flashpoint of o-xylene is significantly higher at 115°F.
3. What are some interferences during the determination of xylene by the tube method?
   Styrene, vinyl acetate, toluene, ethyl benzene and acetaldehyde are indicated at different sensitivities during the determination of xylene in the 10 to 400 ppm range.
4. Which of the three solvents are the most common?
   Meta-xylene and para-xylene are the most common and the most readily available.
5. Of the three xylene isomers, which compounds are used for pharmaceutical synthesis and to prepare vitamins?
   Ortho-xylene and para-xylene are primarily used for synthesis and preparation of vitamins and pharmaceutical compounds.

REFERENCES

1. *Hawley's Condensed Chemical Dictionary,* 11th Ed., rev'd by N. Irving Sax and Richard J. Lewis (New York, N.Y.: Van Nostrand Reinhold, 1987).
2. *Dräger Tube Handbook,* 11th Ed. (Lübeck: Dräger Sicherheitstechnik GmbH., 1998).

CHAPTER 6
NOISE AND THERMAL POLLUTION

THERMAL POLLUTION

Thermal pollution can be defined as heat added to a water body or to air, and combined with various other circumstances to produce deleterious conditions. Thermal pollution can impact the environment in many different ways. However, not all increases in heat in the environment create thermal pollution, but they may be good for the ecology.

Thermal pollution of air can be a problem for residents who live in a valley region surrounded by mountains when trapped heat and air pollutants are present. Thermal pollution in air is evident when heavy traffic passes by on a hot day in a city crowded with people. The heating cycle in the atmosphere depends on weather, wind conditions, evaporation rates, pollution types and quantities, and terrain. Thermal pollution in air impacts the rate of reaction of air pollutants and, to some extent, the type of chemical reactions that will occur.

Thermal pollution of water can have a significant effect on specific types of fish and other species. Most fish have a critical temperature range for survival. For example, trout, northern pike, and walleye survive best in cold waters. Spawning and egg development are inhibited at temperatures above 47 to 48°F. Perch and most bass will not develop at temperatures above 83 to 84°F, while catfish will continue to grow in water environments above 90°F. Many fish kills are the result of higher water temperatures and high "Biochemical Oxygen Demand" (BOD) in the summer months.

Higher water temperatures increase the volatilization of water and can modify weather conditions. When the atmosphere becomes saturated with moisture, the resulting rain can produce flooding and damage. At a higher water temperature, the cycle of moisture in the air appears to become more extreme and cause more damage as the water floods areas. Also, shorelines become decimated or move inland.

Increases in water temperature increase enzyme activity and bacterial growth and metabolism to a maximum level, and then a sudden decrease in activity and growth is observed as temperature continues to increase. Enzymes are destroyed by being denatured at temperatures slightly higher than the optimum.

Thermal pollution can dictate the type of bacteria that live and thrive in a specific ecosystem. Psychrophilic bacteria have temperature optima below 20°C, whereas mesophilic bacteria display temperature optima between 20°C and 45°C. Thermophilic bacteria have a temperature optima above 45°C. Some bacteria are able to grow at 0°C, while others are able to grow at 75 to 80°C. Different types of bacteria can influence the type of plant growth, life and decay processes.

The formation of distinct layers within nonflowing bodies of water is the result of differences in temperature and density. During the summer months, solar radiation and pollution heat the surface layer (epilimnion) which has a lower density and floats on the bottom layer (hypolimnion). This phenomenon is called *thermal stratification.*[1] Layers can have significant temperature differences due to different sources of thermal pollution and solar radiation.

When layers form, they behave independently and have different chemical and biological properties.

Heating of the top layer of water can promote algae growth and the dissolution of compounds that do not normally dissolve in colder water. At higher temperatures, the dissolved oxygen content in water can significantly decrease which will impact life. Dissolved oxygen content in water at 0°C is approximately two times the amount of dissolved oxygen in water at 33°C (92°F). At higher temperatures, bacterial action or biodegradable organic material may cause the water to become anaerobic (lacking dissolved oxygen).[1]

During the autumn, the top water layer cools and the temperatures of the two layers equalize. The disappearance of thermal stratification causes the entire body of water to overturn and completely mix in the spring. However, changes in the original composition of the top layer may cause significant changes in the water as a whole and modify the requirements for water treatment.

Petroleum refining, electric power generation, and many other industries depend on water to dissipate heat from the industrial processes. Power plants generating electricity from coal, oil and nuclear resources dispose of about two-thirds of the energy that they *use*. This energy is converted into heat and disposed of into the environment. Water is normally withdrawn from a water body and passed through the cooling system to absorb heat energy, and then returned to its source. The output of the heated water warms the water body and impacts aquatic life. An electric-producing plant using up to 20 to 30% of the total water in a stream

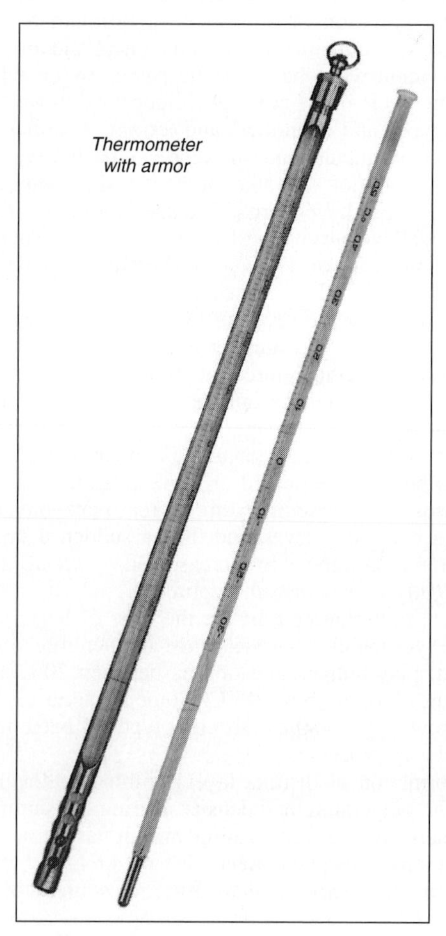

will release a significant amount of heat which will impact the ecology of the stream when the water is released directly back into the stream. In addition, extreme temperature changes will occur in the body of water during power plant startup and shut down. The sudden change in water temperatures can cause serious damage to a variety of aquatic organisms.

In addition, the removal of trees and other vegetation that shade the water in streams can contribute to the thermal pollution problem. High population centers can impact the water temperature and condition of a stream by adding a variety of pollutants. As water temperature rises, the type of fish and life in the stream will change.

Thermal pollution is usually limited to those events caused by humans. However, some natural occurrences such as volcanoes are significant contributors to the problem. Building fires and forest fires, regardless of the cause, can add a considerable amount of thermal pollution to the environment. The clearing of land by burning trees, stumps and other materials can also contribute to the thermal pollution issue. During specific years, these last events probably contribute more to the thermal pollution problem per year than all of the other issues.

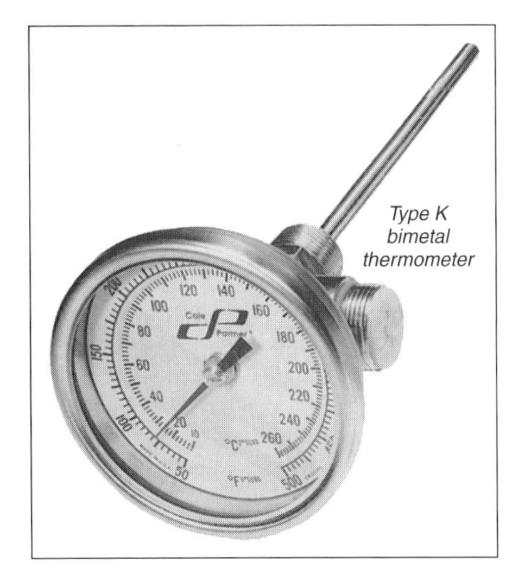

QUESTIONS AND ANSWERS

1. What is thermal pollution?
   Thermal pollution can be defined as heat added to a water body or air to produce deleterious conditions.
2. Do all increases in heat cause thermal pollution?
   Definitely not! Many heat changes are attributed to seasonal changes. Much of the heat generated in society is dissipated and not harmful.
3. How does thermal pollution impact a stream?
   Thermal pollution will dictate the type of life in the stream. The dissolved oxygen concentration in cold water is significantly higher than the dissolved oxygen content in warm water. Trout, northern pike and walleye survive in cold waters. The development and growth of perch and small mouth bass will basically cease at temperatures near 83°F. Catfish will survive in warm waters above 90°F. Fish kills are normal processes in high water temperatures and high BOD levels during the summer months. Thermal pollution also impacts the types of bacteria that will survive and function in a variety of environ-

ments. Also, thermal pollution will increase water evaporation rates and chemical reactions of pollutants in the atmosphere. Thermal pollution can directly impact the amount of rain and weather patterns.

4. What is thermal stratification?
Thermal stratification is the formation of two distinct layers within nonflowing bodies of water attributed to differences in temperature and density. A top layer of water with lower density is warmed during the summer months and floats on top of the bottom layer. The layers can scissor back and forth and display minimal amounts of mixing unless agitated or seasons change.

REFERENCES

1. Stanley E. Manahan, *Environmental Chemistry,* 6th Ed. (Boca Raton, Fla.: Lewis Publishers, 1994).
2. *Hawley's Condensed Chemical Dictionary,* 11th Ed., rev'd by N. Irving Sax and Richard J. Lewis (New York, N.Y.: Van Nostrand Reinhold, 1987).

NOISE POLLUTION

Acoustical energy is called "sound" when its presence is desired and "noise" when its presence is not desired.[1] In many cases, the interpretation of what we hear as sound or noise depends on the attitude of the listener. However, noise can cause deafness, annoyance, and interference with speech patterns. Additional evidence appears to indicate that noise can cause structural damage to the ear system. Loud music or noise can also interfere with the safety of individuals in vehicles and on the job.

Sound is a form of energy that can be transmitted through solids, liquids and gases. Sound is absorbed by some materials and dissipated as heat, while other materials reflect sound. Protection from sound or noise can be provided with helmets, muffs, non-disposable ear plugs, and disposable ear plugs. Cotton balls are considered to offer very poor protection.

Sound energy impacting a surface exerts a pressure which can be measured.[1] Sound level meters measure sound pressure in decibels (*dB 'A'*). The terminology decibels and Bels are used interchangeably where 10 decibels (db) = 1 Bel.

"Pitch" as applied to sound is a sensation which is related to the number of pressure fluctuations striking the eardrum.[1] "Frequency" is the count per second of these fluctuations expressed as hertz (cycles per second). Human voices normally contain energy in the 500 to 3000 Hz range. Below 500 Hz, the sensitivity of the human ear declines significantly. The human ear does not respond to low frequencies as well.

The speech communication criteria are defined as SIL db (Speech Interference Level in decibels) for a voice level located a specific distance from the receiver or listener. For example, a SIL db of 45 would be a normal voice at 10 feet from a person.[1] This would be a relaxed conversation within a private office or conference room. At the other extreme, a SIL db of 75 would consist of very loud voices at 1 foot or shouting at 2 to 3 feet. Also, a SIL db of 65 consists of a raised voice at 2 feet, a very loud voice at 4 feet, and shouting at 8 feet.

Modern sound level meters measure sound in the frequency range of 20 to 20,000 Hz. Acoustic energy is normally audible in this frequency range. Frequencies above the audible range are known as ultrasonics, whereas those below the audible range are called infrasonics. At times, a few individuals can hear noises beyond the specific limits established above, although most individuals cannot hear all of the noises in the whole frequency range. A number of family pets can hear frequencies beyond the normal human range.

In general, the acceptable level of noise in factories where people are exposed to the noise for the whole of each working day is 90 dB 'A.' Typical noise levels for industrial processes range in the area of 80 to 130 db 'A.' Rock concerts and other types of entertainment provided in totally enclosed facilities can display high levels of noise. Some of the higher noise levels may require earplugs or devices to prevent hearing loss. The deterioration of hearing is usually associated with the injurious effect of noise, together with the normal effect of increasing age. However, variations in noise exposure can cause fluctuations in hearing.

The range of sound powers encountered during everyday events is quite extensive. For instance, a jet engine has about 10^{13} times more sound power than a soft whisper.[1] In industry, pneumatic devices such as chisels and trimming equipment often provide the highest levels of noise measured in decibels (db). Earplugs and protective devices should be faithfully worn when this type of equipment is operating.

Noise does not impact everyone's hearing equally. The relationship between noise and hearing loss can be characterized in the following manner: a) above a certain level the more noise in the audible range of frequencies, the greater the hearing loss; b) sound energy which lies beyond the audible frequency range is harmless to hearing; and c) within the audible range, low-frequency sound is relatively less damaging than is high-frequency sound.[1] Intermittent loud noise may not be a hazard to most people.

Very high levels of airborne sound can impact the whole body. Low-frequency noise above 150 dB octave sound pressure level can vibrate the head or chest wall.[1] At very high frequencies, the skin can vibrate causing a heating effect. These levels can be experienced by close exposure to jet aircraft engines.

Cities around the country have enacted "Anti-Noise" laws and regulations limiting loud noises from commercial business, industrial, and home. In some cities, excessive neighborhood noises from horn blowing, music, and loud parties can lead to a summons or a visit from the police. Noise levels in dB 'A' of some common sounds are provided below:

Sounds	Noise level[1]
Quiet Rural Area	30
Rainfall	50
Normal Conversation	60
Vacuum Cleaner	70
Average Factory	80
Lawn Mower	90*
Chain Saw	100
Rock Music	110
Earphones (at loud level)	130
Military Rifle	Above 150

*Permanent hearing loss occurs after 8 hours of exposure to sound at this level and above.

A sound level meter consists of a thin circular metal plate and circuitry which transforms the motion of the plate into a direct current voltage. The movement of the thin plate is proportional to the decibels sound pressure. The instruments are calibrated against an internationally accepted value and readings from the dial of the meter are called dB sound pressure level. Sound level meters set to read "overall SPL" measure the total pressure exerted by all sound waves in the range of 20 to 20,000 Hz (i.e., the frequency range of the human ear).[1]

Several types of sound level meters are available with a variety of instrument features to meet OSHA requirements. Sound level meters operate in a variety of ranges and generally have resolution of 0.1 dB. The traceable sound level meter (Fig. N.1) features three decibel ranges including a low decibel range of 30.0 to 80.0 dB, medium decibel range of 50.0 to 100.0 dB, and a high decibel range of 80.0 to 130.0 dB. The instruments can monitor environmental noise pollution, background sound level, traffic, industrial equipment noise, and any excessive noise in an enclosed area. In addition, an acoustical calibrator (Fig. N.2) can be used to calibrate almost any digital sound meter at 94 dB for compliance with OSHA addendum.

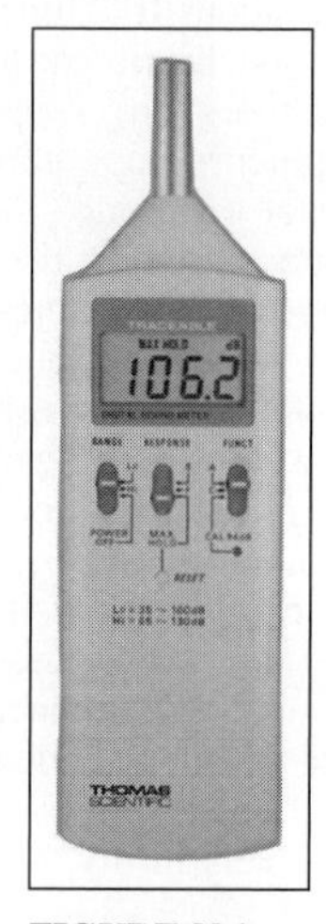

FIGURE N.1

FIGURE N.2

Noise measurements should be made at a height of approximately 3.75 feet (1.2 meters) and 11.75 feet (3.6 meters) away from walls and other reflecting structures.[1] An audiometer can measure the frequency level (Hz or cycles per second) of tone that a person can hear at different levels of loudness or intensity (decibels). Different types of sound level meters demonstrated in Figure N.3 and Figure N.4 can be used to monitor the work environment.

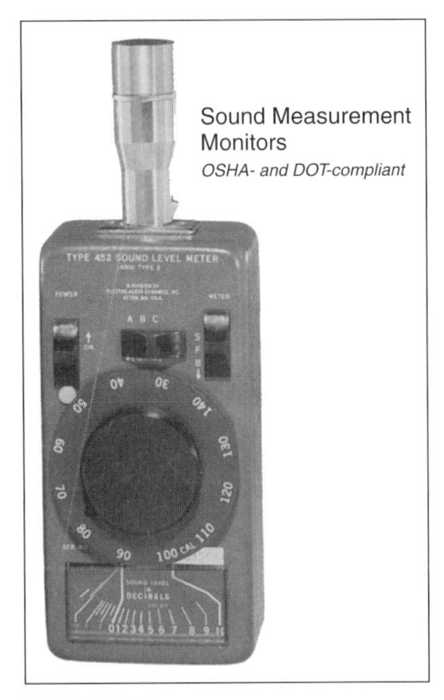

FIGURE N.3

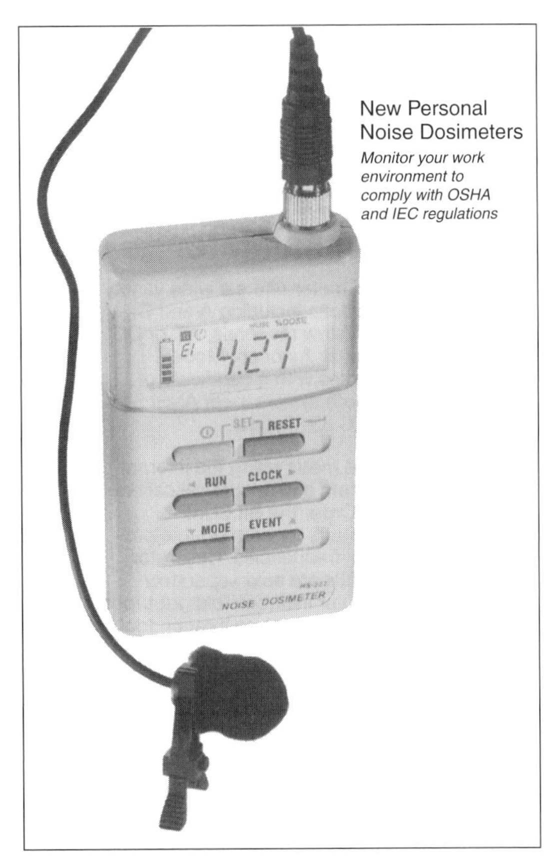

FIGURE N.4

QUESTIONS AND ANSWERS

1. How does noise impact people?
Noise can cause deafness, annoyance, and interfere with speech patterns. Noise can cause structural damage to the ear system. Loud noise can interfere with the safety of people.
2. What range of sound do people hear?
People hear human voices clearly in the range of 500 to 3000 Hz. People can hear sounds within the frequency range of 20 to 20,000 Hz. Most older individuals have difficulty hearing sounds throughout the whole range.
3. What appears to be an acceptable level of noise for people or employees?
For a worker exposed to the noise for the whole working day, an acceptable level of noise in factories is 90 dB 'A.' This same level of "noise" should probably be applied to other activities such as rock concerts inside small closed areas.
4. What are some devices that produce high levels of sound or noise?
Jet engines, rock bands inside of closed areas, pneumatic tools, grinding machines, metal trimming machines, chipping tools, pneumatic chisels (fettling of hard metal castings), and rotary presses.
5. How is a sound meter calibrated?
Place a probe of an acoustical calibrator on the microphone of a digital sound meter, turn on the calibrator and adjust the meter's reading to 94 dB.

REFERENCES

1. G. R. C. Atherley and G. V. Purnell, *Noise Control.* Industrial Safety Handbook, William Handley (ed.) (London, England: McGraw Hill Publishing Co. Ltd., 1969) at 330–346.
2. George Bugliarello, et al., *The Impact of Noise Pollution* (New York, N.Y.: Pergamon Press, 1976).

INDEX

ABOUT THE AUTHORS

Donald A. Drum is professor emeritus at Columbia–Greene Community College in Hudson, New York and also taught at Butler County Community College in Butler, Pennsylvania. As an expert on hazardous materials, he has presented industry seminars and testified during several different types of environmental hearings. He has been a member of the Butler County Hazmat Advisory Team and several professional organizations. Dr. Drum has more than 100 professional publications and has been the recipient of numerous teaching and community awards, including the Chemical Manufacturers Association *Regional Award for Excellence in Chemistry Teaching* and The State University of New York *Chancellor's Award for Excellence in Teaching.*

Shari L. Bauman is a geologist for a major global environmental and engineering consulting firm in Houston, Texas. She holds a Bachelor of Science degree in Geology from Tulane University and a Master of Science in Geochemistry from New Mexico Institute of Mining and Technology. Ms. Bauman specializes in soil and ground water regulatory and compliance projects in the State of Texas. Ms. Bauman has held the position of research assistant and laboratory technician for the X-ray Fluorescence and Instrumental Neutron Activation Analysis Inorganic Chemistry Laboratories and Soils Laboratory, affiliated with New Mexico Institute of Mining and Technology and New Mexico Bureau of Mines and Mineral Resources. She has also held the position of laboratory technician for the Coordinated Instrumentation Facility Laboratory, affiliated with Tulane University. Ms. Bauman has extensive experience with environmental field and laboratory techniques.

Gershon J. Shugar is the author of *The Chemist's Ready Reference Handbook,* Japanese Edition, McGraw-Hill, Tokyo, Japan and the *Chemical Technician's Ready Reference Handbook,* 4th Ed., McGraw-Hill, New York, New York. He is professor emeritus of engineering technologies at Essex County College, Newark, New Jersey. In 1947, Dr. Shugar founded a chemical manufacturing business that became the largest exclusive pearlescent pigment manufacturing company in the United States. In 1968, he was appointed assistant professor of chemistry at Rutgers University, where he taught until his appointment at Essex County College.

ABOUT THE AUTHORS